More information about this series at http://www.springer.com/series/7409

Lecture Notes in Computer Science   11136

Commenced Publication in 1973
Founding and Former Series Editors:
Gerhard Goos, Juris Hartmanis, and Jan van Leeuwen

Denny Vrandečić · Kalina Bontcheva
Mari Carmen Suárez-Figueroa · Valentina Presutti
Irene Celino · Marta Sabou
Lucie-Aimée Kaffee · Elena Simperl (Eds.)

The Semantic Web – ISWC 2018

17th International Semantic Web Conference
Monterey, CA, USA, October 8–12, 2018
Proceedings, Part I

 Springer

Editors
Denny Vrandečić ⓘ
Google
San Francisco, CA
USA

Kalina Bontcheva
University of Sheffield
Sheffield
UK

Mari Carmen Suárez-Figueroa
Universidad Politécnica de Madrid (UPM)
Madrid, Madrid
Spain

Valentina Presutti
National Research Council
Rome, Roma
Italy

Irene Celino ⓘ
Cefriel - Politecnico di Milano
Milan
Italy

Marta Sabou ⓘ
TU Wien
Vienna
Austria

Lucie-Aimée Kaffee ⓘ
University of Southampton
Southampton
UK

Elena Simperl ⓘ
University of Southampton
Southampton
UK

ISSN 0302-9743   ISSN 1611-3349   (electronic)
Lecture Notes in Computer Science
ISBN 978-3-030-00670-9   ISBN 978-3-030-00671-6   (eBook)
https://doi.org/10.1007/978-3-030-00671-6

Library of Congress Control Number: 2018954489

LNCS Sublibrary: SL3 – Information Systems and Applications, incl. Internet/Web, and HCI

This Springer imprint is published by the registered company Springer Nature Switzerland AG
The registered company address is: Gewerbestrasse 11, 6330 Cham, Switzerland

Preface

Now in its 17th year, the ISWC continues to be a focal point of the Semantic Web community. Year after year, it brings together researchers and practitioners from all over the world to present new approaches and findings, share ideas, and discuss experiences. It features a balanced mix of fundamental research, innovative technology, scientific artefacts such as ontologies or benchmarks, and applications that showcase the power of semantics, data, and the Web.

The Web, and all the ideas, technologies, and values that surround it, are at a crossroads. After several decades of growth and prosperity, it is increasingly seen as a means to lock-in customers and their data, spread misinformation, and increase polarization in society. At the same time, there is a palpable sense of excitement as we witness new voices and developments from the community that are fighting this trend in various ways – from more open and transparent forms of scholarly publishing and peer review in some of the workshops featured at the conference to cutting-edge research and applications on topics such as fake news, semantic coherence, and fact checking. Against this background, this year we decided to revive the Blue Sky Ideas track, chaired by Carolina Fortuna and supported by the Computing Community Consortium, to seek visionary ideas and opportunities for research and innovation, which are outside the mainstream topics of the conference.

A child of its times, the 17th ISWC featured a stellar, all-female keynote lineup: Jennifer Golbeck from the University of Maryland talked about human factors in semantic technologies; Vanessa Evers, University of Twente, introduced us to social robotics, an area with interesting applications for the models and technologies developed in our community; while Natasha Noy of Google discussed how we could use semantics to make structured data on the web more accessible and useful for everyone.

This volume contains the proceedings of ISWC 2018, i.e. papers that were peer reviewed and accepted into the main conference program, which covered three tracks: research, resources, and in-use. Altogether, a total of 254 submissions were received, which were evaluated by 486 reviewers. A total of 62 papers were accepted – 39 for the research track, 17 for the resources track, and six for the in-use track. The substantial number of papers in the resources category attests the commitment of the community to sharing and collaboration and to repeatable, reproducible research.

ISWC has an excellent scientific profile – as such, the research track continues to be the most popular venue for submissions. This year the track received overall 167 valid full-paper submissions, which turned into 39 acceptances, leading to an acceptance rate of 23%. We recruited 272 PC members and 67 sub-reviewers, guided by 17 senior PC members. Each paper received at least four reviews, including one from a senior PC member. The papers were assessed for originality, novelty, relevance, and impact of the research contributions, soundness, rigour and reproducibility, clarity and quality of presentation, and grounding in the literature. Each paper was then discussed by the PC chairs and the senior PC members, who helped us reach a consensus.

The resources track promotes the sharing of high-quality information artifacts that have contributed to the generation of novel scientific work. Resources can be datasets, ontologies, vocabularies, ontology design patterns, benchmarks, crowdsourcing designs, software frameworks, workflows, protocols, metrics, among others. The track is becoming demonstratively more and more important to our community as the sharing of reusable resources is key to allowing other researchers to compare new results, reproduce experimental research, and explore new lines of research, in accordance with the FAIR principles for scientific data management. All published resources must address a set of requirements: persistent URI, canonical citation, license specification, to mention a few. This year the track received 55 submissions, of which 17 were accepted (31% acceptance rate), covering a wide range of resource types such as benchmarks, ontologies, datasets, software frameworks, and crowdsourcing designs; a variety of domains such as music, health, education, drama, and audio; and addressing multiple problems such as RDF querying, ontology alignment, linked data analytics, and recommending systems. The reviewing process involved 70 PC members and 9 subreviewers, supported by 8 senior PC members. The average number of reviews per paper were 3.7 (at least three per paper), plus a meta-review provided by a senior PC member. Papers were evaluated based on the availability of the resource, its design and technical quality, impact, and reusability. The review process also included a rebuttal phase and further discussions among reviewers and senior PC members, who provided recommendations. Final decisions were taken following a detailed analysis and discussion of each paper conducted by the program chairs and the senior PC.

The in-use track at ISWC 2018 continued the tradition of demonstrating and learning from the increasing adoption of Semantic Web technologies outside the boundaries of research institutions, by providing a forum for the community to explore the benefits and challenges of applying these technologies in concrete, practical applications, in contexts ranging from industry to government and science. This year, the 32 submissions were reviewed by at least three PC members each and assessed in terms of novelty of the proposed use case or solution, uptake by the target user group, demonstrated or potential impact, as well as the overall soundness and quality. The PC consisted of 43 members. It helped us select 6 papers for acceptance, covering different domains (e.g., healthcare, cultural heritage, industry) and addressing a multitude of research problems (e.g., data integration, collaborative knowledge management, recommendations).

The industry track provides an opportunity for industry adopters to highlight and share the key learnings and new research challenges posed by real-world implementations. This year we had many exciting submissions from small to large companies that are making revealing leaps forward in science and engineering by using and adopting semantic technologies, web of data sources, and knowledge graphs. Each short submission was reviewed by at least three PC members. We accepted 14 out 27 abstracts that showcased a wide range of real-world industrial strength applications. The submissions were assessed in terms of the impact of semantics as a competitive differentiator in industry and discussions on the business value, experiences, insights, as well obstacles that stand in the way of large-scale adoption of semantic technologies.

The main conference program was complemented by presentations from the journal, industry, and posters and demos tracks, as well as the Semantic Web Challenge and a panel on future trends in knowledge graphs.

The conference included a variety of events appreciated by the community, which created more opportunities to present and discuss emerging ideas, network, learn, and mentor. Thanks to Amrapali Zaveri and Elena Demidova, the workshops and tutorials program includes a mix of established topics such as ontology matching and ontology design patterns alongside newer ones that reflect the commitment of the community to innovate and help create systems and technologies that people want and deserve, including re-decentralizing the Semantic Web, augmenting intelligence with humans in the loop, and a perspective workshop discussing open issues and trends. Application-centric workshops range from statistics to science to healthcare. The tutorials covered topics such as ontology modeling, crowdsourcing methods and metrics, RDF data validation and visualization, as well as knowledge graph machine learning and applications.

The conference also included a Doctoral Consortium track, which was chaired by Lalana Kagal and Sabrina Kirrane. The DC afforded PhD students from the Semantic Web community the opportunity to share their research ideas in a critical but supportive environment, where they received feedback from senior members of the community. This year the Program Committee accepted 12 papers for presentation at the event, while a total of 18 students were selected to participate in the DC poster and demo session. All student participants were paired with mentors from the PC who provided guidance on improving their research, producing slides, and giving presentations.

The program was complemented by activities put together by Bo Fu and Anisa Rula as student coordinators, who secured funding for travel grants, managed the grant application process, and organized the mentoring lunch alongside other informal opportunities for students and other newcomers to get to know the community.

Posters and demos are one of the most vibrant parts of every ISWC. This year, the track was chaired by Marieke van Erp and Medha Atre. It included 40 demos and 39 posters selected from a total of 95 submissions. A minute madness session offered time to those who wanted to take to the stage to present a brief preview of their poster or demo to generate interest in the work.

The Semantic Web Challenge has now been a part of ISWC for 15 years. Started as an open challenge to provide a forum for new and prestigious applications of Semantic Web technologies, and seconded by a challenge for scalability with the Billion Triple Challenge since 2003, the challenge was reanimated in 2017 with a new direction, with fixed datasets, and objective measures allowing for direct comparison of challenge entries. The 2018 challenge used a partly public, partly private knowledge graph about company networks owned by Thomson Reuters, and participants were asked to predict supply chain relations between those companies, using both knowledge in the graph itself as well as external sources. The best solutions were presented and discussed at the conference, both in a dedicated plenary session as well as during the poster session.

Delivering a conference is so much more than assembling a program. An event of the scale and complexity of ISWC requires the help, resources, and time of hundreds of people, organizers of satellite events, reviewers, volunteers, and sponsors. We are very grateful to our local team at Stanford University, who have expertly managed conference facilities, accommodation, registrations, the website, and countless other details. They made the conference a place we want to be every year and helped us grow this exciting scientific community.

Our thanks also go to Maribel Acosta, our tireless publicity chair – she played a critical role in ensuring that all conference activities and updates were communicated and promoted across mailing lists and on social media. Oana Inel was the metadata chair this year – her work made sure that all relevant information about the conference was available in a format that could be used across applications, continuing a tradition established at this conference many years ago. We are especially thankful to our proceedings chair, Lucie-Aimée Kaffee, who oversaw the publication of this volume alongside a number of CEUR proceedings for other tracks.

Sponsorship is crucial to the realization of the conference in its current form. We had a highly committed trio of sponsorship chairs, Annalisa Gentile, Maria Maleshkova, and Laura Koesten, who went above and beyond to find new ways to engage with sponsors and promote the conference to them. Thanks to them, the conference now features a social program that is almost as exciting as the scientific one – including a jam session accompanying the posters and demos presented on the second day of the conference and a bike ride from San Jose to Asilomar, the venue of the conference. Our special thanks go to the Semantic Web Science Association (SWSA) for their continuing support and guidance and to the organizers of the conference from 2017 and 2016, who were a constant inspiration, role model, and source of practical knowledge.

August 2018

Denny Vrandečić
Kalina Bontcheva
Mari Carmen Suárez-Figueroa
Valentina Presutti
Irene Celino
Marta Sabou
Lucie-Aimée Kaffee
Elena Simperl

Organization

Organizing Committee

General Chair

Elena Simperl — University of Southampton, UK

Local Chair

Rafael Gonçalves — Stanford University, USA

Research Track Chairs

Denny Vrandečić — Google, Mountain View, USA
Kalina Bontcheva — University of Sheffield, UK

Resources Track Chairs

Mari Carmen Suárez-Figueroa — Universidad Politécnica de Madrid, Spain
Valentina Presutti — Italian National Research Council, Italy

In-Use Track Chairs

Irene Celino — Cefriel, Italy
Marta Sabou — Vienna University of Technology, Austria

Workshop and Tutorial Chairs

Amrapali Zaveri — Maastricht University, Netherlands
Elena Demidova — L3S Research Center in Hannover, Germany

Poster and Demo Track Chairs

Medha Atre — Indian Institute of Technology, India
Marieke van Erp — Knaw Humanities Cluster, Netherlands

Journal Track Chairs

Abraham Bernstein — University of Zurich, Switzerland
Pascal Hitzler — Wright State University, Dayton, USA
Steffen Staab — University of Koblenz-Landau, Germany

Industry Track Chairs

Vanessa Lopez — IBM Research, Dublin, Ireland
Kavitha Srinivas — Rivet Labs, USA

Doctoral Consortium Chairs

Sabrina Kirrane	Vienna University of Economics and Business, Austria
Lalana Kagal	MIT CSAIL, USA

Semantic Web Challenge Chairs

Heiko Paulheimt	University of Mannheim, Germany
Axel Ngonga	University of Paderborn, Germany
Dan Bennett	Enterprise Data Services at Thomson Reuters, USA

Proceedings Chair

Lucie-Aimée Kaffee	University of Southampton, UK

Metadata Chair

Oana Inel	Vrije Universiteit Amsterdam, Netherlands

Sponsorship Chairs

Laura Koesten	Open Data Institute, UK
Maria Maleshkova	Karlsruhe Institute of Technology, Germany
Annalisa Gentile	IBM Research, San Jose, USA

Student Coordinators

Bo Fu	California State University, Long Beach, USA
Anisa Rula	University of Milano-Bicocca, Italy

Publicity Chair

Maribel Acosta	Karlsruhe Institute of Technology, Germany

Program Committee

Senior Program Committee – Research Track

Christian Bizer	University of Mannheim
Kai-Uwe Sattler	TU Ilmenau
Paul Groth	Elsevier Labs
Stefan Schlobach	Vrije Universiteit Amsterdam
Thomas Lukasiewicz	University of Oxford
Steffen Staab	Institut WeST, University Koblenz-Landau and WAIS, University of Southampton
Pascal Hitzler	Wright State University
Harith Alani	The Open University
Alessandro Moschitti	Qatar Computing Research Institute
Claudia d'Amato	University of Bari
Vojtěch Svátek	University of Economics, Prague

Sören Auer   TIB Leibniz Information Center for Science
                                   and Technology and University of Hannover
Oscar Corcho   Universidad Politécnica de Madrid
Abraham Bernstein   University of Zurich
Lalana Kagal   MIT
David Martin   Nuance Communications
Jose Manuel Gomez Perez   Expert System Iberia

Program Committee – Research Track

Maribel Acosta   Karlsruhe Institute of Technology
Alessandro Adamou   The Open University
Nitish Aggarwal   IBM
Harith Alani   The Open University
Panos Alexopoulos   Textkernel B.V.
Jose Julio Alferes   Universidade NOVA de Lisboa
Marjan Alirezaie   Orebro University
José Luis Ambite   University of Southern California
Renzo Angles   Universidad de Talca
Grigoris Antoniou   University of Huddersfield
Manuel Atencia   Univ. Grenoble Alpes and Inria
Ioannis N. Athanasiadis   Wageningen University
Sören Auer   TIB Leibniz Information Center for Science
                                   and Technology and University of Hannover
Nathalie Aussenac-Gilles   IRIT CNRS
Franz Baader   TU Dresden
Payam Barnaghi   University of Surrey
Valerio Basile   University of Turin
Zohra Bellahsene   LIRMM
Michael K. Bergman   Cognonto Corporation
Abraham Bernstein   University of Zurich
Elisa Bertino   Purdue University
Leopoldo Bertossi   Carleton University
Christian Bizer   University of Mannheim
Fernando Bobillo   University of Zaragoza
Kalina Bontcheva   The University of Sheffield
Alex Borgida   Rutgers University
Mihaela Bornea   IBM
Loris Bozzato   Fondazione Bruno Kessler
Adrian M. P. Brasoveanu   MODUL Technology GmbH
Charalampos Bratsas   Aristotle University of Thessaloniki
John Breslin   NUI Galway
Carlos Buil Aranda   Universidad Técnica Federico Santa María
Gregoire Burel   The Open University
Elena Cabrio   Université Côte d'Azur, CNRS, Inria, I3S, France
Andrea Calì   University of London, Birkbeck College

David Carral   TU Dresden
Gerard Casamayor   Universitat Pompeu Fabra
Davide Ceolin   Vrije Universiteit Amsterdam
Ismail Ilkan Ceylan   University of Oxford
Pierre-Antoine Champin   LIRIS, Université Claude Bernard Lyon1
Gong Cheng   Nanjing University
Christian Chiarcos   Universität Frankfurt am Main
Michael Cochez   Fraunhofer
Pieter Colpaert   Ghent University
Simona Colucci   Politecnico di Bari
Olivier Corby   Inria
Oscar Corcho   Universidad Politécnica de Madrid
Luca Costabello   Accenture Labs
Fabio Cozman   University of São Paulo
Isabel Cruz   University of Illinois at Chicago
Philippe Cudre-Mauroux   U. of Fribourg
Bernardo Cuenca Grau   University of Oxford
Claudia d'Amato   University of Bari
Aba-Sah Dadzie   The Open University
Enrico Daga   The Open University
Florian Daniel   Politecnico di Milano
Laura M. Daniele   TNO - Netherlands Organization for Applied Scientific
                          Research
Victor de Boer   Vrije Universiteit Amsterdam
Jeremy Debattista   Trinity College Dublin
Thierry Declerck   DFKI GmbH
Jaime Delgado   Universitat Politècnica de Catalunya
Daniele Dell'Aglio   University of Zurich
Emanuele Della Valle   Politecnico di Milano
Elena Demidova   L3S Research Center
Ronald Denaux   University of Leeds, UK
Dennis Diefenbach   Université Jean Monet
Stefan Dietze   GESIS - Leibniz Institute for the Social Sciences
Ying Ding   Indiana University Bloomington
Mauro Dragoni   Fondazione Bruno Kessler - FBK-IRST
Michel Dumontier   Maastricht University
Henrik Eriksson   Linköping University
Vadim Ermolayev   Zaporozhye National University
Jérôme Euzenat   Inria and Univ. Grenoble Alpes
James Fan   HelloVera.ai
Nicola Fanizzi   Università degli studi di Bari "Aldo Moro"
Anna Fensel   Semantic Technology Institute (STI) Innsbruck,
                          University of Innsbruck
Alberto Fernandez   CETINIA, University Rey Juan Carlos
Javier D. Fernández   Vienna University of Economics and Business
Sebastien Ferre   Université de Rennes 1

Besnik Fetahu   L3S Research Center
Tim Finin   University of Maryland, Baltimore County
Lorenz Fischer   Sentient Machines
Fabian Flöck   GESIS Cologne
Antske Fokkens   Vrije Universiteit Amsterdam
Muriel Foulonneau   Luxembourg Institute of Science and Technology
Flavius Frasincar   Erasmus University Rotterdam
Fred Freitas   Universidade Federal de Pernambuco (UFPE)
Adam Funk   University of Sheffield
Aldo Gangemi   Università di Bologna and CNR-ISTC
Daniel Garijo   Information Sciences Institute
Anna Lisa Gentile   IBM
Jose Manuel Gomez Perez Expert System Iberia
Rafael S. Gonçalves   Stanford University
Gregory Grefenstette   IHMC and Biggerpan, Inc.
Paul Groth   Elsevier Labs
Tudor Groza   The Garvan Institute of Medical Research
Cathal Gurrin   Dublin City University
Christophe Guéret   Accenture
Peter Haase   metaphacts
Armin Haller   Australian National University
Harry Halpin   World Wide Web Consortium
Karl Hammar   Jönköping University
Andreas Harth   University Erlangen-Nuremberg
Oktie Hassanzadeh   IBM
Tom Heath   Open Data Institute
Johannes Heinecke   Orange Labs
Andreas Herzig   IRIT-CNRS
Pascal Hitzler   Wright State University
Aidan Hogan   DCC, Universidad de Chile
Laura Hollink   Vrije Universiteit Amsterdam
Matthew Horridge   Stanford University
Katja Hose   Aalborg University
Andreas Hotho   University of Wuerzburg
Geert-Jan Houben   Delft University of Technology
Wei Hu   Nanjing University
Eero Hyvönen   Aalto University
Yazmin A. Ibanez-Garcia Institute of Information Systems, TU Wien
Luis Ibanez-Gonzalez   University of Southampton
Oana Inel   Vrije Universiteit Amsterdam
Mustafa Jarrar   Birzeit University
Ernesto Jimenez-Ruiz   The Alan Turing Institute
Clement Jonquet   University of Montpellier - LIRMM
Lucie-Aimée Kaffee   University of Southampton
Lalana Kagal   MIT
Martin Kaltenboeck   Semantic Web Company

Mark Kaminski	University of Oxford
Pavan Kapanipathi	IBM T.J. Watson Research Center
Md. Rezaul Karim	Fraunhofer FIT, Germany
Tomi Kauppinen	Aalto University School of Science
Takahiro Kawamura	Japan Science and Technology Agency
Mayank Kejriwal	Information Sciences Institute
Carsten Keßler	Aalborg University Copenhagen
Prashant Khare	Knowledge Media Institute, Open University, UK
Haklae Kim	Samsung Electronics
Sabrina Kirrane	Vienna University of Economics and Business - WU Wien
Matthias Klusch	DFKI
Matthias Knorr	Universidade NOVA de Lisboa
Stasinos Konstantopoulos	NCSR Demokritos
Roman Kontchakov	Birkbeck, University of London
Jacek Kopecky	University of Portsmouth
Adila A. Krisnadhi	Wright State University and Universitas Indonesia
Udo Kruschwitz	University of Essex
Tobias Kuhn	Vrije Universiteit Amsterdam
Benedikt Kämpgen	Empolis Information Management GmbH
Patrick Lambrix	Linköping University
Steffen Lamparter	Siemens AG, Corporate Technology
Agnieszka Lawrynowicz	Poznan University of Technology
Danh Le Phuoc	TU Berlin
Chengkai Li	University of Texas at Arlington
Juanzi Li	Tsinghua University
Nuno Lopes	TopQuadrant, Inc.
Chun Lu	Université Paris-Sorbonne and Sépage
Markus Luczak-Roesch	Victoria University of Wellington
Thomas Lukasiewicz	University of Oxford
Carsten Lutz	Universität Bremen
Alexander Löser	Beuth Hochschule für Technik Berlin
Frederick Maier	Institute for Artificial Intelligence
David Martin	Nuance Communications
Trevor Martin	University of Bristol
Mercedes Martinez-Gonzalez	University of Valladolid
Miguel A. Martinez-Prieto	University of Valladolid
Wolfgang May	Universitaet Goettingen
Diana Maynard	The University of Sheffield
Franck Michel	Université Côte d'Azur, CNRS, I3S
Nandana Mihindukulasooriya	Universidad Politécnica de Madrid
Riichiro Mizoguchi	Japan Advanced Institute of Science and Technology
Marie-Francine Moens	Katholieke Universiteit Leuven
Pascal Molli	University of Nantes - LS2N

Gabriela Montoya	Aalborg University
Federico Morando	Nexa Center for Internet and Society at Politecnico di Torino
Alessandro Moschitti	Qatar Computing Research Institute
Paul Mulholland	The Open University
Raghava Mutharaju	GE Global Research
Lionel Médini	LIRIS lab., University of Lyon
Ralf Möller	University of Luebeck
Hubert Naacke	Sorbonne Université, UPMC, LIP6
Axel-Cyrille Ngonga Ngomo	Paderborn University
Andriy Nikolov	metaphacts GmbH
Lyndon Nixon	MODUL Technology GmbH
Leo Obrst	MITRE
Francesco Osborne	The Open University
Raul Palma	Poznan Supercomputing and Networking Center
Matteo Palmonari	University of Milano-Bicocca
Jeff Z. Pan	University of Aberdeen
Rahul Parundekar	Toyota Info-Technology Center
Bibek Paudel	University of Zurich
Heiko Paulheim	University of Mannheim
Tassilo Pellegrini	University of Applied Sciences St. Pölten
Silvio Peroni	University of Bologna
Catia Pesquita	LaSIGE, Universidade de Lisboa
Reinhard Pichler	Vienna University of Technology
Emmanuel Pietriga	Inria
Giuseppe Pirrò	Institute for High Performance Computing and Networking (ICAR-CNR)
Dimitris Plexousakis	Institute of Computer Science, FORTH
Mike Pool	Goldman Sachs Group
Livia Predoiu	University of Oxford
Cédric Pruski	Luxembourg Institute of Science and Technology
Yuzhong Qu	Nanjing University
Dnyanesh Rajpathak	General Motors
Dietrich Rebholz-Schuhmann	Insight Centre for Data Analytics
Georg Rehm	DFKI
Achim Rettinger	Karlsruhe Institute of Technology
Martin Rezk	Rakuten
German Rigau	IXA Group, UPV/EHU
Carlos R. Rivero	Rochester Institute of Technology
Giuseppe Rizzo	ISMB
Marco Rospocher	Fondazione Bruno Kessler
Camille Roth	Sciences Po
Marie-Christine Rousset	University of Grenoble Alpes
Ana Roxin	University of Burgundy, UMR CNRS 6306

Harald Sack   FIZ Karlsruhe, Leibniz Institute for Information
                            Infrastructure & KIT Karlsruhe
Sherif Sakr   The University of New South Wales
Angelo Antonio Salatino   The Open University
Muhammad Saleem   AKSW, University of Leizpig
Cristina Sarasua   University of Zurich
Felix Sasaki   Lambdawerk
Bahar Sateli   Concordia University
Kai-Uwe Sattler   TU Ilmenau
Vadim Savenkov   Vienna University of Economics and Business (WU)
Marco Luca Sbodio   IBM
Johann Schaible   GESIS - Leibniz Institute for the Social Sciences
Bernhard Schandl   mySugr GmbH
Ansgar Scherp   Kiel University and ZBW – Leibniz Information Center
                            for Economics, Kiel, Germany
Marvin Schiller   Ulm University
Stefan Schlobach   Vrije Universiteit Amsterdam
Claudia Schon   Universität Koblenz-Landau
Marco Schorlemmer   Artificial Intelligence Research Institute, IIIA-CSIC
Lutz Schröder   Friedrich-Alexander-Universität Erlangen-Nürnberg
Daniel Schwabe   PUC-Rio
Erich Schweighofer   University of Vienna
Giovanni Semeraro   University of Bari
Juan F. Sequeda   Capsenta Labs
Luciano Serafini   Fondazione Bruno Kessler
Saeedeh Shekarpour   University of Dayton
Gerardo Simari   Universidad Nacional del Sur and CONICET
Elena Simperl   University of Southampton
Hala Skaf-Molli   Nantes University
Sebastian Skritek   Vienna University of Technology
Monika Solanki   University of Oxford
Dezhao Song   Thomson Reuters
Steffen Staab   Institut WeST, University Koblenz-Landau and WAIS,
                            University of Southampton
Yannis Stavrakas   Institute for the Management of Information Systems
Armando Stellato   University of Rome, Tor Vergata
Audun Stolpe   Norwegian Defence Research Establishment (FFI)
Umberto Straccia   ISTI-CNR
Markus Strohmaier   RWTH Aachen University and GESIS
Heiner Stuckenschmidt   University of Mannheim
Jing Sun   The University of Auckland
York Sure-Vetter   Karlsruhe Institute of Technology
Vojtěch Svátek   University of Economics, Prague
Marcin Sydow   PJIIT and ICS PAS, Warsaw
Mohsen Taheriyan   Google
Hideaki Takeda   National Institute of Informatics

Kerry Taylor	Australian National University and University of Surrey
Annette Ten Teije	Vrije Universiteit Amsterdam
Kia Teymourian	Boston University
Dhavalkumar Thakker	University of Bradford
Allan Third	The Open University
Krishnaprasad Thirunarayan	Wright State University
Ilaria Tiddi	The Open University
Thanassis Tiropanis	University of Southampton
Konstantin Todorov	LIRMM, University of Montpellier
David Toman	University of Waterloo
Nicolas Torzec	Yahoo
Yannick Toussaint	LORIA
Sebastian Tramp	eccenca GmbH
Cassia Trojahn	UT2J & IRIT
Raphaël Troncy	EURECOM
Jürgen Umbrich	Vienna University of Economy and Business (WU)
Joerg Unbehauen	University of Leipzig
Jacopo Urbani	Vrije Universiteit Amsterdam
Herbert Van De Sompel	Los Alamos National Laboratory, Research Library
Jacco van Ossenbruggen	CWI & VU University Amsterdam
Ruben Verborgh	Ghent University – imec
Serena Villata	CNRS - Laboratoire d'Informatique, Signaux et Systèmes de Sophia-Antipolis
Denny Vrandečić	Google
Domagoj Vrgoc	Pontificia Universidad Católica de Chile
Simon Walk	Graz University of Technology
Kewen Wang	Griffith University
Zhichun Wang	Beijing Normal University
Paul Warren	Knowledge Media Institute, Open University, UK
Grant Weddell	University of Waterloo
Erik Wilde	CA Technologies
Cord Wiljes	CITEC, Bielefeld University
Gregory Todd Williams	Hulu
Jiewen Wu	Institute for InfoComm Research, A*STAR
Yong Yu	Shanghai Jiao Tong University
Fouad Zablith	American University of Beirut
Ondřej Zamazal	University of Economics, Prague
Benjamin Zapilko	GESIS - Leibniz Institute for the Social Sciences
Sergej Zerr	L3S Research Center
Qingpeng Zhang	City University of Hong Kong
Ziqi Zhang	Sheffield University
Jun Zhao	University of Oxford

Additional Reviewers – Research Track

Abdel-Qader, Mohammad
Acar, Erman
Akrami, Farahnaz
Allavena, Davide
Angelidis, Iosif
Annane, Amina
Badenes-Olmedo, Carlos
Bakhshandegan Moghaddam, Farshad
Batsakis, Sotiris
Bhardwaj, Akansha
Bilgin, Aysenur
Biswas, Russa
Borgwardt, Stefan
Braun, Tanya
Calleja, Pablo
Chaloux, Julianne
Charalambidis, Angelos
Chaves-Fraga, David
Cheatham, Michelle
Chen, Jiaoyan
Daniel, Ron
Deng, Shumin
Ding, Jiwei
Dudas, Marek
Espinoza, Paola
Frommhold, Marvin
Hildebrandt, Marcel
Hogan, Aidan
Jimenez, Damian
Jouanot, Fabrice
Khosla, Megha
Kilias, Torsten
Kondylakis, Haris

Koopmann, Patrick
Kostylev, Egor V.
Koutraki, Maria
Krieg-Brückner, Bernd
Krishnamurthy, Gangeshwar
Kritikos, Kyriakos
Mireles, Victor
Mogadala, Aditya
Moodley, Kody
Musto, Cataldo
Padhee, Swati
Palumbo, Enrico
Piao, Guangyuan
Priyatna, Freddy
Revenko, Artem
Ribeiro De Azevedo, Ryan
Ringsquandl, Martin
Rodrigues, Cleyton
Rodriguez Muro, Mariano
Rosso, Paolo
Schneider, Rudolf
Shimizu, Cogan
Siciliani, Lucia
Silvestre Vilches, Jorge
Sun, Zequn
Tachmazidis, Ilias
Thoma, Steffen
Umbrich, Jürgen
Volk, Martin
Wielemaker, Jan
Ziebelin, Danielle
Özcep, Özgür Lütfü

Senior Program Committee – Resources Track

Maria Esther Vidal   Universidad Simon Bolivar
Valentina Tamma   University of Liverpool
Anna Lisa Gentile   IBM
Steffen Lohmann   Fraunhofer
Aidan Hogan   DCC, Universidad de Chile
Serena Villata   CNRS - Laboratoire d'Informatique, Signaux et
                         Systèmes de Sophia-Antipolis
Jorge Gracia   University of Zaragoza
Stefan Dietze   L3S Research Center

Program Committee – Resources Track

Muhammad Intizar Ali	Insight Centre for Data Analytics, National University of Ireland, Galway
Ghislain Auguste Atemezing	Mondeca
Mattia Atzeni	University of Cagliari
Maurizio Atzori	University of Cagliari
Elena Cabrio	Université Côte d'Azur, CNRS, Inria, I3S
Timothy Clark	University of Virginia
Francesco Corcoglioniti	University of Trento
Daniele Dell'Aglio	University of Zurich
Emanuele Della Valle	Politecnico di Milano
Stefan Dietze	GESIS - Leibniz Institute for the Social Sciences
Ying Ding	Indiana University Bloomington
Mauro Dragoni	Fondazione Bruno Kessler - FBK-IRST
Mohnish Dubey	University of Bonn
Fajar J. Ekaputra	Vienna University of Technology
Diego Esteves	University of Bonn
Stefano Faralli	University of Mannheim
Mariano Fernández López	Universidad San Pablo CEU
Jesualdo Tomás Fernández-Breis	Universidad de Murcia
Aldo Gangemi	Università di Bologna and CNR-ISTC
Anna Lisa Gentile	IBM
Claudio Giuliano	Fondazione Bruno Kessler
Jose Manuel Gomez-Perez	ExpertSystem
Alejandra Gonzalez-Beltran	University of Oxford
Rafael S Gonçalves	Stanford University
Jorge Gracia	University of Zaragoza
Alasdair Gray	Heriot-Watt University
Tudor Groza	The Garvan Institute of Medical Research
Amelie Gyrard	Kno.e.sis - Ohio Center of Excellence in Knowledge-enabled Computing
Pascal Hitzler	Wright State University
Robert Hoehndorf	King Abdullah University of Science and Technology
Rinke Hoekstra	University of Amsterdam
Aidan Hogan	DCC, Universidad de Chile
Antoine Isaac	Europeana and VU University Amsterdam
Ernesto Jimenez-Ruiz	The Alan Turing Institute
Simon Jupp	European Bioinformatics Institute
Tomi Kauppinen	Aalto University School of Science
Christoph Lange	University of Bonn and Fraunhofer IAIS
Agnieszka Lawrynowicz	Poznan University of Technology
Alejandro Llaves	Fujitsu Laboratories of Europe
Steffen Lohmann	Fraunhofer

Phillip Lord	Newcastle University
Markus Luczak-Roesch	Victoria University of Wellington
Maria Maleshkova	Karlsruhe Institute of Technology
Fiona McNeill	Heriot Watt University
Nandana Mihindukulasooriya	Universidad Politécnica de Madrid
Raghava Mutharaju	GE Global Research
Giulio Napolitano	Fraunhofer Institute and University of Bonn
Vinh Nguyen	National Library of Medicine, NIH
Andrea Giovanni Nuzzolese	University of Bologna
Alessandro Oltramari	Bosch Research and Technology Center
Raul Palma	Poznan Supercomputing and Networking Center
Bijan Parsia	The University of Manchester
Silvio Peroni	University of Bologna
María Poveda-Villalón	Universidad Politécnica de Madrid
Valentina Presutti	CNR - Institute of Cognitive Sciences and Tecnologies
Mariano Rico	Universidad Politécnica de Madrid
German Rigau	IXA Group, UPV/EHU
Giuseppe Rizzo	ISMB
Edna Ruckhaus	Universidad Politécnica de Madrid
Anisa Rula	University of Milano-Bicocca
Michele Ruta	Politecnico di Bari
Satya Sahoo	Case Western Reserve University
Cristina Sarasua	University of Zurich
Stefan Schlobach	Vrije Universiteit Amsterdam
Jodi Schneider	University of Illinois Urbana Champaign
Stefan Schulte	Vienna University of Technology
Hamed Shariat Yazdi	University of Bonn
Elena Simperl	University of Southampton
Mari Carmen Suárez-Figueroa	Universidad Politécnica de Madrid
Valentina Tamma	University of Liverpool
Krishnaprasad Thirunarayan	Wright State University
Cassia Trojahn	UT2J & IRIT
Raphaël Troncy	EURECOM
Maria Esther Vidal	Universidad Simon Bolivar
Natalia Villanueva-Rosales	University of Texas at El Paso
Serena Villata	CNRS - Laboratoire d'Informatique, Signaux et Systèmes de Sophia-Antipolis
Fouad Zablith	American University of Beirut
Amrapali Zaveri	Maastricht University
Jun Zhao	University of Oxford

Additional Reviewers – Resources Track

Bader, Sebastian
Daquino, Marilena
Ebrahimi, Monireh
García-Silva, Andrés
Heling, Lars
Kismihók, Gábor
Nayyeri, Mojtaba

Nechaev, Yaroslav
Pham, Thu Le
Sadeghi, Afshin
Shimizu, Cogan
Weller, Tobias
Zhou, Lu

Program Committee – In-Use Track

Irene Celino	Cefriel
Marco Comerio	Cefriel
Oscar Corcho	Universidad Politécnica de Madrid
Philippe Cudre-Mauroux	University of Fribourg
Mathieu D'Aquin	Insight Centre for Data Analytics, National University of Ireland, Galway
Brian Davis	Insight Centre for Data Analytics, Galway
Stefan Dietze	GESIS - Leibniz Institute for the Social Sciences
Mauro Dragoni	Fondazione Bruno Kessler - FBK-IRST
Achille Fokoue	IBM
Daniel Garijo	Information Sciences Institute
Anna Lisa Gentile	IBM
Jose Manuel Gomez-Perez	ExpertSystem
Rafael S Gonçalves	Stanford University
Paul Groth	Elsevier Labs
Tudor Groza	The Garvan Institute of Medical Research
Christophe Guéret	Accenture
Peter Haase	metaphacts
Lucie-Aimée Kaffee	University of Southampton
Tomi Kauppinen	Aalto University School of Science
Elmar Kiesling	Vienna University of Technology
Craig Knoblock	University of Southern California
Freddy Lecue	Accenture Labs
Vanessa Lopez	IBM
Vassil Momtchev	Ontotext AD
Andriy Nikolov	metaphacts GmbH
Francesco Osborne	The Open University
Jeff Z. Pan	University of Aberdeen
Artem Revenko	Semantic Web Company GmbH
Giuseppe Rizzo	ISMB
Dumitru Roman	SINTEF
Marta Sabou	Vienna University of Technology
Harald Sack	FIZ Karlsruhe, Leibniz Institute for Information Infrastructure and KIT Karlsruhe

Juan F. Sequeda	Capsenta Labs
Elena Simperl	University of Southampton
Dezhao Song	Thomson Reuters
Thomas Steiner	Google
Simon Steyskal	Siemens AG Austria
Anna Tordai	Elsevier B.V.
Raphaël Troncy	EURECOM
Josiane Xavier Parreira	Siemens AG Österreich

Additional Reviewers – In-Use Track

Bakhshandegan Moghaddam, Farshad	Nikolov, Nikolay
Chen, Jiaoyan	Smirnova, Alisa
Fawei, Biralatei	Tietz, Tabea
García-Silva, Andrés	Türker, Rima
Li, Chenxi	Zernichow, Bjørn Marius Von
Lutov, Artem	Zhao, Yuting
Mireles, Victor	

Sponsors

Platinum Sponsors

ELSEVIER

https://www.elsevier.com/

IBM **Research**

http://www.ibm.com/

https://franz.com/

Gold Sponsors

https://data.world/

metaphacts

http://www.metaphacts.
com/

ontotext

https://ontotext.com/

ORACLE

http://www.oracle.com/

 THOMSON REUTERS®

https://www.thomsonreuters.
com/

http://videolectures.net/

*i*Novex

https://inovexcorp.com/

Bronze Sponsors

https://www.google.com/

Contributors

https://www.springer.com/   https://capsenta.com/   https://www.iospress.nl/

Student Travel Award Sponsors

https://www.nsf.gov/   http://swsa.semanticweb.org/

Doctoral Consortium

https://www.journals.elsevier.
com/artificial-intelligence

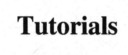

Tutorials

ISWC 2018 Workshop and Tutorial Chairs' Welcome

Besides the main technical program, ISWC 2018 hosts a selection of workshops and tutorials on a range of emerging and established topics. The key areas addressed by the workshop and tutorial programme include core Semantic Web technologies such as knowledge graphs and scalable knowledge base systems, ontology design and modelling, semantic deep learning and statistics, and well as novel applications of semantic technologies to audio and music, IoT, robotics, healthcare, social media and social good topics. Furthermore, several events address the topics on the interface of Semantic Web technologies and humans, including visualization and interaction paradigms for Web Data as well as crowdsourcing applications. The workshops and tutorials provide a setting for focused, intensive scientific exchange among researchers and practitioners in a variety of formats.

The decision on acceptance of workshops and tutorial proposals was made on the basis of their overall quality and their appeal to a reasonable fraction of the Semantic Web community while also targeting diversity of the programme. Overall, we received 31 workshop and tutorial proposals, of which 8 were accepted as full-day events and 17 as half-day events. The full workshop and tutorials programme is available at: http://iswc2018.semanticweb.org/workshops-tutorials.

We would like to take this opportunity to thank the workshop and tutorial organizers for their invaluable and inspiring contributions to the ISWC 2018 programme. We look forward to seeing you in Monterey!

March 2018

Elena Demidova
Amrapali Zaveri
Workshop & Tutorial Chairs

Methods and Tools for Modular Ontology Modeling

Karl Hammar[1], Pascal Hitzler[2], Cogan Shimizu[2],
and Md Kamruzzaman Sarker[2]

[1] Department of Computer Science and Informatics,
Jönköping University, Sweden
karl.hammar@ju.se
[2] Data Semantics Lab, Wright State University, USA
{pascal.hitzler,shimizu.5,sarker.3}@wright.edu

Ontology design patterns and other methods for modular ontology engineering have recently experienced a revival, and several new promising tools and techniques have been presented. The use of methods for modular ontology development and these newly developed tools and technologies promise simpler ontology development and management, in turn furthering increased adoption of ontologies and ontology-based tech, both within and outside of the semantic web academic environment. This workshop intends to spread the word about these method and tooling improvements beyond "the usual crowd"of pattern developers and researchers, for the benefit of the Semantic Web research community as a whole.

This full-day tutorial targets ontology designers, data publishers, and software developers interested in employing semantic technologies and ontologies. We present the state-of-the-art in terms of methods and tools, exemplifying their usage in several real-world cases. We then tutor the attendees on the use of three sets of related tooling for modular ontology development, allowing them to try out leading-edge software that they might otherwise have missed, under the supervision of the tools' main developers. We expect that at the end of the day, the attendees will have developed the ability to independently and with confidence develop ontologies in a modular fashion, using the tools and techniques showcased in this tutorial.

Validating RDF Data Tutorial

Jose Emilio Labra Gayo[1] and Iovka Boneva[2]

[1] University of Oviedo, Spain
labra@uniovi.es
[2] Univ. Lille - CRIStAL, F-59000 Lille, France
iovka.boneva@univ-lille1.fr

RDF promises a distributed database of repurposable, machine-readable data. Although the benefits of RDF for data representation and integration are indisputable, it has not been embraced by everyday programmers and software architects who care about safely creating and accessing well-structured data. Semantic web projects still lack some common tools and methodologies that are available in more conventional settings to describe and validate data. In particular, relational databases and XML have popular technologies for defining data schemas and validating data which had no analog in RDF.

Two technologies have been proposed for RDF validation: Shape Expressions (ShEx) and Shapes Constraint Language (SHACL).

ShEx was designed as an intuitive and human-friendly high level language for RDF validation in 2014 [4]. ShEx 2.0 has recently been proposed by the W3C ShEx community group [3].

SHACL was proposed by the Data Shapes Working Group and accepted as a W3C Recommendation in July 2017 [1].

In this tutorial we will present both ShEx and SHACL using examples, presenting the rationales for their designs, a comparison of the two, and some example applications. The contents of the tutorial will be complemented by the *Validating RDF Data* book [2] written by the presenters.

References

1. Knublauch, H., Kontokostas, D.: Shapes Constraint Language (SHACL). W3C Proposed Recommendation, June 2017
2. Labra Gayo, J.E., Prud'hommeaux, E., Boneva, I., Kontokostas, D.: Validating RDF Data. Morgan & Claypool (2017)
3. Prud'hommeaux, E., Boneva, I., Labra Gayo, J.E., Kellog, G.: Shape Expressions Language 2.0. W3C Community Group Report, Apr 2017
4. Prud'hommeaux, E., Labra, J.E., Solbrig, H.: Shape expressions: an RDF validation and transformation language. In: 10th International Conference on Semantic Systems, Sept 2014

Hybrid Techniques for Knowledge-Based NLP - Knowledge Graphs Meet Machine Learning and All Their Friends

Jose Manuel Gomez-Perez and Ronald Denaux

Expert System, Madrid, Spain
{jmgomez,rdenaux}@expertsystem.com

Many different artificial intelligence techniques can be used to explore and exploit large document corpora that are available inside organizations and on the Web. While natural language is symbolic in nature and the first approaches in the field were based on symbolic and rule-based methods, like ontologies, semantic networks and knowledge bases, many of the most widely used methods are currently based on statistical approaches. Each of these two main schools of thought in natural language processing, knowledge-based and statistical, have their limitations and strengths and there is an increasing trend that seeks to combine them in complementary ways to get the best of both worlds. This tutorial covers the foundations and modern practical applications of knowledge-based and statistical methods and techniques as well as their combination for the exploitation of large document corpora. Following a practical and hands-on approach, the tutorial tries to address a number of fundamental questions to achieve this goal, including: (i) how can machine learning extend previously captured knowledge explicitly represented as knowledge graphs in cost-efficient and practical ways, (ii) what are the main building blocks and techniques enabling such hybrid approach to natural language processing, (iii) how can structured and statistical knowledge representations be seamlessly integrated, (iv) how can the quality of the resulting hybrid representations be inspected and evaluated, and (v) how can this improve the overall quality and coverage of our knowledge graphs. The tutorial will first focus on the foundations that can be used to this purpose, including knowledge graphs and word embeddings, and will then show how these techniques can be effectively combined in NLP tasks (and other data modalities in addition text) related to research and commercial projects where the instructors currently participate.

Building Enterprise-Ready Knowledge Graph Applications in the Cloud

Peter Haase[1] and Michael Schmidt[2]

[1] metaphacts GmbH, 69190 Walldorf, Germany
ph@metaphacts.com
[2] Amazon Web Services, Seattle, WA, USA
schmdtm@amazon.com

Knowledge Graphs are a powerful tool that changes the way we do data integration, search, analytics, and context-sensitive recommendations. Consisting of large networks of entities and their semantic relationships, they have been successfully utilized by the large tech companies, with prominent examples like the Google Knowledge Graph and Wikidata, which makes community-created knowledge freely accessible. Cloud computing has fundamentally changed the way that organizations build and consume IT resources, enabling services to be provisioned on-demand in a pay-as-you-go model. Building Knowledge Graphs in the cloud makes it easy to leverage their powerful capabilities quickly and cost effectively.

In this tutorial, we cover the fundamentals of building Knowledge Graphs in the cloud. In comprehensive hands-on exercises we will cover the end-to-end process of building and utilizing an open Knowledge Graph based on high-quality Linked Open Data sets, covering all aspects of the Knowledge Graph life cycle including enterprise-ready data management, integration and interlinking of sources, authoring, exploration, querying, and search. The hands-on examples will be performed using prepared individual student accounts set up in the AWS cloud, backed by an RDF/SPARQL graph database service with an enterprise Knowledge Graph application platform deployed on top.

Crowdsourcing with CrowdTruth
Harnessing Disagreement in Human Interpretation for Ambiguity-Aware Machine Intelligence

Lora Aroyo[1], Anca Dumitrache[1], Oana Inel[1], and Chris Welty[2]

[1] Vrije Universiteit Amsterdam, Netherlands
lora.aroyo@gmail.com, anca.dumitrache@gmail.com,
oana.inel@gmail.com
[2] Google Research, New York, USA
cawelty@gmail.com
http://crowdtruth.org

In this tutorial, we introduce the *CrowdTruth methodology* for crowdsourcing ground truth by harnessing and interpreting inter-annotator disagreement. CrowdTruth is a widely used crowdsourcing methodology[1] adopted by industrial partners and public organizations, e.g. Google, IBM, New York Times, The Cleveland Clinic, Crowdynews, The Netherlands Institute for Sound and Vision, Rijksmuseum, and in a multitude of domains, e.g. AI, news, medicine, social media, cultural heritage, social sciences. The central characteristic of CrowdTruth is *harnessing the diversity in human interpretation* to capture the wide range of opinions and perspectives, and thus, provide more reliable and realistic real-world annotated data for training and evaluating machine learning components. Unlike other methods, we do not discard dissenting votes, but incorporate them into a *richer and more continuous representation of truth*. The goal of this tutorial is to introduce the Semantic Web audience to a *novel approach to crowdsourcing* that takes advantage of the diversity of opinions (human semantics) inherent to the Web. We believe it is quite timely, as methods that deal with disagreement and diversity in crowdsourcing have become increasingly popular. Creating this more *complex notion of truth* contributes directly to the larger discussion on *how to make the Web more reliable, diverse and inclusive*.

---

[1] http://crowdtruth.org.

Challenges and Opportunities with Big Linked Data Visualization

Laura Po

"Enzo Ferrari" Engineering Department, University of Modena
and Reggio Emilia, Italy
laura.po@unimore.it
http://www.dbgroup.unimo.it/laurapo

The Linked Data Principles defined by Tim-Berners Lee promise that a large portion of Web Data will be usable as one Big interlinked RDF database. Today, we are assisting at a staggering growth in the production and consumption of Linked Open Data (LOD). In this scenario, it is crucial to provide intuitive tools for researchers, domain experts, but also businessmen and citizens to view and interact with increasingly large datasets. Visual analytics integrates the analytic capabilities of the computer and the abilities of the human analyst, allowing novel discoveries and empowering individuals to take control of the analytical process.

This tutorial aims to identify the challenges and opportunities in the representation of Big Linked Data by reviewing some current approaches for exploring and visualizing LOD sources. First, we introduce the problem of finding relevant sources in catalogues of thousands of datasets, we present the issues related to the understanding and exploration of unknown sources. We list the difficulties to visualize large datasets in static or dynamic form. We focus on the practical use of LOD/ RDF browsers and visualization toolkits and examine the support at big scale. In particular, we experience the exploration of some LOD datasets by performing searches of growing complexity. At last, we sketch the main open research challenges with Big Linked Data visualization. By the end of the tutorial, the audience will be able to get started with their own experiments on the LOD Cloud, to select the most appropriate tool for a defined type of analysis and they will be aware of the open issues that remain unsolved in the scenario of the exploration of Big Linked Data.

Contents–Part I

Contents–Part II

In-Use Track

Research Track

Fine-Grained Evaluation of Rule- and Embedding-Based Systems for Knowledge Graph Completion

Christian Meilicke[(⊠)], Manuel Fink, Yanjie Wang,
Daniel Ruffinelli, Rainer Gemulla, and Heiner Stuckenschmidt

Research Group Data and Web Science, University of Mannheim,
Mannheim, Germany
`christian@informatik.uni-mannheim.de`

Abstract. Over the recent years, embedding methods have attracted increasing focus as a means for knowledge graph completion. Similarly, rule-based systems have been studied for this task in the past. What is missing so far is a common evaluation that includes more than one type of method. We close this gap by comparing representatives of both types of systems in a frequently used evaluation protocol. Leveraging the explanatory qualities of rule-based systems, we present a fine-grained evaluation that gives insight into characteristics of the most popular datasets and points out the different strengths and shortcomings of the examined approaches. Our results show that models such as TransE, RESCAL or HolE have problems in solving certain types of completion tasks that can be solved by a rule-based approach with high precision. At the same time, there are other completion tasks that are difficult for rule-based systems. Motivated by these insights, we combine both families of approaches via ensemble learning. The results support our assumption that the two methods complement each other in a beneficial way.

1 Introduction

Knowledge graph completion or link prediction refers to the task of predicting missing information in a knowledge graph. A knowledge graph is a graph where a node represents an entity and an edge is annotated with a label that denotes a relation. A directed edge from s to o labelled with r corresponds to a triple $\langle s, r, o \rangle$. Such a triple can be understood as the fact that subject s is in relation r to object o. As a logical formula we write $r(s, o)$. Often knowledge graphs are created automatically from incomplete data sources that do not fully capture the real relations between the entities. The goal of knowledge graph completion is to use the existing knowledge to find these correct missing links without adding any wrong information. The current evaluation practice estimates model performance by the model's ability to complete incomplete triples like $\langle s, r, ? \rangle$ or $\langle ?, r, o \rangle$ derived from a known fact $\langle s, r, o \rangle$. The task in this case consists of generating a candidate ranking for the empty position that minimizes the amount of wrong suggestions ranked above the correct ones.

© Springer Nature Switzerland AG 2018
D. Vrandečić et al. (Eds.): ISWC 2018, LNCS 11136, pp. 3–20, 2018.
https://doi.org/10.1007/978-3-030-00671-6_1

Recently, a new family of models for knowledge graph completion has received increasing attention. These models are based on embedding the knowledge graph into a low dimensional space. A prominent example is TransE [2], where both nodes (entities) and edge labels (relations) are mapped to vectors in $\mathbb{R}^n$. Other examples include RESCAL [9], TransH [16], TransG [17], DistMult [18], HolE [8] or ProjE [13]. Once the embeddings have been computed, they can be leveraged to generate a candidate ranking for the missing entity of a completion task. Over the last years many different models have been proposed that follow this principle.

In contrast, rule-based approaches learn logical formulas that are the explicit representation of statistical regularities and dependencies encoded in the knowledge graph. To predict candidates for incomplete triples, the learned rules are applied to rank candidates based on the confidence of the rules that fired. Works that focus on embeddings usually do not compare the proposed models with rule-based methods and vice versa. In this paper, we do not present a substantially novel method for knowledge graph completion. Instead, we apply AMIE [4], an existing system for learning rules, as well as our own approach called RuleN to this problem. The development of RuleN is mainly inspired by the idea of using a very simple mechanism that can be completely described in the paper. In our experiments, we have found that on the datasets commonly used for the evaluation of embedding based models, both systems are highly competitive. Among the many different embedding-based models for which results have been reported over the recent years (see [6,12]), only few exceptions performed better.

In a rule-based approach each generated candidate comes with an explanation in terms of the rule that generated this candidate. With the help of these explanations, we analyze the datasets commonly used for the evaluation of embeddings by partitioning their test set. Each subset is associated with the type of the rule which generated the correct test triple with high confidence, e.g., a *symmetry* or *subsumption* rule. This analysis sheds light on the characteristics and difficulty of these datasets. Based on this partitioning, we compare the performance of various rule- and embedding-based approaches (RESCAL [9], TransE [2] and HoleE [8]) on a fine-grained level. Our results show that a large fraction of the test cases is covered by simple rules that have a high confidence. These test cases can be solved easily by a rule-based approach, while the embedding models generate clearly inferior results.

There is also a fraction of test cases that is hard for rule-based approaches. We use the method from [15] to learn an ensemble including both types of approaches. Our results show that the ensemble can achieve better results than the top-performing approach on each dataset used in our experiments. This confirms our findings that both families of approaches are strong on different types of completion tasks, which can be leveraged by the ensemble.

2   Related Work

Within this section, we first discuss methods for learning rules. We continue with approaches that use observed features, which correspond to certain types of rules, to learn a model. Note that there is no clear distinction between the first and the second group of approaches. Finally, we explain latent feature models that are based on the idea of using embeddings and we give some details on the three models we used in our experiments.

Regarding rule-based methods for relational learning, Quickfoil [19] is a highly scalable ILP algorithm that mines first order rules for given target relations. Quickfoil is in principle designed to learn rules that strictly hold. While it also tolerates a small amount of noise, i.e., it can also learn rules even though there are some negative examples in the given knowledge base, it cannot learn rules with a low confidence. However, these rules are also important for ranking the candidates of a knowledge completion task. In many cases, we may not have a strict rule, but only weak evidence.

AMIE [4] is an approach for learning rules that is similar to our approach introduced in the next section as RuleN. It has a different language bias, as explained in more detail in Sect. 3.1. The main difference is that AMIE computes the confidence based on the whole knowledge graph, while our approach will compute an approximation that is based on selecting a random sample. It can be expected that AMIE is complete and that the confidences of AMIE are precise. This is not the case for RuleN. However, due to the underlying sampling mechanism RuleN might be able to mine longer path rules. We use AMIE in our experiments as an alternative approach for learning rules.

The path ranking algorithm [7] (PRA) is based on the idea of using random walks to find characteristic paths that frequently occur next to links of a target relation. These paths are used as features in a matrix where each row corresponds to a pair of entities. By including negative examples generated under the Closed World Assumption, a logistic regression is performed on the matrix to train a classifier. The classifier for a relation can then be used to predict the likelihood of the target relation between two given entities based on the surrounding path features. The rule bodies in RuleN correspond to the paths in PRA. While PRA puts a lot of emphasis on learning how to combine the path features with machine learning, RuleN is simpler in this regard. It uses the path features in a more conservative way for which it approximates the significance of individual paths more thoroughly. A more expressive extension of PRA is presented in [5], where the authors extract further sub-graph features besides paths.

In [10], Niepert proposes Gaifman Models. Gaifman Models are a way of sampling small subgraphs from a knowledge graph in order to learn a model that uses first order rules as features. One of the main differences is that the set of features, which needs to be defined prior to learning the model, comprises all possible rules of a certain type. Contrary to this, RuleN stores only those rules for which we found at least one positive example during sampling. In the experiments presented in [10] the authors use all path features of length 1 and path features of length 2 that use only one relation in the rule body (e.g., rules that express transitivity of a relation), which corresponds to a subset of the rules that AMIE or RuleN can learn.

Another approach that uses observed features has been proposed in [14]. As feature set the authors use path features of length 1 and features that reflect how probable it is for a certain entity to appear in subject/object position of a certain relation. The latter correspond to the constant rules of RuleN. The authors show that such a model can score surprisingly well on the commonly used datasets, which motivates them to propose the FB15k-237 dataset that we will consider in our experiments. The results are compared against several approaches that are based on embeddings. This analysis (observed vs. latent features) is similar to our evaluation effort. However, we use AMIE and RuleN to learn rules that are more expressive than the feature sets used in [14] and [10] without the need for negative examples. Furthermore, we perform a more fine-grained evaluation based on the distinction between different types of completion tasks.

It has already been argued that a simple rule-based approach restricted to learning inverse relations can achieve state-of-the-art results on WN18 and FB15k [3]. Our evaluation extends these findings by partitioning the "easy" test triples into detailed categories, which allow fine-grained insight into the performance of different systems. Also, the Inverse Model in [3] is too simple to represent the state-of-the-art performance of rule-based systems on FB15-237.

In contrast to methods which exploit observed features or rules, latent feature models learn representations of the entities and relations from the knowledge base in a low-dimensional space, such that the structure of the knowledge base is represented in this latent space. These learned representations are known as the *embeddings* of the entities and relations, respectively. The models provide a score function $f(s, r, o)$ which for a given triple $\langle s, r, o \rangle$ reflects the model's confidence in the truthfulness of the triple. Based on this, potential candidates for a given query $\langle s, r, ? \rangle$ can be ranked.

Our comparisons in this work focus on bilinear models, which have been successful in the standard benchmarks for this task. RESCAL [9] is a factorization-based bilinear model. It represents entities as vectors $\mathbf{a}_i \in \mathbb{R}^n$, relations as matrices $\mathbf{R}_k \in \mathbb{R}^{n \times n}$ and has a score function $f(s, r, o) = \mathbf{a}_s^T \mathbf{R}_r \mathbf{a}_o$. HolE [8] represents entities as vectors $\mathbf{a}_i \in \mathbb{R}^n$, relations as vectors $\mathbf{r}_k \in \mathbb{R}^n$ and has a score function $f(s, r, o) = \mathbf{r}_r^T (\mathbf{a}_s \star \mathbf{a}_o)$, where $\star$ refers to the circular correlation between $\mathbf{a}_s$ and $\mathbf{a}_o$. TransE is a translation-based model, which represents entities as vectors $\mathbf{a}_i \in \mathbb{R}^n$, relations as vectors $\mathbf{r}_k \in \mathbb{R}^n$ and has a score function $f(s, r, o) = \|\mathbf{a}_s + \mathbf{r}_r - \mathbf{a}_o\|_2^2$.

3   A Simple Rule-Based Approach

In this work, we are interested in understanding which types of rules help in knowledge base completion and can be applied successfully to the datasets currently used for evaluating state of the art methods. For this goal, we developed our own rule-based system RuleN that is simple enough to be described in detail within this work. It is based on learning the types of rules defined in Sect. 3.1 with a sampling strategy described in Sect. 3.2. In Sect. 3.3 we explain how to apply the learned rules to rank the candidates for a given completion task.

3.1   Types of Rules

Let r and s refer to relations, x and y to variables that quantify over entities, and let a be a constant that refers to an entity. RuleN supports the following types of rules:

$$r(x_1, x_{n+1}) \leftarrow s_1(x_1, x_2) \wedge \ldots \wedge s_n(x_n, x_{n+1}) \qquad (P_n)$$
$$r(x, a) \leftarrow \exists y \; r(x, y) \qquad (C)$$

We call rules of type P_n with $n \geq 1$ path rules. Given two entities x_1 and x_{n+1} that are connected by an r-edge, a path rule describes an alternative path that leads from x_1 to x_{n+1}. Note that a path in this sense may also contain edges implicitly given by the inverse relations, e.g. $s_3^{-1}(x_3, x_4)$ corresponds to $s_3(x_4, x_3)$. Type C rules are rules with a constant in the head of the rule. The language bias introduced by these rule types is similar to that of existing systems such as PRA [7] and AMIE [4] but there are differences. For example, AMIE does not limit constants to the head of a rule and is in general slightly more expressive. However, it does not learn rules of type C. Concrete examples for some of these rule types are shown in the following. These rules have been generated in the experiments that we report about later.

$$hyponym(x, y) \leftarrow hypernym(y, x) \qquad [0.94] \qquad (1)$$
$$celebrityBreakup(x, y) \leftarrow celebrityMarriage(x, y) \qquad [0.08] \qquad (2)$$
$$producedBy(x, z) \leftarrow sequel(x, y) \wedge producedBy(y, z) \qquad [0.55] \qquad (3)$$
$$language(x, English) \leftarrow \exists y \; language(x, y) \qquad [0.64] \qquad (4)$$

Rule 1 and 2 are examples of type P_1. The latter depicts the fact that 8% of celebrity marriages in that dataset ended in divorce. Rule 3 is an example for type P_2. Rule 4 is an example for rule type C that captures that in 64% of the cases, the spoken language of a person is English.

3.2   Learning Rules

For a given rule R, let $h(R) = r(x, y)$ denote its head and $b(R)$ denote its body. As defined in [4], the head coverage is the number of $h(R) \wedge b(R)$ groundings that can be found in the given knowledge graph, divided by the number of $h(R)$ groundings. A head coverage close to 100% suggests that the rule can be used to propose candidates for most completion tasks of relation r. The confidence of a rule is defined as the number of $h(R) \wedge b(R)$ groundings divided by the number of $b(R)$ groundings. Confidence tells us how likely it is that a candidate proposal generated by this rule is correct.

To learn rules for a target relation r, RuleN utilizes a twofold sampling approach instead of a complete search. We first explain the learning of path rules of maximum length n. Given a target relation r, we need to find rule bodies $b(R)$ for $r(x_1, x_{n+1}) \leftarrow b(R)$ that result in helpful rules. The straightforward approach is to look at all triples $\langle a, r, b \rangle$ in the training set and determine all possible paths up to length n between a and b each time using an iterative deepening depth-first search. Using these paths as body for the rule, the confidence can be calculated in a second step. To speed up this rule finding step, it is only performed for k (=sample size) triples $\langle a, r, b \rangle$. Each rule that is constructed this way has a head coverage >0. Moreover, the higher the head coverage of a rule, the more likely it is to be found. For example, a rule with a head coverage of 0.01 will be found for $k = 100$ with a probability $\approx 63.4\%$. This illustrates that the procedure can miss rules with a low head coverage.

We apply a similar approach for C rules. Given a target relation r, we randomly pick k facts $\langle a, r, b \rangle$. For each of these facts, we create the rules $r(x, b) \leftarrow r(x, y)$ and $r(a, y) \leftarrow r(x, y)$. An example is Rule 4.

In a second step, we compute the confidence of path rules by randomly sampling true body groundings. We then approximate the factual confidence by dividing the number of groundings for which the head is also true by the total number of groundings sampled for the body. With respect to a C rule, we simply pick a sample of r facts and count how often we find a or b in subject and object position.

3.3   Applying Rules

Given a completion task $\langle a, r, ? \rangle$, we select all rules with r in their head. Suppose that we have learned four relevant rules as shown in Table 1. For each of the three path rules, we look up all body groundings in the given KB where we replace x by a, collecting all possible values of the variable y. For the constant rule, the body is implicitly true when using the rule to make a prediction for the object position, so it is not checked. What this simply means is that the rule always predicts the constant c when asked for the object position of r, independent of the subject.

Table 1. Four relevant rules for the completion task $\langle a, r, ? \rangle$ resulting in the ranking $\langle g(0.81), d(0.81), e(0.23), f(0.23), c(0.15) \rangle$.

Rule	Type	Confidence	Result
$r(x, y) \leftarrow s(y, x)$	P_1	0.81	$\{d, g\}$
$r(x, y) \leftarrow r(y, x)$	P_1	0.70	$\emptyset$
$r(x, y) \leftarrow t(x, z) \wedge u(z, y)$	P_2	0.23	$\{e, f, g\}$
$r(x, c) \leftarrow \exists y\; r(x, y)$	C	0.15	$\{c\}$

A rule can generate one candidate (fourth row), several candidates (first and third row), or no candidate (second row). There are different ways to aggregate the results generated by the rules. As a basis, we choose the most robust approach. We define the final score of an entity as the maximum of the confidence scores of all rules that generated this entity. If a candidate has been generated by more than one rule, we use the amount of these rules as a secondary sorting attribute among candidates with the same (maximum) score. Hence g is ranked before d in the given example. Combining confidences of multiple rules for the same candidate in a more sophisticated way is difficult due to unknown probabilistic dependencies between rules. For example, we found that an aggregation based on multiplication distorts the results (e.g., when two rules of which one subsumes the other fire simultaneously), leading to worse predictions.

4   Experimental Results

Within our experiments we focussed mainly on the three datasets that have been extensively used to evaluate embedding-based models for knowledge graph completion: the WordNet dataset WN18 described in [1], the FB15k dataset, which is a subset of FreeBase, described in [2] and FB15k-237, which has been designed in [14] as a harder and more realistic variant of the FB15k dataset. FB15k-237 is also known as FB15KSelected. We published additional evaluation results for WN18RR, which is a harder variant of WN18 without inverse relations proposed in [3] online at http://web.informatik.uni-mannheim.de/RuleN/. The web page contains also the RuleN code and other relevant material.

First, we computed results for the two rule-based systems AMIE and RuleN. Our results imply that rule-based systems are competitive and that it is easy to determine settings for them which yield good results. Next, we divided the datasets into partitions to perform a fine-grained evaluation including TransE, RESCAL and HolE, as well as AMIE and RuleN. Finally, we evaluated an ensemble of these five systems showing that this is a way to leverage the strengths of both approaches.

We followed the evaluation protocol proposed in [2]. Each dataset consists of a training, validation and test set which are used for training, hyperparameter tuning and evaluation respectively. Each triple $\langle s, r, o \rangle$ from the test set results in two completion tasks $\langle ?, r, o \rangle$ and $\langle s, r, ? \rangle$ that are used to query the systems for a ranked list of entities for the placeholder. hits@k is the fraction of completion tasks for which the removed entity was ranked at least at rank k. We only looked at filtered hits@k, which means that for each completion task, entities other than the removed one which also result in true triples contained in the dataset, are ignored in the ranked list. The filtered mean reciprocal rank MRR is calculated by summing over all completion tasks the reciprocals of the ranks of the removed candidate after filtering.

4.1   Performance of Rule-Based Approaches

Embedding-based models have hyperparameters which need to be optimized on a validation dataset. Rule-based systems also have hyperparameters. However, in our experiments, we found them easy to set for knowledge base completion even without a validation dataset. As these hyperparameters are typically a mechanism to tune running time versus completeness of the rule learning process, we simply used the most expressive setting that still finishes within a reasonable time.

The hyperparameters of RuleN are the sample size and the length of the path rules. For AMIE, we focused on thresholds for support, head coverage, and rule length. Furthermore, for both systems, it is possible to disable the mining of rules with constants. In our experiments, we have found that there is indeed a positive correlation between setting the hyperparameters as liberally as possible and the prediction performance. (The only exception to this paradigm resulted in a performance drop of less than 1%.) In Table 2, we show the filtered hits@10 results for increasingly liberal settings for runtimes <10 h on WN18 and FB15k.

RuleN has one sampling parameter that affects the number of mined rules and one that determines the precision of the confidence calculation. We tied both to the same value, which we varied between 50 and 1000. It is interesting to see that there seems to be a limit for the sample size of RuleN above which the performance remained stable and that it was possible to achieve very good results already with a low sample size and consequently a low run time. Note that this enables RuleN to be applicable to very large (in number of entities) knowledge graphs as long as the number of relations is bounded.

Table 2. Impact of different settings on performance of rule-based systems. For RuleN, the number in the *Setting* column denotes the sample size. For AMIE, it shows the values for support (s) and head coverage (hc) used for the mining of the path and constant rules respectively. The length of rules with constants was set individually for AMIE as denoted by the *Rule Type* column.

	Rule type	Setting	FB15k				WN18			
			hits@10	Learn	Apply	#Rules	hits@10	Learn	Apply	#Rules
RuleN	P_{12}	50	.853	1167s	137s	69k	.943	5s	5s	230
	P_{12}	100	.859	2491s	165s	96k	.943	8s	5s	314
	P_{12}	500	.862	6120s	170s	158k	.945	22s	5s	693
	P_{12}	1000	.862	6492s	207s	177k	.945	34s	6s	945
	C	1000	.312	1s	25s	94k	.05	1s	10s	12k
	P_{12}, C	1000	**.875**	6493s	191s	270k	.948	6s	12s	13k
	$P_{12[3]}, C$	1000[100]	.870	49868s	10272s	917k	**.958**	398s	20s	41k
	$P_{123[45]}, C$	1000[100]					**.958**	4103s	151s	54k
AMIE	P_{12}, C_1	s = 0, hc = 0.0/0.01	.858	4889s	1952s	861k	.942	17s	4s	352
	P_{123}, C_2	s = 0, hc = 0.0/0.01					.948	868s	29s	4806

An overview on the results that current state of the art approaches achieve on these two datasets can be found in [12] and [6]. In summary, for WN18 there are only few approaches that achieved a hits@10 score higher than 95%, e.g. 96.4% (Inverse Model [3]), 96.4% (R-GCN+ [11]), 95.5% (ConvE [3]) and 95.3% (IRN [12]). The follow-up approaches scored around 92–95%, while there are still many other approaches that achieved less. For the FB15k dataset there is a higher variance in the results. The best approaches achieved a hits@10 score of 92.7% (IRN [12]) and 88.2% (TransG [17]). However the vast majority could not top a score of 84%. Thus, RuleN and AMIE outperformed the majority of models for which results have been reported on WN18 and FB15k. On FB15k there are only few systems that achieved better results and none of them perform better on WN18. These results show that symbolic representations can compete with and perform sometimes better than many of the approaches that are based on embeddings. This insight is not only supported by the good results of RuleN, but especially by the competitive results of AMIE, which we could use almost out of the box to generate the presented results.

All experiments were performed on a machine with 4 cores at 2394 MHz and 8 GB memory. Even in the most complex setting reported in Table 2, we were able to run the rule-based systems in a few hours on FB15k. Runtimes on FB15k-237 were slightly shorter than those on FB15k as it is a subset of it. For the WN18 dataset, there are competitive settings where we finished in less than a minute, including learning and prediction. It would take much longer on this hardware setting to train the embedding-based models to competitive performance. In our experiments, we found that rule-based systems were orders of magnitude faster to train due to the required hyperparameter search of embedding-based models. The training and prediction runtimes for a given hyperparameter setting were comparable to rule-based systems though.

4.2 Dataset Partitioning

In the following we examine each of the datasets in detail. In particular, we analyze which types of rules are relevant to correctly predict the missing information in the test sets of these datasets. For that purpose, we restricted RuleN to learn P_1 and P_2 rules only. We further distinguish between special sub-types of these rules as follows:

- We refer to a rule of form $r(x, y) \leftarrow r(y, x)$ as a **symmetry rule**. An example is $married(x, y) \leftarrow married(y, x)$.
- We refer to a rule of form $r(x, y) \leftarrow s(x, y)$ with $r \neq s$ as an **equivalence rule** if the reverse direction $s(x, y) \leftarrow r(x, y)$ holds also.[1]
- We distinguish in the case of equivalence between **inverse equivalence**, i.e. $r(x, y) \leftarrow s(y, x)$, and plain **equivalence**. An example for an inverse equivalence rule is $hypernym(x, y) \leftrightarrow hyponym(y, x)$.
- We call any P_1 rule that is not a symmetry or (inverse) equivalence rule a **subsumption rule**, e.g., $cityIn(x, y) \leftarrow capitalOf(x, y)$.

We used RuleN with a sample size of 1000 to learn P_1 and P_2 rules for both WN18 and FB15k. Then we removed all rules with a confidence lower than 0.5. We applied this very restrictive set of rules to the completion tasks defined by the test sets. For each completion task we applied all relevant rules in descending order with respect to their confidence. If one of the candidates generated by the

---

[1] We annotate a rule as equivalence rule if it holds in both directions with a *similar confidence*. We said that two confidence values are similar if they do not differ more than 0.05. This is a pragmatic decision, which allows us to define a meaningful category.

rule was the entity replaced with a question mark, we marked the completion task as solved by the type of that specific rule. Note that we did not continue to check the remaining rules. Thus, we annotated each completion task with the type of the most confident rule that could solve the task. This annotation follows the naming convention defined above. If we could not find such a rule, we annotated the task as *Uncovered*. It is not the case that a completion task annotated as *Uncovered* cannot be correctly predicted by a rule-based approach. There is still the possibility that it can be correctly solved by a rule with low confidence or by a rule which is not of type P_1 or P_2.

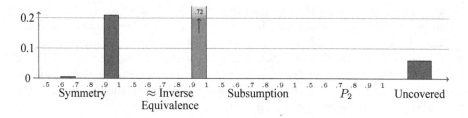

Fig. 1. Rule coverage for the WN18 dataset. We truncated the y-axis; the majority of the test cases are covered by inverse equivalence (72%).

In Fig. 1, we have depicted the results of applying this approach to the WN18 dataset. The dataset has very specific characteristics. Only $\approx 6.12\%$ of the completion tasks fall into the *Uncovered* category. Moreover, the majority of the tasks is covered by equivalence rules (72.5%). Note that we have grouped the rules of each type with respect to their confidence in the ranges from $(0.5, 0.6]$ to $(0.9, 1.0]$. Here, all of the inverse equivalence rules have a confidence higher than 0.9. An example of an equivalence rule that dominates the dataset is Rule 1 (together with its reversed counterpart) that we already presented above. The remaining tasks are covered by symmetry rules. An example for such a rule is $see\_also(x, y) \leftarrow see\_also(y, x)$. Again, most of them are highly confident. It is also interesting to see that subsumption and P_2 rules do not help to detect anything that is not already covered by equivalence or symmetry rules with higher confidence. For that reason, any method that is capable of exploiting equivalence and symmetry should be able to find the correct candidate for $\approx 94\%$ of the test cases.

The results of applying our approach to the FB15k dataset are shown in Fig. 2. For this dataset we observe a heterogeneous set of rules that covers a smaller fraction (still 81.6%) of the tasks in the test set. The dataset is still dominated by equivalence (dark blue) and especially inverse equivalence (light blue) rules. These rules cover around 60% of all completion tasks. However, we find now also subsumption rules (6.8%), that are not equivalence or symmetry rules, and P_2 rules (7.3%). Moreover, the fraction of uncovered tasks (18.4%) is larger compared to WN18, but still rather small.

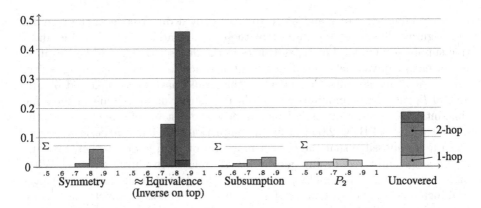

Fig. 2. Rule coverage for the FB15k dataset. Σ shows the total fraction of a specific rule type. (Color figure online)

This time, we also analyzed the uncovered tasks in more detail and further divided it into three subgroups. If such a completion task is based on reconstructing a triple $\langle a, r, b \rangle$, we determined the shortest path between a and b in the training set. In Fig. 2 we distinguish between 1-hop, 2-hop and other test cases (where the shortest path between a and b has a length ≥ 3). Note that for 1-hop and 2-hop test cases there is still a chance that rules of length 1 or 2 can be used to find the correct candidates. However, since these test cases are not in one of the other categories, we know that those rules would have a confidence lower than 50%.

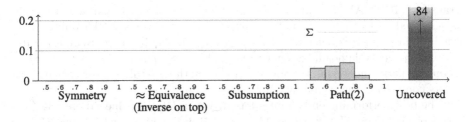

Fig. 3. Rule coverage for the FB15k-237 dataset. 31% of the Uncovered category are ≥ 2-hop testcases, 69% are 2-hop testcases, and none of them are 1-hop testcases.

The high fraction of test cases covered by simple rules could give the impression that WN18 and FB15k are too easy. FB15k-237 has been designed in [14] as a harder variant of the FB15k dataset by making the following two modifications. First, all (inverse) equivalent relations have been removed from the dataset resulting in a knowledge graph with 237 remaining relations. Second, the validation and test sets were changed, such that any triple $\langle x, r, y \rangle$ is removed from it, if there is some other triple $\langle x, s, y \rangle$ or $\langle y, s, x \rangle$ with $s \neq r$ or $s = r$

in the training set, i.e. x and y are connected by a direct edge in the training set. Figure 3 illustrates the impact of these modifications. The first modification suppresses any kind of dependencies in the dataset that would be captured by (inverse) equivalence rules. The second modification is even more aggressive, because it suppresses any dependencies that could have been exploited by any kind of P_1 rule. These modifications result in a harder dataset, while at the same time introducing an unrealistic bias. Suppose the test set of a dataset with the modifications of FB15k-237 contains a completion task like $murdered(?, john)$. Then it is impossible that the correct murderer of $john$ is his brother, his wife, his boss, his employee, or any person directly related to him in any way. What makes this circumstance really problematic is the fact that the training set may well include examples of murders for which there is another direct relationship between the subject and object. Hence, any system that correctly learns this pattern from the examples in the training set will be penalized for it in the common evaluation format, as including directly related entities in the candidate ranking for a test case can only worsen the performance but never improve it. Therefore, results on FB15k-237 need to be taken with a grain of salt, especially if a system makes any use of P_1 rules. Indeed, we found that suppressing all P_1 rules, the performance of AMIE on FB15k-237 actually improved by roughly 2% for hits@10. For the FB15k-237 results presented in this paper, however, we always used the full rule set.

4.3   Fine-Grained Evaluation

In the following, we present results for each of the annotated subsets. We used AMIE and RuleN with the most liberal settings described in Table 2. As approaches that are based on the use of embeddings, we used the methods TransE [2], RESCAL [9], and HolE [8], for which we did a hyperparameter search as described in [15]. The so-found best hyperparameters are available online. The evaluation results are depicted in Tables 3, 4, and 5. The shortcuts Sym, Eq, Sub, and UC in the table headings refer to the subsets *Symmetry, Equivalence, Subsumption* and *Uncovered*. We focus mainly on the FB15k dataset because it covers completion tasks from all subsets.

The best performing embeddings based system (HolE) achieved only 36% in terms of hits@1 on FB15k, while AMIE and RuleN achieved 64.7% and 77.2%. The interesting aspect is not the hits@1 itself, but the pattern that if the rule-based systems presented the correct candidate within the top 10, it was usually on the first position. This is not the case for the embedding models. In the *Symmetry* category, for example, the first candidate of TransE was always wrong. We found that for a completion task like $\langle a, r, ? \rangle$, the highest ranked entity was always a itself. This problem with symmetry was less severe for HolE and RESCAL, however, the tendency is the same.

For the subsets *Equivalence, Subsumption,* and P_2, RuleN and AMIE could not generate results close to 100% anymore. However, they were still significantly ahead in terms of hits@10 and especially hits@1 score. On WN18, HolE was a noteworthy exception as it achieved competitive results to RuleN and AMIE on

Table 3. Fine-grained results for WN18.

	All (100%)		Sym (21.4%)		Eq (72.5%)		UC (6.1%)	
	hits@1	hits@10	hits@1	hits@10	hits@1	hits@10	hits@1	hits@10
AMIE	.872	.948	**1.00**	**1.00**	.904	**1.00**	.047	.166
RuleN	**.945**	**.958**	.999	**1.00**	**.998**	1.00	**.128**	**.325**
HolE	.933	.940	.981	**1.00**	**.998**	.999	.011	.039
RESCAL	.749	.874	.878	.973	.772	.913	.019	.063
TransE	.082	.944	.000	.988	.114	.996	.000	.175
Ensemble	.941	.956	1.00	1.00	.998	1.00	.060	.287

Table 4. Fine-grained results for FB15k (h@k refers to hits@k).

	All (100%)		Sym (7.2%)		Eq (60%)		Sub (6.8%)		P_2 (7.3%)		UC (18.4%)	
	h@1	h@10	h@1	h@10	h@1	h@10	h@1	h@10	h@1	h@10	h@1	h@10
AMIE	.647	.858	.906	.983	.766	.961	.720	.950	.451	**.736**	.205	**.486**
RuleN	**.772**	**.870**	**.992**	1.00	**.940**	**.982**	**.831**	**.954**	**.536**	.724	**.207**	.480
HolE	.366	.706	.046	.936	.484	.811	.505	.814	.179	.438	.127	.339
RESCAL	.267	.600	.126	.768	.308	.638	.333	.645	.288	.546	.158	.416
TransE	.031	.796	.000	.852	.039	.893	.024	.884	.019	.661	.027	.479
Ensemble	.798	.898	.981	1.00	.957	.992	.895	.982	.575	.797	.258	.562

the mentioned subsets. TransE and RESCAL performed worse. If we look at the FB15k *Uncovered* subset, we observed a different pattern. Rule-based and embedding-based approaches performed on a similar level with respect to the hits@10 score.

On FB15k-237, AMIE and RuleN outperformed the other approaches only in the P_2 category. The overall results were slightly below the best performing embedding-based systems RESCAL and TransE as they were superior on the large *Uncovered* subset. These different strengths indicate potential for an ensemble model.

To sum up, some of the approaches that are based on embeddings had rather specific problems with symmetric relations in our experiments. Furthermore, the other subsets that can be covered by highly confident path rules of length one or two, could not be solved reliably by approaches such as TransE, HolE, or RESCAL. This became more obvious when looking at hits@1 instead of looking at hits@10. Overall, we observed rule-based approaches to be more precise. Their top ranked candidate was usually a correct hit (for most categories >50%), while this was not the case for TransE, HolE, or RESCAL. On the other hand, those systems held their ground in test cases that are tough for rule-based systems.

Table 5. Fine-grained results for FB15k-237.

	All (100%)		P$_2$(14%)		UC (86%)	
	hits@1	hits@10	hits@1	hits@10	hits@1	hits@10
AMIE	.174	.409	.437	.656	.131	.368
RuleN	**.182**	.420	**.487**	**.691**	.132	.376
HolE	.096	.291	.166	.337	.085	.283
RESCAL	.167	.418	.342	.546	**.138**	.397
TransE	.106	**.430**	.191	.579	.092	**.405**
Ensemble	.234	.517	.539	.721	.184	.484

4.4   Ensemble Learning

Given that rule-based and embedding-based approaches use unrelated strategies and therefore achieve different results on specific categories, we propose to combine both methods to produce predictions with higher quality. The training time of an ensemble is essentially bottlenecked by the system that requires the most computational effort since models can be built in parallel. Learning the ensemble weights is a negligible effort in comparison. Hence, we feel that this approach is practical given sufficient resources.

We constructed an ensemble that consists of RuleN and AMIE on the one hand and TransE, HolE and RESCAL on the other hand using linear blending to combine these models, as suggested in [15]. The goal is to combine the strength of each model at the relation level. This is in line with our observation that there are relations for which RuleN or AMIE can learn rules with high confidence, while there are also relations where it is not possible to learn such rules. We constructed for each relation a dataset that consisted of all its triples from the training set as well as an equal amount of negative triples obtained by randomly perturbing either subject or object. Then a meta learner (logistic regression) was trained such that the constructed data could be classified correctly, using each individual model's normalized score as input feature.

Learning the weights based on the performance on the training set has its drawbacks. Rule-based systems need access to the training set to infer new knowledge from learned rules. Given this fact, they could trivially replicate all knowledge contained in the training set. To prevent this for each completion task, the triple that defines this task needs to be temporarily suppressed. Embedding-based systems, on the other hand, are trained with the primary goal of remembering the training set as good as possible. To establish equal preconditions, a similar tweak would have to be applied to these systems. However, it is impractical to do so given their latent knowledge representation. Learning the ensemble weights on the validation set, i.e., performing link prediction on unseen data, might be a better alternative. However, in most of the existing works the validation set was used for hyperparameter tuning only. Thus, we refrained from doing so to prevent doubts about the comparability of the results.

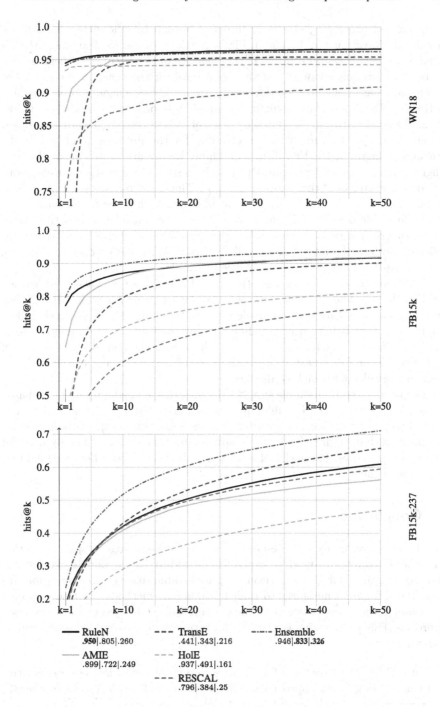

Fig. 4. Hits@k for k = 1 . . . 50 for different systems and ensembles for WN18, FB15k and FB15k-237. Filtered MRRs are shown below the explanation for each approach in the order WN18|FB15k|FB15k-237.

Instead of presenting results in terms of filtered hits@k with a fixed k, we visualized hits@k for $k = 1 \ldots 50$ in Fig. 4. At the bottom, we also added the filtered mean reciprocal rank (MRR).[2] The performance gain of the ensemble over its best performing member system varied between the different datasets. For WN18, it achieved slightly inferior results than the best single approach, which is RuleN. We cannot fully explain the small loss of quality of the ensemble. It should be noted that the characteristics of WN18 heavily reward rule-based systems and that this might be an example for the problem described in the previous paragraph. On FB15k, the ensemble was clearly better than the best single approach, which was again RuleN. The results of the ensemble were about 3 % points better over the whole range of k. The ensemble was even more beneficial on FB15k-237. This supports our assumption that the performance gain of the ensemble over its rule-based member systems correlates with the size of the *Uncovered* fraction of a data set. The high precision of rule-based systems is reflected both in the hits@1 score and the MRR. With the exception of WN18, these scores are further improved by the ensemble.

Additionally, we have analyzed the ensemble weights that have been learned for FB15k-237. The relation *nationality* is an example for which RuleN has high weights. For this relation, RuleN generates many C rules, which reflect the frequency distribution of the different nationalities (most people are from the US, followed by UK, and so on). We have also checked other examples of high weights for rule-based approaches. Most of them were correlated with the existence of rules with high confidence.

The results of our ensemble support the idea that embedding- and rule-based approaches perform well on different types of completion tasks, and that it is fruitful to join predictions of both types of models. This is especially important for datasets that might have less regularities than the datasets usually used for evaluation purposes. For such datasets a combination of both families might be even more beneficial.

5   Conclusion

In this paper, we analyzed rule-based systems for knowledge graph completion on datasets commonly used to evaluate embedding-based models. The generated results allow for a comparison with embedding-based approaches for this task. Besides global measures to rank the different methods, we also classified test cases of the datasets based on the explanations generated by our rule-based approach. This partitioning is available for future works. We gained several interesting insights.

– Both AMIE and RuleN are for the most commonly used datasets competitive to embedding-based approaches. This holds not only with respect to TransE,

---

[2] If a rule-based approach did not rank the candidate, we have set the rank to $n/2$ where n is the set of all entities. This is the average result of randomly ranking the candidates.

RESCAL, or HolE, but still holds for the large majority of the models reported about in [13] and [6]. Only few of these embedding models perform slightly better.

- Rule-based approaches can deliver an explanation for the generated ranking. This feature can be used for a fine-grained evaluation and helps to understand the regularities within and the hardness of a dataset.
- TransE, RESCAL, and HolE have problems in solving specific types of completion tasks that can be solved easily with rule-based approaches. This becomes noticeable in particular when looking solely at the top candidate of the filtered ranking.
- The good results of the rule-based systems are caused by the fact that the standard datasets are dominated by regularities such as symmetry and (inverse) equivalence. FB15k-237 is an exception to this due to the specific way it was constructed.
- It is possible to leverage the outcome of both families of approaches by learning an ensemble. This ensemble achieves better results than any of its members (the WN18 results are a minor deviation).

With this paper, we tried to fill a research gap and shed new light on the insights gained in previous years. Rule-based approaches perform very well and are a competitive alternative to models based on embeddings. For that reason, they should be included as a baseline for the evaluation of knowledge graph completion methods. Moreover, we recommend conducting the evaluation on a more fine-grained level like the one we proposed.

References

1. Bordes, A., Glorot, X., Weston, J., Bengio, Y.: A semantic matching energy function for learning with multi-relational data. Mach. Learn. **94**(2), 233–259 (2014)
2. Bordes, A., Usunier, N., Garcia-Duran, A., Weston, J., Yakhnenko, O.: Translating embeddings for modeling multi-relational data. In: Advances in Neural Information Processing Systems, pp. 2787–2795 (2013)
3. Dettmers, T., Minervini, P., Stenetorp, P., Riedel, S.: Convolutional 2D knowledge graph embeddings. CoRR abs/1707.01476 (2017). http://arxiv.org/abs/1707.01476
4. Galárraga, L.A., Teflioudi, C., Hose, K., Suchanek, F.: AMIE: association rule mining under incomplete evidence in ontological knowledge bases. In: Proceedings of the 22nd International Conference on World Wide Web, pp. 413–422. ACM (2013)
5. Gardner, M., Mitchell, T.M.: Efficient and expressive knowledge base completion using subgraph feature extraction. In: EMNLP, pp. 1488–1498 (2015)
6. Kadlec, R., Bajgar, O., Kleindienst, J.: Knowledge base completion: baselines strike back. arXiv preprint arXiv:1705.10744 (2017)
7. Lao, N., Mitchell, T., Cohen, W.W.: Random walk inference and learning in a large scale knowledge base. In: Proceedings of the Conference on Empirical Methods in Natural Language Processing, pp. 529–539. Association for Computational Linguistics (2011)

8. Nickel, M., Rosasco, L., Poggio, T.A., et al.: Holographic embeddings of knowledge graphs. In: AAAI, pp. 1955–1961 (2016)
9. Nickel, M., Tresp, V., Kriegel, H.P.: A three-way model for collective learning on multi-relational data. In: ICML, vol. 11, pp. 809–816 (2011)
10. Niepert, M.: Discriminative Gaifman models. In: Advances in Neural Information Processing Systems, pp. 3405–3413 (2016)
11. Schlichtkrull, M., Kipf, T.N., Bloem, P., van den Berg, R., Titov, I., Welling, M.: Modeling relational data with graph convolutional networks. In: Gangemi, A., et al. (eds.) ESWC 2018. LNCS, vol. 10843, pp. 593–607. Springer, Cham (2018). https://doi.org/10.1007/978-3-319-93417-4_38
12. Shen, Y., Huang, P.S., Chang, M.W., Gao, J.: Traversing knowledge graph in vector space without symbolic space guidance. arXiv preprint arXiv:1611.04642 (2016)
13. Shi, B., Weninger, T.: ProjE: embedding projection for knowledge graph completion. In: AAAI, vol. 17, pp. 1236–1242 (2017)
14. Toutanova, K., Chen, D.: Observed versus latent features for knowledge base and text inference. In: Proceedings of the 3rd Workshop on Continuous Vector Space Models and their Compositionality, pp. 57–66 (2015)
15. Wang, Y., Gemulla, R., Li, H.: On multi-relational link prediction with bilinear models. In: Association for the Advancement of Artificial Intelligence, AAAI (2018). https://www.aaai.org/ocs/index.php/AAAI/AAAI18/paper/view/16900
16. Wang, Z., Zhang, J., Feng, J., Chen, Z.: Knowledge graph embedding by translating on hyperplanes. In: AAAI, vol. 14, pp. 1112–1119 (2014)
17. Xiao, H., Huang, M., Zhu, X.: TransG: a generative model for knowledge graph embedding. In: Proceedings of the 54th Annual Meeting of the Association for Computational Linguistics (Volume 1: Long Papers), vol. 1, pp. 2316–2325 (2016)
18. Yang, B., Yih, W., He, X., Gao, J., Deng, L.: Embedding entities and relations for learning and inference in knowledge bases. arXiv preprint arXiv:1412.6575 (2014)
19. Zeng, Q., Patel, J.M., Page, D.: QuickFOIL: scalable inductive logic programming. Proc. VLDB Endow. 8(3), 197–208 (2014)

Aligning Knowledge Base and Document Embedding Models Using Regularized Multi-Task Learning

Matthias Baumgartner[1]($\boxtimes$), Wen Zhang[2,3]($\boxtimes$), Bibek Paudel[1]($\boxtimes$),
Daniele Dell'Aglio[1], Huajun Chen[2,3], and Abraham Bernstein[1]

[1] Department of Informatics, University of Zurich, Zurich, Switzerland
{baumgartner,bpaudel,dellaglio,bernstein}@ifi.uzh.ch
[2] College of Computer Science and Technology, Zhejiang University,
Hangzhou, China
{wenzhang2015,huajunsir}@zju.edu.cn
[3] Alibaba-Zhejiang University Joint Institute of Frontier Technologies,
Hangzhou, China

Abstract. Knowledge Bases (KBs) and textual documents contain rich and complementary information about real-world objects, as well as relations among them. While text documents describe entities in freeform, KBs organizes such information in a structured way. This makes these two information representation forms hard to compare and integrate, limiting the possibility to use them jointly to improve predictive and analytical tasks. In this article, we study this problem, and we propose KADE, a solution based on a regularized multi-task learning of KB and document embeddings. KADE can potentially incorporate any KB and document embedding learning method. Our experiments on multiple datasets and methods show that KADE effectively aligns document and entities embeddings, while maintaining the characteristics of the embedding models.

1 Introduction

In recent years, the open data and open knowledge movements gain more and more popularity, deeply changing the Web, with an exponential growth of open and accessible information. Wikipedia is the most successful example of this trend: it is among the top-5 most accessed Web sites and offers more than 40 million articles. Based on Wikipedia, several Knowledge Bases (KBs) have been created, such as FreeBase and DBpedia. Wikidata is a sibling project of Wikipedia which focuses on the construction of a collaboratively edited KB.

It is therefore natural to ask, how precise and complete information can be retrieved from such open repositories. One of the main challenges arising from this question is data integration, where knowledge is usually distributed and

M. Baumgartner, W. Zhang, and B. Paudel contributed equally to this work.

D. Vrandečić et al. (Eds.): ISWC 2018, LNCS 11136, pp. 21–37, 2018.
https://doi.org/10.1007/978-3-030-00671-6_2

complementary, and needs to be combined to get a holistic and common view of the domain. We can envision two common cases where this challenge is relevant: when users need open knowledge from different repositories, and when users need to combine open and private knowledge.

Key to the success of data integration is the *alignment* process, i.e. the combination of *descriptions* that refer to the same real-world object. This is because those descriptions come from data sources that are heterogeneous not only in content, but also in structure (different aspects of an object can be modelled in diverse ways) and format, e.g. relational database, text, sound and images. In this article, we describe the problem of *KB entity-document alignment*. Different from previous studies, we assume that the same real-world object is described as a KB entity and a text document. Note that the goal is not to align an entity with its surface forms, but rather with a complete document. We move a step towards the solution by using existing embedding models for KBs and documents.

A first problem we face in our research is how to enable comparison and contrast of entities and documents. We identify *embedding models* as a possible solution. These models represent each entity in a KB, or each document in a text corpus, by an embedding, a real-valued vector. Embeddings are represented in vector spaces which preserve some properties, such as similarity. Embeddings gained popularity in a number of tasks, such as finding similar entities and predicting new links in KBs, or comparing documents in a corpus.

So far, there are no algorithms to create embeddings starting form descriptions in different formats. Moreover, embeddings generated by different methods are not comparable out of the box. In this study, we ask if *is it possible to represent embeddings from two different models in the same vector space, which (i) brings close embeddings describing the same real world object, and (ii) preserves the main characteristics of the two starting models?*

Our main contribution is KADE, a regularized multi-task learning approach for representing embeddings generated by different models in a common vector space. KADE is a generic framework and it can potentially work with any pair of KB and document embedding models. To the best of our knowledge, our study is the first to present a generic embedding model to deal with this problem. Our experiments show that KADE integrates heterogeneous embeddings based on what they describe (intuitively, embeddings describing the same objects are close), while preserving the semantics of the embedding models it integrates.

2   Related Work

Document Embedding. Paragraph Vector [4], also known as Doc2vec, is a popular document embedding model, based on Word2vec [6]. It has two variants: Distributed Memory and Skip Gram. They represent each document by an embedding such that the learned embeddings of similar documents are close in the embedding space. Document embeddings can also be learned with neural models like CNN and RNN [3,12], topic models, or matrix factorization. Furthermore, pre-trained embeddings from multiple models can be combined in an ensemble using dimensionally reduction techniques [10,20].

KB Embedding. Translation-based embedding models like TransE [1], TransR [5], and TransH [17] have been shown to successfully representing KB entities. Apart from entities, these models also represent each relation by an embedding. They aim to improve the link prediction task and treat each triple as a relation-specific translation. Several other embedding models have since been introduced [8,15]. RDF2Vec [9] uses the same principle as Word2vec in the context of a KB and represents each entity by an embedding. Other type of KB embedding models include neural, tensor and matrix based methods [7].

Combining Text and KB. Several approaches improve KB embeddings by incorporating textual information. [14] exploits lexical dependency paths extracted from entities mentioned in documents. [21] and [16] jointly learn embeddings of words and entities, by using an alignment loss function based on the co-occurrence of entity and words in its descriptions, or Wikipedia anchors, respectively. A similar approach has been used for named entity disambiguation [19]. DKRL [18] treats text descriptions as head and tail entities, and ignores the linguistic similarity while learning KB embeddings.

Further, multiple pre-trained embeddings can be combined into one multimodal space. [13] concatenates embeddings of aligned images, words, and KB entities, then fuses them via dimensionality reduction.

Regularized Multi-Task Learning (MTL). Regularized MTL [2] exploits relatedness between multiple tasks to simultaneously learn their models. It enforces task relatedness by penalizing deviations of individual tasks using regularization.

In contrast to these methods, our goal is to align documents with entities, while at the same time retaining the properties of both document and KB embeddings. Our model is flexible since it does not require a predefined alignment loss function, and learns by means of regularization through task-specific representations of documents and KB. It does not depend on the availability of further linguistic resources like a dependency parser, careful extraction of features like anchor text, or defining a separate alignment loss function. Our solution aims at preserving linguistic and structural properties encoded in the document and KB embeddings respectively, while also representing them in a common space.

3   Preliminaries

Knowledge Bases and Documents. A Knowledge Base $\mathcal{K}$ contains triples $k_j = \{(h, r, t) \mid h, t \in \mathcal{E}; r \in \mathcal{R}\}$, where $\mathcal{E}$ and $\mathcal{R}$ are the sets of entities and relations, respectively. (h, r, t) indicates that the head entity h and tail entity t are related by a relation r. Entities in KBs, such as Freebase and DBPedia, describe real-world objects like places, people, or books.

Text is the most popular way to describe real-world objects. It is usually organised in documents, which delimit the description to a specific object or concept, such as a country or a person. Formally, a document d in a corpus $\mathcal{D}$ is represented as a sequence $\langle w_1, \ldots, w_i, \ldots, w_{n_d} \rangle$, where $|d| = n_d$, and w_i denotes a word in the corpus drawn from a word vocabulary $\mathcal{W}$.

For example, Wikipedia contains a textual document about Mike Tomlin, the head coach of Pittsburgh Steelers, at https://en.wikipedia.org/wiki/Mike_Tomlin (denoted d_{mt}); FreeBase contains a graph-based description of Mike Tomlin, here identified by m.0c5f-j[1], e.g. the triple (m.0c5f-j, current_team_head_coached, m.05tfm) states that *Mike Tomlin* (the head entity) is the *head coach* (the relation) of *Pittsburgh Steelers* (the tail entity, m.05tfm).

We name such different forms of information as *descriptions* of real-world objects, e.g. d_{mt} and $m.0c5f\_j$ are two different descriptions of Mike Tomlin.

Embeddings. Embeddings are dense vectors in a continuous vector space which could be regarded as another representation of descriptions[2]. Embeddings gained popularity in recent years, due to their successful applications [3,6,12].

In this study, we consider two families of embedding models: the *translation-based models* for Knowledge Bases and the *paragraph vector models* for documents. The models of the former family represent KB entities as points in the continuous vector space, and relations as translations from head entities to tail entities. Representative models of the former family are TransE [1], TransH [17] and TransR [5]. TransE defines the score function for a triple (h, r, t) as: $S(h, r, t) = \|\mathbf{h} + \mathbf{r} - \mathbf{t}\|$. TransH and TransR extend TransE to overcome the limited capability of TransE to encode 1-N, N-1 and N-N relations. They define the score function for a triple as: $S(h, r, t) = \|\mathbf{h}_\perp + \mathbf{r} - \mathbf{t}_\perp\|$, in which $\mathbf{h}_\perp$ and $\mathbf{t}_\perp$ are projected head and tail entities. TransH projects entities to relation specific hyperplanes with $\mathbf{w}_r$ as normal vectors, thus $\mathbf{h}_\perp = \mathbf{h} - \mathbf{w}_r^\top \mathbf{h} \mathbf{w}_r$ and $\mathbf{t}_\perp = \mathbf{t} - \mathbf{w}_r^\top \mathbf{t} \mathbf{w}_r$. TransR projects entities from the entity space to the relation space via relation specific matrices $\mathbf{M}_r$, thus $\mathbf{h}_\perp = \mathbf{M}_r \mathbf{h}$ and $\mathbf{t}_\perp = \mathbf{M}_r \mathbf{t}$. Let (h, r, t) be a triple in $\mathcal{K}$, and (h', r', t') be a negative sample, i.e., (h', r', t') is not in $\mathcal{K}$. The margin-based loss function l_{KM} for (h, r, t) is defined as:

$$l_{KM}((h, r, t), (h', r, t')) = \max(0, (\gamma + S(\mathbf{h}, \mathbf{r}, \mathbf{t}) - S(\mathbf{h}', \mathbf{r}', \mathbf{t}'))), \qquad (1)$$

where γ is a margin. The objective of translation-based models is to minimize the margin-based loss function for all triples, i.e., $\mathcal{L}_{KM}(\mathcal{K}) = \sum_{(h,r,t) \in \mathcal{K}} l((h, r, t), (h', r, t'))$, where (h', r', t') is a negative example sampled for (h, r, t) during training.

The paragraph vector models represent documents in a continuous vector space. Examples of models of this family are PV-DM (Distributed Memory) and PV-SG (Skip Gram or DBOW) [4]. Both models learn embeddings for variable-length documents, by training to predict words (in PV-DM) or word-contexts (in PV-SG) in the document. For every document d, the objective of PV-DM is

---

[1] We omit prefixes for the sake of readability.

[2] Throughout this paper, we denote vectors in lowercase bold letters and matrices in uppercase bold letters.

to maximize the average log probability of a target word w_t appearing in a given context c_t, conditioned not only on the context words but also on the document:

$$l_{DM}(d) = \frac{1}{n_d} \sum_{t=1}^{n_d} log\ p(\mathbf{w}_t|\mathbf{d}, \mathbf{c}_t), \qquad (2)$$

where $p(\mathbf{w}_t|\mathbf{d}, \mathbf{c}_t) = \sigma(\mathbf{b} + \mathbf{U}g(\mathbf{d}, \mathbf{c}_t))$, and $\sigma(\cdot)$ is the logistic function, $\mathbf{U}$ and $\mathbf{b}$ are weight and bias parameters, and g is constructed by concatenating (or averaging) its parameters. The PV-SG objective is to maximize l_{DM} defined as:

$$l_{DM}(d) = \frac{1}{n_c} \sum_{t=1}^{n_c} log\ p(\mathbf{c}_t|\mathbf{d}), \qquad (3)$$

where $p(\mathbf{c}_t|\mathbf{d}) = \sigma(\mathbf{b} + \mathbf{U}g(\mathbf{c}_t))$. In this way, both PV-DM and PV-SG capture similarities between documents by using both context and document embeddings to maximize the probability of context or target words. As a result, the embeddings of similar documents are close in the embedding space.

4   Aligning Embedding Models

From Sect. 3, we see that the descriptions of Mike Tomlin in FreeBase and Wikipedia bring complementary information. There is an intrinsic value in considering these two descriptions together, since they offer a more complete view of the person. Embedding models offer a space where descriptions can be contrasted and compared, and it is therefore natural to ask ourselves if we can integrate multiple descriptions by exploiting those models. However, embedding models represent descriptions in different vector spaces. In other words, while two embeddings generated by the same model are comparable, two embeddings generated by different models are not.

We want to study if it is possible to bridge together different embedding models, while preserving the characteristics of the individual models and enabling new operations. This operation, that we name alignment, should take into account two characteristics, namely relatedness and similarity, explained below.

Relatedness. Descriptions of the same real-world object are related. Let $\mathcal{T}$ be a set of descriptions; the document and entity sets ($\mathcal{D}$ and $\mathcal{E}$) are two disjoint subsets of $\mathcal{T}$. We introduce the notion of *relatedness*, to indicate that two descriptions refers to the same real-world object, as a function $rel : \mathcal{T} \mapsto \mathcal{T}$. rel is a symmetric relation, i.e. if $t_i \in \mathcal{T}$ relates to $t_j \in \mathcal{T}$, then the vice versa holds.

Similarity. In addition to relatedness, we intuitively introduce the notion of *similarity*. Let's consider the following example: Pittsburgh Steelers and San Francisco 49ers are two football teams. They are two real-world objects, and in the context of a set of real-world objects, we can define a notion of similarity between them. For example, if we consider all the sport teams in the world, the two teams are similar, since they share several features—they are football teams, they play in the same nation and in the same leagues. However, in the context of NFL, the two teams are less similar—they play on different coasts and have different achievements. From this example, we observe that the notion of similarity can be relation- or context-specific.

Our proposed model is agnostic to the notions of similarity adopted by individual KB and document embedding models. We want our model to be robust to the choice of individual embedding models. In this way, we only loosely interpret the model-specific loss functions and choice of similarity measure.

Alignment. The problem we investigate in our research is of *aligning* the embeddings of related descriptions. It is worth stressing, that alignment captures both, the notion of relatedness and similarity. The goal is to obtain a space where the embeddings of related descriptions are close (i.e. relatedness), while preserving the semantics of their original embeddings (i.e., similarity).

Assumptions. In this study, we make the following assumptions. First, relatedness is defined as a function that relates each entity of $\mathcal{E}$ to a document in $\mathcal{D}$, and vice versa. Formally, we denote this relation with rel_{ed} and it holds: (i) rel_{ed} is injective, (ii) $\forall e \in \mathcal{E}$, $rel_{ed}(e) \in \mathcal{D}$ and (iii) $\forall d \in \mathcal{D}$, $rel_{ed}(d) \in \mathcal{E}$. Based on rel_{ed}, we define the *entity-document relatedness set* as $Q = \{(e, d) \mid rel_{ed}(e) = d \wedge e \in \mathcal{E} \wedge d \in \mathcal{D}\}$, e.g. $(m.0c5f\_j, d_{mt}) \in Q$.

The second assumption is based on the fact that, in real scenarios, it often happens that only some relations between document and entities are known. We therefore assume that the algorithm can access a relatedness set $Q' \subset Q$.

In the following, we abuse the notation and we use d and e to indicate embeddings when it is clear from the context. When not, we use $v(t)$ to indicate the embedding of the description (either entity or document) $t \in \mathcal{T}$.

5   A Regularized Multi-Task Learning Method for Aligning Embedding Models

In this section we present KADE, a framework to align **K**nowledge base **a**nd **D**ocument **E**mbedding models. KADE can be separated into three parts, as shown in Fig. 1: (i) a Knowledge Base embedding model, on the left, where vectors are denoted by circles, (ii) a document embedding model, on the right, where vectors are denoted by squares, and (iii) a regularizer process, in the center.

The construction of the Knowledge Base (or document) embedding model is represented by the arrows with dashed lines, which represent the moving direction for entity (or document) embeddings according to the underlying model. The colors and index numbers in the two models indicate that the vectors are describing the same real-world object, i.e., $(e_i, d_i) \in Q, i \in [1, 4]$.

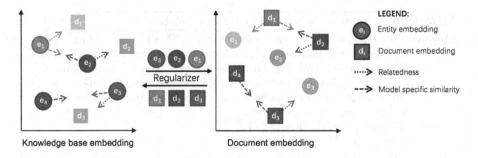

Fig. 1. Illustration of the intuition behind KADE. (Color figure online)

Moreover, (e_1, d_1), (e_2, d_2) and (e_3, d_3) are known related entity-document pairs, i.e., $(e_i, d_i) \in Q', i \in [1, 3]$, while the relatedness information of (e_4, d_4) is not known, i.e., $(e_4, d_4) \in Q$ and $(e_4, d_4) \notin Q'$.

The regularization process builds an embedding space which represents both, the document and the KB entity vectors. There are two regularizers: each of them applies to the training of the document and Knowledge Base embedding, forcing the related document and the entity vectors to be close in the vector space. The regularizer process is shown through the arrows with dotted lines, which represent the moving direction influenced by the related entity-document pairs. This is done by exploiting the information from Q', i.e., the regularizer process cannot use the (e_4, d_4) pair, since it is not in Q'.

Regularizer for the Knowledge Base Embedding Model. We define KADE's objective function $\mathcal{L}^K$ for a Knowledge Base embedding model as follows:

$$
\begin{aligned}
\mathcal{L}^K(\mathcal{D}, \mathcal{K}) = \sum_{\substack{(h,r,t)\in\mathcal{K} \\ (h',r,t')\notin\mathcal{K}}} \Bigg(& l_{KM}((h,r,t),(h',r,t')) \\
& + \lambda_k \Big(\|v(h) - v(rel_{ed}(h))\| + \|v(t) - v(rel_{ed}(t))\| \\
& + \|v(h') - v(rel_{ed}(h'))\| + \|v(t') - v(rel_{ed}(t'))\| \Big) \Bigg),
\end{aligned}
\tag{4}
$$

where l_{KM} is defined in (1), $v(rel_{ed}(e))$ is the embedding corresponding to the document describing e (from the relatedness set), and λ_k is the regularizer parameter for the KB embedding model. If the related entity-document pair is missing in Q', then the regularization term for that entity is zero.

Regularizer for the Document Embedding Model. Similarly, we define the KADE's objective function for a document embedding model as follows:

$$
\mathcal{L}^D(\mathcal{D}, \mathcal{K}) = \sum_{d \in \mathcal{D}} \Big(l_{DM}(d) + \lambda_d \|v(d) - v(rel_{ed}(d))\| \Big)
\tag{5}
$$

Algorithm 1. Iterative learning of embeddings in KADE

Input: A Knowledge Base $\mathcal{K}$ and document corpus $\mathcal{D}$, a KB embedding model
KM and its loss function $\mathcal{L}_{KM}$, a document embedding modem DM and
its loss function $\mathcal{L}_{DM}$, regularizer parameters λ_K and λ_D, relatedness
relation rel_{ed} between entities in $\mathcal{K}$ and documents in $\mathcal{D}$, embedding
dimension k, number of iterations num_iters, loss threshold $\epsilon < 1.0$,
iterations per model $iter\_model$.
Result: Document embeddings $D \in \mathbb{R}^{|\mathcal{D}| \times k}$,
Entity embeddings $E \in \mathbb{R}^{|\mathcal{E}| \times k}$.

1 *Initialize E and D along with other model variables according to KM and DM;*
set $iters = 0$, $iters\_model = 0$,
$loss\_change = km\_loss\_change = dm\_loss\_change = 1$.

2 **while** *$iters \leq num\_iters$ AND $loss\_change > \epsilon$* **do**
3   **for** $i_{km} = 1$; $i_{km} \leq iters\_model$; $i_{km} = i_{km} + 1$ **do**
4    $\mathcal{L}_{KM} \leftarrow$ Calculated according to (1);
5    $\mathcal{L}^K(\mathcal{K}, \mathcal{D}) \leftarrow$ Calculated according to (4) ;
6   Update E and other KM variables using the gradients of $\mathcal{L}^K(\mathcal{K}, \mathcal{D})$;
7   **end**
8   **for** $i_{dm} = 1$; $i_{dm} \leq iters\_model$; $i_{dm} = i_{dm} + 1$ **do**
9    $\mathcal{L}_{DM} \leftarrow$ Calculated according to (2);
10   $\mathcal{L}^D(\mathcal{K}, \mathcal{D}) \leftarrow$ Calculated according to (5);
11   Update D and other DM variables using the gradients of $\mathcal{L}^D(\mathcal{K}, \mathcal{D})$;
12 **end**
13 **if** $iters > 0$ **then**
14   $km\_loss\_change \leftarrow |prev\_km\_loss - \mathcal{L}^K(\mathcal{K}, \mathcal{D})|$;
15   $dm\_loss\_change \leftarrow |prev\_dm\_loss - \mathcal{L}^D(\mathcal{K}, \mathcal{D})|$;
16 **end**
17 $prev\_km\_loss \leftarrow \mathcal{L}^K(\mathcal{K}, \mathcal{D})$;
18 $prev\_dm\_loss \leftarrow \mathcal{L}^D(\mathcal{K}, \mathcal{D})$;
19 $loss\_change = max(km\_loss\_change, dm\_loss\_change)$;
20 $iters \leftarrow iters + 1$;
21 **end**

---

where l_{DM} is the loss function as defined in either (2) or (3), $v(rel_{ed}(d))$ is the
embedding for the entity related to d, and λ_d is the regularizer parameter for
document embedding model. As above, if the related entity-document pair is not
in Q', then the regularization term for the document is set to zero.

Learning Procedure. The learning procedure of KADE is described in Algorithm 1. To learn the complete model, the algorithm trains the document and
the KB embedding models alternately. This is done by using batch stochastic gradient descent, the common training procedure for these kinds of models.
Specifically, one iteration of the learning keeps the KB model fixed and updates
the document model. Within this step, the document model is updated for a

given number of iterations. After that, the KB model is updated by keeping the document model fixed. In a similar way, this step lasts for a given number of iterations. Then the next iterations of KADE proceeds in a similar fashion, until convergence or a fixed number of steps.

6   Experimental Evaluation

Hypotheses. Our experiments aim at verifying three hypotheses. The model learned by KADE retains the characteristics of the document (HP1) and KB (HP2) embedding models, i.e. it does not break the semantics of the document and KB models. Additionally, KADE represents documents and KB entities in the same embedding space such that related documents and entities are close to each other in that space (HP3).

Datasets. We consider three open datasets, which are widely used to benchmark KB embeddings. Their properties are summarized in Table 1. Each of them consists of a Knowledge Base and a document corpus. Related documents are known for all entities, and vice-versa. **FB15k** [1] is derived

Table 1. Dataset sizes

	FB15k	FB40k	DBP50
Entities	14,904	34,609	24,624
Relations	1,341	1,292	351
Total triples	530,663	322,717	34,609
Train triples	472,860	258,175	32,388
Unique words	35,649	50,805	158,921

from the most popular 15,000 entities in Freebase. The text corpus is constructed from the corresponding Wikipedia abstracts. **FB40k**[3] is an alternative to FB15k, containing about 34,000 Freebase entities. It is sparser than FB15k: it includes more entities but fewer relations and triples. The text corpus is derived in the same fashion as for FB15k. **DBP50** is provided by [11]. It is extracted from DBpedia, it has fewer triples than the other datasets, but its documents are longer and its vocabulary is larger. We apply standard pre-processing steps like tokenization, normalization and stopword removal on the documents of the three datasets. The KB triples are split into training and test sets, in a way that each entity and relation in the test set is also present in the training set. The relatedness set Q contains pairs of Freebase/DBpedia entities and documents describing the same real-world object. When not specified, experiments use Q as known relatedness set. Otherwise, we describe how we built $Q' \subset Q$.

Methods. As explained in Sect. 3, for documents, we use two *Paragraph Vector* models: distributed memory (PV-DM) and skip-gram (PV-SG). For KBs, we consider TransE, TransH, and TransR. The method configuration is indicated in parenthesis, e.g. KADE (TransR, PV-SG). We use KADE (TransE, PV-DM) as our reference configuration, denoted by KADE$_{ref}$. When KB or document models are trained on their own, we refer to them as *Independent* models.

---

[3] Available at https://github.com/thunlp/KB2E.

Parameters. The hyperparameters for TransE and TransR are embedding dimension k, learning rate α, margin γ, and a norm. TransH has an additional weight parameter C. KADE further introduces the regularizers λ_k and λ_d. We did a preparatory parameter search to find reasonable values for embedding dimension k and regularizers λ_k, λ_d. The search was limited to KADE$_{\text{ref}}$ on FB15k and resulted in $k = 100$, $\lambda_k = 0.01$, and $\lambda_d = 0.01$. For the remaining parameters, we adopted the values reported by TransE authors [1]: $\alpha = 0.01, \gamma = 1, \text{norm} = \text{L1}$. We used these values for TransH and TransR as well. We adopted $C = 1$ for TransH from [16].

The parameters for the document model are the embedding dimension k, learning rate α, window size w, and number of negative samples n_{neg}. After a preparatory parameter search, we adopted $\alpha = 15, w = 5, neg = 35$.

Experiments are iterated 9600 times[4], in batches of 5000 samples. KADE switches between models every ten training batches (*iters_model* in Algorithm 1).

Implementation. We use our own implementation of TransE, TransH, TransR, since some methods do not provide a reference implementation. To assess our implementations, we ran the link prediction experiments of [1,5,17] with parameters values described above. Results are summarized in Table 2. For each model, the performance of KADE is reasonably close to the originally reported values, although they do not match exactly. These differences are due to parameter settings and implementation details: We used the same parameters for all three methods, while [5,17] optimized such parameters, and we found that slightly different sampling strategies, as implemented in [5], can further improve the result. Ours, as well as other implementations of TransE[5], normalize the embedding vectors during loss computation. This has a positive effect on performance.

It is worth noting that these slight differences in performance do not affect the study of our hypotheses, which is to align document and KB embeddings while retaining the semantics of the original models.

Table 2. Performance of our implementations and published results on KB link prediction. Higher HITS@10 values are better; lower mean rank values are better.

		TransE		TransH		TransR	
		Ours	Bordes [1]	Ours	Wang [17]	Ours	Lin [5]
HITS@10	Raw	0.469	0.349	0.459	0.425	0.443	0.438
	Filtered	0.712	0.471	0.692	0.585	0.656	0.655
Mean rank	Raw	225.410	243.000	234.494	211.000	229.433	226.000
	Filtered	86.304	125.000	97.527	84.000	92.689	78.000

---

[4] This matches the number of training runs over the whole dataset reported in [1].
[5] https://github.com/thunlp/KB2E.

6.1   HP1: KADE Retains the Document Embedding Model

HP1 states that KADE retains the quality of document embedding models. We study HP1 by using the document embeddings as features for binary classifiers[6]. We report the results of $KADE_{ref}$ for FB15k; we had similar results for FB40k.

We first build the category set by retrieving categories from Freebase for each entity in FB15k. Since each entity is related to a textual document through Q, we assign categories to the documents. As a result, we retrieved 4,069 categories, with an average of 46.2 documents per category, a maximum of 14,887 documents[7], and a minimum of one. To have enough positive and negative training examples for each category, we remove categories that belonged to fewer than 10% or more than 50% documents, resulting in 27 categories, with an average of 2,400 documents per category (max: 4,483, min: 1,502). For each category C, we randomly select documents not belonging to C, so that we obtain equal number of documents in positive and negative classes. Next, we randomly assign 70% documents from each class to the training set and the remaining 30% to the testing set. Finally, we train a logistic regression classifier for each category and test the classification accuracy on the testing set. As a result, we train two classifiers for each category (54 classifiers in total): one related to KADE and the other one related to the independent document embedding model PV-DM.

We repeated this procedure five times and report the average result (the standard deviations are small and omitted) in Fig. 2. KADE significantly improves the classification accuracy on all categories. The average classification accuracy from $KADE_{ref}$ is 0.92, while the one from independent training is 0.80. On many

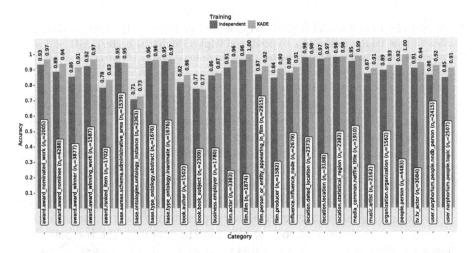

Fig. 2. Accuracy of binary document classification using $KADE_{ref}$ and independently trained document model. The name of Freebase categories and number of documents in each category is listed along the vertical bars.

---

[6] We preferred binary over multi-class classification because it is simpler to explain.

[7] For the class http://rdf.freebase.com/ns/common.topic.

categories, embeddings from KADE achieve an accuracy above 0.95, while the maximum accuracy of embeddings from independent training is 0.89.

The classification accuracy of the document model increases when KADE uses PV-SG rather than PV-DM. The accuracy in the independent case lifts to 88%, with a maximum of 98%. It sill holds that KADE improves over the independent models, even though in many cases the values rougly match.

The results we obtained in this experiment suggest that document embeddings learned by KADE are not worse than the one learned by independent document embedding models. On the contrary, the document embeddings learned by KADE perform better in document classification.

6.2   HP2: KADE Retains the KB Embedding Model

The second hypothesis relates to the ability of KADE to maintain the semantics of the KB embedding model used in the alignment process. We study this hypothesis by performing the link prediction experiment proposed in [1].

Similar to previous studies, for every triple in the test set, we construct corrupted triples by replacing the head (or tail) entity with every other entity in the Knowledge Base. While testing link prediction, we rank all true and corrupted triples according to their scores and get the rank of current test triple.

We report the mean rank (MR) of the test triples and the ratio of test triples in the 10 highest ranked triples (HIT@10). As noted by [1], some corrupted triples be present in the KB. To cope with this, we report *filtered* results, where corrupted triples that are present in the training set are removed.

Table 3. HIT@10 and MR of KADE and independent training. HITS@10 reports the fraction of true triples in the top 10 predicted triples (higher is better). MR indicates the position of the original head (or tail) in the ranking (lower is better).

			FB15k		FB40k		DBP50	
			KADE	Indep.	KADE	Indep.	KADE	Indep.
TransE	HITS@10	Raw	**0.470**	0.469	**0.590**	0.583	**0.440**	0.382
		Filtered	**0.715**	0.712	**0.754**	0.746	**0.469**	0.400
	Mean rank	Raw	**221.257**	225.410	**835.899**	962.244	**1178.849**	2451.215
		Filtered	**82.053**	86.304	**471.996**	598.209	**1130.392**	2403.093
TransH	HITS@10	Raw	0.456	**0.459**	**0.579**	0.571	**0.434**	0.386
		Filtered	0.689	**0.692**	**0.740**	0.731	**0.463**	0.405
	Mean rank	Raw	**233.073**	234.494	**857.130**	1007.940	**1174.965**	2511.550
		Filtered	**96.009**	97.527	**496.215**	649.459	**1126.766**	2463.208
TransR	HITS@10	Raw	0.414	**0.443**	0.554	**0.556**	**0.376**	0.374
		Filtered	0.651	**0.656**	0.705	**0.711**	**0.395**	0.392
	Mean rank	Raw	242.700	**229.433**	**928.089**	929.848	**2414.541**	2430.336
		Filtered	98.620	**92.689**	**561.778**	563.693	**2366.314**	2382.091

Table 3 compares the link prediction performance of KADE$_{ref}$ and independent training over the datasets. KADE$_{ref}$ slightly but consistently outperforms independent training. While there is always an improvement, it is more pronounced on sparser datasets. We conduct the same experiment with TransH and TransR with PV-DM as document model. All other parameters and experimental protocols remain identical. In a few cases KADE embeddings do not outperform the independent embeddings. However, the difference is small (<7%), compared to the differences introduced by the method or dataset.

Also in this case, experiments suggest that HP2 holds, and KADE retains the semantics of the KB embedding models.

6.3   HP3: KADE Aligns KB and Document Embeddings

The first two hypotheses are meant to assess if KADE retains the semantics of the embedding models. Our last hypothesis takes a different perspective, and it aims at verifying if the resulting model is effectively aligning entity and document embeddings. We study this hypothesis in two experiments.

Progression of KADE Training. The first experiment tests if KADE brings document and KB entity embeddings into the same embedding space. For this, we randomly select 100 related entity-document pairs from Q' and call this set Q_R. At different stages of training (i.e. after different numbers of iterations), we take both document and entity embeddings from Q_R. Then, we retrieve the most similar entity embeddings learned by KADE using cosine similarity. In this way, we get two ranked lists of 100 entities, corresponding to a pair $q_R \in Q_R$: one retrieved using the document embedding of q_R as query, and another using the entity embedding of q_R as query. We compare the changes in set overlap of these two ranked lists, and rank correlation as the training of KADE proceeds.

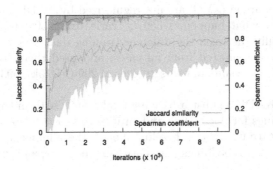

Fig. 3. The process of embedding documents and KB entities in the same space, with KADE$_{ref}$ on FB40k. Shaded areas indicate the standard deviations.

Figure 3 shows the results of this experiment for KADE$_{ref}$ on the FB40k dataset. In the early stages of training, set overlap (measured using Jaccard similarity) as well as rank correlation (measured by the Spearman coefficient) are very low, meaning that the two ranked lists are very different. Both measures show significant improvements as training progresses, showing that KADE is able to align the entities of documents and entities in the same space. We repeated this experiment with the other datasets and we observed a similar behaviour. The same holds when retrieving the document embeddings instead of entity embeddings for query pairs in Q_R.

Alignment Generalization by KADE. This experiment investigates to what extent KADE *generalizes*: can KADE align entity-document pairs that are not available during training, i.e., which are not in Q'? We first introduce the *alignment score* as an evaluation measure, afterwards we present the results.

Given a query document $d \in \mathcal{D}$, the alignment model orders all entities $e \in \mathcal{E}$ with respect to their similarity to d. Let $r_d(e) \in [1, |\mathcal{E}|]$ denote the *rank* of entity e, retrieved for query document d. We define the rank r_e of a document d in an analogous way. A perfect alignment model exhibits $r_d(e) = |\mathcal{E}|$ iff $(e, d) \in Q$, i.e., the related entity is ranked highest out of all possible choices. We further define the normalized version of the ranking measure as $r_d^N(e) = (r_d(e) - 1)/(|\mathcal{E}| - 1) \in [0, 1]$. We define $r_e^N(d)$ analogously.

For a pool of test documents $\mathcal{D}_t \subset \mathcal{D}$ and test entities $\mathcal{E}_t \subset \mathcal{E}$, we average the rankings for all document and entity queries (with respect to their related counterparts) to get the alignment score:

$$AS := 1/2 \left(\frac{1}{|\mathcal{E}_t|} \sum_{e \in \mathcal{E}_t} r_e^N[rel_{ed}(e)] + \frac{1}{|\mathcal{D}_t|} \sum_{d \in \mathcal{D}_t} r_d^N[rel_{ed}(d)] \right)$$

Next, we train KADE with varying sizes of Q' and examine to what extent embeddings of entity-document pairs in $Q \setminus Q'$ are aligned by KADE. Note that embeddings are still computed for all documents and entities, however, the document-entity pairs in $Q \setminus Q'$ are not regularized by KADE.

Figure 4a show what happens when Q' varies from 30% to 98% of Q. To ensure comparability between the different training data sets, we required the same 2% of Q to be omitted from all cases, and used them to calculate the alignment score. This results in test sets sizes of 299 (FB15k), 693 (FB40k), and 493 (DBP50).

We compare KADE against a baseline model built with independently constructed embeddings for documents and entities to show the impact of regularized multi-task learning. In this baseline model, the alignment is achieved by projecting document embeddings onto entity embeddings, and vice-versa, i.e.:

$$\forall e \in \mathcal{E} : e = \sigma \left(rel_{ed}(e) \mathbf{P}_1 + \mathbf{b}_1 \right) \forall d \in \mathcal{D} : d = \sigma \left(rel_{ed}(d) \mathbf{P}_2 + \mathbf{b}_2 \right)$$

The sigmoid function σ allows the model to account for nonlinearity. The projection matrices $\mathbf{P}_{\{1,2\}}$ and biases $\mathbf{b}_{\{1,2\}}$ are estimated from Q'. The model is evaluated on FB15k, with the same test set as used for KADE.

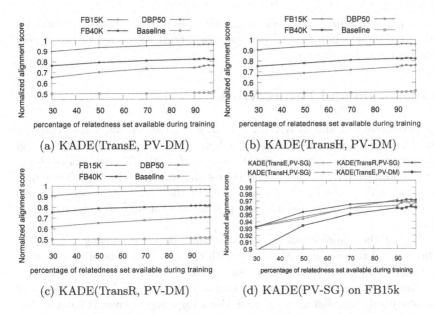

Fig. 4. KADE aligns embeddings of related documents and entities. The x-axis show the percentage of known related pairs ($|Q'|/|Q|$). We measure the alignment score on a fixed set of 2% of Q. In the y-axis, 1 implies that for every query entity (document), the related document (entity) was retrieved as the first result. 0.5 indicates the random baseline.

Figure 4a shows that KADE's performance improves when the size of Q' increases. This effect reflects that embeddings of test pairs are constrained by their neighboring documents and entities.

Further, the performance differs across datasets. FB15k shows the best performance. This can be explained by the fact that we optimized the model parameters based on this dataset. We further explain this result considering the dataset density: higher interconnection allows more coherent embedding construction in the document and entity models.

KADE consistently outperforms the baseline. Although the baseline enhances with more training data, it only narrowly improves over random guessing ($AS = 0.5$). This indicates that independent embedding construction leads to incompatible spaces and highlights the impact of KADE's regularized multi-task learning.

The same experiment is repeated for the other KB or document models. While the method is exchanged, the experimental setup and all parameters are maintained. For KB models (Fig. 4b and c), we observe that the influence of the dataset is much higher than the one of the KB model. This reflects the results from Table 3. For document models, Fig. 4d shows the effect of using PV-SG instead of PV-DM. As baseline, KADE$_{\text{ref}}$ is plotted. Consistent with results from Sect. 6.1, PV-SG outperforms PV-DM, independent of the KB model. The effect is more pronounced if the size of Q' is small.

The experiments in this section assess HP3: KADE aligns the embeddings from different models in a common vector space, built according to the relatedness among the descriptions.

7   Conclusion and Future Work

In this paper, we introduced KADE, a flexible regularized multi-task learning method that represents embedding from heterogeneous models in a common embedding space. We show that KADE can work with different KB or document embedding methods. Our experiments showed that the KADE regularization process does not break the semantics of the underlying document and Knowledge Base embedding models, and in some cases it may even improve their performance.

Looking at the assumptions we took, we think that a promising and important direction of this research is to cope with changes in the set of objects that are considered. In other words, how can KADE cope with description of new objects that are added to the document corpus or the KB? KADE makes use of document embeddings that are pre-trained on a very large corpus. It follows, that the word embeddings which the model learns are general enough for new and unseen documents. This means that KADE might use the already computed document embeddings to find embeddings for new documents. As suggested in [4], it is possible to learn the embeddings for new documents by keeping the network weights and other embeddings constant, and updating the gradients of the document embedding for a few iterations. Similarly, translation-based KB embedding models cannot deal with new entities out of the box. However, we believe it may be possible to learn embeddings for unseen KB entities by using approaches similar to the above one.

Acknowledgements. We would like to thank the SNF Sino Swiss Science and Technology Cooperation Programme program under contract RiC 01-032014, NSFC 61473260/61673338, and the Swiss Re Institute, in particular Axel Mönkeberg, for discussions and financial support.

References

1. Bordes, A., Usunier, N., García-Durán, A., Weston, J., Yakhnenko, O.: Translating embeddings for modeling multi-relational data. In: NIPS, pp. 2787–2795 (2013)
2. Evgeniou, T., Pontil, M.: Regularized multi-task learning. In: KDD, pp. 109–117 (2004)
3. Kim, Y.: Convolutional neural networks for sentence classification. In: EMNLP, pp. 1746–1751. ACL (2014)
4. Le, Q.V., Mikolov, T.: Distributed representations of sentences and documents. In: ICML, vol. 32, pp. 1188–1196. JMLR.org (2014)
5. Lin, Y., Liu, Z., Sun, M., Liu, Y., Zhu, X.: Learning entity and relation embeddings for knowledge graph completion. In: AAAI, vol. 15, pp. 2181–2187 (2015)

6. Mikolov, T., Sutskever, I., Chen, K., Corrado, G.S., Dean, J.: Distributed representations of words and phrases and their compositionality. In: NIPS (2013)
7. Nickel, M., Murphy, K., Tresp, V., Gabrilovich, E.: A review of relational machine learning for knowledge graphs. Proc. IEEE **104**(1), 11–33 (2016)
8. Nickel, M., Rosasco, L., Poggio, T.A.: Holographic embeddings of knowledge graphs. In: AAAI, pp. 1955–1961. AAAI Press (2016)
9. Ristoski, P., Paulheim, H.: RDF2Vec: RDF graph embeddings for data mining. In: Groth, P. (ed.) ISWC 2016. LNCS, vol. 9981, pp. 498–514. Springer, Cham (2016). https://doi.org/10.1007/978-3-319-46523-4_30
10. Saul, L.K., Roweis, S.T.: Think globally, fit locally: unsupervised learning of low dimensional manifolds. JMLR **4**(Jun), 119–155 (2003)
11. Shi, B., Weninger, T.: Open-world knowledge graph completion. CoRR abs/1711.03438 (2017)
12. Tang, D., Qin, B., Liu, T.: Document modeling with gated recurrent neural network for sentiment classification. In: EMNLP, pp. 1422–1432. ACL (2015)
13. Thoma, S., Rettinger, A., Both, F.: Towards holistic concept representations: embedding relational knowledge, visual attributes, and distributional word semantics. In: d'Amato, C., et al. (eds.) ISWC 2017. LNCS, vol. 10587, pp. 694–710. Springer, Cham (2017). https://doi.org/10.1007/978-3-319-68288-4_41
14. Toutanova, K., Chen, D., Pantel, P., Poon, H., Choudhury, P., Gamon, M.: Representing text for joint embedding of text and knowledge bases. In: EMNLP, pp. 1499–1509. ACL (2015)
15. Trouillon, T., Welbl, J., Riedel, S., Gaussier, E., Bouchard, G.: Complex embeddings for simple link prediction. In: ICML, JMLR Workshop and Conference Proceedings, vol. 48, pp. 2071–2080. JMLR.org (2016)
16. Wang, Z., Zhang, J., Feng, J., Chen, Z.: Knowledge graph and text jointly embedding. In: EMNLP, pp. 1591–1601. ACL (2014)
17. Wang, Z., Zhang, J., Feng, J., Chen, Z.: Knowledge graph embedding by translating on hyperplanes. In: AAAI, pp. 1112–1119. AAAI Press (2014)
18. Xie, R., Liu, Z., Jia, J., Luan, H., Sun, M.: Representation learning of knowledge graphs with entity descriptions. In: AAAI, pp. 2659–2665. AAAI Press (2016)
19. Yamada, I., Shindo, H., Takeda, H., Takefuji, Y.: Joint learning of the embedding of words and entities for named entity disambiguation. In: CoNLL, pp. 250–259. ACL (2016)
20. Yin, W., Schütze, H.: Learning word meta-embeddings. In: ACL, vol. 1. ACL (2016)
21. Zhong, H., Zhang, J., Wang, Z., Wan, H., Chen, Z.: Aligning knowledge and text embeddings by entity descriptions. In: EMNLP, pp. 267–272. ACL (2015)

Inducing Implicit Relations from Text Using Distantly Supervised Deep Nets

Michael Glass[1]([✉]), Alfio Gliozzo[1], Oktie Hassanzadeh[1],
Nandana Mihindukulasooriya[2], and Gaetano Rossiello[1,3]

[1] Knowledge Induction and Reasoning Group, IBM Research AI, New York, USA
mrglass@us.ibm.com
[2] Ontology Engineering Group, Universidad Politcnica de Madrid, Madrid, Spain
[3] Department of Computer Science, University of Bari, Bari, Italy

Abstract. Knowledge Base Population (KBP) is an important problem in Semantic Web research and a key requirement for successful adoption of semantic technologies in many applications. In this paper we present Socrates, a deep learning based solution for Automated Knowledge Base Population from Text. Socrates does not require manual annotations which would make the solution hard to adapt to a new domain. Instead, it exploits a partially populated knowledge base and a large corpus of text documents to train a set of deep neural network models. As a result of the training process, the system learns how to identify implicit relations between entities across a highly heterogeneous set of documents from various sources, making it suitable for large-scale knowledge extraction from Web documents. Main contributions of this paper include (a) a novel approach based on composite contexts to acquire implicit relations from Title Oriented Documents, and (b) an architecture for unifying relation extraction using binary, unary, and composite contexts. We provide an extensive evaluation of the system across three different benchmarks with different characteristics, showing that our unified framework can consistently outperform state of the art solutions. Remarkably, Socrates ranked first in both the knowledge base population and attribute validation track at the Semantic Web Challenge at ISWC 2017.

Keywords: Knowledge base population · Deep learning
Distant supervision

1 Introduction

Knowledge Base Population (KBP) is a core problem in Semantic Web research and a key requirement for successful adoption of semantic technologies in many applications. Given a previously defined schema for a knowledge base, the KBP problem consists of acquiring entities and relations from the corpus according to the ontology. The outcome is a knowledge base that can be used to enhance downstream applications such as search engines and business analytics.

© Springer Nature Switzerland AG 2018
D. Vrandečić et al. (Eds.): ISWC 2018, LNCS 11136, pp. 38–55, 2018.
https://doi.org/10.1007/978-3-030-00671-6_3

A common approach to Knowledge Base Population is using Information Extraction (IE) from text, which typically consists of Entity Detection and Linking (EDL) and Relation Extraction (RE) using models that have been pre-trained for the types and relations of interest. The main drawback of supervised IE is that moving to a new domain requires substantial effort. Building a new training set requires reading hundreds, if not thousands, of documents and marking relevant entities and relations in them. This process might take several weeks of work, sometimes providing unsatisfactory results, mostly due to very low recall. A key problem is the existence of implicit relations in text, that occur between entities mentioned across different part of the same document and sometimes across different documents. The majority of supervised IE systems are only able to recognize explicit relations within the same sentence.

In this paper, we present Socrates, a KBP solution that addresses the above problems. Socrates exploits distant supervision [13,20] to minimize domain adaptation cost and is able to identify implicit relations between entities to maximize recall.

Distant supervision can be applied when a partially populated KB for the target schema and a large domain corpus for the target domain are available, providing a cost effective alternative to document level supervision. In a distant supervision approach, the entities and relations in the KB are matched in text to automatically generate training data. The availability of background knowledge can be used to alleviate, if not eliminate, the need of human supervision for domain adaptation. This is a common use case observed particularly in business settings across different industries, including healthcare, finance, customer relationship management, and IT support. For example, in the *ISWC Semantic Web Challenge 2017*[1] on Knowledge Base Population and Validation, Thompson Reuters was interested in extending the public part of the PermID dataset[2] representing popular companies, with information about unseen companies, whose websites were provided as an input.

Different from most IE systems, Socrates enables the recognition of *implicit relations* between entities across different documents, by exploiting the notion of *unary relations*, and across different part of the same document, by leveraging the notion of *composite context sets*, presented in Sect. 5.2. This allows us to substantially increase recall of slot filling queries by capturing implicit information.

In this paper, we present an extensive evaluation of Socrates in three different benchmarks. In Sect. 6.1 we evaluate Socrates on the problem of extending the public part of Thompson Reuters Perm ID with information about new companies. This dataset has been released by the organizers of the *ISWC 2017 Semantic Web Challenge* and enables us to test the *composite context set* approach. Socrates ranked first in both the Knowledge Base Population and Validation tasks of the challenge. In Sect. 6.2, we evaluate the ability to extend a sample of relations in Freebase with information extracted from New York Times articles.

---

[1] http://challenge.semanticweb.org.
[2] https://permid.org.

To this end, we used a standard benchmark [14] that enables a meaningful comparison with state of the art approaches for binary relations, showing significant improvement over the state of the art. Finally, in Sect. 6.3, we evaluate the ability to extend DBPedia with information derived from web crawls [6]. Compared to the previous benchmark, this is a large scale knowledge induction problem involving hundreds of relations and millions of sentences. This setup enables us to test the effectiveness of unary relations. Our results show that unary relations, if combined with binary relations, provide a complementary signal that doubles the recall of the overall process.

The main contributions of this paper are:

- A novel approach based on composite contexts to acquire implicit relations from Title Oriented Documents
- An architecture able to combine binary, unary and composite-context relation extraction.

The rest of the paper is organized as follows: Sect. 2 discusses the existing work on the knowledge base population problem; Sect. 3 presents the Socrates framework and introduces *composite context sets*; Sect. 6 provides a comprehensive evaluation of the Socrates under three different benchmarks, and Sect. 7 draws some conclusions and proposes future work.

2   Related Work

The KBP problem is to induce knowledge graphs from new collections of documents by just providing the schema of the ontology as an input for the system, and no document level annotations for training. As an output the system populates the ontology with new entities and relations identified in text. State of the art approaches for this task [17,18] usually leverage additional examples provided by linked open data to train IE analytics, reducing the need for manual annotations.

Relation extraction using distant supervision has a long history [13,20]. In distant supervision, first mentions of entities from the knowledge base are located in text. When two entities are mentioned in the same sentence that sentence becomes part of the evidence for the relation (if any) between those entities. The set of sentences mentioning an entity pair is used in a machine learning model to predict how the entities are related, if at all. In this work, a novel approach based on unary relations and implicit contexts are presented that is capable of extracting relations even if the two entities do not appear in the same sentence.

Deep learning has been applied to binary relation extraction. Both CNN-based [23] and LSTM-based [21] models have been trained successfully using a sentence as the unit of context. Recently, cross sentence approaches have been explored by building paths connecting the two identified arguments through related entities [24]. These approaches are limited by requiring both entities to be mentioned in a textual context. The context aggregation approaches of state-of-the-art neural models, max-pooling [22] and attention [4,11], do not consider that different contexts may contribute to the prediction in different ways. Instead, the context pooling only determines the degree of a sentence's contribution to the relation prediction. In contrast, the Network-in-Network context aggregation of Socrates can combine textual evidence with different types of contribution to the prediction, not just different degrees.

TAC-KBP[3] is a long running challenge for knowledge base population. Effective systems in these competitions combine many approaches such as rule-based relation extraction, directly supervised linear and neural network extractors, distantly supervised neural network models [25] and tensor factorization approaches to relation prediction. Compositional Universal Schema is an approach based on combining the matrix factorization approach of universal schema [15], with representations of textual relations produced by an LSTM [2]. The rows of the universal schema matrix are entity pairs, and will only be supported by a textual relation if they occur in a sentence together.

Other approaches to relational knowledge induction have used distributed representations for words or entities and used a model to predict the relation between two terms based on their semantic vectors [3]. This enables the discovery of relations between terms that do not co-occur in the same sentence. However, the distributed representation of the entities is developed from the corpus without any ability to focus on the relations of interest. One example of such work is LexNET [19], which developed a model using the distributional word vectors of two terms to predict lexical relations between them (DS_h). The term vectors are concatenated and used as input to a single hidden layer neural network. Unlike our approach to implicit relations, the term vectors are produced by a standard relation-independent model of the term's contexts such as word2vec [12].

3   Socrates Architecture

Socrates is a deep learning based solution for KBP. As an input, Socrates takes a partially populated knowledge graph and extends it with new entities and facts identified from a large collection of documents. Socrates is able to answer slot filling queries about specific entities and does not require additional supervision.

Socrates' architecture is described in Fig. 1. The input of the system is a partially populated KB and a large corpus of text. The output of Socrates is an extended KB, returned as a list of triples with confidence, containing additional facts extracted by the system. Socrates can also be used to validate relations provided as an input. In this case, it returns confidence scores for the input triples by gathering evidence from their textual occurrences.

Optionally, for some entities, Title Oriented Documents (TOD) [5] can be provided as well. These documents are about specific entities, such as the website for a specific company or the Wikipedia page for a music band. TODs are used

---

[3] https://tac.nist.gov/2017/KBP/.

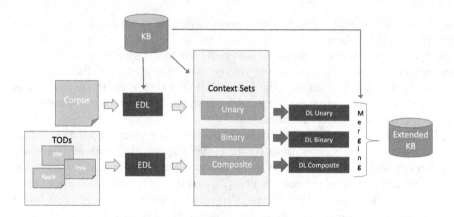

Fig. 1. Socrates architecture

to create composite contexts used to predict relations about the title entity. Socrates does not need manually annotated mentions of entities and relations at all.

At ingestion time, Socrates parses the input document with an Entity Detection and Linking (EDL) system. The goal is to match entity mentions in the corpus to those in the provided KB. EDL is also needed to identify new candidate entities to be added to the KB. A simple option for EDL is gazetteer-based matching. This is effective when labels are provided by the KB, such as in CC-DBP as described in Subsect. 6.3. However, ad-hoc EDL analytics can be provided when working on specific domains, for example to recognize telephone numbers in the ISWC Challenge dataset, or to enable partial match of company names. Although EDL is an interesting research area and might be trained using distant supervision in itself, in this paper we take EDL as a prerequisite to be provided as a pluggable component.

Once EDL is performed, Socrates collects the data needed to train the relation extraction systems. To this aim, it gathers *Context Sets*. Context Sets can be either windows of text, sentences, or composites of multiple parts of a document. Socrates distinguishes three different types of Context Sets:

Binary context sets are contexts containing two different entities. Binary context sets containing two entities in the ontology related by some relations are used as positive examples for those relations. While the negative examples are context sets containing entities not related in the KB.

Unary context sets are contexts containing only one entity, to be used to train a unary-relation extraction system.

Composite context sets are sets of contexts extracted from multiple discontinuous parts of a document. These contexts can support a relation between an entity in the title or section header and another entity mentioned in the body of the document. These are particularly effective for TODs, as described in Subsect. 5.2.

A closer look at the generated training data can provide insight in the value of these three types of context sets. Below are example binary contexts relating an organization to a country. The two arguments are shown in bold. Some contexts where two entities occur together (relevant contexts) will imply a relation between them, while others will not. In the first context, PHILIPPINES and EAGLE CEMENT are not textually related. While in the second context, DYNA MANAGEMENT SERVICES is explicitly stated to be located in BERMUDA.

- The company competes with Holcim **Philippines**, the local unit of Swiss company LafargeHolcim, and **Eagle Cement**, a company backed by diversified local conglomerate San Miguel which is aggressively expanding into infrastructure.
- ... said Richmond, who is vice president of **Dyna Management Services**, a **Bermuda**-based insurance management company.

On the other hand, there are many triples that have no relevant context using binary extraction, but can be supported with unary extraction. JB HI-FI is a company located in AUSTRALIA, (unary relation *hasLocation*:AUSTRALIA). Although "JB Hi-Fi" never occurs together with "Australia" in our corpus, we can gather implicit textual evidence for this relation from its unary relation context sets. Furthermore, even cases where there is a relevant binary context set, the contexts may not provide enough or any textual support for the relation, while the unary context sets might.

- Woolworths, Coles owner Wesfarmers, **JB Hi-Fi** and Harvey Norman were also trading higher.
- **JB Hi-Fi** in talks to buy The Good Guys
- In equities news, protective glove and condom maker Ansell and **JB Hi-Fi** are slated to post half year results, while Bitcoin group is expected to list on ASX.

The key indicators are: "ASX", which is an Australian stock exchange, and the other Australian businesses mentioned, such as Woolworths, Wesfarmers, Harvey Norman, The Good Guys, Ansell and Bitcoin group. There is no strict logical entailment, indicating JB HI-FI is located in Australia, instead there is textual evidence that makes it probable.

Composite context sets can be constructed when the title, section header or document metadata is informative for relation prediction. This is typically true for TODs. In the example below the EDL did not match "TEXAS ELECTRONICS CANADA INC." to TEXAS ELECTRONIQUES CANADA INC. but the title is still part of the context, so both arguments of the possible headquarters PhoneNumber relation are present in the constructed context.

- www.texaselec.com   *Texas Electroniques Canada Inc.*
  East and Latin America. Read more about us TEXAS ELECTRONICS CANADA INC. Tel: **514-842-4431** Toll-free: 1-800-387-9696 Fax: 514-842-8641 E-mail:

– www.texaselec.com contact us   *Texas Electroniques Canada Inc.*
contact form below. Our representatives would be glad to help you! Phone:
514-842-4431 Toll-free: 1-800-387-9696 Fax: 514-842-8641 E-mail:

The core KBP technology used by Socrates is a deep learning based binary
relation extraction system, described in Sect. 4. Variants of this approach are
then used to train unary and composite-context KBP systems, all providing
new triples as an output with associated probabilities. As a final step, Socrates
merges triples generated by all the three techniques as explained in Sect. 5.3.

4   Deep Nets for KBP

Socrates uses all the context sets collected from the corpus to train a deep
learning based relation extraction classifier. To this aim, it feeds them into a
deep neural network, described by Fig. 2. This architecture is largely unchanged
for all three types of context sets.

The sentence-to-vector portion of the neural architecture begins by looking
up the words in a word embedding table. The word embeddings are initialized
with word2vec [12] and updated during training. The position of each word
relative to the entity is also looked up in a position embedding table.

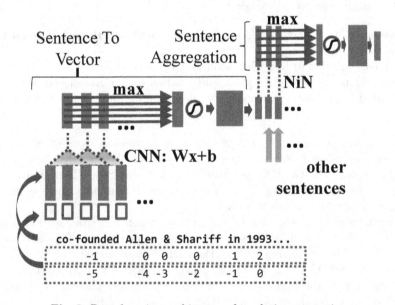

Fig. 2. Deep learning architecture for relation extraction

Formally, the word embedding matrix is $W \in \mathbb{R}^{d_w \times |V|}$ where d_w is the dimen-
sionality of the word embedding and $|V|$ is the size of the vocabulary V. The

position embeddings are $P \in \mathbb{R}^{d_p \times size_p}$ where d_p is the dimensionality of the position embedding and $size_p$ is the number of different relative positions expressible through position embeddings.

For a sentence of length m, the word vector at the ith position, $v_i = [w_i, p_i^{a0}, p_i^{a1}]$, is the concatenation of its word embedding w_i, the position embedding relative to the first argument p_i^{a0} and the position embedding relative to the second argument p_i^{a1}. In the case of unary contexts, only a single argument is used.

A piecewise max-pooled convolution (PCNN) is then applied, with the pieces defined by the position of the argument (or arguments for binary contexts): before the (first) argument, the argument (between the arguments), and after the (second) argument. A fully connected layer then produces the sentence vector representation. This is a refinement of the Neural Relation Extraction (NRE) [11] approach to sentence-to-vector mapping. The fully connected layer over the PCNN is an addition.

Let $v_{i:i+fw}$ indicate the concatenation of word vectors $v_i, v_{i+1}, ..., v_{i+fw}$. The filter matrix is $F \in \mathbb{R}^{fw(d_w+2d_p) \times d_f}$, where fw is the filter width. The position of first and second arguments are indicated by pos_0, pos_1 respectively. The piecewise max-pooled convolution is given below:

$$c_i = tanh(F \cdot v_{i:i+fw} + b_f)$$
$$p_j^{s0} = max_{i \in [0, pos_0]}(c_{i,j})$$
$$p_j^{s1} = max_{i \in [pos_0, pos_1]}(c_{i,j})$$
$$p_j^{s2} = max_{i \in [pos_1, m)}(c_{i,j})$$

The sentence vector x is produced by a fully connected layer over the concatenated outputs of the piecewise max-pool.

$$x = tanh(L_s \cdot [p^{s0}, p^{s1}, p^{s2}] + b_s)$$

The weight matrix for the sentence vector representation is $L_s \in \mathbb{R}^{3d_f \times d_s}$. Dropout is applied on the context vector x.

The sentence vector aggregation portion of the neural architecture uses a Network-in-Network over the sentence vectors. Network-in-Network (NiN) [10] is an approach of 1×1 CNNs to image processing. The width-1 CNN we use for mention aggregation is an adaptation to a set of sentence vectors. The result is max-pooled and put through a fully connected layer to produce the score for each relation. Unlike a maximum aggregation used in many previous works [22] on binary relation extraction, the evidence from many contexts can be combined to produce a prediction. Unlike attention-based pooling also used previously for binary relation extraction [11], the different contexts can contribute to different aspects, not just different degrees. For example, a prediction that a city is in France might depend on the conjunction of several facets of textual evidence linking the city to the French language, the Euro, and Norman history.

Formally, the NiN is a width-1 convolution with filter matrix $A \in R^{d_s \times d_a}$, where d_a is the dimensionality of the resulting context-set vector.

$$a_i = tanh(A \cdot x_i + b_a)$$
$$p_j^a = max_{i \in [0,n)}(a_{i,j})$$

NiN is an optional layer, the alternative is to simply apply the relation prediction to the sentence vector and take the maximum relation prediction over all contexts.

The relation prediction layer has weight matrix $L_r \in \mathbb{R}^{r \times d_a}$ where r is the number of relations. The final relation prediction vector is $sigmoid(L_r \cdot p^a + b_r)$.

The final layer of the network is vector of relation predictions and the intermediate layers are shared. This architecture allows us to efficiently train many relations, while reusing the feature representations in the intermediate layers across relations as a form of transfer learning. The predictions of this network represent the probability for the input entity to belong to each relation.

5   Implicit Relations

In a traditional binary KBP task a triple has a *relevant context set* if the two entities occur at least once together in the corpus - where the notion of 'together' is typically intra-sentential (within a single sentence). To overcome this issue Socrates uses a more aggressive approach to generate context sets that enables us to recognize relations between entities even if they to not occur in the same sentence. In this section we present our solutions to deal with implicit information: unary relations and composite contexts.

5.1   Unary Relations

Unary relations were recently introduced as an approach for gathering implicit knowledge from text [7]. The basic idea is that in many cases relation extraction problems can be reduced to sets of simpler and inter-related unary relation extraction problems. This is possible by providing a specific value to one of the two arguments, transforming the relations into a set of categories. For example, the *livesIn* relation between persons and countries can be decomposed into 195 relations (one relation for each country), including *livesIn*:UNITED_STATES, *livesIn*:CANADA, and so on.

To recognize unary relations we exploit the same deep learning architecture described in Fig. 2, with the only difference that just one entity is marked in the input. Each unary relation is then recognized by a specific neuron in the final layer of the net. A unary relation extraction system is therefore a multi-class, multi-label classifier that takes an entity as input and returns its probability as a slot filler for each relation.

Binary and unary approaches are limited in different important respects. KBP with unary relations can only produce triples when fixing a relation and argument provides a relatively large corpus extension. Triples such as ⟨BARACK_OBAMA *spouse* MICHELLE_OBAMA⟩ cannot be extracted in this way, since neither Barack nor Michelle Obama have a large set of spouses. The limitation of binary relation extraction is that the arguments must occur together. But for many triples, such as those relating to a person's occupation, a film's genre or a company's product type, the second argument is often not given explicitly.

5.2   Composite Contexts

Socrates is also able to process a TOD associated to some input entity and leverage the focus of the TOD as a component of context. TODs are associated to specific entities in the KB and usually contain mostly information related to the entities. In this case, we can work on the assumption that most of the facts expressed in the documents regards the target entity, even though it has not been mentioned explicitly near another entity in the body of the document.

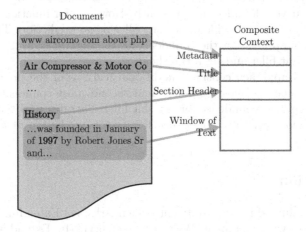

Fig. 3. Construction of composite context

Figure 3 shows the construction of a composite context from a document. The title is an entity in this case, which will always be in-context for any other entity in the document. The document metadata, in this case a URL is also part of any composite context constructed from this document. A section header, if present, is also placed into the context. The main part of the context is a window of text around a mention of an entity.

This enables us to define very effective slot filling strategies for entities where TODs are available. We apply this strategy on the ISWC 2017 KBP challenge, reporting the best performances.

Table 1. Hyperparameters used

Hyperparameter	NYT-FB	CC-DBP(binary)	CC-DBP(unary)	SWC-2017
$size_p$	80	80	80	80
d_w	50	50	50	50
d_p	5	5	5	5
d_s	400	800	400	100
d_a	N/A	N/A	400	16
d_f	1000	3000	1000	3000
fw	3	3	3	3
dropout	0.5	0.5	0.5	0.5

5.3   Final Merger

The relational prediction system considers each prediction for a slot filler, such as phone number or year founded independently. However, for functional relations, the slot filling task is to provide either one filler or no prediction. The simplest approach is to simply assign the confidence of the highest scoring filler as the confidence for that filler and set a threshold.

An improved approach considers additional features of the prediction such as the gap between the most confident and second most confident prediction to determine the final confidence for the slot filler. Socrates uses these features to estimate a more accurate confidence for its top prediction of a functional relation.

6   Evaluation

Socrates was evaluated in three different benchmarks: (a) Extending Thompson Reuters PermID with Company Websites (Sect. 6.1), (b) Extending Freebase with NYT articles (Sect. 6.2), and (c) Extending DBpedia with Web Crawls (Sect. 6.3). The hyperparameters used in these experiments are shown in Table 1.

6.1   Extending Thompson Reuters PermID with Company Websites

Socrates was evaluated against the state of the art KBC tools as part of the *ISWC Semantic Web Challenge 2017*. The challenge consisted of a knowledge graph population task (Task 1) and a knowledge based validation task (Task 2). Detailed task descriptions as well as the training/test datasets are available from the challenge website[4].

In order to apply knowledge induction to the challenge we needed to gather relevant text. We applied an open source crawler to the URLs provided for each test company. Although some websites did not exist, or did not allow crawling,

---

[4] https://iswc2017.semanticweb.org/calls/iswc-semantic-web-challenge-2017/.

we were able to get URLs for over 90% of the companies in the test data. We also crawled websites of 80,000 companies from the training data. The websites were processed with boilerpipe [9] to extract the text.

Semantic Web Challenge 2017 Results. The evaluations for the challenge were performed using the GERBIL Benchmark Framework [16]. The results are shown in Table 2 (Task 1) and Table 3 (Task 2). Two variations of the Socrates system was evaluated in the challenge. The Socrates-KI system is the results from the components operating with unstructured text only, while Socrates is an extension of Socrates-KI results by looking up missing values from three structured data sources: opencorporates.com, crunchbase.com, and usaspending.gov. As it can be seen, the extension with structured data results in only a small improvement in accuracy.

We tuned the confidence thresholds by testing against a subset of the test data with crowdsourced attribute fillers. Rather than select the optimal threshold for this dataset, we probed four to eight possible thresholds for each submission.

Table 2. Attribute prediction

	F1 [%]
Socrates	55.397
Socrates-KI	54.835
Leopard	53.438
Disco	53.315
YellowPage	46.007
Baseline	45.867

Table 3. Attribute validation

	AUC [%]
Socrates-KI	68.014
Leopard	53.088
Baseline	50.000

Document Classification. Because the set of possible countries for a company's headquarters is small, we adopted a document classification approach using logistic regression. For features we used: the bag of words in the company website, the top level domain (TLD) of the website URL, and the bag of countries detected in the location recognition and linking phase. To help correct for the different distribution of countries between train (the public PermID database) and the test data we removed from training any company whose headquarter's country was not in the list of TLDs for test websites.

Attribute Validation. The attribute validation task did not provide the company website URL as a certainty, but instead gave it as a statement to validate. Conversely, the country for the company was provided as a known fact.

We addressed the validation of the website URL by string kernel similarity between the company name and the URL. Since the headquarters country was given as a known fact, we also checked the top-level-domain (TLD) of the website against a TLD to country mapping.

For phone numbers and years we ran our deep learning based extractors over the provided, possibly erroneous, websites. Additionally we checked the country code for the phone number against the known headquarters country.

Detailed Evaluation and Analysis Using Crowdsourcing. To further investigate the performance of our system across different attributes and more deeply analyze the accuracy results, we built our own benchmark using crowd-sourcing over a sample of 2,000 records from the test data. Note that the outcome of crowdsourcing was used *only for the purposed of this evaluation*, i.e., we did not use the outcome as additional training data and we never included any portion of the outcome in our GERBIL submissions for the challenge.

The first interesting observation from the crowdsourcing experience was the difficulty of the task even for humans. We had to make several iterations to design the Mechanical Turk's Human Intelligence Tasks or "HITs" in a way that the outcome had the least noise and the HITs finished in a reasonable amount of time. Interestingly, the level of agreement between the crowd workers were comparable to the accuracy of our automated extraction. For phone numbers, the workers agreed on 730 values (54.6%). For year founded the number of agreements was 935 (69.99%), while for location country the workers agreed on 1,220 values (91.32%).

Using the outcome of crowdsourcing, we evaluated the accuracy of our extractions over different attributes. Table 4 shows the results of this evaluation. Note the low precision and recall values for year and phone number attributes. This is mainly due to the difficulty of finding the right information on the web. An interesting example is that of a company called "Sterigenics" for which the company website states the year founded is 1925, Crunchbase.com has 1978 as the year founded, and the crowd worker provided 2004 as the value. Interestingly, all these values can be seen as correct, as they belong to various subsidiaries and branches of the same company.

Table 4. Results on mechanical turk by attribute

	True positive count	Precision	Recall	F1
Overall results	1880	0.6775	0.4692	0.5544
Phone number	377	0.4883	0.2822	0.3577
Country	1233	0.9229	0.9229	0.9229
Year founded	270	0.4048	0.2022	0.2697

6.2 Extending Freebase with NYT Articles

A standard benchmark for distantly supervised relation extraction was developed by Riedel [14] and used in many subsequent works [8,20,22]. The text of New

York Times was processed with the Stanford NER system and the identified entities linked by name to Freebase. The task is to predict the instances of 52 relations from the sentences mentioning two arguments.

The state-of-the-art for this dataset is NRE's (Neural Relation Extraction) PCNN+ATT model (Piecewise Convolutional Neural Network with Attention) [11]. The binary relation extraction of Socrates is most related to PCNN+ONE, with the incorporation of type information from the entity recognition, an additional fully connected layer before the final max-pooling and an increased number of filters in the sentence-to-vector convolutional layer.

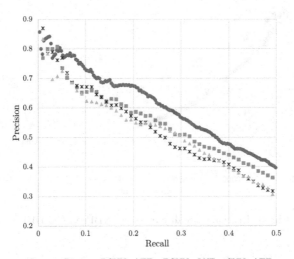

Fig. 4. Precision recall curves for KBP on NYT-FB

Figure 4 shows the performance for Socrates' binary relation extraction on this dataset compared to the models of NRE. Only the binary model is tested on this dataset because the dataset is already processed to the point of context set construction, and only binary contexts are produced. As can be seen from the precision-recall curve, the model of Socrates improves on the state-of-the-art in this standard dataset.

6.3   Extending DBpedia with Web Crawls

We also evaluate on a web-scale knowledge base population benchmark that we called $CC\text{-}DBP^5$. It combines the text of Common Crawl[6] with the triples from 298 frequent relations in DBpedia [1]. Mentions of DBpedia entities are located in text by gazetteer matching of the preferred label.

---

[5] https://github.com/IBM/cc-dbp.
[6] http://commoncrawl.org.

Figure 5 shows the precision-recall curves for unary only, binary only and the combined system. The unary and binary systems alone achieve similar performance. But they are effective at very different triples. This is shown in the large gains from combining these complementary approaches. For example, at 0.5 precision, the combined approach has a recall of more than double (15,750 vs 7,400) compared to binary alone, which represents over 100% relative improvement.

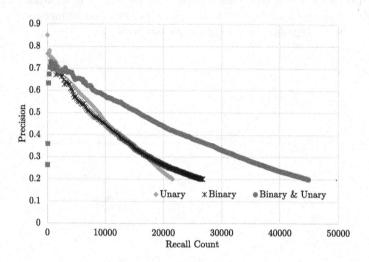

Fig. 5. Precision recall curves for KBP on CC-DBP

We did not identify TODs in common crawl, so we do not use composite contexts for this task. We combine the output of the two systems by, for each triple, taking the highest confidence from each system. We also ran the PCNN+ATT model of NRE on this dataset, but without hyperparameter tuning its performance was very low.

The recall is given as a triple count rather than a percentage. Traditional attempts to measure the recall of KBP systems use the set of all triples explicitly stated in text for the denominator of recall. This is unsuitable for evaluating our approach because the system is able to make probabilistic predictions based on implicit and partial textual evidence, thus producing correct triples outside the classic recall basis.

7   Conclusion and Future Work

Knowledge Base Population is an important research problem in the Semantic Web research and in this paper we presented Socrates, a KBP system able to capture implicit relations in text. To this aim, we introduced the notion of unary context sets and implicit context. Socrates was evaluated in three different benchmarks and we demonstrated that there is a consistent improvement over the state-of-the-art. Our approach is extremely effective and complements existing binary relation extraction methods for KBP. Remarkably, Socrates achieved the best performance on both tasks of the *ISWC Semantic Web Challenge 2017*.

The different approaches to context set construction we have unified in the Socrates system provide complementary sources of textual evidence for the prediction of relations. The binary contexts require no assumptions about the type of document or its structure, but are limited to cases where both arguments of a relation occur together. Unary contexts provide textual evidence for unary relations, but unary relations can only be trained when enough fillers exist for a given relation and fixed argument. Finally, composite contexts still require both arguments to be mentioned in a single document, but by leveraging the document structure we remove the limitation of close co-occurrence.

In future work, we plan to explore the use of more advanced forms of entity detection and linking, including propagating features from the EDL system forward for both unary and binary deep models. In addition we plan to exploit extracted relations as source of evidence to bootstrap a probabilistic reasoning approach, with the goal of leveraging ontological constraints from the KB such as the property domain, range and other axioms. We also plan to develop strategies for integrating the new triples gathered from textual evidence with new triples predicted from existing KB relationships by knowledge base completion.

References

1. Auer, S., Bizer, C., Kobilarov, G., Lehmann, J., Cyganiak, R., Ives, Z.: DBpedia: a nucleus for a web of open data. In: Aberer, K. (ed.) ASWC/ISWC -2007. LNCS, vol. 4825, pp. 722–735. Springer, Heidelberg (2007). https://doi.org/10.1007/978-3-540-76298-0_52

2. Chang, H., et al.: Extracting multilingual relations under limited resources: TAC 2016 cold-start KB construction and slot-filling using compositional universal schema. In: Proceedings of TAC (2016)

3. Drozd, A., Gladkova, A., Matsuoka, S.: Word embeddings, analogies, and machine learning: beyond king - man + woman = queen. In: Proceedings of COLING 2016, the 26th International Conference on Computational Linguistics, pp. 3519–3530 (2016)

4. Feng, X., Guo, J., Qin, B., Liu, T., Liu, Y.: Effective deep memory networks for distant supervised relation extraction. In: Proceedings of the Twenty-Sixth International Joint Conference on Artificial Intelligence, IJCAI 2017, Melbourne, Australia, 19–25 August 2017, pp. 4002–4008 (2017). https://doi.org/10.24963/ijcai.2017/559

5. Ferrucci, D., et al.: Building watson: an overview of the DeepQA project. AI Mag. **31**(3), 59–79 (2010)

6. Glass, M., Gliozzo, A.: A dataset for web-scale knowledge base population. In: Gangemi, A., et al. (eds.) ESWC 2018. LNCS, vol. 10843, pp. 256–271. Springer, Cham (2018). https://doi.org/10.1007/978-3-319-93417-4_17

7. Glass, M., Gliozzo, A.: Discovering implicit knowledge with unary relations. Preprint (2018). https://ibm.box.com/s/31jqgm5xxjixetee4b1upisxdwbtw12r

8. Hoffmann, R., Zhang, C., Ling, X., Zettlemoyer, L., Weld, D.S.: Knowledge-based weak supervision for information extraction of overlapping relations. In: Proceedings of the 49th Annual Meeting of the Association for Computational Linguistics: Human Language Technologies, vol. 1, pp. 541–550. Association for Computational Linguistics (2011)
9. Kohlschütter, C., Fankhauser, P., Nejdl, W.: Boilerplate detection using shallow text features. In: Proceedings of the Third ACM International Conference on Web Search and Data Mining, pp. 441–450. WSDM 2010. ACM, New York, NY, USA (2010). https://doi.org/10.1145/1718487.1718542
10. Lin, M., Chen, Q., Yan, S.: Network in network. arXiv preprint arXiv:1312.4400 (2013)
11. Lin, Y., Shen, S., Liu, Z., Luan, H., Sun, M.: Neural relation extraction with selective attention over instances. In: Proceedings of ACL (2016)
12. Mikolov, T., Sutskever, I., Chen, K., Corrado, G.S., Dean, J.: Distributed representations of words and phrases and their compositionality. In: Burges, C.J.C., Bottou, L., Welling, M., Ghahramani, Z., Weinberger, K.Q. (eds.) Advances in Neural Information Processing Systems 26, pp. 3111–3119. Curran Associates, Inc. (2013)
13. Mintz, M., Bills, S., Snow, R., Jurafsky, D.: Distant supervision for relation extraction without labeled data. In: Proceedings of the Joint Conference of the 47th Annual Meeting of the ACL and the 4th International Joint Conference on Natural Language Processing of the AFNLP: Volume 2-Volume 2, pp. 1003–1011. Association for Computational Linguistics (2009)
14. Riedel, S., Yao, L., McCallum, A.: Modeling relations and their mentions without labeled text. In: Balcázar, J.L., Bonchi, F., Gionis, A., Sebag, M. (eds.) ECML PKDD 2010. LNCS (LNAI), vol. 6323, pp. 148–163. Springer, Heidelberg (2010). https://doi.org/10.1007/978-3-642-15939-8_10
15. Riedel, S., Yao, L., McCallum, A., Marlin, B.M.: Relation extraction with matrix factorization and universal schemas. In: Proceedings of the 2013 Conference of the North American Chapter of the Association for Computational Linguistics: Human Language Technologies, pp. 74–84 (2013)
16. Röder, M., Usbeck, R., Ngomo, A.C.N.: GERBIL-benchmarking named entity recognition and linking consistently. Semant. Web J. (2018). http://www.semantic-web-journal.net/system/files/swj1671.pdf
17. Roth, B., Monath, N., Belanger, D., Strubell, E., Verga, P., McCallum, A.: Building knowledge bases with universal schema: cold start and slot-filling approaches. In: Proceedings of the Eighth Text Analysis Conference (TAC 2015) (2015)
18. Shin, J., Wu, S., Wang, F., De Sa, C., Zhang, C., Ré, C.: Incremental knowledge base construction using deepdive. Proc. VLDB Endow. **8**(11), 1310–1321 (2015)
19. Shwartz, V., Goldberg, Y., Dagan, I.: Improving hypernymy detection with an integrated path-based and distributional method. In: Annual Conference of the Association for Computational Linguistics (ACL), pp. 2389–2398 (2016)
20. Surdeanu, M., Tibshirani, J., Nallapati, R., Manning, C.D.: Multi-instance multi-label learning for relation extraction. In: Proceedings of the 2012 Joint Conference on Empirical Methods in Natural Language Processing and Computational Natural Language Learning, pp. 455–465. Association for Computational Linguistics (2012)
21. Xu, Y., Mou, L., Li, G., Chen, Y., Peng, H., Jin, Z.: Classifying relations via long short term memory networks along shortest dependency paths. In: Proceedings of the 2015 Conference on Empirical Methods in Natural Language Processing, pp. 1785–1794 (2015)
22. Zeng, D., Liu, K., Chen, Y., Zhao, J.: Distant supervision for relation extraction via piecewise convolutional neural networks. In: EMNLP, pp. 1753–1762 (2015)

23. Zeng, D., Liu, K., Lai, S., Zhou, G., Zhao, J.: Relation classification via convolutional deep neural network. In: Proceedings of COLING 2014, the 25th International Conference on Computational Linguistics: Technical Papers, pp. 2335–2344 (2014)
24. Zeng, W., Lin, Y., Liu, Z., Sun, M.: Incorporating relation paths in neural relation extraction. arXiv preprint arXiv:1609.07479 (2016)
25. Zhang, Y., et al.: Stanford at TAC KBP 2016: sealing pipeline leaks and understanding Chinese. In: Proceedings of TAC (2016)

Towards Encoding Time in Text-Based Entity Embeddings

Federico Bianchi(✉), Matteo Palmonari, and Debora Nozza

University of Milano - Bicocca, Viale Sarca 336, Milan, Italy
{federico.bianchi,palmonari,debora.nozza}@disco.unimib.it

Abstract. Knowledge Graphs (KG) are widely used abstractions to represent entity-centric knowledge. Approaches to embed entities, entity types and relations represented in the graph into vector spaces - often referred to as KG embeddings - have become increasingly popular for their ability to capture the similarity between entities and support other reasoning tasks. However, representation of time has received little attention in these approaches. In this work, we make a first step to encode time into vector-based entity representations using a text-based KG embedding model named Typed Entity Embeddings (TEEs). In TEEs, each entity is represented by a vector that represents the entity and its type, which is learned from entity mentions found in a text corpus. Inspired by evidence from cognitive sciences and application-oriented concerns, we propose an approach to encode representations of years into TEEs by aggregating the representations of the entities that occur in event-based descriptions of the years. These representations are used to define two time-aware similarity measures to control the implicit effect of time on entity similarity. Experimental results show that the linear order of years obtained using our model is highly correlated with natural time flow and the effectiveness of the time-aware similarity measure proposed to flatten the time effect on entity similarity.

1 Introduction

Knowledge Graphs (KGs) provide useful abstractions for representing knowledge, with nodes describing real-world entities and entity types, and labeled edges representing relations between entities, between types, and between entities and types. Traditional approaches to represent KGs use graph databases and semantic web technologies based on the RDF model[1]. More recently, complementary models to represent KGs have been proposed, which embed KG elements such as entities, types and relations into vector spaces of fixed dimensionality and learn such representations from large amounts of data [1,6,14,19,21,30,32]. We refer to these models as *KG embeddings*. In KG embeddings, entities are represented by vectors, and efficient geometric operations can support a variety of tasks such as the evaluation of similarity between arbitrary entity pairs.

---

[1] https://www.w3.org/RDF/.

© Springer Nature Switzerland AG 2018
D. Vrandečić et al. (Eds.): ISWC 2018, LNCS 11136, pp. 56–71, 2018.
https://doi.org/10.1007/978-3-030-00671-6_4

Some approaches generate KG embeddings using structured data as a source, e.g., relations occurring in the KG, and are mainly targeted at predictive reasoning tasks such as link prediction [6,14,32]. Other approaches generate the KG embeddings from text corpora using methods similar to the ones used to generate word embeddings [1], under the distributional hypothesis [11]. These models referred to as *text-based KG embeddings* in the following, are mainly targeted at similarity evaluation tasks.

In a previous work, we have presented Typed Entity Embeddings (TEEs) as one of the latter models [2,4]. In TEEs, embeddings of entities and types are generated under the following entity-centric reinterpretation of the distributional hypothesis: *entities and types that appear in similar contexts are similar*. An entity linking algorithm [22] is used to find entity mentions in the corpus, while the KG is used to extract the most specific types of the mentioned entities. Then, based on the co-occurrence of entities and types in the text corpora two vector spaces are generated, one for entities and one for types. The direct sum of the two vector spaces leads to a *typed entities space*. In this space, each entity is represented by the concatenation of its vector in the entity space and the vector of its type. For example, the typed vector of the DBpedia entity dbr:Rome is the concatenation of the vectors generated for dbr:Rome and dbo:City.

Time is an important aspect in knowledge representation and has been extensively studied in the field of qualitative temporal representation and reasoning [15,23,25,31]. In addition, time is essential to human cognition, as people "place events in time, deciding when they occurred, in which order and on what scale, whether that of a lifetime or of a few seconds" [8]. Finally, recent work has investigated temporal word embeddings to study language evolution along time using diachronic corpora [27]. Thus, we believe that encoding time into KG embeddings models is an important research objective.

Our work is inspired by evidence found in cognitive science studies as well as by application-oriented concerns. Time and time perception have been deeply investigated in the cognitive psychology literature. Since it has been observed that "the succession of events is an inherent property of our time perception. Memory is necessary, and the order of these events is fundamental" [26], we may consider textual descriptions of events found in text corpora as a sort of memory, and as a source for learning representations of time. Encoding time into KG embeddings has also several practical applications, in particular when evaluating entity similarity with text-based KG embeddings, where similarity depends on entity co-occurrence in similar contexts. Time can sneak into entity similarity in a way that cannot be controlled, because entities that share a temporal context are more likely to co-occur in the text (we refer to this implicit effect of time on entity similarity as to the *time effect hypothesis*). As a consequence, we may find that the most similar entity to dbr:Winston Churcill is the little-known dbr:Harold McMillian. This makes perfect sense, *if time is considered when evaluating the similarity*, but if we want to compare UK politician by international relevance and fame, rather then by their chronological order, we may prefer to find also more famous prime ministers like dbr:Margaret Thatcher among the most similar entities to dbr:Winston Churcill. If we are able to explicitly incorporate time into KG embeddings, then we can control its effect when evaluating the similarity between entities, boosting or flattening time effect in entity similarity. Such control over similarity is helpful for example in knowledge exploration applications, which we investigated in previous work [3]. Other potential applications can be found in time-aware entity recommendations [18,28] (e.g., find "related contemporary entities" vs. "related entities in the past" vs. "time-independent related entities") and in temporal information retrieval, where it is important to keep track of the time factor.

To the best of our knowledge, in this paper we propose a first approach to make time a first-class citizen in KG embeddings. We use the text-based TEEs model as background and learn explicit representations of temporal entities as part of this model. In particular, we encode representations of years, i.e., we embed regular time periods with a yearly granularity. We build year representations from the textual description of events occurring during each year, which are available in different web sources[2]. We generate year representations by aggregating the representations of the entities that take part in events occurring in those particular years. These representations are then used to define two parametric *time-aware similarity functions*: time-flattening and time-boosting similarities.

In other words, in this paper we tackle the following research challenges: (1) to generate representations of time periods that are inspired by evidence found in cognitive psychology for memory being a fundamental aspect in time representation [26]; (2) to use these representations to control the effect of time over entity similarity for practical applications. The contributions with respect to these objectives can be summarized as follows:

– We learn representations of time periods at a yearly granularity starting from natural language descriptions of events occurring in these periods, showing that, even if the natural time flow is not explicitly encoded into the model, the generated year representations are highly correlated with the natural sequence of years.
– We provide evidence for the time-effect on entity similarity in text-based KG embeddings.
– We propose two parametric time-aware similarity measures to control the time effect in entity similarity.

Our approach to encoding time into text-based KG embeddings is explained in Sect. 2. Experiments to evaluate the time effect, properties of the year representations and the time-aware similarities are discussed in Sect. 3. Related work is discussed in Sect. 4. Conclusions and future work end the paper.

2 Typed Entity Embeddings with Time Periods

We use a minimal definition of Knowledge Graph (KG) as a directed labeled graph, where nodes are *entities* or *types*, and labeled edges represent

---

[2] Examples are Wikipedia pages for years, https://www.onthisday.com/events-by-year.php and https://www.history.com/this-day-in-history.

relations between entities or between types or between entities and types. Relations between types define a sub-type graph (often a hierarchy). However, the relations between the entities are not considered for generating the embeddings in our work. An example of KG, which will be used in the rest of the paper, is DBpedia, where types are classes in the DBpedia Ontology. For simplicity, we assume that given an entity we can determine its minimum type (i.e., its most specific type) using the typing assertions and the sub-type graph. In case an entity has more minimal types none of which is the minimum, different strategies can be considered to select one of them as representative type. In DBpedia the most specific type of entities can be determined by using dedicated resources[3].

A Typed Entity Embedding (TEE) with Time Periods consists in:

- A set of typed entities E, which includes a subset E^τ of temporal periods.
- Two embedding functions; $\phi : E \setminus E^\tau \to \mathbb{R}^k$ and one $\omega : E^\tau \to \mathbb{R}^k$.
- A similarity function $\eta : E \times E \to [0,1]$ constructed as operation over the typed entity vectors.
- A proximity function $\rho : E \to E^\tau$ that allows us to find the most representing time period for a given typed entity.
- A time-aware similarity function $\psi : E \times E \to [0,1]$ that computes the similarity between two typed entities by considering their time distance as a factor.

In the following, we explain each component of the model more in detail.

2.1 TEEs and Their Generation

The ϕ embedding function has been intuitively explained in Sect. 1. For more details we refer to previous work [2,4]. Here we provide some more insights about the generation process using Fig. 1. As a corpus we use a set of documents, each one describing an entity in natural language. As shown in the figure, after the entity mentions are found in a document by an entity linking algorithm, we generate a second document that consists in the sequence of the entities found in the corpus text. This document is transformed into a third document, where entities are replaced by their most specific type, obtaining a sequence of types. As a result, we have two corpora, one for entities and one for types. At this point we run word2vec [16] on each corpus to generate Entity Embeddings (**EE**) and Type Embeddings (**TE**) [4] separately. These two embeddings can have different dimensionality. Finally, for each entity, we concatenate its entity vector with the vector of its most specific type, thus obtaining a typed entity vector, i.e., a Typed Entity Embedding (**TEE**). Entities of the same type (or of similar types) are more likely to be closer to each other in the TEE space built upon this concatenation than in the EE space that consists of entity-only vectors [2]. Given an entity e, we use the bold notation **e** to refer to its typed entity vector.

---

[3] http://wiki.dbpedia.org/services-resources/documentation/datasets#InstanceTyp es.

Observe that since we have generated the typed entity vector space as the direct sum of the entity and the type vector spaces, we can easily drop the type-component in typed entity vectors and use entity vectors without representing the entity types. We will use these simplified entity vectors in the experiments discussed in Sect. 3.1.

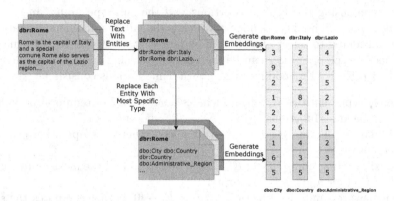

Fig. 1. Entity embedding process: textual content is replaced by entities. Each entity is replaced with its own type. Embeddings can be then generated using word2vec. Finally each entity is concatenated to its own type.

2.2   Encoding Temporal Periods into TEEs

In this work, we consider time as a set of connected time periods E^τ, i.e., a sequence of time periods totally ordered by a relation $<^\tau$. We represent time periods at the year granularity, meaning that each year represents the time period that spans over the year duration. In the following, we describe the function ω, which embeds time periods into $\mathbb{R}^k$.

Our main hypothesis is that discrete periods of times can be embedded in a vector space, where each period is represented by a vector, in such a way that years that are near in time have similar vectors. A second hypothesis that drives our approach is that a period of time, e.g., a year, can be described by the entities that take part in the events that occur during the time period. For example, years in the first half of the 40s are characterized by World War II events and by the entities that had a relevant role in these events. For the experiments conducted in this paper, we consider textual descriptions of events that appear in the Wikipedia pages that describe years[4].

To generate the representation of a year, we extract entities from the corresponding Wikipedia page and compute the average vector of the entity vectors defined in the EE space[5]. In other words, we drop the type component from the

---

[4] E.g., https://en.wikipedia.org/wiki/1943.

[5] Our representation of years is independent from the ϕ embedding function: other embedding algorithms could be used to compute the entities representation.

typed entity vectors, to use a more entity-centric representation (types occur more regularly across years). To generate a TEE-compatible representation of the time periods in the vector space, we concatenate each embedding generated in the EE space with a vector consisting of 0s of the same dimension of type vector in TEE. This process is briefly summarized in Fig. 2. Slight variations of this process are also possible, e.g., different vector aggregation methods, as discussed in Sect. 2.3. Finally, we empirically found that it is better to consider only the "Events" section in year Wikipedia pages for entity extraction because the sections "Births" and "Deaths" produce noisier representations.

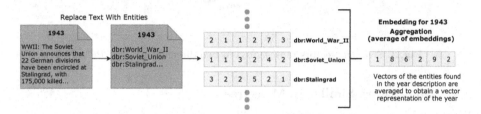

Fig. 2. Year embedding process.

Since non-temporal and temporal entities (i.e., time periods) are embedded in the same space, a comparison between entities of these two kinds is possible. In addition, these representations of time periods are generated using collective knowledge of what has happened during their time. However, since no explicit constraints over time ordering are used in the generation process, a natural question is which relation can be found between the vector-based representations of time periods and the natural time order. In Fig. 3 we show an example of the 2D representation of the years from 1900 to 2015 using PCA. Interestingly, the years seem to follow a natural time order from left to right. A statistical correlation analysis between the one-dimensional projection of years using PCA and their natural order confirms this intuition (Kendall $\tau = \mathbf{0.80}$, Spearman Rank correlation coefficient $= \mathbf{0.94}$).

2.3 Temporal Embeddings Alternative Configurations

There are different ways to use the entities that are found in the year description: (1) considering the entities only one time (i.e., as a set of entities); (2) considering the entities multiple times if they appear more than once (i.e., United States might appear more than one time in the text); (3) Using TF-IDF on the whole year corpus to weight each term by its own TF-IDF score and apply this with (1) and (2). We generated models using all these alternative configurations and we projected the embeddings into 1D using PCA and compared this ordering in 1D with the natural flow time order (i.e., the natural sequence from 1900 to 2015) and we obtained that the model that considers each entity only once (1) is the one that is most correlated to the natural time order. We thus decided to use this configuration to generate temporal embeddings.

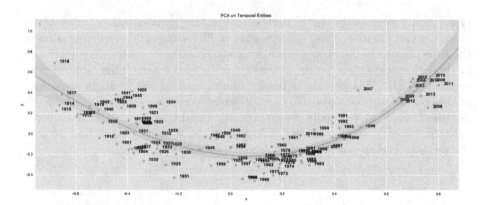

Fig. 3. Average vectors represented in two dimensions using PCA.

2.4   Time-Aware Similarity Measures

We propose a new way of computing similarity that also considers the temporal factor in the embeddings. Given an entity $e \in E$ is it possible to find its most representing year in E^τ by considering the functions defined in the model. To get the most representative year for a given entity we select the most similar year in E^τ to a given entity.

$$\rho(e) = \underset{e^\tau \in E^\tau}{\operatorname{argmax}} \, cos(\phi(e), \omega(e^\tau))$$

We use $\mathbf{e}^\tau$ to denote the vector of $\rho(e)$, i.e., the closer year to a given entity in the vector space.

We can now define two time-aware similarity functions: a **time-flattened similarity** and a **time-boosted similarity**. The time-flattened similarity can be computed using the following formula.

$$\psi(e_1, e_2) = \alpha\eta(\mathbf{e_1}, \mathbf{e_2}) - (1 - \alpha)\eta_n(\mathbf{e_1^\tau}, \mathbf{e_2^\tau})$$

Where $\mathbf{e_1}$, $\mathbf{e_2}$ are the embeddings of the entities e_1, e_2, η is the cosine similarity in the typed entity space, and α is a parameter that can be used to regulate the weight of the time flattening factor. Time flattening is obtained by subtracting the temporal similarity η_n of the most representative temporal periods (i.e., years) of both entities $\mathbf{e_1^\tau}$ and $\mathbf{e_2^\tau}$. The temporal similarity is defined as the cosine similarity between two years in the typed entity space normalized in the interval $[0, 1]$ with a max-min approach, by considering the maximum similarity between two years in the representation and the minimal similarity between two years in the representation. We adopt this normalization to make the year factor of the similarity work as a weight factor.

A time-boosted similarity function can be defined analogously by adapting the formula in such a way that a time-boosting factor is summed to the similarity between the typed entities:

$$\psi(e_1, e_2) = \alpha\eta(\mathbf{e_1}, \mathbf{e_2}) + (1 - \alpha)\eta_n(\mathbf{e_1^\tau}, \mathbf{e_2^\tau})$$

3 Experimental Evaluation

The experiments that we discuss in this section have the following goals: (1) validate the time-effect hypothesis introduced in Sect. 1, that is, that time influences the distribution of entities in vector spaces, and the rationale behind the generation of time period representations from those of the entities that take part in the events that occur during these time periods; (2) evaluate the effectiveness of the time-aware similarity function.

Experimental Settings (for all the Experiments). Our entity embeddings were generated using the DBpedia's abstracts (4M of textual documents) from the 2016 dump[6]. We used the skip-gram algorithm for obtaining the entities embeddings [16]. We used a window of 5 in the algorithm and types and entities are embedded into 100 dimensional vector spaces. We annotated text using DBpedia Spotlight [7]. For the year embeddings we decide to concentrate our experiments on the years from 1900 to 2015. Code and dataset are freely available online so that experiments can be replicated[7].

3.1 Time Effect and Temporal Representations in Text-Based Entity Embeddings

In a first experiment, we evaluate if the time effect can be noticed in text-based entity embeddings, i.e., temporal contexts shared by entities have an effect on their similarity. Then, with a second experiment, we provide evidence that years that are close in time are more likely to have descriptions that share a larger number of entities, thus supporting the rationale behind using entity representations to generate temporal representations. The time effect validated with the first experiment adds even more substance to this idea, since the time effect suggests that time is implicitly encoded in entity representations. As a consequence, temporal representations are generated using entity representations that implicitly encode some temporal characterization. Finally, in a third experiment, we investigate if the space that jointly represents years (temporal entities) and other entities can support entity ordering over time, to further evaluate the quality of the temporal representations and their relation to the representations of the other entities. We remind that some properties of our model, e.g., the correlation between projection on one dimension of temporal entities and natural flow of time, have been discussed in Sect. 2.

Classifying World War I and World War II Battles
Our assumption is that entity embeddings share time-based context, and thus entities that live nearby times are closer to each other in the vector space. We collect battles from World War I and World War II using the list provided by Wikipedia military engagements pages[8,9]. We used a clustering algorithm to

---

[6] http://wiki.dbpedia.org/dbpedia-version-2016-04.
[7] https://github.com/vinid/time-aware.
[8] https://en.wikipedia.org/wiki/List_of_World_War_I_battles.
[9] https://en.wikipedia.org/wiki/List_of_military_engagements_of_World_War_II.

understand if there is an underlying pattern that puts entities in two different groups. We use K-means (number of clusters equal to 2 to cluster the vector representations of the battles in two different groups represented by the two wars). We will evaluate the performance of the clustering algorithm by considering how many years were clustered in the correct group.

Dataset. Our dataset contains 152 battles linked to Wikipedia (and thus DBpedia) from the two different periods 1914–1918 and 1939–1945. 63 battles are from World War I while 89 are from World War II.

Results. In Table 1 we show the confusion matrix that we obtained after clustering the embedded vectors of the battles with K-Means. Out of 152 samples, 146 were correctly associated to the same cluster, while 8 were classified in the wrong one. Accuracy is around 95%. Another interesting result is that the two cluster centroids are closer to the respective war years: the first centroid is near the years of World War I, while the second centroid is close to the years of World War II.

Table 1. Confusion matrix for World War I and II clustering.

Actual Values (n = 152)	Predicted	
	World War I	World War II
World War I	57	6
World War II	2	87

Adherence to Natural Time Order

Our intuition suggests that the descriptions of contiguous years (e.g. 1943 and 1944) share more entities than the descriptions of years that are not contiguous (e.g. 1901 and 1992).

Methodology and Dataset. We collect every possible combination of two and three contiguous years (e.g. 1900–1901, 1901–1902; 1900–1901–1902, 1901–1902–1903) and we compute the average number of shared entities. We compare the values of these two samples with the average values of shared entities of *every* possible combination of two and three years (e.g. 1902–1992, thus considering also noncontiguous years).

Evaluation. Results on the average number of shared entities are reported in Table 2. Pairs and triples of contiguous years have a higher amount of shared entities with respect to noncontinuous years.

We use the Kolmogorov-Smirnov test to detect if the average number of shared entities of contiguous years is statistically different from the respective value for of all the combinations of years. A p-value lower than 0.05 confirms our hypothesis.

Table 2. Average number of shared entities between continuous and non contiguous years.

	Contiguous-2	All-2	Contiguous-3	All-3
Average	55.6	33.5	27.7	12.6
Std	30.3	19.2	13.3	8.22

Relative Ordering of Entities by Time

In this experiment, we show that time actually affects the position of entities in the space.

Dataset. We pick 101 entities from different groups of people and events (United States Presidents, British Prime Ministers, French Presidents, Fifa World-Cup Years, Wars over in 1900, Olympics Events). Entities and groups have been chosen so as to select pairs of entities for which a chronological order can be established upon a reasonably objective criterion (e.g., dbr:Barack Obama is a president elected after dbr:Woodrow Wilson). We acknowledge that ordering people by considering one single feature is a strong assumption. However, being prime minister or president is a very discriminant feature for people.

Methodology. For each entity pair, we compare the manually determined relative order with the order of their most representative years according to our model. The most representative year of an entity is the the closest year to the entity in the vector space. We want to show that given two entities $e_1, e_2 \in E$ such that e_1 is known to chronologically come before e_2, it is likely that $\rho(e_1) >^\tau \rho(e^2)$. For each pair, we also computed the number of *time steps* separating the two entities. This is a measures that indicates a relative distance between two entities: the number of time steps between Barack Obama and George W. Bush is 1, because Bush was the US president before Obama, while between Barack Obama and Bill Clinton it is 2, because Bush was between Obama and Clinton. The same is applied to events like the Fifa World Cup (e.g., the time step between the 2006 world cup and the 2002 world cup is 1).

Results. The accuracy of the relative orders was 70%. For 217 pairs the model was not able to decide a relative ordering since $\rho(e_1) = \rho(e_2)$[10]. In Fig. 4, we show the distribution of correctly relative ordered pairs, incorrect relative ordered pairs and pairs that the model could not order by time steps involved in the relative order. The radius of the points is used to indicate the number of time pairs with a certain time step in each category. It is clear that the farther in time two entities are, the easier it is to determine a correct relative order. Thus, time influences the position of entities in the vector space and the estimation of a relative time order between entities using their representative year is quite accurate.

---

[10] If for the generation of e^τ we consider the average of the nearest 10 years to an entity all the 902 pairs can be compared and the accuracy reaches 92%.

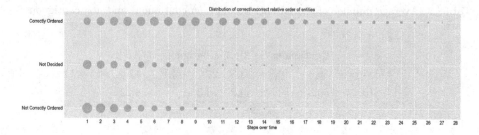

Fig. 4. Distribution of correctly and not correctly ordered pairs with the use of EE.

3.2 Time-Aware Similarity

To test time-aware similarity measures we concentrate on time-flattened similarity for two reasons: it is useful to mitigate the time-effect that we have discussed in the previous experiment and it can be evaluated more objectively using data available in KGs. Defining what is similar when considering the time variable is a challenging task. In this experiment we decided to provide a small-scale experiment on the possible use of the time-flattened similarity by considering reasonably objective orders is time. A time-flattening similarity should reduce the effect that a shared temporal context has on finding similar entities. A time-flattened similarity can be used to find entities that are similar independently from the temporal context they share. For example, given a prime minister, we would expect to find many other prime ministers among its most similar entities if we neglect time, but the time-effect moves many prime ministers down in the ranking with cosine similarity.

Dataset. In line with other experiments done on similarity and relatedness [12] we create a dataset containing entities that are related but distant in time. Given an input entity the task consists in finding similar entities that are distant in time (e.g., given dbr:Barack Obama, a time-flattened similarity should rank in higher position the entities dbr:Theodore Roosevelt and dbr:William Howard Taft). Given a set of 12 prime ministers, the task is therefore to compare the number of prime ministers found in the 5/10 most similar entities retrieved using non-time flattened similarity measures and time-flattening similarity measures. We used prime ministers because this is a salient feature in making entities "ontologically" similar (e.g., very few entities share this feature). We selected 6 entities representing the most recent US presidents (from a list of the most recent 19 presidents) and 6 entities representing the most recent British Prime Ministers (from a list of the most recent 19 prime minister).

Algorithms and Methods. We consider different algorithms to test the time-flattened similarity: we want to test if our model is actually able to retrieve entities that are far in time starting from input entities. We select the 100 nearest entities given an input entity using cosine and order them using time-flattened similarity. We then compute Precision@K and Recall@K. This task is tested on both TEE and the EE model: we will compare standard neighborhood of the input entity (based on cosine similarity) and the time-flattened one (considering the time-flattened similarity).

Baseline. As a Baseline we consider a similarity measure that uses a skip-gram model trained on a corpus that also contains mentions to entity years[11]. Time-flattening in the baseline is computed considering the closest *entity* year to a given entity as the most representative year, similarly to what we do in our similarity. The difference with our representation is that, in this Baseline, the entity year representations are learned by considering the co-occurrence in text as in standard embedding models and do not have an explicit representation generated by a dedicated embedding function like in our model. The tested models are:

- Time-aware Similarity TEE (TATEE), with time-flattened similarity;
- Similarity TEE (STEE) (standard neighborhood with cosine);
- Time-Aware Similarity EE (TAEE), with time-flattened similarity;
- Similarity EE (SEE) (standard neighborhood with cosine);
- Time-flattened similarity Baseline (Baseline).

Experiments on the time-flattened similarity were run with $\alpha = 0.7$.

Results. Table 3 shows the results. The use of a time-flattening factor can improve the retrieval of entities that are distant in time. Models that use types have an advantaged: the tasks, in fact, consists in finding entities that share more or less of the same type. However, the performance of the model that does not use types, but time-flattened (TAEE) is better than the baseline. We can conclude that the use of both the TEE model and time-awareness (time-flattening in this case) allows achieving better performance on this task.

Table 3. Results for time-flattened similarity (0* means small values)

	Precision@5	Precision@10	Recall@5	Recall@10
TATEE	**0.40**	**0.40**	**0.20**	**0.21**
STEE	0.14	0.21	0.07	0.10
TAEE	0.05	0.04	0*	0.01
SEE	0.02	0.02	0*	0.01
Baseline	0.01	0.01	0*	0.01

Qualitative Evaluation. To provide further insights on the behavior of our time-flattening similarity, we discuss an example in details. If we consider the entity *dbr:Winston Churchill* and the top-100 entities more similar to it, recent but popular British prime ministers are found distant in the TEE model when retrieved using plain cosine similarity: *dbr:Tony Blair* and *dbr:Gordon Brown*

---

[11] https://github.com/idio/wiki2vec.

are respectively in the 49th and the 41th position. If we use our time-flattening similarity, the two entities are found respectively at the 16th and 14th position. The nearest entity to *dbr:Winston Churchill* is *dbr:Harold Macmillan* (Member of Churchill government and British prime minister two years after Winston Churchill) if we use plain cosine similarity, and *dbr:Margaret Thatcher* if we use our time-flattened similarity.

Time-Flattening/Time-Boosting. In Table 4, we list the top-10 most similar entities to Barack Obama, when retrieved with time-flattened, plain cosine, and time-boosted similarity. For time-aware similarities we also show differences for different values of α, to show the effect of this parameter (remember that 0.7 was used as value in previous experiments). We believe that this example shows an interesting behavior: removing the time effect with time-flattened similarity pushes old presidents of the United States in higher positions. Otherwise, if we use the time-boosted similarity, members of the Obama government and his rivals during the elections (i.e., John McCain and Mitt Romney) are the ones pushed in higher positions.

Table 4. Time-flattened and time-boosted similarity on the entity Barack Obama.

Time flattened similarity - time ←		Cosine similarity	Time boosted similarity → + Time	
$\alpha = 0.1$	$\alpha = 0.7$		$\alpha = 0.7$	$\alpha = 0.1$
G. Ford	B. Clinton	B. Clinton	B. Clinton	G. Bush
C. Coolidge	Reagan	Reagan	G. Bush	**J. Kerry**
H. Hoover	Carter	G. Bush	Reagan	D. Cheney
Truman	Al Gore	Carter	Kerry	**McCain**
F. Roosevelt	Nixon	Al Gore	D. Cheney	**Biden**
W. Wilson	**G. Ford**	Nixon	McCain	Ron Paul
Eleanor Roosevelt	G. Bush	J. Kerry	Biden	H. Humphrey
D. Eisenhower	**C. Coolidge**	D. Cheney	Carter	**Romney**
W. Harding	T. Kennedy	McCain	Al Gore	C. Powell
G. Cleveland	**H. Hoover**	Biden	Ron Paul	W. Mondale

4   Related Work

Qualitative temporal representation and reasoning is a topic covered by a vast literature in Artificial Intelligence and related fields, for which we refer to several surveys [23,25,31]. Different mathematical models of time such as point-based vs. interval-based, linear time vs. branching time, have been proposed [31]. Models to support reasoning with approximate time intervals have also been proposed [5]. Previous work has surveyed models to represent temporal information in RDF [23] and reason about time in natural language processing [25]. In our work, we use a simple model of time as a sequence of regular time periods and do not tackle logic-based temporal reasoning. Otherwise, none of the previous approaches has addressed the problem of generating temporal representations from texts. Extraction of temporal information from text (see, e.g., [15]) and imputation of temporal validity intervals for RDF triples (see, e.g., [24]), are tasks also very different from the one addressed in this paper.

Many approaches for KG embeddings that consider knowledge graph structure have been introduced in literature [6,14,19–21,30,32]. For example, TransE [6] embeds entities in a space in which for each triple (s, p, o), $\mathbf{s} + \mathbf{p} \approx \mathbf{o}$ holds. All these methods are able to efficiently represent entities and relations of a KG into a vector space, but none of them take explicit steps towards the representation of temporal entities. Other methods have been introduced to represent temporal information in KGs [9,13,29] and have obtained good results in tasks like time-aware link prediction. The main difference with our approach is that we explicitly embed temporal entities inside a vector space.

We used a semantic annotator to extract entities from text. If we consider Wikipedia, text can be replaced with the use of links as done in other models [1]. Our approach can be generalized to any kind of text, even those that do not contain links, such as books or newspapers.

Worth mentioning in this context are the works on temporal word embeddings [27]. These representations are often called diachronic embeddings [10], since they start from collections of documents coming from different periods in times and build and embedding for each of the periods. The study of embeddings at different points in time has shown that words are subject to a shift in meaning that can be quantified using distance measures between different embeddings across the vector space. The main difference between our work and theirs is that we are embedding temporal entities inside a KG, while often the task proposed in other approaches is to study the changes of meaning in words over during time [10]. Following this methodology, a recent work on time-aware entity relatedness that uses word embeddings learned from a collection of documents that spans different time periods has been proposed [17].

5   Conclusions and Future Work

In this paper we have presented an approach to encoding temporal periods into text-based entity embedding models. In particular, we used our previous work to generate Typed Entity Embeddings (TEEs) from textual descriptions of entities and encoded into this model the representations of years. These representations are generated using natural language descriptions of events occurring during the years. To the best of our knowledge, this is the first attempt to explicitly encode time into KG embedding models. In addition, we have defined a parametric time-aware similarity function that can be tuned to boost or flatten the effect of time when computing the entity similarity.

In our experiments we have shown that time has an effect on entity embeddings built from text, thus validating the main hypothesis behind this work. Then we have tested our time-aware similarity function to show that it can capture aspects of similarity that other time-agnostic similarity measures cannot capture. Such a similarity measure can provide novel knowledge exploration methods where time can be factored when finding entities similar to each other.

Our results provide a first contribution to the problem of encoding time into KG embeddings built from text, which poses several challenges that we want to address in future work. So far, we have considered sequences of regular time periods at a yearly granularity. An important challenge would be to consider different granularity levels and, even more important, to study the compositional nature of temporal representations extracted from text. For example, we would like to generate a vector for the 70 s by composing vectors of years 197X. More in general, we would like to investigate how vector-based representations of time periods can be composed so as to provide a soft account of relations between time intervals that are considered in qualitative models of temporal reasoning like Allen algebra.

References

1. Basile, P., Caputo, A., Rossiello, G., Semeraro, G.: Learning to rank entity relatedness through embedding-based features. In: Métais, E., Meziane, F., Saraee, M., Sugumaran, V., Vadera, S. (eds.) NLDB 2016. LNCS, vol. 9612, pp. 471–477. Springer, Cham (2016). https://doi.org/10.1007/978-3-319-41754-7_51
2. Bianchi, F., Palmonari, M.: Joint learning of entity and type embeddings for analogical reasoning with entities. In: NL4AI Workshop, Co-located with the International Conference of the Italian Association for Artificial Intelligence (AI* IA) (2017)
3. Bianchi, F., Palmonari, M., Cremaschi, M., Fersini, E.: Actively learning to rank semantic associations for personalized contextual exploration of knowledge graphs. In: Blomqvist, E., Maynard, D., Gangemi, A., Hoekstra, R., Hitzler, P., Hartig, O. (eds.) ESWC 2017. LNCS, vol. 10249, pp. 120–135. Springer, Cham (2017). https://doi.org/10.1007/978-3-319-58068-5_8
4. Bianchi, F., Soto, M., Palmonari, M., Cutrona, V.: Type vector representations from text: an empirical analysis. In: DL4KGS Workshop, Co-located with the ESWC (2018)
5. Bittner, T.: Approximate qualitative temporal reasoning. Ann. Math. Artif. Intell. **36**(1–2), 39–80 (2002)
6. Bordes, A., Usunier, N., Garcia-Duran, A., Weston, J., Yakhnenko, O.: Translating embeddings for modeling multi-relational data. In: NIPS, pp. 2787–2795 (2013)
7. Daiber, J., Jakob, M., Hokamp, C., Mendes, P.N.: Improving efficiency and accuracy in multilingual entity extraction. In: I-Semantics (2013)
8. Damasio, A.R.: Remembering when. Sci. Am. **287**(3), 66–73 (2002)
9. Esteban, C., Tresp, V., Yang, Y., Baier, S., Krompaß, D.: Predicting the co-evolution of event and knowledge graphs. In: 2016 19th International Conference on Information Fusion (FUSION), pp. 98–105, July 2016
10. Hamilton, W.L., Leskovec, J., Jurafsky, D.: Diachronic word embeddings reveal statistical laws of semantic change. arXiv preprint arXiv:1605.09096 (2016)
11. Harris, Z.S.: Distributional structure. Word **10**(2–3), 146–162 (1954)
12. Hoffart, J., Seufert, S., Nguyen, D.B., Theobald, M., Weikum, G.: Kore: keyphrase overlap relatedness for entity disambiguation. In: CIKM, pp. 545–554. ACM (2012)
13. Jiang, T., et al.: Encoding temporal information for time-aware link prediction. In: EMNLP, pp. 2350–2354 (2016)
14. Lin, Y., Liu, Z., Sun, M., Liu, Y., Zhu, X.: Learning entity and relation embeddings for knowledge graph completion. In: AAAI, pp. 2181–2187 (2015)

15. Ling, X., Weld, D.S.: Temporal information extraction. In: AAAI. vol. 10, pp. 1385–1390 (2010)
16. Mikolov, T., Sutskever, I., Chen, K., Corrado, G.S., Dean, J.: Distributed representations of words and phrases and their compositionality. In: NIPS, pp. 3111–3119 (2013)
17. Mohapatra, N., Iosifidis, V., Ekbal, A., Dietze, S., Fafalios, P.: Time-aware and corpus-specific entity relatedness. In: DL4KGS Workshop, Co-located with the ESWC (2018)
18. Nguyen, T.N., Kanhabua, N., Nejdl, W.: Multiple models for recommending temporal aspects of entities. In: Gangemi, A., et al. (eds.) ESWC 2018. LNCS, vol. 10843, pp. 462–480. Springer, Cham (2018). https://doi.org/10.1007/978-3-319-93417-4_30
19. Nickel, M., Rosasco, L., Poggio, T.A., et al.: Holographic embeddings of knowledge graphs. In: AAAI, pp. 1955–1961 (2016)
20. Nickel, M., Tresp, V., Kriegel, H.P.: A three-way model for collective learning on multi-relational data. In: Proceedings of ICML-11, pp. 809–816 (2011)
21. Ristoski, P., Paulheim, H.: RDF2Vec: RDF graph embeddings for data mining. In: Groth, P., et al. (eds.) ISWC 2016. LNCS, vol. 9981, pp. 498–514. Springer, Cham (2016). https://doi.org/10.1007/978-3-319-46523-4_30
22. Rizzo, G., Troncy, R.: NERD: a framework for unifying named entity recognition and disambiguation extraction tools. In: EACL, pp. 73–76. ACL (2012)
23. Rula, A., Palmonari, M., Harth, A., Stadtmüller, S., Maurino, A.: On the diversity and availability of temporal information in linked open data. In: Cudré-Mauroux, P., et al. (eds.) ISWC 2012. LNCS, vol. 7649, pp. 492–507. Springer, Heidelberg (2012). https://doi.org/10.1007/978-3-642-35176-1_31
24. Rula, A., Palmonari, M., Ngonga Ngomo, A.-C., Gerber, D., Lehmann, J., Bühmann, L.: Hybrid acquisition of temporal scopes for RDF data. In: Presutti, V., d'Amato, C., Gandon, F., d'Aquin, M., Staab, S., Tordai, A. (eds.) ESWC 2014. LNCS, vol. 8465, pp. 488–503. Springer, Cham (2014). https://doi.org/10.1007/978-3-319-07443-6_33
25. Sanampudi, S.K., Kumari, G.V.: Temporal reasoning in natural language processing: a survey. Int. J. Comput. Appl. 1(4), 68–72 (2010)
26. Snaider, J., McCall, R., Franklin, S.: Time production and representation in a conceptual and computational cognitive model. Cogn. Syst. Res. 13(1), 59–71 (2012)
27. Szymanski, T.: Temporal word analogies: identifying lexical replacement with diachronic word embeddings. In: Association for Computational Linguistics, Vancouver, Canada. ACL, August 2017
28. Tran, N.K., Tran, T., Niederée, C.: Beyond time: dynamic context-aware entity recommendation. In: Blomqvist, E., Maynard, D., Gangemi, A., Hoekstra, R., Hitzler, P., Hartig, O. (eds.) ESWC 2017. LNCS, vol. 10249, pp. 353–368. Springer, Cham (2017). https://doi.org/10.1007/978-3-319-58068-5_22
29. Trivedi, R., Dai, H., Wang, Y., Song, L.: Know-Evolve: deep temporal reasoning for dynamic knowledge graphs. In: ICML, pp. 3462–3471 (2017)
30. Trouillon, T., Welbl, J., Riedel, S., Gaussier, É., Bouchard, G.: Complex embeddings for simple link prediction. In: ICML, pp. 2071–2080 (2016)
31. Van Beek, P.: Reasoning about qualitative temporal information. Artif. Intell. 58(1–3), 297–326 (1992)
32. Wang, Z., Zhang, J., Feng, J., Chen, Z.: Knowledge graph embedding by translating on hyperplanes. In: AAAI, pp. 1112–1119 (2014)

Rule Learning from Knowledge Graphs Guided by Embedding Models

Vinh Thinh Ho[1], Daria Stepanova[1(✉)], Mohamed H. Gad-Elrab[1],
Evgeny Kharlamov[2], and Gerhard Weikum[1]

[1] Max Planck Institute for Informatics, Saarbrücken, Germany
`dstepano@mpi-inf.mpg.de`
[2] University of Oxford, Oxford, UK

Abstract. Rules over a Knowledge Graph (KG) capture interpretable patterns in data and various methods for rule learning have been proposed. Since KGs are inherently incomplete, rules can be used to deduce missing facts. Statistical measures for learned rules such as confidence reflect rule quality well when the KG is reasonably complete; however, these measures might be misleading otherwise. So it is difficult to learn high-quality rules from the KG alone, and scalability dictates that only a small set of candidate rules could be generated. Therefore, the ranking and pruning of candidate rules are major problems. To address this issue, we propose a rule learning method that utilizes probabilistic representations of missing facts. In particular, we iteratively extend rules induced from a KG by relying on feedback from a precomputed embedding model over the KG and external information sources including text corpora. Experiments on real-world KGs demonstrate the effectiveness of our novel approach both with respect to the quality of the learned rules and fact predictions that they produce.

1 Introduction

Motivation. Rules are widely used to represent relationships and dependencies between data items in datasets and to capture the underlying patterns in data [1,24]. Applications of rules include health-care [37], equipment diagnostics [16,19], telecommunications [18], and commerce [27]. To facilitate rule construction, a variety of rule learning methods have been developed, see e.g. [8,17] for an overview. Moreover, various statistical measures such as confidence, actionability, and unexpectedness to evaluate the quality of the learned rules have been proposed.

Rule learning has recently been adapted to the setting of Knowledge Graphs (KGs) [9,10,32,36] where data is represented as a graph of entities interconnected via relations and labeled with classes, or more formally as a set of grounded binary and unary atoms typically referred to as facts. Examples of large-scale KGs include Wikidata [33], Yago [30], NELL [21], and Google's KG. Since many KGs are constructed from semi-structured knowledge, such as Wikipedia, or

© Springer Nature Switzerland AG 2018
D. Vrandečić et al. (Eds.): ISWC 2018, LNCS 11136, pp. 72–90, 2018.
https://doi.org/10.1007/978-3-030-00671-6_5

harvested from the Web with a combination of statistical and linguistic methods, they are inherently incomplete [10].

Rules over KGs are of the form *head* ← *body*, where *head* is a binary atom and *body* is a conjunction of, possibly negated, binary or unary atoms. When rules are automatically learned, statistical measures like support and confidence are used to assess the quality of rules. Most notably, the confidence of a rule is the fraction of facts predicted by the rule that are indeed true in the KG. However, this is a meaningful measure for rule quality only when the KG is reasonably complete. For rules learned from largely incomplete KGs, confidence and other measures may be misleading, as they do not reflect the patterns in the missing facts. For example, a KG that knows only (or mostly) male CEOs would yield a heavily biased rule *gender*$(X, male)$ ← *isCEO*(X, Y), *isCompany*(Y), which does not extend to the entirety of valid facts beyond the KG. Therefore, it is crucial that rules can be ranked by a meaningful quality measure, which accounts for KG incompleteness.

Example. Consider a KG about people's jobs, residence and spouses as well as office locations and headquarters of companies. Suppose a rule learning method has computed the following two rules:

$$r_1 : livesIn(X, Y) \leftarrow worksFor(X, Z), hasOfficeIn(Z, Y) \tag{1}$$

$$r_2 : livesIn(Y, Z) \leftarrow marriedTo(X, Y), livesIn(X, Z) \tag{2}$$

The rule r_1 is quite noisy, as companies have offices in many cities, but employees live and work in only one of them, while the rule r_2 clearly is of higher quality. However, depending on how the KG is populated with instances, the rule r_1 could nevertheless score higher than r_2 in terms of confidence measures. For example, the KG may contain only a specific subset of company offices and only people who work for specific companies. If we knew the complete KG, then the rule r_2 should presumably be ranked higher than r_1.

Suppose we had a perfect oracle for the true and complete KG. Then we could learn even more sophisticated rules such as:

$$r_3 : livesIn(X, Y) \leftarrow worksFor(X, Z), hasHeadquarterIn(Z, Y),$$
$$not\ locatedIn(Y, USA)$$

This rule would capture that most people work in the same city as their employers' headquarters, with the USA being an exception (assuming that people there are used to long commutes). This is an example of a rule that contains a negated atom in the rule body (so it is no longer a Horn rule) and has a partially grounded atom with a variable and a constant as its arguments.

Problem. The problem of KG incompleteness has been tackled by methods that (learn to) predict missing facts for KGs (or actually missing relational edges between existing entities). A prominent class of approaches is statistics-based and includes tensor factorization, e.g. [23] and neural-embedding-based models, e.g. [2,22]. Intuitively, these approaches turn a KG, possibly augmented with external sources such as text [38] or log files [29], into a probabilistic representation of its entities and relations, known as *embeddings*, and then predict the likelihood of missing facts by reasoning over the embeddings (see, e.g. [34] for a survey).

These kinds of embeddings can complement the given KG and are a potential asset in overcoming the limitations that arise from incomplete KGs. Consider the following gedankenexperiment: we compute embeddings from the KG and external text sources, that can then be used to predict the complete KG that comprises all valid facts. This would seemingly be the perfect starting point for learning rules, without the bias and quality problems of the incomplete KG. However, this scenario is way oversimplified. The embedding-based fact predictions would themselves be very noisy, yielding also many spurious facts. Moreover, the computation of all fact predictions and the induction of all possible rules would come with a big scalability challenge: in practice, we need to restrict ourselves to computing merely small subsets of likely fact predictions and promising rule candidates.

Approach. In this work we propose a novel approach for rule learning guided by external sources that allows to learn high-quality rules from incomplete KGs. In particular, our method extends rule learning by exploiting probabilistic representations of missing facts computed by embedding models of KGs and possibly other external information sources. We iteratively construct rules over a KG and collect feedback from a precomputed embedding model, through specific queries issued to the model for assessing the quality of (partially constructed) rule candidates. This way, the rule induction loop is interleaved with the guidance from the embeddings, and we avoid scalability problems. Our machinery is also more expressive than many prior works on rule learning from KGs, by allowing non-monotonic rules with negated atoms as well as partially grounded atoms. Within this framework, we devise confidence measures that capture rule quality better than previous techniques and thus improve the ranking of rules.

While enhancing embeddings with precomputed rules or constraints has been studied in several works [14,15,28,35,35], accounting for embeddings in rule construction as we propose, has not been considered before to the best of our knowledge.

Contribution. The salient contributions of our work are as follows:

- We propose a rule learning approach guided by external sources, and show how to learn high-quality rules by utilizing feedback from embedding models.
- We implement our approach and present extensive experiments on real-world KGs, demonstrating the effectiveness of our approach with respect to both the quality of the learned rules and the fact predictions that they produce.

– Our code and data are made available to the research community at https://github.com/hovinhthinh/RuLES.

2 Rule Learning Guided by External Sources

In this section, we first give some necessary preliminaries, then introduce our framework for rule learning guided by external sources, discuss challenges associated with it, and finally propose a concrete instantiation of our framework with embedding models.

2.1 Background

We assume countable sets $\mathcal{R}$ of unary and binary relation names and $\mathcal{C}$ of constants. A *knowledge graph* (KG) $\mathcal{G}$ is a finite set of ground atoms a of the form $p(b, c)$ and $c(b)$ over $\mathcal{R} \cup \mathcal{C}$. With $\Sigma_{\mathcal{G}}$, the *signature* of $\mathcal{G}$, we denote elements of $\mathcal{R} \cup \mathcal{C}$ that occur in $\mathcal{G}$.

We define rules over KGs following the standard approach of non-monotonic logic programs under the answer set semantics [11]. Let $\mathcal{X}$ be a countable set of variables. A *rule* r is of the form *head* ← *body*, where *head*, or *head(r)*, is an atom over $\mathcal{R} \cup \mathcal{C} \cup \mathcal{X}$ and *body*, or *body(r)*, is a conjunction of positive and negative atoms over $\mathcal{R} \cup \mathcal{C} \cup \mathcal{X}$. Finally, $body^+(r)$ and $body^-(r)$ denote the atoms that occur in $body(r)$ positively and negatively respectively; that is, the rule can be written as $head(r) \leftarrow body^+(r), not\ body^-(r)$. A rule is *Horn*, if all head variables occur in the body, and $body^-(r)$ is empty.

We define *execution* of rules with default negation [11] over KGs in the standard way. More precisely, let $\mathcal{G}$ be a KG, r a rule over $\Sigma_{\mathcal{G}}$, and a be an atom over $\Sigma_{\mathcal{G}}$. Then, $r \models_{\mathcal{G}} a$ holds if there is a variable assignment that maps atoms $body^+(r)$ in $\mathcal{G}$ such that it does not map any of the atoms in $body^-(r)$ in $\mathcal{G}$. Then, let $\mathcal{G}_r = \mathcal{G} \cup \{a \mid r \models_{\mathcal{G}} a\}$. Intuitively, $\mathcal{G}_r$ extends $\mathcal{G}$ with edges derived from $\mathcal{G}$ by applying r. Note that to avoid propagating uncertain predictions, given a set of rules R we execute every rule in R on $\mathcal{G}$ independently, i.e., $\mathcal{G}_R = \bigcup_{r \in R} \mathcal{G}_r$. Given additional syntactic restrictions on rules in R, which disallow cycles through negation, consistency is ensured.

2.2 Problem Statement and Proposal of General Solution

Let $\mathcal{G}$ be a KG over the signature $\Sigma_{\mathcal{G}} = (\mathcal{R}_{\mathcal{G}}, \mathcal{C}_{\mathcal{G}})$. A *probabilistic KG* $\mathcal{P}$ is a pair $\mathcal{P} = (\mathcal{G}, f)$ where $f : \mathcal{R}_{\mathcal{G}} \times \mathcal{C}_{\mathcal{G}} \times \mathcal{C}_{\mathcal{G}} \rightarrow [0, 1]$ is a probability function over the facts over $\Sigma_{\mathcal{G}}$. We assume $f(a) = 1$ for each fact $a \in \mathcal{G}$, which is already known to be true.

The goal of our work is to learn rules that not only describe the available graph $\mathcal{G}$ well, but also predict highly probable facts based on the function f. The key questions now are how to define the quality of a given rule r based on $\mathcal{P}$ and how to exploit this quality during rule learning for pruning out unpromising rules.

A quality measure μ for rules over probabilistic KGs is a function $\mu : (r, \mathcal{P}) \mapsto \alpha$, where $\alpha \in [0, 1]$. In order to measure the quality μ of r over $\mathcal{P}$ we propose:

- to measure the quality μ_1 of r over $\mathcal{G}$, where $\mu_1 : (r, \mathcal{G}) \mapsto \alpha \in [0, 1]$,
- to measure the quality μ_2 of $\mathcal{G}_r$ by relying on $\mathcal{P}_r = (\mathcal{G}_r, f)$, where $\mu_2 : (\mathcal{G}', (\mathcal{G}, f)) \mapsto \alpha \in [0, 1]$ for $\mathcal{G}' \supseteq \mathcal{G}$ is the quality of extension $\mathcal{G}'$ of $\mathcal{G}$ over $\Sigma_\mathcal{G}$ given f, and
- to combine the result as the weighted sum.

Formally, we define our hybrid rule quality function $\mu(r, \mathcal{P})$ as follows:

$$\mu(r, \mathcal{P}) = (1 - \lambda) \times \mu_1(r, \mathcal{G}) + \lambda \times \mu_2(\mathcal{G}_r, \mathcal{P}) \tag{3}$$

In this formula μ_1 can be any classical quality measure of rules over the given KG $\mathcal{G}$. Intuitively, $\mu_2(\mathcal{G}_r, \mathcal{P})$ is the quality of $\mathcal{G}_r$ wrt f that allows us to capture the information about facts missing in $\mathcal{G}$ that are relevant for r. The weighting factor λ, we call it *embedding weight*, allows one to choose whether to rely more on the classical measure μ_1 or on the measure μ_2 of the quality of the extension $\mathcal{G}_r$ of r over $\mathcal{G}$.

Challenges. There are several challenges that one faces when realising our approach. First, given an incomplete $\mathcal{G}$, one has to define f such that $(\mathcal{G}, f)$ satisfies the expectations, i.e., reflects well the probabilities of missing facts. Second, one has to define μ_1 and μ_2 that also satisfy the expectations and admit efficient implementation. Finally, the adaptation of existing rule learning approaches to account for the probabilistic function f without the loss of scalability is not trivial. Indeed, materializing f by augmenting $\mathcal{G}$ with all possible probabilistic facts over $\Sigma_\mathcal{G}$ and subsequently applying standard rule learning methods on the obtained graph is not practical. Storing such potentially enormous augmented graph where many probabilistic facts are irrelevant for the extraction of meaningful rules might be simply infeasible.

2.3   Realization of General Solution

We now describe how we addressed the above stated challenges. In this section, we present concrete realizations of f, μ_1 and μ_2, and in Sect. 3 we discuss how we implemented them and adapted within an end-to-end rule learning system.

Realization of the Probabilistic Function f. We propose to define f by relying on embeddings of KGs. Embeddings are low-dimensional vector spaces that represent nodes and edges of KGs and can be used to estimate the likelihood (not necessary probability) of potentially missing binary atoms using a scoring function $\xi : \mathcal{R}_\mathcal{G} \times \mathcal{C}_\mathcal{G} \times \mathcal{C}_\mathcal{G} \to \mathbb{R}$. Examples of concrete scoring functions can be found, e.g., in [34]. Since embeddings per se are not in the focus of our paper, we will not give further details on them and refer the reader to [34] for an overview. Note that our framework is not dependent on a concrete embedding model. What is important for us is that embeddings can be used to construct

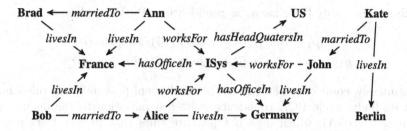

Fig. 1. An example knowledge graph.

probabilistic representations [22] of atoms missing in KGs and we use this to define f.

Consider an auxiliary definition. Given a KG $\mathcal{G}$, and an atom $a = p(s,o)$, the set $\mathcal{G}^s$ consists of a and all atoms a' that are obtained from a by replacing s with a constant from $\Sigma_\mathcal{G}$, except for those that are already in $\mathcal{G}$. Then, given a scoring function ξ, $[\mathcal{G}^s]$ is a list of atoms from $\mathcal{G}^s$ ordered in the descending order. Finally, the *subject rank* [12] of a given ξ, *subject_rank*$_\xi(a)$ is the position of a in $[\mathcal{G}^s]$. Analogously, one can define $[\mathcal{G}^o]$ and the corresponding *object rank* [12] of a given ξ, that is, *object_rank*$_\xi(a)$.

Now we are ready to define the function f for an atom $a \notin \mathcal{G}$ as the average of its subject and object inverted ranks given ξ [12], i.e.:

$$f_\xi(a) = 0.5 \times (1/subject\_rank_\xi(a) + 1/object\_rank_\xi(a))$$

Note that we assume $f_\xi(a) = 1$ for $a \in \mathcal{G}$.

Realization of μ_1. This measure should reflect the descriptive quality of a given rule r with respect to $\mathcal{G}$. There are many classical data mining measures that can be used as μ_1, see, e.g. [10,20,31,41] for μ_1s proposed specifically for KGs.

In this work, we selected the following two measures for μ_1: *confidence* and *PCA confidence* [10], where PCA stands for the partial completeness assumption, that can be defined using rule support, *r-supp*, body support, *b-supp*, and partial body support, *pb-supp*, as follows. Let $r : head \leftarrow body^+, not\ body^-$ be a rule, x be the subject variable of the *head*, and let h denote a *head*'s variable assignment that we with a slight abuse of notation use as a homomorphism on (sets of) atoms. Then,

$$r\text{-}supp(r,\mathcal{G}) = |\{h \mid h(head) \in \mathcal{G}, \exists h' \supseteq h \text{ s.t. } h'(body^+) \in \mathcal{G}, h'(body^-) \notin \mathcal{G}\}|,$$

$$b\text{-}supp(r,\mathcal{G}) = |\{h \mid \exists h' \supseteq h \text{ s.t. } h'(body^+) \in \mathcal{G}, h'(body^-) \notin \mathcal{G}\}|,$$

$$pb\text{-}supp(r,\mathcal{G}) = |\{h \mid \exists h' \supseteq h \text{ s.t. } h'(body^+) \in \mathcal{G}, h'(body^-) \notin \mathcal{G}, \text{and}$$
$$\exists h'' \text{ s.t. } h(x) = h''(x), h''(head) \in \mathcal{G}\}|.$$

Finally, we are ready to define μ_1 as confidence or PCA confidence:

$$\mu_1 = conf(r,\mathcal{G}) = r\text{-}supp(r,\mathcal{G})/b\text{-}supp(r,\mathcal{G}),$$
$$\mu_{1,pca} = conf_{pca}(r,\mathcal{G}) = r\text{-}supp(r,\mathcal{G})/pb\text{-}supp(r,\mathcal{G}).$$

Intuitively, confidence of a rule is the conditional probability of rule's head given its body, while PCA confidence is its generalisation to the open world assumption (OWA), which does not penalize rules that predict facts $p(s,o)$, such that $p(s,o') \notin \mathcal{G}$ for any o'.

Example 1. Consider the KG $\mathcal{G}$ in Fig. 1 and recall the rules r_1 and r_2 from Eqs. (1)–(2). For r_1, we have $conf(r_1,\mathcal{G}) = conf_{pca}(r_1,\mathcal{G}) = \frac{3}{6}$, while for r_2 it holds that $conf(r_2,\mathcal{G}) = conf_{pca}(r_2,\mathcal{G}) = \frac{1}{3}$. If Alice was not known to live in Germany, then $conf_{pca}(r_2, \mathcal{G} \setminus \{livesIn(Alice, Germany)\}) = \frac{1}{2}$. Finally, for the following rule with negation:

$$r_4 : livesIn(Y,Z) \leftarrow marriedTo(X,Y), livesIn(X,Z), not\ researcher(X)$$

stating that married people live together unless one is a researcher, and $\mathcal{G}' = \mathcal{G} \cup \{researcher(bob)\}$, we have $conf(r_4,\mathcal{G}') = conf_{pca}(r_4,\mathcal{G}') = \frac{1}{2}$.   □

Realization of μ_2. There are various ways how one can define the quality $\mu_2(\mathcal{G}_r, \mathcal{P})$ of $\mathcal{G}_r$. A natural candidate to define the quality of $\mathcal{G}_r$ is the probability of $\mathcal{G}_r$, that is, as $\mu_2(\mathcal{G}_r, \mathcal{P}) = \prod_{a \in \mathcal{G}_r} f(a) \times \prod_{a \in (\mathcal{R}_\mathcal{G} \times C_\mathcal{G} \times C_\mathcal{G}) \setminus \mathcal{G}_r} (1 - f(a))$. A disadvantage of such quality measure is that in practice it will be very low, as the product of many (potentially) small probabilities, and thus Eq. 3 will be heavily dominated by $\mu_1(r,\mathcal{G})$. Therefore, we advocate to define $\mu_2(\mathcal{G}_r, \mathcal{P})$ as the average probability of predicted facts in $\mathcal{G}_r$:

$$\mu_2(\mathcal{G}_r, \mathcal{P}) = (\Sigma_{a \in \mathcal{G}_r \setminus \mathcal{G}} f(a))/|\mathcal{G}_r \setminus \mathcal{G}|.$$

Example 2. Consider the KG $\mathcal{G}$ in Fig. 1, and the rules from Eqs. (1)–(2) with their confidence values as presented in Example 1. Suppose that a text-enhanced embedding model produced a relatively accurate estimation of the probabilities of facts over *livesIn* relation. For example, even though there is no direct connection between Germany and Berlin within the graph, relying on the living places of entities similar to John and hidden semantic relations between Germany and Berlin such as co-occurrences in text and other linguistic features, for the fact $a = livesIn(john, berlin)$ we obtained $f(a) = 0.9$, while for $a' = livesIn(john, france)$, a much lower probability $f(a') = 0.09$. These naturally support the predictions of r_2 but not those of r_1.

Generalising this idea, assume that on the whole dataset we get $\mu_2(\mathcal{G}_{r_1}, \mathcal{P}) = 0.1$ and $\mu_2(\mathcal{G}_{r_2}, \mathcal{P}) = 0.8$, where $\mathcal{P} = (\mathcal{G}, f)$. Thus, for $\lambda = 0.5$ we have $\mu(r_1, \mathcal{P}) = (1 - 0.5) \times 0.5 + 0.5 \times 0.1 = 0.3$, while for $\mu(r_2, \mathcal{P}) = (1 - 0.5) \times \frac{1}{3} + 0.5 \times 0.8 \approx 0.57$, resulting in the desired ranking of r_2 over r_1 based on μ.   □

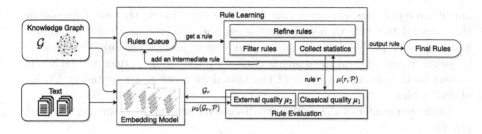

Fig. 2. Overview of our system.

3 Approach Description

In this section, we describe our rule learning system with embedding support. Conceptually, it extends the standard relational association rule learners [10,13] to also take into account the feedback from embedding models through the probabilistic function f.

Following common practice [10] we restrict ourselves to rules that are *closed*, where every variable appears at least twice (moreover, we extract only rules whose Horn part is *closed*), and *safe*, where variables appearing in the negated part also appear in the positive part of the rule.

Overview. The input of the system are a KG, possibly a text corpus, and a set of user specified parameters that are used to terminate rule construction. These parameters include an embedding weight λ, a minimum threshold for μ_1, a minimum rule support $r\text{-}supp$ and other *rule-related* parameters such as a maximum number of positive and negative atoms allowed in r. The KG and text corpus are used to train the embedding model that in turn is used to construct the probabilistic function f. The rules r are constructed in the iterative fashion, starting from the head, by adding atoms to its body one after another until at least one of the termination criteria (that depend on f) is met. In parallel with the construction of the rule r, the quality $\mu(r)$ is computed.

In Fig. 2 we present a high level architecture of our system, where arrows depict information flow between blocks. The *Rule Learning* block constructs rules over the input KG, *Rule Evaluation* supplies it with quality scores μ for rules r, using $\mathcal{G}$ and f, where f is computed by the *Embedding Model* block from $\mathcal{G}$ and text.

We now discuss the algorithm behind the *Rule Learning* block in Fig. 2. Following [10] we model rules as sequences of atoms, where the first atom is the head of the rule and other atoms are its body. The algorithm maintains a priority queue of intermediate rules (see the *Rules Queue* block in Fig. 2). Initially all possible binary atoms appearing in $\mathcal{G}$ are added to the queue with empty bodies. At each iteration, a single rule is selected from the queue. If the rule satisfies the *filtering criteria* (see the *Filer rules* block) which we define below, then the system returns it as an output. If the rule is not filtered, then it is processed with one of the *refinement operators* (see the *Refine rules* block) that we define below that expand the rule with one more atom and produce new rule candidates, which are then pushed into the queue (if not being pushed before). The iterative process is repeated until the queue is empty. All the reported rules will be finally ranked by the decreasing order of the hybrid measure μ, computed in *Collect statistics* block.

In the remainder of the section we discuss refinement operators and filtering criteria.

Refinement Operators. We rely on the following three standard refinement operators [10] that extend rules:

(i) *add a positive dangling atom*: add a binary positive atom with one fresh variable and another one appearing in the rule, i.e., *shared*.

(ii) *add a positive instantiated atom*: add a binary positive atom with one argument being a constant and the other one being a shared variable.

(iii) *add a positive closing atom*: add a binary positive atom with both of its arguments being shared variables.

Additionally, we introduce two more operators to allow negated atoms in rule bodies:

(iv) *add an exception instantiated atom*: add a binary negated atom with one of its arguments being a constant, and the other one being a shared variable.

(v) *add an exception closing atom*: add a binary negated atom to the rule with both of its arguments being shared variables.

These two operators are only applied to closed rules. Moreover, we ensure that the addition of exception atoms to the rule $r : head(r) \leftarrow body^+(r)$, should result in $r' : head(r) \leftarrow body^+(r), not\ body^-(r)$, such that

$$r\text{-}supp(head(r) \leftarrow body^+(r), body^-(r), \mathcal{G}) = 0.$$

Intuitively, we aim at adding exceptions that explain the absence of predictions expected to be in the graph rather then their presence. Thus, the introduced exceptions should not affect the rule support, i.e., $r\text{-}supp(r, \mathcal{G}) = r\text{-}supp(r', \mathcal{G})$.

Filtering Criteria. After applying one of the refinement operators to a rule, a set of candidate rules is obtained. For each candidate rule we first verify that the hybrid measure μ has increased and discard the rule if it has not. Then, we compute its *h-cover* [10] and our novel exception confidence measure *e-conf* that are defined as follows:

$$h\text{-}cover(r, \mathcal{G}) = r\text{-}supp(r, \mathcal{G})/|\{h \mid h(head(r, \mathcal{G})) \in \mathcal{G}\}|,$$
$$e\text{-}conf(r, \mathcal{G}) = conf(r'', \mathcal{G}),$$

where r'' : $body^-(r) \leftarrow body^+(r), not\ head(r)$. If the h-cover and e-conf are below the user specified threshold, then the rule is discarded. Intuitively, h-cover quantifies the ratio of the known true facts that are implied by the rule. In contrast, e-conf is the conditional probability of the exception given predictions produced by the Horn part of r, which helps to disregard insignificant exceptions, i.e., those that explain the absence in $\mathcal{G}$ of only a small fraction of predictions made by $head(r) \leftarrow body^+(r)$, as such exceptions likely correspond to noise. Observe that not all of the filtering criteria are relevant for all rule types. For example, exception confidence is relevant only for non-monotonic rules to ensure the quality of the added exceptions.

Finally, note that by exploiting the embedding feedback, we can now distinguish exceptions from noise. Consider the rule stating that married people live together. This rule can have several possible exceptions, e.g., either one of the spouses is a researcher or he/she works at a company, which has headquarter in the US. Whenever the rule is enriched with an exception, naturally, the support of its body decreases, i.e., the size of $\mathcal{G}_r$ goes down. Relying on our filtering criteria, we aim at adding such negated atoms, that the average quality of $\mathcal{G}_r$ increases, meaning that the introduced negated atoms prevent unlikely predictions.

4   Evaluation

We have implemented our hybrid rule learning approach in Java within a system prototype RuLES, and conducted experiments on a Linux machine with 80 cores and 500 GB RAM. In this section we report the results of our experimental evaluation, which focuses on *(i)* the benefits of our hybrid embedding-based rule quality measure over traditional rule measures; *(ii)* the effectiveness of RuLES against the state-of-art Horn rule learning systems; and *(iii)* the quality of non-monotonic rules learned by RuLES compared to existing methods.

4.1   Experimental Setup

Datasets. We performed experiments on the following two real world datasets:

- *FB15K* [2]: a subset of Freebase with 592K binary facts over 15K entities and 1345 relations commonly used for evaluating KG embedding models [34].
- *Wiki44K*: a dataset with 250K binary facts over 44K entities and 100 relations, which is a subset of Wikidata dataset from December 2014 used in [10].

In the experiments for each incomplete KG $\mathcal{G}$ we need its *ideal* completion $\mathcal{G}^i$ that would give us a gold standard for evaluating our approach and comparing it to others. Since obtaining a real life $\mathcal{G}^i$ is hard, we used the KGs FB15K and Wiki44K as reference graphs $\mathcal{G}^i_{appr}$ that approximate $\mathcal{G}^i$. We then constructed $\mathcal{G}$ by randomly selecting 80% of its facts while preserving the distribution of facts over predicates.

Embedding Models. We experimented with the three state-of-the-art embedding models: TransE [2], HolE [22], and the text-enhanced SSP [38] model. We reuse the implementation of TransE, HolE[1], and SSP[2]. TransE and HolE were trained on $\mathcal{G}$ and SSP on $\mathcal{G}$ enriched with a textual description for each entity extracted from Wikidata. We compared the effectiveness of the models and selected for every KG the best one. Apart from SSP, which showed the best performance on both KGs, we also selected HolE for FB15K and TransE for Wiki44K. Note that in this work as a proof of concept we considered some of the most popular embedding models, but conceptually any model (see [34] for overview) can be used in our system.

Evaluation Metric. To evaluate the learned rules we use the quality of predictions that they produce when applied on $\mathcal{G}$, i.e., the more correct facts beyond $\mathcal{G}$ a ruleset produces, the better it is. We consider two evaluation settings: *closed world* setting (CW) and *open world* setting (OW). In the CW setting, we define the prediction precision of a rule r and a set of rules R as:

$$pred\_prec_{CW}(r) = \frac{|\mathcal{G}_r \cap \mathcal{G}_{appr}^i \setminus \mathcal{G}|}{|\mathcal{G}_r \setminus \mathcal{G}|}, \quad pred\_prec_{CW}(R) = \frac{\sum\limits_{r \in R} pred\_prec_{CW}(r)}{|R|}.$$

In the OW setting, we also take into account the incompleteness of $\mathcal{G}_{appr}^i$ and consider the quality of predictions outside it by performing a random sampling and manually annotating the sampled facts relying on Web resources such as Wikipedia. Thus, we define the OW prediction precision $pred\_prec_{OW}$ for a set of rules R as follows:

$$pred\_prec_{OW}(R) = \frac{|\mathcal{G}' \cap \mathcal{G}_{appr}^i| + |\mathcal{G}' \setminus \mathcal{G}_{appr}^i| \times accuracy(\mathcal{G}' \setminus \mathcal{G}_{appr}^i)}{|\mathcal{G}'|}.$$

where $\mathcal{G}' = \bigcup_{r \in R} \mathcal{G}_r \setminus \mathcal{G}$ is the union of predictions generated by rules in R, and $accuracy(S)$ is the approximated ratio of true facts inside S computed via manual checking of facts sampled from S. Finally, to evaluate the meaningfulness of exceptions in a rule (i.e., negated atoms) we compute the *revision precision*, which according to [32] is defined as the ratio of incorrect facts in the difference between predictions produced by the Horn part of a rule and its non-monotonic version over the total number of predictions in this difference (the higher the revision precision, the better the rule exceptions) computed per ruleset. Formally,

$$rev\_prec_{OW}(R) = 1 - \frac{|\mathcal{G}'' \cap \mathcal{G}_{appr}^i| + |\mathcal{G}'' \setminus \mathcal{G}_{appr}^i| \times accuracy(\mathcal{G}'' \setminus \mathcal{G}_{appr}^i)}{|\mathcal{G}''|}.$$

where $\mathcal{G}'' = \mathcal{G}_H \setminus \mathcal{G}_R$ and H is the set of Horn parts of rules in R. Intuitively, $\mathcal{G}''$ contains facts not predicted by the rules in R but predicted by their Horn versions.

---

[1] https://github.com/mnick/scikit-kge.
[2] https://github.com/bookmanhan/Embedding.

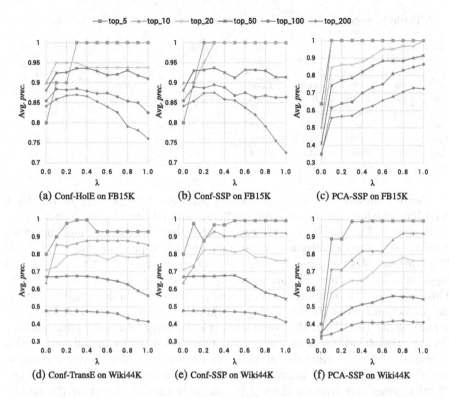

Fig. 3. $pred\_prec_{CW}$ of the $top\text{-}k$ rules with various $embedding\ weights$.

RuLES Configuration. We run RuLES in several configurations where μ_1 is set to either *standard confidence (Conf)* or *PCA confidence (PCA)*, and μ_2 is computed based on either TransE, HolE, or SSP models. Through the experiments the configurations are named as $\mu_1\text{-}\mu_2$ (e.g. Conf-HolE).

4.2   Embedding-Based Hybrid Quality Function

In this experiment we study the effect of using our hybrid embedding-based rule measure μ from Eq. 3 on the rule ranking compared to traditional measures and embedding models independently. We do it by first learning rules of the form $r\ :\ h(X,Z) \leftarrow p(X,Y), q(Y,Z)$ from $\mathcal{G}$ where $r\text{-}supp(r,\mathcal{G}) \geq 10$, $conf(r,\mathcal{G}) \in [0.1, 1)$ and $h\text{-}cover(r,\mathcal{G}) \geq 0.01$. Then, we rank these rules using Eq. 3 with $\lambda \in \{0, 0.1, 0.2, \ldots, 1\}$, $\mu_1 \in \{conf, conf_{pca}\}$ and with μ_2 that is computed by relying on TransE, HolE and SSP. Note that $\lambda = 0$ simulates learning rules using the standard measure μ_1 similar to [10], while $\lambda = 1$ corresponds to ranking rules solely based on the predictions of the embedding models. Configuring λ indirectly allows us to compare our hybrid measure to both traditional measures and quality of embedding models.

Table 1. $pred\_prec_{CW}$ of the *top-k* rules learned using different measures.

top-k	FB15K				Wiki44K			
	Conf ($\lambda = 0$)	PCA ($\lambda = 0$)	Conf-HolE ($\lambda = 0.3$)	Conf-SSP ($\lambda = 0.3$)	Conf ($\lambda = 0$)	PCA ($\lambda = 0$)	Conf-TransE ($\lambda = 0.3$)	Conf-SSP ($\lambda = 0.3$)
5	0.800	0.638	**1.000**	**1.000**	0.800	0.402	**0.995**	0.968
10	0.900	0.506	**1.000**	**1.000**	0.638	0.321	0.863	**0.932**
20	0.900	0.499	0.950	**1.000**	0.712	0.357	0.802	**0.825**
50	0.881	0.410	0.936	**0.937**	0.670	0.352	**0.675**	0.674
100	0.855	0.348	0.885	**0.895**	**0.477**	0.331	0.474	0.474
200	0.842	0.355	0.870	**0.875**	–	–	–	–

Figure 3 shows the average prediction precision $pred\_prec_{CW}$ of the *top-k* rules ranked using our measure μ for different embedding weights λ (*x-axis*). In particular, in Figs. 3a, b, d, and e we observe that combining confidence with any embedding model increases the average prediction precision for $0 \leq \lambda \leq 0.3$. Moreover, we observe the decrease of prediction precision for $0.4 \leq \lambda \leq 1$ and *top-k* rules learned from FB15K when $k \geq 20$ and from Wiki44K when $k \geq 10$. This shows that the combination of μ_1 and μ_2 gives noticeable positive effect on the prediction results. Ranking using hybrid measure with λ around 0.3 achieves better results than both the traditional rule learning and embedding models. On the other hand, for $\mu_1 = conf_{pca}$ the precision increases significantly when combined with embedding models and only decreases slightly for $\lambda = 1$ (Figs. 3c and f). Utilizing $conf_{pca}$ instead of $conf$ as μ_1 in our hybrid measure is less effective, since our training data $\mathcal{G}$ is randomly sampled breaking the partial completeness assumption adopted by the PCA confidence.

Table 1 compactly summarizes the average prediction precision of *top-k* rules ranked by the standard rule measures and our μ for the best value of $\lambda = 0.3$ and highlights the effect of using the better embedding model (text-enhanced vs standard). We observe that the accuracy of a utilized embedding model is naturally propagated to the accuracy of the rules that we obtain using our hybrid ranking measure μ. This demonstrates that the use of a better embedding model positively effects the quality of learned rules.

4.3 Horn Rule Learning

In this experiment, we compare RuLES under Conf-SSP configuration (with embedding weight $\lambda = 0.3$) with the state-of-art Horn rule learning system AMIE. We used the default AMIE-PCA configuration with $conf_{pca}$ and AMIE-Conf with $conf$ measures respectively. For a fair comparison, we set the two configurations of AMIE and our system to generate rules with at most three positive atoms and filtered them based on minimum confidence of 0.1, head coverage of 0.01 and rule support of 10 in case of FB15K and 2 in case of Wiki44K. We then filtered out all rules with $conf(r, \mathcal{G}) = 1$, as they do not produce any predictions.

Table 2 shows the number of facts (see the *Facts* column) predicted by the set R of *top-k* rules in the described settings and their prediction precision

Table 2. $pred\_prec_{OW}$ of the *top-k* rules generated by RuLES and AMIE.

top-k	FB15K						Wiki44K					
	AMIE-PCA		AMIE-Conf		RuLES		AMIE-PCA		AMIE-Conf		RuLES	
	Facts	*Prec.*	*Facts*	*Prec.*	*Facts*	*Prec.*	*Facts*	*Prec.*	*Facts*	*Prec.*	*Facts*	*Prec.*
20	1029	0.28	82	0.63	44	1.00	185	0.73	91	0.95	3291	0.98
50	1716	0.43	190	0.74	186	0.92	47099	0.10	3594	0.95	6154	0.88
100	3085	0.65	255	0.78	539	0.80	56831	0.20	13870	0.83	13253	0.82
200	10586	0.62	1210	0.83	1205	0.88	82288	0.39	19538	0.72	20408	0.73
500	40050	0.51	2702	0.75	7124	0.95	219264	0.35	124836	0.23	128256	0.48

Table 3. $pred\_prec_{OW}$ of the *top-k* rules generated by NeuralLP and RuLES.

top-k	Family-NeuralLP		Family-Conf-TransE	
	Facts	*Prec.*	*Facts*	*Prec.*
10	3709	0.72	4201	0.68
20	8821	0.53	6957	0.72
30	11337	0.49	9368	0.71
40	14662	0.46	11502	0.72
50	18768	0.40	14547	0.62

$pred\_prec_{OW}(R)$ (see the *Prec.* column). The size of the random sample outside $\mathcal{G}^i_{appr}$ is 20. We can observe that on FB15K, RuLES consistently outperforms both AMIE configurations. The *top-20* rules have the highest precision difference (outperforming AMIE-PCA and AMIE-Conf by 72% and 37% respectively). This is explained by the fact that the hybrid embedding quality penalizes rules with higher number of false predictions. For Wiki44K, RuLES is capable of achieving better precision in most of the cases. Notably, for the *top-20* rules RuLES predicted significantly more facts then competitors yet with a high precision.

In Table 3, we compare RuLES with the recently developed NeuralLP system [40]. For this we utilized the Family dataset used by NeuralLP with 28K facts over 3K entities and 12 relations. Starting from the *top-20* rules RuLES is capable of achieving significantly better precision. For the *top-10* rules the precision of NeuralLP is slightly better, but RuLES predicts many more facts.

More experiments and analysis on different datasets are provided in the technical report at https://github.com/hovinhthinh/RuLES.

4.4 RuLES for Exception-Aware Rule Learning

In this experiment, we aim at evaluating the effectiveness of RuLES for learning exception-aware rules. First, consider in Table 4 examples of such rules learned by RuLES over Wiki44K dataset. The first rule r^1 says that a person is a citizen of the country where his alma mater is located, unless it is a research institution, since most researchers in universities are foreigners. The second rule r^2 states

Table 4. Example rules with exception generated by RuLES.

r^1: $nationality(X, Y) \leftarrow graduated\_from(X, Z), in\_country(Z, Y), \textbf{not } research\_uni(Z)$
r^2: $scriptwriter\_of(X, Y) \leftarrow preceded\_by(X, Z), scriptwriter\_of(Z, Y), \textbf{not } tv\_series(Z)$
r^3: $noble\_family(X, Y) \leftarrow spouse(X, Z), noble\_family(Z, Y), \textbf{not } chinese\_dynasties(Y)$

Table 5. $pred\_prec_{OW}$ (left) and rev_prec_{OW} (right) of the *top-k* rules learned by RUMIS and RuLES.

top-k	FB15K				Wiki44K			
	RUMIS		RuLES		RUMIS		RuLES	
	Facts	Prec.	Facts	Prec.	Facts	Prec.	Facts	Prec.
20	672	0.95	34	0.97	5844	0.93	5640	0.93
50	1797	0.94	158	0.99	8585	0.83	13333	0.84
100	2672	0.94	434	0.99	21081	0.76	25265	0.81
200	4103	0.87	1155	0.96	50957	0.51	43677	0.67
500	13439	0.76	5466	0.90	-	-	-	-

top-k	FB15K				Wiki44K			
	RUMIS		RuLES		RUMIS		RuLES	
	Facts	Prec.	Facts	Prec.	Facts	Prec.	Facts	Prec.
20	76	0.70	111	0.68	63	0.47	81	0.94
50	126	0.51	435	0.74	191	0.28	611	0.69
100	183	0.43	680	0.76	543	0.49	1698	0.79
200	310	0.30	1112	0.87	4861	0.40	3175	0.80
500	1155	0.53	3760	0.59	-	-	-	-

that the scriptwriter of some artistic work is also the scriptwriter of its sequel unless it is a TV series, which actually reflects the common practice of having several screenwriters for different seasons. Additionally, r^3 encodes that someone belonged to a noble family if his/her spouse is also from the same noble family, excluding the Chinese dynasties.

To quantify the quality of RuLES in learning non-monotonic rules, we compare the Conf-SSP configuration of RuLES (with embedding weight $\lambda = 0.3$) with RUMIS [32], which is a revision-based non-monotonic rule learning system, which extracts rules of the form $r : h(X, Z) \leftarrow p(X, Y), q(Y, Z), not\ E$, where E is either $e(X, Z)$ or $e(X)$. For a fair comparison we restricted RuLES to learn rules of the same form. We configured both systems setting the minimum rule support threshold to 10 and exception confidence for RuLES to 0.05. To enable the systems to learn rules with exceptions of the form $e(X)$, we enriched our KGs with *types* from original Freebase and Wikidata KGs.

Table 5 (left) reports the number of predictions produced by a rule set R of *top-k* non-monotonic rules learned by both systems as well as their precision $pred\_prec_{OW}(R)$ with a sample of 20 prediction outside $\mathcal{G}^i_{appr}$. The results show that RuLES consistently outperforms RUMIS on both datasets. For Wiki44K, and $k \in \{50, 100\}$, the *top-k* rules produced by RuLES predicted more facts than those induced by the competitor achieving higher overall precision. Regarding the number of predictions, the converse holds for the FB15K KG; however, the rules learned by RuLES are still more accurate.

To evaluate the quality of the chosen exceptions, we compare the $rev\_prec_{OW}(R)$ with a sample of 20 predictions. Observe that in Table 5 (right), rules induced by RuLES prevented the generation of more facts than RUMIS. In all of the cases apart from *top-20* for FB15K, our system managed to remove a larger fraction of erroneous predictions. For Wiki44K, RuLES consistently performs twice as good as RUMIS. In conclusion, the guidance from the embedding model exploited in our system gives us hints on which among the possible exception candidates likely correspond to noise.

5 Related Work

Inductive Logic Programming (ILP) addresses the problem of rule learning from data. In its probabilistic setting, given a set of probabilistic examples for grounded atoms and a target predicate p, the task is to learn rules for predicting probabilities of atoms for p [5,25,26]. which quickly grows to sizes that ILP methods cannot handle.

A recently proposed differentiable ILP framework [7] has advantages over traditional ILP in its robustness to noise and errors in the underlying data. However, [7] requires negative examples, which in our case are hard to get due to the large KG size. Moreover, [7] is memory-expensive as authors admit, and cannot scale to the size of modern KGs.

Unsupervised relational association rule learning systems such as [10,13] induce logical rules from the data by mining frequent patterns and casting them into rules. In the context of KGs [3,10,32] such approaches address the incompleteness of KGs by exploiting sophisticated measures over the original graph, possibly enhanced with a schema [6] or constraints on the number of missing edges [31]. However, these methods do not tap any unstructured information like we do. Indeed, our hybrid embedding-based measure allows us to conveniently account for unstructured information implicitly via embeddings as well as making use of various graph-based rule metrics.

Exploiting embedding models for rule learning is a new research direction that has recently gained attention [39,40]. To the best of our knowledge, existing methods are purely statistics-based, i.e., they reduce the rule learning problem to algebraic operations on neural-embedding-based representations of a given KG. The work [39] constructs rules by modeling relation composition as multiplication or addition of two relation embeddings. The authors of [40] propose a differentiable system for learning models defined by sets of first-order rules that exploits a connection between inference and sparse matrix multiplication [4]. However, existing approaches pose strong restrictions on target rule patterns, which often prohibit learning interesting rules, e.g. non-chain-like or exception-aware ones, which we support.

Another line of work concerns enhancing embedding models with rules and constraints, e.g. [14,15,28,35]. While our direction is related, we pursue a different goal of leveraging the feedback from embeddings to improve the quality of the learned rules. To the best of our knowledge, this idea has not been considered in any prior work.

6 Conclusion

We presented a method for learning rules that may contain negated atoms from KGs that dynamically exploits feedback from a precomputed embedding model.

Our approach is general in that any embedding model can be utilized including text-enhanced ones, which indirectly allows us to harness unstructured web sources for rule learning. We evaluated our approach with various configurations on real-world datasets and observed significant improvements over state-of-the-art rule learning systems.

An interesting future direction is to extend our work to more complex non-monotonic rules with higher-arity predicates, aggregates and existential variables or disjunctions in rule heads, which is challenging due to inevitable scalability issues.

Acknowledgements. This work was partially supported by the EPSRC projects DBOnto, MaSI³ and ED³.

References

1. Agrawal, R., Imieliński, T., Swami, A.: Mining association rules between sets of items in large databases. SIGMOD Rec. **22**(2), 207–216 (1993)
2. Bordes, A., Usunier, N., García-Durán, A., Weston, J., Yakhnenko, O.: Translating embeddings for modeling multi-relational data. In: Proceedings of NIPS, pp. 2787–2795 (2013)
3. Chen, Y., Goldberg, S., Wang, D.Z., Johri, S.S.: Ontological pathfinding. In: SIGMOD (2016)
4. Cohen, W.W.: Tensorlog: a differentiable deductive database. CoRR, abs/1605.06523 (2016)
5. Corapi, D., Sykes, D., Inoue, K., Russo, A.: Probabilistic rule learning in nonmonotonic domains. In: Leite, J., Torroni, P., Ågotnes, T., Boella, G., van der Torre, L. (eds.) CLIMA 2011. LNCS (LNAI), vol. 6814, pp. 243–258. Springer, Heidelberg (2011). https://doi.org/10.1007/978-3-642-22359-4_17
6. d'Amato, C., Staab, S., Tettamanzi, A.G., Minh, T.D., Gandon, F.: Ontology enrichment by discovering multi-relational association rules from ontological knowledge bases. In: SAC, pp. 333–338 (2016)
7. Evans, R., Grefenstette, E.: Learning explanatory rules from noisy data. J. Artif. Intell. Res. **61**, 1–64 (2018)
8. Fürnkranz, J., Kliegr, T.: A brief overview of rule learning. In: Bassiliades, N., Gottlob, G., Sadri, F., Paschke, A., Roman, D. (eds.) RuleML 2015. LNCS, vol. 9202, pp. 54–69. Springer, Cham (2015). https://doi.org/10.1007/978-3-319-21542-6_4
9. Gad-Elrab, M.H., Stepanova, D., Urbani, J., Weikum, G.: Exception-enriched rule learning from knowledge graphs. In: Groth, P., et al. (eds.) ISWC 2016. LNCS, vol. 9981, pp. 234–251. Springer, Cham (2016). https://doi.org/10.1007/978-3-319-46523-4_15
10. Galarraga, L., Teflioudi, C., Hose, K., Suchanek, F.M.: Fast rule mining in ontological knowledge bases with AMIE+. VLDB J. **24**(6), 707–730 (2015)
11. Gelfond, M., Lifschitz, V.: The stable model semantics for logic programming. In: ICLP/SLP, pp. 1070–1080. MIT Press (1988)
12. Glorot, X., Bordes, A., Weston, J., Bengio, Y.: A semantic matching energy function for learning with multi-relational data. CoRR, abs/1301.3485 (2013)

13. Goethals, B., Van den Bussche, J.: Relational association rules: getting WARMER. In: Hand, D.J., Adams, N.M., Bolton, R.J. (eds.) Pattern Detection and Discovery. LNCS (LNAI), vol. 2447, pp. 125–139. Springer, Heidelberg (2002). https://doi.org/10.1007/3-540-45728-3_10
14. Guo, S., Wang, Q., Wang, L., Wang, B., Guo, L.: Jointly embedding knowledge graphs and logical rules. In: EMNLP (2016)
15. Guo, S., Wang, Q., Wang, L., Wang, B., Guo, L.: Knowledge graph embedding with iterative guidance from soft rules. CoRR, abs/1711.11231 (2017)
16. Kharlamov, E., et al.: Semantic rules for machine diagnostics: execution and management. In: CIKM, pp. 2131–2134 (2017)
17. Kotsiantis, S., Kanellopoulos, D.: Association rules mining: a recent overview. GESTS Int. Trans. CS Eng. **32**(1), 71–82 (2006)
18. Mannila, H., Toivonen, H., Verkamo, A.I.: Discovering frequent episodes in sequences. In: KDD 1995 (1995)
19. Mehdi, G., et al.: Semantic rule-based equipment diagnostics. In: d'Amato, C., Fernandez, M., Tamma, V., Lecue, F., Cudré-Mauroux, P., Sequeda, J., Lange, C., Heflin, J. (eds.) ISWC 2017. LNCS, vol. 10588, pp. 314–333. Springer, Cham (2017). https://doi.org/10.1007/978-3-319-68204-4_29
20. Duc Tran, M., d'Amato, C., Nguyen, B.T., Tettamanzi, A.G.B.: Comparing rule evaluation metrics for the evolutionary discovery of multi-relational association rules in the semantic web. In: Castelli, M., Sekanina, L., Zhang, M., Cagnoni, S., García-Sánchez, P. (eds.) EuroGP 2018. LNCS, vol. 10781, pp. 289–305. Springer, Cham (2018). https://doi.org/10.1007/978-3-319-77553-1_18
21. Mitchell, T., et al.: Never-ending learning. In: AAAI (2015)
22. Nickel, M., Rosasco, L., Poggio, T.A.: Holographic embeddings of knowledge graphs. In: AAAI (2016)
23. Nickel, M., Tresp, V., Kriegel, H.-P.: A three-way model for collective learning on multi-relational data. In: ICML (2011)
24. Piatetsky-Shapiro, G.: Discovery, analysis, and presentation of strong rules. In: Knowledge Discovery in Databases, pp. 229–248. AAAI/MIT Press (1991)
25. Raedt, L.D., Dries, A., Thon, I., den Broeck, G.V., Verbeke, M.: Inducing probabilistic relational rules from probabilistic examples. In: IJCAI, pp. 1835–1843. AAAI Press (2015)
26. Raedt, L.D., Thon, I.: Probabilistic rule learning. In: ILP (2010)
27. Ras, Z.W., Wieczorkowska, A.: Action-rules: how to increase profit of a company. In: Zighed, D.A., Komorowski, J., Żytkow, J. (eds.) PKDD 2000. LNCS (LNAI), vol. 1910, pp. 587–592. Springer, Heidelberg (2000). https://doi.org/10.1007/3-540-45372-5_70
28. Rastogi, P., Poliak, A., Durme, B.V.: Training relation embeddings under logical constraints. In: KG4IR (2017)
29. Ringsquandl, M., et al.: Event-enhanced learning for knowledge graph completion. In: ESWC (2018)
30. Suchanek, F.M., Kasneci, G., Weikum, G.: Yago: a core of semantic knowledge. In: Proceedings of WWW, pp. 697–706 (2007)
31. Tanon, T.P., Stepanova, D., Razniewski, S., Mirza, P., Weikum, G.: Completeness-aware rule learning from knowledge graphs. In: d'Amato, C. (ed.) ISWC 2017. LNCS, vol. 10587, pp. 507–525. Springer, Cham (2017). https://doi.org/10.1007/978-3-319-68288-4_30
32. Tran, H.D., Stepanova, D., Gad-Elrab, M.H., Lisi, F.A., Weikum, G.: Towards non-monotonic relational learning from knowledge graphs. In: ILP, pp. 94–107 (2016)

33. Vrandecic, D., Krötzsch, M.: Wikidata: a free collaborative knowledgebase. CACM **57**(10), 78–85 (2014)
34. Wang, Q., Mao, Z., Wang, B., Guo, L.: Knowledge graph embedding: a survey of approaches and applications. IEEE Trans. Knowl. Data Eng. **29**(12), 2724–2743 (2017)
35. Wang, Q., Wang, B., Guo, L.: Knowledge base completion using embeddings and rules. In: IJCAI (2015)
36. Wang, Z., Li, J.: RDF2Rules: Learning rules from RDF knowledge bases by mining frequent predicate cycles. CoRR, abs/1512.07734 (2015)
37. Wojtusiak, J.: Rule learning in healthcare and health services research. In: Dua, S., Acharya, U.R., Dua, P. (eds.) Machine Learning in Healthcare Informatics. ISRL, vol. 56, pp. 131–145. Springer, Heidelberg (2014). https://doi.org/10.1007/978-3-642-40017-9_7
38. Xiao, H., Huang, M., Meng, L., Zhu, X.: SSP: semantic space projection for knowledge graph embedding with text descriptions. In: AAAI (2017)
39. Yang, B., Yih, W., He, X., Gao, J., Deng, L.: Embedding entities and relations for learning and inference in knowledge bases. CoRR, abs/1412.6575 (2014)
40. Yang, F., Yang, Z., Cohen, W.W.: Differentiable learning of logical rules for knowledge base reasoning. In: NIPS, pp. 2316–2325 (2017)
41. Zupanc, K., Davis, J.: Estimating rule quality for knowledge base completion with the relationship between coverage assumption. In: WWW 2018, pp. 1073–1081 (2018)

A Novel Ensemble Method for Named Entity Recognition and Disambiguation Based on Neural Network

Lorenzo Canale[1,2], Pasquale Lisena[1], and Raphaël Troncy[1(✉)]

[1] EURECOM, Sophia Antipolis, France
{canale,lisena,troncy}@eurecom.fr
[2] Politecnico di Torino, Turin, Italy

Abstract. Named entity recognition (NER) and disambiguation (NED) are subtasks of information extraction that aim to recognize named entities mentioned in text, to assign them pre-defined types, and to link them with their matching entities in a knowledge base. Many approaches, often exposed as web APIs, have been proposed to solve these tasks during the last years. These APIs classify entities using different taxonomies and disambiguate them with different knowledge bases. In this paper, we describe Ensemble Nerd, a framework that collects numerous extractors responses, normalizes them and combines them in order to produce a final entity list according to the pattern (surface form, type, link). The presented approach is based on representing the extractors responses as real-value vectors and on using them as input samples for two Deep Learning networks: ENNTR (Ensemble Neural Network for Type Recognition) and ENND (Ensemble Neural Network for Disambiguation). We train these networks using specific gold standards. We show that the models produced outperform each single extractor responses in terms of micro and macro F1 measures computed by the GERBIL framework.

1 Introduction

A crucial task in knowledge extraction from textual document consists in the two complementary tasks of Named Entity Recognition (NER) and Named Entity Disambiguation (NED), achieving the goal of assigning to parts of text (tokens) respectively a type – from a pre-defined taxonomy – and a unique identifier – normally in the form of URI – that points univocally to the referred entity in a given knowledge base. The combination of these two tasks is often abbreviated with the acronym NERD [5,6]. The current state of the art offers an interesting number of NERD extractors. Some of them can be trained by a developer on his own corpus, while other ones are only accessible as black-box services exposed via web APIs offering a limited number of parameters.

In terms of NER, each service provides generally its own taxonomy of named entity types which can be recognised. While they all provide support for three major types (person, organization, location), they largely differ for more fine-grained types which makes hard their comparison and combination. In terms

© Springer Nature Switzerland AG 2018
D. Vrandečić et al. (Eds.): ISWC 2018, LNCS 11136, pp. 91–107, 2018.
https://doi.org/10.1007/978-3-030-00671-6_6

of NED, each extractor can potentially disambiguate entities against specific knowledge bases (KB), but in practice, they mostly rely on popular ones, namely DBpedia, Wikidata, Freebase or YAGO. For this reason, comparing and merging the results of these extractors require some post-processing tasks that typically rely on mappings between those KBs. This task is however simpler than the type alignment, because of the large presence of `owl:sameAs` links between the different KBs.

In this paper, we present **Ensemble Nerd**, a multilingual ensemble method that combines the responses of different NERD extractors. This method relies on a real-value vectorial representation as input samples for two Deep Learning networks, ENNTR (Ensemble Neural Network for Type Recognition) and ENND (Ensemble Neural Network for Disambiguation). The networks provide models for performing type alignment and named entity linking to a knowledge base. This strategy is evaluated against some well-known gold standards, showing that the output of the ensemble outperforms the results of single extractors.

This work aims to answer the following research questions: Can we define an ensemble method that combines the extractors responses in order to create a new more powerful extractor? Is it possible to define an ensemble method that avoids a type alignment step or that computes it automatically, without any human intervention? Which ensemble method should be adopted to exploit all the collected information? Considering that extractors return list of named entities – together with the type and the disambiguation link of each of them –, how this data can be numerically represented? Can we better understand which features contribute more to improve the ensemble output response? How dependant is this feature selection of the corpora, language, entity types and what is the influence of the KB?

The remainder of this paper is organised as follows: Sect. 2 describes some related work. Section 3 details how we represent the extractors responses, while Sect. 4 presents the core of the ensemble method. An evaluation is proposed in Sect. 5, while conclusion and and future work are discussed in Sect. 6.

2   State of the Art

Ensemble methods for the NER and NED tasks have already largely been studied in the literature. The **NERD** framework [5,6] allows to compare and evaluate some of the most popular named entity extractors. It can analyse any textual resource published on the web and to extract the named entities that are detected, typed and disambiguated by various named entity extractor APIs. For overcoming the different type taxonomies, the authors designed the *NERD ontology* which provides a set of mappings between these various classifications and consequently makes possible an evaluation of the quality of each extractor. This task was originally a one time modeling exercise: the authors manually mapped the different taxonomies to the NERD ontology.

NERD-ML, a machine learning approach developed on top of the NERD framework, combines the responses of single extractors applying alternatively three different algorithms: Naive Bayes (NB), k-Nearest Neighbours (k-NN) and Support Vector Machines (SVM) [6,11]. It is a more sophisticated and robust approach that uses machine learning inductive techniques for passing from the output type of single extractors to the right entity type in a normalized types set, i.e. the NERD Ontology [7]. **FOX** [9,10] is a framework that relies on ensemble learning by integrating and merging the results of four NER tools: the **Stanford Named Entity Recognizer** [3], the **Illinois Named Entity Tagger** [4], the **Ottawa Baseline Information Extraction** (Balie) and the **Apache OpenNLP Name Finder**. FOX compares the performance of these tools for a small set of classes namely LOCATION, ORGANIZATION and PERSON. For achieving this goal, the entity types of each NER tools is mapped to these three classes. Given any input text t, FOX processes t with each of the n tools it integrates. The result of each tool T_i is a piece of annotated text t_i, in which either a specific class or zero (not belonging to the label of a named entity) is assigned to each token. The tokens in t are then represented as vectors of length n and are used for getting the final type. The author demonstrates that a Multi-Layer Perceptron (MLP) gets the best results among a pool of 15 different algorithms [9].

3   Feature Engineering for NERD

Ensemble Nerd currently integrates a set of 8 extractors shown in Table 1. An extractor can belong to the set T (extractors that perform NER task) or to the set U (extractors that perform NED task). Currently, *TextRazor* is the only one in both sets: $T \cap U = \{TextRazor\}$. All these extractors relies on Wikidata, Wikipedia or DBpedia for entity disambiguation.

Each extractor produces a list of named entities as response for a specific input text. From this output, we generate 4 different kinds of feature.

1. Surface form features. They are strictly related to the text used to extract named entity. The input text is split into tokens and a word embedding

Table 1. Extractor included in Ensemble Nerd. ✓ indicates that the extractor supports the action (type recognition or named entity disambiguation)

Extractor	Type recognition	NE disambiguation
AlchemyAPI	✓	✗
DandelionAPI	✗	✓
DbSpotlight	✗	✓
TextRazor	✓	✓
Babelfy	✗	✓
MeaningCloud	✓	✗
ADEL	✓	✗
OpenCalais	✓	✗

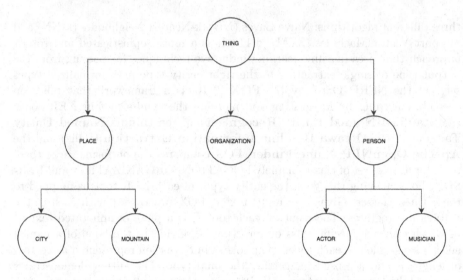

Fig. 1. Example of type taxonomy for a generic extractor.

representation is assigned to each of them. We consider also the stop words, assigning also to them a real-value vectorial representation. The word vectors are computed using *fastText* [1]. We define s^x as the real-valued vector associated to a specific token x:

$$s^x = \left[s_p^x | s_c^x \right], dim(s^x) = 400 \tag{1}$$

where $|$ (pipe) is the concatenation operator and *dim* is the vector dimension.

s_p^x, $dim(s_p^x) = 300$, consists in the token embedding computed using the Wikipedia pre-trained *fastText* models released by the authors. The model changes depending on the language used in the text, since all localised Wikipedia have been used to train language specific models.

s_c^x, $dim(s_c^x) = 100$, is the token embedding computed when training *fastText* directly on a particular textual corpus – i.e. the one for which we want to perform the NERD tasks. This means that s_c^x does not vary depending on the language but on the gold standard itself.

2. Type features. Each extractor $e \in T$ has its own type taxonomy o which is a taxonomy of a maximum depth L. In the following, we consider a simple example of an taxonomy o with just a 2 levels hierarchy (Fig. 1):

1. Level 1 includes three types: PLACE, ORGANIZATION and PERSON.
2. Level 2 includes four types: CITY and MOUNTAIN (subtypes of PLACE) and ACTOR and MUSICIAN (subtypes of PERSON).

We name C_i the number of different types inside the level i (e.g. $C_1 = 3$). We infer a one-hot encoding representation for each level as shown in Table 2.

For a generic type τ in the last layer (e.g. ACTOR), the features vector v_τ consists in the concatenation of the one-hot representation of each type founded

Table 2. Representation of types through one-hot encoding.

Level 1		Level 2	
Type	Representation	Type	Representation
PERSON	001	ACTOR	0001
ORGANIZATION	010	MUSICIAN	0010
PLACE	100	CITY	0100
		MOUNTAIN	1000

on the walk from the root to the leaf associate to τ. The features vector for ACTOR is therefore 0010001, where the first three values 001 derive from PERSON and the last four values 0001 derive from ACTOR. Hence, we can state that $dim(v_\tau) = \sum_i^L C_i$. If the extractor $e \in T$ returns a type that is not the last level in the hierarchy, as PERSON, we fill the missing vector positions with 0. The features vector v_{PERSON} associated to PERSON is thus 0010000. This mechanism is extensible to any taxonomy. However the $dim(v_\tau)$ is different for each extractor, depending on the taxonomy that it uses.

This procedure can be extended also to extractors that do not perform NER. A generic extractor e, where $e \in U \wedge e \notin T$, returns a link for each entity. Following the interlinks between KBs, we can always obtain an entity in Wikidata. The type of the entity would be the class of this entity in Wikidata, which is the value of the property *instance of (P31)*[1]. Entities might possess multiple types and for this reason they are represented through K-hot encoding.

For a **typed named entity** w^t with the format (surface form, type), the type feature vector $v_e^{w^t}$ is computed for the extractor e where $e \in U \vee e \in T$. $dim(v_e^{w^t})$ varies accordingly to the considered extractor. In fact, we get a real-value numerical type representation without a type alignment phase. For this reason, the number of dimensions that forms the type features vector depends on the the number of types in the extractor taxonomy.

3. Entity features. These features represent the similarity between two Wikidata entities w_1 and w_2, as a vector of 5 dimensions. The first four dimensions correspond to **semantic knowledge**:

1. the first dimension $S_{uri}(w_1, w_2)$ indicates if the compared entities share the same URI with a Boolean;
2. the second dimension provides the string similarity between the labels l_{w_1} and l_{w_2} associated to the compared entities:

$$S_{Lev}(w_1, w_2) = max(1 - d_{Lev}(l_{w_1}, l_{w_2})/\beta, 0), \beta = 8$$

where $d_{Lev}(l_{w_1}, l_{w_2})$ is the **Levenshtein distance** between the compared strings and β is a constant equals to the number of maximum differences after which the similarity is saturated to 0.

---

[1] https://www.wikidata.org/wiki/Property:P31.

3. the third dimension $S_{TfIdf}(w_1, w_2)$ represents the **TF-IDF Cosine Similarity** between the abstracts associated to the compared entities. This dimension represents a textual knowledge as in [12];

4. the fourth dimension $S_{occ}(w_1, w_2)$. value indicates if the compared entities share the same *occupation (P106)*.[2] This property is specific for entities of type PERSON: this Wikidata class has no other subclasses, as opposed to the other types. For this reason this similarity dimension greatly helps in the disambiguation of people with similar names but different professions. $S_{occ}(w_1, w_2)$ is set to 1 when the two entities referred to people that have the same profession, and 0 otherwise (different profession or not a PERSON).

The fifth and last dimension of the vector represents the structural similarity as in [12]. We define a property set P, containing three properties: *subclass of (P279)*[3], *instance of (P31)*[4], and *part of (P361)*[5]. A subgraph G is extracted from Wikidata selecting all the triples in which a property in P appears. We define the distance d_{w_1,w_2} between two generic entities w_1 and w_2 as the shortest path length that links w_1 and w_2 in G. Then, we compute the maximum distance between two nodes in the graph G, defining it as d_{max}. We assess the structural similarity between w_1 and w_2 as:

$$S_{stc}(w_1, w_2) = -\frac{d_{w_1,w_2}}{d_{max}} + 1$$

The total similarity between w_1 and w_2 can be expressed as:

$$S(\boldsymbol{w_1}, \boldsymbol{w_2}) = [S_{uri}(w_1, w_2), S_{Lev}(w_1, w_2), S_{TfIdf}(w_1, w_2), S_{occ}(w_1, w_2), S_{stc}(w_1, w_2)] \tag{2}$$

The choice of representing the similarity between two entities as a real-value vectors rather than using an entity embedding is in line with our goal of representing how the extractors differ in the prediction rather than directly representing an entity. This approach avoids to compute embeddings on the whole Wikidata KB. We rely on interlinks between KBs for guaranteeing that we can always compare Wikidata entities. This causes the risk that no Wikidata entity exists for the source one, i.e. because the information is not present. However, this case is very rare (Table 3) in all the considered benchmarks in the evaluation, thanks to the reliance of all the involved extractors on Wikidata, Wikipedia or DBpedia, which containing similar information. This would become a limit when using different KBs (e.g. thematic ones), not fully interlinkable to Wikidata and for which a loss in information should be taken in account.

4. Score features. Some extractors return scores representing either the confidence or the saliency for each named entity. For each extractor $e \in K$, w^k is

---

[2] https://www.wikidata.org/wiki/Property:P106.
[3] https://www.wikidata.org/wiki/Property:P279.
[4] https://www.wikidata.org/wiki/Property:P31.
[5] https://www.wikidata.org/wiki/Property:P361.

Table 3. Coverage of matching against Wikipedia of disambiguated entity in the ground truth.

Extractor	Disambiguation KB	WD coverage
Dandelion	Wikipedia	99%
DBSpotlight	DBpedia Fr	98%
TextRazor	Wikidata	100%
Babelfy	DBpedia	100%

a **named entity score** with the format (surface form, scores). We define $v_e^{w^k}$ as the features vector representing the scores for w^k and the extractor e. $dim(v_e^{w^k})$ depends on the considered extractors, more precisely on the number of scores returned by it.

4   Ensemble NERD: ENNTR and ENND

Our experimental ensemble method relies on two Neural Networks that receive in input the features described in the previous Section. We respectively name them with the acronyms **Ensemble Neural Network for Type Recognition (ENNTR)** and **Ensemble Neural Network for Disambiguation (ENND)**. For both networks, the hyper parameter optimization was done using Grid Search.

These networks architectures come after a series of previous experiments that involved LSTM and BiLSTM, receiving a complete vector including all the features as input sample. A really slow training, the ease of network overfitting to the sample input, and huge difference in dimensionality (and so in impact to the results) between the different features were some of the reasons for which we have abandoned these approaches.

Ensemble Neural Network for Type Recognition (ENNTR). We consider a generic ground truth GT formed by N textual fragments (e.g. sentences), such that we can split each fragment in tokens. X_i is the ordered list of tokens for fragment i. Concatenating the lists X_i, we get a list X, that is the ordered list of tokens for the whole corpus. We call x a generic token in X.

GT associates a type in a taxonomy o_{Gt} to each token x. We identify the neural network target as Y_t. The number of samples in Y_t is equal to the total number of tokens: $dim(Y_t) = dim(X)$. The neural network goal is to assign the right type to each token and its architecture is represented in Fig. 2.

ENNTR has an output layer O formed by $H = card(o_{GT})$ neurons, where $card(o_{GT})$ is the number of different types (or cardinality) in o_{GT}. As a consequence, each value returned by a neuron in the output layer corresponds to the probability that a token x belongs to a specific type. Hence, each target sample y_t is a vector formed by H values, where each value corresponds to a type and a neuron. In Fig. 2, we are assuming that $H = 4$.

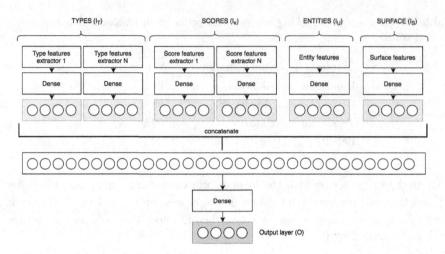

Fig. 2. ENNTR architecture

ENNTR presents many input layers. Using the same notation used in Sect. 3, T is the set of extractors that return type information, K is the set of extractors that return score information, U is the set of extractors that perform disambiguation. Defining I as the set of input layers of ENNTR, we can identify four different types of input layer depending on the kind of features being input.

$$I = I_T \cup I_K \cup I_U \cup I_S$$

$$|I| = |I_T| + |I_K| + |I_U| + |I_S| = |T \cup U| + |K| + 1 + 1$$

All the input layers works at token level, so that the features at entity level defined in Sect. 3 requires a transformation to token-level. The surface form of an entity w (e.g. *Barack Obama*) can be tokenised, producing the list of tokens X_w (e.g. *[Barack, Obama]*). The feature vector of token x is equal to the one of an entity w if x is a token in X_w. Otherwise it is equal to a padding vector $\boldsymbol{d}$, of the same dimension and containing only 0 values.

In particular, I_T receives in input a type features vector $\boldsymbol{t}_e^x$, computed like:

$$\boldsymbol{t}_e^x = \begin{cases} \boldsymbol{v}_e^{w^t} & if \ x \in X_{w^t} \\ \boldsymbol{d}_t & if \ x \notin X_{w^t} \end{cases} \tag{3}$$

$$\boldsymbol{d}_t = [0, ..., 0], dim(\boldsymbol{d}_k) == dim(\boldsymbol{v}_e^{w^t})$$

Similarly, I_K receives in input a type features vector $\boldsymbol{k}_e^x$, computed like:

$$\boldsymbol{k}_e^x = \begin{cases} \boldsymbol{v}_e^{w^k} & if \ x \in X_{w^k} \\ \boldsymbol{d}_k & if \ x \notin X_{w^k} \end{cases} \tag{4}$$

$$\boldsymbol{d}_k = [0, ..., 0], dim(\boldsymbol{d}_k) == dim(\boldsymbol{v}_e^{w^k})$$

The Wikidata entity u_e^x for the token x is:

$$u_e^x = \begin{cases} u_e^{w^u} & if\ x \in X_{w^u} \\ NAN & if\ x \notin X_{w^u} \end{cases} \tag{5}$$

The layers I_U receive in input the entity features vector $\boldsymbol{u}^x$, computed for a token x as:

$$\boldsymbol{u}^x = [S(u_1^x, u_1^x), S(u_1^x, u_2^x), ..., S(u_P^x, u_P^x)]$$

Finally, the input layers I_S receive the surface features vector $\boldsymbol{s}^x$ without any further transformation.

Each input layer I_n is fully connected with a layer M_n. M_n, like O, is composed by H neurons, where H is the number of types in the ground truth. The activation of neurons in M_n is linear.

In this first part of the network, each I_n – composed by a different number of neurons depending on the related features vector – is mapped on H neurons in M_n. This avoids that the neural network privileges features vectors with higher dimension – it happens directly concatenating different features vectors. This part of the network can be considered as an **alignment block** since it automatically map the types between the extractors and the ground truth taxonomy. This is pretty similar to the *Inductive Entity Typing Alignment* work described in [7], with the difference that the alignment step is learned by a fully connected layer. Differently from previous works [9,10], the approach does not need any preliminary alignment and recognition, because they are part of the same network.

The last part of the network is the **ensemble block**. M_k layers are concatenated forming a new layer R. $|o_{GT}|$ is the number of types in the ground truth, $|I|$ the number of input layers and $|P|$ the number of neurons in R:

$$|P| = |o_{GT}| \cdot |I|$$

R is fully connected to the output layer O. The activation of the neurons in O is linear. This means that ENNTR finally consists in a linear combinations of features: the key is the way in which the features are generated and entered in the network. The values v_h of the H output neurons in O correspond to the probability that a given type is correct. We take the highest value v_{max} between them and if it is greater than a threshold θ, we set the type related to its neuron as the predicted one. The final output of the ensemble method is a list of predicted type l_p for each token x. In a final step, sequences of token which belong to the same type are merged to a single entity, similarly to [9,10].

Ensemble Neural Network for Disambiguation (ENND). We consider a ground truth GT, similar to the one seen for ENNTR, that this time associates a Wikidata entity identifier (URI) to each token. We identify the target as Y_d.

The ENND architecture is represented in Fig. 3. Differently from related work, the goal of the network would not be to directly predict the right disambiguated entity, but to determine if the predicted entity by an extractor e, where $e \in U$, is correct or not. For this reason, the number of samples in

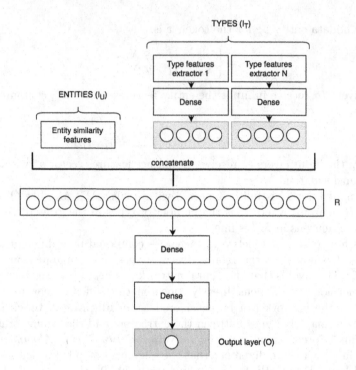

Fig. 3. ENND architecture

target Y_d is not equal to the number of tokens. For each token x, each extractor e returns a predicted entity u_e^x: we call C_x the set of predicted entities for the token x, and v_x the correct entity; $|C_x| \leq |U|$ because more extractors could predict the same entity. For each candidate $c_{x,j} \in C_x$, where $0 < j \leq |C_x|$, we generate a target sample $y_d \in Y_d$:

$$y_d = \begin{cases} 1 \ if \ c_{x,j} = v_x \\ 0 \ if \ c_{x,j} \neq v_x \end{cases}$$

The output layer O contains a single neuron that should converge to y_d. The O activation is a sigmoid. Naming I the set of input layers of ENND, two different types of input can be identified depending on the kind of features.

$$I = I_U \cup I_T$$

$$|I| = |I_U| + |I_T| = 1 + |T \cup U|$$

The entity similarity features enter through I_U. We define $c_{x,j}$ as a candidate entity for the token x. For each target sample y_d, we compute a similarity features sample $u^{x,j}$ as:

$$\boldsymbol{u_{x,j}} = [\boldsymbol{S}(c_{x,j}, u_1^x)|\boldsymbol{S}(c_{x,j}, u_2^x)|...|\boldsymbol{S}(c_{x,j}, u_R^x)] \ where \ R = card(U)$$

$$dim(\boldsymbol{u_{x,j}}) = dim(\boldsymbol{S}(\boldsymbol{w_1}, \boldsymbol{w_2})) \cdot card(U)$$

The input layers I_T receive in input the the type feature vector t_e^w, computed with the same method used for ENNTR. I_T layers are fully connected to the layers M_n as in ENNTR. M_n is formed by H neurons, where H is an hyperparameter, set to 4 during our experiment. As for ENNTR, the M_n activation is linear.

After this step, the I_U layer and the M_k layers are concatenated in a new layer R. In this layer, some neurons represent the type information, some other the entity features. This combination aims to exploit the fact that some extractors better disambiguate on certain types. The number of neurons in R is equal to $dim(\boldsymbol{u}_{x,j}) + |T \cup U| \cdot H$.

The last part of the network is composed by two dense layers[6] and the output layer O discussed before. The activation functions of the dense layers cannot be a *softmax* function since the number of candidates – and so is the number of neurons in the output layer – is variable according to each specific token. We so opted for the **Scaled Exponential Linear Units (selu)**:

$$selu(x) = \lambda \begin{cases} x & if \ x > 0 \\ \alpha e^x - \alpha \ if \ x \leq 0 \end{cases}$$

The loss function used to train the network is the Mean Square Error, that gives slightly better results and similar training time if compared to MSE.

The neural network goal is to determine the probability that an entity candidate is right. In fact, for each sample, we get an output value that corresponds to this probability. $o_{x,j}$ corresponds to the output value of the input sample associated to the candidate entity j for token x. We select the candidate associated with the highest value $o_{x,max}$ among all output values $\{o_{x,1}, o_{x,2}, ..., o_{x,card(C_x)}\}$. Defining a threshold τ_d, if $o_{x,max} > \tau_d$, we can select as predicted entity for token x the one related to $o_{x,max}$. Otherwise, we consider that the token x is not part of a named entity. This process of **candidate selection** returns the list z_p of predicted Wikidata entities identifiers at token level. In a final step, sequences of tokens which belong to the same Wikidata entity identifiers are merged to a single entity. A_p represents the predicted corpus of annotated fragments.

5   Experiment and Evaluation

We developed an implementation of the two neural networks using Keras.[7] In order to make our approach comparable with the state of the art, our evaluation relies on well-known corpora and metrics, which have been already applied to related work. Moreover, we evaluate our approach on a new gold standard that we provide to the community.

---

[6] A *dense layer* is a layer fully connected to the previous one.

[7] The source code is available at https://github.com/D2KLab/ensemble-nerd, together with the documentation for accessing the live demo at http://enerd. eurecom.fr.

- **OKE2016:** annotated corpus of English textual resources, created for the 2016 OKE Challenge. The types set contains 4 different tags.[8] This ground truth disambiguates the entities using DBpedia. The ensemble technique we use for scoring is averaging, but not boosting or bagging.
- **AIDA/CoNLL:** English corpus and contains assignments of entities to the mentions of named entities, linked to DBpedia. This dataset does not infer types for NEs and can only be used for evaluating NED.
- **NexGenTV corpus:**[9] dataset composed of 77 annotated fragments of transcripts from politician television debates in French.[10] Each fragment lasts in average 2 min. The corpus is split in 64 training and 13 test samples. The list of types includes 13 different labels.[11] Entities are disambiguated through Wikidata.

Table 4. OKE2016 corpus NER Evaluation

	Token based			Entity based		
	fsc	pre	rec	fsc	pre	rec
adel	0.87	0.88	0.87	0.84	0.85	0.83
alchemy	0.79	0.93	0.68	0.88	0.92	0.86
babelfy	0.66	0.88	0.7	0.74	0.79	0.7
dandelion	0.64	0.89	0.51	0.78	0.83	0.75
dbspotlight	0.59	0.75	0.49	0.6	0.77	0.52
meaning cloud	0.59	0.91	0.44	0.72	0.78	0.69
opencalais	0.56	0.97	0.39	0.69	0.71	0.68
textrazor	0.74	0.86	0.65	0.77	0.81	0.74
ensemble	**0.91**	**0.91**	**0.91**	**0.94**	**0.95**	**0.92**
ensemble ($I = I_T$)	**0.88**	**0.91**	**0.85**	**0.88**	**0.92**	**0.84**
ensemble ($I = I_S$)	**0.50**	**0.53**	**0.47**	**0.50**	**0.52**	**0.48**
ensemble ($I = I_U$)	**0.44**	**0.47**	**0.41**	**0.43**	**0.43**	**0.43**
ensemble ($I = I_K$)	**0.37**	**0.40**	**0.34**	**0.38**	**0.40**	**0.36**

Type Recognition. For each gold standard GT, two different kinds of score are computed. The *token based* scores have been used in [9,10]. From GT, a list of target types l_t with dimension $|X|$ is extracted. We can obtain from ENNTR the list of predicted types l_p. For each type t_{GT} in GT, we compute precision

---

[8] PERSON, ORGANIZATION, PLACE, ROLE.

[9] http://enerd.eurecom.fr/data/training_data/nexgen_tv_corpus/.

[10] The debates are in the context of the 2017 French presidential election.

[11] PERSON, ORGANIZATION, GEOGRAPHICAL POINT, TIME, TIME INTERVAL, NUMBER, QUANTITY, OCCURRENCE, EVENT, INTELLECTUAL WORK, ROLE, GROUP OF HUMANS and OCCUPATION.

$Precision(l_t, l_p, t_{GT})$, recall $Recall(l_t, l_p, t_{GT})$ and F1 score $F1(l_t, l_p, t_{GT})$. Then, we compute micro averaged measures $Precision_{micro}(l_t, l_p)$, $Recall_{micro}(l_t, l_p)$ and $F1_{micro}(l_t, l_p)$ [8].

The *entity based* scores follow the definition of precision and recall coming from the **MUC-7 test scoring** [2]. Given A_t and A_p as the annotated fragment in GT, the computed measures are $Precision_{brat}(A_t, A_p)$, $Recall_{brat}(A_t, A_p)$ and $F1_{brat}(A_t, A_p)$.

The computed scores for OKE2016 and NexGenTv corpora are reported in Tables 4 and 5. The tables show also the same metrics applied to single extractors, after that their output types have been mapped to the ones of GT through the alignment block of ENNTR. For both token and entity scores, the ensemble method outperforms the single extractors for all metrics.

Table 5. NexGenTv corpus NER evaluation

	Token based			Entity based		
	fsc	pre	rec	fsc	pre	rec
adel	0.68	0.84	0.57	0.75	0.83	0.7
alchemy	0.80	0.83	0.77	0.87	0.97	0.81
babelfy	0.55	0.83	0.41	0.65	0.74	0.59
dandelion	0.26	0.69	0.16	0.51	0.69	0.42
dbspotlight	0.48	0.75	0.34	0.5	0.61	0.45
meaning cloud	0.82	0.88	0.77	0.8	0.87	0.76
opencalais	0.58	0.81	0.45	0.81	0.9	0.76
textrazor	0.81	0.89	0.74	0.75	0.8	0.72
ensemble	**0.94**	**0.97**	**0.91**	**0.92**	**0.98**	**0.87**
ensemble $(I = I_T)$	**0.87**	**0.91**	**0.83**	**0.89**	**0.93**	**0.85**
ensemble $(I = I_S)$	**0.54**	**0.58**	**0.50**	**0.53**	**0.56**	**0.50**
ensemble $(I = I_U)$	**0.47**	**0.49**	**0.45**	**0.46**	**0.47**	**0.45**
ensemble $(I = I_K)$	**0.40**	**0.42**	**0.38**	**0.39**	**0.40**	**0.38**

In order to identify the most impacting features in the obtained results, ENTTR has been sequentially adapted and retrained in order to receive in input only a specific kind of features, i.e. only I_T, I_K, I_U or I_S. The tokens based scores for these new trained networks reveals that the type features I_T are the only ones that, used alone as input, continue to make ENTRR outperforming single extractors, as can be expected given the type recognition goal. The other feature kinds, while having a lower impact, are still improving the final results when combined in the ensemble.

Entity Linking. We evaluate the entity linking for both OKE2016, AIDA/CoNLL and NexGenTv corpora using the GERBIL framework[12] and in particular micro and macro scores for the experiment type "Disambiguate to Knowledge Base" (D2KB). The computed scores are reported in Tables 6 and 7; the ensemble method outperforms again the single extractors that it integrates for all metrics. As for type recognition, we repeated the experiment using only a specific kind of features, in order to show the feature impact. In such case, the most influential features are the entity ones I_U. However, the impact of type features I_T is still crucial because its absence reduce drastically the improvement of the ensemble method with respect to the single extractors.

Tables 8 and 9 compare the NED extractors presented on GERBIL with our ensemble. For OKE2016, PBOH is the only tool which obtains a better score However this extractors reaches very low scores for AIDA/CoNLL, while our

Table 6. GERBIL Micro scores on OKE2016, NexGenTV and AIDA/CoNLL corpus

	OKE2016			NEXGEN			AIDA		
	fsc	pre	rec	fsc	pre	rec	fsc	pre	rec
babelfy	0.54	0.64	0.47	0.51	0.51	0.51	0.66	0.70	0.62
dandelion	0.59	0.77	0.48	0.34	0.50	0.26	0.45	0.66	0.34
dbspotlight	0.39	0.53	0.30	0.38	0.29	0.54	0.47	0.65	0.36
textrazor	0.53	0.78	0.40	0.61	0.55	0.69	0.62	0.57	0.53
ensemble	**0.66**	**0.88**	**0.52**	**0.69**	**0.70**	**0.64**	**0.68**	**0.79**	**0.60**
ensemble ($I = I_U$)	**0.59**	**0.80**	**0.47**	**0.59**	**0.60**	**0.58**	**0.55**	**0.60**	**0.50**
ensemble ($I = I_T$)	**0.41**	**0.45**	**0.38**	**0.42**	**0.47**	**0.38**	**0.48**	**0.52**	**0.45**

Table 7. GERBIL Macro scores on OKE2016, NexGenTV and AIDA/CoNLL corpus

	OKE2016			NEXGEN			AIDA		
	fsc	pre	rec	fsc	pre	rec	fsc	pre	rec
babelfy	0.54	0.65	0.47	0.51	0.52	0.51	0.60	0.65	0.57
dandelion	0.59	0.76	0.49	0.35	0.50	0.27	0.43	0.52	0.37
dbspotlight	0.39	0.52	0.32	0.38	0.29	0.55	0.45	0.63	0.37
textrazor	0.54	0.77	0.42	0.61	0.54	0.71	0.57	0.78	0.45
ensemble	**0.65**	**0.86**	**0.53**	**0.67**	**0.69**	**0.64**	**0.68**	**0.76**	**0.61**
ensemble ($I = I_U$)	**0.59**	**0.77**	**0.48**	**0.59**	**0.59**	**0.59**	**0.55**	**0.59**	**0.51**
ensemble ($I = I_T$)	**0.42**	**0.44**	**0.40**	**0.41**	**0.42**	**0.40**	**0.49**	**0.51**	**0.47**

---

[12] GERBIL is a general Linked Data benchmarking that offers an easy-to-use web-based platform for the agile comparison of annotators using multiple datasets and uniform measuring approaches.

Table 8. GERBIL scores on OKE2016

	Micro scores			Macro scores		
	fsc	pre	rec	fsc	pre	rec
agdistis	0.50	0.50	0.50	0.52	0.52	0.52
aida	0.49	0.63	0.41	0.5	0.64	0.42
dexter	0.44	0.92	0.29	0.43	0.81	0.31
fox	0.48	0.77	0.35	0.47	0.69	0.37
freme ner	0.31	0.57	0.21	0.26	0.27	0.25
kea	0.64	0.67	0.61	0.63	0.66	0.61
pboh	0.69	0.69	0.69	0.69	0.69	0.69
ensemble	**0.66**	**0.88**	**0.52**	**0.65**	**0.86**	**0.53**

Table 9. GERBIL scores on AIDA-CoNLL

	Micro scores			Macro scores		
	fsc	pre	rec	fsc	pre	rec
agdistis	0.58	0.58	0.58	0.59	0.59	0.59
aida	0.00	0.00	0.00	0.00	0.00	0.00
dexter	0.51	0.76	0.38	0.47	0.75	0.36
fox	0.57	0.63	0.51	0.56	0.64	0.51
freme ner	0.38	0.62	0.27	0.29	0.30	0.27
kea	0.60	0.65	0.56	0.59	0.63	0.56
pboh	0.00	0.00	0.00	0.00	0.00	0.00
ensemble	**0.68**	**0.79**	**0.60**	**0.68**	**0.76**	**0.61**

ensemble still continues to have good performances. For the NexGenTV dataset, we cannot compare the other NERD extractors because the majority of them perform NED only for the English language.

6   Conclusion and Future Work

In this paper, we presented two multilingual ensemble methods which combine the responses of web services (extractors) performing Named Entity Recognition and Disambiguation. The method relies on two Neural Networks that outperform the single extractors respectively in NER and NED tasks. Furthermore, the NER network allows to avoid the manually type alignment between the type taxonomies of each extractor and the ground truth taxonomy. We demonstrated the importance of the features generation for the success of these ensemble methods. In terms of NER, the type features play most of the work in the ensemble. For the NED task, while entity features have the greater impact, only a combination with type features really improve the effectiveness of the ensemble method with respect to single extractor predictions.

As future work, we plan to enhance the input feature set with Part of Speech tags features that would be assigned to each token. We also aim to vary the neural network architecture, and in particular, we are planning to replace the dense layer receiving the surface features with a BiLSTM, which would also take in consideration the context in which the tokens are sequentially appearing. Finally, all the neural networks models have been trained when all extractors APIs were reachable. A training that involves some samples which simulates the extractors failures and unavailability would make the network models more robust to API failures.

Acknowledgements. This work has been partially supported by the French National Research Agency (ANR) within the ASRAEL project (ANR-15-CE23-0018), the French Fonds Unique Interministériel (FUI) within the NexGen-TV project and the European Union's Horizon 2020 research and innovation programme via the project MeMAD (GA 780069).

References

1. Bojanowski, P., Grave, E., Joulin, A., Mikolov, T.: Enriching word vectors with subword information. arXiv preprint arXiv:1607.04606 (2016)
2. Chinchor, N.: Appendix B: MUC-7 test scores introduction. In: Seventh Message Understanding Conference (MUC-7), Fairfax, Virginia, USA (1998)
3. Finkel, J.R., Grenager, T., Manning, C.: Incorporating non-local information into information extraction systems by Gibbs sampling. In: 43rd Annual Meeting on Association for Computational Linguistics (ACL), Ann Arbor, Michigan, USA, pp. 363–370 (2005)
4. Ratinov, L., Roth, D.: Design challenges and misconceptions in named entity recognition. In: 13th Conference on Computational Natural Language Learning (CoNLL), Boulder, Colorado, USA, pp. 147–155, June 2009
5. Rizzo, G., Troncy, R.: NERD: a framework for unifying named entity recognition and disambiguation extraction tools. In: 13th Conference of the European Chapter of the Association for Computational Linguistics (EACL), Avignon, France, pp. 73–76 (2012)
6. Rizzo, G., van Erp, M., Troncy, R.: Benchmarking the extraction and disambiguation of named entities on the semantic web. In: 9th International Conference on Language Resources and Evaluation (LREC), Reykjavik, Iceland (2014)
7. Rizzo, G., van Erp, M., Troncy, R.: Inductive entity typing alignment. In: 1st International Workshop on Linked Data for Information Extraction (LD4IE), Riva del Garda, Italy (2014)
8. Sebastiani, F.: Machine learning in automated text categorization. ACM Comput. Surv. **34**(1), 1–47 (2002)
9. Speck, R., Ngonga Ngomo, A.-C.: Ensemble learning of named entity recognition algorithms using multilayer perceptron for the multilingual web of data. In: 9th International Conference on Knowledge Capture (K-CAP), Austin, TX, USA (2017)

10. Speck, R., Ngonga Ngomo, A.-C.: Ensemble learning for named entity recognition. In: Mika, P. (ed.) ISWC 2014 Part I. LNCS, vol. 8796, pp. 519–534. Springer, Cham (2014). https://doi.org/10.1007/978-3-319-11964-9_33

11. van Erp, M., Rizzo, G., Troncy, R.: Learning with the web: spotting named entities on the intersection of NERD and machine learning. In: 3rd International Workshop on Making Sense of Microposts (#MSM), Concept Extraction Challenge, Rio de Janeiro, Brazil (2013)

12. Zhang, F., Yuan, N.J., Lian, D., Xie, X., Ma, W.-Y.: Collaborative knowledge base embedding for recommender systems. In: 22nd ACM SIGKDD International Conference on Knowledge Discovery and Data Mining (KDD), San Francisco, California, USA, pp. 353–362 (2016)

EARL: Joint Entity and Relation Linking
for Question Answering over Knowledge
Graphs

Mohnish Dubey[1,2(✉)], Debayan Banerjee[1],
Debanjan Chaudhuri[1,2], and Jens Lehmann[1,2]

[1] Smart Data Analytics Group (SDA), University of Bonn, Bonn, Germany
{dubey,chaudhur,jens.lehmann}@cs.uni-bonn.de, debayan@uni-bonn.de
[2] Fraunhofer IAIS, Bonn, Germany
jens.lehmann@iais.fraunhofer.de

Abstract. Many question answering systems over knowledge graphs
rely on entity and relation linking components in order to connect the
natural language input to the underlying knowledge graph. Traditionally,
entity linking and relation linking have been performed either as dependent sequential tasks or as independent parallel tasks. In this paper, we
propose a framework called EARL, which performs entity linking and
relation linking as a joint task. EARL implements two different solution
strategies for which we provide a comparative analysis in this paper:
The first strategy is a formalisation of the joint entity and relation linking tasks as an instance of the Generalised Travelling Salesman Problem
(GTSP). In order to be computationally feasible, we employ approximate GTSP solvers. The second strategy uses machine learning in order
to exploit the connection density between nodes in the knowledge graph.
It relies on three base features and re-ranking steps in order to predict
entities and relations. We compare the strategies and evaluate them on
a dataset with 5000 questions. Both strategies significantly outperform
the current state-of-the-art approaches for entity and relation linking.

Keywords: Entity linking · Relation linking · GTSP
Question answering

1 Introduction

Question answering over knowledge graphs (KGs) is an active research area con-
cerned with techniques that allow obtaining information from knowledge graphs
based on natural language input. Specifically, Semantic Question Answering
(SQA) as defined in [8] is the task of users asking questions in natural language
(NL) to which they receive a concise answer generated by a formal query over a
KG.

Semantic question answering systems can be a fully rule based systems [4]
or end-to-end machine learning based systems [19]. The main challenges faced

© Springer Nature Switzerland AG 2018
D. Vrandečić et al. (Eds.): ISWC 2018, LNCS 11136, pp. 108–126, 2018.
https://doi.org/10.1007/978-3-030-00671-6_7

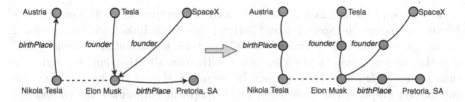

Fig. 1. An excerpt of the subdivision knowledge graph for the example question "Where was the founder of Tesla and Space X born?". Note that both entities and relations are nodes in the graph.

in SQA are (i) entity identification and linking, (ii) relation identification and linking, (iii) query intent identification and (iv) formal query generation.

Some QA systems have achieved good performance on simple questions [11], i.e. those questions which can be answered by linking to at most one relation and at most one entity in the KG. Recently, the focus has shifted towards complex questions [30], comprising of multiple entities and relations.

Usually, all entities and relations need to be correctly linked to the knowledge graph in order to generate the correct formal query and successfully answer the question of a user. Hence, it is crucial to perform the linking process with high accuracy and this is a major bottleneck for the widespread adoption of current SQA systems. In most entity linking systems [12,26], disambiguation is performed by looking at other entities present in the input text. However, in the case of natural language questions (short text fragments) the number of other entities for disambiguation is not high. Therefore, it is potentially beneficial to consider entity and relation candidates for the input questions in combination, to maximise the usable evidence for the candidate selection process. To achieve this, we propose EARL (Entity and Relation Linker), a system for jointly linking entities and relations in a question to a knowledge graph. EARL treats entity linking and relation linking as a single task and thus aims to reduce the error caused by the dependent steps.

EARL uses the knowledge graph to jointly disambiguate entity and relations: It obtains the context for entity disambiguation by observing the relations surrounding the entity. Similarly, it obtains the context for relation disambiguation by looking at the surrounding entities. The system supports multiple entities and relations occurring in complex questions. EARL implements two different solution strategies: The first strategy is a formalisation of the joint entity and relation linking tasks as an instance of the Generalised Travelling Salesman Problem (GTSP). Since the problem is NP-hard, we employ approximate GTSP solvers. The second strategy uses machine learning in order to exploit the connection density between nodes in the KG. It relies on three base features and re-ranking steps in order to predict entities and relations. We compare the strategies and evaluate them on a dataset with 5000 questions. Both strategies outperform the current state-of-the-art approaches for entity and relation linking.

Let us consider an example to explain the underlying idea: *"Where was the founder of Tesla and SpaceX born?"*. Here, the entity linker needs to perform disambiguation for the keyword "Tesla" between the scientist "Nikola Tesla" and the car company "Tesla Motors". EARL uses all other entities and relations (*SpaceX, founder, born*) present in the query. It does this by analysing the subdivision graph of the knowledge graph fragment containing the candidates for relevant entities and relations. While performing the joint analysis (Fig. 1), EARL detects that there is no likely combination of candidates, which supports the disambiguation of "Tesla" as "Nikola Tesla", whereas there is a plausible combination of candidates for the car company "Tesla Motors".

Overall, our contributions in this paper are as follows:

1. The framework EARL, where GTSP solver or Connection Density can be used for joint linking of entities and relations (Sect. 4).
2. A formalisation of the joint entity and relation linking problem as an instance of the Generalised Travelling Salesman (GTSP) problem (Sect. 4.2).
3. An implementation of the GTSP strategy using approximate GTSP solvers.
4. A "Connection Density" formalisation and implementation of the joint entity and relation linking problem as a machine learning task (Sect. 4.3).
5. An adaptive E/R learning module, which can correct errors occurring across different modules (Sect. 4.3).
6. A comparative analysis of both strategies - GTSP and connection density (Table 2).
7. A fully annotated version of the 5000 question LC-QuAD data-set, where entity and relations are linked to the KG.
8. A large set of labels for DBpedia predicates and entities covering the syntactic and semantic variations.[1]

The paper is organised into the following sections: (2) Related Work outlining some of the major contributions in entity and relation linking used in question answering; (3) Problem Statement, where we discuss the problem in depth and our hypotheses for the solution; (4) the architecture of EARL including preprocessing steps followed by (i) a GTSP solver or (ii) a connection density approach; (5) Evaluation, with various evaluation criteria and results; (6) Discussion; and (7) Conclusion.

2 Related Work

The entity and relation linking challenge has attracted a wide variety of solutions over time. Linking natural language phrases to DBpedia resources, Spotlight [12] breaks down the process of entity spotting into four phases. It identifies the entity using a list of surface forms and then generates DBpedia resources candidates. It then disambiguates the entity based on surrounding context. AGDISTIS [26] follows the inherent structure of the target knowledge base more closely to solve

---

[1] Dataset available at https://github.com/AskNowQA/EARL.

Table 1. State of the art for Entity and Relation linking in question answering

Linking approach	QA system	Advantage	Disadvantage
Sequential	[2,4,21]	-Reduces candidate search space for Relation Linking -Allows schema verification	-Relation Linking information cannot be exploited in Entity Linking process - Errors in Entity Linking cannot be overcome
Parallel	[16,27,28]	- Lower runtime - Re-ranking of Entities possible based on Relation Linking	- Entity Linking process cannot use information from Relation Linking process and vice versa - Does not allow schema verification
Joint (with limited candidate set)	[1,30]	- Potentially high accuracy - Reduces error propagation - Better disambiguation - Allows schema verification - Allows re-ranking	- Complexity increase - Larger search space

the problem. Being a graph-based disambiguation system, AGDISTIS performs disambiguation based on the hop-distance between the candidates for the entities in a given text, where multiple entities are present. Babelfy [13] uses word sense disambiguation for entity linking. On the other hand, S-MART [29] is often appropriated as an entity linking system over Freebase resources. It generates multiple regression trees and then applies sophisticated structured prediction techniques to link entities to resources.

As relation linking is generally considered to be a problem-specific task, only a few general purpose relation linking systems are in use. Iterative bootstrapping strategies for extracting RDF resources from unstructured text have been explored in BOA [5] and PATTY [15]. It consists of natural language patterns corresponding to relations present in the knowledge graph. Word embedding models are also frequently used to overcome the linguistic gap for relation linking. RelMatch [20] improves the accuracy of the PATTY dataset for relation linking. There are tools such as ReMatch [14] which uses wordnet similarity for relation linking.

Many QA systems use an out-of-the-box entity linker, often one of the afore-mentioned ones. These tools are not tailor-made for questions and are instead trained on large text corpora, typically devoid of questions. This may create several problems as questions do not span over more than one sentence, thereby rendering context-based disambiguation relatively ineffective. Further, graph based systems rely on the presence of multiple entities in the source text and disambiguate them based on each other. This becomes difficult when dealing with questions, as they seldom consist of multiple entity.

Thus, to avoid the issues mentioned, a variety of approaches have been employed for entity and relation linking for question answering. Semantic parsing-based systems such as AskNow [4] and TBSL [25] first link the entities and generate a list of candidate relations based on the identified resources. They use several string and semantic similarity techniques to finally select the correct entity and relation candidates for the question. In these systems, the process of relation linking depends on linking the entities. Generating entity and relation candidates has also been explored by [30], which uses these candidates to create staged query graphs, and later re-ranks them based on textual similarity between the query and the target question, computed by a Siamese architecture-based neural network. There are some QA systems such as Xser [28], which performs relation linking independent of entity linking. STAGG [30] takes the top 10 entities given by the entity linker and tries to build query-subgraph chains corresponding to the question. This approach considers a ranked list of entity candidates from the entity linker and chooses the best candidate based on the query subgraph formed. Generally, semantic parsing based systems treat entity and relation linking as separate tasks which can be observed in the generalised pipeline of Frankenstein [21] and OKBQA www.okbqa.org/.

3   Overview and Preliminaries

3.1   Overview and Research Questions

As discussed previously, in question answering the tasks of entity and relation linking are performed either sequentially or in parallel. In sequential systems, usually the entity linking task is performed first, followed by relation linking. As a consequence, information in the relation linking phase cannot be exploited during entity linking in this case. In parallel systems, entity and relation linking are performed independently. While this is efficient in terms of runtime performance, the entity linking process cannot benefit from further information obtained during relation linking and vice versa. We illustrate the advantages and disadvantages of both approaches, as well as the systems following them, in Table 1. Our main contribution in this paper is the provision of a system, which takes candidates for entity and relation linking as input and performs a joint optimisation selecting the best combination of entity and relation candidates.

Postulates. We have three postulates, which we want to verify based on our approach:

H1: Given candidate lists of entities and relations from a question, the correct solution is a cycle of minimal cost that visits exactly one candidate from each list.

H2: Given candidate lists of entities and relations from a question, the correct candidates exhibit relatively dense and short-hop connections among themselves in the knowledge graph compared to wrong candidate sets.

H3: Jointly linking entity and relation leads to higher accuracy compared to performing these tasks separately.

We will re-visit all of these postulates in the evaluation section of the paper.

3.2   Preliminaries

We will first introduce basic notions from graph theory:

Definition 1 (Graph). *A (simple, undirected) graph is an ordered pair $G = (V, E)$ where V is a set whose elements are called* vertices *and E is a set of pairs of vertices which is called* edges.

Definition 2 (Knowledge Graph). *Within the scope of this paper, we define a* knowledge graph *as a labelled directed multi-graph. A labelled directed multi-graph is a tuple $KG = (V, E, L)$ where V is a set called* vertices, *L is a set of* edge labels *and $E \subseteq V \times L \times V$ is a set of ordered triples.*

It should be noted that our definition of knowledge graphs captures basic aspects of RDF datasets as well as property graphs [6]. The knowledge graph vertices represent entities and the edges represent relationships between those entities (Fig. 2).

Definition 3 (Subdivision Graph). *The subdivision graph [24] $S(G)$ of a graph G is the graph obtained from G by replacing each edge $e = (u, v)$ of G by a new vertex w_e and 2 new edges (u, w_e) and (v, w_e).*

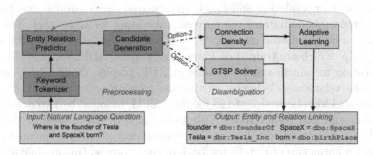

Fig. 2. EARL architecture: In the disambiguation phase one may choose either Connection Density or GTSP. In cases where training data is not available beforehand GTSP works better.

4   EARL

In general, entity linking is a two step process. The first step is to identify and spot the span of the entity. The second step is to disambiguate or link the entity to the knowledge graph. For linking, the candidates are generated for the spotted span of the entity and then the best candidate is chosen for the linking. These two steps are similarly followed in standard relation linking approaches. In our approach, we first spot the spans of entities and relations. After that, the (disambiguation) linking task is performed jointly for both entities and relations.

In this section we first discuss the step of span detection of entity and relation in natural language question and candidate list generation. We perform the disambiguation by two different approaches, which are discussed later in this section.

4.1 Candidate Generation Steps

4.1.1 Shallow Parsing

Given a question, extract all keyword phrases out. EARL uses SENNA [3] as the keyword extractor. We also remove stop words from the question at this stage. In example question "Where was the founder of Tesla and SpaceX born?" we identify $<founder, Tesla, SpaceX, born>$ as our keyword phrases.

4.1.2 E/R Prediction

Once keyword phrases are extracted from the questions, the next step in EARL is to predict whether each of these is an entity or a relation. We use a character embedding based long-short term memory network (LSTM) to do the same. The network is trained using labels for entity and relation in the knowledge graph. For handling out of vocabulary words [17], and also to encode the knowledge graph structure in the network, we take a multi-task learning approach with hard parameter sharing. Our model is trained on a custom loss given by:

$$\mathcal{E} = (1 - \alpha) * \mathcal{E}_{BCE} + \alpha * \mathcal{E}_{ED} \tag{1}$$

where, $\mathcal{E}_{BCE}$ is the binary cross entropy loss for the learning objective of a phrase being an entity or a relation and $\mathcal{E}_{Ed}$ is the squared euclidian distance between the predicted embedding and the correct embedding for that label. The value of α is empirically selected as 0.25. We use pre-trained label embeddings from RDF2Vec [18] which are trained on knowledge graphs. RDF2Vec provides latent representation for entities and relations in RDF graphs. It efficiently captures the semantic relatedness between entities and relations.

We use a hidden layer size of 128 for the LSTM, followed by two dense layers of sizes 512 and 256 respectively. A dropout value of 0.5 is used in the dense layers. The network is trained using Adam optimizer [9] with a learning rate of 0.0001 and a batch size of 128. Going back to the example, this module identifies "founder" and "born" as *relations*, "Tesla" and "SpaceX" as *entities*.

4.1.3 Candidate List Generation

This module retrieves a candidate list for each keyword identified in the natural language question by the shallow parser. To retrieve the top candidates for a keyword we create an Elasticsearch[2] index of URI-label pairs. Since EARL requires an exhaustive list of labels for a URI in the knowledge graph, we expand the labels. We used Wikidata labels for entities which are in same-as relation in the knowledge base. For relations we require labels which were semantically equivalent (such as writer, author) for which we took synonyms from the Oxford Dictionary API[3]. To cover grammatical variations of a particular label, we added inflections from fastText[4]. We avoid any bias held towards or against popular entities and relations.

---

[2] https://www.elastic.co/products/elasticsearch.
[3] https://developer.oxforddictionaries.com/.
[4] https://fasttext.cc/.

The output of these pre-processing steps are (i) set of keywords from the question, (ii) every keyword is identified either as relation or entity, (iii) for every keyword there is a set of candidate URIs from the knowledge graph.

4.2 Using GTSP for Disambiguation

At this point we may use either a GTSP based solution or Connection Density (later explained in Sect. 4.3) for disambiguation. We start with the formalisation for GTSP based solution.

The entity and relation linking process can be formalised via spotting and candidate generation functions as follows: Let S be the set of all strings. We assume that there is a function $spot : S \rightarrow 2^S$ which maps a string s (the input question) to a set $\mathcal{K}$ of substrings of s. We call this set $\mathcal{K}$ the *keywords* occurring in our input. Moreover, we assume there is a function $cand_{KG} : \mathcal{K} \rightarrow 2^{V \cup L}$ which maps each keyword to a set of candidate node and edge labels for our knowledge graph $G = (V, E, L)$. The goal of joint entity and relation linking is to find combinations of candidates, which are closely related. How closely nodes are related is modelled by a cost function $cost_{KG} : (V \cup L) \times (V \cup L) \rightarrow [0, 1]$. Lower values indicate closer relationships. According to our first postulate, we aim to encode graph distances in the cost function to reward those combinations of entities and relations, which are located close to each other in the input knowledge graph. To be able to consider distances between both relations and entities, we transform the knowledge graph into its subdivision graph (see Definition 3). This subdivision graph allows us to elegantly define the distance function as illustrated in Fig. 4.

Given the knowledge graph KG and the functions *spot*, *cand* and *cost*, we can cast the problem of joint entity and relation linking as an instance of the Generalised Travelling Salesman (GTSP) problem: We construct a graph G with $V = \bigcup_{k \in K} cand(k)$. Each node set $cand(k)$ is called a cluster in this vertex set. The GTSP problem is to find a subset $V' = (v_1, \dots, v_n)$ of V which contains exactly one node from each cluster and the total cost $\sum_{i=1}^{n-1} cost(v_i, v_{i+1})$ is minimal with respect to all such subsets. Please note that in our formalisation of the GTSP, we do not require V' to be a cycle, i.e. v_1 and v_n can be different. Moreover, we do not require clusters to be disjoint, i.e. different keywords can have overlapping candidate sets.

Figure 3 illustrates the problem formulation. Each candidate set for a keyword forms a cluster in the graph. The weight of each edge in this graph is given by the cost function, which includes the distance between the nodes in the subdivision graph of the input knowledge graph as well as the confidence scores of the candidates. The GTSP requires the solution to visit one element per cluster and minimises the overall distance.

Approximate GTSP Solvers. In order to solve the joint entity and relation linking problem, the corresponding GTSP instance needs to be solved. Unfortunately, the GTSP is NP-hard [10] and hence it is intractable. However, since

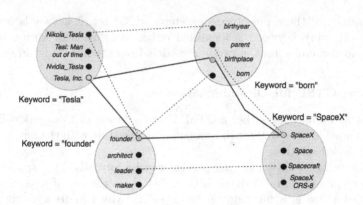

Fig. 3. Using GTSP for disambiguation: The bold line represents the solution offered by the GTSP solver. Each edge represents an existing connection in the knowledge graph. The edge weight is equal to the number of hops between the two nodes in the knowledge graph. We also add the index search ranks of the two nodes the edges connect to the edge weight when solving for GTSP.

GTSP can be reduced to standard TSP, several polynomial approximation algorithms exist to solve GTSP. The state-of-the-art approximate GTSP solver is the Lin–Kernighan–Helsgaun algorithm [7]. Here, a GTSP instance is transformed into standard asymmetric TSP instances using the Noon-Bean transformation. It allows the heuristic TSP solver LKH to be used for solving the initial GTSP. Among LKH's characteristics, its use of 1-tree approximation for determining a candidate edge set, the extension of the basic search step, and effective rules for directing and pruning the search contribute to its efficiency.

While a GTSP based solution would be suitable for solving the joint entity and relation linking problem, it has the drawback that it can only provide the best candidate for each keyword given the list of candidates. Most approximate GTSP solutions do not explore all possible paths and nodes and hence a comprehensive scoring and re-ranking of nodes is not possible. Ideally, we would like to go beyond this and re-rank all candidates for a given keyword. This would open up new opportunities from a QA perspective, i.e. a user could be presented with a sorted list of multiple possible answers to select from.

4.3   Using Connection Density for Disambiguation

As discussed earlier, once the candidate list generation is achieved, EARL offers two independent modules for the entity and relation linking. In the previous Subsect. 4.2 we discussed one approach using GTSP. In this subsection we will discuss the second approach for disambiguation using Connection Density, which works as an alternative to the GTSP approach. We have also compared the two methods in Table 2.

Table 2. Comparison of GTSP based approach and Connection density for Disambiguation

GTSP	Connection Density
Requires no training data	Requires data to train the XGBoost classifier
The approximate GSTP LKH solution is only able to return the top result as not all possible paths are explored	Returns a list of all possible candidates in order of score
Time complexity of LKH is $\mathcal{O}(nL^2)$ where n = number of nodes in graph, L = number of clusters in graph of	Time complexity is $\mathcal{O}(N^2L^2)$ where N = number of nodes per cluster, L = number of clusters in graph
Relies on identifying the path with minimum cost	Depends on identifying dense and short-hop connections

4.3.1   Formalisation of Connection Density

For identified keywords in a question we have the set $\mathcal{K}$ as defined earlier. For each keyword K_i we have list L_i which consists of all the candidate uris generated by text search. We have n such candidate lists for each question given by, $\mathcal{L} = \{L_1, L_2, L_3, ..., L_n\}$. We consider a probable candidate $c_m^i \in L_i$, where m is the total number of candidates to be considered per keyword, which is the same as the number of items in each list.

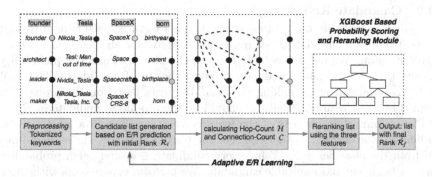

Fig. 4. Connection Density with example: The dotted lines represent corresponding connections between the nodes in the knowledge base.

The hop distance $dKGhops(c_i^k, c_j^o) \in \mathbb{Z}^+$ is number of hops between c_i^k and c_j^o in the subdivision knowledge graph. If the shortest path from c_i^k and c_j^o requires the traversal of h edges then $dKGhops(c_i^k, c_j^o) = h$.

Connection Density is based on the three features: Text similarity based initial Rank of the List item ($\mathcal{R}_i$) Connection-Count ($\mathcal{C}$) and Hop-Count ($\mathcal{H}$).

Initial Rank of the List ($\mathcal{R}_i$), is generated by retrieving the candidates from the search index via text search. This is achieved in the preprocessing steps as mentioned in the Sect. 4. Further, to define $\mathcal{C}$ we introduce $dConnect$.

$$dConnect(c_i^k, c_j^o) = \begin{cases} 1 & \text{if } dKGhops(c_i^k, c_j^o) \leqslant 2 \\ 0 & \text{otherwise} \end{cases} \quad (2)$$

The Connection-Count $\mathcal{C}$ for an candidate c, is the number of connections from c to candidates in all the other lists divided by the total number n of keywords spotted. We consider nodes at hop counts of greater than 2 disconnected because nodes too far away from each other in the knowledge base do not carry meaningful semantic connection to each other.

$$\mathcal{C}(c_i^k) = 1/n \sum_{o|o \neq k} \sum_{j=1}^{j=m} dConnect(c_i^k, c_j^o) \quad (3)$$

The Hop-Count $\mathcal{H}$ for a candidate c, is the sum of distances from c to all the other candidates in all the other lists divided by the total number of keywords spotted.

$$\mathcal{H}(c_i^k) = 1/n \sum_{o|o \neq k} \sum_{j=1}^{j=m} dKGhops(c_i^k, c_j^o) \quad (4)$$

4.3.2   Candidate Re-ranking

$\mathcal{H}, \mathcal{C}$ and $\mathcal{R}_i$ constitute our feature space $\mathcal{X}$. This feature space is used to find the most relevant candidate given a set of candidates for an identified keyword in the question. We use a machine learning classifier to learn the probability of being the most suitable candidate $\bar{c}^i$ given the set of candidates. The final list $\mathcal{R}_f$ is obtained by re-ranking the candidate lists based on the probability assigned by the classifier. Ideally, $\bar{c}^i$ should be the top-most candidate in $\mathcal{R}_f$.

The training data consists of the features $\mathcal{H}, \mathcal{C}$ and $\mathcal{R}_i$ and a label 1 if the candidate is the correct, 0 otherwise. For the testing, we apply the learned function from the classifier f on $\mathcal{X}$ for every candidate $\in c_i$ and get a probability score for being the most suitable candidate. We perform experiments with three different classifiers, namely extreme gradient boosting(xgboost), SVM (with a linear kernel) and logistic regression to re-rank the candidates. The experiments are done using a 5-fold cross-validation strategy where, for each fold we train the classifier on the training set and observe the mean reciprocal rank (MRR) of $\bar{c}^i$ on the testing set after re-ranking the candidate lists based on the assigned probability. The average MRR on 5-fold cross-validation for the three classifiers are 0.905, 0.704 and 0.794 respectively. Hence, we use xgboost as the final classifier in our subsequent experiments for re-ranking.

4.3.3 Algorithm

We now present a pseudo-code version of the algorithm to calculate the two features: Connection Density algorithm is used for finding hop count and connection count for each candidate node. We then pass these features to a classifier for scoring and ranking This algorithm (Algorithm 1 Connection Density) has a time complexity given by $\mathcal{O}(N^2 L^2)$ where N is the number of keywords and L is the number of candidates for each keyword.

Algorithm 1. Connection Density

---

   function: ConnectionDensity( )
   input   : $\mathcal{L}$, with n number of keywords // **an array of arrays**
   output : Hop-Count $\mathcal{H}$, Connection-Count $\mathcal{C}$
1 dConnectCounter = { } // **Count for connections from and to each node**
2 dHopCounter = { } // **Similarly hop counts for each node**
3 **foreach** $L_a \in \mathcal{L}$ **do**
4   **foreach** $c_i^a \in L_a$ **do**
5         $dConnectCounter[c_i^a] = 0$ // **Initialising the dictionary**
6         $dHopCounter[c_i^a] = 0$

7 **foreach** $(L_a, L_b) \in \mathcal{L}$ **do**
8   **foreach** $c_i^a \in L_a$ **do**
9   **foreach** $c_j^b \in L_b$ **do**
10   **if** $dKGhops(c_i^a, c_j^b) <= 2$ **then**
11               $dConnectCounter[c_i^a] += 1$
12               $dConnectCounter[c_j^b] += 1$
13            $dHopCounter[c_i^a] += \mathrm{dKGhops}(c_i^a, c_j^b)$
14            $dHopCounter[c_j^b] += \mathrm{dKGhops}(c_i^a, c_j^b)$

15 **foreach** $(c_i, score) \in dConnectCounter$ **do**
16      $\mathcal{C}(c_i) = dConnectCounter(c_i)/n$ // **Normalisation with respect to**
         number of keywords spotted
17 **foreach** $(c_i, score) \in dHopCounter$ **do**
18      $\mathcal{H}(c_i) = dHopCounter(c_i)/n$
19 **return** (Hop-Count $\mathcal{H}$, Connection-Count $\mathcal{C}$)

4.4 Adaptive E/R Learning

EARL uses a series of sequential modules with little to no feedback across them. Hence, the errors in one module propagate down the line. To trammel this, we implement an adaptive approach especially for curbing the errors made in the pre-processing modules. While conducting experiments, it was observed that most of the errors are in the shallow parsing phase, mainly because of grammatical errors in LC-QuAD which directly affects the consecutive E/R prediction and candidate selection steps. If the E/R prediction is erroneous, it will search in a

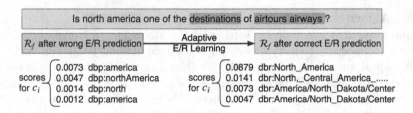

Fig. 5. Adaptive E/R learning

wrong Elasticsearch index for probable candidate list generation. In such a case none of the candidates $\in c^i$ for a keyword would contain $\bar{c}^i$ as is reflected by the probabilities assigned to c^i by the re-ranker module. If the maximum probability assigned to c^i is less than a very small threshold value, empirically chosen as 0.01, we re-do the steps from ER prediction after altering the original prediction. If the initial assigned probability is *entity*, we change it to *relation* and vice-versa, example Fig. 5. This module is empirically evaluated in Table 5.

5   Evaluation

Data Set: LC-QuAD [23] is the largest complex questions data set available for QA over KGs. We have annotated this data set to create a gold label data set for entity and relation linking, i.e. each question now contains the correct KG entity and relation URIs with their respective text spans in the question. This annotation was done in a semi-automated process and subsequently manually verified. The annotated dataset of 5000 questions is publicly available at https://figshare.com/projects/EARL/28218.

5.1   Experiment 1: Comparison of GTSP, LKH and Connection Density

Aim: We evaluate hypotheses (**H1** and **H2**) that the connection density and GTSP can be used for joint linking task. We also evaluate the LKH approximation solution of GTSP for doing this task. We compare the time complexity of the three different approaches.

Results: Connection density results in a similar accuracy as that of an exact GTSP solution with a better time complexity (see Table 3). Connection density has worse time complexity than approximate GTSP solver LKH if we assume the best case of equal cluster sizes for LKH. However, it provides a better accuracy. Moreover, the average time taken in EARL using connection density (including the candidate generation step) is 0.42 s per question. Further observing Table 3, we can see that the brute force GTSP solution and Connection Density have similar accuracy, but the brute force GTSP solution has exponential time complexity. The approximate solution LKH has polynomial run time, but its accuracy drops

Table 3. Empirical comparison of Connection Density and GTSP: n = number of nodes in graph; L = number of clusters in graph; N = number of nodes per cluster; top K results retrieved from ElasticSearch.

Approach	Accuracy (K = 30)	Accuracy (K = 10)	Time complexity
Brute Force GTSP	0.61	0.62	$\mathcal{O}(n^2 2^n)$
LKH - GTSP	0.59	0.58	$\mathcal{O}(nL^2)$
Connection Density	0.61	0.62	$\mathcal{O}(N^2 L^2)$

compared to the brute force GTSP solution. Moreover, from a question answering perspective the ranked list offered by the Connection Density approach is useful since it can be presented to the user as a list of possible correct solutions or used by subsequent processing steps of a QA system. Hence, for further experiments in this section we used the connection density approach.

5.2 Experiment 2: Evaluating Joint Connectivity and Re-ranker

Aim: Evaluating the performance of Connection Density for predicting the correct entity and relation candidates from a set of possible E-R candidates. Here we evaluate hypothesis **H2**, the correct candidates exhibit relatively dense and short-hop connections.

Table 4. Evaluation of joint linking performance

Value of k	$\mathcal{R}_f$ based on $\mathcal{R}_i$	$\mathcal{R}_f$ based on $\mathcal{C}, \mathcal{H}$	$\mathcal{R}_f$ based on $\mathcal{R}_i, \mathcal{C}, \mathcal{H}$
$k = 10$	0.543	0.689	0.708
$k = 30$	0.544	0.666	0.735
$k = 50$	0.543	0.617	**0.737**
$k = 100$	0.540	0.534	0.733
$k^* = 10$	0.568	0.864	**0.905**
$k^* = 30$	0.554	0.779	0.864
$k^* = 50$	0.549	0.708	0.852
$k^* = 50$	0.545	0.603	0.817

Metrics: We use the mean reciprocal rank of the correct candidate $\bar{c}^i$ for each entity/relation in the query. From the probable candidate list generation step, we fetch a list of top candidates for each identified phrase in a query with a k value of 10, 30, 50 and 100, where k is the number of results from text search for each keyword spotted. To evaluate the robustness of our classifier and features we perform two tests. (i) On the top half of Table 4 we re-rank the top k candidates returned from the previous step. (ii) On the bottom half of Table 4 we artificially insert the correct candidate into each list to purely test re-ranking abilities of our system (this portion of the table contains k^* as the number of items in each candidate list). We inject the correct uris at the lowest rank (see k^*), if it was not retrieved in the top k results from previous step.

Results: The results in Table 4 depict that our algorithm is able to successfully re-rank the correct URIs if the correct ones are already present. In case correct URIs were missing in the candidate list, we inserted URIs artificially as the last candidate. The MRR then increased from 0.568 to 0.905.

5.3   Experiment 3: Evaluating Entity Linking

Aim: To evaluate the performance of EARL with other state-of-the-art systems on the entity linking task. This also evaluates our hypothesis **H3**.

Metrics: We are reporting the performance on accuracy. Accuracy is defined by the ratio of the correctly identified entities over the total number of entities present.

Result: EARL performs better entity linking than the other systems (Table 5), namely Babelfy, DBpediaSpotlight, TextRazor and AGDISTIS + FOX (limited to entity types - LOC, PER, ORG). We conducted this test on the LC-QuAD and QALD-7 dataset[5]. The value of **k** is set to 30 while re-ranking and fetching the most probable entity.

Table 5. Evaluating EARL's Entity Linking performance

System	Accuracy LC-QuAD	Accuracy - QALD
FOX [22] + AGDISTIS [26]	0.36	0.30
DBpediaSpotlight [12]	0.40	0.42
TextRazor[a]	0.52	0.53
Babelfy [13]	0.56	0.56
EARL without adaptive learning	0.61	0.55
EARL with adaptive learning	**0.65**	**0.57**

[a]https://www.textrazor.com/.

5.4   Experiment 4: Evaluating Relation Linking

Aim: Given a question, the task is to the perform relation linking in the question. This also evaluates our hypothesis **H3**.

Metrics: We use the same accuracy metric as in the Experiment 3.

---

[5] https://project-hobbit.eu/challenges/qald2017/.

Results: As reported in Table 6, EARL outperforms other approaches we could run on LC-QuAD and QALD. The large difference in accuracy of relation-linking over LC-QuAD over QALD, is due to the face that LC-QuAD has 82% questions with more than one relation, thus detecting relation phrases in the question was difficult.

Table 6. Evaluating EARL's Relation Linking performance

System	Accuracy LC-QuAD	Accuracy - QALD
ReMatch [14]	0.12	0.31
RelMatch [20]	0.15	0.29
EARL without adaptive learning	0.32	0.45
EARL with adaptive learning	**0.36**	**0.47**

6 Discussion

Our analysis shows that we have provided two tractable (polynomial with respect to the number of clusters and the elements per cluster) approaches of solving the joint entity and relation linking problem. We experimentally achieve similar accuracy as the exact GTSP solution with both LKH-GTSP and Connection Density with better time complexity, which allows us to use the system in QA engines in practice. It must be noted that one of the salient features of LKH-GTSP is that it requires no training data for the disambiguation module while on the other hand Connection Density performs better given training data for its XGBoost classifier. While the system was tested on DBpedia, it is not restricted to a particular knowledge graph.

There are some limitations: The current approach does not tackle questions with hidden relations, such as "How many shows does HBO have?". Here the semantic understanding of the corresponding SPARQL query is to count all TV shows (*dbo:TelevisionShow*) which are owned by (*dbo:company*) the HBO (*dbr:HBO*). Here *dbo:company* is the hidden relation which we do not attempt to link. However, it could be argued that this problem goes beyond the scope of relation linking and could be better handled by the query generation phase of a semantic QA system.

Another limitation is that EARL cannot be used as inference tool for entities as required by some questions. For example Taikonaut is an astronaut with Chinese nationality. The system can only link taikonaut to *dbr:Astronaut*, but additional information can not be captured. It should be noted, however, that EARL can tackle the problem of the "lexical gap" to a great extent as it uses synonyms via the grammar inflection forms.

Our approaches of LKH-GTSP and Connection Density both have polynomial and approximately similar time complexities. EARL with either Connection Density or LKH-GTSP can process a question in a few hundred milliseconds on a standard desktop computer on average. The result logs, experimental setup and source code of our system are publicly available at: https://github.com/AskNowQA/EARL.

7   Conclusions and Future Work

Here we propose EARL, a framework for joint entity and relation linking. We provided two strategies for joint linking - one based on reducing the problem to an instance of the Generalised Travelling Salesman problem and the other based on a connection density based machine learning approach. Our experiments on QA benchmarks resulted in accuracies which are significantly above the results of current state-of-the-art approaches for entity and relation linking. In future, we will improve the candidate generation phase to ensure that a higher proportion of correct candidates are retrieved.

Acknowledgement. This work is supported by the funding received from the EU H2020 projects WDAqua (ITN, GA. 642795) and HOBBIT (GA. 688227).

References

1. Berant, J., Chou, A., Frostig, R., Liang, P.: Semantic parsing on freebase from question-answer pairs. In: EMNLP, vol. 2, p. 6 (2013)
2. Both, A., Diefenbach, D., Singh, K., Shekarpour, S., Cherix, D., Lange, C.: Qanary – a methodology for vocabulary-driven open question answering systems. In: Sack, H., Blomqvist, E., d'Aquin, M., Ghidini, C., Ponzetto, S.P., Lange, C. (eds.) ESWC 2016. LNCS, vol. 9678, pp. 625–641. Springer, Cham (2016). https://doi.org/10.1007/978-3-319-34129-3_38
3. Collobert, R., Weston, J., Bottou, L., Karlen, M., Kavukcuoglu, K., Kuksa, P.: Natural language processing (almost) from scratch. J. Mach. Learn. Res. **12**(Aug), 2493–2537 (2011)
4. Dubey, M., Dasgupta, S., Sharma, A., Höffner, K., Lehmann, J.: AskNow: a framework for natural language query formalization in SPARQL. In: Sack, H., Blomqvist, E., d'Aquin, M., Ghidini, C., Ponzetto, S.P., Lange, C. (eds.) ESWC 2016. LNCS, vol. 9678, pp. 300–316. Springer, Cham (2016). https://doi.org/10.1007/978-3-319-34129-3_19
5. Gerber, D., Ngomo, A.-C.N.: Bootstrapping the linked data web. In: 1st Workshop on Web Scale Knowledge Extraction@ ISWC, vol. 2011 (2011)
6. Gubichev, A., Then, M.: Graph pattern matching: do we have to reinvent the wheel? In: Proceedings of Workshop on GRAph Data. ACM (2014)
7. Helsgaun, K.: Solving the equality generalized traveling salesman problem using the Lin-Kernighan-Helsgaun algorithm. Math. Program. Comput. **7**, 269–287 (2015)
8. Höffner, K., Walter, S., Marx, E., Usbeck, R., Lehmann, J., Ngonga Ngomo, A.-C.: Survey on challenges of question answering in the semantic web. Semant. Web **8**(6), 895–920 (2017)
9. Kingma, D., Ba, J.: Adam: A method for stochastic optimization. arXiv preprint arXiv:1412.6980 (2014)

10. Laporte, G., Mercure, H., Nobert, Y.: Generalized travelling salesman problem through n sets of nodes: the asymmetrical case. Discrete Appl. Math. **18**(2), 185–197 (1987)

11. Lukovnikov, D., Fischer, A., Lehmann, J., Auer, S.: Neural network-based question answering over knowledge graphs on word and character level. In: Proceedings of the 26th International Conference on World Wide Web, pp. 1211–1220 (2017)

12. Mendes, P.N., Jakob, M., García-Silva, A., Bizer, C.: DBpedia spotlight: shedding light on the web of documents. In: Proceedings of the 7th International Conference on Semantic Systems, pp. 1–8. ACM (2011)

13. Moro, A., Raganato, A., Navigli, R.: Entity linking meets word sense disambiguation: a unified approach. Trans. Assoc. Comput. Linguist. (2014)

14. Mulang, I.O., Singh, K., Orlandi, F.: Matching natural language relations to knowledge graph properties for question answering. In: Proceedings of the 13th International Conference on Semantic Systems, pp. 89–96. ACM (2017)

15. Nakashole, N., Weikum, G., Suchanek, F.: Patty: a taxonomy of relational patterns with semantic types. In: Proceedings of the EMNLP 2012, pp. 1135–1145. Association for Computational Linguistics (2012)

16. Park, S., Kwon, S., Kim, B., Lee, G.G.: ISOFT at QALD-5: hybrid question answering system over linked data and text data. In: CLEF (Working Notes) (2015)

17. Pinter, Y., Guthrie, R., Eisenstein, J.: Mimicking word embeddings using subword RNNs. In: EMNLP, pp. 102–112 (2017)

18. Ristoski, P., Paulheim, H.: RDF2Vec: RDF graph embeddings for data mining. In: Groth, P., et al. (eds.) ISWC 2016. LNCS, vol. 9981, pp. 498–514. Springer, Cham (2016). https://doi.org/10.1007/978-3-319-46523-4_30

19. Serban, I.V., et al.: Generating factoid questions with recurrent neural networks: the 30m factoid question-answer corpus. arXiv preprint arXiv:1603.06807 (2016)

20. Singh, K., et al.: Capturing knowledge in semantically-typed relational patterns to enhance relation linking. In: Proceedings of the Knowledge Capture Conference, p. 31. ACM (2017)

21. Singh, K., et al.: Why reinvent the wheel: let's build question answering systems together. In: Proceedings of the 2018 World Wide Web Conference on World Wide Web, pp. 1247–1256. International World Wide Web Conferences Steering Committee (2018)

22. Speck, R., Ngonga Ngomo, A.-C.: Ensemble learning for named entity recognition. In: Mika, P., et al. (eds.) ISWC 2014. LNCS, vol. 8796, pp. 519–534. Springer, Cham (2014). https://doi.org/10.1007/978-3-319-11964-9_33

23. Trivedi, P., Maheshwari, G., Dubey, M., Lehmann, J.: LC-QuAD: a corpus for complex question answering over knowledge graphs. In: d'Amato, C., et al. (eds.) ISWC 2017. LNCS, vol. 10588, pp. 210–218. Springer, Cham (2017). https://doi.org/10.1007/978-3-319-68204-4_22

24. Trudeau, R.J.: Introduction to Graph Theory (corrected, enlarged republication. ed.) (1993)

25. Unger, C., Bühmann, L., Lehmann, J., Ngonga Ngomo, A.-C., Gerber, D., Cimiano, P.: Template-based question answering over RDF data. In: Proceedings of the 21st International Conference on World Wide Web, pp. 639–648. ACM (2012)

26. Usbeck, R., et al.: AGDISTIS - graph-based disambiguation of named entities using linked data. In: Mika, P., et al. (eds.) ISWC 2014. LNCS, vol. 8796, pp. 457–471. Springer, Cham (2014). https://doi.org/10.1007/978-3-319-11964-9_29

27. Veyseh, A.P.B.: Cross-lingual question answering using common semantic space. In: TextGraphs@ NAACL-HLT, pp. 15–19 (2016)

28. Xu, K., Zhang, S., Feng, Y., Zhao, D.: Answering natural language questions via phrasal semantic parsing. In: Zong, C., Nie, J.Y., Zhao, D., Feng, Y. (eds.) Natural Language Processing and Chinese Computing. CCIS, vol. 496, pp. 333–344. Springer, Heidelberg (2014). https://doi.org/10.1007/978-3-662-45924-9_30
29. Yang, Y., Chang, M.-W.: S-mart: novel tree-based structured learning algorithms applied to tweet entity linking. In: ACL 2015 (2015)
30. Yih, W.-T., Chang, M.-W., He, X., Gao, J.: Semantic parsing via staged query graph generation: question answering with knowledge base. In: Proceedings of the 53rd ACL Conference, vol. 1, pp. 1321–1331 (2015)

TSE-NER: An Iterative Approach
for Long-Tail Entity Extraction
in Scientific Publications

Sepideh Mesbah[(⊠)], Christoph Lofi, Manuel Valle Torre, Alessandro Bozzon,
and Geert-Jan Houben

Delft University of Technology, Mekelweg 4, 2628 CD Delft, The Netherlands
{s.mesbah,c.lofi,m.valletorre,a.bozzon,g.j.p.m.houben}@tudelft.nl

Abstract. Named Entity Recognition and Typing (NER/NET) is a
challenging task, especially with long-tail entities such as the ones found
in scientific publications. These entities (e.g. "WebKB", "StatSnowball")
are rare, often relevant only in specific knowledge domains, yet important
for retrieval and exploration purposes. State-of-the-art NER approaches
employ supervised machine learning models, trained on expensive type-
labeled data laboriously produced by human annotators. A common
workaround is the generation of labeled training data from knowledge
bases; this approach is not suitable for long-tail entity types that are, by
definition, scarcely represented in KBs. This paper presents an iterative
approach for training NER and NET classifiers in scientific publications
that relies on minimal human input, namely a small seed set of instances
for the targeted entity type. We introduce different strategies for training data extraction, semantic expansion, and result entity filtering. We
evaluate our approach on scientific publications, focusing on the long-tail
entities types *Datasets*, *Methods* in computer science publications, and
Proteins in biomedical publications.

1 Introduction

The growth of domain-specific knowledge available as digital text demands
more effective methods for querying, accessing, and exploring document collections. Scientific publications are a compelling example: online digital libraries
(e.g. IEEE Xplore) contain hundreds of thousands documents; yet, the available retrieval functionality is often limited to keyword/faceted search on *shallow*
meta-data (e.g. title, terms in abstract). A query like *retrieve the publications
that used a social media dataset for food recipe recommendation* is bound to
return unsatisfactory results[1].

Named entities, obtained through an analysis of a document's content, are an
effective way to achieve better retrieval and exploration capabilities. Automatic
Named Entity Recognition and Typing (NER/NET) is essential to unlock and

---

[1] https://scholar.google.de/scholar?q=publications+using++social+media+datasets
+for+food+recipes+recommendation.

© Springer Nature Switzerland AG 2018
D. Vrandečić et al. (Eds.): ISWC 2018, LNCS 11136, pp. 127–143, 2018.
https://doi.org/10.1007/978-3-030-00671-6_8

mine the knowledge contained in digital libraries, as most smaller domains lack the resources for manual annotation work.

To perform well, state-of-the-art NER/NET methods [3,4,11] either require comprehensive domain knowledge (e.g. to specify matching rules), or rely on a large amount of human-labeled training data for machine learning. Both solutions are expensive and time-consuming.

A cheaper alternative is to generate labeled training data by obtaining existing instances of the targeted entity type from Knowledge Bases (KBs) [3]. This of course requires that the desired entity type is well-covered in the KB.

Problem Statement. While achieving impressive performance with high-recall named entities (e.g. locations and age) [11], generic NER/NETs show their limits with domain-specific and long-tail entity types. Consider the following sentence: *"We evaluated the performance of SimFusion+ on the WebKB dataset"*. Despite *WebKB*[2] being a popular dataset in the Web research community, generic NERs (e.g. Textrazor[3]) mistype it as an `Organization` instead of the domain-specific entity type `Dataset`. The entity *SimFusion+* of type `Software` is missed completely.

Literature [20,26,27] shows that training of *domain-specific* NER/NETs is still an open challenge for two main reasons: (1) the *long-tail* nature of such entity types, both in existing knowledge bases *and* in the targeted document collections [22]; and (2) the high cost associated with the creation of hand-crafted rules, or human-labeled training datasets for supervised machine learning techniques. Few approaches addressed these problems by relying on bootstrapping [27] or Entity Expansion [3,11] techniques, achieving promising performance. However, how to train high-performance *long-tail* Entity Extraction and Typing with minimal human supervision remains an open research question.

Original Contribution. We contribute TSE-NER, an iterative approach for training NER/NET classifiers for long-tail entity types that exploits Term and Sentence Expansion, extensively expanding on [16]. TSE-NER relies on minimal human input – a seed set of instances of the targeted entity type. We introduce different strategies for training data extraction, semantic expansion, and result entity filtering. Different combinations of these strategies allow to tune the technique for either higher recall or higher precision scenarios.

We performed extensive evaluations comparing to state-of-the-art methods, and assess several sentence expansion and term filtering strategies. As our core use case, we focus on 15,994 data science publications from 10 conference series with the *Dataset* (e.g. `Imagenet`) and data processing *Methods* (e.g. `LSTM`) long-tail entity types. We show that our approach is able to consistently outperform state-of-the-art low-cost supervision methods, even with small amount of training information: with a seed set of 100 entities, our approach can achieve precision up to 0.91 when tuned for precision, and recall up to 0.41 when tuned for recall, or 0.77 and 0.30 for a balanced setting. When applied in an iterative

---

[2] http://www.cs.cmu.edu/~WebKB/.

[3] https://www.textrazor.com/.

fashion, our approach can achieve comparable performance with an initial seed set of only 5 entities. We show that sentence expansion and filtering strategies can provide a spectrum of performance profiles, suitable for different retrieval applications such as search (high precision) and exploration (high recall).

To study the performance of TSE-NER across scientific domains, we processed 4,525 biomedical publications focusing on *Protein* (e.g. Myoglobin) entity type. Evaluation on the Craft corpus [2] shows that TSE-NER can achieve performance comparable to existing dictionary-based systems, and obtain precision up to 0.40 and recall up to 0.28 with just 25 seed terms. TSE-NER is implemented in the *SmartPub* platform [17]; its source code is available on the companion Website [18], and its application shown in the video screencast at the following address: https://youtu.be/zLLMwOT5sZc.

Outline. The remainder of the paper is organized as follows. In Sect. 2 we cover related work. Section 3 presents our approach, and describes alternative data expansion and entity filtering strategies. The experimental setup and results are presented in Sect. 4. Section 5 concludes.

2   Related Work

A considerable amount of literature published in recent years addressed the *deep* analysis of text. Common approaches for *deep* analysis of publications rely on techniques such as bootstrapping [27], word-frequency analysis [25], probabilistic methods like Latent Dirichlet Allocation [8], etc. In contrast to current research [25] which limits the analysis of a publication's content to its title, abstract, references, and authors, we extract entity instances from the much richer full text. In addition, our method does not rely on existing knowledge bases [20,23] and it is not based on selecting the most frequent keywords [25]. More recent research [26] used both corpus-level statistics and local syntactic patterns of scientific publications to identify entities of interest. Our method uses only a small set of seed names (i.e 5–100), and automatically trained distributed word representations to train a NER in iterative steps (i.e. 2–3).

Entity Instances Extraction. Named Entity Recognition (NER) has been applied to identify both entity types of general interest (e.g. Person, Location, Cell, Brand, etc.) as well as for specific domains (e.g., medicine or other domain where resources for training a NER are easily available). NERs rely on different approaches such as dictionary-based, rule-based, machine-learning [26] or hybrid (combination of rule based and machine learning) [29] techniques. Despite its high accuracy, a major drawback of dictionary-based approaches is that they require an exhaustive dictionary of domain terms, which are expensive to create and many smaller domains lack the resources to do so. The same holds for rule-based techniques, which rely on formal languages to express rules and require comprehensive domain knowledge and time to create.

Bootstrapping and Entity Set Expansion. Most current NERs are based on Machine Learning techniques, which require a large corpus of labeled training text [9]. Again, the high costs of data annotation is one of the main challenges in adopting specialized NER for rare entity types in specialized domains [26]. In recent years, many attempts have been made to reduce annotation costs. Active learning techniques have been proposed, asking users to annotate a small part of a text for machine learning methods [7].

Transfer learning techniques [21] use the knowledge gained from one domain and apply it to a different but related named entity type. Co-training [1] starts with a small amount of manually annotated supervised training data and attempt to increase the amount of annotated data. In contrast to previous work, we are not dependent on manually annotated supervised training data [1]; we do not require a large training corpus [21] for transfer learning; also, our approach differs from works on high-recall entity extractors (e.g. with regular expression extractors) for detecting entity types such as location and age [11].

Entity Set Expansion is a technique finding similar entities to a given small set of seed entities [3,6,11]. Bootstrapping [27] is another approach similar to our method that uses seed terms and extracts features such as unigrams, bigrams, left unigram, closest verb, etc. These are used to annotate more concept mentions which leads to extracting new features. This step operates in an iterative fashion until no new features are detected. Our approach is inspired by Entity Set Expansion and bootstrapping, but relies on different expansion strategies and does not require concepts already being available in knowledge bases [3].

3   Approach

The TSE-NER (Term and Sentence Expansion) approach for domain-specific long-tail entity recognition is organized in five steps, as shown in Fig. 1. ❶ An initial set of seed terms is used to identify a set of sentences used as initial *training data* (Sect. 3.1). ❷ *Expansion* strategies can be used to expand the set of initial seed terms, and the *training data* sentences (Sect. 3.2). ❸ The *Training Data Annotation* step annotates the training data using the (possibly expanded) seed terms set (Sect. 3.3). ❹ A new Named Entity Recognizer (NER) is *trained* using the annotated training data, and the newly trained NER is applied on the corpus to detect a candidate set of entities (Sect. 3.4). ❺ The *Filtering* step refines the set candidate entities set, to improve the quality of outputted *Verified Terms* set (Sect. 3.5).

Fig. 1. Overview of the domain-specific long-tail named entities recognition approach.

TSE-NER operates under the hypothesis that there are recurring patterns in the mentions of domain-specific named entities, and that they appear in similar contexts. If this hypothesis holds, by training a classifier on the texts containing the entities, we are able to extract the instances of the entity type of interest. The process can be iterated, by repeating the first step using the newly detected terms as seeds to generate new training data. We rely on the following concepts (some are only relevant for the evaluation, and could be omitted in setups where evaluation is not necessary). The companion website [18] provides a complete unified algorithm covering the TSE-based NER training workflow.

Known Entity Terms $T_{all} := T_{seed} \cup T_{test}$: This represents a manually created set of instances of the entity type for which a NER classifier is to be trained. In this work, we split this set into a set of seed terms T_{seed} used for training, and test terms T_{test} used for evaluation purposes. In a real-life scenario not requiring a formal evaluation, of course only the seed terms would be necessary. T_{seed} may be small. In this work we consider seed sets $5 \leq |T_{seed}| \leq 100$. Creating T_{seed} is the only manual input required for NER training in our approach.

Document Corpus $D_{all} := \{d_1, ..., d_{|D|}\}$: This is the complete document corpus available to our system. Parts of it can potentially be used for training, others for testing. Each document is considered to be a sequence of sentences.

All Sentences $S_{all} := \{s | s \in d \wedge d \in D_{all}\}$: This represents all sentences of the whole document corpus. Each sentence is considered to be a sequence of terms.

Test Sentences $S_{test} := \bigcup_{t \in T_{test}} \{s | s \in S_{all} \wedge t \in s\}$: These are all sentences containing any term from the test set, and they need to to be excluded from any training in order to ensure the validity of our later evaluations, resulting in the set of **Development Sentences** $S := S_{all} \setminus S_{test}$.

In the following, we introduce the iterative version of our approach, representing the current iteration number as i whereas initially $i = 0$. Each iteration i uses its own term list T_i, which initially is $T_0 \subseteq T_{seed}$ (the size of the subset of T_{seed} depends on the desired use case, as discussed in Sect. 4.3).

3.1 Training Data Extraction

As a first step, a set of training data sentences S_i for the current iteration is created by extracting suitable sentences from S. At this stage, this is realized by selecting all sentences containing any of the seed terms. Therefore, S_i provides examples of the positive classification class as they are guaranteed to contain a desired entity instance. To better capture the usage context of the seed entity, we also extract surrounding sentences in the text: $S_i := \bigcup_{t \in T_i} \{s | s \in S \wedge (t \in s \vee t \in successor(s) \vee t \in predecessor(s))\}$.

3.2 Expansion

The small size of the seed term set T_{seed} has two obvious shortcoming that can greatly hinder the accuracy and recall of the trained NERs: (1) the amount of training data sentences S_i is limited; and (2) there are only few examples of mentions of the entity instances of the given type. In addition, the *generalization* capability of the NER for identifying new named entities can also be affected: an insufficient amount of positive examples can lead to entities of the targeted type being labeled negatively; while the extraction of sentences in the training data that are related to seed terms will cause a shortage of negative examples. To account for these issues, we designed two *expansion* strategies.

Term Expansion (TE). Term Expansion is designed to increase the number of known instances of the desired entity type before training the NER. An expanded set of entities will provide more positive examples in the training data, thus ideally improving the precision of the NER. In scientific documents, it is common for domain-specific named entities to be in close proximity, e.g. to enumerate alternative solutions, or list technical artifacts. The *Term Expansion* (TE) strategy is therefore designed to test and exploit this hypothesis.

We introduce the interface $expandTerms(terms_s)$, with $terms_s \subseteq terms_i$. While many different implementations for this interface are possible, in this work we use *semantic similarity*: terms which are semantically similar to terms in the seed list should be included in the expansion. For example, given the dataset seed terms Clueweb and cim-10, the expansion should add similar terms like trec-2005.

We exploit the distributional hypothesis [10] stating that terms frequently occurring in similar context are semantically related, using the popular *word2vec* implementation of skip-n-gram word embeddings [19]. In essence, *word2vec* embeds each term of a large document corpus into low-dimensional vector space (100 dimensions in our case), and the cosine distance between two vectors has been shown to be a high-quality approximation of semantic relatedness [14]. In our implementation, we trained the *word2vec* model on the whole development sentence collection S, as described in [19], learning all uni- and bigram word vectors of all terms in the corpus. Then, in its most basic version, we select all terms from all sentences, and cluster them with respect to their embedding vectors using K-means clustering. Silhouette analysis is used to find the optimal number k of clusters. Finally, clusters that contain at least one of the seed terms are considered to (only) contain entities the same type (e.g *Dataset*).

Initial experiments have shown that this naive approach is slow, and that it can potentially introduce many false positives due to (1) the large number of considered terms, and (2) the sometimes faulty assumption that all terms in cluster are indeed similar as *word2vec* relatedness is not always reliable for similarity measurements [14].

Algorithm 1. TE using Semantic Relatedness

function EXPANDTERMS($terms_s$)

$T_{entity} := \{t | t \in s \land s \in S \land isEntity(t)\}$
$\qquad\qquad\qquad\qquad\qquad$ ▷ All entities in S

$clusters := cluster(word2vec(T_{entity}))$
$\qquad\qquad\qquad\qquad\qquad$ ▷ Cluster the embeddings

$clusters_{correct} := \{c | c \in clusters \land t \in terms_s \land t \in c\}$

$\qquad\qquad$ ▷ Select clusters containing any initial term

return $\bigcup_{c \in clusters_{correct}}$

end function

To improve, in the following we only consider terms which are likely to be named entities by using NLTK entity detection to obtain a list of all entities E_{all} contained in S^4. This results in the Algorithm 1.

Sentence Expansion (SE). A second (optional) measure to increase the size and variety of the training set is the *Sentence Expansion* (SE) strategy. It addresses the problem of the over-representation of positive examples resulting from selecting only sentences with instances of the desired type (see Sect. 3.1). The goal is to include negatives sentences not containing instances of the desired type, but are still very similar in semantics and vocabulary.

We rely on *doc2vec* document embeddings [13], a variant of *word2vec*, to learn vector representations of all sentences. For each sentence in S, we use the cosine distance to discover the most similar sentences filtered to those not containing any known instance of the targeted type. As such sentences might contain an unknown instance of that type, we always combine SE with term expansion to minimize the risk of accidentally mislabeling them as negative examples.

3.3   Training Data Annotation

The annotation of training data from the (expanded) seed terms is performed automatically, with no human intervention. After obtaining an (expanded) set of instances T_i (the current term list) and training sentences S_i, we annotate each term $A_{T_i} := annotate_{T_i}(S_i)$ in all training sentences if they are a positive instance of the targeted entity type, i.e. if the term $\in T_i$. Using A_{T_i}, any state-of-the-art supervised NER can be trained.

3.4   NER Training

For training a new NER_i, we used the Stanford NER tagger[5] to train a Conditional Random Field (CRF) model. As the focus of this paper is the process of training data generation, we do not consider additional algorithms. CRF has shown to be an effective technique on different NER tasks [12]; the goal of CRF is to learn the hidden structure of an input sequence. This is done by defining a set of feature functions (e.g. word features, current position of the word labels of the nearby word), assigning them weights and transforming them to a probability to detect the output label of a given entity. The features used in the training of the model are listed in the companion website. After a NER for the current iteration N_i is trained, it is used to annotate the whole development corpus S, i.e. $A_{NER_i} := annotate_{NER_i}(S)$. All positively annotated terms are considered newly discovered instances of our desired type.

---

[4] NLTK entity detection is based on grammatical context. It does not perform any typing, and due to it's simplicity, has high recall values.

[5] https://github.com/dat/stanford-ner.

3.5   Filtering

After applying the NER to the development corpus, we obtain a list of new candidate terms. As our process relied on several steps which might have introduced noise and false positives (like the expansion steps, but also the NER itself), the goal of this last (optional) step is to filter out candidate terms that are unlikely of the targeted type using a set of external heuristics with different assumptions:

Wordnet + Stopwords (WS) Filtering. In the domain-specific language of scientific documents, it is common for named entities to be "proper" of that domain (like `Simlex-999`), or to be expressed as acronyms (like `Clueweb`, `SVM`, `RCV`). In this strategy, named entities are assumed to be not relevant if they are part of the "common" English language, either as proper nouns (e.g. *software, database, figure*), or a Stopwords (e.g. *on, at*). This is achieved by performing lookup operations in WordNet[6] and in common lists of stopwords[7]. As both sources focus on general English language, only domain-specific terms should be preserved.

Similar Terms (ST) Filtering. In order to distinguish between different entity types that pertain to a given domain (e.g. `SVM` is of type `Method`, while `Clueweb` is of type `Dataset`), this filtering strategy employs an approach similar to the one used in the *Term Expansion* (TE) strategy. The idea is to cluster entities based on their embedding feature using K-means clustering, and keep all the entities that appear in the cluster that contains a seed term.

Pointwise Mutual Information (PMI) Filtering. This filtering strategy adopts a semantic similarity measure derived from the number of times two given keywords appear together in a *sentence* in our corpus. The heuristic behind this filter is vaguely inspired by Hearst Patterns [24], as we manually compile a list of context terms/patterns CX which likely indicate the presence of an instance of our desired class (e.g., "we evaluate on x" typically indicates a dataset). Unlike the other filters, it does increase the manual resource costs for training.

Given a set of candidate entities CT_i and the context term set CX, we measure the PMI between them using $\log \frac{N(ct,cx)}{N(ct)N(cx)}$ with $ct \in CT_i \wedge cx \in CX$, and $N(ct,cx)$ being the number of sentences in which both a candidate entity (ct) and a given keyword (t) occur (analogously, $N(ct)$ counts the number of occurrences of ct). Finally, candidate terms are filtered and excluded if their PMI value is below a given threshold value.

Knowledge Base Lookup (KBL) Filtering. Our target are long-tail domain-specific entities, i.e. entities that are not part of existing knowledge bases. Named entities that could be linked to a knowledge base could be assumed incorrect, and therefore amenable to exclusion from the final named entity set. In the KBL approach we exclude the entities that have a reference in the DBpedia.

---

[6] http://wordnet.princeton.edu/.
[7] http://www.nltk.org/book/ch02.html.

Ensemble (EN) Filtering. Different filtering strategies are likely to remove different named entities. To reduce the likelihood of misclassification, the *Ensemble* (EN) filtering strategy combines the judgment of multiple filtering strategies, to preserve candidate entities that are considered correct by one or more strategy. Intuitively, if each strategy makes different errors, then a combination of the filters' judgment can reduce the total error. We preserve the entities that are passed through two out of three selected filtering strategies.

4   Evaluation

This section reports on an empirical evaluation to assess the performance of the approach (and its variants) described in Sect. 3, and the ability to utilize it for long-tail named entity recognition. Sect. 4.1 describes the experimental set-up, followed by the results (Sect. 4.2), and their discussion (Sect. 4.3).

4.1   Experimental Setup

Corpora. Our main evaluation, shown in the following sections, is performed on the data science (15,994 papers from 10 conference series) domain. To assess the performance of TSE-NER in other scientific domains, at the end of the section we describe an experiment over 4,525 publications from 10 biomedical journals. The full description of the corpora is described in the companion Web site [18]. Publications are processed using GROBID [15], to extract a structured full-text representation of their content.

Long Tail Entity Types Selection. Scientific publications contain a large quantity of long-tail named entities. Focusing on the data science domain, we address the entity types *Dataset* (i.e. dataset presented or used in a publication), and *Methods* (i.e. algorithms – novel or pre-existing – used to create/enrich/analyze a dataset). Both entities types are scarcely represented in existing knowledge bases[8]. To evaluate the performance of our approach, we create a set of 150 seed instances T_{all} for each targeted type, collected public from public websites[9].

For each type, 50 of those are selected as test terms for that type T_{test}, while 100 are used as seed terms T_{seed}.

Evaluation Dataset. As discussed in Sect. 3, in the training process all test sentences S_{test} (i.e. sentences mentioning terms in T_{test}) in the corpus D_{all} are removed. For evaluation, we manually created a type-annotated test set: for

---

[8] In DBPedia, the type `dbo:database` features 989 instances, but mostly related to biology, economy, and history. The type `dbo:software` contain names of several algorithms, but the list is clearly incomplete.

[9] For instance: https://github.com/caesar0301/awesome-public-datasets. The full list of seed entity instances, as well as the list of sources are available on the companion Website.

each test term, we select all sentences in which they are contained including any adjacent sentence, forming the set of annotated sentences $S_{annotated} := \cup_{t \in T_{test}}\{s|s \in S_{test} \wedge (t \in s \vee t \in successor(s) \vee t \in predecessor(s))\}$. An expert annotator labeled each term as an instance of the target type to create the test annotation set used for evaluation $A_{test} := annotate_{expert}(S_{annotated})$.

Details of statistics on sentences used for training and testing can be found in the companion Web site. For training, depending on the seed set size between 5 and 100, we used between 198 and 2863 sentences for the *dataset* entity type and 617 to 18545 sentences for the *Method* entity type.

Algorithm 2. Evaluation Protocol

---

function EVALUATE(*seed_size*)

$T \subseteq_{seed\_size} T_{seed}$

$NER_{final} := longtailTrain(T, S_{all})$

$A_{final} := annotate_{NER_{final}}(S_{annotated})$

$result := analyze(A_{final}, A_{test})$

end function

For testing 50 seed terms were used for both dataset (i.e. 3149 sentences) and method (i.e. 1097 sentences) entity type. The evaluation protocol is described in Algorithm 2, where the *seed_size* values can be initialized with different values. Our analysis was not limited to the 50 test seed terms, we further evaluated 200 entities recognized by TSE-NER via a pooling technique.

4.2   Results

For a given entity type (*Dataset* and *Method*), we test the performance with differently sized seed sets and expansion strategies to create the training data for generating the NER model, and different filtering strategies. We report the performance of the basic WS, PMI, and EN strategies, plus a combination of the WS, ST, and KBL strategies, as listed in Table 1. The complete evaluation results for all the seed set size and the filtering techniques can be found in the companion Web site. We investigate iterative performance, and results on the manually annotated test from the previous section.

Tables 1 and 2 summarize the performance achieved for *Dataset* and *Method* entity types. In Table 2, the *No Expansion* and *Term Expansion* figures for the *Method* type are omitted for brevity's sake. Our approach is able to achieve excellent precision [89% – 91%] with both entity types, and good recall (up to 41%) with the *Dataset* type. The lower recall obtained with the *Method* type can be explained with the greater diversity (in terms of n-grams and use of acronyms) of method names.

The expansion strategies lead to an average +200% (SE – *Dataset*) and +300% (TE – *Dataset*) increase in recall, thus demonstrating their effectiveness for generalization. On average, filtering decrease recall, but with precision improvements up to +20% (PM – *Method*). These are promising figures, considering the minimal human supervision involved in the training of the NERs. We can also show the different trade-offs our approach can strike: different configurations of filtering and expansion lead to different results with respect to precision and recall values, allowing for example a high-precision slightly-lower recall setup for a digital library, and a higher recall lower precision setup for a Web retrieval system.

Expansion Strategies. Expansion strategies increase the size and variety of training datasets, thus improving the precision and recall. Both strategies achieve the expected results, although with different performance increase: compared to *NE* strategy, both TE and SE achieve a considerable performance boost ($\mu = +190\%$) for recall, but at cost of lower precision ($\mu = -8.7\%$). We account the better recall performance of TE to the contextual similarity (and proximity) of named entities of the same type in technical documents (e.g. Gov2, Robust04, ClueWeb and Wt10g). The precision decrease in TE can be accounted to treating some terms incorrectly as positive instances due to their presence in the same embedding clusters as the seed terms (see also Sect. 3.2). The SE strategy shows lower recall ($\mu = +210\%$ over NE), but with less precision loss ($\mu = -5.2\%$ than NE). We account this positive behaviour to the presence of more quality negative examples, helping to maintain the generalization capabilities of the NER, while refining the quality of its recognition.

Filtering Strategies. We observe no significant improvement in precision with the WS filtering approach. Manual inspection of results reveal that most of the false positives are already domain-specific terms (e.g. Pagerank, Overcite for *Dataset*, and NDCG for *Method*) which are not included in Wordnet, but that are of the wrong type. SS slightly increases the precision by keeping only the entities that appear in the same cluster as the seed names; however, this comes at a cost,

Table 1. *Dataset* entity type: precision/recall/F-score on evaluation dataset. Legend: *NE* – No Expansion; *TE* – Term Expansion; *SE* – Sentence Expansion; *NF* – No Filtering; *WS* – Wordnet + StopWords; *SS* – Similar Terms + WS; *KS* – Knowledge Base Lookup + SS; *PM* – Point-wise Mutual Information; *EN* – Ensemble.

Strategy	#S	NF	WS	SS	KS	PM	EN
NE	5	.83/.05/.10	.84/.04/.08	.86/.03/.07	.75/.01/.01	.90/.04/.09	.86/.04/.08
	25	.84/.08/.16	.83/.07/.13	.86/.07/.13	.78/.01/.03	.91/.08/.15	.85/.07/.13
	100	.85/.15/.26	.85/.13/.22	.87/.13/.22	.82/.03/.07	.91/.13/.24	.86/.12/.22
TE	5	.76/.14/.25	.78/.13/.22	.79/.11/.20	.74/.04/.09	.83/.13/.23	.80/.13/.22
	25	.72/.29/.42	.73/.28/.40	.75/.27/.40	.73/.17/.28	.77/.27/.40	.75/.27/.40
	100	.69/.41/.51	.70/.39/.50	.71/.38/.50	.71/.28/.40	.74/.38/.50	.72/.38/.50
SE	5	.83/.07/.14	.84/.06/0.12	.86/.05/.10	.82/.01/.02	.91/.07/.13	.86/.06/.11
	25	.81/.22/.35	.80/.18/.29	.83/.17/.29	.77/.04/.08	.89/.20/.33	.82/.18/.29
	100	.77/.30/.43	.77/.24/.37	.80/.23/.36	.78/.07/.13	.86/.26/.40	.79/.24/.37

Table 2. *Method* entity type: precision/recall/F-score. Legend as in Table 1.

Strategy	#S	NF	WS	SS	KS	PM	EN
SE	5	.76/.04/.08	.77/.03/.07	.77/.01/.01	.84/.01/.01	.86/.01/.03	.84/.03/.05
	25	.77/.14/.24	.77/.12/.21	.79/.09/.16	.87/.05/.09	.86/.05/.09	.85/.09/.17
	100	.68/.15/.25	.67/.14/.23	.65/.12/.20	.84/.07/.13	.85/.05/.10	.83/.10/.19

as the recall is also penalized by the exclusion of entities of interest that are in other clusters. KB excludes popular entities that are contained in the knowledge base (e.g. `Wordnet`, `Dailymed`), but also some rare entities that are mistyped.

For instance, the *Dataset* entities `Ratebeer`[10] or `Jester` can be retrieved from DBpedia using the lookup search, although the result points to another entity. This is a clear limitation with the adopted lookup technique, which could be avoided with a more precise implementation of the lookup function. PMI usually gets the highest precision; the strategy proved effective in removing false positives, but penalizes recall by excluding entities that do not appear with the words in the context list CX. For instance, `Unigene` (*Dataset*) often appears in with the term *data source*, which is not in our context list and thus filtered out. The EN strategy keeps only the entities that are preserved by two out of three (WS, KB and PMI) filtering strategies. While reducing the number of false positives, this proves to be too restrictive; for instance *Dataset* names such as `Yelp`, `Twitter`, `Foursquare` and `Nasdaq` are removed by both the WS and KB strategies.

Seed Set Size. We randomly initialize $T \subseteq T_{seed}$ with $|T| = 5, 10, 25, 50, 100$ (see Algorithm 2). We execute the evaluation cycle 10 times for each size of T, and again vary expansion and filtering strategies. The recall performance sharply increase with the number of seeds term ($\mu = +340\%$ from 5 to 100 seeds): this is due to the increase in the number of sentences available for NER training, and is an expected behaviour. The decrease in precision is an average of -6% from 5 to 100 seeds, with an average value of -5.1% for *Dataset* and -6.9% for *Methods*. Noteworthy are the good performance with as little as 5 seed entities (*Datasets*: 0.25 F-score with TE strategy and no filtering).

Iterative NER Training. Figure 2 shows the result of the iterative NER training using Sentence Expansion with 5 seeds. We report the results with the PMI (*Dataset*) and EN (*Methods*) filtering, as they are the ones offering the most balanced performance in both precision and recall. Despite the small initial seed seed, it is possible to achieve precision and recall comparable to the ones obtained with an initial set of 100 seeds in only 2 iterations.

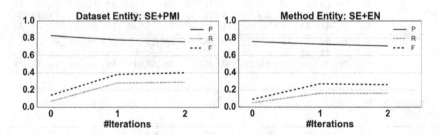

Fig. 2. Dataset (L) and Method (R) entity: iterative NER training using 5 initial seeds.

---

[10] http://lookup.dbpedia.org/api/search/KeywordSearch?QueryClass=&QueryString =ratebeer.

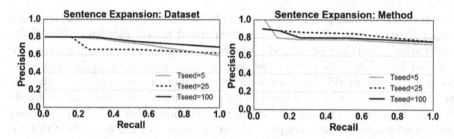

Fig. 3. *Dataset* (L) and *Method* (R): precision and recall for ranked top 10, 25, 50, 100 and 200 entities, varying seeds sizes.

Analysis of Recognized Entities. To widen the scope of our evaluation, we extended our result analysis beyond the 150 named entities in T_{all}. We manually investigated up-to-now unknown named entities which have been recognized by the NER after training. We applied a method inspired by the pooling technique typically used in information retrieval research: given a list of seed terms T_{seed} of a given type, and a list of recognized potential filtered terms FT of an yet unknown type, the idea is to rank the items in the list of candidate terms FT according to their embedding similarity to the items in the seed set T_{seed} and collect the top K. As a result, the obtained precision and recall measurements are only approximate values. The similarity is measured based on the cosine similarity between the *word2vec* embedding vectors. Each entity in the lists has been manually checked by an expert. Figure 3 shows the precision and recall of the top $K = 10$, 25, 50, 100, and 200 retrieved entities using the SE approach. As in the previous experiment, we used the PMI and EN filtering strategies respectively for *Dataset* and *Method* types. Precision performance are consistently high at all level of recall. Note that we randomly selected $T \subseteq T_{seed}$ with —T—=5,25,100 seed terms and used them to train the NER using the SE strategy. Variations in precision performance in Fig. 3 are therefore accountable on the initial seed term used in each configuration (seed terms might bring in more false positives).

The *Dataset* entities mslr-web10 (a benchmark collection for learning to rank method) and ace2004 (ACE 2004 Multilingual Training Corpus); and *Method* entities such as *TimedTextTank* and *StatSnowball* are a sample of extracted entities. More examples can be found in the companion website. Some examples of incorrect detected entities are due to ambiguous nature of the sentence. Consider the following sentence: *"The implementation of scikitlearn toolkit was adopted for these methods"*, since it is similar to a sentence that contains a method entity, the entity *scikitlearn* was detected as a method although its a software library. In another sentence: *"The Research Support Libraries Programme (RSLP) Collection Description Project developed a model."*, RSPL (a project) was detected as a dataset due to its surrounding words (e.g. *collection, libraries*).

Comparison with State-of-the-Art. We compared our method with: (1) the BootStrapping (BS) based concept extraction approach [27], a commonly used state-of-the-art technique in scientific literature; the experiments where executed with the code and the parameters (k, n, t) to (2000, 200, 2) provided in [27], and with 100 seeds. And, (2) improved and expanded Hearst Pattern (HP) [24] for automatically building or extending knowledge bases extracting type-instance relations e.g., X such as Y as in "we used datasets such as twitter". Intuitively, the performance of BS decreases with less number of seed terms. For the HP we kept type-instance pairs related to dataset or method (i.e. the context words in CX). Experiments on our evaluation dataset shown that TSE-NER achieved better performance in terms of precision/recall/fscore for the *dataset* entity type (0.77/0.30/0.43) compared to *BS* (0.08/0.13/0.10) and *HP* (0.92/0.15/0.27) as well as for the *method* entity (*TSE-NER*: 0.68/0.15/0.25, *BS*: 0.11/0.32/0.16, *HP*: 0.64/0.04/0.07). The high precision and low recall in *HP* is explained by the limited set of *HP* patterns. We infer that different expansion strategies augment the performance of our technique compared to the *BS* which just relies on features such as unigrams, bigrams, closest verb, etc. Finally we also evaluated the performance of traditional supervised annotation. The supervised approach can achieve precision/recall/f-score of 0.82/0.35/0.49 for *dataset* entity type and 0.70/0.17/0.28 for *method* entity type using training data from 100 seeds.

Biomedical Domain. To test the performance of TSE-NER on another domain, we processed 4,525 biomedical publications from 10 journals focusing on the *Protein* entity type. The seed terms were selected from the protein ontology.[11] We excluded test terms appearing in the Craft corpus [2] (a manually annotated corpus containing 67 full-text biomedical journals) and kept only those with references in the publications (see companion site). We randomly initialized $T \subseteq T_{seed}$ with $|T| = 5, 25, 100$ and employed the SE strategy and a simple WS filtering. The evaluation cycle has been executed 10 times for each size of T, and results are averaged. TSE-NER can achieve precision/recall/f-score of 0.57/0.08/0.14 using 5 seeds, 0.40/0.28/0.32 using 25 seeds, and 0.38/0.46/0.41 with 100 seeds. The latter results are comparable to extensive dictionary-based systems [28] (0.44/0.43/0.43) [5] (0.57/0.57/0.57) where existing ontologies in the biomedical domain are used for matching *Protein* entities of the text.

4.3   Discussion

The design goal of the TSE-NER approach was minimizing the training costs in scenarios where the targeted entity types are rare, and little to no resources (for manual annotations) are available. In these cases, relying on dictionaries or knowledge-bases is not feasible, and common techniques like supervised learning cannot be applied. We believe to have successfully reached that goal, as we could show that even with small seed lists T_{seed} with little as 5 or 25 terms, high-precision NERs could be trained.

Nonetheless, this ease-of-training comes at a price: recall values are low, and are unlikely to be able to compete with known much more elaborately trained NERs for popular types. However, by selecting different configurations for filtering and expansion, recall can be moderately improved at the cost of precision.

---

[11]   http://obofoundry.org/ontology/pr.html.

Also, the effectiveness of such changes of configurations seems to slightly differ between the *Dataset* and *Method* entity types. As a result, we cannot identify one clear best configuration as TSE-NER seems to benefit from some entity type-specific tuning. However, this also provides some flexibility to tune with respect to different quality and application requirements.

Furthermore, some of our underlying assumptions, heuristics and implementation choices, are designed as a simplistic prove-of-concepts, and deserve further discussion and refinement. As an example, consider WS WordNet filtering: we assumed domain-specific named entities would not be part of common English language. While this is true for many relevant domain-specific entities, several datasets (for instance) do indeed carry common names like the `census dataset`. For a production system, more complex implementations and tailored crafting is necessary for reaching better performance values. Another restriction is related to the core heuristics found in the term and sentence expansion, where we assume that similar types of entities occur in similar contexts – which is not necessarily always the case.

Threats To Validity. Our evaluation has been performed on an extensive document corpus, covering two distinctively different domains. However, we focused only on a limited set of entity types. The hypothesis described in Sect. 3 hold for *Datasets*, *Methods*, and *Proteins*, but further experiments are needed for other entity types in the same domains (e.g. *Software*) or in other domains. Despite the good performance achieved, it could already be noted that even between those three types, no single TSE-NER configuration is clearly the best. In order to obtain a complete understanding of the full capabilities, limitations, and trade-offs of our approach, more studies addressing additional domains and entity types are necessary.

5   Conclusion

We presented a novel approach for the extraction of domain-specific long-tail entities from scientific publications. A limiting factor in this scenario is the lack of resources and/or available explicit knowledge to allow for established NER training techniques. We explored techniques able to limit the reliance on human supervision, resulting in an iterative approach that requires only a small set of seed terms of the targeted type. Our core contributions, in addition to the overall approach, are a set of expansion strategies exploiting semantic relatedness between terms to increase the size and labelling quality of the generated training dataset, as well as several filtering techniques to control the noise.

In our evaluation, we could show that we can reach a precision of up to 0.91, or a recall of up to 0.41 – a good result considering the very cheap training costs. Furthermore, we could show that recall can be traded for more precision to a moderate extend by changing the configuration of our NER training process.

For future work, additional evaluation addressing more domains and entity types is of importance to better understand the range of applicability of our approach. Also, many of our currently still simplistic heuristics and implementation choices can benefit from (domain-specific) improvement and optimization.

References

1. Agerri, R., Rigau, G.: Robust multilingual named entity recognition with shallow semi-supervised features. Artif. Intell. **238**, 63–82 (2016)
2. Bada, M., et al.: Concept annotation in the craft corpus. BMC bioinf. **13**(1), 161 (2012)
3. Brambilla, M., Ceri, S., Della Valle, E., Volonterio, R., Acero Salazar, F.X.: Extracting emerging knowledge from social media. In: International Conference on World Wide Web, pp. 795–804 (2017)
4. Derczynski, L., Nichols, E., van Erp, M., Limsopatham, N.: Results of the WNUT2017 shared task on novel and emerging entity recognition. In: Proceedings of the 3rd Workshop on Noisy User-Generated Text, pp. 140–147 (2017)
5. Funk, C., et al.: Large-scale biomedical concept recognition: an evaluation of current automatic annotators and their parameters. BMC bioinf. **15**(1), 59 (2014)
6. García-Pablos, A., Cuadros, M., Rigau, G.: W2VLDA: almost unsupervised system for aspect based sentiment analysis. Expert Syst. Appl. **91**, 127–137 (2018)
7. Goldberg, S., Wang, D.Z., Grant, C.: A probabilistically integrated system for crowd-assisted text labeling and extraction. J. Data Inf. Qual. (JDIQ) **8**(2), 10 (2017)
8. Griffiths, T.L., Steyvers, M.: Finding scientific topics. Proc. Nat. Acad. Sci. **101**(suppl 1), 5228–5235 (2004)
9. Habibi, M., Weber, L., Neves, M., Wiegandt, D.L., Leser, U.: Deep learning with word embeddings improves biomedical named entity recognition. Bioinformatics **33**(14), i37–i48 (2017)
10. Harris, Z.: Distributional structure. Word **10**, 146–162 (1954)
11. Kejriwal, M., Szekely, P.: Information extraction in illicit web domains. In: International Conference on World Wide Web, pp. 997–1006 (2017)
12. Lafferty, J., McCallum, A., Pereira, F.C.: Conditional random fields: probabilistic models for segmenting and labeling sequence data. In: International Conference on Machine Learning, vol. 951, pp. 282–289 (2001)
13. Le, Q., Mikolov, T.: Distributed representations of sentences and documents. In: International Conference on Machine Learning (ICML-14), pp. 1188–1196 (2014)
14. Lofi, C.: Measuring semantic similarity and relatedness with distributional and knowledge-based approaches. Inf. Media Tech. **10**(3), 493–501 (2015)
15. Lopez, P.: GROBID: combining automatic bibliographic data recognition and term extraction for scholarship publications. In: Agosti, M., Borbinha, J., Kapidakis, S., Papatheodorou, C., Tsakonas, G. (eds.) ECDL 2009. LNCS, vol. 5714, pp. 473–474. Springer, Heidelberg (2009). https://doi.org/10.1007/978-3-642-04346-8_62
16. Mesbah, S., Fragkeskos, K., Lofi, C., Bozzon, A., Houben, G.-J.: Semantic annotation of data processing pipelines in scientific publications. In: Blomqvist, E., Maynard, D., Gangemi, A., Hoekstra, R., Hitzler, P., Hartig, O. (eds.) ESWC 2017. LNCS, vol. 10249, pp. 321–336. Springer, Cham (2017). https://doi.org/10.1007/978-3-319-58068-5_20
17. Mesbah, S., Lofi, C., Bozzon, A., Houben, G.-J.: SmartPub: a platform for long-tail entity extraction from scientific publications. In: The Web Conference (2018)
18. Mesbah, S., Lofi, C., Bozzon, A., Houben, G.-J.: TSE-NER companion page (2018). https://sites.google.com/view/iswc2018/
19. Mikolov, T., Sutskever, I., Chen, K., Corrado, G.S., Dean, J.: Distributed representations of words and phrases and their compositionality. In: Advances in Neural Information Processing Systems, pp. 3111–3119 (2013)

20. Osborne, F., de Ribaupierre, H., Motta, E.: TechMiner: extracting technologies from academic publications. In: Blomqvist, E., Ciancarini, P., Poggi, F., Vitali, F. (eds.) EKAW 2016. LNCS (LNAI), vol. 10024, pp. 463–479. Springer, Cham (2016). https://doi.org/10.1007/978-3-319-49004-5_30
21. Qu, L., Ferraro, G., Zhou, L., Hou, W., Baldwin, T.: Named entity recognition for novel types by transfer learning. In: EMNLP (2016)
22. Reinanda, R., Meij, E., de Rijke, M.: Document filtering for long-tail entities. In: Proceedings of the 25th ACM International on Conference on Information and Knowledge Management, pp. 771–780. ACM (2016)
23. Sateli, B., Witte, R.: What's in this paper?: Combining rhetorical entities with linked open data for semantic literature querying. In: International Conference on World Wide Web, pp. 1023–1028 (2015)
24. Seitner, J., et al.: A large database of hypernymy relations extracted from the web. In: LREC (2016)
25. Shubankar, K., Singh, A., Pudi, V.: A frequent keyword-set based algorithm for topic modeling and clustering of research papers. In: 2011 3rd Conference on Data Mining and Optimization (DMO), pp. 96–102. IEEE (2011)
26. Siddiqui, T., Ren, X., Parameswaran, A., Han, J.: FacetGist: collective extraction of document facets in large technical corpora. In: International Conference on Information and Knowledge Management, pp. 871–880. ACM (2016)
27. Tsai, C.-T., Kundu, G., Roth, D.: Concept-based analysis of scientific literature. In: International Conference on Information Knowledge Management. ACM (2013)
28. Tseytlin, E., Mitchell, K., Legowski, E., Corrigan, J., Chavan, G., Jacobson, R.S.: Noble-flexible concept recognition for large-scale biomedical natural language processing. BMC bioinf. **17**(1), 32 (2016)
29. Tuarob, S., Bhatia, S., Mitra, P., Giles, C.L.: Algorithmseer: a system for extracting and searching for algorithms in scholarly big data. IEEE Trans. Big Data **2**(1), 3–17 (2016)

An Ontology-Driven Probabilistic Soft Logic Approach to Improve NLP Entity Annotations

Marco Rospocher[✉]

Fondazione Bruno Kessler – IRST, Via Sommarive 18, 38123 Trento, Italy
`rospocher@fbk.eu`

Abstract. Many approaches for Knowledge Extraction and Ontology Population rely on well-known Natural Language Processing (NLP) tasks, such as Named Entity Recognition and Classification (NERC) and Entity Linking (EL), to identify and semantically characterize the entities mentioned in natural language text. Despite being intrinsically related, the analyses performed by these tasks differ, and combining their output may result in NLP annotations that are implausible or even conflicting considering common world knowledge about entities. In this paper we present a Probabilistic Soft Logic (PSL) model that leverages ontological entity classes to relate NLP annotations from different tasks insisting on the same entity mentions. The intuition behind the model is that an annotation likely implies some ontological classes on the entity identified by the mention, and annotations from different tasks on the same mention have to share more or less the same implied entity classes. In a setting with various NLP tools returning multiple, confidence-weighted, candidate annotations on a single mention, the model can be operationally applied to compare the different annotation combinations, and to possibly revise the tools' best annotation choice. We experimented applying the model with the candidate annotations produced by two state-of-the-art tools for NERC and EL, on three different datasets. The results show that the joint "a posteriori" annotation revision suggested by our PSL model consistently improves the original scores of the two tools.

1 Introduction

The problem of identifying and semantically characterizing the entities mentioned in a natural language text has been extensively investigated over the years. Several Natural Language Processing (NLP) tasks have been defined and investigated. Some of them, such as Named Entity Recognition and Classification (NERC) and Entity Linking (EL), directly tackle the problem of recognizing the entities in a text, characterizing them according to some predefined categories (NERC) or disambiguating them with respect to a reference Knowledge Base (EL). Other tasks, though conducting different analyses than explicitly identifying entities, may also contribute to their characterization: an example is Semantic

© Springer Nature Switzerland AG 2018
D. Vrandečić et al. (Eds.): ISWC 2018, LNCS 11136, pp. 144–161, 2018.
https://doi.org/10.1007/978-3-030-00671-6_9

Role Labeling (SRL), the task of identifying the role (e.g., seller, buyer, goods) of words, and thus also entities, in a sentence.

Several tools have been proposed to effectively perform these tasks. However, despite the good performances on the single tasks, when combining them, as for instance in Knowledge Extraction frameworks (e.g., NewsReader [1], PIKES [2]), the output of these tools may result in unlikely or even contradictory information. Consider for instance the sentence "Lincoln is based in Michigan.". Here, the entity mention"Lincoln" refers to the company "Lincoln Motor Company".[1] However, using two state-of-the-art NLP tools, one for NERC (Stanford NER[2]) and one for EL (DBpedia Spotlight[3]), the first correctly identifies "Lincoln" as an organization, while the second wrongly links it to the DBpedia entity corresponding to"Abraham Lincoln". As another example, on the sentence "San Jose is one of the strongest hockey team.", the NERC tool wrongly identifies the mention"San Jose" as a location, while the EL one correctly links it to the entity "San Jose Sharks".[4]

In this paper we present **PSL4EA**, a novel approach based on Probabilistic Soft Logic (PSL) that, leveraging ontological background knowledge, enables relating the entity annotations produced by different NLP tools on the same entity mentions, and to assess their coherence. In a nutshell, given the mention of an entity in a text, the proposed PSL model enables:

1. to express the ontological entity classes of the background knowledge likely implied by the involved annotations; and,
2. to assess the coherence of the annotations, as the extent to which they share the same implied ontological entity classes.

If available, information on the confidence of the tools on the provided annotations can be included in the model, and it is taken in consideration when assessing the coherence of the annotations. As a consequence, if the considered tools provide multiple *candidate* annotations— i.e., alternative annotations on the same mention, weighted with a confidence score—the model can be applied to select the combination of annotations (one for each tool) that maximizes the annotation coherence in light of their confidences, possibly overruling the best candidate choices of the tools.

We present the creation of the model for a concrete scenario involving NERC and EL annotations, leveraging YAGO [3] as background ontological knowledge. To assess the effectiveness of the approach, we applied the model on the candidate annotations produced by two state-of-the-art tools for NERC (Stanford NER [4]) and EL (DBpedia Spotlight [5]), on three reference evaluation datasets (AIDA CoNLL-YAGO [6], MEANTIME [7], TAC-KBP [8]), showing experimentally that the joint annotation revision suggested by the model consistently

---

[1] https://en.wikipedia.org/wiki/Lincoln_Motor_Company (last accessed on April 1, 2018).

[2] http://nlp.stanford.edu:8080/corenlp/ (last accessed on April 1, 2018).

[3] http://demo.dbpedia-spotlight.org/ (last accessed on April 1, 2018).

[4] https://en.wikipedia.org/wiki/San_Jose_Sharks (last accessed on April 1, 2018).

improves the scores of the considered tools. We also discuss how to extend the model to (entity) annotations beyond NERC and EL.

While PSL was previously applied [9] for Knowledge Graph Identification (i.e., deriving a knowledge graph from triples automatically extracted from text), to the best of our knowledge this is the first work exploiting this powerful framework, with ontological knowledge, to assess the coherence and to improve NLP entity annotations. Differently from other approaches that have investigated jointly trained NERC and EL models (e.g., [10,11]), **PSL4EA** works "a posteriori" on the annotations for the considered tasks, leveraging ontological knowledge. This makes the approach applicable to many existing NLP tools for entity annotation.

The paper is structured as follows. Section 2 briefly recaps the main aspects of Probabilistic Soft Logic. Section 3 presents our novel, ontology-driven PSL approach for jointly assessing the coherence and revising NLP annotations. Section 4 reports the empirical assessment of using **PSL4EA** to improve the performances of Stanford NER and DBpedia Spotlight on three reference datasets for NERC and EL. Section 5 discusses some aspects of the proposed approach, including the extension to other (entity) annotation types (e.g., Semantic Role Labeling). Section 6 compares with relevant related works, while Sect. 7 concludes.

2   Background on Probabilistic Soft Logic

Probabilistic Soft Logic (PSL) [12] is a powerful, general-purpose probabilistic programming language that enables users to specify rich probabilistic models over continuous variables. It is a statistical relational learning framework that uses first-order logic to compactly define Markov networks, and comes with methods for performing efficient probabilistic inference for the resulting models. Differently from other related works, variables in PSL are continuous in the range [0, 1] rather than binary.

A PSL program consists of a PSL model and some data. A PSL model is composed of a set of weighted if-then, first-order logic rules, such as:

$$1.2 : \mathsf{WorksFor}(b, c) \,\&\, \mathsf{BossOf}(b, e) \rightarrow \mathsf{WorksFor}(e, c) \tag{1}$$

stating that employees are likely to work for the same company as their boss. Here: 1.2 is the *weight* of the rule; b, c, and e are universally-quantified *variables*; WorksFor and BossOf are *predicates*; $\mathsf{WorksFor}(b, c)$ is an *atom*; the part on the left of the arrow is called *body*, while the part on the right is named *head*. The *grounding* of a rule is the substitution of variables in the rule's atoms with constants (e.g., the ground atom $\mathsf{WorksFor}(B, C)$ results by assigning constants B and C to variables b and c), and ground atoms take a *soft-truth value* in the range [0, 1].

To compute soft-truth values for logical formulas, PSL adopts *Lukasiewicz t-norm* and *co-norm* to provide a relaxation of the logical conjunction ($\wedge$), disjunction ($\vee$) and negation($\neg$). Let I (*interpretation*) be an assignment of soft-truth values to ground atoms, and let a_1 and a_2 be two ground atoms, we have:

$$I(a_1) \wedge I(a_2) = max\{I(a_1) + I(a_2) - 1, 0\}$$
$$I(a_1) \vee I(a_2) = min\{I(a_1) + I(a_2), 1\} \tag{2}$$
$$\neg I(a_1) = 1 - I(a_1)$$

Given a rule r, with body r_b and head r_h, r is said to be *satisfied* if and only if $I(r_b) \leq I(r_h)$. For instance, with $I(\mathsf{WorksFor}(B,C)) = 0.6$, $I(\mathsf{BossOf}(B,E)) = 0.6$ and $I(WorksFor(E,C)) = 0.5$, rule (1) is satisfied. Otherwise, PSL defines a *distance to satisfaction* $d(r) = max\{0, I(r_b) - I(r_h)\}$, capturing how far a rule is from being satisfied. For instance, with $I(\mathsf{WorksFor}(B,C)) = 0.8$, $I(\mathsf{BossOf}(B,E)) = 0.9$ and $I(\mathsf{WorksFor}(E,C)) = 0.3$, rule (1) has a distance to satisfaction equal to 0.4.

By leveraging the distance to satisfaction, PSL defines a *probability distribution*

$$f(I) = \frac{1}{Z} \exp\left[-\sum_{r \in R} w_r d(r)^p \right] \tag{3}$$

over interpretations, where Z is a normalization constant, w_r is the weight of rule r, R is the set of all rules, and $p \in \{1, 2\}$ identifies a linear or quadratic loss function.

Different inference tasks can be investigated on a PSL program. One relevant for this paper is Most Probable Explanation (MPE) inference and corresponds to finding the overall interpretation with the maximum probability (i.e., the most likely soft-truth values of unknown ground atoms) given a set of known ground atoms. That is, the interpretation that minimizes the distance to satisfaction by trying to satisfy all rules as much as possible.

3   A PSL Model for NERC and EL

In this section, we outline **PSL4EA** (*PSL for Entity Annotations*), the PSL model we propose to jointly assess the coherence, and possibly revise, the entity annotations produced for some NLP tasks. We present the approach focusing on the two typical NLP tasks for entity annotation,[5] namely:

- **Named Entity Recognition and Classification (NERC)**: the task of labeling mentions in a text that refer to named things such as persons, organizations, etc., and choosing their type according to some predefined categories (e.g., PER, ORG);
- **Entity Linking (EL)**: the task of aligning an entity mention in a text to its corresponding entity in a Knowledge Base (e.g., YAGO [3], DBpedia [13]).

The approach is based on the assumption that, given the mention of a named entity in a text, the entity can be typed with all its ontological classes[6] defined in a given Knowledge Base K, our ontological background knowledge.

---

[5] The extension to other types of entity annotations is discussed later in Sect. 5.

[6] Typically, an entity is typed with many ontological classes, cf. rdf:type assertions from YAGO on http://dbpedia.org/page/Lincoln_Motor_Company (last accessed on April 1, 2018).

We discuss the general case where we have multiple alternative annotations (*candidates*) for each task on the same mention. That is, given a mention M, and assuming to have n_N NERC and n_E EL candidates on M, we indicate with $A_1^N, \ldots, A_{n_N}^N$ and $A_1^E, \ldots, A_{n_E}^E$ the NERC and EL candidates, while $w(M, A_j^i)$ indicates the confidence score assigned to annotation A_j^i on mention M.

The PSL model comprises two parts: the first one exploiting the relation between NLP annotations and ontological classes from the background knowledge; and, the second one capturing the coherence of the NLP annotations via these ontological classes.

3.1 Classes Implied by NLP Annotations

The intuition behind this part of the model is that given an annotation for an entity mention, if this annotation is compatible with some ontological classes of the background knowledge, then the ontological classes characterizing the entity should be among them.

Given a mention M and a NERC annotation A_i^N, we define the rule:

$$w(M, A_i^N) : \mathsf{Ann}_N(M, A_i^N) \,\&\, \mathsf{ImpCl}_N(A_i^N, c) \to \mathsf{ClAnn}_N(M, A_i^N, c) \qquad (4)$$

where:

- $\mathsf{Ann}_N(x, y)$ relates a mention x to a NERC annotation y. The grounding of the predicate has value 1 if the mention is annotated with that NERC type, 0 otherwise;
- $\mathsf{ImpCl}_N(x, y)$ captures to which extent seeing a certain NERC annotation x implies that the entity is typed with the ontological class y. This quantity can be learned from gold data (see Sect. 3.1);
- $\mathsf{ClAnn}_N(x, y, z)$ captures that mention x corresponds to an entity that is instance of class z due to annotation y.

For the first two predicates, the soft-truth value of the atoms is known (input data), while the value for the ground atoms of ClAnn_N has to be determined by the model. Furthermore, the rule is partly grounded, i.e., the only variable is the ontological class c. Given a mention M on which we have n_N NERC candidates, we have n_N such rules, one for each candidate, weighted according to the corresponding confidence score.

Similarly, given a mention M and an EL annotation A_i^E, we define the rule:

$$w(M, A_i^E) : \mathsf{Ann}_E(M, A_i^E) \,\&\, \mathsf{ImpCl}_E(A_i^E, c) \to \mathsf{ClAnn}_E(M, A_i^E, c) \qquad (5)$$

where $\mathsf{Ann}_E(x, y)$, $\mathsf{ImpCl}_E(x, y)$, $\mathsf{ClAnn}_E(x, y, z)$ are defined analogously to the NERC case. Again, note that we have n_E such rules.

Determining ImpCl_N and ImpCl_E. $\mathsf{ImpCl}_N(x, y)$ captures the "likelihood" that a certain NERC annotation implies an ontological class. The higher the soft-truth value for a given NERC type x and ontological class y, the higher are the chances that if an entity mention is NERC annotated with x, than the entity is an instance of class y. To determine $\mathsf{ImpCl}_N(x,y)$ we assume the availability of a gold standard corpus G where each entity mention is annotated with both (i) its NERC type and (ii) all its ontological classes from the background knowledge, or, alternatively, an annotation deterministically alignable to them (e.g., an EL annotation, with the entity typed according to the ontological classes). We then use G as data for another PSL program, with rules:

$$1.0 : \mathsf{Gold}_N(m,t) \,\&\, \mathsf{ImpCl}_N(t,c) \rightarrow \mathsf{Gold}_C(m,c)$$
$$1.0 : \mathsf{Gold}_N(m,t) \,\&\, \neg\mathsf{ImpCl}_N(t,c) \rightarrow \neg\mathsf{Gold}_C(m,c) \tag{6}$$

where $\mathsf{Gold}_N(m,t)$ is 1 if mention m is annotated with t in G, and 0 otherwise, while $\mathsf{Gold}_C(m,c)$ is 1 if c is one of the ontological classes of the entity denoted by the mention m, and 0 otherwise. That is, the soft-truth values of the ground atoms of Gold_C and Gold_N are known, while the value for the ground atoms of ImpCl_N has to be determined by this specific model. Note that two rules are used in (6): they respectively account for the cases where mentions, NERC annotated with a type t, are annotated (i) also with class c, and (ii) not with class c, so to properly capture the "likelihood" that a NERC type implies some classes but not others.

The model has to estimate ImpCl_N for all possible NERC types and ontological classes. While all possible NERC types are typically occurring in G, some very specific class c of the background knowledge K may be observed few times (or even not at all) in it. However, especially for coarse-grain NERC types such as the classical 4-type (PER, ORG, LOC, MISC) model, there is little benefit in considering rarely observed, very specific ontological classes. We thus restrict our attention to popular classes, those observed at least $\bar{n}$ times (an hyperparameter of our approach) in G, typically general classes in the class taxonomy, filtering out any remaining class in K.

For EL, if the entities in the target EL Knowledge Base and the background knowledge K are aligned,[7] the soft-truth value of the ImpCl_E atoms can be deterministically obtained via such alignment: $\mathsf{ImpCl}_E(x,y)$ has soft-truth value 1 if y is one of the ontological classes of the entity z corresponding to x in the alignment, 0 otherwise.[8]

3.2 Annotation Coherence via Classes

The second part of the PSL model puts in relation the predicates ClAnn_N and ClAnn_E via ontological classes:

$$w_1 : \mathsf{ClAnn}_N(m,t,c)\,\&\,\mathsf{ClAnn}_E(m,e,c) \rightarrow \mathsf{Ann}_{PSL}(m,t,e)$$
$$w_2 : \mathsf{ClAnn}_N(m,t,c)\,\&\,\neg\mathsf{ClAnn}_E(m,e,c) \rightarrow \neg\mathsf{Ann}_{PSL}(m,t,e) \tag{7}$$
$$w_3 : \neg\mathsf{ClAnn}_N(m,t,c)\,\&\,\mathsf{ClAnn}_E(m,e,c) \rightarrow \neg\mathsf{Ann}_{PSL}(m,t,e)$$

---

[7] This clearly includes the special case where the EL Knowledge Base is actually K.

[8] This assumes that K contains complete information about entity classes (closed-world assumption), which usually holds for the most general classes in the class taxonomy.

where Ann_{PSL} is the predicate we use to estimate the coherence of a couple of NERC and EL candidate annotations on a given mention. The intuition here is that a NERC and an EL annotation implying the same classes[9] from the ontological background knowledge are likely to be coherent, and thus the soft-truth value of the corresponding Ann_{PSL} atom should be higher than when the annotations imply different classes. Note that these rules are not grounded. Rule weights w_1, w_2, w_3 are hyperparameters of our approach: the higher their values, the stronger the satisfaction of those rules — and hence coherence enforcement — is accounted for during inference.

Note that the two parts of the model have one important distinctive feature: for the actual construction of the model, the first part is *dynamic*, in the sense that the (partially-grounded) rules are instantiated based on the actual annotations and confidence scores available, while the second part is *static*, with rules involving only variables (and no constants) and thus defined once for all.

$0.9 : \text{Ann}_N(\text{L}, \text{ORG}) \ \& \ \text{ImpCl}_N(\text{ORG}, c) \rightarrow \text{ClAnn}_N(\text{L}, \text{ORG}, c)$

$0.1 : \text{Ann}_N(\text{L}, \text{PER}) \ \& \ \text{ImpCl}_N(\text{PER}, c) \rightarrow \text{ClAnn}_N(\text{L}, \text{PER}, c)$

$0.5 : \text{Ann}_E(\text{L}, \text{A. Lincoln}) \ \& \ \text{ImpCl}_E(\text{A. Lincoln}, c) \rightarrow \text{ClAnn}_E(\text{L}, \text{A. Lincoln}, c)$

$0.3 : \text{Ann}_E(\text{L}, \text{Lincoln MC}) \ \& \ \text{ImpCl}_E(\text{Lincoln MC}, c) \rightarrow \text{ClAnn}_E(\text{L}, \text{Lincoln MC}, c)$

$0.2 : \text{Ann}_E(\text{L}, \text{Lincoln UK}) \ \& \ \text{ImpCl}_E(\text{Lincoln UK}, c) \rightarrow \text{ClAnn}_E(\text{L}, \text{Lincoln UK}, c)$

$0.9 : \text{Ann}_N(\text{M}, \text{LOC}) \ \& \ \text{ImpCl}_N(\text{LOC}, c) \rightarrow \text{ClAnn}_N(\text{M}, \text{LOC}, c)$

$0.05 : \text{Ann}_N(\text{M}, \text{PER}) \ \& \ \text{ImpCl}_N(\text{PER}, c) \rightarrow \text{ClAnn}_N(\text{M}, \text{PER}, c)$

$0.05 : \text{Ann}_N(\text{M}, \text{ORG}) \ \& \ \text{ImpCl}_N(\text{ORG}, c) \rightarrow \text{ClAnn}_N(\text{M}, \text{ORG}, c)$

$0.9 : \text{Ann}_E(\text{M}, \text{Michigan}) \ \& \ \text{ImpCl}_E(\text{Michigan}, c) \rightarrow \text{ClAnn}_E(\text{M}, \text{Michigan}, c)$

$0.1 : \text{Ann}_E(\text{M}, \text{U. of Michigan}) \ \& \ \text{ImpCl}_E(\text{U. of Michigan}, c) \rightarrow \text{ClAnn}_E(\text{M}, \text{U. of Michigan}, c)$

$10 : \text{ClAnn}_N(m, t, c) \ \& \ \text{ClAnn}_E(m, e, c) \rightarrow \text{Ann}_{PSL}(m, t, e)$

$10 : \text{ClAnn}_N(m, t, c) \ \& \ \neg\text{ClAnn}_E(m, e, c) \rightarrow \neg\text{Ann}_{PSL}(m, t, e)$

$10 : \neg\text{ClAnn}_N(m, t, c) \ \& \ \text{ClAnn}_E(m, e, c) \rightarrow \neg\text{Ann}_{PSL}(m, t, e)$

Fig. 1. Instantiation of the PSL model for the sentence "Lincoln is based in Michigan."

Figure 1 shows an example of instantiation of the model on the sentence "Lincoln is based in Michigan.", with two mentions $m_1 = $ Lincoln and $m_2 = $ Michigan (shortened for compactness to L and M, respectively), and assuming to have two NERC (ORG [0.9], PER [0.1]) and three EL (A. Lincoln [0.5], Lincoln MC [0.3], Lincoln UK [0.2]) confidence-weighted candidates on the first, and three NERC (LOC [0.9], PER [0.05], ORG [0.05]) and two EL (Michigan [0.9], U. of Michigan [0.1]) confidence-weighted candidates on the second.

---

[9] Note that, for a given grounding of m, t and e, the value of Ann_{PSL} results from the contribution of several classes c.

The PSL model is further complemented with negative priors, i.e., additional rules stating that by default all open ground atoms (i.e., whose value has to be determined by the model) of investigated predicates ($ClAnn_N$, $ClAnn_E$, Ann_{PSL}) have 0 soft-truth value.

By running MPE inference on the model, we can compute the soft-truth value of all the ground atoms of Ann_{PSL}. Intuitively, the higher this value, the more likely a NERC annotation and an EL annotation are coherent on the given mention, with the combination of candidates scoring the highest value being the best NERC and EL annotation for the model, in light of their original confidence scores and the ontological knowledge.

By comparing the soft-truth value of the resulting Ann_{PSL} ground atoms with a threshold value θ (an hyperparameter of our approach), we can decide to which extent to rely on the prediction of the model, especially when revising (and possibly overruling) the best-choice candidate annotations proposed by some NERC and EL tools.

4 Evaluation

We conduct an evaluation, in a scenario where both NERC and EL analyses are run, to show that our PSL approach, leveraging some ontological background knowledge and applied "a posteriori" on the confidence-weighted candidate annotations returned by a NERC tool and a EL tool, suggests better annotations than the highest score ones independently returned by the given tools. The data used by the PSL model (including the soft-truth values for $ImpCl_N$ and $ImpCl_E$ ground atoms), the evaluation package (excluding copyrighted dataset material), and additional result tables are available on the **PSL4EA** web-folder.[10]

4.1 Background Knowledge and Tools

As background knowledge we use YAGO [3]. We materialize, applying RDF_{pro} [14], all the inferable classes for an entity based on the YAGO TBox (e.g., subclass axioms), obtaining class information for 6,016,695 entities taken from a taxonomy of 568,255 classes.

To produce the NERC and EL annotations, we exploit two state-of-the-art tools:

- **Stanford NER** [4]: a reference tool for NERC. We use Stanford NER with the traditional CoNLL 2003 model consisting of 4 NERC types: Location (LOC), Person (PER), Organization (ORG), and Miscellaneous (MISC). By default, Stanford NER returns the best NERC labeling of a sentence, but it can be instructed to provide many alternative weighted NERC labelings of a sentence, from which it is possible to derive NERC candidates (and their confidences) for a mention;

---

[10] http://pikes.fbk.eu/psl4ea.html.

– **DBpedia Spotlight** [5]: a reference tool for EL that uses DBpedia [13] as target knowledge base. Via its *candidates* service, DBpedia Spotlight can be instructed to return ten EL candidates (and their confidences) for a given mention.

4.2 Datasets

To verify the capability of our approach to generalize over different annotated data, we use three distinct datasets in our evaluation. They consist of textual documents together with gold-standard annotations, both for NERC and EL:[11]

– **AIDA CoNLL-YAGO** [6]: it consists of 1,393 English news articles from Reuters, hand-annotated with named entity types (PER, ORG, LOC, MISC) and YAGO2 entities (and Wikipedia page URLs). It is organized in three parts: **eng.train** (946 docs), **eng.testa** (216 docs), **eng.testb** (231 docs);
– **MEANTIME** [7]: it consists of 480 news articles from Wikinews, in four languages. In our evaluation, we only use all the 120 articles of the English section. The dataset includes manual annotations (limited to the first 5 sentences of the articles) for named entity types (only PER, ORG, LOC) and DBpedia entities;
– **TAC-KBP** [8]: it consists of 2,231 English documents (news article, newsgroup and blog posts, forum discussions).
For each document, it is known that all the mentions of one or a few *query* entities can be linked to a certain Wikipedia page and to a specific NERC type (only PER, ORG, LOC), thus giving rise to a (partially) annotated gold standard for NERC and EL.

4.3 Research Question and Evaluation Measures

We address the following research question:

Does the ontology-driven PSL4EA a posteriori joint revision of Stanford NER and DBpedia Spotlight annotations improve their NERC and EL performances?

In investigating this research question, we remark that by construction the PSL model relies on the mentions detected by the NLP tools used, so the model may revise the NERC types and/or the EL entities proposed by the tools, but does not alter other aspects such as the mention span (i.e., the textual tokens that constitute the mention). As such, meaningful measures for our evaluation are the following ones, typically adopted in NERC and EL evaluation campaigns:

---

[11] We choose these datasets, among many available ones for NERC and for EL as they have both NERC and EL annotations that can be used to evaluate the improvement on both tasks.

- **type**: a mention is counted as correct if it has the same span and NERC type as a gold annotation. It is the measure used in the CoNLL2003 NER evaluation, and corresponds to **strong_typed_mention_match** in the TAC-KBP official scorer;[12]
- **link**: a mention is counted as correct if it has the same span and EL entity as a gold annotation. It corresponds to **strong_link_match** in the TAC-KBP official scorer;
- **type+link**: an entity mention is counted as correct if it has the same span, NERC type, and EL entity as a gold annotation. It corresponds to **strong_typed_link_match** in the TAC-KBP official scorer.

For evaluating the performance on these measures, we use the standard metrics, namely precision (P), recall (R), and F_1, computed using the TAC-KBP official scorer on the predicted and gold standard annotations as follow:

- true positives (TP) = predicted annotations, in the gold standard;
- false positives (FP) = predicted annotations, not in the gold standard;
- false negatives (FN) = gold standard annotations, not predicted;
- $P = \frac{TP}{TP+FP}$, $R = \frac{TP}{TP+FN}$ and $F_1 = \frac{2 \cdot P \cdot R}{P+R}$.

4.4 Evaluation Procedure

We use AIDA **eng.train** as the gold standard G for determining ImpCl_N — Table 1 provides, for each NERC type, an overview of the YAGO classes of the top 10 soft-truth value ground atoms of ImpCl_N — while ImpCl_E is deterministically obtained directly via the DBpedia-YAGO alignment. We use AIDA **eng.testa** to optimize the **PSL4EA** model hyperparameters (cf. Sect. 3), namely $\bar{n}$ (=200),[13] w_1, w_2, w_3 (=10.0), and θ (=0.2). We adopt the quadratic loss function (cf. Eq. (3)).

All datasets are preprocessed in order to use entity URIs from the same version of DBpedia (namely, 2016-04) as the used DBpedia Spotlight version. In particular, the Wikipedia URLs in AIDA and TAC-KBP are aligned to the 2016-04 DBpedia URIs via DBpedia's 'Redirects', 'Revision URIs', and 'Wikipedia Links' datasets.

The experiment is conducted comparing the metric scores for the considered measures in two settings, without (*standard*) and with (*with PSL4EA*) the contribution of the **PSL4EA** model: in the *standard* setting we annotate the documents of the three corpora directly using the highest confidence score NERC type and EL entity proposed by Stanford NER and DBpedia spotlight; instead, in the *with PSL4EA* setting, the **PSL4EA** model picks, among all the confidence-weighted candidate annotations returned by the tools on the same

---

[12] https://github.com/wikilinks/neleval (last accessed on April 1, 2018).
[13] With $\bar{n} = 200$, the background knowledge used in the model is reduced to 214 YAGO classes.

Table 1. Top 10 YAGO classes for each NERC type according to the soft-truth value (in parentheses) of ImpCl_N ground atoms learned from AIDA eng.train.

NERC type	YAGO classes
PER	PhysicalEntity100001930 (.991), CausalAgent100007347 (.988), Object100002684 (.963), YagoLegalActorGeo (.963), Whole100003553 (.962), YagoLegalActor (.961), LivingThing100004258 (.960), Organism100004475 (.960), Person100007846 (.960), WikicatLivingPeople (.850)
ORG	YagoPermanentlyLocatedEntity (.945), Abstraction100002137 (.945), YagoLegalActorGeo (.938), YagoLegalActor (.925), Group100031264 (.924), SocialGroup107950920 (.923), Organization108008335 (.914), Association108049401 (.642), Club108227214 (.637), Unit108189659 (.340)
LOC	YagoPermanentlyLocatedEntity (.986), YagoLegalActorGeo (.967), PhysicalEntity100001930 (.909), Object100002684 (.907), YagoGeoEntity (.905), Location100027167 (.889), Region108630985 (.883), District108552138 (.866), AdministrativeDistrict108491826 (.865), Country108544813 (.524)
MISC	YagoPermanentlyLocatedEntity (.843), YagoLegalActorGeo (.679), PhysicalEntity100001930 (.614), Object100002684 (.609), YagoGeoEntity (.591), Location100027167 (.572), Region108630985 (.571), AdministrativeDistrict108491826 (.568), District108552138 (.568), Country108544813 (.549)

mention, the ⟨NERC type, EL entity⟩ combination with the highest soft-truth value for Ann_{PSL}.[14]

We remark that our approach is not a complete NER+EL solution on its own but relies on annotations provided by NERC and EL tools (e.g., Stanford NER and DBpedia Spotlight as in the considered experiment), revised "a posteriori" using ontological knowledge. Therefore, in line with the investigated research question, we focus our study on comparing the scores between the two aforementioned settings, rather than analyzing the absolute scores obtained, which inherently depend also on the performances of the tools providing the candidate annotations (i.e., changing the tools would likely results in different overall P, R, and F_1 scores).

Furthermore, as some datasets are only partially annotated (e.g., TAC-KBP), in the paper we focus the evaluation only on the mentions detected by the tools (i.e., annotated with NERC and/or EL) — which we recall are the same in both settings — that are in the gold standard, in order to better compare performances across the different datasets, and to avoid obtaining scores, namely P and F_1,

---

[14] If the highest soft-truth value on a mention is below the threshold θ, the approach falls back to the best NERC and EL candidate annotations suggested by the tools on it.

overly biased by *FP* in both settings. For completeness, scores considering all mentions returned by the tools as well as macro-averaged variants (by document, by NERC type) are provided on the web-folder.

4.5   Results and Discussion

Table 2 reports precision, recall, and F_1 (micro-averaged) for the evaluation measures on all the datasets, for both settings considered.

For all the metrics computed over the three datasets, the scores are consistently higher in the *with PSL4EA* setting than in the *standard* one, with improvements ranging from .004 to .032. Most of the improvements (24 out of 27) are statistically significant ($p < 0.05$) according the Approximate Randomization test. Similar outcomes (cf. **PSL4EA** web-folder for all the detailed data) are observed when:

- considering all mentions returned by the tools (rather than just those in the gold standard): improvements ranging from .003 to .025;
- macro-averaging by document: improvements ranging from .003 to .029;
- macro-averaging by NERC type: improvements ranging from .003 to .020.

Improvements for **type+link** (from .010 to .032), besides being all statistically significant, are always higher than the ones for the other two measures (**type** and **link**), thus confirming that the model is particularly effective in proposing, for a given mention, the correct ⟨NERC, EL⟩ annotation combination among the available candidates.

Table 2. Precision, recall, and F_1 scores for **type**, **link**, and **type+link** measures for both settings on the three datasets (number of gold standard mentions in parentheses). Score differences (*with PSL4EA − standard*) are reported, with statistical significance ones marked in bold.

		type			link			type+link		
		P	R	F_1	P	R	F_1	P	R	F_1
	standard	.943	.875	.908	.662	.652	.656	.634	.625	.630
AIDA (5616)	with **PSL4EA**	.947	.879	.912	.670	.659	.665	.646	.635	.640
	Δ	**.004**	**.004**	**.004**	**.008**	**.007**	**.009**	**.012**	**.010**	**.010**
	standard	.882	.695	.777	.703	.556	.621	.635	.502	.561
MEANTIME (792)	with **PSL4EA**	.902	.711	.795	.714	.564	.630	.667	.527	.589
	Δ	**.020**	**.016**	**.018**	.011	.008	.009	**.032**	**.025**	**.028**
	standard	.911	.652	.760	.401	.423	.412	.367	.386	.376
TAC-KBP (4969)	with **PSL4EA**	.925	.662	.772	.408	.430	.419	.384	.404	.394
	Δ	**.014**	**.010**	**.012**	**.007**	**.007**	**.007**	**.017**	**.018**	**.018**

Analyzing more in detail the results, it is worth remarking that the model used for the evaluation, while trained only on AIDA **eng.train**, performs reasonably well also on the other two datasets, as confirmed by the substantially higher scores for the *with PSL4EA* setting over the *standard* one, with statistical significant improvements in most of the cases. This may suggest that the instantiated model generalizes well over different document collections, something we plan to further confirm with additional experiments in future work.

Summing up, the results on multiple datasets show that exploiting the **PSL4EA** model to "a posteriori" revise the annotations provided by Stanford NER and DBpedia Spotlight allows to consistently improve their NERC and EL scores, and thus we can positively answer our research question.

5   Discussion

Peculiarity of the PSL4EA model with respect to other PSL applications. PSL has been applied for different structural relational learning tasks, including the distillation of a Knowledge Graph from candidate relation triples extracted from text [9]. In that work, the authors encode the confidence score of extracted relation triples as the soft-truth value of the corresponding atoms, instead of rule weights like in **PSL4EA**. We experimented also with such configuration for the NERC and EL joint annotation revision setting, achieving however worse performances than modeling confidences as rule weights.

Applicability to Other NERC and EL Tools. In the experiments discussed in Sect. 4, we applied **PSL4EA** to jointly revise the NERC and EL annotations produced by Stanford NER and DBpedia Spotlight. However, we remark that **PSL4EA** works on NERC and EL candidate *annotations*, and thus its applicability is not limited only to those specific tools. Indeed, the model used for the evaluation can be applied as-is to any couple of NERC and EL tools provided that: (i) the NERC tool annotates with the 4-type CoNLL2003 NERC categories (or its popular 3-type version omitting MISC); and, (ii) the EL tool annotates with DBpedia URIs. Clearly, the model can be adapted to other NERC categories and EL reference Knowledge Bases, revising $ImpCl_N$ and $ImpCl_E$.

Implementation and Performances. We implemented the **PSL4EA** approach used in the evaluation as a Java module[15] of PIKES [2], an open-source knowledge extraction framework exploiting several NLP analyses, including NERC (via Stanford NER) and EL (via DBpedia Spotlight). For the PSL inference, we use the open-source Java PSL software [12].[16] In details, the module (i) builds a PSL model and data dynamically for each named entity mention having both NERC and EL annotations, (ii) performs MPE inference, and (iii) saves the results in the PIKES output. Computationally, the performances of the module are roughly comparable to the annotation costs.[17]

Extension to Other Types of Entity Annotations. In Sect. 3 we presented an ontology-driven PSL model for assessing the coherence and jointly revising

---

[15] To be distributed with the next PIKES release.

[16] https://github.com/linqs/psl.

[17] Note that substantial improvements of running time performances can be achieved with further engineering and optimization, out-of-scope for the purposes of this work.

NERC and EL annotations. That model can be extended to other typologies of annotations, that may involve (named) entities. Here we briefly discuss some ideas on how these additional annotations could contribute to the model, leaving the actual development of the model (and its evaluation) to future work.

Semantic Role Labeling (SRL) is the task of finding the semantic role of each argument of each (verbal or nominal) predicate in a sentence. For instance, in the sentence "Sergio Mattarella is the president of Italy", "president" evokes a *Leadership* frame (according to FrameNet [15]), and has two arguments, "Sergio Mattarella" (with role *Leader*) and "Italy" (with role *Governed*). Clearly, role annotations may contribute to further characterize entities, and, similarly to NERC and EL, they may imply some ontological classes. For instance, a *Leader* role annotation is more likely to occur on the mention of an entity of type "Leader109623038" in YAGO than an entity of type "Airplane102691156". We can thus think to include role annotations in **PSL4EA** with rules similar to the ones for NERC and EL:

$$w(M, A_i^R) : \mathsf{Ann}_R(M, A_i^R) \,\&\, \mathsf{ImpCl}_R(A_i^R, c) \rightarrow Cl\mathsf{Ann}_R(M, A_i^R, c) \qquad (8)$$

where predicate ImpCl_R, capturing the ontological classes implied by role annotations, can be learned from data as described in Sect. 3.1.[18] However, to more precisely handle SRL annotations, the PSL model should be further extended to capture the fact that role annotations on different mentions (e.g., the *Leader* on "Sergio Mattarella" and the *Governed* on "Italy" in the example considered) but originating from the same predicate have to be related (i.e., selecting one candidate on one mention may affect the candidates on the others). Furthermore, the addition of the SRL annotations requires the extension of the rules ensuring the annotation coherence — cf. (7).

Another typology of annotation that may extend the **PSL4EA** model is entity coreference, i.e., the task of identifying that two or more mentions in a text refer to the same entity. Coreference should instruct the model to propagate the same annotations on all coreferring mentions, as suggested by the following rule for two coreferring mentions:

$$w_C(M_1, M_2) : \mathsf{Ann}_{PSL}(M_1, t, e) \,\&\, \mathsf{Coref}(M_1, M_2) \rightarrow \mathsf{Ann}_{PSL}(M_2, t, e) \qquad (9)$$

where $\mathsf{Coref}(M_1, M_2)$ and $w_C(M_1, M_2)$ capture the coreference annotation and its confidence.

6   Related Work

We briefly overview some literature works related to our contribution.

PSL Application to Knowledge Extraction and NLP. Probabilistic Soft Logic has been applied for some information extraction and NLP tasks. In [9] the

---

[18] A dataset to derive such information is presented in [16], where FrameNet frame elements (i.e., roles) are related to "compatible" WordNet synsets, which in turns can be directly mapped to YAGO classes.

authors apply PSL for Knowledge Graph Identification (KGI), that is the task of distilling a knowledge graph from the noisy output (subject-predicate-object triples) of information extractors (cf. also later in this section). The approach combines different strategies (e.g., entity classification, relational link prediction) together with constraints from existing ontologies. In [17] PSL is used to combine logical and distributional representations of natural-language meaning for the task of semantic textual similarity (STS). In [18] PSL is exploited to classify events mentioned in text leveraging event-event associations and fine-grained entity types. In [19] PSL is applied for the lexical inference problem, i.e., to guess unknown word meaning by leveraging linguistic and contextual features.

In our work PSL is applied to assess the coherence and revise entity annotations, exploiting ontological background knowledge. We are not aware of other works applying PSL to specifically improve NLP annotations.

NLP Annotation Improvement. Some previous works have tackled the problem of improving the performances of some NLP tasks by leveraging or combining related analyses, focusing mainly on NERC and EL. In some works, one NLP analysis is used to influence the performance of another NLP task, in a pipeline, one-direction fashion. For instance, in [10,20] named entities are firstly recognized (NERC) and used to influence the entity disambiguation step (EL). Joint models for multiple tasks, in particular for NERC and EL, have also been developed, applying different techniques such as re-ranking mechanisms [21], conditional random field (CRF) extensions [22], semi-Markov structured linear classifiers [23], and probabilistic graphical models [11]. In [24], a joint model implemented as a structured CRF has been proposed, where NERC and EL analyses are complemented by coreference information.

Our work differs from all these approaches under several aspects. First, our approach is not a complete joint NERC and EL solution, but it works a posteriori on produced candidate annotations. This makes our approach applicable to many existing NERC and EL approaches as-is (i.e., without re-training their models or changing their implementations) granted they provide confidence-weighted candidate annotations. Second, it does not impose a directionality on the influence between the considered tasks, like in approaches such as [10,20]. Third, our approach stands out for the central role of the ontological background knowledge, exploited as "interlingua" to assess the coherence of the annotations from different NLP tasks. This is similar to the approach adopted by **JPARK** [25], where a pure probabilistic model — derived from some conditional independence assumptions, and leveraging class sets rather than individual class contributions like in **PSL4EA** — is used to revise entity annotations.

Knowledge Graph Construction. Approaches for Knowledge Graph construction from text (e.g., Google's Knowledge Vault [26] and DeepDive [27]) have tackled the problem of determining the correctness of large sets of potentially noisy subject-predicate-object triples, obtained via information *extractors* from various types of content (e.g., documents, tables). Some of these works exploit ontological knowledge to constrain the selection of the extracted candidate triples. In NELL (Never-Ending Language Learning) [28], ontological constraints (e.g., a person cannot be a city) are used to filter the extracted triples. In other works, ontological knowledge is integrated directly in a probabilistic model, together with the confidence values of extractor candidates, such as in [29] (exploiting Markov Logic Networks) and the previously discussed PSL approach in [9]. Instead, a MAX-SAT algorithm is proposed in [30], to select high confidence triples that maximize the number of satisfied ontological constraints.

Our work differs from all these approaches and it is not directly comparable with them. To begin with, our approach works at the level of NLP annotations, rather than triples typically returned by relation extractors, and aims at improving the coherence of these annotations on a given mention, rather filtering extracted triples in order to be compliant with or to maximize the given set of ontological constraints. Furthermore, in all these approaches the relation extractors are aligned by construction with the relations and classes of the ontology used for constraining the triple selection, while in our work determining the ontological knowledge classes likely implied by the annotations is part of the problem and encoded into the PSL model.

7 Conclusions

In this paper we presented **PSL4EA**, an approach based on Probabilistic Soft Logic that, leveraging ontological background knowledge, aims at improving the joint annotation of entity mentions by NLP tools, for tasks such as NERC and EL. NLP annotations for different tasks are mapped to ontological classes of a common background knowledge, then exploited to jointly assess the annotation coherence. Given confidence-weighted candidate annotations by multiple NLP tools for different tasks on the same textual entity mention, **PSL4EA** can be operationally applied to jointly revise the best annotation choices performed by the tools, in light of the coherence of the candidate annotations via the ontological knowledge.

We developed the approach for NERC and EL, leveraging YAGO as ontological background knowledge. We experimented with the model on the NERC and EL candidate annotations provided by two state-of-the-art tools, Stanford NER and DBpedia Spotlight, on three distinct reference datasets. The results show the capability of **PSL4EA** to jointly improve their annotations, as confirmed by the higher scores on all measures and metrics when applying the model.

As discussed in the paper, our future work mainly aims at concretely extending the proposed model to other NLP annotations than NERC and EL, starting with SRL and entity coreference. Furthermore, for the NERC and EL scenario, we plan to experiment with different training sets, possibly produced by combining different datasets, in order to further improve the generality and representativeness of the model obtained using the training part of the AIDA CoNLL-YAGO dataset.

Acknowledgments. The author would like to thank Dr. Francesco Corcoglioniti for some useful suggestions and fruitful discussions while developing the idea.

References

1. Vossen, P., et al.: NewsReader: using knowledge resources in a cross-lingual reading machine to generate more knowledge from massive streams of news. Knowl.-Based Syst. **110**, 60–85 (2016). https://doi.org/10.1016/j.knosys.2016.07.013
2. Corcoglioniti, F., Rospocher, M., Aprosio, A.P.: Frame-based ontology population with PIKES. IEEE Trans. Knowl. Data Eng. **28**(12), 3261–3275 (2016)
3. Hoffart, J., Suchanek, F.M., Berberich, K., Weikum, G.: YAGO2: a spatially and temporally enhanced knowledge base from wikipedia. Artif. Intell. **194**, 28–61 (2013)
4. Finkel, J.R., Grenager, T., Manning, C.: Incorporating non-local information into information extraction systems by Gibbs sampling. In: Proceedings of ACL 2005, pp. 363–370 (2005)
5. Daiber, J., Jakob, M., Hokamp, C., Mendes, P.N.: Improving efficiency and accuracy in multilingual entity extraction. In: Proceedings of I-Semantics (2013)
6. Hoffart, J., et al.: Robust disambiguation of named entities in text. In: Proceedings of EMNLP 2011 (2011)
7. Minard, A.L., et al.: MEANTIME, the newsreader multilingual event and time corpus. In: Proceedings of LREC 2016 (2016)
8. Ji, H., Grishman, R., Dang, H.: Overview of the TAC2011 knowledge base population track. In: TAC 2011 Proceedings Papers (2011)
9. Pujara, J., Miao, H., Getoor, L., Cohen, W.: Knowledge graph identification. In: Alani, H. (ed.) ISWC 2013. LNCS, vol. 8218, pp. 542–557. Springer, Heidelberg (2013). https://doi.org/10.1007/978-3-642-41335-3_34
10. Stern, R., Sagot, B., Béchet, F.: A joint named entity recognition and entity linking system. In: Proceedings of HYBRID 2012, pp. 52–60 (2012)
11. Nguyen, D.B., Theobald, M., Weikum, G.: J-NERD: joint named entity recognition and disambiguation with rich linguistic features. TACL **4**, 215–229 (2016)
12. Bach, S.H., Broecheler, M., Huang, B., Getoor, L.: Hinge-loss Markov random fields and probabilistic soft logic. J. Mach. Learn. Res. (JMLR) **18**(109), 1–67 (2017)
13. Lehmann, J., et al.: Dbpedia - a large-scale, multilingual knowledge base extracted from Wikipedia. Semant. Web **6**(2), 167–195 (2015)
14. Corcoglioniti, F., Rospocher, M., Mostarda, M., Amadori, M.: Processing billions of RDF triples on a single machine using streaming and sorting. In: Proceedings of the 30th Annual ACM Symposium on Applied Computing. SAC 2015, pp. 368–375. ACM (2015)
15. Baker, C.F., Fillmore, C.J., Lowe, J.B.: The Berkeley FrameNet Project. In: Proceedings of ACL 1998, pp. 86–90 (1998)
16. Tonelli, S., Bryl, V., Giuliano, C., Serafini, L.: Investigating the semantics of frame elements. In: ten Teije, A. (ed.) EKAW 2012. LNCS (LNAI), vol. 7603, pp. 130–143. Springer, Heidelberg (2012). https://doi.org/10.1007/978-3-642-33876-2_13
17. Beltagy, I., Erk, K., Mooney, R.J.: Probabilistic soft logic for semantic textual similarity. In: Proceedings of the 52nd Annual Meeting of the Association for Computational Linguistics (ACL-14), Baltimore, MD, pp. 1210–1219 (2014)
18. Liu, S., Liu, K., He, S., Zhao, J.: A probabilistic soft logic based approach to exploiting latent and global information in event classification. In: AAAI, pp. 2993–2999. AAAI Press (2016)
19. Wang, W.C., Ku, L.W.: Identifying Chinese lexical inference using probabilistic soft logic. In: 2016 IEEE/ACM International Conference on Advances in Social Networks Analysis and Mining (ASONAM), pp. 737–743, August 2016

20. Plu, J., Rizzo, G., Troncy, R.: A hybrid approach for entity recognition and linking. In: Gandon, F., Cabrio, E., Stankovic, M., Zimmermann, A. (eds.) SemWebEval 2015. CCIS, vol. 548, pp. 28–39. Springer, Cham (2015). https://doi.org/10.1007/978-3-319-25518-7_3
21. Sil, A., Yates, A.: Re-ranking for joint named-entity recognition and linking. In: Proceedings of CIKM 2013, pp. 2369–2374 (2013)
22. Luo, G., Huang, X., Lin, C.Y., Nie, Z.: Joint named entity recognition and disambiguation. In: Proceedings of EMNLP 2015, pp. 879–888 (2015)
23. Leaman, R., Lu, Z.: TaggerOne: joint named entity recognition and normalization with semi-Markov models. Bioinformatics 32(18), 2839–2846 (2016)
24. Durrett, G., Klein, D.: A joint model for entity analysis: coreference, typing, and linking. TACL 2, 477–490 (2014)
25. Rospocher, M., Corcoglioniti, F.: Joint posterior revision of NLP annotations via ontological knowledge. In: Proceedings of the Twenty-Seventh International Joint Conference on Artificial Intelligence, IJCAI-18, pp. 4316–4322 (2018). https://doi.org/10.24963/ijcai.2018/600
26. Dong, X., et al.: Knowledge vault: a web-scale approach to probabilistic knowledge fusion. In: Proceedings of ACM KDD 2014, pp. 601–610 (2014)
27. De Sa, C., et al.: DeepDive: declarative knowledge base construction. SIGMOD Rec. 45(1), 60–67 (2016)
28. Mitchell, T., et al.: Never-ending learning. In: Proceedings of AAAI-15 (2015)
29. Jiang, S., Lowd, D., Dou, D.: Learning to refine an automatically extracted knowledge base using markov logic. In: Proceedings of ICDM 2012, pp. 912–917 (2012)
30. Suchanek, F.M., Sozio, M., Weikum, G.: SOFIE: a self-organizing framework for information extraction. In: WWW 2009 (2009)

Ontology Driven Extraction of Research Processes

Vayianos Pertsas[1(✉)], Panos Constantopoulos[1,2],
and Ion Androutsopoulos[1,2]

[1] Department of Informatics, Athens University of Economics and Business,
Athens, Greece
{vpertsas,panosc,ion}@aueb.gr
[2] Digital Curation Unit, IMSI - Athena Research Centre, Athens, Greece

Abstract. We address the automatic extraction from publications of two key concepts for representing research processes: the concept of research activity and the sequence relation between successive activities. These representations are driven by the Scholarly Ontology, specifically conceived for documenting research processes. Unlike usual named entity recognition and relation extraction tasks, we are facing textual descriptions of activities of widely variable length, while pairs of successive activities often span multiple sentences. We developed and experimented with several sliding window classifiers using Logistic Regression, SVMs, and Random Forests, as well as a two-stage pipeline classifier. Our classifiers employ task-specific features, as well as word, part-of-speech and dependency embeddings, engineered to exploit distinctive traits of research publications written in English. The extracted activities and sequences are associated with other relevant information from publication metadata and stored as RDF triples in a knowledge base. Evaluation on datasets from three disciplines, Digital Humanities, Bioinformatics, and Medicine, shows very promising performance.

Keywords: Ontology population · Information extraction
Machine learning methodologies · Linked data

1 Introduction

The steep increase of scientific publications in every major discipline [1] makes it increasingly difficult for experts to maintain an overview of their domain, increases the risk of missing new work or reinventing solutions, and makes it harder to relate ideas from different domains [2]. This situation could be significantly alleviated by supporting queries such as: find all papers that address a given problem; how was the problem solved; which methods are employed by whom in addressing particular tasks; etc. Answering queries like these essentially requires access to information about research processes. Such information could be compiled interactively, or automatically extracted from research publications, finally offered in a structured form suitable for

D. Vrandečić et al. (Eds.): ISWC 2018, LNCS 11136, pp. 162–178, 2018.
https://doi.org/10.1007/978-3-030-00671-6_10

supporting semantic queries. It is to be noted that search engines widely used by researchers, such as Google Scholar[1], Scopus[2] or Semantic Scholar[3], mostly leverage article metadata, while knowledge expressed in the actual text is only exploited in a shallow manner mostly by matching query terms to documents [3].

Understanding and encoding the knowledge contained in research articles is a complex task which poses several challenges. For instance, in order to extract the context of the research reported in an article (who is involved, what are their interests, affiliations, etc.), information from the metadata of the article must be extracted, analyzed and mapped onto a schema, so that activities, entities etc. extracted from the text of the article can be placed in the right context. Furthermore, the actual text of publications needs to be processed in order for activities, entities, and more generally concepts relevant to the documentation of research processes to be identified, extracted and associated according to predefined relation types of the same schema.

In this paper we address the problem of automatically extracting from publications, in the English language, two key concepts for representing research processes: the concept of *research activity* and the *sequence relation* between successive activities. We associate the information extracted from the texts of the articles with relevant information previously extracted from the articles' metadata or other digital repositories, and publish the resulting information in the form of RDF triples adhering to Linked Data standards. We consider these to be the first steps towards populating an ontology specifically designed for modeling research processes and practices [4], thus generating a research process documentation knowledge base.

Research activities and sequence relations manifest themselves in texts in ways that need to be specifically taken into account in order to achieve satisfactory extraction performance. For example, unlike usual named entities (e.g., persons, locations), research activities have textual descriptions of widely variable length, while pairs of successive (in time) activities often span multiple sentences, unlike simpler relation extraction tasks. We engineered several task-specific features exploiting the semantic context of the ontology being populated, syntactic dependencies of words and other syntactic structure information, which we combined with word embeddings. The latter are dense vector representations of words that can be produced in an unsupervised manner from unlabeled corpora and have proved instrumental in many Natural Language Processing (NLP) tasks in the past years [5, 6]. We actually employ three kinds of embeddings: word embeddings, part-of-speech (POS) tag embeddings, and dependency embeddings, all pre-trained for the domain of research processes, following the example of [7] where the first two kinds were combined.

We developed and compared several sliding window classifiers[4], thus exploring the activity and sequence extraction tasks along three dimensions:

(1) *Processing granularity*. We tested the effectiveness of classification at three levels of granularity: token-, sentence- and chunk-based classification.

---

[1] https://scholar.google.com/.

[2] https://www.elsevier.com/solutions/scopus.

[3] https://www.semanticscholar.org.

[4] Our software and data will be available at: http://nemo.dcu.gr/resources/.

(2) *Feature space.* The usual NLP practices were extended with the special task-specific features we developed and we assessed their effectiveness.

(3) *Machine learning (ML) method.* We developed classifiers employing Logistic Regression (LR) [8], linear Support Vector Machines (SVM) [9], and Random Forests (RF) [10], as well as a two-stage pipeline combination.

The performance of these classifiers was evaluated with datasets from three different disciplines: Digital Humanities, Bioinformatics, and Medicine. We measured Precision, Recall and F1 scores in token- and entity-based evaluations with very promising results, indicating the potential for creating a reliable research process knowledge base. The results also confirmed the contribution of the specially designed features in achieving that performance. We view the methods presented in this paper as strong baselines for extending our work to extracting other entities and relations describing research processes (e.g. goals, methods employed, propositions, etc.), and for experimenting with other classifiers (e.g., CRFs [11]), especially deep learning-based ones (e.g., RNNs, CNNs [12]) when larger datasets become available.

The rest of this paper proceeds as follows: in Sect. 2 we present related work and explain how our task is different; in Sect. 3 we describe the methodology and experimental setup; in Sect. 4 we discuss the evaluation experiments and their results; and we conclude in Sect. 5.

2 Related Work

To the best of our knowledge, the task of extracting variable-length textual descriptions of research activities from publications, and associating them on the basis of sequential order as inferred from the text, has not been addressed in previous work. That said, however, information extraction (IE) from scientific papers has attracted a lot of interest over the past years, as testified by the recent creation of a challenge on Scientific Information Extraction (ScienceIE) [3], the ACL RD-TEC Reference Dataset for Terminology Extraction and Classification [13], or domain-specific competitions such as BioCreAtIve[5]. Recent works deal with the extraction of key-phrases denoting tasks, scientific methods and materials from research documents [14, 15], the association of the extracted entities with Linked Data [16–18], or the recognition of biomedical entities such as genes [19, 20]. They use features based on surface form, POS tags, or word embeddings and they employ classifiers such as SVMs, CRFs or neural networks, to extract key-phrases and named entities from text, as well as binary lexical semantic relations (synonym-of, hyponym-of).

In [21], key-phrases denoting the "Focus", "Technique" and "Domain" of the articles are identified on the basis of syntactic patterns matched via rules to the dependency tree of each sentence in article abstracts. In [22], rule-based methods are employed in understanding the dynamics preceding the creation of new topics. In [23], sentences from abstracts in the domains of clinical trials and biomedicine are classified

---

[5] http://biocreative.sourceforge.net/index.html.

in categories, such as introduction, purpose, method, results and conclusion, using various bag-of-words or bag-of-n-grams representations.

A specialized system for extracting specific elements from legal contracts [7] uses sliding window classifiers and handcrafted features combined with word and POS tag embeddings to extract contract elements such as title, date, signatories' names, etc.

In [24, 25], portions of text mentioning specific papers are extracted and relations to the corresponding citations are generated using rule based approaches or features that deal mainly with the surface form or structural aspects of text (e.g., they examine the existence of specific POS tags or lexical terms that indicate references, other citations, etc. in the current or previous sentence). In [26], authors and organizations are identified in scientific papers via CRFs using features that mainly deal with token surface form (lower/upper case, presence in gazetteers, font size, etc.) or structural text characteristics (appearance in sections/paragraphs, first word in line, etc.). The extracted entities are then interrelated by further extracting the *hasAffiliation* property. For that, an SVM with Gaussian kernel is used with features related to the author affiliation markers and the distance of extracted strings.

In other works related to action sequencing, such as [27], the authors create abstractions of action sentences based on a predefined template and then cluster those abstractions together based on a functional similarity measure. In [28], the authors use deep reinforcement learning, in order to extract sequences of labeled actions from sentences; each action is represented by arguments constructed from the verb and its object (e.g., cook (rice)) and the sequencing relations can be selected or eliminated based on their type (i.e., optional, exclusive or essential). In [29], the authors use a predefined list of names to map their action descriptions and interpret them as action sequences, or to generate navigational action descriptions using an encoder-aligner-decoder structure. Unlike the above methods, we identify and associate actions that are not expressed by single words or mapped to a fixed template or list of names. Instead, in our work actions have complex textual representations of variable length and cannot be labeled with words from a name-list. Moreover, we are not confined to deriving sequence relations from single lexical keywords. Instead, sequence relations are inferred from a combination of the actual textual context of activities along with structural properties of the text (e.g., relative positions of the entities in the texts).

In all of the approaches reviewed above, IE from text is addressed using either rules or ML methods based on features that handle mainly the surface form of words disregarding other information, such as attributes derived from syntactic dependencies or more complex syntactic patterns. ML methods of that kind perform inadequately in extracting research activities from text, as suggested by the evaluation of our baseline method that uses similar features. This behavior can be attributed to the following characteristics of the task at hand:

- Research activities are entities manifesting themselves only by their textual description and not by any specific nomenclature. Furthermore, their textual description does not follow any specific surface form.
- The textual chunks representing research activities can be of arbitrary length. This has been observed to exceed 50 tokens, which is significantly higher than the lengths of entity names in common Named Entity Recognition (NER).

- Unlike other NER tasks, the surface form of the tokens inside the textual description of a research activity can vary so much, that it is insignificant for the purpose of extracting activities.
- Contrary to common NER tasks, where the extracted entities cover only a small portion of a sentence, research activities may cover almost entire sentences, and even more than one sentence. Here we restrict our investigation to activities contained in one sentence each. Multi-sentence activities are usually composed of smaller ones. The hierarchical decomposition of those composite activities eventually leads to simple single-sentence ones.
- Sequence relations between activities cannot be detected solely from lexical cues in the text. Other attributes of the activities, including their relative position in text, actual textual description, etc., are also employed to improve classification.

The main contributions of the work reported here are:

- The way we address the complexity of the particular task by combining information from the ontology (e.g., available relation types, constraints on their domain and range), task-specific embeddings of words, POS tags, and syntactic dependencies, features detecting special syntactic sequences of words and their order of appearance in texts, specialized features dealing with lexico-syntactic patterns, as opposed to just word surface form, currently employed in other works related to extracting knowledge from scientific literature.
- The proposed methods are applicable to any scientific domain, since no domain-specific lexica or training corpora are required, and they are demonstrated with test sets from three disciplines, capturing a variety of writing styles.
- Our methods yield higher performance compared to common NER or rule-based solutions, as evidenced by comparing to the baselines, which is notable especially considering the fact that the limited sizes of the datasets we had available do not allow for more sophisticated ML approaches (such as deep learning methods).

Furthermore, we show how –based on the semantics from an ontology, specifically designed to represent research processes [4] - information extracted from text can be associated with knowledge from article metadata and other sources (such as ORCID[6]) as part of creating a comprehensive research process knowledge base.

3   Setup and Methodology

We use as schema for research process knowledge bases the Scholarly Ontology (SO) [4], a domain-independent ontology of scholarly/scientific work. A specialization, in fact precursor, of SO already applied to the domain of Digital Humanities is the NeDiMAH Methods Ontology (NeMO) [30]. A brief overview of SO core concepts is given in the following section. For a full account see [4].

---

[6] https://orcid.org.

3.1   Conceptual Framework: The Scholarly Ontology

Figure 1 shows the core concepts and relations in SO. The rationale behind the ontology is to support documenting *"who* does *what, when,* and *how"* in a given scholarly domain. The ontology is built around the central notion of *activity* and combines three perspectives: the *agency* perspective, concerning actors and intentionality; the *procedure* perspective, concerning the intellectual framework and organization of work; and the *resource* perspective, concerning the material and immaterial objects consumed, used or produced in the course of activities.

Activity concerns real events that have occurred in the form of intentional acts carried out by actors. The instances of the *Activity* class are real processes with specific results, as opposed to those of the *Method* class, which are specifications, procedures, or recipes for carrying out activities so as to address specific goals. Sequence and composition of activities are represented by the *follows* and *partOf* relations respectively. *Actor* instances are entities capable of performing intentional acts that they can be accounted or referenced for. They can participate in activities, actively or passively, in one or more roles. Subclasses of *Actor* are the classes *Person* and *Group*, representing individual persons and collective entities respectively. Further specializations of *Group* are the classes *Organization* and *ResearchTeam*. *ContentItem* comprises information resources, regardless of their physical carrier, in human readable form (with images, tables, articles, bibliographic references, etc. being specializations of *ContentItem* class). *Assertion* includes all kinds of assertions in the scholarly domain and captures the intellectual essence of scholarly activity, comprising propositions resulting from activities and can be *supportedBy* evidence provided by content items. Finally, the class *Topic* comprises thematic keywords which function as tags expressing the subject of methods, the topic of content items, research interests of actors, etc.

In this paper, we focus on extracting from text and automatically populating two key concepts of the ontology: (i) *Activity*, a unary predicate denoting research processes

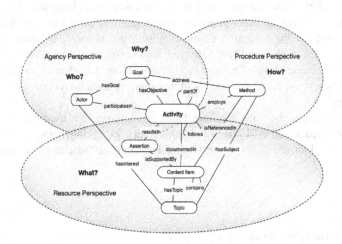

Fig. 1. Scholarly ontology core

such as a biological experiment, an archeological excavation, an anthropological or medical study, etc. and (ii) *follows*, a binary predicate denoting the sequence relation between two successive activities. Figure 2 shows an example of textual chunks representing research activities -highlighted- and their sequence relations.

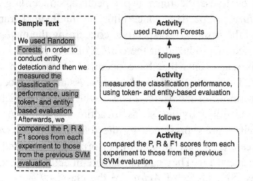

Fig. 2. Activities and sequential relations

3.2   The Dataset

An unlabeled dataset obtained from 50,000 open-access research papers was used in order to create embeddings. The dataset consisted of approximately 10,000,000 sentences after metadata cleaning and parsing using spaCy[7], yielding 300,000,000 tokens and eventually a vocabulary of approx. 1,000,000 unique words (types). Word, part-of-speech tag (POS) and dependency (DEP) embeddings were generated from the above. Specifically: 100-dimensional word embeddings were produced using the Gensim implementation of word2vec[8] (skip-gram model); 25-dimensional POS embeddings were produced by replacing each token by its corresponding POS tag before running word2vec; and 25-dimensional DEP embeddings were produced by replacing each token by the label of the (unique) arc linking the token to its head in the dependency tree. Our experiments with other general-purpose, publicly available embeddings, such as those trained on the Common Crawl corpus using GloVe[9], or those trained on Wikipedia articles with word2vec, showed inferior performance compared to our domain-specific embeddings. This can be attributed to the fact that our embeddings are trained exclusively on scholarly articles, thus capturing the idiosyncrasies of scholarly writing styles.

To train and evaluate our machine learning methods, we used research articles randomly selected using APIs from publishers such as Springer and Elsevier, or by scraping online journals such as the Digital Humanitites Quarterly. To annotate the

---

[7] https://spacy.io/.

[8] https://radimrehurek.com/gensim/.

[9] https://nlp.stanford.edu/projects/glove/.

dataset with ground truth, we used human annotators, appropriately trained in the use of SO. Guidelines and examples were provided to the annotators.

The training set, comprising texts from 50 research articles covering 9 research domains, was annotated by two post-graduate students. Three annotation trials (one article annotated by both annotators per trial, followed by discussion) were initially performed. Inter-annotator agreement was 81% kappa statistic, measured on 5 articles annotated by both annotators at the end of the annotation trials. Subsequently, the remaining articles were annotated by one annotator each. The annotation of the training set yielded approx. 1,000 sequence relations and 1,700 activities comprising approx. 31,000 tokens. For hyper-parameter tuning we used 3-fold cross-validation.

For testing, we used articles from three disciplines, Digital Humanities (DH), Bioinformatics (BIOINF) and Medicine (MED), to expose our classifiers, trained on a generic set, to a wide variety of writing styles. Three test sets, 15 articles per discipline, were annotated by two expert -per discipline- annotators. The annotators were trained on 5 articles per discipline, annotated by both annotators, with discussion after anno-tating each article. Inter-annotator agreement was 81%, 83% and 85% kappa for DH, BIOINF and MED, respectively, for the fifth article of each discipline. The remaining articles were annotated by one annotator each. For each test set, human annotation produced approx. 600 activities containing approx. 10,000 tokens. Concerning sequence relations, human annotation produced approx. 200 relations for DH, 500 for BIOINF, and 600 for MED. The differences in the numbers can be attributed to the granularity of activities and the writing style prevalent in each research field.

3.3 Extracting Research Activities

Seven sliding window classifiers (SWC) and a two-stage pipeline classifier were implemented for extracting research activities (Table 1). They all perform token-based classification by examining each token t and its surrounding tokens in a fixed-size window, and classifying t as positive if it is part of a phrase expressing a research activity, or negative otherwise. The size of the window was set at 30 tokens around t (a total of $30 + 30 + 1 = 61$ tokens) following hyper-parameter tuning. Zero-padding was used to represent tokens exceeding the sentence boundary. Each window of tokens was turned into a feature vector representing the token t being classified. We experimented with Logistic Regression, linear Support Vector Machines and Random Forests, with different feature specifications as detailed below. We use the notation M.E.F or M.E. F to denote the resulting classifiers, where M denotes the learning method used, E the embeddings and F the special features.

The first and second classifiers, **LR.WP.B** and **SVM.WP.B**, use Logistic Recres-sion (LR) and linear SVM respectively, while they both employ 139 features: 125 derived from the 100- and 25-dimensional vectors of the word and POS embeddings (WP), and another 14 binary hand-crafted features labeled "basic" (B) that deal with the surface form of tokens. Of those features, 7 capture specific token surface forms (title, capitalized, digit, punctuation mark, etc.), while the other 7 determine whether the token's lexical form indicates neighboring activities. For example, words that indicate sequencing of events ('first', 'afterwards', 'finally', etc.), specialization ('concretely', 'specifically', etc.), causality ('for', 'to', etc.,), etc. The total number of features in the window is: $61 \times 139 = 8,479$.

The third and fourth classifiers, **LR.WPD.BS** and **SVM.WPD.BS**, differ from the first two in that they extend the embeddings-related features with 25 originating from DEP embeddings (WPD) and the special features with 10 binary "smart" features (BS) related to special syntactic structures. The latter are meant to capture the inclusion of a token in patterns suggesting activities, either directly, such as sub-sentences with verb in past tense and subject in first person (e.g.: "we performed stylistic analysis"), or indirectly, such as sub-sentences with causal modifiers indicating goals of neighboring activities (e.g. "[ACT: performed stylistic analysis], in order [GOAL: to recognize each characteristic]"). The total number of features is now $61 \times (139 + 25 + 10) = 10,614$.

The fifth classifier, **RF.PD.BS**, employs Random Forests (RF) and uses 51 one-hot features representing POS tags and 71 one-hot features representing DEP tags (PD), rather than embeddings. It also uses the same binary features (14 "basic" and 10 "smart") as the third and fourth classifiers. The total number of features in the sliding window is $61 \times (14 + 10 + 57 + 71) = 9,272$.

The sixth and seventh classifiers, **LR.PD.S.BS** and **SVM.PD.S.BS,** are like the third and fourth with the difference that: (a) they omit features related to word embeddings, and (b) they account for the syntactic sequence (S) of words, i.e., the sequence from the syntactic dependency of the word to its head and the head of its head, thus encoding joint information for 3 tokens instead of just one. As an example of such a syntactic sequence, consider in the first sentence of Fig. 2, the word "conduct", with its syntactic head "order" and the syntactic head of its head "in". The total number of features in the sliding window is now $61 \times (10 + 14 + 50) \times 3 = 13,542$.

In addition to the above classifiers we implemented a two-stage pipeline (see Fig. 3). The first classifier, **SVM.WPD.BS,** is trained on all the sentences of the training set, as before, but now performs sentence classification instead of token classification, i.e., detects only the existence of research activities in the sentence without identifying their boundaries. For the first classifier, each sentence is represented using averaged word/POS/DEP embeddings of the contained tokens. This produces a vector of 100 or 25 features derived from the 100-dimensional word embeddings or the 25-dimensional POS or DEP embeddings respectively, keeping the number of features/dimensions independent from the actual number of tokens in the sentence. In addition, we used 14 binary features for representing the existence or absence inside the sentence of the -previously described- special syntactic patterns and lexical forms that provide indirect activity identifiers. For the second classifier, we used **SVM.PD.S.BS**, but now trained only on sentences containing at least one research activity. This performs token-based classification and determines the boundaries of the chunks describing research activities in the sentences classified as positives by the first classifier. The intuition behind the pipeline is that, by splitting the task into two simpler sub-tasks, each separate classifier will achieve high enough accuracy for their concatenation to produce better results, which was proven correct in the evaluation.

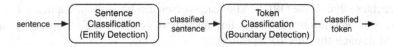

Fig. 3. Activity extraction pipeline

3.4 Extracting Sequence Relations

Extracting sequence relations requires examining all plausible activity pairs. For every pair of extracted activities, the text chunk bounded by these two entities, [*act1*, ..., *act2*], is treated as expressing a candidate sequence relation. A maximum chunk length of 500 tokens, set during hyper-parameter tuning, serves to restrict the search to a reasonable set of candidates excluding pairs of too distant entities unlikely to be sequential, yet including pairs of entities from neighboring paragraphs or sections with reasonable chance of being related. A classifier then determines whether the bounding activities of the chunk satisfy the property *follows*.

Each chunk is represented using averaged word/POS/DEP embeddings of the tokens in the chunk together with 11 special features: 5 that examine certain structural properties of the chunks (*act1* and *act2* are in the same sentence/adjacent sentences/same paragraph; other entities intervene; the chunk contains conjuncts, like the word "and", syntactically associated with tokens inside the boundary entities); 3 for *act1* and 3 for *act2* that examine the entire sentence(s) containing each one of them in order to capture possible sequence indicators (e.g. the words *"then"* and *"Afterwards"* in Fig. 2) referring to *act1* and *act2*, even when they are not inside the chunk bounded by *act1* and *act2* or the individual chunks representing *act1* and *act2* respectively.

We implemented three classifiers for extracting sequence relations between activities. The first sequence extractor, **LR.WPD.B**, uses Logistic Regression and 161 features per chunk: 100 features for the averaged word embeddings of the tokens in the chunk, 25 for the averaged POS, 25 for the averaged DEP embeddings, 5 for structural chunk properties and 6 for sequence indicators, as discussed above. The second extractor, **SVM.WPD.B**, uses the same features, but with a linear SVM. The third extractor, **RF.PD.B**, uses Random Forests (RF) and the per-dimension sum of the one-hot encodings of the POS and DEP tags of each token in the chunk. We also experimented with the average and the TF-IDF-weighted average of the encodings, but without better results in either case.

3.5 Background Context Integration and URI Creation

Having extracted research activities and their sequence relations, we attach to them contextual information obtained from the metadata of the publications. Specifically, we have created mappings that currently support the association of article metadata from two major publishers (Springer and Elsevier) with relevant SO classes such as participants in the research processes (the authors of the paper), their interests (author keywords) and their personal information (affiliations, email, etc.), the *ContentItem* that they are documented in (the research articles), etc. We also provide integration through API with ORCID, a non-for-profit organization for assigning unique, persistent IDs to researchers, so that (i) the ORCID id of each person can be used for duplicate detection and (ii) additional information regarding related projects, funding or biography can be retrieved through the ORCID repository.

The research process knowledge base is created by encoding the extracted information as RDF triples adhering to Linked Data principles and the RDFS[10] and NIF[11] models. For entities with a proper name, such as *Persons*, *Organizations*, *Articles* and *Topics*, their URIs are derived by combining the namespace of the knowledge base, the entity type according to SO, and a unique id provided by the entity name (such as ORCID id or email for persons, article id, topic name, etc.). For activities and sequence relations, URIs are generated by combining the namespace of the knowledge base, the entity type according to SO, the source of extraction (publication id) and the two offsets identifying the boundaries of the extracted entity inside the text, thus ensuring that each URI is unique. A small excerpt of the knowledge base is shown in Fig. 4. Based on our measurements, information extracted from 50 articles translates roughly to 100,000 triples, this being highly dependent on the writing style and the discipline. Indicative running times (on a PC with an Intel i7, 16 GB RAM) for the entire process are approx. 100 s/article.

```
1    <?xml version="1.0" encoding="UTF-8"?>
2    <rdf:RDF
3        xmlns:ns1="http://dcu.gr/ontologies/so/instances/"
4        xmlns:ns2="http://dcu.gr/ontologies/so/"
5        xmlns:ns3="http://persistence.uni-leipzig.org/nlp2rdf/ontologies/nif-core#"
6        xmlns:rdf="http://www.w3.org/1999/02/22-rdf-syntax-ns#"
7        xmlns:rdfs="http://www.w3.org/2000/01/rdf-schema#"
8    >
9        <rdf:Description rdf:about="http://dcu.gr/ontologies/so/instances/ContentItem-S1873506111000596/Activity#offset_13533_13533">
10           <ns1:isDocumentedIn rdf:resource="http://dcu.gr/ontologies/so/instances/ContentItem-Elsevier-S1873506111000596"/>
11           <ns3:referenceContext>http://api.elsevier.com/content/article/pii/S1873506111000596</ns3:referenceContext>
12           <rdf:type rdf:resource="http://dcu.gr/ontologies/so/Activity"/>
13           <ns1:hasParticipant rdf:resource="http://dcu.gr/ontologies/so/instances/Person-ORCID-0000-0001-9512-4708"/>
14           <ns2:follows rdf:resource="http://dcu.gr/ontologies/so/instances/ContentItem-S1873506111000596/Activity#offset_13262_13262"/>
15           <ns3:beginIndex>13533</ns3:beginIndex>
16           <ns3:endIndex>13533</ns3:endIndex>
17           <rdfs:label>performed token- and entity-based evaluation</rdfs:label>
18       </rdf:Description>
```

Fig. 4. Excerpt from the produced RDF triples

4 Evaluation

In general, metadata association has exhibited very good performance since it relies solely on pre-constructed mappings between fixed schemas. Few isolated incidents (lower than 1%) of improper association were due to errors in XML/HTML tags in the article (e.g., an empty or misplaced bracket) and can be treated with additional escape rules as part of the general debugging process.

Regarding the information extraction from text, we evaluated the performance of all the classifiers by measuring Precision, Recall and F1 scores. After window-size selection and hyper-parameter tuning using 3-fold cross-validation on the training set, all the classifiers were trained on the entire training set. As previously mentioned, we

---

[10] https://www.w3.org/TR/rdf-schema/.

[11] http://persistence.uni-leipzig.org/nlp2rdf/.

used three different test sets, DH, BIOINF, and MED, presumably representing different writing styles, as well as their combination (ALL Test Set).

Approximate Randomization Tests (ART) [32] between every classifier and the relevant baseline were carried out to ensure the statistical significance of the tests. Classifiers were grouped in zones of statistically similar results (shown by dividing lines in Tables 1 and 3) and ARTs were run on every combination of methods from different zones in order to ensure that the difference between any two measurements is statistically significant given our test sets. The Bonferroni correction was used to adjust the threshold (p-value) from the default 0.05 to 0.00625 for activity extraction and 0.0125 for sequence relation extraction, since we compared more than two systems. All pair combinations gave probabilities below the above thresholds in ARTs, therefore all the results shown are statistically significant.

4.1 Research Activity Extraction Evaluation

The evaluation of activity extraction methods involves comparing classifier results against a reference standard produced by human annotators on the basis of Precision, Recall and F1 scores calculated as usual[12]. In addition, we compare the classifiers with a "baseline" method, similar to those commonly used in NER tasks [7], with a smaller sliding window of 15 tokens (7 left, the central token t, 7 right), 100 features for word embeddings, 25 features for POS embeddings and 14 "basic" binary features for surface form representation, in total $15 \times (125 + 14) = 2,085$ features. The baseline uses a linear SVM trained on the same training set, as this has proved experimentally to perform slightly better than LR and RF. Two groups of comparisons are made: token-based and entity-based.

In token-based evaluation, a *true positive* (TP) is a token correctly classified as part of a chunk representing a research activity, a *false positive* (FP) is a token incorrectly classified as part of a research activity, and a *false negative* (FN) is a token incorrectly classified as non-part of a research activity. Results of the token-based evaluation for each test set are shown in Table 1. Regarding the pipeline classifier which consists of a sentence- and a token-based classifier in tandem, detailed per stage and aggregate performance results are shown in Table 2. The aggregate scores of the pipeline are also shown in Table 1 for comparison with the other methods.

The **Pipeline** classifier achieved the highest scores on every test set and criterion. The aggregate performance of the pipeline is inferior to that of the individual stages (see Table 2) due to error propagation, since the sentences that are wrongly classified in the first classifier are fed as input into the second. The baseline, on the other hand, performed worse than all the other classifiers on every test set and criterion. This can mainly be attributed to two factors: (a) the difference in the size of the sliding window (as indicated from the performance increase between the baseline and the **SVM.WP.B**); and (b) the use of the DEP embeddings and the "smart" features. Moreover, word embeddings do not add much to the overall improvement of the classification, as suggested by the performance of the **RF.PD.BS**, **LR.PD.S.BS** and **SVM.PD.S.BS**

---

[12] $P = \frac{TP}{TP+FP}, R = \frac{TP}{TP+FN}, F1 = \frac{2*P*R}{P+R}.$

Table 1. Token-based evaluation

	DH test set			BIOINF test set			MED test set			ALL test set		
	P	R	F1	P	R	F1	P	R	F1	P	R	F1
Baseline	0.54	0.30	0.38	0.76	0.50	0.60	0.76	0.62	0.69	0.72	0.50	0.59
1 LR.WP.B	0.62	0.44	0.52	0.79	0.59	0.68	0.79	0.66	0.72	0.75	0.58	0.65
2 SVM.WP.B	0.60	0.50	0.54	0.80	0.66	0.72	0.78	0.68	0.73	0.74	0.63	0.68
3 LR.WPD.BS	0.78	0.76	0.77	0.83	0.81	0.82	0.88	0.83	0.85	0.84	0.80	0.82
4 SVM.WPD.BS	0.76	0.80	0.78	0.83	0.83	0.83	0.87	0.85	0.86	0.83	0.83	0.83
5 RF.PD.BS	0.79	0.80	0.80	0.85	0.83	0.84	0.89	0.83	0.86	0.85	0.82	0.83
6 LR.PD.S.BS	0.77	0.79	0.78	0.82	0.83	0.83	0.88	0.88	0.88	0.83	0.84	0.84
7 SVM.PD.S.BS	0.79	0.82	0.80	0.84	0.84	0.84	0.89	0.89	0.89	0.85	0.85	0.85
8 SVM-Pipeline	**0.83**	**0.82**	**0.82**	**0.87**	**0.89**	**0.88**	**0.90**	**0.93**	**0.92**	**0.87**	**0.89**	**0.88**

Table 2. Pipeline evaluation

	DH test set			BIOINF test set			MED test set			ALL test set		
Entity identification:	P	R	F1	P	R	F1	P	R	F1	P	R	F1
SVM.WPD.BS	0.90	0.89	0.89	0.96	0.94	0.95	0.96	0.96	0.96	0.94	0.93	0.94
Boundary detection:												
SVM.PD.S.BS	0.92	0.89	0.90	0.92	0.95	0.94	0.95	0.96	0.95	0.93	0.94	0.93
Pipeline:												
SVM-Pipeline	0.83	0.82	0.82	0.87	0.89	0.88	0.90	0.93	0.92	0.87	0.89	0.88

classifiers, as word embeddings can be replaced by other contextual information regarding the syntactic sequence of tokens. Therefore, the distinctive features of the methods we developed prove to contribute significantly to the performance of research activity extraction.

In entity-based evaluation, each maximal sequence of consecutive positive tokens is considered as a research activity ("entity"). Ideally, an entity is correctly predicted by a classifier only if it matches 100% with one annotated by humans, counting as errors even the slightest deviations. In practice, a close match suffices, especially in cases where the extracted entities are very long. A threshold of 86% was automatically selected by averaging the Levenshtein distances of a sample of 100 pairs of overlapping strings (a predicted and a gold entity in each pair) for which the annotators indicated that the overlap was sufficient. This translated roughly into a difference of 1-5 tokens (including punctuation marks) at the boundaries of each entity. Consequently, in entity-based evaluation a true positive (TP) is a predicted string that matches a reference standard string by at least 86%; a false positive (FP) is an un-matched predicted string; and a false negative (FN) is an un-matched reference standard string. Results of the entity-based evaluation are shown in Table 3.

The **RF.PD.BS** and **Pipeline** classifiers compete for best performance in the case of entity-based evaluation with similar results on most test sets. The baseline again performs worse than all other methods. Performance results in entity-based evaluation are

Table 3. Entity-based evaluation

		DH test set			BIOINF test set			MED test set			ALL test set		
		P	R	F1	P	R	F1	P	R	F1	P	R	F1
	Baseline	0.16	0.30	0.20	0.23	0.60	0.34	0.28	0.76	0.40	0.23	0.60	0.33
1	LR.WP.B	0.48	0.60	0.53	0.56	0.72	0.63	0.52	0.74	0.62	0.53	0.70	0.60
2	SVM.WP.B	0.42	0.64	0.50	0.54	0.76	0.63	0.51	0.78	0.62	0.50	0.74	0.60
3	LR.WPD.BS	0.58	0.80	0.67	0.57	0.80	0.66	0.58	0.82	0.68	0.58	0.80	0.67
4	SVM.WPD.BS	0.54	0.82	0.65	0.55	0.79	0.65	0.57	0.83	0.67	0.55	0.81	0.66
5	LR.PD.S.BS	0.59	0.82	0.69	0.58	0.78	0.66	0.60	0.84	0.70	0.59	0.81	0.68
6	SVM.PD.S.BS	0.61	0.83	0.70	0.62	0.76	0.68	0.61	0.83	0.70	0.61	0.80	0.70
7	RF.PD.BS	**0.68**	0.79	**0.73**	**0.66**	0.78	**0.72**	**0.66**	0.83	**0.74**	**0.67**	0.80	**0.73**
8	SVM-Pipeline	0.64	**0.83**	**0.73**	0.62	**0.84**	**0.72**	0.60	**0.86**	0.71	0.62	**0.85**	0.72

inferior to those in token-based evaluation. Error analysis showed that this can be attributed to tokens occurring in entity chunks incorrectly classified as not being research activities; this causes the split of the original entity into smaller ones, in turn producing additional errors (1 FN for the undetected original entity and 1 FP for each smaller entity). Consider, for instance, the second sentence in Fig. 2. Had the classifier produced 0 for the token "to" inside the sentence, the activity "compared the P, R and F1 scores from the previous experiment to those from the SVM evaluation" would have been split into two smaller entities: "compared the P, R and F1 scores from the previous experiment" and "those from the SVM evaluation". Since each of the new smaller entities matches the original by less than 86%, the resulting misclassification would give 2 FPs for the smaller activities and 1 FN for the original.

The performance decrease in entity evaluation was found to vary among domains. Indeed, a 6.9% average decrease in F1 scores was observed with the DH test set, while the decrease was 14.5% with the BIOINF test set, and 15.9% with MED. Error analysis indicates that this can be attributed mainly to the differences in writing style. For example, in the DH test set, the research activity entities were found to have smaller size and contain fewer "error prone" tokens (such as acronyms or formulas) that could cause individual token misclassification and thus split of the entity chunk.

4.2 Sequence Relation Extraction Evaluation

The evaluation of sequence relation extraction methods involves comparing the predicted relations among the reference standard entities in each test set with those produced by the human annotators on the basis of Precision, Recall and F1 scores calculated as usual. A true positive (TP) is a chunk [act1, ..., act2] for which the classifier correctly predicted the follows (act2, act1) property; a false positive (FP) is a chunk for which follows (act2, act1) was incorrectly predicted; and a false negative (FN) is a chunk for which follows (act2, act1) incorrectly failed to be predicted. Classifier performance is also compared with that of a simple baseline method that assigns a sequence relation to all adjacent activities in a paragraph and activities connected by sequence cue words (e.g., "then", "subsequently"). Results are shown in Table 4.

Table 4. Relation extraction evaluation

	DH test set			BIOINF test set			MED test set			ALL test set		
	P	R	F1	P	R	F1	P	R	F1	P	R	F1
Baseline	0.62	0.72	0.67	0.65	0.89	0.76	0.59	0.92	0.72	0.62	0.88	0.72
1 LR(WPD)E-AVG-B	**0.87**	0.90	**0.88**	0.85	0.58	0.69	**0.94**	0.69	0.80	0.87	0.77	0.82
2 SVM(WPD)E-AVG-B	0.80	0.93	0.86	0.83	0.65	0.73	0.91	0.75	0.82	0.84	0.80	0.84
3 RF(PD)1H-SUM-B	0.81	**0.93**	0.87	**0.87**	**0.85**	**0.86**	**0.94**	**0.90**	**0.92**	**0.88**	**0.89**	**0.89**

In sequence relation extraction, **RF.PD.B** performed best in BIOINF, MED and overall (F1: 0.86, 0.92 and 0.89 respectively), while for the DH test set the forerunner was **LR.WPD.B** (F1: 0.88). Error analysis suggests that misclassifications are mostly due to adjacent sentences containing multiple activities, a situation more frequent in DH and BIOINF. For example, consider the excerpt: "[*act1:* Two-thirds of the extracted bootstrap samples were used for constructing the model] and then [*act2:* the other one-third were used for testing]. To calculate variable importance, we first [*act3:* put down the out-of-bag cases] and [*act4:* counted the number of votes cast for the correct class], and then [*act5:* randomly permuted the values of variable *root j* in the out-of-bag cases]". One classifier associated *act3* and *act4* of the second sentence with the last activity of the first sentence (*act2*), and another associated the first entity of the second sentence (*act3*) with each entity in the first (*act1, act2*). These predicted associations are treated as wrong because, by definition, *follows* only holds for *immediately successive* activities, with no others in between. Classifiers also tended to fail to detect activity sequences in texts where activities were sparse (e.g., no adjacent paragraphs with at least one activity each), probably because of the large size of text between activities and the structure (not adjacent paragraphs).

5   Conclusion

We addressed the automatic extraction from the text of publications of two core elements of research processes, research activities and their sequence relations, as a basic step towards populating research process knowledge bases complying to an ontology for research process documentation, the Scholarly Ontology (SO). We showed that the complexity of the task demands more complex feature engineering than usual NER tasks. We implemented and tested several sliding window classifiers employing features specifically designed to deal with particular lexical, syntactic, structural and semantic aspects of textual context. Alternative implementations were compared using linear SVMs, Logistic Regression, and Random Forests, as well as a two-stage pipeline classifier specifically configured for the task of activity extraction.

The classifiers were evaluated against a reference standard produced by human annotators, with three different test sets from three domains (Digital Humanities, Bioinformatics and Medicine) and very promising results: overall F1 score 0.88 for research activity extraction in token-based evaluation and 0.73 in entity-based evaluation, and 0.89 for sequence relation extraction. The classifiers were also compared with simpler baselines which were configured without the special features of this work and with smaller sliding window size closer to those used in common NER tasks. The baseline classifiers were consistently inferior in both activity and sequence relation extraction, an additional evidence in support of the effectiveness of the special features and window width we employed. We also showed how contextual information from article metadata and other sources such as ORCID can be associated with the extracted entities according to the Scholarly Ontology and stored as RDF triples adhering to Linked Data standards.

Future work includes extracting further concepts for documenting research processes according to the Scholarly Ontology, such as goals, research questions, propositions, methods, etc., along with their corresponding relations (such as *partOf, employs, hasObjective*, etc.) and experimenting with more complex classifiers (e.g. CNNs or RNNs [12]) when additional larger training datasets become available.

References

1. Bornmann, L., Mutz, R.: Growth rates of modern science: a bibliometric analysis based on the number of publications. J. Assoc. Inf. Sci. Technol. Technol. **66**, 2215–2222 (2015)
2. Renear, A.H., Palmer, C.L.: Strategic reading, ontologies, and the future of scientific publishing. Science **325**, 828–832 (2009)
3. Augenstein, I., Das, M., Riedel, S., Vikraman, L., McCallum, A.: SemEval 2017 Task 10: ScienceIE, pp. 546–555 (2017)
4. Pertsas, V., Constantopoulos, P.: Scholarly ontology: modelling scholarly practices. Int. J. Digit. Libraries. **18**, 173–190 (2017)
5. Levy, O., Goldberg, Y.: Linguistic regularities in sparse and explicit word representations. In: CoNLL, pp. 171–180 (2014)
6. Levy, O., Goldberg, Y., Dagan, I.: Improving distributional similarity with lessons learned from word embeddings. Trans. ACL **3**, 211–225 (2015)
7. Chalkidis, I., Michos, A., Androutsopoulos, I.: Extracting contract elements. In: ICAIL, pp. 19–28, London (2017)
8. McCullagh, P., Nelder, J.A.: Generalized Linear Models, Chapman and Hall London – New York (1983). 261 S
9. Cristianini, N., Shawe-Taylor, J.: An introduction to support vector machines and other kernel-based learning methods. Cambridge University Press, Cambridge (2000). ISBN 0-521-78019-5
10. Breiman, L.: Random forests. Mach. Learn. **45**, 5–32 (2001)
11. Lafferty, J., McCallum, A., Pereira, F.C.N.: Conditional random fields: Probabilistic models for segmenting and labeling sequence data. In: ICML 2001. vol. 8, pp. 282–289 (2001)
12. Goldberg, Y.: A primer on neural network models for natural language processing. J. Artif. Intell. Res. **57**, 345–420 (2016)
13. QasemiZadeh, B., Schumann, A.-K.: The ACL RD-TEC 2.0: a language resource for evaluating term extraction and entity recognition methods. In: LREC, pp. 1862–1868 (2016)
14. Lee, L.-H., Lee, K.-C., Tseng, Y.-H.: The NTNU System at SemEval-2017 Task 10: extracting keyphrases and relations from scientific publications using multiple CRFs. In: 11th International Workshop on SemEval-2017, pp. 950–954 (2017)
15. Luan, Y., Ostendorf, M., Hajishirzi, H.: Scientific Information Extraction with Semi-supervised Neural Tagging, pp. 2631–2641. arXiv:1708.06075 (2017)

16. Sateli, B., Witte, R.: What's in this paper? Combining rhetorical entities with linked open data for semantic literature querying. In: ICWWW ACM, pp. 1023–1028 (2015). https://doi.org/10.1145/2740908.2742022
17. Osborne, F., de Ribaupierre, H., Motta, E.: TechMiner: extracting technologies from academic publications. In: Blomqvist, E., Ciancarini, P., Poggi, F., Vitali, F. (eds.) EKAW 2016. LNCS (LNAI), vol. 10024, pp. 463–479. Springer, Cham (2016). https://doi.org/10.1007/978-3-319-49004-5_30
18. Sateli, B., Witte, R.: Semantic representation of scientific literature: bringing claims, contributions and named entities onto the Linked Open Data cloud. PeerJ Comput. Sci. **1**, e37 (2015)
19. Song, Y., Yi, E., Kim, E., Lee, G.G., Park, S.J.: POSBIOTM-NER: a machine learning approach for bio-named entity recognition (2004). Doi=10.1.1.101.1165
20. Plake, C., et al.: A support vector classifier for gene name recognition. In: BioCreAtIvE Workshop, Granada, Spain, pp. 1–5 (2004)
21. Gupta, S., Manning, C.: Analyzing the dynamics of research by extracting key aspects of scientific papers. In: IJCNLP, pp. 1–9 (2011)
22. Salatino, A.A., Osborne, F., Motta, E.: How are topics born? Understanding the research dynamics preceding the emergence of new areas. PeerJ Comput. Sci. **3**, e119 (2017)
23. Ruch, P., et al.: Using argumentation to extract key sentences from biomedical abstracts. Int. J. Med. Inform. **76**, 195–200 (2007)
24. Di Iorio, A., Nuzzolese, A.G., Peroni, S.: Towards the automatic identification of the nature of citations. In: CEUR Workshop Proceedings, pp. 63–74 (2013)
25. Athar, A., Teufel, S.: Context-enhanced citation sentiment detection. In: NAACL HLT 2012, pp. 597–601 (2012)
26. Do, H.H.N., Chandrasekaran, M.K., Cho, P.S., Kan, M.-Y.: Extracting and matching authors and affiliations in scholarly documents. In: ACM/IEEE-CS - JCDL 2013, pp. 219–228 (2013)
27. Lindsay, A., Read, J., Ferreira, J.F., Hayton, T., Porteous, J., Gregory, P.: Framer: planning models from natural language action descriptions. In: ICAPS, pp. 434–442 (2017)
28. Feng, W., Zhuo, H.H., Kambhampati, S.: Extracting Action Sequences from Texts Based on Deep Reinforcement Learning. arXiv:1803.02632 (2018)
29. Mei, H., Bansal, M., Walter, M.R.: Listen, Attend, and Walk: Neural Mapping of Navigational Instructions to Action Sequences. arXiv:1506.04089 (2015)
30. Pertsas, V., Christodoulou, T., Dallas, C., Constantopoulos, P., Papachristopoulos, L., Hughes, L.: Contextualized integration of digital humanities research: using the NeMO ontology of digital humanities methods. In: Digital Humanities 2016: Conference Abstracts, pp. 161–163. Jagiellonian University & Pedagogical University (2016)
31. Yeh, A.: More accurate tests for the statistical significance of result differences. In: COLING. vol. 2, pp. 947–953 (2000)

Enriching Knowledge Bases
with Counting Quantifiers

Paramita Mirza[1]([⊠]), Simon Razniewski[1], Fariz Darari[2],
and Gerhard Weikum[1]

[1] Max Planck Institute for Informatics, Saarbrücken, Germany
{paramita,srazniew,weikum}@mpi-inf.mpg.de
[2] Universitas Indonesia, Depok, Indonesia
fariz@cs.ui.ac.id

Abstract. Information extraction traditionally focuses on extracting relations between identifiable entities, such as ⟨*Monterey, locatedIn, California*⟩. Yet, texts often also contain Counting information, stating that a subject is in a specific relation with a number of objects, without mentioning the objects themselves, for example, *"California is divided into 58 counties"*. Such counting quantifiers can help in a variety of tasks such as query answering or knowledge base curation, but are neglected by prior work.

This paper develops the first full-fledged system for extracting counting information from text, called *CINEX*. We employ distant supervision using fact counts from a knowledge base as training seeds, and develop novel techniques for dealing with several challenges: *(i)* non-maximal training seeds due to the incompleteness of knowledge bases, *(ii)* sparse and skewed observations in text sources, and *(iii)* high diversity of linguistic patterns. Experiments with five human-evaluated relations show that CINEX can achieve 60% average precision for extracting counting information. In a large-scale experiment, we demonstrate the potential for knowledge base enrichment by applying CINEX to 2,474 frequent relations in Wikidata. CINEX can assert the existence of 2.5M facts for 110 distinct relations, which is 28% more than the existing Wikidata facts for these relations.

1 Introduction

Motivation. General-purpose knowledge bases (KBs) like Wikidata, DBpedia or YAGO [1, 31, 35] find increasing use in applications such as question answering, entity search or document enrichment, and their automated construction from Internet sources has been greatly advanced. So far, information extraction (IE) to this end has focused on fully qualified subject-predicate-object (SPO) facts such as ⟨*Monterey, locatedIn, California*⟩. However, texts often contain only counting information: the number of objects that stand in a specific relation with a certain entity, without mentioning the objects themselves. Examples are: *"California is divided into 58 counties"*, *"Clint Eastwood directed more than twenty movies"* or *"Trump has three sons and two daughters"*.

© Springer Nature Switzerland AG 2018
D. Vrandečić et al. (Eds.): ISWC 2018, LNCS 11136, pp. 179–197, 2018.
https://doi.org/10.1007/978-3-030-00671-6_11

This kind of knowledge can be codified into an extension of existentially quantified formulas known in AI and logics as *counting quantifiers (CQs)*: they assert the existence of a specific number of SPO triples without fully knowing the triples themselves. Counting information can substantially extend the scope and value of knowledge bases. First, they allow accurate answers for queries that involve counts (e.g., number of counties per US state) or existential quantifiers (e.g., directors who made at least 5 movies). Second, an important use case is KB curation [8,34]. KBs are notoriously incomplete, contain erroneous triples, and are limited in keeping up with the pace of real-world changes. Counting information helps to identify gaps and inaccuracies. For example, knowing the exact number of counties in California or a lower bound for the number of films directed by Eastwood are important cues to complete and enrich a KB.

State-of-the-Art and Challenges. The predominant approach to extracting facts for KB population is distant supervision, using seeds for the SPO triples of interest (e.g., [21,32]). The seeds are usually taken from an initial KB or are manually compiled. Spotting the seeds in a text corpus (e.g., *Clint Eastwood, directed* and *Gran Torino*) then allows learning patterns for relations (e.g., *"director of"* or *"⟨someone⟩'s masterpiece"*), which in turn lead to observing new fact candidates. This methodology is known as the pattern-relation duality principle [2].

Distant supervision is a natural approach for extracting counting information as well: the cardinality of distinct O arguments for a given SP pair, $n := |\{O \mid SPO \in KB\}|$, serves as a seed for the counting assertion, $\langle S, P, \exists n \rangle$. However, it is more challenging than traditional SPO-fact extraction and needs to cope with several issues:

(1) *Non-maximal seeds:* Unlike for SPO-fact extraction, the incompleteness of KBs not only leads to a reduction in the number of seeds, but to seeds that systematically underestimate the count of facts that are valid in reality. For example, a KB that knows only a subset of Trump's children, say three out of five, leads to a non-maximal seed that may reward spurious patterns like *"owns three golf resorts"* at the cost of patterns like *"his five children"*. Even worse, KBs often have complete blanks on certain relations, e.g., not knowing any of Eastwood's movies despite labeling his occupation as *film director* and *film producer* (https://www.wikidata.org/wiki/Q43203).

(2) *Sparse and skewed observations:* For many relations, counting information is expressed in text in a sparse and highly skewed way. For example, the non-existence of children is rarely mentioned. For musicians, the first Grammy someone has won often has more mentions than later ones, hence giving undue weight to the pattern *"his/her first award"*. The number of members in a music band is often around four, which makes it hard to learn patterns for very large or very small bands.

(3) *Linguistic diversity:* Counting information can be expressed in a variety of linguistic forms like
   (i) *explicit numerals* as cardinal numbers (e.g., *"has five children"*),
   (ii) *lower bounds* via ordinal numbers (e.g., *"her third husband"*),

(iii) *number-related noun phrases* such as 'twins' or *'quartet'*,

(iv) *existence-proving articles* as in *"has a child"*,

(v) *non-existence adverbs* such as *'never'* and *'without'*.

Open IE methods [18] cannot cope with these challenges. For example, the sentence *"Trump has five children"* would typically result in the triple ⟨ *Trump, has, five children*⟩, failing to recognize that 'five' is a numeric modifier of 'children'. On the other hand, IE methods with pre-specified relations for KB population (e.g., NELL [23]) capture relevant O values only for few relations specified to have numeric literals as their range, such as *numberofkilledinbombing* or *earthquakecasualtiesnumber* (http://rtw.ml.cmu.edu/rtw/kbbrowser/).

Approach and Contributions. In this paper, we develop the first full-fledged system for Counting Information Extraction, called CINEX. Our method is based on machine learning for sequence labeling, judiciously designed to cope with the outlined challenges. We leverage distant supervision from fact counts in a given KB, but devise special techniques to handle non-maximal seeds, sparseness and skew in observing count information in text, and linguistic diversity of patterns. We counter non-maximal seeds (Challenge 1) by relaxing matching conditions for numbers higher than KB counts, and by reducing the training to popular, more complete entities. Sparseness and skew (Challenge 2) are addressed by discounting uninformative numbers using entropy measures. Linguistic variance (Challenge 3) is handled by careful consolidation of detected mentions. We devise both a traditional feature-based conditional random field (CRF) and a bi-directional LSTM-CRF model using TensorFlow, finding that both perform roughly comparable, although the traditional approach is more robust when dealing with noisy training data.

The salient original contributions of this paper are:

- The methodology of our extraction system, CINEX.
- An empirical evaluation with five manually annotated relations, showing 60% precision on average.
- An application and large-scale experimental study of CINEX on 2,474 frequent relations of Wikidata, showing that counting information can extend the SPO facts in Wikidata for 110 distinct relations by 28%.
- Code and data made available to the research community on Github.[1]

The remainder of this paper is structured as follows. In Sect. 2 we specify the scope of counting quantifiers and discuss the incompleteness of KBs, using Wikidata as a reference point. Section 3 presents our methodology for extracting counting information at large scale, which we then detail in Sects. 4 and 5. Section 6 gives experimental results on the quality of our extraction method, with a particular focus on how CINEX can enrich the Wikidata KB in Sect. 6.4. Section 7 discusses related work.

---

[1] https://github.com/paramitamirza/CINEX.

2   Counting Information in Knowledge Bases

Counting quantifiers for a KB with SPO triples are statements on a subset of the SPO arguments. We focus on the dominant case of quantification of O arguments for a given SP pair. We write counting statements as $\langle S, P, \exists n \rangle$, where S is the subject, P is the predicate and n is a natural number (including zero). For instance, the statement that President Garfield has 7 children would be written as $\langle Garfield, hasChild, \exists 7 \rangle$. In the OWL description logics, this statement is written as:

```
ClassAssertion(ObjectExactCardinality(7 :hasChild) :Garfield)
```

Wikidata. To illustrate how today's KBs deal with counting information, we briefly discuss the case of Wikidata, presumably the world's largest and best curated publicly available KB. Wikidata already contains counting relations for a few topics such as *numberOfChildren*, *numberOfSeasons* (of a TV series), or *numberOfHouseholds* (of an administrative entity). This information can coexist with fully qualified SPO facts. Regarding children, for example, Wikidata knows 4 out of the 7 children of President Garfield by name, and knows that he had 7 in total (see Fig. 1). However, the *numberOfChildren* predicate is asserted for only 0.2% of persons in Wikidata so far. Even the *child* property is asserted for only 2.2% of persons, creating uncertainty about whether the others have no children or whether Wikidata does not know about them.

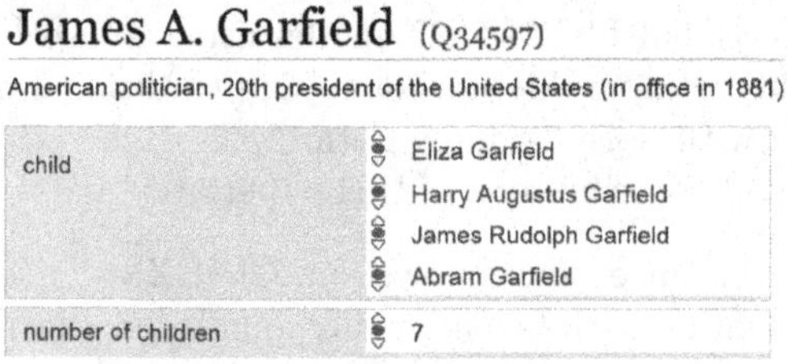

James A. Garfield (Q34597)

American politician, 20th president of the United States (in office in 1881)

child		Eliza Garfield
		Harry Augustus Garfield
		James Rudolph Garfield
		Abram Garfield
number of children		7

Fig. 1. SPO facts and counting information in Wikidata.

Counting information is beneficial for search and question answering, for example to answer *"Which US presidents were married twice?"* We analyzed the number of questions in the TREC 2003, 2004 and 2007 QA test datasets [4], and found that 5% to 10% of the questions (typically starting with *"How many"*) fall into this category.

Potential for KB Enrichment. To quantitatively assess the gap in Wikidata, for which counting information can contribute to KB enrichment, we had one expert read the Wikipedia articles of 200 randomly selected people, with the task of comparing the text-borne counting information on the *hasChild* relation with the explicitly stated children names. The expert was instructed to look at two kinds of cues: *(i) explicit numerals* expressing counting information, *(ii) counting names* of children mentioned in the article. We compare these numbers against *(iii)* the *Wikidata SPO triples* for the person's *hasChild* predicate. Note that approach (ii) corresponds to what standard IE aims to achieve (i.e., extracting full triples and then counting).

We found that counting information via numerals allows the discovery of children counts for 12% of all test entities, while names of children are only mentioned for 7%, and Wikidata contains facts about children for only 2.5%. As for the total number of children, counting information asserts the existence of *twice* as many children, i.e., 0.35 children per person, as spotting and counting children names (0.18), and even *eleven* times more than Wikidata currently knows of (0.03).

3   System Overview

The CINEX system aims to solve the following problem:

Problem 1 (Counting Quantifier Extraction). *Given a text about a subject S, and a predicate P, the task of counting quantifier (CQ) extraction is to determine the number of objects with which S stands in relation regarding P.*

For instance, given the sentence *"Trump has three sons and two daughters"*, the output for the predicate *numberOfChildren* should be 5.

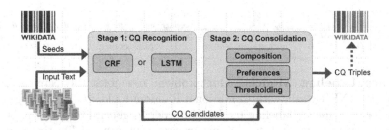

Fig. 2. Overview of the CINEX system.

Figure 2 gives a pictorial overview of the system architecture of CINEX. We split the overall task into two main components: the recognition of counting information and the consolidation of intermediate results into the final output of counting quantifiers. These components are presented in Sects. 4 and 5, respectively.

CINEX utilizes seeds from Wikidata in a judicious way in order to train a model for CQ recognition, using one of two options: a conditional random field (CRF) or a bidirectional LSTM neural network. When applied to new text, the output of the recognition model is a set of CQ candidates, which are often fairly noisy, though. Subsequently, the second stage of CINEX – CQ consolidation – cleans and aggregates the counting information and produces the final output of CINEX. The resulting CQ triples could potentially be added to a knowledge base such as Wikidata.

4   Counting Quantifier Recognition

The first stage of CINEX aims to recognize counting information in text, this way collecting a pool of CQ candidates for further cleaning and consolidation. We cast the CQ recognition into a sequence labeling task, operating on a per-sentence basis and learned separately for each predicate P. We are interested in counting information for a given subject-predicate (SP) pair and assume that the subject is already identified by the sentence context (e.g., the main entity featured in a document, like a Wikipedia article about S or S's homepage on the Web). Furthermore, we assume that the input sentence is pre-processed by detecting terms that indicate counting information: cardinals, ordinals and number-related terms (numterms).

Task 1 (Counting Quantifier Recognition). *Given a sentence about subject S and predicate P containing at least one cardinal, ordinal or number-related term (numterm), the task of CQ recognition is to label each token of the sentence with one of the following tag: (i)* COUNT, *for denoting a CQ mention, (ii)* COMP, *for denoting compositional cues and (iii)* O, *for others.*

The following shows an example:

sentence	Jolie brought her	twins	,	one	daughter	and	three	adopted children to the gala .
pre-processed	Jolie brought her	NUMTERM	,	CARDINAL	daughter	and	CARDINAL	adopted children to the gala .
output tags	O O O	COUNT	COMP	COUNT	O	COMP	COUNT	O O O O O O

Sequence Labeling Models. Our problem resembles the Named Entity Recognition (NER) task, with Conditional Random Fields (CRFs) being a typical choice of sequence labeling models. In order to generalize patterns beyond specific numeric values/tokens, we pre-process sentences to lift these specific tokens into placeholders *cardinal, ordinal* and *numeric term (numterm)*. For instance, the sentence *"Donald Trump has three children from his first wife."* becomes *"Donald Trump has* CARDINAL *children from his* ORDINAL *wife."*

CINEX learns one sequence labeling model for each predicate of interest (e.g., with separate models for children and spouses). We have devised solutions based on two sequence labeling methods:

1. *Feature-based model.* We constructed a CRF-based sequence classifier using CRF++ [14] with n-gram features (up to pentagrams), taking into account lemmas and placeholders (e.g., {*Trump, have,* CARDINAL, *child, from*}) instead of the original tokens.

2. *Neural model.* We adopt the bidirectional LSTM-CRF architecture proposed in [15] using TensorFlow, presently the state-of-the-art method for sequence-to-sequence learning, to build our sequence labeling model. The neural architecture takes into account words, placeholders and character embeddings to represent the input sequence. The neural model should be able to exploit, for example, that word embeddings for *'children'*, *'daughters'* and *'sons'* are close to each other in the embedding space. Furthermore, word embeddings for out-of-vocabulary words such as *'ennealogy'* can be generated via character embeddings, recovering similarity to e.g. *'pentalogy'*.

Incompleteness-Aware Distant Supervision. We employ distant supervision to generate training data, as common in relation extraction [3,21,32]. Given a knowledge base (KB) relation P, for each entity S in the KB that appears as the subject of P, we retrieve *(i)* the *triple count* $|\langle S, P, *\rangle|$ from the KB and *(ii)* sentences about S containing *candidate mentions*, e.g., cardinal numerals. Candidate mentions that are equal to or representing the triple count will be labelled with the tag COUNT denoting counting quantifier mentions, i.e., as positive examples. Otherwise, candidate mentions will be labeled with the O tag, i.e., as negative examples, like any other non-candidate mentions (e.g., non-numerals). We built separate training data for each relation P of interest.

Incomplete information from the KB used as the ground truth may negatively affect the quality of training data resulting from the distant supervision approach. To mitigate the effect that KB incompleteness has on training data quality, we investigated filtering the ground truth based on *subject popularity*, according to the number of stored KB triples for that subject, which is also highly correlated with other popularity measures like PageRank or Wikipedia article length. For example, for 10 random entities from the 99th, 90th and 80th percentile w.r.t. popularity, the mean difference between Wikidata children counts and a manually established ground truth from Wikipedia is 0.8, 1.5 and 2.4, respectively. Assuming that popularity and completeness are correlated in general, we can thus trade training data quantity for quality by disregarding less popular entities during training.

Candidate counts that are higher than the KB count are normally considered as not expressing the object count for the relation of interest, i.e., as negative training examples. *But this can also happen to mentions that actually express the correct count*, when the KB is incomplete and only knows counts lower than the correct one. Our remedy is to treat mentions higher than KB counts neither as positive nor as negative examples, but to simply exclude them from the training set. However, there is the need to maintain enough negative examples; otherwise, the classifier would get overly optimistic. For this purpose we utilize upper bound information of triple counts specific to each relation, i.e., the triple count at 99th percentile (e.g., 3 for number of spouses), as found in the KB. A higher count mention will then still be treated as a negative example if it is deemed to be impossible to represent count information for the relation in question.

Furthermore, the more frequent a certain number occurs in a text, the more probable it is to occur in various contexts. As a way to give the classifier less noisy training examples, we ignore sentences that contain count mentions of numbers that have a low entropy in the given text, even when they represent the actual object count. This way we ensure that the models only learn from correct number mentions in the right context.

Linguistic Diversity. As mentioned in the introduction, there are several ways to express count information in natural language text, cardinals and ordinals being only the most obvious ones.

Number-Related Terms. We exploited the *relatedTo* relation in ConceptNet [29] for collecting around 1,200 terms related to numbers. The terms are split into two groups, those having Latin/Greek prefixes[2] and those not having them. For the first group, we generated a list of Latin/Greek prefixes (e.g., *quadr-*) and a list of possible suffixes (e.g., *-plets*). When generating training data, a term with Latin/Greek affixes was labeled with the positive COUNT tag if its prefix matched the triple count. For feature-based models we also replaced such terms in the input with placeholders NUMTERM appended with their Latin/Greek suffixes, while we use the original tokens for neural models.

From the second group we manually selected 15 terms that were especially strongly associated with specific counts (e.g., *twins*, *dozen*). During preprocessing, these terms are then either replaced with corresponding terms/phrases containing cardinal numbers, e.g., *thrice* → *three times* and *a dozen* → *twelve*, or replaced with corresponding Latin/Greek suffix placeholders (e.g. NUMTERM-PLETS for *twins*).

Indefinite Articles. Indefinite articles (i.e., 'a', 'an') are similar to the ordinal *first* insofar as they can express the existence of at least one object. We initially planned to treat them this way, yet due to their overwhelming frequency our classifiers could not cope with them. Thus we now disregard them in the training stage and only consider them as candidate mentions when applying the learned models, by replacing them with the CARDINAL placeholder, and treating them as the mention *one*.

Compositionality. To account for compositional mentions occurring in one sentence, we introduce an extra label, *compositionality tag* (COMP), for the sequence labeling models. During training data generation, we identify consecutive candidate tokens with label COUNT such that *(i)* the sum of their values is equal to the triple count and *(ii)* there exist *compositional cues* (commas and 'and') in between, which are then tagged with the COMP label.

---

[2] http://phrontistery.info/numbers.html.

5 Counting Quantifier Consolidation

Once tokens expressing counting or compositionality information have been identified, these need to be consolidated into a single prediction for the number of objects.

Task 2 (Counting Quantifier Consolidation). *For a given subject S and predicate P, the input to this second stage is a set of token lists, where each token list consists of words/numbers and their corresponding input and output labels (i.e., cardinal, ordinal, numterm, count or comp) and at least one token is tagged cardinal, ordinal or numterm. The desired output is a single number for the counting quantifier for S and P, that is, the correct number of objects for S and P.*

For example, for the pair $\langle AngelinaJolie, hasChild \rangle$, the following token lists may have been detected (annotated as <u>counting information</u> and [compositional cues], with confidences as subscripts):

l_1: *Angelina has a grand total of <u>six</u>$_{0.4}$ children together: <u>three</u>$_{0.3}$ biological [and]$_{0.6}$ <u>three</u>$_{0.5}$ adopted.*

l_2: *The arrival of the <u>first</u>$_{0.5}$ biological child of Jolie and Pitt caused an excited flurry with fans.*

l_3: *On July 12, 2008, she gave birth to <u>twins</u>$_{0.8}$: <u>a</u>$_{0.1}$ son, Knox Léon, [and]$_{0.5}$ <u>a</u>$_{0.2}$ daughter, Vivienne Marcheline.*

We use the following algorithm to consolidate the counting quantifier (CQ) candidates from these labeled token lists.

Algorithm 1 (Mention Consolidation)

1. Sum up compositional mentions. *Mentions having compositional cues in between are summed up, and their confidence score is set to the highest confidence score of the mentions.*
2. Select prediction per type. *For multiple mentions of type cardinal and number-related term, only the mention with the highest confidence is retained if it is above a certain threshold, with compositional mentions treated like cardinals. For ordinals, we always select the highest ordinal available in the candidate pool, regardless of the confidence scores.*
3. Rank mention types. *In the last step, the final prediction is chosen based on the preference $n_{cardinal} \gg n_{numterm} \gg n_{ordinal} \gg n_{article}$, i.e., whenever a cardinal mention exists, it is returned as final answer, otherwise a number-related term, ordinal or article.*

In the example above, in the first step, the two mentions of <u>three</u> in s_1 are summed up to one mention <u>6</u>$_{0.5}$, and the two indefinite articles in s_3 are combined into <u>2</u>$_{0.2}$. In the second step, <u>6</u>$_{0.5}$ is chosen as highest-confidence cardinal, <u>twins</u>$_{0.8}$ as highest ranking numterm (with numerical value 2), and <u>first</u>$_{0.5}$ as highest ranking ordinal. In the last step, the cardinal <u>6</u>$_{0.5}$ or the term <u>twins</u>$_{0.8}$ is chosen as final prediction, depending on whether the confidence threshold is below 0.5 or not.

Confidence Scores. We interpret *marginal probabilities* given by CRFs, i.e., the probability of a token labeled with a certain tag resulting from forward-backward inference, as the *confidence scores* of identified mentions. When a CRF layer is not applied on top of the neural models, the probabilities are simply given by the *softmax* output layer.

Count Zero. We so far only considered counting information for counts greater than zero. Reliably recognizing subjects without objects is difficult for two reasons, (i) because reliable training data is even harder to come by, and (ii) because the count zero is neither expressed via cardinals nor ordinals or indefinite articles. We thus consider count zero only in passing, focusing on two especially frequent ways to express it: *(i)* determiners *'no'* and *'any'* (used in negation) and *(ii)* non-existence-proving adverbs *'without'* and *'never'*. We approach their labeling in a manner similar to the identification of count information via indefinite articles, i.e., not using the count quantifier cues for training but considering them when applying the models.

We performed text preprocessing beforehand to ensure that the non-existence cues can be discovered by the learned models. This preprocessing step includes transforming sentences containing *'not-any'*, *'never'* and *'without'* into sentences containing *'no'* and *'0'*, for example:

They didn't have any children	→	*They have no children*
He has never been married	→	*He has been married 0 times*
The marriage was without children	→	*The marriage was with no children.*

Finally, textual occurrences of *'no'* and *'0'* are replaced with CARDINAL and treated as count zero.

6   Experiments

6.1   Experimental Setup

Dataset. We chose Wikidata as our source KB and Wikipedia pages about given subject entities as our source text for the distant supervision approach.[3] While some Wikidata properties are self-explanatory, like *child* or *spouse*, some others are overloaded, i.e., used in highly diverse domains with different semantics depending on the type of the subject entities, e.g. *has part*. Thus, we define *relations* in our experiments as pairs of a Wikidata subject type/class and a Wikidata property. We focus on five diverse relations (listed in Table 1 under the *Relation* column) using the four Wikidata properties already used in [22], but using two specific Wikidata classes for the overloaded *has part* property, i.e., *series of creative works* and *musical ensemble*. We use four sets of entities for training and evaluation:

---

[3] Both in their version as of March 20, 2017.

Table 1. Number of Wikidata instances as subjects (#Subject) of each relation in the training set.

Wikidata subject class	Wikidata property	Relation	#Subjects
series of creative works (Q7725310)	has part (P527)	containsWork	642
musical ensemble (Q2088357)	has part (P527)	hasMember	8,901
admin. territ. entity (Q56061)	contains admin. territ. entity (P150)	containsAdmin	6,266
human (Q5)	child (P40)	hasChild	40,145
human (Q5)	spouse (P26)	hasSpouse	45,261

1. *Training set*: For each relation, all subject entities with an English Wikipedia page that have at least one object in Wikidata, except those used for development and testing (counts are shown in Table 1).
2. *Manual test set*: 200 entities per relation randomly chosen from the training set (i.e., have at least one object).
3. *Automated test set*: 200 of the 10% most popular entities per relation removed from the training set (i.e., have at least one object).
4. *Zero-count test set*: 64 and 168 entities for the *hasChild* and *hasSpouse* relations, respectively, which are entities in Wikidata having child (P40) and spouse (P26) properties set to the special value *no-value*.

For the *manual test set* we manually annotated mentions in text that correspond to counting quantifiers, and established the correct object count from Wikipedia. The *automated test set* is used for parameter tuning of the neural models, and as silver standard for evaluating our system beyond the 5 gold-annotated relations. For evaluating zero-count quantifier detection, we use two relations for which manually created data from Wikidata is available.

Hyperparameters. We set 0.1 as the confidence score threshold in the mention consolidation task (Sect. 5), after experimenting with varying values. For training the neural models, we employed Adam [12] with a learning rate of 0.001. Using stochastic gradient descent (SGD) with a gradient clipping of 5.0 as reported in [15] results in worse performance. The LSTM network uses a single layer with 300 dimensions. The hidden dimension of the forward and backward character LSTMs are 100. We set the dropout rate to 0.5. We also use GloVe pre-trained embeddings [26] to initialize our lookup table.

6.2   Evaluation

Evaluation Scheme. We evaluate our system, CINEX (Counting Information Extraction), on quantifier recognition, quantifier consolidation, and on the end-to-end task with the following metrics:

We use *precision*, *recall* and *F1-score* to evaluate how well the system can identify counting information in a given text. For entities for which the system recognized at least one counting quantifier (CQ) candidate, we then measure *precision* in choosing the correct final CQ. Finally, we evaluate the system for

Table 2. Performance of CINEX on recognizing counting quantifier mentions, with different architectures and in comparison with the baseline. Highest F1-score per relation in boldface.

Relation	Baseline [22]			CINEX								
				CRF			biLSTM			biLSTM-CRF		
	P	R	F1	P	R	F1	P	R	F1	P	R	F1
containsWork	22.4	24.0	23.1	61.9	29.3	**39.8**	61.1	19.6	29.6	54.9	28.9	37.8
hasMember	1.5	4.3	2.2	55.7	56.5	**56.1**	38.2	18.8	25.2	35.9	33.3	34.6
containsAdmin	51.1	64.3	57.0	72.5	82.9	77.3	78.4	82.9	80.6	78.7	84.3	**81.4**
hasChild	6.4	49.4	11.4	54.5	44.4	**49.0**	33.9	11.7	17.4	26.1	14.8	18.9
hasSpouse	1.9	12.1	3.3	58.2	67.2	**62.4**	20.4	36.2	26.1	27.1	32.8	29.7

Table 3. Performance of CINEX-CRF on recognizing counting quantifier mentions, per mention type. *Numt.* stands for number-related terms, *Art.* for indefinite articles. Baseline comparison is only for cardinals (highest F1-score per relation in boldface).

Relation	Baseline [22]			CINEX-CRF (per type)								
	Cardinals			Cardinals			Numt. + Art.			Ordinals		
	P	R	F1	P	R	F1	P	R	F1	P	R	F1
containsWork	22.4	77.8	**34.8**	60.0	18.3	28.1	53.1	98.1	68.9	77.6	19.9	31.7
hasMember	1.5	25.0	2.9	50.0	33.3	**40.0**	55.7	64.2	59.6	100	25.0	40.0
containsAdmin	51.1	64.3	57.0	84.1	82.9	**83.5**	0	0	0	0	0	0
hasChild	6.4	72.7	11.8	75.6	56.9	**64.9**	24.3	100	39.1	7.7	2.3	3.5
hasSpouse	1.9	87.5	3.7	76.9	90.9	**83.3**	0	0	0	85.3	63.0	72.5

the end-to-end task in terms of *coverage*, i.e., for how many subject entities the system can extract correct object counts from text, and *Mean Absolute Error (MAE)*, to understand how much system predictions deviate from the truth.

Quantifier Recognition. We report in Table 2 the performance results of different architectures w.r.t. precision, recall and F1-score. We also compare our system with the best performing method for extracting cardinals reported in [22] as baseline. As one can see, feature-based CRF models are the most robust sequence labeling approach across relations for this task, although the neural models achieve higher F1-score with 3.3 percentage point difference for *containsAdmin*. Adding a CRF layer on top of bidirectional LSTM models improves performance across relations, although this architecture still fails to beat the feature-based CRF models in most cases. We conjecture that this is due to neural models being much more prone to overfitting to noisy distantly supervised training data. Still, both feature-based and neural models consistently outperform the baseline by a large margin, in particular w.r.t. precision.

In Table 3 we split this analysis further by mention type. This provides a more fair comparison with the baseline that only considers cardinal numbers. Still, CINEX-CRF achieves a higher precision on all relations, and a higher F1-

Table 4. Performance of CINEX-CRF in consolidating counting quantifier mentions w.r.t. precision (P), coverage (Cov) and MAE. *Numt.* stands for number-related terms, *Art.* for articles. Results per type show contribution ($Contr$) to overall output and precision of individual types.

Relation	Baseline [22]			CINEX-CRF			CINEX-CRF (per type)					
							Cardinals		Numt. + Art.		Ordinals	
	P	Cov	MAE	P	Cov	MAE	P	Contr	P	Contr	P	Contr
containsWork	42.0	29.0	3.7	**49.2**	**29.0**	**2.6**	55.0	33.9	62.5	40.7	20.0	25.4
hasMember	11.8	6.0	3.8	**64.3**	**18.0**	**1.2**	62.5	28.6	65.0	71.4	0	0
containsAdmin	51.8	14.5	7.3	**78.6**	**22.0**	**1.7**	85.7	87.5	33.3	10.7	0	1.8
hasChild	37.0	**22.0**	2.2	**50.0**	19.5	2.3	67.3	70.5	6.3	20.5	14.3	9.0
hasSpouse	26.8	11.0	1.3	**58.1**	**12.5**	**0.5**	75.0	18.6	43.8	37.2	63.2	44.2
hasZeroChild				92.3	18.8	-						
hasZeroSpouse				71.9	13.7	-						

score on 4 out of 5. We also see variety within the mention types and relations, ordinals for instance being well picked up for *hasSpouse*, but badly for *hasChild*.

Quantifier Consolidation. Table 4 shows the performance of CINEX-CRF, our best performing system for recognizing counting information, on the consolidation and end-to-end task. We report the results broken down per mention type, as well as in overall.

Table 5. Examples of correct and incorrect predictions by CINEX-CRF.

	Relation	Subject	#O	Predicted counting quantifiers	
Correct	containsWork	The Heroes of Olympus	5	The Heroes of Olympus is a <u>pentalogy</u> of adventure...	5
	hasMember	Siria	2	The music <u>duo</u> Siria is composed of...	2
	containsAdmin	Gusevsky District	5	...was subdivided into <u>one</u> urban settlement and <u>four</u> rural settlements.	5
	hasChild	Hanna Neumann	5	Four of her <u>five</u> children became mathematicians...	5
	hasSpouse	Hannelore Schroth	3	Her <u>third</u> marriage to a lawyer produced a son...	3
Incorrect	containsWork	Scandal (TV series)	7	...this season was split into two runs, the first consisting of <u>ten</u> episodes.	10
	hasMember	Ladysmith Black Mambazo	9	...Mazibuko (the eldest of the <u>six</u> brothers) joined Mambazo...	6
	containsAdmin	Cottbus	4	Cottbus has <u>a</u> football team called FC Energie Cottbus...	1
	hasChild	Barack Obama	2	The couple's <u>first</u> daughter, Malia Ann, was born on July 4, 1998.	1
	hasSpouse	Ruth Williams Khama	1	...and <u>twins</u> Anthony and Tshekedi were born in Bechuanaland...	2

In predicting counting quantifiers through recognizing cardinals in text, CINEX-CRF achieves 55–85% precision. This is a considerable improvement (up to 48.9 percentage points) compared to the baseline [22]. Although the baseline yields a comparable coverage, its low precision suggests that it has difficulties to pick up correct context and produces some matches only by chance.

Number-related terms and articles are beneficial in improving coverage particularly for *containsWork* and *hasMember*, yet produce low precision results for *hasChild*, possibly due to spurious indefinite articles frequently identified as counting quantifiers. Overall, taking compositionality as well as mention types

Table 6. KB enrichment potential for 40 relations, showing only relations with accuracy (Acc) >50% and coverage (Cov) >5%.

Wikidata subject class	Wikidata property	P	Cov	#Existing facts	#Missing facts	KB increase
duo	has part	88.9	26.7	561	51	9.1%
rock band	has part	78.6	18.3	1,148	187	16.3%
band	has part	70.2	16.5	9,342	3,905	41.8%
township of China	contains admin	100.0	63.0	7,254	19	0.3%
municipality with town privileges	contains admin	100.0	13.7	3,343	25	0.7%
amphoe (subdivision of Thailand)	contains admin	98.0	63.2	6,226	1,032	16.6%
town in China	contains admin	97.8	29.0	38,894	377	1.0%
canton of France (until 2015)	contains admin	97.2	38.5	9,191	189	2.1%
county of China	contains admin	89.5	35.7	22,401	236	1.1%
District of China	contains admin	88.9	35.6	11,828	170	1.4%
municipality of the Czech Republic	contains admin	76.9	5.0	8,279	184	2.2%
fictional human	child	100.0	9.1	327	141	43.1%
race horse	child	87.0	27.4	1,800	1,742	96.8%
mythological Greek character	child	85.7	21.4	624	44	7.1%
human biblical figure	child	66.7	16.7	274	42	15.3%
human	child	58.8	28.5	73,527	117,942	160.4%
human	spouse	61.4	17.5	50,373	48,778	96.8%
		Total (over all 40)		**224,216**	**173,256**	**77.3%**

other than cardinals into account improve both accuracy and coverage of the system, with MAE of not more than 2.6 across relations. The performance of CINEX-CRF on predicting non-existence of objects is reported in the last two rows of Table 4. We obtain a high accuracy of 92.3% for *hasChild* and 71.9% for *hasSpouse*.

Qualitative Analysis. Table 5 lists notable examples of correct and incorrect predictions. Errors for *hasMember* and *hasSpouse* are sometimes caused by wrongly labelled mentions that are related instead with other relations, e.g., musical ensemble members and siblings. For some relations, understanding the fine-grained types of subject entities may help in choosing the correct context of counting quantifiers. For instance, a *TV series* consists of *seasons* while a specific season of the series contains *episodes*.

Notable is also the low precision of ordinals shown in Table 4. A main reason is that ordinals only reliably express lower bounds (see e.g. fourth incorrect example). If one considers ordinals as correct whenever they are not higher than the true count, the reported precision scores increase from 14.3–63.2% to 85.7–89.5%.

6.3   KB Enrichment Potential

In this section we return to our original goal of enlarging the number of facts known to exist. We investigate the potential of CINEX on 40 relations, by focusing on the 4 previously used Wikidata properties, but looking at the up to 10 most frequent subject classes of entities using each property. For each relation, we then perform automated evaluation of CINEX as described in Sect. 6.1. In Table 6, we report relations for which CINEX-CRF gave precision >0.5 and coverage >0.05. For each relation we report the number of existing facts in Wikidata, and the existence of how many more facts we can infer from the counting quantifiers. For instance, we can derive the existence of 160.4% more children relationships than currently stored. In sum, CINEX is able to identify the existence of 173K more facts than Wikidata currently knows, thus increasing the existential knowledge of Wikidata for these 40 relations by 77.3%.

We also applied CINEX to all human entities to find out how many subjects are found to have no objects w.r.t. the *hasChild* and *hasSpouse* relations, finding 1,648 instances for children and 557 for spouses. These assertions increase the existing known zero cases in Wikidata for both relations by a factor of 25.8 and 3.3, respectively.

Table 7. Classes along with relations for which count information could be retrieved best.

Human	Creative works	Admin. territorial	Musical ensemble	Organization	Transport. facility
occupation	nominated for	contains settlement	has part	subsidiary	connecting line
employer	genre	contains admin. territorial	nominated for	founded by	adjacent station
influenced by	cast member	capital of	record label	-	-
award received	screenwriter	member of	award received	-	-
child	voice actor	sister city	genre	-	-

6.4   Count Information Across KB Relations

So far we only evaluated CINEX on four manually chosen Wikidata properties. In this section we investigate to which extent counting quantifiers are present for arbitrary relations, and to which extent they can be extracted by CINEX.

To this end, we collected all Wikidata properties that were interesting, i.e., were not asserted to be single-value[4], had a functionality degree (#*subjects*/#*triples*) of less than 0.98 [10], and were used by at least 500 subjects, obtaining 267 properties in total. For each of these properties, we identified the 10 most frequent entity classes used as subjects, resulting in a total of 2,474 relations. For each relation, we then performed automated evaluation of CINEX as described in Sect. 6.1, finding 110 relations for which CINEX gave precision >50% and coverage >5%.

---

[4] Properties having the constraint https://www.wikidata.org/wiki/Q19474404.

Among the frequent classes (grouped by theme) of subjects for which we can mine counting quantifiers from the corresponding Wikipedia pages are: *human* (including twin, fictional human, biblical figure and mythological Greek character), *creative works* (e.g., film, television series), *administrative territorial entity* (e.g., country, municipality), *musical ensemble* (e.g., band, duo), *organization* (e.g., business enterprise, nonprofit organization) and *transportation facility* (e.g., metro station, train station). We show in Table 7 the top 5 Wikidata properties for each mentioned subject type. Other notable relations include: <*battle, participant*>, <*human spaceflight, crew member*> and <*star, child astronomical body*>.

In terms of KB enrichment, CINEX was able to extract a total of 851K counting quantifier facts, which in turn state the existence of 2.5M facts not yet asserted for these 110 <*Wikidata class, Wikidata property*> pairs. These existential facts, provided on Github, increase the number of facts known to exist for these relations by 28.3%.

7   Related Work

Knowledge bases have seen a rise of attention in recent years. Aside from a few manual efforts like Wikidata, the construction of these knowledge bases is usually done via automated information extraction, focusing either on structured data (DBpedia [1], YAGO [31]), or on unstructured contents from the web. For the latter, directions include extracting arbitrary facts without predefined schema, called Open IE [6,19,23], and extracting triples based on well-defined knowledge base relations [13,25,33], in which the distant supervision approach is widely used [3,21,32]. The idea of distant supervision is to use facts from an existing KB in order to label sentences as positive/negative training samples, depending on whether the entities from the existing facts occur in them or not. A major challenge for distant supervision is knowledge base incompleteness: If the KB used for labeling the training data misses facts, candidates may wrongly be classified as negative samples, reducing the quality of the learning process. Approaches to mitigate this effect include heavily under-sampling the negative evidence [27,33], to learn only from positive samples [20], or to use heuristics in selecting negative samples [9,10], yet these do not help with potentially wrong seed counts.

Most works on information extraction focus on relations that link entities, like ⟨*Trump, presidentOf, USA*⟩, or that store String or measurement values. Counting quantifiers have received comparably little attention. Numbers, a major construct for expressing counts, were investigated mostly in the context of temporal information, e.g. to enrich facts with timestamps/durations [16,30], or in the context of quantities and measures like ⟨*MtEverest, height, 8848mt*⟩ [11,17,24,28]. In contrast, terms that express counting quantifiers are either extracted incorrectly by state-of-the-art Open-IE systems, or not at all. While NELL, for instance, knows 13 relations about the number of casualties and injuries in disasters, they all contain only seed facts and no learned facts. In [22], which we use as baseline for our experiments, we have proposed a single-stage process for identifying numbers that express relation counts. Yet, we there only consider explicit cardinals and do not tackle training data incompleteness nor compositionality, thus achieving only moderate precision and coverage.

While a few counting qualifier predicates such as *number of children*, *number of seasons* (of a TV series) or *number of households* (of a territory) already exist in Wikidata, it should be noted that a proper interpretation of counting quantifiers requires to go beyond the standard open-world assumption of the Semantic Web, as they allow to infer negative information. Appropriate models require to combine open-world and closed-world reasoning, as does for instance the local closed-world assumption [5,7].

8   Conclusions

We have proposed to enrich KBs with counting quantifiers, and discussed the challenges that set counting quantifier extraction apart from standard information extraction. In particular, we showed that it is imperative to consider the compositionality of counts, and their expression in non-numeric form. We have shown that our system, CINEX, can extract counting quantifiers with 60% average precision on five relations, and when applied to a large set of relations, it is possible to extend the number of facts known to exist in 110 of them by 28%. We believe that the extraction of counting quantifiers opens interesting avenues for tasks such as question answering, information extraction or KB curation. Our data and code are available at https://github.com/paramitamirza/CINEX.

References

1. Auer, S., Bizer, C., Kobilarov, G., Lehmann, J., Cyganiak, R., Ives, Z.: DBpedia: a nucleus for a web of open data. In: Aberer, K., et al. (eds.) ASWC/ISWC -2007. LNCS, vol. 4825, pp. 722–735. Springer, Heidelberg (2007). https://doi.org/10.1007/978-3-540-76298-0_52
2. Brin, S.: Extracting patterns and relations from the World Wide Web. In: WebDB (1998)
3. Craven, M., Kumlien, J., et al.: Constructing biological knowledge bases by extracting information from text sources. In: ISMB (1999)
4. Dang, H.T., Kelly, D., Lin, J.J.: Overview of the TREC 2007 question answering track. TREC **7**, 63 (2007)
5. Darari, F., Nutt, W., Pirrò, G., Razniewski, S.: Completeness statements about RDF data sources and their use for query answering. In: Alani, H., et al. (eds.) ISWC 2013. LNCS, vol. 8218, pp. 66–83. Springer, Heidelberg (2013). https://doi.org/10.1007/978-3-642-41335-3_5
6. Del Corro, L., Gemulla, R.: ClausIE: clause-based open information extraction. In: WWW (2013)
7. Denecker, M., Cortés-Calabuig, A., Bruynooghe, M., Arieli, O.: Towards a logical reconstruction of a theory for locally closed databases. ACM Trans. Database Syst. **35**(3) (2010)

8. Dong, X.L., et al.: From data fusion to knowledge fusion. PVLDB **7**(10), 881–892 (2014)
9. Dong, X.L., et al.: Knowledge vault: a web-scale approach to probabilistic knowledge fusion. In: KDD (2014)
10. Galárraga, L., Teflioudi, C., Hose, K., Suchanek, F.M.: Fast rule mining in ontological knowledge bases with AMIE+. VLDB J. **24**(6), 707–730 (2015)
11. Ibrahim, Y., Riedewald, M., Weikum, G.: Making sense of entities and quantities in web tables. In: CIKM (2016)
12. Kingma, D., Ba, J.: Adam: a method for stochastic optimization. arXiv:1412.6980 (2014)
13. Koch, M., Gilmer, J., Soderland, S., Weld, D.S.: Type-aware distantly supervised relation extraction with linked arguments. In: EMNLP (2014)
14. Kudo, T.: CRF++: Yet another CRF toolkit (2005). https://sourceforge.net/projects/crfpp/
15. Lample, G., Ballesteros, M., Subramanian, S., Kawakami, K., Dyer, C.: Neural architectures for named entity recognition. In: NAACL (2016)
16. Ling, X., Weld, D.S.: Temporal information extraction. In: AAAI (2010)
17. Madaan, A., Mittal, A., Mausam, G.R., Ramakrishnan, G., Sarawagi, S.: Numerical relation extraction with minimal supervision. In: AAAI (2016)
18. Mausam: Open information extraction systems and downstream applications. In: IJCAI (2016)
19. Mausam, Schmitz, M., Soderland, S., Bart, R., Etzioni, O.: Open language learning for information extraction. In: EMNLP (2012)
20. Min, B., Grishman, R., Wan, L., Wang, C., Gondek, D.: Distant supervision for relation extraction with an incomplete knowledge base. In: HLT-NAACL (2013)
21. Mintz, M., Bills, S., Snow, R., Jurafsky, D.: Distant supervision for relation extraction without labeled data. In: ACL/IJCNLP (2009)
22. Mirza, P., Razniewski, S., Darari, F., Weikum, G.: Cardinal virtues: extracting relation cardinalities from text. In: ACL 2017 (Short Papers) (2017)
23. Mitchell, T.M., et al.: Never-ending learning. In: AAAI (2015)
24. Neumaier, S., Umbrich, J., Parreira, J.X., Polleres, A.: Multi-level semantic labelling of numerical values. In: Groth, P., et al. (eds.) ISWC 2016. LNCS, vol. 9981, pp. 428–445. Springer, Cham (2016). https://doi.org/10.1007/978-3-319-46523-4_26
25. Palomares, T., Ahres, Y., Kangaspunta, J., Ré, C.: Wikipedia knowledge graph with DeepDive. In: ICWSM (2016)
26. Pennington, J., Socher, R., Manning, C.D.: GloVe: global vectors for word representation. In: EMNLP (2014)
27. Riedel, S., Yao, L., McCallum, A.: Modeling relations and their mentions without labeled text. In: Balcázar, J.L., Bonchi, F., Gionis, A., Sebag, M. (eds.) ECML PKDD 2010. LNCS (LNAI), vol. 6323, pp. 148–163. Springer, Heidelberg (2010). https://doi.org/10.1007/978-3-642-15939-8_10
28. Saha, S., Pal, H., Mausam: Bootstrapping for numerical open IE. In: ACL (2017)
29. Speer, R., Havasi, C.: Representing general relational knowledge in ConceptNet 5. In: LREC (2012)
30. Strötgen, J., Gertz, M.: Heideltime: high quality rule-based extraction and normalization of temporal expressions. In: SemEval Workshop (2010)
31. Suchanek, F.M., Kasneci, G., Weikum, G.: YAGO: a core of semantic knowledge. In: WWW (2007)
32. Suchanek, F.M., Sozio, M., Weikum, G.: SOFIE: a self-organizing framework for information extraction. In: WWW (2009)

33. Surdeanu, M., Tibshirani, J., Nallapati, R., Manning, C.D.: Multi-instance multi-label learning for relation extraction. In: ACL (2012)
34. Tan, C.H., Agichtein, E., Ipeirotis, P., Gabrilovich, E.: Trust, but verify: predicting contribution quality for knowledge base construction and curation. In: WSDM (2014)
35. Vrandečić, D., Krötzsch, M.: Wikidata: a free collaborative knowledgebase. In: CACM (2014)

QA4IE: A Question Answering Based Framework for Information Extraction

Lin Qiu[1]([✉]), Hao Zhou[1], Yanru Qu[1], Weinan Zhang[1], Suoheng Li[2],
Shu Rong[2], Dongyu Ru[1], Lihua Qian[1], Kewei Tu[3], and Yong Yu[1]

[1] Shanghai Jiao Tong University, Shanghai, China
{lqiu,kevinqu,maxru,qianlihua,yyu}@apex.sjtu.edu.cn,
{zhou1998,wnzhang}@sjtu.edu.cn
[2] Yitu Tech, Shanghai, China
{suoheng.li,shu.rong}@yitu-inc.com
[3] ShanghaiTech University, Shanghai, China
tukw@shanghaitech.edu.cn

Abstract. Information Extraction (IE) refers to automatically extracting structured relation tuples from unstructured texts. Common IE solutions, including Relation Extraction (RE) and open IE systems, can hardly handle cross-sentence tuples, and are severely restricted by limited relation types as well as informal relation specifications (e.g., free-text based relation tuples). In order to overcome these weaknesses, we propose a novel IE framework named QA4IE, which leverages the flexible question answering (QA) approaches to produce high quality relation triples across sentences. Based on the framework, we develop a large IE benchmark with high quality human evaluation. This benchmark contains 293K documents, 2M golden relation triples, and 636 relation types. We compare our system with some IE baselines on our benchmark and the results show that our system achieves great improvements.

1 Introduction and Background

Information Extraction (IE), which refers to extracting structured information (i.e., relation tuples) from unstructured text, is the key problem in making use of large-scale texts. High quality extracted relation tuples can be used in various downstream applications such as Knowledge Base Population [16], Knowledge Graph Acquisition [22], and Natural Language Understanding. However, existing IE systems still cannot produce high-quality relation tuples to effectively support downstream applications.

1.1 Previous IE Systems

Most of previous IE systems can be divided into Relation Extraction (RE) based systems [27,51] and Open IE systems [3,8,36].

Early work on RE decomposes the problem into Named Entity Recognition (NER) and relation classification. With the recent development of neural networks (NN), NN based NER models [18,26] and relation classification models [48]

© Springer Nature Switzerland AG 2018
D. Vrandečić et al. (Eds.): ISWC 2018, LNCS 11136, pp. 198–216, 2018.
https://doi.org/10.1007/978-3-030-00671-6_12

show better performance than previous handcrafted feature based methods. The recently proposed RE systems [33,52] try to jointly perform entity recognition and relation extraction to improve the performance. One limitation of existing RE benchmarks, e.g., NYT [34], Wiki-KBP [23] and BioInfer [31], is that they only involve 24, 19 and 94 relation types respectively comparing with thousands of relation types in knowledge bases such as DBpedia [4,6]. Besides, existing RE systems can only extract relation tuples from a single sentence while the cross-sentence information is ignored. Therefore, existing RE based systems are not powerful enough to support downstream applications in terms of performance or scalability.

On the other hand, early work on Open IE is mainly based on bootstrapping and pattern learning methods [2]. Recent work incorporates lexical features and sentence parsing results to automatically build a large number of pattern templates, based on which the systems can extract relation tuples from an input sentence [3,8,36]. An obvious weakness is that the extracted relations are formed by free texts which means they may be polysemous or synonymous and thus cannot be directly used without disambiguation and aggregation. The extracted free-text relations also bring extra manual evaluation cost, and how to automatically evaluate different Open IE systems fairly is an open problem. Stanovsky and Dagan [41] try to solve this problem by creating an Open IE benchmark with the help of QA-SRL annotations [10]. Nevertheless, the benchmark only involves 10K golden relation tuples. Hence, Open IE in its current form cannot provide a satisfactory solution to high-quality IE that supports downstream applications.

There are some recently proposed IE approaches which try to incorporate Question Answering (QA) techniques into IE. Levy et al. [21] propose to reduce the RE problem to answering simple reading comprehension questions. They build a question template for each relation type, and by asking questions with a relevant sentence and the first entity given, they can obtain relation triples from the sentence corresponding to the relation type and the first entity. Roth et al. [35] further improve the model performance on a similar problem setting. However, these approaches focus on sentence level relation argument extractions and do not provide a full-stack solution to general IE. In particular, they do not provide a solution to extract the first entity and its corresponding relation types before applying QA. Besides, sentence level relation extraction ignores the information across sentences such as coreference and inference between sentences, which greatly reduces the information extracted from the documents.

1.2 QA4IE Framework

To overcome the above weaknesses of existing IE systems, we propose a novel IE framework named QA4IE to perform document level general IE with the help of state-of-the-art approaches in Question Answering (QA) and Machine Reading Comprehension (MRC) area.

The input of QA4IE is a document D with an existing knowledge base K and the output is a set of relation triples $R = \{e_i, r_{ij}, e_j\}$ in D where e_i and e_j are two individual entities and r_{ij} is their relation. We ignore the adverbials and only consider the entity pairs and their relations as in standard RE settings. Note that we process the entire document as a whole instead of processing individual sentences separately as in previous systems. As shown in Fig. 1, our QA4IE framework consists of four key steps:

1. Recognize all the candidate entities in the input document D according to the knowledge base K. These entities serve as the first entity e_i in the relation triples R.
2. For each candidate entity e_i, discover the potential relations/properties as r_{ij} from the knowledge base K.
3. Given a candidate entity-relation or entity-property pair $\{e_i, r_{ij}\}$ as a query, find the corresponding entity or value e_j in the input document D using a QA system. The query here can be directly formed by the word sequence of $\{e_i, r_{ij}\}$, or built from templates as in [21].
4. Since the results of step 3 are formed by free texts in the input document D, we need to link the results to the knowledge base K.

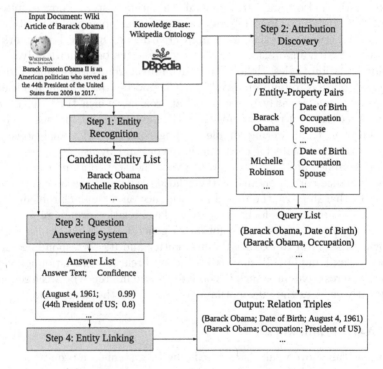

Fig. 1. An overview of our QA4IE framework.

This framework determines each of the three elements in relation triples step by step. Step 1 is equivalent to named entity recognition (NER), and state-of-the-art NER systems [25,26] can achieve over 0.91 F1-score on CoNLL'03 [43], a well-known NER benchmark. For attribution discovery in step 2, we can take advantage of existing knowledge base ontologies such as Wikipedia Ontology to obtain a candidate relation/property list according to NER results in step 1. Besides, there is also some existing work on attribution discovery [20, 49] and ontology construction [9] that can be used to solve the problem in step 2. The most difficult part in our framework is step 3 in which we need to find the entity (or value) e_j in document D according to the previous entity-relation (or entity-property) pair $\{e_i, r_{ij}\}$. Inspired by recent success in QA and MRC [37, 46, 47], we propose to solve step 3 in the setting of SQuAD [32] which is a very popular QA task. The problem setting of SQuAD is that given a document $\tilde{D}$ and a question q, output a segment of text a in $\tilde{D}$ as the answer to the question. In our framework, we assign the input document D as $\tilde{D}$ and the entity-relation (or entity-property) pair $\{e_i, r_{ij}\}$ as q, and then we can get the answer a with a QA model. Finally in step 4, since the QA model can only produce answers formed by input free texts, we need to link the answer a to an entity e_j in the knowledge base K, and the entity e_j will form the target relation triple as $\{e_i, r_{ij}, e_j\}$. Existing Entity Linking (EL) systems [28, 38] directly solve this problem especially when we have high quality QA results from step 3.

As mentioned above, step 1, 2 and 4 in the QA4IE framework can be solved by existing work. Therefore, in this paper, we mainly focus on step 3. According to the recent progress in QA and MRC, deep neural networks are very good at solving this kind of problem with a large-scale dataset to train the network. However, all previous IE benchmarks [41] are too small to train neural network models typically used in QA, and thus we need to build a large benchmark. Inspired by WikiReading [12], a recent large-scale QA benchmark over Wikipedia, we find that the articles in Wikipedia together with the high quality triples in knowledge bases such as Wikidata [45] and DBpedia can form the supervision we need. Therefore, we build a large scale benchmark named QA4IE benchmark which consists of 293K Wikipedia articles and 2M golden relation triples with 636 different relation types.

Recent success on QA and MRC is mainly attributed to advanced deep learning architectures such as attention-based and memory-augmented neural networks [5, 42] and the availability of large-scale datasets [11, 13] especially SQuAD. The differences between step 3 and SQuAD can be summarized as follows. First, the answer to the question in SQuAD is restricted to a continuous segment of the input text, but in QA4IE, we remove this constraint which may reduce the number of target relation triples. Second, in existing QA and MRC benchmarks, the input documents are not very long and the questions may be complex and difficult to understand by the model, while in QA4IE, the input documents may be longer but the questions formed by entity-relation (or entity-property) pair are much simpler. Therefore, in our model, we incorporate Pointer Networks [44] to adapt to the answers formed by any words within the document in any order as well as Self-Matching Networks [47] to enhance the ability on modeling longer input documents.

1.3   Contributions

The contributions of this paper are as follows:

1. We propose a novel IE framework named QA4IE to overcome the weaknesses of existing IE systems. As we discussed above, the problem of step 1, 2 and 4 can be solved by existing work and we propose to solve the problem of step 3 with QA models.
2. To train a high quality neural network QA model, we build a large IE benchmark in QA style named QA4IE benchmark which consists of 293K Wikipedia articles and 2 million golden relation triples with 636 different relation types.
3. To adapt QA models to the IE problem, we propose an approach that enhances existing QA models with Pointer Networks and Self-Matching Networks.
4. We compare our model with IE baselines on our QA4IE benchmark and achieve a great improvement over previous baselines.
5. We open source our code and benchmark for repeatable experiments and further study of IE.[1]

2   QA4IE Benchmark Construction

This section briefly presents the construction pipeline of QA4IE benchmark to solve the problem of step 3 as in our framework (Fig. 1). Existing largest IE benchmark [41] is created with the help of QA-SRL annotations [10] which consists of 3.2K sentences and 10K golden extractions. Following this idea, we study recent large-scale QA and MRC datasets and find that WikiReading [12] creates a large-scale QA dataset based on Wikipedia articles and WikiData relation triples [45]. However, we observe about 11% of QA pairs with errors such as wrong answer locations or mismatch between answer string and answer words. Besides, there are over 50% of QA pairs with the answer involving words out of the input text or containing multiple answers. We consider these cases out of the problem scope of this paper and only focus on the information within the input text.

Therefore, we choose to build the benchmark referring the implementation of WikiReading based on Wikipedia articles and golden triples from Wikidata and DBpedia [4,6]. Specifically, we build our QA4IE benchmark in the following steps.

Dump and Preprocessing. We dump the English Wikipedia articles with Wikidata knowledge base and match each article with its corresponding relation triples according to its title. After cleaning data by removing low frequency tokens and special characters, we obtain over 4M articles and 18M triples with over 800 relation types.

---

[1] Our source code and benchmark datasets can be found at https://github.com/SJTU-lqiu/QA4IE.

Clipping. We discard the triples with multiple entities (or values) for e_j (account for about 6%, e.g., a book may have multiple authors). Besides, we discard the triples with any word in e_j out of the corresponding article (account for about 50%). After this step, we obtain about 3.5M articles and 9M triples with 636 relation types.

Incorporating DBpedia. Unlike WikiData, DBpedia is constructed automatically without human verification. Relations and properties in DBpedia are coarse and noisy. Thus we fix the existing 636 relation types in WikiData and build a projection from DBpedia relations to these 636 relation types. We manually find 148 relations which can be projected to a WikiData relation out of 2064 DBpedia relations. Then we gather all the DBpedia triples with the first entity is corresponding to one of the above 3.5M articles and the relation is one of the projected 148 relations. After the same clipping process as above and removing the repetitive triples, we obtain 394K additional triples in 302K existing Wikipedia articles.

Distillation. Since our benchmark is for IE, we prefer the articles with more golden triples involved by assuming that Wikipedia articles with more annotated triples are more informative and better annotated. Therefore, we figure out the distribution of the number of golden triples in articles and decide to discard the articles with less than 6 golden triples (account for about 80%). After this step, we obtain about 200K articles and 1.4M triples with 636 relation types.

Query and Answer Assignment. For each golden triple $\{e_i, r_{ij}, e_j\}$, we assign the relation/property r_{ij} as the query and the entity e_j as the answer because the Wikipedia article and its corresponding golden triples are all about the same entity e_i which is unnecessary in the queries. Besides, we find the location of each e_j in the corresponding article as the answer location. As we discussed in Sect. 1, we do not restrict e_j to a continuous segment in the article as required in SQuAD. Thus we first try to detect a matched span for each e_j and assign this span as the answer location. Then for each of the rest e_j which has no matched span, we search a matched sub-sequence in the article and assign the index sequence as the answer location. We name them **span-triples** and **seq-triples** respectively. Note that each triple will have an answer location because we have discarded the triples with unseen words in e_j and if we can find multiple answer locations, all of them will be assigned as ground truths.

Dataset Splitting. For comparing the performance on span-triples and seq-triples, we set up two different datasets named QA4IE-SPAN and QA4IE-SEQ. In QA4IE-SPAN, only articles with all span-triples are involved, while in QA4IE-SEQ, articles with seq-triples are also involved. For studying the influence of the article length as longer articles are normally more difficult to model by LSTMs, we split the articles according to the article length. We name the set of articles with lengths shorter than 400 as S, lengths between 400 and 700 as M, lengths greater than 700 as L. Therefore we obtain 6 different datasets named QA4IE-SPAN-S/M/L and QA4IE-SEQ-S/M/L. A 5/1/5 splitting of train/dev/test

Table 1. Detailed statistics of QA4IE benchmark.

		S	M	L	Total
SPAN	# Docs	52898	29352	65124	147374
	# Triples	342361	195944	457509	995814
SEQ	# Docs	52559	29188	64385	146132
	# Triples	341820	196138	457033	994991
	# Seq-triples	46521	27176	57507	131204
	%Seq-triples	13.61	13.86	12.58	13.19

Table 2. Comparison between existing IE benchmarks and QA benchmarks. The first two are IE benchmarks and the rest four are QA benchmarks.

Dataset	Source	#Docs	#Triples/queries	Remarks
QA4IE benchmark	Wikipedia/ WikiData/ DBpedia	293K	2M	Automatical generation
Open IE [41]	WSJ/Wikipedia	3.2K	10K	Generated from QA-SRL annotations
Zero-Shot Benchmark [21]	Wikipedia/WikiData	N/A	30M	Sentence level docs, only 120 relation types
WikiReading [12]	Wikipedia/WikiData	4.7M	18.58M	11% errors, 50% out of document answers
SQuAD [32]	Wikipedia	536	100K	Crowdsourced, span answers only
CNN/Daily mail [11]	CNN/Daily mail	300K	1.4M	Semi-synthetic cloze-style query
CBT [13]	Children's book	688K	688K	Semi-synthetic cloze-style query

sets is performed. The detailed statistics of QA4IE benchmark are provided in Table 1.

We further compare our QA4IE benchmark with some existing IE and QA benchmarks in Table 2. One can observe that QA4IE benchmark is much larger than previous IE and QA benchmarks except for WikiReading and Zero-Shot Benchmark. However, as we mentioned at the beginning of Sect. 2, WikiReading is problematic for IE settings. Besides, Zero-Shot Benchmark is a sentence-level

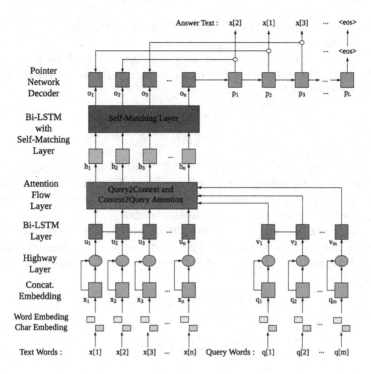

Fig. 2. An overview of our QA model.

dataset and we have described the disadvantage of ignoring information across sentences at Sect. 1.1. Thus to our best knowledge, QA4IE benchmark is the largest document level IE benchmark and it can be easily extended if we change our distillation strategy.

3 Question Answering Model

In this section, we describe our Question Answering model for IE. The model overview is illustrated in Fig. 2. The input of our model are the words in the input text $x[1], \ldots, x[n]$ and query $q[1], \ldots, q[n]$. We concatenate pre-trained word embeddings from GloVe [30] and character embeddings trained by Char-CNN [17] to represent input words. The $2d$-dimension embedding vectors of input text $x_1, \ldots, x_n$ and query $q_1, \ldots, q_n$ are then fed into a Highway Layer [40] to improve the capability of word embeddings and character embeddings as

$$
\begin{aligned}
g_t &= \text{sigmoid} \left(W_g x_t + b_g \right) \\
s_t &= \text{relu} \left(W_x x_t + b_x \right) \\
u_t &= g_t \odot s_t + \left(1 - g_t \right) \odot x_t.
\end{aligned}
\tag{1}
$$

Here $W_g, W_x \in \mathbb{R}^{d \times 2d}$ and $b_g, b_x \in \mathbb{R}^d$ are trainable weights, u_t is a d-dimension vector. The function relu is the rectified linear units [19] and $\odot$

is element-wise multiply over two vectors. The same Highway Layer is applied to q_t and produces v_t.

Next, u_t and v_t are fed into a Bi-Directional Long Short-Term Memory Network (BiLSTM) [14] respectively in order to model the temporal interactions between sequence words:

$$
\begin{aligned}
u_t' &= \text{BiLSTM}(u_{t-1}', u_t) \\
v_t' &= \text{BiLSTM}(v_{t-1}', v_t).
\end{aligned}
\tag{2}
$$

Here we obtain $\mathbf{U} = [u_1', \ldots, u_n'] \in \mathbb{R}^{2d \times n}$ and $\mathbf{V} = [v_1', \ldots, v_m'] \in \mathbb{R}^{2d \times m}$. Then we feed $\mathbf{U}$ and $\mathbf{V}$ into the attention flow layer [37] to model the interactions between the input text and query. We obtain the $8d$-dimension query-aware context embedding vectors $h_1, \ldots, h_n$ as the result.

After modeling interactions between the input text and queries, we need to enhance the interactions within the input text words themselves especially for the longer text in IE settings. Therefore, we introduce Self-Matching Layer [47] in our model as

$$
\begin{aligned}
o_t &= \text{BiLSTM}(o_{t-1}, [h_t, c_t]) \\
s_j^t &= w^T \tanh(W_h h_j + \tilde{W}_h h_t) \\
\alpha_i^t &= \exp(s_i^t) / \Sigma_{j=1}^n \exp(s_j^t) \\
c_t &= \Sigma_{i=1}^n \alpha_i^t h_i.
\end{aligned}
\tag{3}
$$

Here $W_h, \tilde{W}_h \in \mathbb{R}^{d \times 8d}$ and $w \in \mathbb{R}^d$ are trainable weights, $[h, c]$ is vector concatenation across row. Besides, α_i^t is the attention weight from the tth word to the ith word and c_t is the enhanced contextual embeddings over the tth word in the input text. We obtain the $2d$-dimension query-aware and self-enhanced embeddings of input text after this step. Finally we feed the embeddings $\mathbf{O} = [o_1, \ldots, o_n]$ into a Pointer Network [44] to decode the answer sequence as

$$
\begin{aligned}
p_t &= \text{LSTM}(p_{t-1}, c_t) \\
s_j^t &= w^T \tanh(W_o o_j + W_p p_{t-1}) \\
\beta_i^t &= \exp(s_i^t) / \Sigma_{j=1}^n \exp(s_j^t) \\
c_t &= \Sigma_{i=1}^n \beta_i^t o_i.
\end{aligned}
\tag{4}
$$

The initial state of LSTM p_0 is o_n. We can then model the probability of the tth token a^t by

$$
\begin{aligned}
\text{P}(a^t | a^1, \ldots, a^{t-1}, \mathbf{O}) &= (\beta_1^t, \beta_2^t, \ldots, \beta_n^t, \beta_{n+1}^t) \\
\text{P}(a_i^t) &\triangleq \text{P}(a^t = i | a^1, \ldots, a^{t-1}, \mathbf{O}) = \beta_i^t.
\end{aligned}
\tag{5}
$$

Here β_{n+1}^t denotes the probability of generating the "eos" symbol since the decoder also needs to determine when to stop. Therefore, the probability of generating the answer sequence $\mathbf{a}$ is as follows

$$
\text{P}(\mathbf{a} | \mathbf{O}) = \prod_t \text{P}(a^t | a^1, \ldots, a^{t-1}, \mathbf{O}).
\tag{6}
$$

Given the supervision of answer sequence $\mathbf{y} = (y_1, \ldots, y_L)$, we can write down the loss function of our model as

$$L(\theta) = -\sum_{t=1}^{L} \log P(a_{y_t}^t).$$ (7)

Table 3. Comparison of QA models on SQuAD datasets. We only include the single model results on the dev set from published papers.

	Dev Set
Span model	EM/F1
LR baseline [32]	40.0/51.0
Match-LSTM [46]	64.1/73.9
BiDAF [37]	67.7/77.3
R-Net [47]	71.1/79.5
MEMEN [29]	71.0/80.4
M-Reader + RL [15]	72.1/81.6
SAN [24]	**76.2/84.1**
Sequence model	
Match-LSTM (Seq) [46]	54.4/68.2
Our model	**61.7/72.5**

To train our model, we minimize the loss function $L(\theta)$ based on training examples.

4   Experiments

4.1   Experimental Setup

We build our QA4IE benchmark following the steps described in Sect. 2. In experiments, we train and evaluate our QA models on the corresponding train and test sets while the hyper-parameters are tuned on dev sets. In order to make our experiments more informative, we also evaluate our model on SQuAD dataset [32].

The preprocessing of our QA4IE benchmark and SQuAD dataset are all performed with the open source code from [37]. We use 100 1D filters with width 5 to construct the CharCNN in our char embedding layer. We set the hidden size $d = 100$ for all the hidden states in our model. The optimizer we use is the AdaDelta optimizer [50] with an initial learning rate of 2. A dropout [39] rate of 0.2 is applied in all the CNN, LSTM and linear transformation layers in our model during training. For SQuAD dataset and our small sized QA4IE-SPAN/SEQ-S datasets, we set the max length of input texts as 400 and a mini-batch size of 20. For middle sized (and large sized) QA4IE datasets, we set the max length as 700 (800) and batch size as 7 (5). We introduce an early stopping in training process after 10 epochs. Our model is trained on a GTX 1080 Ti GPU and it takes about 14 h on small sized QA4IE datasets. We implement our model with TensorFlow [1] and optimize the computational expensive LSTM layers with LSTMBlockFusedCell[2].

4.2 Results in QA Settings

We first perform experiments in QA settings to evaluate our QA model on both SQuAD dataset and QA4IE benchmark. Since our goal is to solve IE, not QA, the motivation of this part of experiments is to evaluate the performance of our model and make a comparison between QA4IE benchmark and existing datasets. Two metrics are introduced in the SQuAD dataset: Exact Match (EM) and F1-score. EM measures the percentage that the model prediction matches one of the ground truth answers exactly while F1-score measures the overlap between the prediction and ground truth answers. Our QA4IE benchmark also adopts these two metrics.

Table 3 presents the results of our QA model on SQuAD dataset. Our model outperforms the previous sequence model but is not competitive with span models because it is designed to produce sequence answers in IE settings while baseline span models are designed to produce span answers for SQuAD dataset.

Table 4. Comparison of QA models on 6 datasets of our QA4IE benchmark. The BiDAF model cannot work on our SEQ datasets thus the results are N/A.

Model	SPAN-S	SPAN-M	SPAN-L	SEQ-S	SEQ-M	SEQ-L
	EM/F1	EM/F1	EM/F1	EM/F1	EM/F1	EM/F1
BiDAF [37]	88.89/90.89	82.37/85.04	68.00/70.29	N/A	N/A	N/A
Match-LSTM [46]	85.88/88.21	79.19/82.05	66.87/70.44	89.60/91.95	83.57/87.40	62.64/68.98
Our model	**91.53/93.19**	**86.04/88.65**	**70.86/74.51**	**91.20/93.04**	**85.52/88.43**	**71.96/76.11**

The comparison between our QA model and two baseline QA models on our QA4IE benchmark is shown in Table 4. For training of both baseline QA models,[3] we use the same configuration of max input length as our model and tune the rest of hyper-parameters on dev sets. Our model outperforms these two baselines on all 6 datasets. The performance is good on S and M datasets but worse for longer documents. As we mentioned in Sect. 4.1, we set the max input

---

[2] https://www.tensorflow.org/api_docs/python/tf/contrib/rnn/
LSTMBlockFusedCell.

[3] The code of BiDAF is from https://github.com/allenai/bi-att-flow.
The code of Match-LSTM is from https://github.com/fuhuamosi/MatchLstm.

length as 800 and ignore the rest words on L datasets. Actually, there are 11% of queries with no answers in the first 800 words in our benchmark. Processing longer documents is a tough problem [7] and we leave this to our future work.

Table 5. Model ablation on QA4IE-SEQ-S. The first line is our original model and each of the following lines is the original model with a component ablated.

	EM/F1	Training hours
Our original model	91.20/93.04	14
− Char embedding	89.78/91.76	14
− Highway	90.04/91.97	14
− Self Matching	89.55/91.60	10
− LSTMBlockFusedCell		90

To study the improvement of each component in our model, we present model ablation study results in Table 5. We do not involve Attention Flow Layer and Pointer Network Decoder as they cannot be replaced by other architectures with the model still working. We can observe that the first three components can effectively improve the performance but Self Matching Layer makes the training more computationally expensive by 40%. Besides, the LSTMBlockFusedCell works effectively and accelerates the training process by 6 times without influencing the performance.

4.3 Results in IE Settings

In this subsection, we put our QA model in the entire pipeline of our QA4IE framework (Fig. 1) and evaluate the framework in IE settings. Existing IE systems are all free-text based Open IE systems, so we need to manually evaluate the free-text based results in order to compare our model with the baselines. Therefore, we conduct experiments on a small dataset, the dev set of QA4IE-SPAN-S which consists of 4393 documents and 28501 ground truth queries.

Our QA4IE benchmark is based on Wikipedia articles and all the ground truth triples of each article have the same first entity (i.e. the title of the article). Thus, we can directly use the title of the article as the first entity of each triple without performing step 1 (entity recognition) in our framework. Besides, all the ground truth triples in our benchmark are from knowledge base where they are disambiguated and aggregated in the first place, and therefore step 4 (entity linking) is very simple and we do not evaluate it in our experiments.

A major difference between QA settings and IE settings is that in QA settings, each query corresponds to an answer, while in the QA4IE framework, the QA model take a candidate entity-relation (or entity-property) pair as the query and it needs to tell whether an answer to the query can be found in the input text. We can consider the IE settings here as performing step 2 and then step 3 in the QA4IE framework.

In step 2, we need to build a candidate query list for each article in the dataset. Instead of incorporating existing ontology or knowledge base, we use a simple but effective way to build the candidate query list of an article. Since we have a ground truth query list with labeled answers of each article, we can add all the neighboring queries of each ground truth query into the query list. The neighboring queries are defined as two queries that co-occur in the same ground truth query list of any articles in the dataset. We transform the dev set of QA4IE-SPAN-S above by adding neighboring queries into the query list. After this step, the number of queries grows to 426336, and only 28501 of them are ground truth queries labeled with an answer.

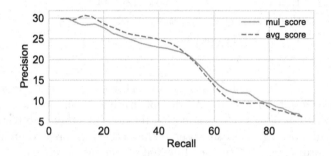

Fig. 3. Precision-recall curves with two confidence scores on the dev set of QA4IE-SPAN-S.

In step 3, we require our QA model to output a confidence score along with the answer to each candidate query. Our QA model produces no answer to a query when the confidence score is less than a threshold δ or the output is an "eos" symbol. For the answers with a confidence score $\geq \delta$, we evaluate them by the EM measurement with ground truth answers and count the true positive samples in order to calculate the precision and recall under the threshold δ. Specifically, we try two confidence scores calculated as follows:

$$\text{Score}_{\text{mul}} = \prod_{t=1}^{L} P(a_{i_t}^t), \ \text{Score}_{\text{avg}} = \sum_{t=1}^{L} P(a_{i_t}^t)/L, \tag{8}$$

where $(a_{i_1}^1, \ldots, a_{i_L}^L)$ is the answer sequence and $P(a_i^t)$ is defined in Eq. (5).$\text{Score}_{\text{mul}}$ is equivalent to the training loss in Eq. (7) and $\text{Score}_{\text{avg}}$ takes the answer length into account.

The precision-recall curves of our framework based on the two confidence scores are plotted in Fig. 3. We can observe that the EM rate we achieve in QA settings is actually the best recall (91.87) in this curve (by setting $\delta = 0$). The best F1-scores of the two curves are 29.97 (precision = 21.61, recall = 48.85, $\delta = 0.91$) for $\text{Score}_{\text{mul}}$ and 31.05 (precision = 23.93, recall = 44.21, $\delta = 0.97$) for $\text{Score}_{\text{avg}}$. $\text{Score}_{\text{avg}}$ is better than $\text{Score}_{\text{mul}}$, which suggests that the answer length should be taken into account.

Table 6. Results of three Open IE baselines on the dev set of QA4IE-SPAN-S.

	Open IE 4	Stanford IE	ClauseIE
#Extracted triples	32309	120147	75078
#After filtering	487	467	554
#True positive	403	301	133

We then evaluate existing IE systems on the dev set of QA4IE-SPAN-S and empirically compare them with our framework. Note that while [21] is closely related to our work, we cannot fairly compare our framework with [21] because their systems are in the sentence level and require additional negative samples for training. [35] is also related to our work, but their dataset and code have not been published yet. Therefore, we choose to evaluate three popular Open IE systems, Open IE 4 [36], Stanford IE [3] and ClauseIE [8].

Since Open IE systems take a single sentence as input and output a set of free-text based triples, we need to find the sentences involving ground truth answers and feed the sentences into the Open IE systems. In the dev set of QA4IE-SPAN-S, there are 28501 queries with 44449 answer locations labeled in the 4393 documents. By feeding the 44449 sentences into the Open IE systems, we obtain a set of extracted triples from each sentence. We calculate the number of true positive samples by first filtering out triples with less than 20% words overlapping with ground truth answers and then asking two human annotators to verify the remaining triples independently. Since in the experiments, our framework is given the ground-truth first entity of each triple (the title of the corresponding Wikipedia article) while the baseline systems do not have this information, we ask our human annotators to ignore the mistakes on the first entities when evaluating triples produced by the baseline systems to offset this disadvantage. For example, the 3rd case of ClauseIE and the 4th case of Open IE 4 in Table 7 are all labeled as correct by our annotators even though the first entities are pronouns. The two human annotators reached an agreement on 191 out of 195 randomly selected cases.

The evaluation results of the three Open IE baselines are shown in Table 6. We can observe that most of the extracted triples are not related to ground truths and the precision and recall are all very low (around 1%) although we have already helped the baseline systems locate the sentences containing ground truth answers.

4.4 Case Study

In this subsection, we perform case studies of IE settings in Table 7 to better understand the models and benchmarks. The baseline Open IE systems produce triples by analyzing the subjects, predicates and objects in input sentences, and thus our annotators lower the bar of accepting triples. However, the analysis on semantic roles and parsing trees cannot work very well on complicated input

Table 7. Case study of three Open IE baselines and our framework on dev set of QA4IE-SPAN-S, the results of baselines are judged by two human annotators while the results of our framework are measured by Exact Match with ground truth. The triples in red indicate the wrong cases.

Input Sentence	Ground Truth Triple	Open IE 4	Stanford IE	ClauseIE	Ours
Dieter Kesten was born on 9 June 1914 at Gelsenkirchen.	(Dieter Kesten; date of birth; 9 June 1914)	(Dieter Kesten; was born; on 9 June 1914 at Gelsenkirchen)	(Dieter Kesten; was born on; 9 June 1914)	(Dieter Kesten; was born; on 9 June 1914)	(Dieter Kesten; date of birth; 9 June 1914)
Hamilton died on 2 March 1625 at Whitehall, London, from a fever and was buried in the family mausoleum at Hamilton, on 2 September of that year.	(James Hamilton; date of death; 2 March 1625)	(Hamilton; died; on 2 March 1625 at Whitehall)	(Hamilton; died on; 2 September) (Hamilton; died on; 2 March 1625)	(Hamilton; died on; 2)	(James Hamilton; date of death; 2 March 1625)
She attended Texas A&M University, where she swam for the Texas A&M Aggies swimming and diving team in National Collegiate Athletic Association (NCAA) competition from 2011 to 2014.	(Breeja Larson; member of sports team; Texas A&M Aggies)	(She; attended; Texas A&M University)	(She; attended; M University)	(She; attended; Texas A&M University) (she; swam; for the Texas A&M Aggies swimming and diving team)	(Breeja Larson; member of sports team; Texas A&M Aggies)
His grave and memorial are at Balbeggie Churchyard, St. Martin's, near Perth, Scotland.	(John Simpson; place of death; St. Martin's)	(His grave and memorial; are; at Balbeggie Churchyard, St. Martin's, near Perth)	(Perth; near Churchyard is; St. Martin's)	(Balbeggie Churchyard near Perth Scotland; is; St. Martin's)	(John Simpson; place of death; Balbeggie Churchyard)
He served in the British Army and was wounded in World War I.	(William Dobbie; conflict; World War I)	(He; was wounded; in World War I)	(He; was wounded in; World War I)	(He; was wounded; in World War I)	(William Dobbie; conflict; World War I)

sentences like the 2nd and the 3rd cases. Besides, the baseline systems can hardly solve the last two cases which require inference on input sentences.

Our framework works very well on this dataset with the QA measurements EM = 91.87 and F1 = 93.53 and the IE measurements can be found in Fig. 3. Most of the error cases are the fourth case which is acceptable by human annotators. Note that our framework takes the whole document as the input while the baseline systems take the individual sentence as the input, which means the experiment setting is much more difficult for our framework.

4.5   Human Evaluation on QA4IE Benchmark

Finally, we perform a human evaluation on our QA4IE benchmark to verify the reliability of former experiments. The evaluation metrics are as follows:

Triple Accuracy is to check whether each ground truth triple is accurate (one cannot find conflicts between the ground truth triple and the corresponding article) because the ground truth triples from WikiData and DBpedia may be incorrect or incomplete.

Contextual Consistency is to check whether the context of each answer location is consistent with the corresponding ground truth triple (one can infer from the context to obtain the ground truth triple) because we keep all matched answer locations as ground truths but some of them may be irrelevant with the corresponding triple.

Triple Consistency is to check whether there is at least one answer location that is contextually consistent for each ground truth triple. It can be calculated by counting the results of Contextual Consistency.

We randomly sample 25 articles respectively from the 6 datasets (in total of 1002 ground truth triples with 2691 labeled answer locations) and let two human annotators label the Triple Accuracy for each ground truth triple and the Contextual Consistency for each answer location. The two human annotators reached an agreement on 131 of 132 randomly selected Triple Accuracy cases and on 229 of 234 randomly selected Contextual Consistency cases. The human evaluation results are shown in Table 8. We can find that the Triple Accuracy and the Triple Consistency is acceptable while the Contextual Consistency still needs to be improved. The Contextual Consistency problem is a weakness of distant supervision, and we leave this to our future work.

Table 8. Human evaluation on QA4IE benchmark.

	SPAN-S	SPAN-M	SPAN-L	SEQ-S	SEQ-M	SEQ-L	Total
Triple accuracy	98.8%	96.9%	98.1%	97.1%	96.2%	97.8%	97.5%
	161/163	154/159	159/162	170/175	152/158	181/185	977/1002
Contextual consistency	78.6%	65.1%	70.3%	75.4%	73.9%	82.4%	74.6%
	195/248	239/367	494/703	230/305	264/357	586/711	2008/2691
Triple consistency	93.3%	87.4%	91.4%	92.0%	92.4%	92.4%	91.5%
	152/163	139/159	148/162	161/175	146/158	171/185	917/1002

5   Conclusion

In this paper, we propose a novel QA based IE framework named QA4IE to address the weaknesses of previous IE solutions. In our framework (Fig. 1), we divide the complicated IE problem into four steps and show that the step 1, 2 and 4 can be solved well enough by existing work. For the most difficult step 3, we transform it to a QA problem and solve it with our QA model. To train this QA model, we construct a large IE benchmark named QA4IE benchmark that consists of 293K documents and 2 million golden relation triples with 636 different relation types. To our best knowledge, our QA4IE benchmark is the largest document level IE benchmark. We compare our system with existing best IE baseline systems on our QA4IE benchmark and the results show that our system achieves a great improvement over baseline systems.

For the future work, we plan to solve the triples with multiple entities as the second entity, which is excluded from problem scope in this paper. Besides, processing longer documents and improving the quality of our benchmark are all challenging problems as we mentioned previously. We hope this work can provide new thoughts for the area of information extraction.

Acknowledgements. W. Zhang is the corresponding author of this paper. The work done by SJTU is sponsored by National Natural Science Foundation of China (61632017, 61702327, 61772333) and Shanghai Sailing Program (17YF1428200).

References

1. Abadi, M., et al.: Tensorflow: a system for large-scale machine learning. In: OSDI, vol. 16, pp. 265–283 (2016)
2. Agichtein, E., Gravano, L.: Snowball: extracting relations from large plain-text collections. In: Proceedings of the Fifth ACM Conference on Digital Libraries, pp. 85–94. ACM (2000)
3. Angeli, G., Premkumar, M.J.J., Manning, C.D.: Leveraging linguistic structure for open domain information extraction. In: ACL, vol. 1, pp. 344–354 (2015)
4. Auer, S., Bizer, C., Kobilarov, G., Lehmann, J., Cyganiak, R., Ives, Z.: DBpedia: a nucleus for a web of open data. In: Aberer, K. (ed.) ASWC/ISWC -2007. LNCS, vol. 4825, pp. 722–735. Springer, Heidelberg (2007). https://doi.org/10.1007/978-3-540-76298-0_52
5. Bahdanau, D., Cho, K., Bengio, Y.: Neural machine translation by jointly learning to align and translate. In: International Conference on Learning Representations (ICLR) (2015)
6. Bizer, C., Lehmann, J., Kobilarov, G., Auer, S., Becker, C., Cyganiak, R., Hellmann, S.: Dbpedia-a crystallization point for the web of data. Web Semant. Sci. Serv. Agents World Wide Web **7**(3), 154–165 (2009)
7. Chen, D., Fisch, A., Weston, J., Bordes, A.: Reading Wikipedia to answer open-domain questions. In: ACL, vol. 1, pp. 1870–1879 (2017)
8. Del Corro, L., Gemulla, R.: ClausIE: clause-based open information extraction. In: Proceedings of International Conference on World Wide Web, pp. 355–366 (2013)
9. Gupta, R., Halevy, A., Wang, X., Whang, S.E., Wu, F.: Biperpedia: an ontology for search applications. Proc. VLDB Endow. **7**(7), 505–516 (2014)
10. He, L., Lewis, M., Zettlemoyer, L.: Question-answer driven semantic role labeling: using natural language to annotate natural language. In: EMNLP, pp. 643–653 (2015)
11. Hermann, K.M., et al.: Teaching machines to read and comprehend. In: Advances in Neural Information Processing Systems, pp. 1693–1701 (2015)
12. Hewlett, D., et al.: WikiReading: a novel large-scale language understanding task over Wikipedia. In: ACL, vol. 1, pp. 1535–1545 (2016)
13. Hill, F., Bordes, A., Chopra, S., Weston, J.: The goldilocks principle: reading children's books with explicit memory representations. arXiv preprint arXiv:1511.02301 (2015)
14. Hochreiter, S., Schmidhuber, J.: Long short-term memory. Neural Comput. **9**, 1735–1780 (1997)
15. Hu, M., Peng, Y., Qiu, X.: Reinforced mnemonic reader for machine comprehension. CoRR, abs/1705.02798 (2017)
16. Ji, H., Grishman, R.: Knowledge base population: successful approaches and challenges. In: ACL, pp. 1148–1158 (2011)
17. Kim, Y., Jernite, Y., Sontag, D., Rush, A.M.: Character-aware neural language models. In: AAAI, pp. 2741–2749 (2016)
18. Lample, G., Ballesteros, M., Subramanian, S., Kawakami, K., Dyer, C.: Neural architectures for named entity recognition. In: Proceedings of NAACL-HLT, pp. 260–270 (2016)

19. Le, Q.V., Jaitly, N., Hinton, G.E.: A simple way to initialize recurrent networks of rectified linear units. arXiv preprint arXiv:1504.00941 (2015)
20. Lee, T., Wang, Z., Wang, H., Hwang, S.W.: Attribute extraction and scoring: a probabilistic approach. In: 29th International Conference on Data Engineering, pp. 194–205 (2013)
21. Levy, O., Seo, M., Choi, E., Zettlemoyer, L.: Zero-shot relation extraction via reading comprehension. In: CoNLL, pp. 333–342 (2017)
22. Lin, Y., Liu, Z., Sun, M., Liu, Y., Zhu, X.: Learning entity and relation embeddings for knowledge graph completion. In: AAAI, vol. 15, pp. 2181–2187 (2015)
23. Ling, X., Weld, D.S.: Fine-grained entity recognition. In: AAAI (2012)
24. Liu, X., Shen, Y., Duh, K., Gao, J.: Stochastic answer networks for machine reading comprehension. arXiv preprint arXiv:1712.03556 (2017)
25. Luo, G., Huang, X., Lin, C.Y., Nie, Z.: Joint entity recognition and disambiguation. In: EMNLP, pp. 879–888 (2015)
26. Ma, X., Hovy, E.: End-to-end sequence labeling via bi-directional LSTM-CNNs-CRF. In: ACL, vol. 1, pp. 1064–1074 (2016)
27. Mintz, M., Bills, S., Snow, R., Jurafsky, D.: Distant supervision for relation extraction without labeled data. In: ACL, pp. 1003–1011 (2009)
28. Moro, A., Raganato, A., Navigli, R.: Entity linking meets word sense disambiguation: a unified approach. TACL 2, 231–244 (2014)
29. Pan, B., Li, H., Zhao, Z., Cao, B., Cai, D., He, X.: MEMEN: multi-layer embedding with memory networks for machine comprehension. arXiv preprint arXiv:1707.09098 (2017)
30. Pennington, J., Socher, R., Manning, C.: GloVe: global vectors for word representation. In: EMNLP, pp. 1532–1543 (2014)
31. Pyysalo, S., Ginter, F., Heimonen, J., Björne, J., Boberg, J., Järvinen, J., Salakoski, T.: BioInfer: a corpus for information extraction in the biomedical domain. BMC Bioinform. 8(1), 50 (2007)
32. Rajpurkar, P., Zhang, J., Lopyrev, K., Liang, P.: SQuAD: 100,000+ questions for machine comprehension of text. In: EMNLP, pp. 2383–2392 (2016)
33. Ren, X., et al.: CoType: joint extraction of typed entities and relations with knowledge bases. In: Proceedings of the 26th International Conference on World Wide Web, pp. 1015–1024 (2017)
34. Riedel, S., Yao, L., McCallum, A.: Modeling relations and their mentions without labeled text. In: Balcázar, J.L., Bonchi, F., Gionis, A., Sebag, M. (eds.) ECML PKDD 2010. LNCS (LNAI), vol. 6323, pp. 148–163. Springer, Heidelberg (2010). https://doi.org/10.1007/978-3-642-15939-8_10
35. Roth, B., Conforti, C., Poerner, N., Karn, S., Schütze, H.: Neural architectures for open-type relation argument extraction. arXiv preprint arXiv:1803.01707 (2018)
36. Schmitz, M., Bart, R., Soderland, S., Etzioni, O., et al.: Open language learning for information extraction. In: EMNLP, pp. 523–534 (2012)
37. Seo, M., Kembhavi, A., Farhadi, A., Hajishirzi, H.: Bidirectional attention flow for machine comprehension. arXiv preprint arXiv:1611.01603 (2016)
38. Shen, W., Wang, J., Han, J.: Entity linking with a knowledge base: issues, techniques, and solutions. IEEE Trans. Knowl. Data Eng. 27(2), 443–460 (2015)
39. Srivastava, N., Hinton, G., Krizhevsky, A., Sutskever, I., Salakhutdinov, R.: Dropout: a simple way to prevent neural networks from overfitting. J. Mach. Learn. Res. 15(1), 1929–1958 (2014)
40. Srivastava, R.K., Greff, K., Schmidhuber, J.: Highway networks. arXiv preprint arXiv:1505.00387 (2015)

41. Stanovsky, G., Dagan, I.: Creating a large benchmark for open information extraction. In: EMNLP, pp. 2300–2305 (2016)
42. Sukhbaatar, S., Weston, J., Fergus, R., et al.: End-to-end memory networks. In: Advances in Neural Information Processing Systems, pp. 2440–2448 (2015)
43. Tjong Kim Sang, E.F., De Meulder, F.: Introduction to the CoNLL-2003 shared task: language-independent named entity recognition. In: Proceedings of NAACL-HLT, pp. 142–147 (2003)
44. Vinyals, O., Fortunato, M., Jaitly, N.: Pointer networks. In: Advances in Neural Information Processing Systems, pp. 2692–2700 (2015)
45. Vrandečić, D., Krötzsch, M.: Wikidata: a free collaborative knowledgebase. Commun. ACM **57**(10), 78–85 (2014)
46. Wang, S., Jiang, J.: Machine comprehension using match-LSTM and answer pointer. arXiv preprint arXiv:1608.07905 (2016)
47. Wang, W., Yang, N., Wei, F., Chang, B., Zhou, M.: Gated self-matching networks for reading comprehension and question answering. In: ACL, vol. 1, pp. 189–198 (2017)
48. Xu, K., Feng, Y., Huang, S., Zhao, D.: Semantic relation classification via convolutional neural networks with simple negative sampling. In: EMNLP, pp. 536–540 (2015)
49. Yahya, M., Whang, S., Gupta, R., Halevy, A.: ReNoun: fact extraction for nominal attributes. In: EMNLP, pp. 325–335 (2014)
50. Zeiler, M.D.: ADADELTA: an adaptive learning rate method. arXiv preprint arXiv:1212.5701 (2012)
51. Zeng, D., Liu, K., Chen, Y., Zhao, J.: Distant supervision for relation extraction via piecewise convolutional neural networks. In: EMNLP, pp. 1753–1762 (2015)
52. Zheng, S., Wang, F., Bao, H., Hao, Y., Zhou, P., Xu, B.: Joint extraction of entities and relations based on a novel tagging scheme. In: ACL, vol. 1, pp. 1227–1236 (2017)

Constructing a Recipe Web
from Historical Newspapers

Marieke van Erp[1]($\boxtimes$), Melvin Wevers[1], and Hugo Huurdeman[2]

[1] KNAW Humanities Cluster, DHLab, Amsterdam, The Netherlands
{marieke.van.erp,melvin.wevers}@dh.huc.knaw.nl
[2] Universiteit van Amsterdam, Amsterdam, The Netherlands
h.c.huurdeman@uva.nl

Abstract. Historical newspapers provide a lens on customs and habits of the past. For example, recipes published in newspapers highlight what and how we ate and thought about food. The challenge here is that newspaper data is often unstructured and highly varied. Digitised historical newspapers add an additional challenge, namely that of fluctuations in OCR quality. Therefore, it is difficult to locate and extract recipes from them. We present our approach based on distant supervision and automatically extracted lexicons to identify recipes in digitised historical newspapers, to generate recipe tags, and to extract ingredient information. We provide OCR quality indicators and their impact on the extraction process. We enrich the recipes with links to information on the ingredients. Our research shows how natural language processing, machine learning, and semantic web can be combined to construct a rich dataset from heterogeneous newspapers for the historical analysis of food culture.

Keywords: Natural language processing · Information extraction
Food history · Digitised newspapers · Digital humanities

1 Introduction

There is no dearth of structured recipes available online (cf. Epicurious, Foodnetwork.com).[1] Recipes can also be found in non-structured form in digitized newspapers and magazines. Because of their diachronic nature, these recipes can offer valuable insights into the evolution of food customs, making them of particular interest to historians and ethnologists. However, their lack of structure and varying OCR quality make it more difficult to identify, extract, and use these recipes for analysis. In this paper, we present our work on extracting and enriching recipes from a collection of Dutch historical newspapers (1945–1995).

Scholars in the humanities and social sciences approach what, how, when, where, and why we consume as constitutive signifiers of national and local identities [1]. Diachronic analyses of recipes offer insights into changes in food culture, shedding light on "analyses of everyday culture, the changing foundations

---

[1] http://www.epicurious.com, http://www.foodnetwork.com.

© Springer Nature Switzerland AG 2018
D. Vrandečić et al. (Eds.): ISWC 2018, LNCS 11136, pp. 217–232, 2018.
https://doi.org/10.1007/978-3-030-00671-6_13

of nations in a globalising world, and of food and drink as subjects of objects of consumption within the dynamic material worlds of late capitalism and late modernity [2]." Van Otterloo argues that the perception of a typical Dutch food culture formed during the 1950s. She also claims that it is difficult to get an overview of all the ideas and perceptions of food and the consumption of food. Computational approaches, however, are able to process large amounts of data and can possibly extract a more comprehensive overview of developments of ideas associated with food.

Newspapers function as transceivers; they are both producer and messenger of public discourse [3,4]. In other words, newspapers both reflect and shape prevailing ideas and tastes in particular periods. Recipes have been part of newspapers at least since the late nineteenth century. In addition, newspapers also contain reports containing views on daily life and customs in a national context. This information regularly appeared in recipes, offering an understanding of food cultures of the past. On the whole, this makes newspapers an invaluable source for studies of food culture.

However, recipes in digitised newspapers are not easily accessible. For instance, a query using the search term 'recept' (recipe) not only retrieves articles containing food recipes, but also recipes for homemade remedies and articles mentioning doctor's prescriptions—the same word in Dutch. Furthermore, not all articles that include recipes include the term 'recipe.' Due to noise introduced in the digitisation process and the diachronic language variation, standard information extraction methods perform poorly on such data. This paper addresses these issues and presents our method and experiments for (1) automatically identifying recipes in newspapers using a classification algorithm, (2) classifying the recipes using a multi-label classifier, (3) extracting ingredients, quantities and units using automatically extracted lexicons, and (4) linking the ingredients to information on their origins. In all steps, we investigate methods for which we can automatically generate training data (via distant supervision) or automatically extracted lexicons from domain-specific and generic resources. This approach also lowers the threshold to transfer the approaches to other domains. Furthermore, we evaluate the quality of the OCR and of our extraction process. All annotations, including the OCR quality indicators, are made available, enabling researchers to gain insights into the quality of the extracted information.

Our contributions are twofold: (1) a distant supervised method for extracting, structuring and enriching recipes from newspapers; (2) a dataset consisting of 27,411 historical recipes extracted from Dutch newspapers (1945–1995), which can be used for further research.

Our software, experiments and data can be found at: https://github.com/DHLab-nl/historical-recipe-web. Due to copyright restrictions, the text from the newspaper articles is not included, but can be retrieved via the document IDs.

The remainder of this paper is structured as follows. In Sect. 2, we discuss the background and related work. In Sect. 3, we describe the datasets used in this work. Our extraction, structuring and enrichment pipeline and evaluation are described in Sect. 4. Statistics on our historical recipes dataset are presented in Sect. 5. We discuss strengths and limitations of our approach in Sect. 6 and conclude with directions for future work in Sect. 7.

2 Related Work

The food domain has recently gained some attention in the AI community as a versatile application domain. Various recipe databases are available for research purposes, such as [5] and [6]. These can, for example, be used for the construction of recipe workflows containing specific actions to be carried out [7,8].

Recipe extraction and classification is clearly a multilingual research domain, as [8,9] show by taking Japanese and Italian as their domains, respectively. [10] enrich German recipes with category tags. We apply this type of tag classification in Subsect. 4.3, but we amend the feature selection to fit our dataset. Closer to our work is the extraction of ingredients and quantities and units from recipes such as presented in [9] and [11]. However, the main difference with our work is that their corpora are digital-born and thus not affected by fluctuations in text quality from the digitisation process as our corpus is (which also holds for [5–8]).

We take inspiration from [9] concerning the use of different lexicons for the extraction of ingredients (see Subsect. 4.4). The fluctuation in digitisation quality of our corpus affects our options for the application of standard natural language processing tools. There is some work on information extraction from noisy OCR data, such as [12] who investigate the impact of error rates from different OCR engines in a Named Entity Recognition (NER) task using a dictionary, regular expressions, a Maximum Entropy Markov Model, a CRF, and a combination of the approaches. For Dutch ingredient, quantity and unit extraction, there is no training data available as there is for NER. Therefore, we focus on dictionary and regular expression-based methods for that part of our research.

In the Semantic Web domain, the two main dedicated food datasets we found were Open Food Facts[2], an open collaborative database containing information about food in English and French and Foodpedia, a linked dataset containing Russian food products [13]. Although there are dedicated recipe vocabularies such as the BBC Food Ontology[3], and the Food Ontology[4], the number of datasets using those is limited, not open, or not easily findable.

Some examples of analyses that rich food datasets can provide can be found in [14], which presents an exploratory interface for comparing 487 chocolate chip cookie recipes collected from the Web. Restaurant menus also provide a window into social status as a linguistic analysis of 6,511 restaurant menus by [15] shows. They found that more expensive restaurants use longer and more foreign words. As different newspapers target different audiences, our dataset may also provide such insights, but the core goal of this research paper is to investigate the extent to which distant supervised methods can be used to identify, classify, and structure recipes from a historical newspaper corpus.

---

[2] https://world.openfoodfacts.org/data.

[3] https://www.bbc.co.uk/ontologies/fo.

[4] http://data.lirmm.fr/ontologies/food.

3   Data

Using Optical Layout Recognition, pages have been segmented into separate articles, available as images and OCR'ed text. The quality of the digitised text varies throughout the corpus. The age and quality of the original material are important determinants of the ability of the software to recognise the text; hence, older newspapers contain more errors than more recent papers.

The National Library of the Netherlands allows researchers to access data through an API and selected parts of the corpus are available as downloadable data dumps.[5] Access to the source material enables more substantial analyses, which are not possible on resources that are solely accessible through web search interfaces such as the Library of Congress' Chronicling America Corpus.[6]

In addition to our dataset of newspapers, we used a corpus of structured recipes to bootstrap the extraction of ingredients and to train a multi-label classifier to tag the historical recipes. This additional corpus consists of approximately 16,000 recipes from *Allerhande*, the recipe resource from the oldest and one of the largest Dutch supermarket chains.[7] Its recipes have been marked up with schema.org information[8] as well as tags, nutritional information, the source of publication, and ratings.

Data Selection. We selected four recently-digitised newspapers because of their higher OCR quality. These newspapers are the liberal *NRC handelsblad* (1970–1994), the social-democratic Amsterdam-based newspaper *Het Parool* (1946–1995), the Catholic *Volkskrant* (1950–1995) and the Protestant newspaper *Trouw* (1950–1995). Table 1 details the descriptive statistics of our dataset.[9]

Apart from their higher OCR quality, the historical period represented by the selected newspapers is of particular interest for research into Dutch food culture. The period after the Second World War exhibited rapid modernisation and industralisation. The recipes might show how these processes affected food culture and perceptions of cooking within households. The Netherlands also welcomed people from its former colonies Indonesia and Surinam as well as migrant workers from Morocco and Turkey. These migrant communities introduced new recipes and styles of cooking to the Netherlands. We argue that a dataset of historical recipes and their descriptions can be used to better understand how these cuisines were perceived and appropriated in the Netherlands [1, 16–18].

---

[5] Due to copyright restrictions, a user agreement is required for newspapers published after 1876.

[6] https://chroniclingamerica.loc.gov/.

[7] https://www.ah.nl/allerhande/.

[8] http://schema.org.

[9] Note that the decreased type-token ratio for the *NRC* suggests that the OCR quality in this newspaper is probably the lowest. Of these four newspapers, *NRC* was digitised first, which might explain the lower OCR quality.

Table 1. Statistics of the four selected historical newspapers: number of pages, number of articles, the number of unique tokens (types), the number of tokens in total (tokens) and token to type ratio (TTR)

	Pages	Articles	Types	Tokens	TTR
Parool	14,194	2,380,697	23,651,078	612,036,106	0.039
Volkskrant	13,628	2,248,652	28,616,758	744,275,792	0.038
NRC	7,199	947,198	11,735,250	489,397,816	0.024
Trouw	13,891	2,578,731	24,520,472	656,941,631	0.037

4   Constructing the Historical Recipe Web

In our workflow, we first generate lists of ingredients, recipe tags, and recipe descriptions from the structured recipe background dataset (Allerhande). We use this dataset to train a recipe tag classifier (described in Subsect. 4.3) and to bootstrap an ingredient and quantities and units extractor (described in Subsect. 4.4). The first step includes the detection of historical recipes using a seed list and the training of a recipe classifier based on historical recipes (describe in Subsect. 4.1). Then, we tag the historical recipes using our tag classifier and we extract the ingredient and quantify information from them. Finally, we enrich the set of structured historical recipes by linking the ingredients to DBpedia, recovering their scientific name, if available, and linking the ingredients to the Global Biodiversity Information Facility to obtain their origin.

4.1   Recipe Identification

From the four newspapers, we selected articles that include the tokens 'recept' or 'recepten' *and* one of the following tokens: 'gram, kilogram, pond, keuken, koken, kook, bakken, eetlepel, gerecht, theelepel, snijden' (recipe, recipes, gram, kilogram, pound, kitchen, cooking, cook, baking, tablespoon, dish, teaspoon, cut). We then manually annotated which of these articles were actually recipes (Table 2). Some recipes are part of a larger article describing an entire menu. In such cases, we treated the article as a single recipe.

Table 2. Results of recipe annotation from seed tokens

	Correct	False	Total
Volkskrant	1,526	796	2,322
Parool	1,481	971	2,452
Trouw	2,568	926	3,494
NRC	1,913	753	2,666
Total	7,488	3,466	10,954

Table 3. f_1, precision, and recall of the recipe classifier

	f_1	Precision	Recall
articles	0.97	0.96	0.97
recipes	0.95	0.96	0.95

Next, we created a training set of the articles annotated as recipes, articles falsely extracted as recipes, and 24,000 articles randomly selected from the four newspapers bar the articles annotated as recipes. This dataset was used to train a recipe classifier. After removing the search terms used for the initial query to improve the performance of the classifier, we transformed the text into a TF-IDF feature space based on unigrams and bigrams. On this feature space, we trained three classifiers: a multinomial Naive Bayes, a Support Vector Machine (SVM) with Stochastic Gradient Descent (SGD), and a Linear Support Vector Classification using cross-validated randomized search on hyperparameters. The latter scored the best with an accuracy score of 0.96 using a 5-fold cross validation (see Fig. 3 for precision, recall, and f_1 scores) (Table 3).

After applying the trained classifier to the four sets of newspaper articles, the number of recipes found increased drastically, especially for earlier periods, yielding 27,411 articles of which we have a high confidence that they are recipes. Using the classifier resulted in an almost six-fold increase over the initial seed list (see Fig. 1).

4.2   OCR Quality of the Recipe Dataset

While the Delpher newspaper data was digitised and OCR'ed relatively recently, the OCR quality is not perfect. To get an indication of the OCR quality, we performed a lexicon-based OCR quality check developed at the Dutch Language Institute.[10] This method checks what proportion of tokens present in an article occurs in a range of historical lexicons.[11] Most OCR software will give an indication of the certainty of its decisions by attaching a score to a document or batch of documents. However, these scores often give an indication of the errors at the character level, while for our purpose, it is more useful to know how many words (or tokens) are correct in a text, as information extraction techniques do not read as easily over character errors than humans do.

Figure 2 shows the results of this measure on the different newspapers (left) and per 5-year interval (right). Fortunately, the majority of the texts scores about 80%, although there is some difference between the newspapers and the different time periods. The scores are also provided in the historical recipe dataset, such that researchers can choose to exclude articles with a lower OCR score.

---

[10] https://ivdnt.org/the-dutch-language-institute.
[11] https://www.digitisation.eu/tools-resources/language-resources/historical-and-nam ed-entities-lexica-of-dutch/.

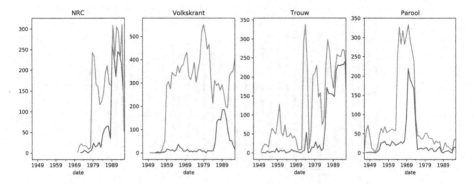

Fig. 1. Retrieved articles using seed list (blue) and using classifier (orange) (Color figure online)

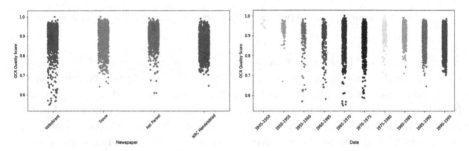

Fig. 2. Lexicon-based OCR quality indicators per newspaper (left) and per five-year period (right)

4.3   Tag Classifier

To categorise the recipes, we trained a multi-label classifier using the tags associated with recipes in the Allerhande dataset. Recipes in the Allerhande dataset are tagged with one, two, or three tags drawn from a set of 69 tags. These tags either indicate the type of dish (Thai, American, Italian), type of diet (Vegetarian, Healthy, Lactose-free), occasion (Christmas, Easter), or style of cooking (Grilling, Baking, Oven, Fast, Budget).

After initial training of the classifier on all tags, we removed tags with an accuracy score < 0.1, tags occurring in fewer than fifty recipes, and those that were specific to the Allerhande set such as 'advertorial' and 'wat eten we vandaag' (what's for dinner today). Also, we grouped together similar tags, such as 'healthy' and 'slim', and 'without meat/fish' and 'vegetarian'. These steps resulted in a set of 36 tags.

As input variables, we used the title, description, and cooking instruction fields from the Allerhande set. From this text we removed the names of tags to make the classifier less sensitive to the presence of these words. After converting the text into a TF-IDF feature space with an ngram range of (1, 5), we trained a OneVsRest Classifier balanced Linear SVC. The overall accuracy score of the

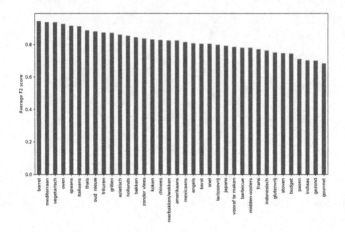

Fig. 3. Accuracy scores per tag of tag classifier on Allerhande dataset

classifier is 0.75. The Hamming Loss is 0.014, and the average F2 test score: 0.82. Figure 3 shows the scores per tag based on the Allerhande training set.

Subsequently, we applied the tag classifier to the annotated recipes extracted from the historical newspapers. Figure 4 shows the number of tags found in this dataset. In the bar chart, we find that a small set of tags were found relatively often, while others were infrequently found, or not at all. This suggests that some tags are quite specific to the Allerhande data and do not generalize quite well. On the other hand, tags such as 'vegetarian', 'italian', 'asian', and the more specific 'thai', 'grilling', and 'deep frying' were found with high accuracy in historical recipes.

For evaluation, we constructed a dataset of 100 recipes for every tag and 100 recipes that were not tagged. If tags appeared in fewer than 100 recipes, we selected all these recipes, for the other cases we took a sample. The tagged set included 1,197 recipes. We manually annotated recipes with the tags: 'italian', 'vegetarian', and 'asian'. These tags occurred relatively often and were easier to tag since they were less ambiguous than for instance, 'budget'. For these tags, the tagger scored relatively well (Table 4). During manual tagging, we also noticed that recipes tagged as 'asian' did not receive the more specific tags 'japanese', 'indonesian', 'chinese', or 'thai', even though they were described as such. The low recall for 'vegetarian' partly stems from the fact that in the Allerhande desserts, while often vegetarian, are almost never tagged as such. We annotated these recipes as 'vegetarian'. An interesting find was also that a recipe described as 'vegetarian' in a newspaper article was not tagged as 'vegetarian' by our tagger. Here the classifier was actually correct, since the recipe used chicken and trasi, a spice paste made of fermented shrimp. This perhaps suggests a changing concept of vegetarian food.

Table 4. Evaluation of tagger on historical recipes

	Precision	Recall	Accuracy	f_1
Asian	0.97	0.72	0.95	0.83
Italian	0.83	0.84	0.96	0.84
Vegetarian	0.78	0.45	0.78	0.57

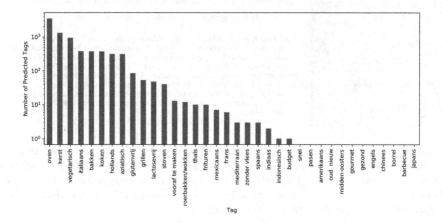

Fig. 4. Frequency of tags found in historical recipes

4.4 Ingredient and Quantity Extraction

Figure 5 illustrates some of the difficulties in extracting information from a digitised newspaper source. As the scan of the newspaper page shows (left), some of the text on the right-hand side is difficult to read because of the fold of the newspaper, resulting in gaps or misrecognised characters in the OCR output (top right). We have annotated the ingredients that do not contain any errors in blue, potential ingredients contained in strings with OCR errors in pink, and quantities in green. Interestingly, not all ingredients are precisely quantified, such as 'a pinch' (literally 'a knife's point' in Dutch). This makes it difficult to, for example, compute the nutritional value of the dish, even if the OCR was perfect and all ingredients and quantities could be recognised correctly.

To evaluate the ingredient and quantity extraction, we selected a random stratified sample from the recipe set created using seed list in Subsect. 4.1. The sample consists of 100 articles (1.35% of the set) following the same distribution over newspapers and time periods.

Ingredients, quantities and units in the sampled recipes were annotated using the Recogito annotation tool.[12] Furthermore, ingredients that contained OCR errors were marked separately to gain insight into the proportion of ingredients affected by these errors. Three annotators contributed to the gold standard. Six

---

[12] http://recogito.pelagios.org/.

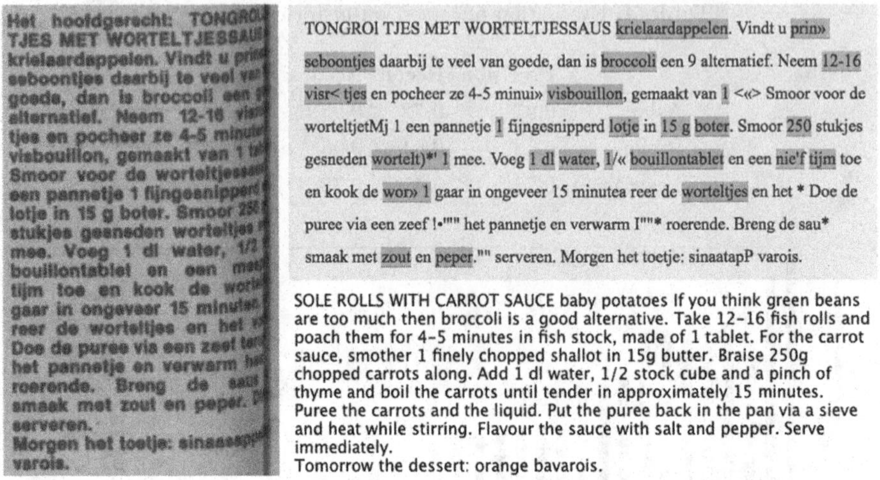

Fig. 5. Example of a newspaper recipe scan, its resulting OCR'ed text, marked up with ingredients that our approach should be able to recognise (blue), potential ingredients (pink) and quantities (green) as well as the recipe's English translation. Source: NRC Handelsblad 24 April 1988, page 20, https://resolver.kb.nl/resolve?urn=KBNRC01: 000029338:mpeg21:a0179 (Color figure online)

articles were annotated by all three annotators for which we computed Krippendorff's alpha to measure inter-annotator agreement, yielding a score of 0.85 [19]. Overall, the agreement is high, but we do see disagreement on whether or not parentheses are included and for the OCR category particularly it is unclear when a garbled-up word starts and ends. For example, in one instance Annotator 1 annotated *j ° ar,anen'* and Annotator 2: *°ar,anen'*.[13]

Ingredient Extraction. Many of the recipes do not follow a structured format where the ingredients are presented at the start of the article (as web-based recipes or formal cookbooks usually do). Segmenting the articles into 'ingredient' and 'description' paragraphs is therefore not an option. Experiments with standard NLP tools to identify noun chunks and part-of-speech tags are not robust against the OCR variation in our corpus. Therefore, ingredients are extracted using a dictionary-based tagger. We generate several ingredients lists inspired by [9]. In that work, a domain specific resource was used to bootstrap ingredients from AGROVOC[14] and a combination of three generic resources based on WordNet [20]. As a Dutch version of AGROVOC does not exist, we used the Allerhande corpus to generate a list of unique ingredients consisting of 2,723 food stuffs ranging from 'uien' (onions) to 'Ben Jerry's Cinnamon Buns ijs' (Ben Jerry's Cinnamon Buns ice cream).

---

[13] The article actually stated '4 bananen'.
[14] http://aims.fao.org/vest-registry/vocabularies/agrovoc-multilingual-agricultural-thesaurus.

We compared the Allerhande list to lists of ingredients from two generic datasets: Dutch DBpedia and Open Dutch Wordnet. From DBpedia, we select resources in the categories 'food' and 'lists of food'.[15] After excluding some categories (e.g. "List of Belgian Beers", which contained a fair few mentions of breweries), 2,642 potential ingredients remained. Singular nouns were automatically expanded with plural forms using the pattern library.[16] This yielded a total of 4,110 ingredients. From Open Dutch WordNet, we selected lexemes with the superclass 'Food' or 'Plant', yielding 1,602 entries, which were also automatically expanded to their plural forms, added up to 3,204 ingredients.

Table 5 presents the results of four types of ingredients extraction: (1) exact match using the entire list of ingredients; (2) exact match using only ingredients harvested from DBpedia; (3) exact match using ingredients harvested from WordNet; and (4) exact match using the combined lists (AH-DBP-WN). In an effort to tackle spelling variations and OCR errors, we experimented with fuzzy matching, but this only decreased performance by introducing more noise and no gains in recall.

Some ingredients may be mentioned several times in the recipe but we only note each ingredient once, thus performing a type analysis rather than a token analysis. Our gold standard contains 1,538 ingredients without OCR errors and the annotators identified 150 strings denoting ingredients containing OCR errors.

Error Analysis. The low recall stems from insufficient coverage of the ingredient lists, but simply adding ingredients would not yield 100% recall as there is also variation in parts of ingredients, e.g. 'brandneteltopjes' (tips of nettles) or 'kabeljauwkoppen' (cod heads). Furthermore, recipes occassionally mention ingredients by referring to a brand name, e.g. 'Delfiatablet' (a brand of butter), or by describing a foreign foodstuff, e.g. 'warka-vellen' (Moroccan phyllo).

Errors in precision stem from noise in the lexicons. For example, the Allerhande ingredients list contains 'aardappelsalade' (potato salad) and 'chocolade-cake' (chocolate cake), whereas in newspaper article this is the name of the final product. The annotators were instructed to only annotate the base ingredient and not its shape. For example, in 'kokend water' (boiling water) only 'water' is annotated. This decision was made to keep close to unprocessed ingredients and not have to account for variant such as chopped, diced, sliced, grated, etcetera. These variants, however, do occur in the Allerhande ingredients list. In addition to names of dishes, the DBpedia and WordNet lists contain cooking actions such as 'fruiten' (sautée) and other related terms such as 'dier' (animal), 'blikje' (can), and 'ingrediënten' (ingredients). This notwithstanding, our setup is to test the extent to which automatically harvested lexicons can be used for ingredient extraction. Some cleanup would improve the precision, but for the recall an automatically bootstrapped lexicon, or a statistical method will probably yield better results.

---

[15] http://nl.dbpedia.org/resource/Categorie:Voedsel; http://nl.dbpedia.org/resource/
   Categorie:Lijsten_van_voedsel. The resources typed with dbo:Food are mostly beers.
[16] https://www.clips.uantwerpen.be/pages/pattern-nl.

Table 5. Results of ingredients extraction from recipes. 'Clean ingredients' denotes results on ingredients without OCR errors, 'With OCR errors' denotes results including OCR errors. The number of correct items is the same for both sets as no new mentions from the set of OCR errors was retrieved.

	Clean ingredients				With OCR errors		
	Precision	Recall	f_1	Correct	Precision	Recall	f_1
Allerhande	0.70	0.65	0.67	998	0.70	0.59	0.64
DBpedia	0.60	0.33	0.47	513	0.60	0.30	0.45
WordNet	0.62	0.50	0.56	764	0.62	0.45	0.54
AH-DBP-WN	0.56	0.75	0.66	1,154	0.56	0.68	0.62

Quantity and Unit Extraction. Quantities and units are extracted using a regular expression that utilises a list of 91 units generated from the Allerhande dataset containing terms such as 'kilogram' and 'liter', but also 'pakje' (package) and 'pot' (jar). The units were pluralised automatically yielding 182 instances. The matcher checks for an occurrence of one or more digits followed by a unit or a digit followed by an ingredient. This quantity extraction method correctly identified 312 units with a precision of 0.74, a recall of 0.51, and an F_1 of 0.62.

Error Analysis. Precision errors are often caused by half matches, e.g. recognition of '4 pot' (4 jar) where the full annotation states '4 potten'. Part-of-speech tagging might resolve some of these problems, if the available taggers can be made more robust in dealing with OCR errors. The case for recall is more complex. On the one hand, the units lexicon can be expanded with variations on, for instance, pieces, wine glasses, layers, and tea cups. However, we also found some quite poetic variations on quantities and units expressions, such as 'een paar royale slagen met de pepermolen' (a couple of generous twists on the pepper grinder), 'een niet kinderachtige hoeveelheid' (a not childish amount), and 'een snuf snuf' (a sniff sniff). The use of these variants might be distinctive of particular historical periods.

Table 6. Results of ingredient to DBpedia linking

	Precision	Recall	f_1	Unique	Scientific	dbpedia-en
String match	95.56 (280)	10.77 (293)	53.17	293	37	293
Spotlight	85.45 (1,034)	44.47 (1,210)	64.96	438	76	397

4.5   Linking Recipe Elements

The food on our plates is often sourced from all corners of the world. To gain an insight into the different localities from which our ingredients originated, we linked items in our Allerhande ingredients list to the Global Biodiversity

Information Facility (GBIF).[17] This resource gives information about different species and their native range. To establish these links, we first collected an ingredient's scientific name from DBpedia, which was then queried in GBIF to obtain its origin. In this step, we also created links between our ingredient list and DBpedia, through which we also obtained links to the English DBpedia. We use two different approaches to generate these links: a simple string match and DBpedia spotlight [21]. The resulting links (Table 6) were judged by one annotator.

Error Analysis. The precision on the string match is naturally quite high, only in cases where ingredient names are ambiguous this fails. DBpedia Spotlight has more trouble, as it has a higher coverage. It, for example, links 'salsa' to salsa dancing instead of the sauce. Its increase in recall over the string match method is thanks to its access to synonyms such as "Zwaardherik" for 'Rucola' (arugula). There are still quite some ingredients for which no link was found. Some are quite surprising, such as the lack of a link for 'aardbeien' (strawberries), but for ingredients such as 'Amelander verse sladressing' (Amelander fresh salad dressing) or 'kippenbouillontablet' (chicken stock cube) this is not surprising. For many of the processed food items, such as cheese, there is no scientific name and corresponding GBIF entry. There are other interesting sources to relate these to, such as consumer price indices, but we leave this for future work.

5   Dutch Historical Recipe Web

Our extracted and enriched historical recipes dataset of over 27k recipes and over 365k ingredients can for example be used to investigate ingredient combinations in different time periods or popular tags in different newspapers. Table 7 shows the statistics of our recipes dataset.

It should be noted that the newspaper dataset does not include all published newspapers, so any comparative or proportional analyses derived from the newspaper corpus or our dataset will have to take this into account. Recipes may be repeated, but differences in OCR performance makes detecting the same recipe not trivial.

Table 7. Statistics of Dutch historical recipe web

	Recipes	Tags	Ingredients	Quantities	DBpedia	Scientific	GBIF
Parool	4,440	5,221	46,685	11,620	2,423	277	170
Volkskrant	13,270	16,962	185,872	56,626	7,395	730	349
NRC	3,764	4,943	59,717	17,738	1,850	282	142
Trouw	5,937	7,353	72,859	21,880	3,232	368	168
Total	27,411	34,479	365,133	107,864	14,900	1,657	829

---

[17] https://www.gbif.org/.

6   Discussion

In this paper, we focused on distant supervision approaches to detect and classify recipes from newspapers; to extract ingredients, quantities and units; and to add links to external datasets. The obtained scores show that for the identification and classification tasks, this works quite well, as the recipes from our seed datasets generalise well over to the newspapers dataset.

For the more fine-grained extraction and enrichment, i.e. the ingredients, quantities, units and external links, there are clear limitations to using available lexicons and resources. Although ingredients are not as infinite a set as, for example, named entities, our newspaper dataset shows enough variation to affect the performance of the approach. As the OCR quality affects standard natural language processing tools, such as part-of-speech tagging or noun chunking, it is difficult to bootstrap patterns from the dataset to grow the lexicons. Solutions can be sought in (a) only working with those articles that obtain a high OCR score, (b) cleaning up the OCR, or (c) training NLP systems to deal with noisy text. In our dataset, the OCR lexical coverage scores are provided, so researchers can choose to only use those articles in their analyses. Correcting the OCR is difficult, in particular with images that are already difficult to read for humans, but some tools are becoming available such as PICCL.[18]

7   Conclusions and Future Work

We presented a distant supervised method and experiments to construct a recipe dataset from historical newspapers. To the best of our knowledge, we are the first to combine natural language processing, machine learning, and semantic web for information extraction from noisy OCR data. Our evaluations show that articles denoting recipes can be identified with an F_1 score of 0.96, tags can be assigned with F_2 scores between 0.57 and 0.84, ingredients can be identified with an F_1 of 0.67, quantities and units with an F_1 of 0.62 and link with an F_1 of 0.64. These results leave room for improvement, but the approach does not require manually labeled training data. The resulting 27,411 recipes can be used by (humanities) researchers interested in food culture to more easily access relevant sources. We will continue to expand this dataset with additional newspapers and time periods and explore diachronic lexicons and machine learning methods to improve the classification and extraction.

The lexicon-based method that was used for the ingredients and quantities and units extraction is limited by the scope of available lexicons and cleanliness. The Allerhande lexicon, which was derived from schema.org ingredient elements, shows that such markup allows flexibility on the content provider's side, but makes it difficult to repurpose, for example, to use as an ingredients lexicon. Furthermore, the coverage of the Dutch DBpedia in the food domain was also lower and less well-structured than expected.

---

[18] https://github.com/LanguageMachines/PICCL.

We have also assessed the impact of OCR errors in the newspapers corpus by providing an indication of the article's lexical coverage and by annotating OCR problems in the ingredients lists in our evaluation dataset. The use of PICCL and other methods will be investigated to improve the quality of the sources.

As our method relies on distant supervision and automatically extracted lexicons, it can easily be ported to other domains to construct similar datasets from (historical) newspapers or magazines such as sport reports or music reviews.

The dataset, software and experiments described in this paper can be found at: https://github.com/DHLab-nl/historical-recipe-web

Acknowledgements. The authors thank the National Library of the Netherlands for making available the newspaper collection for research purposes as well as for organising the HackaLOD hackathon with Rijksmuseum and Netwerk Digitaal Erfgoed where this project got started. We thank Jesse de Does for the OCR quality measure, Marten Postma and Emiel van Miltenburg for querying Open Dutch WordNet, and Richard Zijdeman for fruitful discussions on the dataset concept. No Hawaiian pizzas were consumed during the writing of this paper.

References

1. van Otterloo, A.H.: Eten en eetlust in Nederland, 1840–1990: een historisch-sociologische studie. B. Bakker, Amsterdam (1990)
2. Wilson, T.M. (ed.): Food, Drink and Identity In Europe. European studies. Rodopi, Amsterdam (2006)
3. Schudson, M.: The Power of News. Harvard University Press, Cambridge (1982)
4. Marchand, R.: Advertising the American Dream: Making Way for Modernity, 1920–1940. University of California Press, Berkeley (1985)
5. Harashima, J., Ariga, M., Murata, K., Ioki, M.: A large-scale recipe and meal data collection as infrastructure for food research. In: Proceedings of the Tenth International Conference on Language Resources and Evaluation (LREC 2016). European Language Resources Association (ELRA), Paris, May 2016
6. Tasse, D., Smith, N.A.: SOUR CREAM: toward semantic processing of recipes. Technical report CMU-LTI-08-005, Carnegie Mellon University, Pittsburgh, PA (2008)
7. Maeta, H., Sasada, T., Mori, S.: A framework for recipe text interpretation. In: Proceedings of the 2014 ACM International Joint Conference on Pervasive and Ubiquitous Computing, Adjunct Publication. ACM, pp. 553–558 (2014)
8. Mori, S., Maeta, H., Yamakata, Y., Sasada, T.: Flow graph corpus from recipe texts. In: Proceedings of the Ninth International Conference on Language Resources and Evaluation (LREC 2014). European Language Resources Association (ELRA), Reykjavik, May 2014
9. Mazzei, A.: On the lexical coverage of some resources on Italian cooking recipes. In: Proceedings of CLiC-it 2014, First Italian Conference on Computational Linguistics, pp. 254–259 (2014)
10. Kicherer, H., Dittrich, M., Grebe, L., Scheible, C., Klinger, R.: What you use, not what you do: automatic classification of recipes. In: Frasincar, F., Ittoo, A., Nguyen, L.M., Métais, E. (eds.) NLDB 2017. LNCS, vol. 10260, pp. 197–209. Springer, Cham (2017). https://doi.org/10.1007/978-3-319-59569-6_22

11. Greene, E.: Extracting structured data from recipes using conditional random fields. The New York Times Open Blog (2015)
12. Packer, T.L., et al.: Extracting person names from diverse and noisy OCR text. In: Proceedings of the Fourth Workshop on Analytics for Noisy Unstructured Text Data, pp. 19–26. ACM (2010)
13. Kolchin, M., Chistyakov, A., Lapaev, M., Khaydarova, R.: FOODpedia: Russian food products as a linked data dataset. In: Gandon, F., Guéret, C., Villata, S., Breslin, J., Faron-Zucker, C., Zimmermann, A. (eds.) ESWC 2015. LNCS, vol. 9341, pp. 87–90. Springer, Cham (2015). https://doi.org/10.1007/978-3-319-25639-9_17
14. Chang, M., Hare, V.M., Kim, J., Agrawala, M.: RecipeScape: mining and analyzing diverse processes in cooking recipes. In: Proceedings of the 2017 CHI Conference Extended Abstracts on Human Factors in Computing Systems, pp. 1524–1531. ACM (2017)
15. Jurafsky, D., Chahuneau, V., Routledge, B., Smith, N.: Linguistic markers of status in food culture: Bordieu's distinction in a menu corpus. J. Cult. Anal. (2016). http://culturalanalytics.org/2016/10/linguistic-markers-of-status-in-food-culture-bourdieus-distinction-in-a-menu-corpus/
16. Schuyt, K., Taverne, E.: Dutch Culture in a European Perspective: 1950, Prosperity and Welfare. Palgrave Macmillan, Basingstoke (2004)
17. Hoving, I., Dibbits, H., Schrover, M., eds.: Cultuur en migratie in Nederland. Veranderingen in het Alledaagse, 1950–2000. Sdu Uitgevers, The Hague (2005)
18. Schot, J., Rip, A., Lintsen, H. (eds.): Technology and the Making of the Netherlands: The Age of Contested Modernization, 1890–1970. MIT Press, Cambridge (2010)
19. Krippendorff, K.: Computing krippendorff's alpha-reliability (2011)
20. Postma, M., van Miltenburg, E., Segers, R., Schoen, A., Vossen, P.: Open Dutch WordNet. In: Proceedings of the Eight Global WordNet Conference, Bucharest, Romania (2016)
21. Daiber, J., Jakob, M., Hokamp, C., Mendes, P.N.: Improving efficiency and accuracy in multilingual entity extraction. In: Proceedings of the 9th International Conference on Semantic Systems (I-Semantics) (2013)

Structured Event Entity Resolution
in Humanitarian Domains

Mayank Kejriwal[✉], Jing Peng, Haotian Zhang, and Pedro Szekely

USC Information Sciences Institute, Marina Del Rey, USA
kejriwal@isi.edu

Abstract. In domains such as humanitarian assistance and disaster relief (HADR), events, rather than named entities, are the primary focus of analysts and aid officials. An important problem that must be solved to provide situational awareness to aid providers is automatic clustering of sub-events that refer to the same underlying event. An effective solution to the problem requires judicious use of both domain-specific and semantic information, as well as statistical methods like deep neural embeddings. In this paper, we present an approach, AugSEER (Augmented feature sets for Structured Event Entity Resolution), that combines advances in deep neural embeddings both on text and graph data with minimally supervised inputs from domain experts. AugSEER can operate in both online and batch scenarios. On five real-world HADR datasets, AugSEER is found, on average, to outperform the next best baseline result by almost 15% on the cluster purity metric and by 3% on the F1-Measure metric. In contrast, text-based approaches are found to perform poorly, demonstrating the importance of semantic information in devising a good solution. We also use sub-event clustering visualizations to illustrate the qualitative potential of AugSEER.

Keywords: Events · Structured event resolution
Hybrid embeddings · Crisis informatics
Humanitarian and disaster relief · Clustering

1 Introduction

As the devastating consequences of recent disasters such as Hurricanes Irma and Harvey illustrate, effective mobilizing of resources and personnel is an important problem, with technology playing an increasingly important role, both in taking preventive action (e.g., evacuations) and dealing with the disaster's aftermath [6], [8]. The impact of disasters, and other events with a humanitarian dimension, is global: according to the 2016 Human Development report [7], conflicts, disasters and natural resources constitute key global concerns, with more than 21.3 million people (roughly the population of Australia) being affected by the refugee crisis alone. Technology can play an important role in alleviating this suffering by equipping HADR analysts with *situational awareness* [20]. Situational awareness

© Springer Nature Switzerland AG 2018
D. Vrandečić et al. (Eds.): ISWC 2018, LNCS 11136, pp. 233–249, 2018.
https://doi.org/10.1007/978-3-030-00671-6_14

is a broad notion, involving analytics that can cover text, sentiments, entities and spatio-temporal information. Examples include *entity-centric search* and aggregate sentiment analyses that help pinpoint emerging hotspots [10]. In some cases, *posthoc analysis* also needs to be conducted, perhaps by performing batch analytics on newswire or social media collected over a time interval.

For a continuously deployed system to conduct even basic *event-centric* analysis, at global scales of space and over arbitrary periods of time, the *structured event entity resolution* (SEER) problem needs to be solved. Along with named entities, HADR ontologies (whether simple or complex), also include *event entities* as first-class citizens. Event entities tend to be semi-structured objects that are sometimes extracted from documents, but (in the HADR space) can also be entire document fragments. This is especially the case when considering heterogeneous corpora such as specialized newswire (e.g. an article describing a single incident or event), social media and SMS. Events can span multiple days, week or in some cases (such as the Syrian refugee crisis), years. For posthoc analysis (the batch mode), users input their own heterogeneous corpus, usually collected over a multi-year period of time, and desire semi-automatic non-overlapping event clustering as a first step. In this sense, each data item is a 'sub-event', and a collection of sub-events represent a 'resolved' event.

Adequately solving the SEER problem involves several challenges not completely addressed by modern or classic text classification and clustering approaches. First, in addition to being relatively robust to errors, a good SEER system must handle the *topical flux* (more generally, called *concept drift*) that an evolving event exhibits across documents, space and time, often in unprecedented ways. As an example, consider the case of the Haiti earthquake in 2010. In an initial set of documents describing this disaster, the topics were primarily along the lines of earthquakes and landslides. In later documents, the key issues were humanitarian aid, politics and an unfortunate Cholera outbreak due to waste mismanagement by rescuers. Experiments described later show that topic modeling methods (or more recently, document embeddings) yield poor performance by themselves as they are not able to deduce that all of these circumstances relate to the same situation, namely a localized disaster in Haiti that has its origins in the earthquake.

The case above suggests that, barring large quantities of training data, a multi-pronged i.e. *statistical-semantic* approach may be necessary to address the SEER problem. In this paper, we present AugSEER, an approach that can judiciously accommodate both domain expertise and recent advances in neural representation learning to respond to users in online and batch modes. AugSEER is continuously running and minimally supervised. It interfaces directly both with a Neon engine that powers an interactive GUI, and with a NoSQL database that stores a knowledge graph of both named and event entities, (translated and original) texts, and NLP analytics such as sentiment analysis (Fig. 1). The GUI and the overall system (called THOR[1]) is already undergoing user studies with

---

[1] Text-enabled Humanitarian Operations in Real-time.

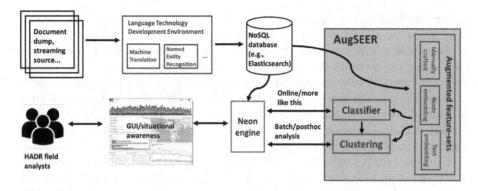

Fig. 1. A schematic of the overall HADR situational awareness system (THOR) within which AugSEER (the focus in this paper) is embedded.

real-world analysts, and is able to incorporate NLP outputs from independent state-of-the-art systems.

Contributions. We introduce and model the Structured Event Entity Resolution (SEER) problem, motivated by rapid mobilization of resources in the HADR domain. To the best of our knowledge, SEER is a difficult, socially consequential AI challenge not addressed by existing work. Second, we present AugSEER, which uses a hybrid combination of feature sets, both manually defined and automatically constructed using neural vector space embeddings, to address the SEER problem in both online and batch modes. AugSEER supports the online *more like this* mode by framing the SEER problem as a probabilistic binary classification task. To support the batch setting (e.g., for posthoc analyses), AugSEER uses a combination of classification and spectral clustering. AugSEER is also *minimally supervised*, being able to achieve reasonably accurate results using 30% (or fewer) training labels. To the best of our knowledge, this is the first application to demonstrate empirical utility from combining feature subspaces in a manner that has not been attempted in prior work on neural embeddings. Third, we rigorously evaluate multiple aspects of AugSEER on five HADR datasets encompassing diverse events, using clustering and classification metrics in tandem with visualizations.

2   Related Work

Feature embeddings have become popular in the AI and knowledge discovery communities in recent years, with vector space embeddings developed for words, sentences, documents, nodes in networks and graphs, particularly knowledge graphs, along with embeddings of the entire graph itself. Many recent models either adapt or extend the skip-gram model, used first for word2vec [13], or in the case of knowledge graph embeddings, surveyed by [21], use hand-crafted energy functions to optimize performance on applications such as triples ranking.

Other similar kinds of graph embeddings have also been proposed in the broader community (see [1] for a recent synthesis).

Our work is different from the above for several reasons. First, none of the embedding papers cited above attempt to combine manual features with graph and text-based feature embeddings in an effort to improve performance as well as allow the domain expert (in an unusual domain like HADR) to exert a level of control over the machine learning process. In general, AI research in the HADR domain has been limited; far more attention has been paid instead to good *data management* techniques [6], [8]. As [8] describe, only a handful of free systems exist for powerful HADR analytics, and none cover the SEER problem. Examples of specific work in HADR, but with much narrower scope than this paper, include 'social sensing' of earthquakes [18], and location extraction [9], both on Twitter data. To the best of our knowledge, no existing HADR system has fully leveraged recent advances in neural embeddings.

Second, existing work on entity resolution and linking is typically limited to resolving *atomic* entities like persons or organizations [4]. In contrast, we are attempting to resolve an entire event, which is a complex data structure with auxiliary information sets like words and entities. To the best of our knowledge, this is the first paper that presents a minimally supervised approach for addressing the SEER problem in a socially consequential domain like HADR.

We also note that, in contrast with graph-theoretic communities, the NLP community majorly focuses on text-centric techniques for a similar problem, namely *event co-reference* resolution [15], [11]. Events in the NLP community tend to be strictly typed according to a shallow schema, and are extracted from documents with corresponding information such as actors and dates. In contrast, our techniques make no such assumptions, since they are unrealistic in HADR. For example, a news article may discuss an event several hours or days after it strikes, while social media could be instantaneous. Often, location information is not available, and many document fragments that our approach takes as input may not even be 'events' in the NLP sense. Most importantly, we are clustering entire semi-structured objects, and not just sentences or triggers that are embedded within a larger textual context. This makes the problem more challenging, and as we describe later, text-only methods perform poorly in many cases.

3   Structured Event Entity Resolution (SEER)

We assume a set of *situation frames*, where a situation frame is intuitively defined as the finest-grained unit of data collected in that HADR problem domain. A situation frame may include such artifacts as SMS messages, intelligence fragments or even social media. Many NLP tasks are performed at the level of situation frames, following which the outputs (such as named entities) are used to enrich the situation frame further. A simple, but representative illustration, of this enrichment and the various artifacts involved, can be seen in Fig. 2. In particular, the situation frame is itself part of an *event ontology*, which captures the core elements of the analyses[2].

Given a set of situation frames, the SEER problem can be defined as inferring (whether automatic or not) *Same Event* relationships between situation frames. The *Same Event* relationship is currently assumed to have equivalence class (i.e. reflexive, symmetric and transitive) semantics, although future work may relax the transitivity assumption. Given these assumptions, each connected component (in the knowledge sub-graph where situation frames are nodes, and edges exist between frames if they are part of the same event) is called a resolved event cluster. The ultimate goal of a *batch* SEER system is to recover such clusters from a given dump of situation frames. In an (alternative) *online* setting, also called *more like this*, users (typically interactively) select a single situation frame as query, sometimes preceded by keyword search, and desire related frames that provide more insight into the broader event described by the query.

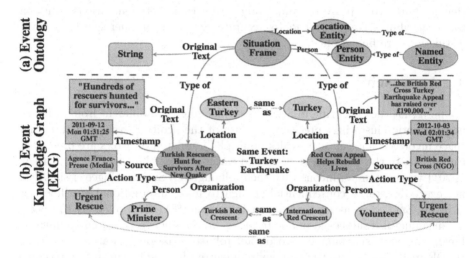

Fig. 2. A schematic illustrating the key representational details of the event ontology and event knowledge graph (EKG) for supporting solutions to the SEER problem.

While the online and batch modes are related, there are several challenges in solving either one. First, raw HADR frames are not only highly heterogeneous in terms of information content and quality, but in low-resource regions of the world (where such technology would have maximal impact), come in a computationally under-studied language like Uighyur [5]. As a first step, machine translation (MT) algorithms have to be executed to automatically translate the text into English [19]. The resulting translated text is noisy, because MT algorithms for such languages are not as well developed as for English. Next, because

---

[2] Although not fully described herein, the ontology is quite rich in practice, and includes inferential elements like sentiments and offsets (for extraction provenance).

named entities are important, both for SEER and for situational analytics, a named entity recognition system has to be executed [3], following which, *duplicate* named entities have to be *resolved*. However, highly accurate, automatic *entity resolution* [4] is far from solved, despite decades of research from the AI community.

Finally, because each disaster event tends to be *unique* compared to other disaster events, building a representative training set, and automating the solution completely using static machine learning modules, is also difficult. An example illustrating how text, topics and entities are collectively important, but can also naïvely interact to give misleading results, is in the case of the earthquake in Turkey in 2011. Around the time the earthquake struck Turkey, the country was also dealing with the Syrian refugee crisis. Frames describing either crisis tended to have similar statistical, entity and word profiles. For example, aid agencies, like the UN, or governmental entities like the Turkish army, were common to both crises. In the next section, we describe AugSEER, which is an approach that attempts to capture the important interactions between various situation frame attributes that can lead to accurate *Same Event* inference even when distinctions are fine-grained.

4   Approach

We note that the clustering in SEER is challenging (and different from ordinary non-semantic clustering) precisely because of the arbitrary scales of time and space involved, since at such scales, multiple, unrelated disasters are present in the corpus. In the example we described earlier, the earthquake that hit Turkey in 2011 was contemporaneous with the (still ongoing) Syrian refugee crisis. Also, not every disaster is consequential enough to make international headlines, or is in an English-speaking region. An important HADR problem is to generate meaningful results even in *low-resource, minimally supervised* environments. Ideally, an analyst would like to obtain robust situational awareness on each HADR-relevant event in such an environment with little technical expertise.

In order to learn good representations for addressing the HADR-specific challenges of the SEER problem, AugSEER relies heavily on an *augmented feature set* that relies on recent advances in latent space embedding models (both for text and graphs [2], [1]) as well as on a small set of similarity features that captures the intuitions of domain experts. More details are provided below.

Manually Crafted Features. Domain experts, who have studied the HADR problem over several years, understand that the text alone does not adequately convey all relevant information about an event to statistical methods. Instead, one must also rely on auxiliary *information sets*, such as extracted entities. Based on initial data exploration and feature engineering, we devised ten real-valued feature functions (Table 1), where each feature function is a similarity function that applies to some information set of a *pair* (D_1, D_2) of situation frames.

We consider three similarity functions, namely cosine similarity on TFIDF, cosine similarity on latent space embeddings derived using the paragraph2Vec

Table 1. Manually crafted feature descriptions. Each feature is computed on a pair of situation frames (D_1 and D_2).

Name	Description
$TFIDF_{\{W,E\}}$	The respective cosine similarities based on bag-of-words and bag-of-entities TFIDF representations of the text fields of D_1 and D_2
$TFIDF_{avg}$	The average of $TFIDF_W$ and $TFIDF_E$
$DV_{\{W,E,avg\}}$	Same as above, except using text embedding (rather than TFIDF) representations [2].
$JAC_{\{L,O,P\}}$	The Jaccard similarity between the {location, organization, person} extracted entity sets of D_1 and D_2
JAC_{all}	The Jaccard similarity between the set of all entities extracted from D_1 and D_2

algorithm [2], and Jaccard similarity. We consider two information sets, namely the set of entities extracted from each frame, and the tokens in the text. In the case of Jaccard similarity, we do not consider the text as an information set, but we do consider finer-grained sets like *differently typed* entities. Descriptions are provided in Table 1.

Importantly, unlike the (subsequently described) node embedding and text embedding features, the manually crafted features are computed for the texts in each *pair* of situation frames. This makes the features inherently more suited to the *more like this* online setting than to the clustering setting, to which their application and scalability is not obvious.

Node Embedding Features. Entities play an important role in the HADR domain, as many key events revolve around a specific set of persons, locations and organizations, some of which might be *latent* (i.e. not explicitly mentioned in the text). On the other hand, some entities might be wrongly extracted or typed due to imperfections in the underlying extraction system. Features relevant to explicitly extracted entities can be captured by the manual features. However, those features cannot capture latent information, and are also not good at distinguishing which entities might prove to be more important to the problem at hand. Instead, to capture the special nature of entities, we construct an *undirected entity-SF bipartite graph* from the corpus by (1) assigning a unique node in the *situation frame (SF) layer* to each frame D, and (2) assigning a unique node in the *entity layer* to the pair (E, T), where T is the type (e.g. person) of an entity E extracted from the text of at least one situation frame. Edges in this bipartite network are created by linking an entity node to each frame node from which the entity was extracted.

Next, we execute a model inspired by the skip-gram based DeepWalk algorithm on the constructed network to obtain an embedding for each situation frame and each entity [17]. DeepWalk was originally designed for learning node representations in unweighted social networks like YouTube and Flickr. In this paper, we use its philosophy for learning entity-centric frame embeddings by first sampling nodes from the bipartite graph and initiating a constant number of random walks from the node; and then treating each random walk like a list of tokens that can be embedded using skip-gram. More details on the skip-gram model and on DeepWalk may be found in the respective papers [13], [17]. We denote the entity-centric node embedding of D, obtained through the procedure described above, as D_N (boldface indicating vectorization). Note that, because of connectivity and co-occurrence information about extracted entities across the corpus, entities that have not been explicitly extracted can also influence D_N owing to the continuous representations learned by DeepWalk in a dense real-valued vector space.

Text Embedding Features. Finally, to capture statistical signals in the text, we use skip-gram based document embeddings (also called Paragraph2Vec or PV) first described in [2]. Specifically, we tokenize the machine-translated (if in a foreign language) text of a situation frame using a standard set of delimiters, convert all words to lower-case, and feed each list of tokens to the PV algorithm. For a frame D, we denote the text embedding feature vector as D_T. These embeddings are also used in computing DV features in Table 1 for frame pairs.

4.1   Classification and Clustering

AugSEER supports the SEER problem both in batch and online settings. The latter is a pairing problem, whereby a domain expert uses the system in a *more like this* manner by first specifying a situation frame as input and then expecting the system to retrieve other situation frames (possibly with other constraints specified in the GUI, like keywords or entities, but not discussed herein) that refer to the same underlying event. In AugSEER, we frame this as a *probabilistic binary classification problem* on pairs of frames, whereby the pair should have higher probability of a positive label if they represent two sub-events resolving to the same underlying event.

In a supervised setting, given a labeled set of positive and negative pairs, we construct an augmented feature vector for a pair (D_1, D_2) by (1) computing the ten manual features on the pair, (2) concatenating the node embedding feature vectors of D_1 and D_2, and (3) concatenating the text embedding feature vectors of D_1 and D_2. The final feature vector is itself a concatenated combination of all three feature sets. A classifier C is trained using the labeled data, and applied on the test data. Based on these scores (i.e. the positive class probability output by C per test item), a ranked list of relevant situation frames can be interactively shown to the HADR domain expert using the system.

In a supervised *batch* setting, the user inputs a document dump into THOR and expects clusters of situation frames, such that each cluster describes an event. As Fig. 1 illustrates, the documents first undergo processing through various components (e.g., NLP components like entity recognition and machine translation) that precede THOR. While clustering can generally be either supervised or unsupervised, it is supervised in this case because a user has specific cluster semantics (and granularity) in mind. If this were not the case, one could also achieve a 'good' clustering by executing a topic modeling algorithm like LDA. In early trials, this was found to yield poor results in terms of capturing events, due to topical flux within event clusters; see, for example, the case of the Haiti earthquake in the introduction.

Instead, AugSEER combines *spectral clustering* with the classification scheme described earlier in a supervised setting [14]. Given a set $\mathcal{D}$ of frames, the input to AugSEER is a $|\mathcal{D} \times \mathcal{D}|$ affinity matrix. We assume training sets T_P and T_N respectively of positive and negative pairs, exactly like with classification. As a first step, we train the classifier $\mathcal{C}$ on the training sets. For efficiency reasons, we use either the (concatenated) node embedding or text embedding feature representations (not both) and we do not use the manual features[3]. The second step is to construct a *symmetric* affinity matrix $\mathcal{A}$ as follows. For a cell $\mathcal{A}_{ij}$ in the matrix indexed by (i, j), we use the following assignment function:

$$
\mathcal{A}_{ij} = \begin{cases} 1 & \text{if } (D_i, D_j) \in T_P \\ 0 & \text{if } (D_i, D_j) \in T_N \\ \mathcal{C}(D_i, D_j) & \text{if } (D_i, D_j) \notin T_P \cup T_N \end{cases} \tag{1}
$$

We assume that the classifier $\mathcal{C}$ outputs the probability of the statement $(D_i, D_j) \in T_P$. Note that spectral clustering, like many other well-known clustering algorithms like k-Means, requires the desired number of clusters as a hyperparameter. Because AugSEER is a tunable system designed to assist users in exploring events (not in giving final single-point outputs), we allow the user to set this number, but also provide guidance through validation. In evaluations, this value is set at the number of clusters in the ground-truth, both for AugSEER and baselines.

5 Experiments

AugSEER has been in development for almost a year, and several evaluations have been conducted. We evaluate the algorithmic potential of AugSEER on the SEER task, both quantitatively and through qualitative visualizations.

5.1 Datasets

We evaluate AugSEER on five HADR datasets described in Table 2. Each dataset is derived from real-world disasters, of which details were publicly published on, and scraped from, the *Relief Web Processed* portal[4]. The datasets describe different HADR categories and are quite diverse in their information content. In addition, we also consider a *global* dataset that combines the information in Datasets 1–5. We use this dataset both for exploring the generalization potential of the system, as well as the loss in performance when we do not combine

---

[3] A more technical reason is that we can visualize the representations this way using an algorithm like t-SNE [12], as we illustrate subsequently.

[4] http://reliefweb.int/.

Table 2. Dataset (and gold standard) details. *pos.* stands for *positively labeled.* Column 5 separately breaks down PER/ORG/LOC entity mentions. The average number of frames per cluster in datasets 1–5 are 3, 11, 4, 6 and 5 resp.

ID	Dominant themes	Unique frames	Unique pos. pairs	Unique entity mentions	Unique words	Clusters
1	Floods	234	535	972/1,069/1,398	13,108	74
2	Earthquakes/landslides	424	11,425	1,559/1,855/1,394	18,735	38
3	Cyclones/hurricanes	101	276	372/534/440	7,479	25
4	Disease-related/tropical	135	1,401	508/513/434	8,495	21
5	Miscellaneous	461	5,117	1,554/1,512/1,576	18,689	85

feature sets into an ensemble. Note that, to ensure a fair evaluation, the machine translation and named entity recognition outputs are already provided by the program for each situation frame in the datasets, in addition to the (not used) original, non-translated text.

Negatively labeled pairs for the classification task were generated as follows. Using each frame D in the corpus as a 'query', we computed a ranked list of all other frames in the corpus using a simple bag-of-words approach on the translated text. We computed the rank of the last frame D_i that describes the same event as D. All documents between rank 1 and i not describing the same event as D were paired with D and assigned a negative label. After computing such pairs using all documents as queries, and removing duplicate pairs, we sampled about 400,000 negative pairs (20x the total number of positive pairs in Datasets 1–5) as the negatively labeled evaluation corpus, shared among Datasets 1–5, as described subsequently.

5.2   Preliminaries

We simulate the *more like this* use-case by using each frame in an event cluster as a query, and by framing the problem of 'pairing' the query frame with relevant sub-event frames as a binary classification task (described earlier in Sect. 4.1).

Parameter Tuning. We used the Python sklearn library implementations for Random Forest (RF) and Logistic Regression (LogReg) classifiers, and for spectral clustering. The gensim package in Python was used both for paragraph2Vec, as well as the word2vec model that feeds into the DeepWalk node embedding. The best hyperparameters for LogReg were found using the LogisticRegressionCV class in sklearn that uses cross-validation (on the training set), and using grid search with cross-validation for RF.

Training Protocol. Training percentages vary with the experiment as described later, but training is always *balanced.* Namely, once $|T_D|$ is fixed for a given experiment, we sample $|T_N|$ pairs from the large negative pairs corpus described earlier in Sect. 5.1. The rest of the corpus is always used for testing. Because of sampling, all experiments are conducted over ten trials, and averages are reported. We use the unpaired (two sample) Student t-test for computing statistical significance of the best performance against the *next* best alternative.

Metrics. Like with other entity resolution scenarios, *precision, recall* and their *F1-Measure* metrics on the positive class are used to report classification accuracy. For evaluating clustering, we use both the *cluster purity* and *F1-Measure* metrics. Given a cluster where each data item (i.e. a frame) has a label (withheld during clustering), in this case the underlying event that the frame is a part of, we compute cluster purity by taking the ratio of the number of frames having the majority label divided by the cluster size. F1-Measure can be computed by using the set of all pairs of frames sharing a cluster as the set of positives, and comparing against the known set of true positives to obtain the precision and recall (and by extension, their F1-Measure), similar to classification. We note that for all metrics, the higher the score, the better the performance.

5.3 Baselines

AugSEER involves a number of different interacting components both in classification and clustering settings. To illustrate that many of these components are jointly necessary for achieving good performance, we considered a range of competitive alternatives. We note that, because the SEER problem has not been studied in detail in the research literature (see Sect. 2), especially in the HADR domain, there are no direct SEER baselines available.

Classification. We consider three alternative feature-sets (or combinations) as baselines: only the manual features (M), only the DeepWalk features on the bipartite entity-frame network (N), and a combination of the two (MN). We also consider the PV text embedding baseline (T) in isolation, along with other text-only baselines like bag-of-words and topic models (using LDA), but all text-only baselines consistently under-performed the alternatives described above by significant margins. The full system includes all three feature sets (MNT).

Clustering. We tried several alternate clustering models, including Gaussian mixture models and agglomerative clustering, and found the latter to work best. We use both average (*agg-avg*) and complete (*agg-c*) linking when performing agglomerative clustering. Results are reported separately for node embedding and text embedding features. We also use *unsupervised* spectral clustering using node embeddings in a cosine similarity affinity space (*spec-N*) as a baseline, to investigate the effects of supervision in AugSEER's model of supervised spectral clustering. We also explored using the latter with TFIDF representations, but performance significantly declined, and we do not report those results herein.

5.4 Results

Four different sets of quantitative experiments, described below, were conducted to test the online and batch potential of AugSEER.

Experiment 1. For the very first set of experiments, we drew on standard findings that focus primarily on text and textual contexts, whether using embeddings or bag-of-words baselines. We considered both the classification and clustering settings, and describe the latter here (results were consistent for both). First, we built a supervised affinity matrix in the manner described in Sect. 4.1, using text embeddings, followed by spectral clustering. Across ten trials, the F1-Measure was only 10.73%, while cluster purity was higher at 71.6%. We also used cosine similarity to build an unsupervised affinity matrix, and while F1 was better for TFIDF (21.48%), the F1 for text embeddings was only 9.24%, almost 1.5% lower than for the supervised setting. Compared to the results described later, these results illustrate the non-viability of using text only, whether in low or high dimensional spaces, for addressing the challenges of SEER in the HADR domain. Alternatives like topic models, as well as alternate choices of word embeddings (e.g., PV vs. fastText), did not yield significant differences.

Experiment 2. For the second set of experiments, we tested the performance of AugSEER by using 30% and 15% of the positive samples in the *global* dataset for (balanced) training, and the rest for testing. We used both the Logistic Regression and Random Forest classifiers (with best hyperparameters determined using cross validation) with all the baseline feature sets mentioned earlier. The average best[5] F1-Measures over ten trials are reported in Table 3.

Table 3. F1-Measure results on the global dataset. MNT is the full feature set ensemble implemented in AugSEER.

Classifier (Training %)	MNT	MN	M	N
LogReg (30%)	**0.4982**	0.4924	0.4185	0.2570
LogReg (15%)	**0.5120**	0.5075	0.4439	0.2847
RF (30%)	0.7725	**0.7737**	0.4165	0.7729
RF (15%)	**0.7423**	0.7296	0.4359	0.7155

To test how the performance varied by the disaster theme, we used 30% of each dataset in Table 2 for training, and the other 70% for testing (over 10 trials). While we do not reproduce the full table herein, an absolute F1-measure improvement, using RF, was achieved by AugSEER (MNT) in the range of 0.8–18% for all five parts over the next best baseline (MN). We note that these results far outperform the text-only results[6] presented in Experiment 1.

Of the results in Table 3, RF (15%) and LogReg (15%) are significant at the 99% and 90% levels respectively. In other cases, there is no significant difference between MN and MNT. This provides some indication that all three feature

---

[5] By best, we mean that we chose the classifier threshold for all systems such that F1-measure achieved by that system was maximized in that trial at that threshold.

[6] Using average best F1 reporting and the 30% training methodology.

Table 4. *Precision/recall/F1-Measure* scores testing generalization of AugSEER (MNT and MN). All results are statistically significant at the 99% confidence level. LogRef (MNT), which is all bold, performs uniformly worse than LogReg (MN), omitted here due to space.

Training Dataset	Test Datasets	RF (MNT)	RF (MN)	LogReg (MN)
1	2+3+4+5	0.223/0.494/0.307	**0.320/0.494/0.393**	**0.228/0.296/0.275**
2	1+3+4+5	0.587/0.150/0.238	**0.614/0.163/0.258**	**0.272/0.228/0.248**
3	1+2+4+5	0.294/0.474/0.363	**0.339/0.475/0.395**	**0.271/0.300/0.284**
4	1+2+3+5	0.166/**0.457**/0.243	**0.205/0.390/0.268**	**0.257/0.205/0.228**
5	1+2+3+4	0.186/0.483/0.268	**0.209/0.506/0.296**	**0.156/0.225/0.184**

Table 5. Cluster *purity* scores using either *node embeddings/text embeddings* in a cosine similarity space (agg-*) or affinity matrix (AugSEER), except spec-N (only node embedding results reported).

ID	agg-av	agg-c	AugSEER	spec-N
1	0.611/ 0.415	0.633/ 0.402	**0.671/ 0.633**	0.556
2	0.718/ 0.384	0.723/ 0.410	**0.920/ 0.880**	0.678
3	0.644/ 0.426	0.634/ 0.416	**0.792/ 0.822**	0.624
4	0.748/ 0.496	0.704/ 0.578	**0.963/ 0.852**	0.644
5	0.639/ 0.475	0.641/ 0.456	**0.755/ 0.592**	0.522

sets have merit, with the effects more dramatic for Logistic Regression than for Random Forest. Overall, the node embedding feature vectors D_N are found to be especially instrumental, illustrating the importance of entities, both latent and explicit, for the SEER task. The good absolute performance of RF over LogReg, even after cross-validation, provides further evidence for the importance of robust feature combinations. Additionally, RF is able to generalize without overfitting, when given more training data (unlike LogReg, which clearly starts overfitting in the 30% setting, compared to the 15% setting). We tried other classifiers like SVM, and found that they underperformed RF as well.

Experiment 3. We isolate the generalization ability of different feature sets in a setting resembling *transfer learning* [16]. We used one of the datasets in Table 2 as positively labeled training data, and the others for testing. We used the same negatively labeled dataset described in Sect. 5.1 for all experiments. Maximal performance was found to be achieved across all settings with balanced training. This resulted in five training/testing paradigms. We report the average (over ten trials[7]) best F1-Measure achieved, along with corresponding precision and recall

---

[7] Because of balanced training, we had to randomly sample the negative training set; the positive set remained constant per trial.

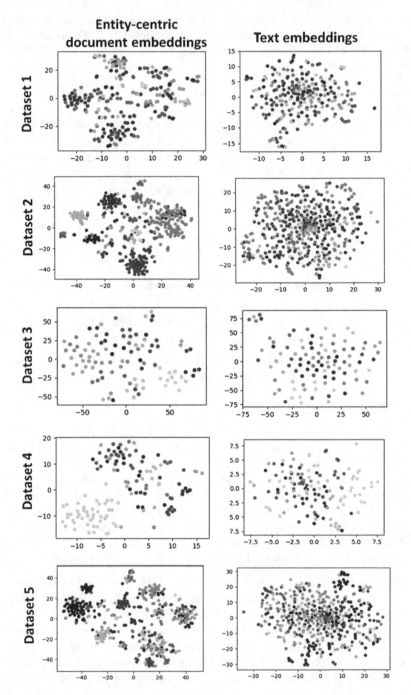

Fig. 3. t-SNE visualizations of all datasets using node/text embeddings (for visual purposes, the same color is sometimes re-used to represent different events). Dimensions have no intrinsic meaning in t-SNE [12].

in Table 4, limiting results to only the two best performing systems, which were always MN and MNT.

Similar to Table 3, we find that the two feature combinations perform similarly, but the trend is reversed. The text embeddings, which *weakly* increased the power of the classifier in the first experiment, have negative influence in this experiment. This experiment offers a cautionary lesson in naively transferring text embeddings, even in domains that seem somewhat similar (every dataset is from an HADR domain). If the data in the training phase does not sufficiently represent the test data (true in this experiment, but not the previous experiment), text embeddings can reduce F1-Measure by as much as 5%.

Experiment 4. We evaluate AugSEER in the batch/posthoc analysis setting. Using 30% positively labeled pairs in a (balanced training) supervised setting, and the RF classifier, we test AugSEER's performance against the agglomerative clustering baselines (using both average and complete link functions) as well as unsupervised spectral clustering (spec-N). In all cases, AugSEER outperforms rival methods on the cluster purity metric by a considerable margin[8], both when using node and text embeddings. When using the F1-Measure metric, a similar trend is observed, but with narrower improvements (3% average improvement, rather than the 15% achieved using cluster purity). In the next section, we use visualizations to emphasize that the latent space model and representation that AugSEER employs for entities has considerable influence on performance.

Visualization Experiments. Visualization is an important function in AugSEER as it is primarily a cognitive system designed to facilitate rapid situational awareness in both military and civilian situations. All visualizations described in this section employ the unsupervised t-SNE algorithm [12]. In an actual deployment, we use THOR (Fig. 1) for an interactive interface. Figure 3 shows that clusters for all datasets achieve an intuitive separation into different events when using the entity-document node embedding representation, but not the text embedding representation, supporting the hypothesis that entities and semantics are fundamental in addressing SEER challenges.

6 Conclusion

This paper presented AugSEER, a statistical-semantic approach for addressing structured event-entity resolution. AugSEER supports a combination of graph and text embeddings, and manually devised feature sets to achieve 77% highest F1-Measure on a challenging classification problem, using only 30% labeled training data. Similar results are achieved in the clustering scenario. AugSEER has also been implemented into a broader HADR system called THOR (Fig. 1) that is designed to ingest noisy NLP outputs and assist HADR field analysts in real-time in low-resource environments[9].

---

[8] All results in Table 5 are statistically significant at the 99% level, except AugSEER node embedding results on Dataset 1.

[9] THOR was recently demonstrated in an academic venue also: https://www2018. thewebconf.org/program/demos-track/.

Acknowledgements. The authors gratefully acknowledge the ongoing support and funding of the DARPA LORELEI program, and the aid of our partner collaborators and users in providing detailed analysis. The views and conclusions contained herein are those of the authors and should not be interpreted as necessarily representing the official policies or endorsements, either expressed or implied, of DARPA, AFRL, or the U.S. Government.

References

1. Cai, H., Zheng, V.W., Chang, K.: A comprehensive survey of graph embedding: problems, techniques and applications. In: IEEE Transactions on Knowledge and Data Engineering (2018)
2. Dai, A.M., Olah, C., Le, Q.V.: Document embedding with paragraph vectors. arXiv preprint arXiv:1507.07998 (2015)
3. Finkel, J.R., Manning, C.D.: Nested named entity recognition. In: Proceedings of the 2009 Conference on Empirical Methods in Natural Language Processing, vol. 1, pp. 141–150. Association for Computational Linguistics (2009)
4. Getoor, L., Machanavajjhala, A.: Entity resolution: theory, practice & open challenges. Proc. VLDB Endow. **5**(12), 2018–2019 (2012)
5. Hahn, R.F.: Modern uighur language research in China: four recent contributions examined. Cent. Asiat. J. **30**(1/2), 35–54 (1986)
6. Hristidis, V., Chen, S.-C., Li, T., Luis, S., Deng, Y.: Survey of data management and analysis in disaster situations. J. Syst. Softw. **83**(10), 1701–1714 (2010)
7. Jahan, S.: Human Development Report 2016: Human Development for Everyone. United Nations Development Programme (UNDP), New York (2016)
8. Li, T., et al.: Data-driven techniques in disaster information management. ACM Comput. Surv. (CSUR) **50**(1), 1 (2017)
9. Lingad, J., Karimi, S., Yin, J.: Location extraction from disaster-related microblogs. In: Proceedings of the 22nd International Conference on World Wide Web, pp. 1017–1020. ACM (2013)
10. Liu, B., Zhang, L.: A survey of opinion mining and sentiment analysis. In: Aggarwal, C., Zhai, C. (eds.) Mining Text Data. Springer, Boston (2012). https://doi.org/10.1007/978-1-4614-3223-4_13
11. Lu, J., Ng, V.: Joint learning for event coreference resolution. In: Proceedings of the 55th Annual Meeting of the Association for Computational Linguistics (Volume 1: Long Papers), vol. 1, pp. 90–101 (2017)
12. Maaten, Lvd, Hinton, G.: Visualizing data using t-SNE. J. Mach. Learn. Res. **9**, 2579–2605 (2008)
13. Mikolov, T., Sutskever, I., Chen, K., Corrado, G.S., Dean, J.: Distributed representations of words and phrases and their compositionality. In: Advances in neural information processing systems, pp. 3111–3119 (2013)
14. Ng, A.Y., Jordan, M.I., Weiss, Y.: On spectral clustering: analysis and an algorithm. In: Advances in Neural Information Processing Systems, pp. 849–856 (2002)
15. Ng, V.: Machine learning for entity coreference resolution: a retrospective look at two decades of research. In: AAAI, pp. 4877–4884 (2017)
16. Pan, S.J., Yang, Q.: A survey on transfer learning. IEEE Trans. Knowl. Data Eng. **22**(10), 1345–1359 (2010)
17. Perozzi, B., Al-Rfou, R., Skiena, S.: Deepwalk: online learning of social representations. In: Proceedings of the 20th ACM SIGKDD International Conference on Knowledge Discovery and Data Mining, pp. 701–710. ACM (2014)

18. Sakaki, T., Okazaki, M., Matsuo, Y.: Earthquake shakes twitter users: real-time event detection by social sensors. In: Proceedings of the 19th International Conference on World Wide Web, pp. 851–860. ACM, 2010
19. Vandeghinste, V., Schuurman, I., Carl, M., Markantonatou, S., Badia, T.: METIS-II: machine translation for low resource languages. In: Proceedings of LREC 2006 (2006)
20. Verma, S., et al.: Natural language processing to the rescue? extracting "situational awareness" tweets during mass emergency. In: ICWSM (2011)
21. Wang, Q., Mao, Z., Wang, B., Guo, L.: Knowledge graph embedding: a survey of approaches and applications. IEEE Trans. Knowl. Data Eng. **29**(12), 2724–2743 (2017)

That's Interesting, Tell Me More! Finding Descriptive Support Passages for Knowledge Graph Relationships

Sumit Bhatia[1][(✉)], Purusharth Dwivedi[2], and Avneet Kaur[2]

[1] IBM Research AI, Delhi, India
sumitbhatia@in.ibm.com
[2] IIIT Delhi, Delhi, India
{purusharth14081,avneet14027}@iiitd.ac.in

Abstract. We address the problem of finding descriptive explanations of facts stored in a knowledge graph. This is important in high-risk domains such as healthcare, intelligence, etc. where users need additional information for decision making and is especially crucial for applications that rely on automatically constructed knowledge graphs where machine-learned systems extract facts from an input corpus and working of the extractors is opaque to the end-user. We follow an approach inspired from information retrieval and propose a simple, yet effective and efficient solution that takes into account passage level as well as document level properties to produce a ranked list of passages describing a given input relation. We test our approach using Wikidata as the knowledge base and Wikipedia as the source corpus and report results of user studies conducted to study the effectiveness of our proposed model.

1 Introduction

Knowledge Graphs are becoming increasingly important in knowledge and data management applications as they afford a semantic structure to the underlying data. They form crucial components of modern web search engines, state-of-the-art question answering systems such as IBM Watson, and are used in a variety of applications in domains as diverse as healthcare [28], finance [33], media [16], cybersecurity [21], etc. Entities are the fundamental units of knowledge graphs and are often presented to users as a result of a search query, or are used in applications such as exploratory search where users can search about entities of interest and browse their important relationships [19]. For various critical applications such as exploring interactions between genes and drugs [14], intelligence applications [36], etc., users may want some additional description or supporting evidence that provides some explanation of the relationship presented to them in order to build confidence in their decision making process. Even for generic information search or browsing activities the entities and relationships presented to the user may be unknown to her and thus, she may not be able to fully appreciate the relevance of information presented to her by the system.

© Springer Nature Switzerland AG 2018
D. Vrandečić et al. (Eds.): ISWC 2018, LNCS 11136, pp. 250–267, 2018.
https://doi.org/10.1007/978-3-030-00671-6_15

As an example, consider the relationship triple $<H.\ R.\ McMaster,\ military\_rank,$ $Lieutenant\ General>$ and its following description as extracted by our proposed approach (Sect. 3).

> *...In February 2014, Defense Secretary Chuck Hagel nominated McMaster for Lieutenant General and in July 2014, McMaster pinned on his third star when he began his duties as Deputy Commanding General of the Training and Doctrine Command and Director of TRADOCs Army Capabilities Integration Center. Army Chief of Staff General Martin Dempsey remarked in 2011 that McMaster was "probably our best Brigadier General. McMaster made Times list of the 100 most influential people in the world in April 2014...*

An end-user who does not know that Mr. McMaster is a US Army officer may find the above fact much more useful when presented with the accompanying supporting description rather than presenting the fact alone. It may also help build his trust and confidence in the system. In fact, it has been found that in scenarios where users are dealing with uncertain information, use of natural language descriptions helps in the decision making process [15]. In web search engines, usefulness of small text snippets to improve end-user experience is well studied [10]. Likewise, in context of scientific digital libraries, it has been found that accompanying figures, tables, etc. with small textual descriptions helps users in judging their importance [5,35]. Therefore, we posit that providing users with small textual explanations of the relationships may help their understanding, build their confidence in the system and help them in accomplishing their intended tasks. We believe that such a capability is even more crucial for systems that rely on knowledge graphs that are constructed automatically [29], especially using deep neural networks [37] where interpretability is a big issue.

In this work, we describe *a probabilistic method based on language models to extract supporting passages from an underlying text corpus that provide descriptive explanations of a knowledge graph relationship* (Sect. 3). Given an input relationship, our model takes into account passage-level and document-level evidence to rank different passages in the order of their relevance to the input relationship. Previous works on explainability of knowledge graph data have mainly focused on explaining how two entities in a graph may be related and the *explanations* are often in the form of a set of common entities or paths connecting the two entities [1,12,31] and thus, suffer from the same issues as discussed above. Efforts on generating textual descriptions of relationships have also focused mainly on template based methods where given a set of facts and an underlying text corpus, different templates are learned that could be used for representing the relationship [2,43]. For example, for relations of type $<X, dateOfBirth, Y>$, sentence templates such as "X was born on Y" are learned. However, such sentences offer textual *representations* of the input relationship rather than a *supporting explanation* which is the main focus of our work. We propose an approach that is simple, effective, and unsupervised, and thus, can be easily adopted by different systems. We implemented and evaluated our approach using Wikipedia as our background text corpus and Wikidata as our knowledge base and results obtained through user studies conducted to study the effectiveness of our proposed techniques are encouraging (Sect. 4). The query and result sets generated by this work are also being made available for the community. We also discuss the strengths and limitations of our proposed approach and lay down directions for future work (Sect. 5).

2   Related Work

We provide a brief overview of related work categorized under two broad categories. First we provide an overview of most relevant papers that have looked at generating small textual descriptions of results in different search scenarios such as web search, academic search, etc. Next, we focus on works that have addressed the problem of explaining relatedness between knowledge graph entities through both graph-based and textual summaries.

2.1   Supporting Search Results with Textual Descriptions

User studies conducted by Tombros and Sanderson [41] have shown that in document retrieval systems, presenting users with short textual summaries describing the retrieved documents help them judge the importance and utility of the results much better and faster. Likewise, in Web Search Engines, it is a common practice to present results along with a small textual summary or *snippet* extracted from the web page [42] and the positive influence of snippets on end-user experience and behavior is well studied [10]. Metzger et al. [26] proposed a semantic aware document-retrieval method that transforms a given keyword query into RDF statements, and ranks documents based on their relevance to the statements. Further, the sentences matching RDF statements in the documents are extracted and presented as snippets to the user [11]. In context of academic search engines such as CiteSeer and Google Scholar, Bhatia and Mitra [6] studied the problem of generating small descriptions of *document-elements* (figures, tables, and pseudo-codes) present in academic papers to help users quickly decide their importance without having to read the whole paper. Similarly, snippets have been found useful for XML search systems [20] and ontology search systems [30] where small textual descriptions have helped users select the most suitable results for their information needs.

2.2   Explaining Knowledge Graph Relationships

Graph-Based Approaches: On receiving an entity query, Web search engines such as Google, Bing, etc. often show a list of related entities on the search page or in a separate entity box populated by information derived from the underlying knowledge base. However, it is not always apparent to the users how the suggested entities are connected to the input entity. Fang et al. [12]

describe their system *REX* that takes as input two knowledge graph entities and produces a ranked list of relationships between the two entities efficiently. Bhatia et al. [3] proposed a relationship ranking function that takes into account features such as entity popularity, affinity between the input entities and strength of different relationships between them. Pirrò [31] considered the problem of explaining how two entities in a knowledge graph might be related as a sub-graph finding problem where the sub-graph consists of nodes and edges in the set of paths between the two input entities. Thus, the explanation of the relatedness between two entities is provided by means of shared entities and relationships between them. Aggarwal et al. [1] considered the task of explaining relationships between two entities as a path-ranking problem and propose a scoring mechanism to identify informative and discriminative paths.

Text Based Approaches: In context of web search where the systems present entities as part of search results, Blanco and Zaragoza [8] studied the problem of finding support sentences for explaining why an output entity is considered relevant to the original ad-hoc text query by the user. Saldanha et al. [34] addressed the problem of generating descriptions of lesser known companies and describe a template based approach to create such descriptions by generating sentences from RDF triples found in DBPedia and Freebase about the company. These sentences are generated by utilizing the RDF triples and corresponding Wikipedia sentences for known companies and learning templates such as "<company> was founded by <founder>". Voskarides et al. [44] describe a learning to rank based sentence extraction and ranking method to find human readable descriptions of a relationship between two knowledge graph entities. Their follow-up work [43] tackles the problem using a template based approach. For a given relationship type, they identify representative sentences describing some of the relationship instances and then generating textual description of other instances of the same relationship type by selecting a suitable template and filling it with appropriate entities. Such template based approaches requires manual construction of templates for each relationship type that may be difficult for many practical applications. For example, Wikidata contains more than 1600 unique relationships types, DBPedia contains more than 2800 relationship types. The problem is exacerbated in domain specific knowledge graphs where domain knowledge is required for generating appropriate templates. Further, machine learning of such templates or other learning based methods require significant amount of training data and it may not always be feasible due to lack of such data and thus, may only be useful for a few specific relationship types.

3   Proposed Approach

Let us consider a relationship $\mathcal{R} = <s, r, t>$ in a knowledge Graph $\mathcal{K}$ where s and t correspond to the source and target nodes (entities), respectively, and r is the relationship edge label. Let P be the set of passages extracted from an underlying text corpus[1]. We wish to rank the passage $p \in P$ based on the probability that it contains a descriptive explanation of $\mathcal{R}$. Mathematically, having observed the relationship $\mathcal{R}$, we are interested in computing the probability that passage p is relevant to $\mathcal{R}$, i.e., $P(p|\mathcal{R})$. By application of Bayes' Theorem, we have:

$$P(p|\mathcal{R}) = \frac{P(p) \times P(\mathcal{R}|p)}{P(\mathcal{R})} \propto P(p) \times P(\mathcal{R}|p) \tag{1}$$

Here, $P(\mathcal{R})$ in the denominator has been ignored as it will be same for all the passages $p \in P$. The component $P(p)$ can be interpreted as the prior probability of the passage p being of interest. Note that this prior is independent of the relationship (query) and can be used to model certain domain specific characteristics based on the application requirements. For example, in a medical domain application, passages coming from a peer-reviewed article can be assigned a higher prior than passages coming from a non-authoritative article. In this work, we are focused on the general performance of the framework and hence, we assume a uniform prior as is common in information retrieval [25, Chap. 12] and thus, $P(p)$ can also be ignored for ranking purposes. With these assumptions and assuming conditional independence of three components of the relationship $\mathcal{R}$ (namely, s, r, and t), Eq. 1 reduces as follows.

$$P(p|\mathcal{R}) \propto \underbrace{P(s|p) \times P(t|p)}_{\substack{\text{entity} \\ \text{probability}}} \times \underbrace{P(r|p)}_{\substack{\text{relationship} \\ \text{probability}}} \tag{2}$$

Here, $P(s|p)$ and $P(t|p)$ represent the probability of observing mentions of source and target entities, s and t, respectively in the passage p. Likewise, $P(r|p)$ represents the probability that relation label r is being described in passage p. In order to compute these probabilities, we adapt the query likelihood model based on multinomial unigram language model [25] that computes probability of generating a query given a text document. We can treat each passage in P as our source document and compute the probabilities of generating the entities s, t and relation r as specified in Eq. 2. Note that the names of entities s and t and relationship label r consist of multiple individual words and assuming conditional independence of terms, we can simplify Eq. 2 as follows.

$$P(p|\mathcal{R}) \propto \prod_{w \in S \cup T \cup R} P(w|p), \tag{3}$$

Here, S, and T are the sets of terms in names of source entity s and target entity t, respectively, and R is the set of terms representing the relationship r. Note that relationship labels in knowledge graphs are often created like variable names (*bornOn, citizen_of*, etc.) that are generally not used in standard written vocabulary. Further, a given relationship may be described by different

---

[1] Given a text corpus, there are multiple ways of extracting passages and the approach for ranking these passages is independent of the way passages are extracted. We detail our choice of passage extraction method in the section on experiments (Sect. 4).

synonymous terms (occupation, profession, etc.). Therefore, to account for these variations, R can be constructed by using a set of synonyms representing a given relationship type. In this work, we have chosen relationship label aliases provided by Wikidata to obtain a set of terms that could be used for representing a given relationship type. For example, for the label *date of birth*, the list of aliases as provided by Wikidata[2] includes *born on, birthday, DOB,* etc. We note that depending upon the application at hand, different domain specific synonyms can also be used for this purpose.

Another important consideration is that a typical passage is only a few sentences long. As a result, a given passage alone may not have sufficient information to reliably approximate the probability of observing a term from the passage due to data sparsity issues. The probabilities are over estimated for the terms that are present in the passage and are under estimated for the terms that are not present in the passage. This is especially exacerbated in case of entity names (nouns) that are often mentioned as corresponding pronouns (his, her, she, etc.). As a result, a highly useful passage may get a very low score if the entity of interest is mentioned by its pronoun in the passage. Likewise, it is possible that a non-relevant passage may get a very high score because of multiple occurrences of just one or two terms in the passage. In order to account for such imbalances, the probability estimations are smoothed by adding document and collection level statistics. Consequently, the unigram language model of passage p is then modeled as a mixture of passage, document, and collection (corpus) language models, respectively, as follows:

$$P(w|p) = P(w|\Theta_{MM}) \tag{4}$$
$$= \lambda_1 \underbrace{P(w|\Theta_p)}_{\substack{\text{passage-level} \\ \text{evidence}}} + \lambda_2 \underbrace{P(w|\Theta_d)}_{\substack{\text{document-level} \\ \text{evidence}}} + \lambda_3 \underbrace{P(w|\Theta_c)}_{\substack{\text{collection-level} \\ \text{evidence}}} \tag{5}$$

where, $\lambda_1 + \lambda_2 + \lambda_3 = 1$. We set $\lambda_1 = 0.6, \lambda_2 = \lambda_3 = 0.2$ for our experiments. The values are chosen to give relatively more weight to passage level evidence and use document and collection level evidence as normalizing factors.

Modeling the entity probabilities and smoothing as just described serves multiple objectives. First, it helps overcome the sparsity problem due to the short length of the passage. Second, the document level evidence gives a higher score to passages that come from documents that talk more about the entities involved in input relationship. Thus, passages coming from documents that are majorly about the involved entities are given a higher weight by the ranking function described in Eq. 5. Also note that such a formulation also addresses the problem of co-reference resolution [17] to some extent and can be interpreted as a probabilistic variant of the heuristic used by Wu and Weld [46] that replaces most frequent pronouns in Wikipedia article with article title. Lastly, the collection level evidence is also important as it plays the role of a reference or *background*

---

[2] Details of *date of birth* relationship label (also called as property in Wikidata): https://www.wikidata.org/wiki/Property:P569.

language model and provides term weighing similar to inverse document frequency (IDF) [48].

The individual probabilities in Eq. 5 can be computed by using the statistics from passage, document, and collection as follows:

$$\text{Passage Evidence: } P(w|\theta_p) = \frac{count(w,p) + 1}{|p| + |V|} \tag{6}$$

$$\text{Document Evidence: } P(w|\theta_d) = \frac{count(w,d) + 1}{|d| + |V|} \tag{7}$$

$$\text{Collection Evidence: } P(w|\theta_c) = \frac{count(w,c)}{|C|} \tag{8}$$

Here, V is the vocabulary of the corpus and $|\cdot|$ indicates the size of the set. Note that we have added the constant one in Eqs. 6 and 7 to prevent zero probabilities for terms that may not be present in the respective passage or document. Further, the denominators are chosen so that the sum of probabilities over the entire vocabulary is one. Also note that the additive factor is not required in the collection model as all the terms in the vocabulary are present in the collection by definition.

4 Experimental Evaluation

4.1 Data Description

In this section we discuss the dataset used in our experiments and how the queries and relevance judgments were obtained. The resulting resources (queries, results, and relevance judgments, and parameters used) are being made available to the community through our git repository[3].

Relationship Queries: We need relationship triples of the form <s,r,t> that will constitute our query relationships for which the supporting passages need to be retrieved from the underlying corpus. In order to create such a query set, we selected titles of the top 25 most viewed pages[4] each for the months of January–April, 2017. From these 100 (25 for each month) page titles, we retained only those that correspond to named entities by manually filtering out titles like *List of Black Mirror episodes, Deaths in 2017*, etc. That gave us a total of 80 unique entities. Next, we used Wikidata[5] as our knowledge base and retrieved all relationships of the entities selected previously using the SPARQL end-points provided by Wikidata. From all these retrieved relationships, we manually filtered out the relationships that were not in English language, were of type *instance of* and *subclass of*, and, where the target entity was not a named entity. This resulted in a final set of 1250 unique relationship triples from which we selected 150 triples at random as our final relationship query set that was used in subsequent experiments.

---

[3] https://github.com/sumit-research/kg-support-passages.
[4] https://en.wikipedia.org/wiki/Wikipedia:Top_25_Report.
[5] https://www.wikidata.org/wiki/Wikidata:Main_Page.

Source Corpus and Passages: We chose Wikipedia[6] as our underlying corpus. There are multiple ways to extract a set of passages given a text corpus such as utilizing the document structure and paragraph or section markers present in the documents itself. However, the passages thus extracted are usually very long, often running into tens of sentences. Further, while such paragraph or section markers are available for well-structured corpora such as Wikipedia, they may not always be available for different source documents. More importantly, such long passages may be detrimental to the end-user experience as they consume valuable screen real estate and reading them requires significant additional efforts from users. Another option is to use text segmentation methods such as TextTiling [18] that segment the input text into topically coherent passages. However, such approaches require significant pre-processing efforts, especially for large corpora (few millions of documents) often encountered in real world applications. In practice, simple (and *fast*) segmentation of input text into fixed length, overlapping passages using a sliding window approach is found to be equally effective [4,22,39,40], if not better, and is the approach we also take. Use of overlapping passages is also encouraged as it reduces the chances of relevant information getting split between two consecutive passages [9]. Therefore, we split the input text of each document into overlapping passages of three consecutive sentences using a sliding window of size three as suggested by Spangler et al. [38]. This resulted in about 80.5 million extracted passages that constitute our source set of passages (set P in Sect. 3). Note that one drawback of such a pre-processing is that multiple overlapping passages containing a highly relevant sentence can all appear in top positions in the final ranked list, thereby artificially boosting the proportion of relevant passages and at the same time, causing a degraded user experience due to repetitive results. Therefore, we perform a post-processing step where such repetitions are detected and only the highest scoring passage is retained and the rest of the overlapping passages are removed from the final ranked list.

In order to compute the different passage, document, and collection based statistics, we used the Indri toolkit provided by the Lemur project[7]. The toolkit offers capabilities to query and index a collection of documents, and APIs to compute term statistics required for language model based computations described in our ranking function (Eq. 5). Specifically, we created two indexes using Indri – a *passage index* of all the extracted passages to compute passage level statistics and an *article index* of all the Wikipedia articles (about 5.34 million articles) to compute document and collection level statistics. A standard stopword list provided by the Onix text retrieval toolkit[8] was used to filter out common stop words and stemming was performed using Porter's [32]. The parameter files used for creating and querying the indexes can be found in our git repository.

---

[6] Specifically, we used the dump of 20th April, 2017.

[7] https://www.lemurproject.org/.

[8] http://www.lextek.com/manuals/onix/stopwords1.html.

Baseline Methods: We use the inference network based generative passage retrieval algorithm implemented in Indri [27] as our first baseline method (*Inf. N/w*). This is a state-of-the-art passage retrieval method and is often chosen as a baseline for various research tasks related to passage retrieval [45,47]. Given a query, this method finds documents that are relevant for the query and then extracts specific continuous portions of text from the documents that are highly relevant for the query. Given an input relationship tuple <s,r,t>, the input query to Indri consists of all the terms in source and target entity names and relationship description. Next, as our second baseline method, we add query expansion [25, Chap. 9] on top of the first baseline by using relationship aliases (as described in Sect. 3). We denote this baseline as *Inf. N/w + Rel.Exp* in subsequent discussions. Note that for passage retrieval, Indri requires passage length as an input parameter. For comparison purposes, we specify the length of passages to be returned by Indri as 600 characters as this is the average length of passages extracted by our proposed approach. Further, note that due to fixed length of the passages retrieved, it is possible that the retrieved passaged are often truncated and thus, have incomplete sentences. In order to overcome this shortcoming, such truncated sentences are completed in a post-processing step so that the extracted passages are well-formed.

4.2   Effectiveness Evaluation

In order to study the effectiveness of our proposed approach for finding high quality descriptive passages, we selected a random set of 50 relationship triples from the set of 150 triples described above. For each of these 50 triples, we retrieved the top five passages from the corpus ranked by our ranking function (Eq. 5). We also obtained five passages for each relationship triple by the two baseline methods. This resulted in 695 unique <query,passage> pairs. Note that this number is less than 750 (50 queries × 3 methods × 5 passages) because in some cases, the same passage was retrieved by multiple methods.

Next, we took the help of three human evaluators to evaluate the quality and correctness of the passages retrieved by the baselines and our proposed method. The evaluators were advanced graduate students in Computer Science, not associated with the project, and had good command of the English language. The evaluators worked independently of each other and were compensated monetarily for their efforts.

For each of the 50 queries used in this study, all the extracted passages for that query by the three methods were presented to the evaluators in a randomized order and they were not informed which passage was retrieved by which method. The evaluators were asked to rate each passage on three point scale – 0 if the passage is incorrect, irrelevant or not at all useful; 1 if the passage contains the relationship but is only partially relevant and does not provide a good explanation; and 2 if the passage is correct and highly relevant and provides a good explanation. All three evaluators provided their judgments for all the 695 <query, passage> pairs.

Table 1. Distribution of the labels assigned by the three evaluators. There were a total of 695 <query, passage> pairs and each evaluator provided judgments for all 695 pairs. Last column reports the results after combining all the judgments where the final rating of a <query, passage> pair was decided after taking the majority vote.

	Evaluator 1	Evaluator 2	Evaluator 3	Final
Non-relevant	406	438	444	449
Partially-relevant	41	11	43	12
Relevant	248	246	208	234

Inter Annotator Agreement: We used Fleiss' Kappa coefficient [13] to measure the agreement between the three evaluators. The value of Kappa coefficient was computed to be 0.67, indicating substantial agreement. For 695 <query, passage> pairs, all three evaluators agreed on the label 545 times, two evaluators provided the same label 130 times and for 20 pairs, all three evaluators provided different ratings. In case of conflict, the final label for a <query, passage> pair was decided by the majority vote and 20 pairs where all three evaluators disagreed were assigned a label of 0 (irrelevant). Table 1 provides further details about the distribution of evaluations provided by the three evaluators.

Results: Table 2 compares the three approaches by using precision, precision at rank 1 ($P@1$), and mean reciprocal rank (MRR). While precision measures how many of the passages extracted by each method are relevant, $P@1$ and MRR measure the ability of the respective methods to identify a relevant passage as the top-ranked passage. This is important because in real world applications, due to limited screen real estate and to minimize users' efforts, we want to present the best results at the top position. As can be observed, the proposed approach achieves a $P@1$ of 0.86 compared to 0.251 and 0.165 for the baseline methods. Similar out-performance is observed in the case of MRR values. Further, we note that the proposed approach achieves an overall precision of 0.727 compared to 0.156 and 0.088 for the baselines. Next, for a fine-grained analysis, Table 3 provides the distribution of passages marked as irrelevant, partially relevant, and highly relevant for the three approaches. We note that for the proposed approach, only about 16% of the passages were found to be irrelevant by the evaluators compared to about 80% for the baseline approaches. These results indicate not only that the proposed approach is able to retrieve a lot more relevant passages describing the input query relationship (as indicated by precision), it is also able to offer relevant results at top positions (as indicated by $P@1$ and MRR values).

A surprising observation from these results is the poor performance of the *Inf. N/w + Rel. Exp.* baseline method, even when compared with the plain *Inf. N/w* method. Aliases of relationship labels were incorporated in order to enable the *Inf. N/w* method to identify passages where variations of relationship terms are used. However, on analysis of the retrieved passages, we observed that addition of the alias terms led to retrieval of many passages that talked about the relationship label in general. For example, for the query <Mariah Carey, spouse, Nick Cannon>, the following passage is retrieved that talks about the concept of *spouse* in general.

> *Wife: Intro A wife is a female partner in a continuing marital relationship. A wife may also be referred to as a spouse, which is a gender-neutral term. The term continues to be applied to a woman who has separated from her partner, and ceases to be applied to such a woman only when her marriage has come to an end, following a legally recognized divorce or the death of her spouse. On the death of her partner, a wife is referred to as a widow, but not after she is divorced from her partner.*

Evidently, the above passage contains multiple mentions of different aliases of the *spouse of* relationship[9] and thus, this passage got a very high score. This example illustrates the strength of the proposed approach that avoids such a dominance of certain terms in the passage by incorporating the document and collection level evidences in the ranking function (Eq. 5) that assigns lower score to passages from documents that contain little or no information about the entities involved in the relationship.

4.3   Preference Evaluation

In this section, we describe the experiment conducted to study the preferences of end-users when passages extracted by different approaches are presented to them side by side. We chose only the *Inf. N.w* baseline for comparison with the proposed approach due to its superior performance compared with the other baseline method. For this experiment, we used the full set of 150 relationship tuples (Sect. 4.1) and recruited 3 undergraduate computer science students that were not associated with this project and were compensated monetarily for their efforts. For each query, the top scored passages extracted by the baseline and our proposed approach were presented to the evaluators side by side and they were asked to chose from one of the following four options: *(i)* both passages are equally good/useful, *(ii)* both passages are equally bad, *(iii)* passage on

Table 2. Performance of the baseline methods and proposed approach as measured by precision, P@1, and MRR.

	P@1	Precision	MRR
Inf. N/w	0.251	0.156	0.272
Inf. N/w + Rel. Exp.	0.165	0.088	0.144
Proposed approach	**0.860**	**0.727**	**0.805**

---

[9] Aliases of spouse include husband, wife, married to, consort, partner, marry, marriage, partner, married, wedded to, wed, and life partner. https://www.wikidata.org/wiki/Property:P26.

Table 3. Distribution of judgment labels for the baseline and proposed approach. Note that the total number of passages for proposed approach is 246 instead of 250 because some passages appeared for more than one query.

	No. of passages marked as		
	Irrelevant	Partially-relevant	Highly relevant
Inf. N/w	198	4	43
Inf. N/w + Rel. Exp.	210	3	32
Proposed method	41	5	159

the left offer a better description, and *(iv)* passage on the right offers a better description. Note that the order in which the passages were presented to the evaluators was randomized and they were not informed of the method that produced a specific passage. Each evaluator provided preference judgments for 50 relationships. The results are summarized in Table 4. As can be seen from the results, for an overwhelming majority of the time, all the evaluators preferred the passages extracted by the proposed approach. Overall, more than 50% of the times, passages retrieved by the proposed approach were preferred (73 out of 150) whereas the passages retrieved by the baseline method was preferred only 11 times.

5 Discussions

In this section, we provide some representative examples to illustrate the strengths and weaknesses of our proposed approach and discuss possible future directions of research. Consider the relationship <*John Cena, nickname, The Protoype*>, for which the passages as produced by the baseline and our proposed approach are as follows.

> **Baseline:** *A prototype is something that is representative of a category of things, or an early engineering version of something to be tested. Prototype may also refer to: Automobiles. Citroën Prototype C, a range of vehicles*

Table 4. Comparative evaluations provided by the three evaluators when presented with top passages from the baseline and proposed approach side by side. Each evaluator provided evaluations for 50 relationship queries.

	Both not useful	Both equally useful	Baseline	Proposed
Evaluator 1	14	13	3	20
Evaluator 2	6	17	2	25
Evaluator 3	10	6	6	28
Total	30	36	11	73

created by Citroën from 1955 to 1956 Citroën Prototype Y, a project of
replacement of the Citroën Ami studied by Citroën in the early seventies
Daytona Prototype, a sports ca.

Proposed approach: *In 2001, Cena signed a developmental contract with*
the WWF and was assigned to its developmental territory Ohio Valley
Wrestling (OVW). During his time there, Cena wrestled under the ring
name The Prototype and held the OVW Heavyweight Championship for
three months and the OVW Southern Tag Team Championship (with Rico
Constantino) for two months. Throughout 2001, Cena would receive four
tryouts for the WWF main roster, as he wrestled multiple enhancement
talent wrestlers on both WWF house shows and in dark matches before
WWF television events.

Note that the first passage contains multiple occurrences of the word *pro-*
totype which is also a less frequent word in the corpus, and thus was highly
ranked by the baseline approach. On the other hand, the passage produced by
the proposed approach is able to correctly identify a good passage even though
it only had one occurrence of *prototype*. One reason for this passage getting a
very high score is the document level component of the ranking function (Eq. 5).
This passage comes from the Wikipedia article about *John Cena* and thus, its
score was boosted by the document-evidence component.

Error Analysis: By further analyzing the passages extracted by the proposed
approach and feedback from the evaluators, we observed two major characteristics of the passages that were not rated as relevant by the evaluators. In the first
category, while the extracted passage does talk about the entities involved, it
does not provide any description of the relationship specified in the query. Consider the following passage for the relationship <*Alan Comes, employer, Fox*
News>.

...Goldlines television advertising includes cable networks such as CNN,
CNBC, Fox News, History International and Fox Business. Goldline has
also been the sponsor of the shows of a number of conservative radio and
television hosts, including The American Advisor, and The Glenn Beck
Program, The Laura Ingraham Show, The Fred Thompson Show, The
Huckabee Report, The Lars Larson Show, The Monica Crowley Show, The
Mark Levin Show, and The Alan Colmes Show. In 2009, Goldline incor-
rectly labeled Glenn Beck as a paid spokesman on its website which raised
concerns with his employer, Fox News, which prohibit such a relationship;
they later corrected it to radio sponsor...

This passage got a high score by the proposed scoring function because it talks
about Fox News and Alan Comes and the originating document also has other
mentions of Fox News. However, it does not provide any description about the
employment of Alan Comes at Fox News. Instead, it provides a lot of unnecessary
information to the user.

The other type of passages that were not judged relevant by the evaluators were the ones that made an indirect reference to the relationship query. Consider the following passage for the query < *Warren Beatty, occupation, Film Producer*>.

> *In 1994, Astin directed and co-produced (with his wife, Christine Astin) the short film Kangaroo Court, which received an Academy Award nomination for Best Live Action Short Film. Astin continued to appear in films throughout the 1990s, including the Showtime science fiction film Harrison Bergeron (1995), the Gulf War film Courage Under Fire (1996), and the Warren Beatty political satire Bulworth (1998). After The Goonies, Astin appeared in several more films, including the Disney made-for-TV movie, The B. R. A*

Here again, the passage contains a lot of unnecessary information and only contains a fleeting reference to Warren Beatty and the movie Bulworth. There is no explicit mention here that Warren Beatty is a film producer and thus, the evaluators did not find this passage to be very informative.

Directions for Future Work: In the present work, we focused on relationship triples between two named entities. It will be interesting to extend the proposed models to triples where the target is a data value instead of a named entity (e.g. <*Burj Khalifa, height, 828 metres*>). This is a challenging problem as the information in text could be present in multiple formats (numbers, text, etc.) as well as in different units. Another aspect of our proposed approach that merits further research is handling of negations. For example, consider the relationship < $X, spouseOf, Y$ > and a sentence, X *is not wife of* Y. Such a sentence will also be considered a relevant sentence by our method even though it offers negative evidence of the fact under consideration. However, handling negations in text is a hard problem and is an active area of research [23]. One related interesting application of our proposed approach that is worth exploring further is in *fact checking* systems such as DeFacto [24] where the users could query for supporting evidence for facts presented to them and can evaluate if the information shown to them is correct.

Another direction for future work is to combine the proposed approach with existing methods for entity search and recommendation [7] and path ranking [1,12,31], and offer textual descriptions for how two entities in the knowledge graph may be related. Such techniques will be useful for discovery and exploratory search based applications and may improve end-user experience by offering human readable explanations of systems' graphical output.

6   Conclusions

We studied the problem of providing descriptive explanations for relationships in a knowledge graph and described a probabilistic method for ranking passages derived from an input corpus in order of their relevance to the input relationship. The proposed method is simple, effective, and outperformed state-of-the-art baseline methods in user studies conducted for evaluating the effectiveness of our proposed approach. We presented some representative examples to illustrate the strengths and weaknesses of our approach and provided directions for future work.

References

1. Aggarwal, N., Bhatia, S., Misra, V.: Connecting the dots: explaining relationships between unconnected entities in a knowledge graph. In: Sack, H., Rizzo, G., Steinmetz, N., Mladenić, D., Auer, S., Lange, C. (eds.) ESWC 2016. LNCS, vol. 9989, pp. 35–39. Springer, Cham (2016). https://doi.org/10.1007/978-3-319-47602-5_8

2. Althoff, T., Dong, X.L., Murphy, K., Alai, S., Dang, V., Zhang, W.: Timemachine: timeline generation for knowledge-base entities. In: Proceedings of the 21th ACM SIGKDD International Conference on Knowledge Discovery and Data Mining, pp. 19–28. ACM (2015)

3. Bhatia, S., Goel, A., Bowen, E., Jain, A.: Separating wheat from the chaff – a relationship ranking algorithm. In: Sack, H., Rizzo, G., Steinmetz, N., Mladenić, D., Auer, S., Lange, C. (eds.) ESWC 2016. LNCS, vol. 9989, pp. 79–83. Springer, Cham (2016). https://doi.org/10.1007/978-3-319-47602-5_17

4. Bhatia, S., He, B., He, Q., Spangler, S.: A scalable approach for performing proximal search for verbose patent search queries. In: Proceedings of the 21st ACM International Conference on Information and Knowledge Management, pp. 2603–2606. ACM (2012)

5. Bhatia, S., Lahiri, S., Mitra, P.: Generating synopses for document-element search. In: Proceedings of the 18th ACM Conference on Information and Knowledge Management, CIKM 2009, pp. 2003–2006. ACM, New York (2009)

6. Bhatia, S., Mitra, P.: Summarizing figures, tables, and algorithms in scientific publications to augment search results. ACM Trans. Inf. Syst. **30**(1), 3:1–3:24 (2012)

7. Bhatia, S., Vishwakarma, H.: Know thy neighbors, and more! Studying the role of context in entity recommendation. In: Proceedings of the 29th ACM Conference on Hypertext and Social Media, HT 2018, ACM, New York (2018)

8. Blanco, R., Zaragoza, H.: Finding support sentences for entities. In: Proceedings of the 33rd International ACM SIGIR Conference on Research and Development in Information Retrieval, pp. 339–346. ACM (2010)

9. Callan, J.P.: Passage-level evidence in document retrieval. In: Croft, B.W., van Rijsbergen, C.J. (eds.) SIGIR 1949. Springer, London (1994)

10. Clarke, C.L., Agichtein, E., Dumais, S., White, R.W.: The influence of caption features on clickthrough patterns in web search. In: Proceedings of the 30th Annual International ACM SIGIR Conference on Research and Development in Information Retrieval, pp. 135–142. ACM (2007)

11. Elbassuoni, S., Hose, K., Metzger, S., Schenkel, R.: ROXXI: reviving witness documents to explore extracted information. Proc. VLDB Endowment **3**(1–2), 1589–1592 (2010)

12. Fang, L., Sarma, A.D., Yu, C., Bohannon, P.: REX: explaining relationships between entity pairs. Proc. VLDB Endowment **5**(3), 241–252 (2011)

13. Fleiss, J.L.: Measuring nominal scale agreement among many raters. Psychol. Bull. **76**(5), 378 (1971)
14. Fokoue, A., Sadoghi, M., Hassanzadeh, O., Zhang, P.: Predicting drug-drug interactions through large-scale similarity-based link prediction. In: Sack, H., Blomqvist, E., d'Aquin, M., Ghidini, C., Ponzetto, S.P., Lange, C. (eds.) ESWC 2016. LNCS, vol. 9678, pp. 774–789. Springer, Cham (2016). https://doi.org/10.1007/978-3-319-34129-3_47
15. Gkatzia, D., Lemon, O., Rieser, V.: Natural language generation enhances human decision-making with uncertain information. In: Proceedings of the 54th Annual Meeting of the Association for Computational Linguistics, ACL 2016, 7–12 August 2016, Berlin, Germany, vol. 2 (2016). Short Papers
16. Gutiérrez-Cuellar, J., Gómez-Pérez, J.M.: Havas 18 labs: a knowledge graph for innovation in the media industry. In: Polleres, A., Castro, A.G., Benjamins, R. (eds.) International Semantic Web Conference (Industry Track), CEUR Workshop Proceedings, vol. 1383 (2014)
17. Hajishirzi, H., Zilles, L., Weld, D.S., Zettlemoyer, L.: Joint coreference resolution and named-entity linking with multi-pass sieves. In: Proceedings of the 2013 Conference on Empirical Methods in Natural Language Processing, pp. 289–299 (2013)
18. Hearst, M.A.: Texttiling: segmenting text into multi-paragraph subtopic passages. Comput. Linguis. **23**(1), 33–64 (1997)
19. Heim, P., Lohmann, S., Stegemann, T.: Interactive relationship discovery via the semantic web. In: Aroyo, L., et al. (eds.) ESWC 2010. LNCS, vol. 6088, pp. 303–317. Springer, Heidelberg (2010). https://doi.org/10.1007/978-3-642-13486-9_21
20. Huang, Y., Liu, Z., Chen, Y.: Query biased snippet generation in xml search. In: Proceedings of the 2008 ACM SIGMOD International Conference on Management of Data, pp. 315–326. ACM (2008)
21. Iannacone, M., et al.: Developing an ontology for cyber security knowledge graphs. In: Proceedings of the 10th Annual Cyber and Information Security Research Conference, CISR 2015, pp. 12:1–12:4. ACM, New York (2015)
22. Khalid, M.A., Verberne, S.: Passage retrieval for question answering using sliding windows. In: Coling 2008: Proceedings of the 2nd workshop on Information Retrieval for Question Answering, pp. 26–33. Association for Computational Linguistics (2008)
23. Konstantinova, N., De Sousa, S.C., Díaz, N.P.C., López, M.J.M., Taboada, M., Mitkov, R.: A review corpus annotated for negation, speculation and their scope. In: LREC, pp. 3190–3195 (2012)
24. Lehmann, J., Gerber, D., Morsey, M., Ngonga Ngomo, A.-C.: DeFacto - deep fact validation. In: Cudré-Mauroux, P., et al. (eds.) ISWC 2012. LNCS, vol. 7649, pp. 312–327. Springer, Heidelberg (2012). https://doi.org/10.1007/978-3-642-35176-1_20
25. Manning, C.D., Raghavan, P., Schütze, H.: Introduction to Information Retrieval. Cambridge University Press, New York (2008)
26. Metzger, S., Elbassuoni, S., Hose, K., Schenkel, R.: S3K: seeking statement-supporting top-k witnesses. In: Proceedings of the 20th ACM International Conference on Information and Knowledge Management, pp. 37–46. ACM (2011)
27. Metzler, D., Croft, W.: Combining the language model and inference network approaches to retrieval. Inf. Process. Manag. **40**(5), 735–750 (2004)
28. Nagarajan, M., et al.: Predicting future scientific discoveries based on a networked analysis of the past literature. In: KDD, 2015, pp. 2019–2028 (2015)
29. Niu, F., Zhang, C., Ré, C., Shavlik, J.W.: Deepdive: web-scale knowledge-base construction using statistical learning and inference. VLDS **12**, 25–28 (2012)

30. Penin, T., Wang, H., Tran, T., Yu, Y.: Snippet generation for semantic web search engines. In: Domingue, J., Anutariya, C. (eds.) ASWC 2008. LNCS, vol. 5367, pp. 493–507. Springer, Heidelberg (2008). https://doi.org/10.1007/978-3-540-89704-0_34
31. Pirrò, G.: Explaining and suggesting relatedness in knowledge graphs. In: Arenas, M., et al. (eds.) ISWC 2015. LNCS, vol. 9366, pp. 622–639. Springer, Cham (2015). https://doi.org/10.1007/978-3-319-25007-6_36
32. Porter, M.F.: An algorithm for suffix stripping. Program **14**(3), 130–137 (1980)
33. Ruan, T., Xue, L., Wang, H., Hu, F., Zhao, L., Ding, J.: Building and exploring an enterprise knowledge graph for investment analysis. In: Groth, P., et al. (eds.) ISWC 2016. LNCS, vol. 9982, pp. 418–436. Springer, Cham (2016). https://doi.org/10.1007/978-3-319-46547-0_35
34. Saldanha, G., Biran, O., McKeown, K., Gliozzo, A.: An entity-focused approach to generating company descriptions. In: The 54th Annual Meeting of the Association for Computational Linguistics, p. 243 (2016)
35. Sandusky, R.J., Tenopir, C.: Finding and using journal-article components: Impacts of disaggregation on teaching and research practice. J. Assoc. Inf. Sci. Technol. **59**(6), 970–982 (2008)
36. Sheth, A., Aleman-Meza, B., Arpinar, I.B., Bertram, C.: Semantic association identification and knowledge discovery for national security applications. J. Database Manag. **16**(1), 33 (2005)
37. Socher, R., Chen, D., Manning, C.D., Ng, A.: Reasoning with neural tensor networks for knowledge base completion. In: Advances in Neural Information Processing Systems, pp. 926–934 (2013)
38. Spangler, S., Kreulen, J.T., Lessler, J.: Generating and browsing multiple taxonomies over a document collection. J. Manag. Inf. Sys. **19**(4), 191–212 (2003)
39. Tiedemann, J.: Comparing document segmentation strategies for passage retrieval in question answering. In: Proceedings of the Conference on Recent Advances in Natural Language Processing (RANLP07), vol. 1 (2007)
40. Tiedemann, J., Mur, J.: Simple is best: experiments with different document segmentation strategies for passage retrieval. In: Coling 2008: Proceedings of the 2nd Workshop on Information Retrieval for Question Answering, pp. 17–25. Association for Computational Linguistics (2008)
41. Tombros, A., Sanderson, M.: Advantages of query biased summaries in information retrieval. In: Proceedings of the 21st Annual International ACM SIGIR Conference on Research and Development in Information Retrieval, pp. 2–10. ACM (1998)
42. Turpin, A., Tsegay, Y., Hawking, D., Williams, H.E.: Fast generation of result snippets in web search. In: Proceedings of the 30th Annual International ACM SIGIR Conference on Research and Development in Information Retrieval, pp. 127–134. ACM (2007)
43. Voskarides, N., Meij, E., de Rijke, M.: Generating descriptions of entity relationships. In: Jose, J.M., et al. (eds.) ECIR 2017. LNCS, vol. 10193, pp. 317–330. Springer, Cham (2017). https://doi.org/10.1007/978-3-319-56608-5_25
44. Voskarides, N., Meij, E., Tsagkias, M., de Rijke, M., Weerkamp, W.: Learning to explain entity relationships in knowledge graphs. In: ACL, vo. 1. pp. 564–574 (2015)
45. Wang, M., Si, L.: Discriminative probabilistic models for passage based retrieval. In: Proceedings of the 31st Annual International ACM SIGIR Conference on Research and Development in Information Retrieval, pp. 419–426. ACM (2008)

46. Wu, F., Weld, D.S.: Open information extraction using Wikipedia. In: Proceedings of the 48th Annual Meeting of the Association for Computational Linguistics, pp. 118–127. ACL 2010 (2010)

47. Yang, H., Callan, J., Si, L.: Knowledge transfer and opinion detection in the TREC 2006 blog track. In: TREC (2006)

48. Zhai, C., Lafferty, J.: The dual role of smoothing in the language modeling approach. In: Proceedings of the Workshop on Language Models for Information Retrieval (LMIR) 2001 (2001)

Exploring RDFS KBs Using Summaries

Georgia Troullinou[1][(✉)], Haridimos Kondylakis[1], Kostas Stefanidis[2],
and Dimitris Plexousakis[1]

[1] ICS-FORTH, Heraklion, Greece
{troulin,kondylak,dp}@ics.forth.gr
[2] University of Tampere, Tampere, Finland
kostas.stefanidis@uta.fi

Abstract. Ontology summarization aspires to produce an abridged version of the original data source highlighting its most important concepts. However, in an ideal scenario, the user should not be limited only to static summaries. Starting from the summary, s/he should be able to further explore the data source requesting more detailed information for a particular part of it. In this paper, we present a new approach enabling the dynamic exploration of summaries through two novel operations *zoom* and *extend*. Extend focuses on a specific subgraph of the initial summary, whereas zoom on the whole graph, both providing granular information access to the end-user. We show that calculating these operators is NP-complete and provide approximations for their calculation. Then, we show that using extend, we can answer more queries focusing on specific nodes, whereas using global zoom, we can answer overall more queries. Finally, we show that the algorithms employed can efficiently approximate both operators.

1 Introduction

The recent explosion of the Web of Data and the associated Linked Open Data (LOD) initiative have led to an enormous amount of widely available RDF datasets [6]. These datasets often have extremely complex schemas, which are difficult to comprehend, limiting the exploitation potential of the information they contain. As a result, there is an increasing need to develop methods and tools that facilitate the quick understanding and exploration of these data sources [9,19].

To this direction, many approaches focus on generating ontology summaries [21,24,25,29]. Ontology summarization [30] is defined as the process of distilling knowledge from an ontology in order to produce an abridged version. Although generating summaries is an active field of research, most of the works focus only on identifying the most important nodes, exploit limited semantic information or produce static summaries, limiting the exploration and the exploitation potential of the information they contain. In addition, although exploration operators over summaries have already been identified as really useful (e.g. [15]), the available approaches so far are limited, expanding only the hierarchy and the connections

© Springer Nature Switzerland AG 2018
D. Vrandečić et al. (Eds.): ISWC 2018, LNCS 11136, pp. 268–284, 2018.
https://doi.org/10.1007/978-3-030-00671-6_16

of selected nodes [11]. As a result, there is an increasing need to develop methods and tools in order to facilitate the understanding and exploration of various data sources, through exploration operators on summaries.

Consider for example that we would like to get a quick view of the DBpedia version 3.8 shown in Fig. 1(a). By visualizing the graph of the schema, it is difficult to understand the contents of the KB. Even if we highlight the most representative nodes (the red ones), according to some importance measure (e.g. Betweenness) the problem persists. Now consider selecting the top-k most representative nodes and connecting them. The result is shown in Fig. 1(b). Here, we can better understand the contents of the DBpedia v3.8. However, still the user might find the presented information overwhelming and s/he would like to see less information, focusing only on the top-10 nodes. Ideally, s/he should be able to zoom-in and zoom-out at will in the presented graph to understand the contents at a selected granularity level. More than this, s/he might want to have more detailed information not only on the whole schema graph but on a selected subset of it. This could happen by selecting some nodes, requesting more details on those. Those details could be offered in terms of showing other nodes dependent on the selected ones as shown in Fig. 1(b) (green nodes). Although exploration operators over summaries have already been identified as useful (e.g. [15]), the available approaches are limited, expanding only the hierarchy and the connections of the selected nodes.

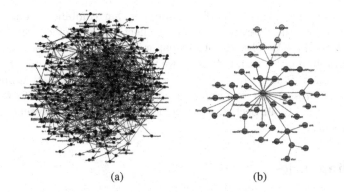

(a)   (b)

Fig. 1. The DBpedia 3.8 schema graph (a) and a schema summary (b) generated using [17]. (Color figure online)

Motivated by the lack of an effective method to explore KBs starting from summaries, we have developed RDFDigest+. RDFDigest+ is a system that transparently and efficiently handles exploratory operations on large KBs. In its core, it employs an algebra where two operators are treated as first-class citizens in various exploration scenarios. Our algebra contains the *extend* and the *zoom* operators with particular semantics. Extend focuses on a specific subgraph of the initial summary, whereas zoom on the whole graph, both providing granular information access to the end-user.

More specifically, in this paper, we focus in RDFS ontologies and demonstrate an efficient and effective method to enable exploration of RDFS KBs, using schema summaries that can be extended and zoomed according to user selections. Our contributions are the following:

- We present RDFDigest+, a novel system that is able to generate summaries, enabling further exploration using zoom and extend operations.
- Summary generation is a two-steps process. First, all schema nodes are ranked according to various measures, and then, the top-k selected nodes are linked using edges that introduce the minimum number of additional nodes over the initial schema graph.
- Over these generated summaries, we enable zoom-in and zoom-out operations to get granular information, adding more important nodes or removing existing ones from the generated summary.
- In addition, through the extend operator, we allow selecting a subset of the presented nodes to visualize other dependent nodes.
- We provide algorithms for calculating the aforementioned operators on a given schema graph and we show that the problem is NP-complete. To this end, we provide effective and efficient approximations as well.
- We demonstrate the added value of these operators, evaluating summary's ability to answer the most-frequent real users queries, and we show that the approximate algorithms proposed can efficiently approximate both operators.

To our knowledge, this is the first approach that combines summaries with both zoom and extend operations, enabling effectively and efficiently the granular exploration of a KB. The rest of this paper is structured as follows: In Sect. 2, we present preliminaries and, in Sect. 3, we provide more details on schema summarization. Then, in Sect. 4, we introduce our ontology exploration operations. In Sect. 5, we present our experimental evaluation and, in Sect. 6, we discuss related work. Finally, in Sect. 7, we conclude this paper and present directions for further work.

2   Preliminaries

In this paper, we focus on RDFS KBs, as RDFS is among the widely-used standards for publishing and representing data on the Web. Our approach handles OWL ontologies as well, considering however only the RDFS part of these ontologies. The representation of knowledge in RDF is based on triples of the form (subject, predicate, object). RDF datasets have attached semantics through RDFS [1], a vocabulary description language. Representation of RDF data is based on three disjoint and infinite sets of *resources*, namely: URIs ($\mathcal{U}$), literals ($\mathcal{L}$) and blank nodes ($\mathcal{B}$). We impose typing on resources, so we consider three disjoint sets of resources: classes ($\mathbf{C} \subseteq \mathcal{U} \cup \mathcal{B}$), properties ($\mathbf{P} \subseteq \mathcal{U}$), and individuals ($\mathbf{I} \subseteq \mathcal{U} \cup \mathcal{B}$). The set $\mathbf{C}$ includes all classes, including RDFS classes and XML datatypes (e.g., xsd:string, xsd:integer). The set $\mathbf{P}$ includes all properties, except rdf:type, which connects individuals with the classes they are instantiated under. The set **I** includes all individuals, but not literals. In addition, our approach adopts the unique name assumption, i.e. resources identified by different URIs are different.

Here, we will follow an approach similar to [26], which imposes a convenient graph-theoretic view of RDF data that is closer to the way the users perceive their datasets. As such, we separate between the schema and the instances of an RDFS KB, represented in separate graphs (G_S and G_I, respectively). The schema graph contains all classes and the properties the classes associated with (via the properties domain/range specification); multiple domains/ranges per property are allowed, by having the property URI be a label on the edge, via a labeling function λ, rather than the edge itself. The instance graph contains all individuals, and the instantiations of schema properties; the labeling function λ applies here as well for the same reasons. Finally, the two graphs are related via the τ_c function, which determines the class(es) each individual is instantiated under.

Definition 1 (RDFS KB). *An RDFS KB is a tuple $V = \langle G_S, G_I, \lambda, \tau_c \rangle$, where:*

- *G_S is a labelled directed graph $G_S = (V_S, E_S)$ such that V_S, E_S are the nodes and edges of G_S, respectively, and $V_S \subseteq \mathbf{C} \cup \mathcal{L}$.*
- *G_I is a labelled directed graph $G_I = (V_I, E_I)$ such that V_I, E_I are the nodes and edges of G_I, respectively, and $V_I \subseteq \mathbf{I} \cup \mathcal{L}$.*
- *A labelling function $\lambda : E_S \cup E_I \mapsto 2^{\mathbf{P}}$ determines the property URI that each edge corresponds to (properties with multiple domains/ranges may appear in more than one edge).*
- *A function $\tau_c : \mathbf{I} \mapsto 2^{\mathbf{C}}$ associating each individual with the classes that it is instantiated under.*

In the following, we will write $p(v_1, v_2)$ to denote an edge e in G_S, where $v_1, v_2 \in V_S$, or G_I, where $v_1, v_2 \in V_I$, from node v_1 to node v_2, such that, $\lambda(e) = p$. In addition, for brevity, we will call schema node a node $s \in V_S$, class node a node $c \in \mathbf{C} \cap V_S$, and instance node a node $i \in \mathbf{I} \cap V_I$. A path from a node v_s to v_i, denoted by $path(v_s \rightarrow v_i)$, is the finite sequence of edges, which connect a sequence of nodes, starting from v_s and ending at v_i. The length of a path, denoted by $dpath(v_s \rightarrow v_i)$, is the number of the edges that exist in that path. Finally, having a schema graph G_S, the closure of G_S, denoted by $Cl(G_S)$, contains all triples that can be inferred from G_S using inference. From now on, when we use G_S, we will mean $Cl(G_S)$ for reasons of simplicity, unless stated otherwise. This is to ensure that the result will be the same, independent of the number of inferences applied on an input schema graph G_S.

3 Schema Summarization

Schema summarization aims to highlight the most representative concepts of a schema, preserving important information and reducing the size and the complexity of the whole schema. Central questions to summarization are (i) how to rank the schema nodes according to an importance measure, and (ii) how to link the top-k ones in order to produce a valid sub-schema graph.

3.1 Identifying Important Nodes in RDFDigest+

To identify the most important nodes, RDFDigest+ employs a variety of centrality measures like Degree, Bridging Centrality, Harmonic Centrality, Radiality, Ego Centrality and Betweenness [17]. As [17] shows, among these measures, Betweenness produces summaries with a better quality. In addition, in this paper we explore for the first time to this purpose, PageRank and HITS, two additional well-known centrality measures [5]. Specifically, the importance measures (IM) we are going to explore for our experiments, for selecting the top-k most important nodes are the following:

- *Betweenness (BE)*. The number of the shortest paths from all nodes to all others that pass through a node.
- *PageRank (PR)*. This centrality measure assigns a score based on node's connections, and their connections. PageRank takes link direction and weight into account so links can only pass influence in one direction, and pass different amounts of influence.
- *HITS (HT)*. HITS algorithm is based on the idea that in the Web, and in all document collections which can be represented by directed networks, there are two types of important nodes: hubs and authorities. Hubs are nodes which point to many nodes of the type considered important. Authorities are these important nodes.

Independently of the importance measure (IM) selected, since those measures have been developed for generic graphs, we adapt them to be used for RDFS graphs. To achieve that we first normalize each measure IM on a scale of 0 to 1:

$$normal(IM(v)) = \frac{IM(v) - \min(IM(G_S))}{\max(IM(G_S)) - \min(IM(G_S))} \qquad (1)$$

where $IM(v)$ is the importance value of a node v in G_S, and $min(IM(G_S))$ is the minimum and $max(IM(G_S))$ is the maximum importance value in G_S.

Similarly, we normalize the number of instances (InstV) that belong to a schema node. As such, the *adapted importance measure* (AIM) of each node is the sum of the normalized values of the importance measures and the instances.

$$AIM(v) = normal(IM(v)) + normal(InstV(v)) \qquad (2)$$

Next, let $TOP_k^{AIM}(V)$ be the function that returns the top-k nodes of an RDFS KB V, according to the selected adapted importance measure (AIM) - for brevity we will use $TOP_k(V)$ independently of the importance measure selected.

Overall, our system is flexible enough to enable the uninterrupted addition of new importance measures by adding new function calls. The diverse set of importance measures offered, enable exploring RDFS KBs according to the way users perceive importance, offering many alternatives and enhancing the exploration abilities of our system.

3.2   Linking Important Nodes

Having a way to rank the schema nodes of an RDFS KB according to the perceived importance, we then focus on selecting the paths that link those nodes, aiming to produce a valid sub-schema graph. As the main problem of previous approaches [17,26] was the introduction of many additional nodes (besides the top-k ones), in this paper, we focus on selecting the paths that introduce the minimum number of additional nodes to the final summary graph. As such, we model the problem of linking the most important nodes as a variation of the well-known *Graph Steiner-Tree problem (GSTP)* [27]. The corresponding algorithm targets at minimizing the additional nodes introduced for connecting the top-k most important nodes [17]. However, the problem is NP-hard, and as such approximation algorithms should be used for large datasets.

3.3   Summary Schema Graph

Having identified ways for locating important nodes and, in turn, for connecting them, we define next the summary schema graph as follows:

Definition 2 (Summary Schema Graph of size n). *Let $V = \langle G_S, G_I, \lambda, \tau_c \rangle$ be an RDFS KB. A summary schema graph of size n for V is a connected schema graph $G'_S = (V'_S, E'_S)$, $G'_S \subseteq Cl(G_S)$, with:*

- *$V'_S = TOP_k(V) \cup V_{ADD}$,*
- *$\forall v_i, v_j \in TOP_k(V), \exists path(v_i \rightarrow v_j) \in G'_S$,*
- *V_{ADD} represents the nodes in the summary used only to link the nodes in $TOP_k(V)$,*
- *$\nexists$ summary schema graph $G''_S = (V''_S, E''_S)$ of size n for V, such that, $|V''_S| < |V'_S|$.*

4   Exploration Through Summaries

Getting the summaries, users can better understand the contents of a KB. However, still the user might find the presented information overwhelming and he/she may like to see less information, focusing for example, only on the top-10 nodes (zoom) or requesting more detailed information for a specific subgraph of the summary (extend).

4.1   The Extend Operator

The extend operator gets as input a subgraph of the schema graph and identifies other nodes that are depending on the selected nodes. Dependence has not only to do with distance, but with additional parameters, including importance. Like TF-IDF, the basic hypothesis here is that the greater the influence of a property on identifying a corresponding instance is, the less times it is repeated, or in other words, infrequent properties are more informative than frequent ones.

This way, we define the dependence between two classes as a combination of their cardinality closeness (defined in the sequel), the adapted importance measures (AIM) of the classes and the number of edges appearing in the path connecting these two classes. So, dependence is defined as:

$$Dependence(u, v) = \frac{AIM(u) - \sum_{i \in Y} \frac{AIM(i)}{CC((i-1),i)}}{dpath(u \to v)} \tag{3}$$

where the cardinality closeness CC is defined for a pair of classes as the number of distinct edges over the number of all edges between them. Formally:

Definition 3 (Cardinality Closeness). *Let c_k, c_s be two adjacent schema nodes and $u_i, u_j \in G_I$ such that $\tau_c(u_i) = c_k$ and $\tau_c(u_j) = c_s$. The cardinality closeness of $p(c_k, c_s)$, namely the $CC(p(c_k, c_s))$, is defined as:*

$$CC(p(c_k, c_s)) = \frac{1 + |c|}{|c|} + \frac{DistinctV(p(u_i, u_j))}{Instances(p(u_i, u_j))} \tag{4}$$

where $|c|, c \in C \cap V_S$, is the number of nodes in the schema graph, $DistinctV(p(u_i, u_j))$ is the number of distinct $p(u_i, u_j)$ and $Instances(p(u_i, u_j))$ is the number of $p(u_i, u_j)$. When there are no instances, $Instances(p(u_i, u_j)) = 1$ and $DistinctV(p(u_i, u_j)) = 0$.

As we move away from a node, the dependence becomes smaller by calculating the differences of AIM across a selected path in the graph. We penalize additionally dependence dividing by the distance of the two nodes. The highest the dependence of a path, the more appropriate is the first node to represent the final node of the path. Also note that $Dependence(u, v)$ is different than $Dependence(v, u)$, since the dependence of a more important node towards a less important node is higher than the other way around, although, they share the same cardinality closeness. To identify the dependent nodes of a selected node, we use the function $dependend(u_i, range, number\_of\_nodes)$ that returns at most $number\_of\_nodes$ nodes depending on u_i with a distance at most $range$.

The *extend* operator takes into account a particular subgraph of a summary schema graph, and is defined as follows:

Definition 4 (Extend operator). *Let $G'_S = (V'_S, E'_S)$ be the summary schema graph of an RDFS KB $V = \langle G_S, G_I, \lambda, \tau_c \rangle$. The extend operator, i.e., $extend(G_e)$, takes as input a subgraph $G_e = (V_e, E_e)$ of G'_S, $G_e \subseteq G'_S$, and returns a connected schema graph $G'_e = (V'_e, E'_e)$, $V_e \subseteq V'_e$, for which:*

- *$G'_e \subseteq Cl(G_S)$,*
- *$V'_e \backslash V_e = V_d \cup V_{ADD'}$, where V_d includes, $\forall v_i \in V_e$, all nodes v_j, such that, $dependend(v_j, range, number\_of\_nodes) = v_i$, and $V_{ADD'}$ the nodes that link the nodes in V_d with the other summary nodes,*
- *$\forall v_i \in V_d \cup TOP_k(V), \exists path(v_x \to v_y) \in G'_e$,*
- *$\nexists G''_e = extend(G_e) = (V''_e, E''_e)$, such that, $|V''_e| < |V'_e|$.*

Algorithm 1 presents the extend algorithm. The algorithm identifies the dependent nodes (lines 2–5) using the depencence function. Due to lack of space, the detailed description of the algorithm used for locating the *dependent* nodes is omitted, however abstractly, it starts from u_i and calculate the dependence of the adjacent nodes expanding progressively the range until it reaches the *number_of_nodes*. Next, the algorithm tries to link the top-k nodes using the Steiner-Tree algorithm (line 6). However, as the Steiner-Tree algorithm is NP-complete, our problem is NP-complete as well.

Algorithm 1. Extend

    Input $G'_S = (V'_S, E'_S)$ the summary schema graph of G_S, $G_e = (V_e, E_e)$ the selected summary schema subgraph

    Output $G'_e = (V'_e, E'_e)$ the result schema graph

1: **procedure** EXTEND
2:     $V'_e = V'_S$
3:     **for** each v_i in V_e **do**
4:         $V'_e = V'_e \cup dependent(v_i, range, number\_of\_nodes)$
5:     **end for**
6:     Calculate E'_e using the Steiner-Tree algorithm over G_S with the nodes in V_e as terminals
7: **end procedure**

Two optimizations that we explore in this work are the following:

CHINS. CHINS is an approximation of the Steiner-Tree algorithm [27] proved to have a worst case bound of 2, i.e., $Z_T/Z_{opt} \leq 2 \cdot (1 - l/|Q|)$, where Z_T and Z_{opt} denote the objective function values of a feasible solution and an optimal solution respectively, Q the set of nodes to be linked (for the extend operator the top-k nodes and the selected dependent ones) and l a constant [3]. The algorithm proceeds as follows:

1. Start with a partial solution consisting of a single selected node.
2. While the solution does not contain all selected nodes do find the nearest nodes $u* \in V_t$ and $p*$ being a top-k node not in V_t.

As such, for each node to be linked, the algorithm has to visit at worst the whole set of nodes and edges of the graph, and the corresponding complexity is $O(Q \cdot |V + E|)$. CHINS has been proved to offer an optimal trade-off between quality of the generated summaries and execution time [17], when used for generating summaries.

Shortest Paths. CHINS starts from a single node extending one by one the set of selected nodes. However, having the nodes in the summary already, there is no need to start from the first node. As such, another approximation could be to start with the nodes already available in the summary and then proceed to step 2 of CHINS. The algorithm for each one of the $|Q \backslash TOP_K(V)|$ nodes needs at worst to visit the whole graph. This way, the worst-case complexity of the algorithm is $O(|Q \backslash TOP_K(V)| \cdot |V + E|)$.

Dependent Paths. In order to calculate the dependence between the selected nodes and the ones introduced by the *dependent* functions, the visited paths can be recorded and use these, already visited paths for connecting the selected nodes with the original summary. So, in this approximation, instead of finding the shortest path between the existing summary and each dependent node, we calculate the shortest path between the extended and the dependent node, which is already calculated in the previous step (the *dependent* function). The complexity remains the same with the previous algorithm ($O(|Q \backslash TOP_K(V)| \cdot |V + E|)$), since only the $|Q \backslash TOP_K(V)|$ nodes are considered sequentially for linking them to the existing summary.

4.2   The Zoom Operator

In this section, we focus on zooming operations, by exploiting the schema graph as a whole. That is, we introduce the *zoom-out* and *zoom-in* operators to produce more detailed or coarse summary schema graphs. To this end, we consider the n' schema nodes with the highest importance in G_S, where n' can be either greater than n, for achieving a *zoom-out*, or smaller than n, for achieving a *zoom-in*, where n represents the number of the most important nodes in a given summary.

Definition 5 (Zoom-out operator). *Let* $G'_S = (V'_S, E'_S)$ *be the summary schema graph of size* n *of an RDFS KB* $V = \langle G_S, G_I, \lambda, \tau_c \rangle$. *The zoom-out operator* $zoom_{out}(G'_S, n')$, *with* $n' > n$, *returns a connected schema graph* $G'_{zo} = (V'_{zo}, E'_{zo})$, *for which:*

- $G'_{zo} \subseteq Cl(G_S)$,
- $V'_{zo} = V'_S \cup TOP \cup V_{ADD}$, *where* $TOP = TOP_{n'}(V) \backslash V'_S$,
- $\forall v_i \in TOP, \exists v_j \in V'_S$, *such that,* $\exists path(v_i \rightarrow v_j) \in G'_{zo}$,
- V_{ADD} *represents the nodes in* G'_{zo} *used only to link the nodes in* TOP,
- $\nexists G''_z o = zoom_{out}(G'_S, n') = (V''_{zo}, E''_{zo})$, *such that,* $|V''_z o| < |V'_z o|$.

Definition 6 (Zoom-in operator). *Let* $G'_S = (V'_S, E'_S)$ *be the summary schema graph of size* n *of an RDFS KB* $V = \langle G_S, G_I, \lambda, \tau_c \rangle$. *The zoom-in operator* $zoom_{in}(G'_S, n')$, *with* $n' < n$, *returns a connected schema graph* $G'_{zi} = (V'_{zi}, E'_{zi})$, *for which:*

- $G'_{zi} \subseteq G'_S$,
- $V'_{zi} = TOP_{n'}(V) \cup V_{ADD}$,
- V_{ADD} *represents the nodes in* G'_{zi} *used only to link the nodes in* $TOP_{n'}(V)$,
- $\nexists G''_{zi} = zoom_{in}(G'_S, n') = (V''_{zi}, E''_{zi})$, *such that,* $|V''_{zi}| < |V'_{zi}|$.

The simplest approach for zooming-in/out, is to calculate from scratch the $TOP_{n'}(V)$ and then to use the Steiner-Tree algorithm from scratch to link the selected nodes. However, since we already have an existing summary as a basis for our zoom operations, we explore the following approximations.

Zoom-In. Remove the nodes in $TOP_n(V) \backslash TOP_{n'}(V)$ and their connections without recalculating the Steiner-Tree algorithm for $TOP_{n'}(V)$ – this might leave additional nodes in the resulting summary.

Zoom-Out - CHINS. Add the nodes in $TOP_{n'}(V) \backslash TOP_n(V)$ and link them with the existing summary, using the CHINS approximation algorithm.

Zoom-Out - Shortest Paths. Add the nodes in $TOP_{n'}(V) \backslash TOP_n(V)$ and link them with the existing summary, using the *Shortest Paths* approximation algorithm.

5 Evaluation and Implementation

To evaluate our approach, we use the version 3.8 of DBpedia[1], which is consisted of 359 classes, 1323 properties and more that 2.3M instances, and offers an interesting use-case for exploration. To identify the quality of our approach, we use a query log containing 50K user queries provided by the DBpedia SPARQL endpoint for the corresponding DBpedia version. Our goal is to assess the percentage of the queries that can be answered solely by using the generated schema summary along with the corresponding instances, i.e. the *coverage* of the queries from a schema summary.

Having a summary, we can calculate for each query the percentage of the classes and properties that are included in the summary. A class/property appears within a query either directly or indirectly. Directly when the said class/property appears within a triple pattern of the query. Indirectly for a class is when the said class is the type of an instance or the domain/range of a property that appear in a triple pattern of the query. Indirectly for a property is when the said property is the type of an instance. Having the percentages of the classes and properties included in the summary, the query coverage is the weighted sum of these percentages. As our summaries are node-based (they are generated based on the top-k most important nodes; in zoom we add/remove important nodes; in extend we add the dependent nodes) the weight on the nodes is larger than the one on the properties (for our experiments we used 0.8 for nodes and 0.2 for edges).

5.1 Quality - Evaluating the Zoom Operator

In this section, we evaluate the quality of the zoom-out operator. To do that we start from a summary containing 10% of the initial schema graph, and we zoom-out progressively by 10%, until we reach the 40% of the schema graph.

---

[1] http://wiki.dbpedia.org/.

Having the *coverage* of each query, we can calculate the average coverage for all queries in our log. In essence, an average coverage of 70% means that on average the 70% of the queries in the query log can be answered only using the summary accompanied with its corresponding instances. As when zooming-out, the next more important nodes are added to the summary, we expect that the average coverage of all queries should grow accordingly. The results are shown in Fig. 2, whereas the actual improvement is shown in Fig. 3. As we can observe, indeed as the percentage of the summary increases, more queries are covered by the result summary. In addition, HITS and Betweenness perform better, competing each other in all cases. Specifically, HITS presents a more stable behavior with the best coverage from the smallest zoom-out percentage, while Betweenness performs better from the 20% zoom-out and on. PageRank is always worse than HITS and Betweenness. As a baseline we added the Random bar as well, where we randomly select nodes from the schema graph (connecting them with the corresponding measure). Even if some-times randomly adding more nodes improves a bit the results, overall, this is the approach with the worst performance, clearly showing the benefits of our approach. Regarding the actual improvement, we observe that CHINS and Shortest Paths return results of the same quality, with Shortest Paths being slightly better in some cases. In this sense, Betweenness appears to be the most stable measure with improvements around 35% to 45%, while PageRank shows a good improvement, around 35%, for cases in which a 40% zoom-out is performed. Due to space limitations, we omit the results of the zoom-in operator that presents similar behavior.

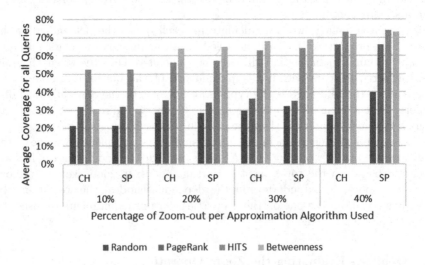

Fig. 2. Zooming-out using various centrality measures and approximation algorithms CHINS (CH) and Shortest Paths (SP).

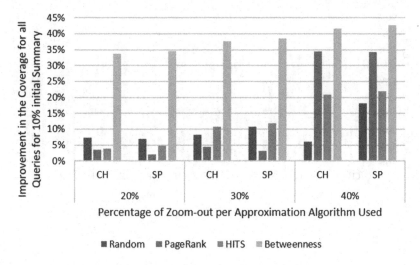

Fig. 3. Improvement on zooming-out using various centrality measures and approximation algorithms CHINS (CH) and Shortest Paths (SP).

5.2   Quality - Evaluating the Extend Operator

Next, we evaluate the extend operator. To do that, we start again from a summary containing 10% of the initial schema graph, and we extend progressively requesting to extend 10% of the available nodes in the summary, until we reach 40% of the initial summary schema graph being extended.

As now we are interested in getting information relevant to particular selected nodes, and not for the whole schema graph, we calculate the average coverage for the queries including only classes from the selected part to be extended. In this case, an average coverage of 70% means that on average the 70% of the queries in the query log, including one of the extended nodes, can be answered only using the summary accompanied by its corresponding instances. As when more nodes related to the extended ones, are added to the summary, we expect that the average coverage of those queries should grow accordingly. The results are shown in Fig. 4, whereas the actual improvement is shown in detail in Fig. 5.

Overall, we observe here that indeed the more nodes we extend, the more "local" queries are covered. In addition, the Shortest Paths algorithm provides the best results in all cases, followed by CHINS. This is reasonable since the Shortest Paths algorithm targets at identifying the shortest path between the dependent nodes and the available summary, and as such, it prioritizes nodes closest to the ones to be extended. On the other hand, the Dependent paths algorithm does a minimum effort trying to connect the dependent nodes to the existing summary and this has a direct effect on the quality of the produced summary. PageRank presents the best coverage, on average around 68% to 78%, while HITS follows with coverage around 65% to 73%. In turn, Betweenness has a coverage around 59% to 72%, while, as expected, Random presents the worst behavior with coverage from 35% to 40%. Overall, even if PageRank has the best performance, we observe that Betweenness has the best improvement.

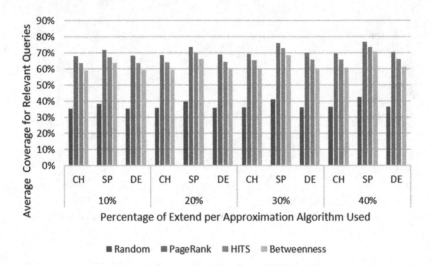

Fig. 4. Extend using HITS and Betweenness, and the approximation algorithms random (RA), CHINS (CH), Shortest Paths (SP) and Dependent (DE).

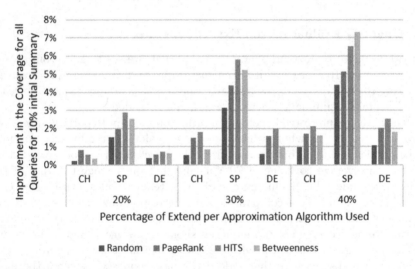

Fig. 5. Improvement on extending using HITS and betweenness, and the approximation algorithms random (RA), CHINS (CH), Shortest Paths (SP) and Dependent (DE).

5.3   The RDFDigest+ System

All aforementioned measures and algorithms are available online on the RDFDigest+ system[2], a novel system that enables effective and efficient RDFS KB exploration using summaries. An instance of RDFDigest+ is shown in Fig. 6.

---

[2] http://rdfdigest.ics.forth.gr.

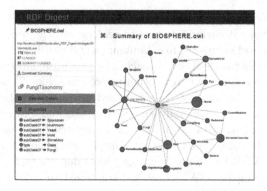

Fig. 6. The RDFDigest+ system.

Users can upload their own datasets, and RDFDigest+ produces a visual summary identifying and linking the most important nodes in the KB. In the presented summary graph, the size of a node depends on its importance. By clicking on a node, additional metadata (e.g. the number of instances, and the connected properties and instances) are provided to enhance the ontology understanding. Further exploration of the data source is allowed by clicking on the details (on the left) of the selected class and properties. When clicked, its instances and connections appear in a pop-up window. In addition, exploration of the data source is allowed by double-clicking on a node to extend the summary on that specific node. Besides a specific node, a whole area can be selected, requesting more detailed information to be presented regarding the selected nodes. The summary can be zoomed-in and zoomed-out in order to present more detailed or more generic information regarding the whole summary. Finally, the user is able to download the summary as a valid RDFS document.

6   Related Work

According to [20], an effective ontology exploration system should provide a number of core functionalities, such as providing a high level overview of the data, zooming in specific parts of the data and filtering out irrelevant parts.

Ontology Visualization Systems. Towards this direction, toolkits like Protege [16], TopBraid Composer [2] and Neon [8], include visualization plug-ins using the node-link diagram paradigm to represent entities in an ontology and their taxonomy to domain relationships. In addition, many plug-ins, like OwlViz in Protege and Graph View in TopBraid, allow navigating the ontology hierarchy by expanding and hiding nodes.

SpaceTree [18] follows the node-link paradigm as well, but is able to maximize the nodes on display by assessing the available display space. It also avoids clutter by utilizing informative preview icons giving the user an idea of the size and shape of the corresponding subtrees. CropCircles [28] on the other hand, uses geometric containment as an alternative to classing node-link displays sacrificing space to make it easier for users to understand the topological relations in an ontology. Hybrid solutions, like Jambalaya [23] and Knoocks [12], combine containment-based and node-link approaches by providing alternative integrated views of the two paradigms, whereas other approaches, like [7], are based on the notion of distorting the view of the presented graph to combine context and focus. The node on focus is usually the central one and the rest of the nodes are presented around it, reduced in size until they reach a point that they are no longer visible. Finally, WebVOWL [14] implements the Visual Notation for OWL Ontologies (VOWL) by providing graphical depictions for elements of the Web Ontology Language (OWL) that are combined to a force-directed graph layout representing the ontology.

However, all aforementioned approaches in essence, use geometric techniques to provide the necessary abstraction, such as hyperbolic or force-directed graphs, geometric containment or miniature sub-trees. However, we argue that an ideal visualization approach should start with the most important elements of the ontology allowing then progressively the users to explore other less important areas.

Ontology Summarization Systems. Besides pure ontology visualization systems, ontology summarization systems have adopted as well zooming functionalities. An example is KC-Viz [15], which focuses on the key concepts of the ontology based on psycholinguistic criteria. Our system on the other hand, allows users to select multiple measures for identifying importance. KC-Viz provides a set of navigation and visualization mechanisms, including flexible zooming into and hiding of specific parts of an ontology. However, this work is limited in selectively expanding the hierarchy and the connections of selected nodes, whereas in our case besides zooming, we also visualize dependent nodes enabling further exploration of the data source.

[13] supports zoom, filter, details-on-demand, relate, history and extract operations using hierarchical connected circles to provide overview, indented trees to relate different concepts and node-links for filtering and details on-demand, enabling the users to choose the level of semantic zoom. However, the operations performed are not formalized, the corresponding algorithms are not presented and an evaluation is completely missing from the aforementioned work.

[10] proposes a tool that supports three visual exploration options. The first one, named *landmark view*, provides an overview of the class (property) taxonomy giving only representative classes in the hierarchy - selected automatically by a set of statistics measures and user preferences. Then, a user can further explore a specific area by extending (or collapsing) branches. The *local view* displays the full hierarchy of a set of classes (properties) whereas *the axiom view*, provides information about a selected class and its connectivity in the ontology. Compared to our work, this approach is limited mostly on hierarchical structures.

7   Conclusions

In this paper[3] we present a novel platform enabling KB exploration operations over summaries. We introduce the zoom and extend operations, focusing on the number of important nodes of the generated summary, and on getting more detailed information for selected schema summary nodes, respectively. We explore various approximation algorithms showing that we can calculate efficiently the aforementioned operations without sacrificing the quality of the result summary. In fact, we show that the Shortest Paths algorithm provides an optimal trade-off between efficiency and quality.

To the best of our knowledge RDFDigest+ is currently the only system enabling such exploration operations over summaries. As future work, we intent to enable KB exploration at the instance level as well, going from schema summaries to instance summaries, enabling zoom and extend operations both as schema and instance level, or exploiting big data frameworks to speed the summarization process [4]. Moreover, given the dynamically evolving datasets we handle, users are often interested in the state of affairs on previous versions of the datasets, along with their corresponding summaries. To address this need, archiving policies [22] typically store adequate deltas between versions, which are generally small, but this would create the overhead of generating versions at query time. As a direct extension of our system, we will study the trade-offs involved when focusing on archiving dynamic RDF summaries.

References

1. RDF Schema 1.1. http://www.w3.org/TR/rdf-schema/. Accessed Apr 2018
2. TopBraid Composer. https://www.topquadrant.com/tools/ide-topbraid-compos er-maestro-edition/. Accessed Oct 2017
3. Du, D.-Z., Smith, J.M., Rubinstein, J.H. (eds.): Advances in Steiner Trees. Kluwer Academic Publishers, Dordrecht (2000)
4. Agathangelos, G., Troullinou, G., Kondylakis, H., Stefanidis, K., Plexousakis, D.: RDF query answering using apache spark: review and assessment. In: ICDE (2018)
5. Boldi, P., Vigna, S.: Axioms for centrality. Internet Math. **10**(3–4), 222–262 (2014)
6. Christophides, V., Efthymiou, V., Stefanidis, K.: Entity Resolution in the Web of Data. Synthesis Lectures on the Semantic Web: Theory and Technology. Morgan & Claypool Publishers, San Rafael (2015)
7. de Souza, K.X.S., dos Santos, A.D., Evangelista, S.R.M.: Visualization of ontologies through hypertrees. In: CLIHC (2003)
8. Erdmann, M., Waterfeld, W.: Overview of the neon toolkit. In: Ontology Engineering in a Networked World, pp. 281–301 (2012)
9. Fafalios, P., Iosifidis, V., Stefanidis, K., Ntoutsi, E.: Multi-aspect entity-centric analysis of big social media archives. In: TPDL (2017)
10. Jiao, Z.L., Liu, Q., Li, Y., Marriott, K., Wybrow, M.: Visualization of large ontologies with landmarks. In: GRAPP and IVAPP (2013)
11. Kondylakis, H., Troullinou, G., Stefanidis, K., Plexousakis, D.: Beyond summaries for ontology exploration. ERCIM News **2018**(113) (2018)

---

[3] The work was partially supported by the TEKES Finnish project Virpa D. and by the iManageCancer EU project (H2020, #643529).

12. Kriglstein, S., Wallner, G.: Knoocks - a visualization approach for OWL lite ontologies. In: CISIS (2010)
13. Kuhar, S., Podgorelec, V.: Ontology visualization for domain experts: a new solution. In: International Conference on Information Visualisation, IV (2012)
14. Lohmann, S., Link, V., Marbach, E., Negru, S.: WebVOWL: web-based Visualization of Ontologies. In: Lambrix, P. (ed.) EKAW 2014. LNCS (LNAI), vol. 8982, pp. 154–158. Springer, Cham (2015). https://doi.org/10.1007/978-3-319-17966-7_21
15. Motta, E., Peroni, S., Li, N., d'Aquin, M.: Kc-Viz: a novel approach to visualizing and navigating ontologies. In: EKAW (2010)
16. Musen, M.A.: The protégé project: a look back and a look forward. AI Matters 1(4), 4–12 (2015)
17. Pappas, A., Troullinou, G., Roussakis, G., Kondylakis, H., Plexousakis, D.: Exploring importance measures for summarizing RDF/S KBs. In: ESWC (2017)
18. Plaisant, C., Grosjean, J., Bederson, B.B.: SpaceTree: supporting exploration in large node link tree, design evolution and empirical evaluation. In: InfoVis (2002)
19. Roussakis, Y., Chrysakis, I., Stefanidis, K., Flouris, G., Stavrakas, Y.: A flexible framework for understanding the dynamics of evolving RDF datasets. In: Arenas, M., et al. (eds.) ISWC 2015. LNCS, vol. 9366, pp. 495–512. Springer, Cham (2015). https://doi.org/10.1007/978-3-319-25007-6_29
20. Shneiderman, B.: The eyes have it: a task by data type taxonomy for information visualizations. In: IEEE Symposium on Visual Languages (1996)
21. Peroni, S., Motta, E., d'Aquin, M.: Identifying key concepts in an ontology, through the integration of cognitive principles with statistical and topological measures. In: Domingue, J., Anutariya, C. (eds.) ASWC 2008. LNCS, vol. 5367, pp. 242–256. Springer, Heidelberg (2008). https://doi.org/10.1007/978-3-540-89704-0_17
22. Stefanidis, K., Chrysakis, I., Flouris, G.: On designing archiving policies for evolving RDF datasets on the web. In: Yu, E., Dobbie, G., Jarke, M., Purao, S. (eds.) ER 2014. LNCS, vol. 8824, pp. 43–56. Springer, Cham (2014). https://doi.org/10.1007/978-3-319-12206-9_4
23. Storey, M.D., Noy, N.F., Musen, M.A., Best, C., Fergerson, R.W., Ernst, N.A.: Jambalaya: an interactive environment for exploring ontologies. In: IUI (2002)
24. Troullinou, G., Kondylakis, H., Daskalaki, E., Plexousakis, D.: RDF digest: efficient summarization of RDF/S KBs. In: Gandon, F., Sabou, M., Sack, H., d'Amato, C., Cudré-Mauroux, P., Zimmermann, A. (eds.) ESWC 2015. LNCS, vol. 9088, pp. 119–134. Springer, Cham (2015). https://doi.org/10.1007/978-3-319-18818-8_8
25. Troullinou, G., Kondylakis, H., Daskalaki, E., Plexousakis, D.: RDF digest: ontology exploration using summaries. In: ISWC (2015)
26. Troullinou, G., Kondylakis, H., Daskalaki, E., Plexousakis, D.: Ontology understanding without tears: the summarization approach. Semant. Web 8(6), 797–815 (2017)
27. Voß, S.: Steiner's problem in graphs: heuristic methods. Discrete Appl. Math. 40(1), 45–72 (1992)
28. Wang, T.D., Parsia, B.: CropCircles: topology sensitive visualization of OWL class hierarchies. In: Cruz, I. (ed.) ISWC 2006. LNCS, vol. 4273, pp. 695–708. Springer, Heidelberg (2006). https://doi.org/10.1007/11926078_50
29. Wu, G., Li, J., Feng, L., Wang, K.: Identifying potentially important concepts and relations in an ontology. In: Sheth, A., et al. (eds.) ISWC 2008. LNCS, vol. 5318, pp. 33–49. Springer, Heidelberg (2008). https://doi.org/10.1007/978-3-540-88564-1_3
30. Zhang, X., Cheng, G., Qu, Y.: Ontology summarization based on RDF sentence graph. In: WWW (2007)

What Is the Cube Root of 27?
Question Answering Over CodeOntology

Mattia Atzeni and Maurizio Atzori[✉]

Math/CS Department, University of Cagliari,
Via Ospedale 72, 09124 Cagliari, CA, Italy
m.atzeni38@studenti.unica.it, atzori@unica.it

Abstract. We present an unsupervised approach to process natural language questions that cannot be answered by factual question answering nor advanced data querying, requiring instead ad-hoc code generation and execution. To address this challenging task, our system, AskCO, performs language-to-code translation by interpreting the natural language question and generating a SPARQL query that is run against *CodeOntology*, a large RDF repository containing millions of triples representing Java code constructs. The query retrieves a number of Java source code snippets and methods, ranked by AskCO on both syntactic and semantic features, to find the best candidate, that is then executed to get the correct answer. The evaluation of the system is based on a dataset extracted from StackOverflow and experimental results show that our approach is comparable with other state-of-the-art proprietary systems, such as the closed-source WolframAlpha computational knowledge engine.

Keywords: Question answering over linked data
Natural language programming · Semantic parsing · Machine reading
Language-to-code

1 Introduction

Question Answering over Linked Data and ontologies allows leveraging structured data and Natural Language Processing to give a precise answer to the input provided by the end user. However, most of the information available in the Web is organized in the form of unstructured or semi-structured data, thereby being difficult to be automatically processed by such approaches. A paradigmatic example is represented by massive open source code repositories, where source code is not readily available to be queried as Linked Open Data, despite the great potential for the development of computational knowledge engines capable of leveraging this impressive amount of information. To overcome this issue, we have recently introduced CodeOntology[1] [1,2], as a resource aimed at allowing the adoption of the Semantic Web technology stack within the domain of

---

[1] http://codeontology.org.

© Springer Nature Switzerland AG 2018
D. Vrandečić et al. (Eds.): ISWC 2018, LNCS 11136, pp. 285–300, 2018.
https://doi.org/10.1007/978-3-030-00671-6_17

software development and engineering. CodeOntology consists of two main contributions *(i)* an ontology modeling object-oriented code constructs and *(ii)* a parser which is capable of analyzing Java source code and serializing it into RDF triples. CodeOntology also includes a dataset containing millions of RDF triples extracted from OpenJDK [3].

Following this research line, in this paper we introduce an algorithmic approach that addresses the task of Natural Language Programming by employing CodeOntology for Question Answering over source code. Hence, we target a Question Answering problem where the answer to the input question is not directly available in the data, but the dataset contains the information that is needed to *compute* the correct answer. This challenging task is accomplished by performing an unsupervised semantic parsing of natural language utterances into a Java source code, which can be automatically executed to retrieve the answer to the input question.

We discuss two approaches: *(i)* a fast coarse-grained approach which only supports natural language commands corresponding to the invocation of a single method, and *(ii)* a fine-grained approach which is based on dependency parsing and is capable of tagging substrings of the input question with entities from CodeOntology, thereby supporting the execution of more complex expressions, involving the invocation of multiple methods. Within the coarse-grained approach, we propose a simple technique to rank entities available in CodeOntology (specifically, Java methods), based on syntactic and semantic features. On the one hand, the first approach is aimed at providing a natural language interface to Java source code, focusing on applications for developers, such as Computer Assisted Coding tools pluggable within IDEs. Hence, we assume that the user can specify a description of the method to be invoked and the actual arguments. These arguments can be of any arbitrary type, including user-defined classes. On the other hand, the fine-grained approach is aimed at providing a computational knowledge engine for Question Answering and other end-user applications, such as speech-driven tools like Amazon Alexa. Hence, we assume the input is a single natural language question and the actual arguments are provided within the question as literals, thereby limiting the type of the parameters that the user can effectively specify.

Experimental results are based on a dataset extracted from StackOverflow and show that our approach is comparable with state-of-the-art systems, such as the proprietary closed-source WolframAlpha computational knowledge engine. Thus, the main contributions of this work are:

- we introduce an unsupervised approach capable of mapping natural language utterances into Java source code, by leveraging the possibility of extracting Linked Data from any Java project;
- we propose a technique to rank entities from CodeOntology (Java methods) based on syntactic and semantic features;
- we provide a dataset derived from simple questions extracted from Stack-Overflow, to evaluate the performances of our system[2].

---

[2] available at: https://doi.org/10.6084/m9.figshare.6071663.

We remark that, while the paper focuses on OpenJDK methods only, the resulting system, that we called AskCO, is general enough to be applied with any custom set of Java repositories.

2 Related Work

Natural language represents certainly one of the easiest ways to interact with a computer for humans. In Question Answering over Linked Data (QALD), indeed, natural language questions are translated into SPARQL queries to find factual information or more advanced statistics from, e.g., datacubes [4]. Although falling in the area of QALD, our work focuses on questions for which no answer can be found by only querying a repository, since the correct answer needs to be computed by generating and executing code.

In this sense, this work resembles more closely related approaches to natural language querying in software engineering. A large body of work has been done to allow software engineers to manage information about large software systems. For instance, LaSSIE [5] was a prototype tool which made use of a frame-based description language, as well as explicit knowledge representation and reasoning, to address the problem of discovering and learning new information about an existing system. LaSSIE was also embedded with a simple natural language interface based on a taxonomy of the domain and on a lexicon, which included the words known to the system. This work has inspired several more recent research projects, such as [6], where Semantic Web technologies have been applied to support *guided-input natural language* queries concerning static source code information. The presented approach allows importing knowledge about the evolution of a software system into a RHDB (Release History Database), which is augmented with ontological information on source code. Although similar to our work, the expressiveness of this approach is in fact limited by the kind of questions it supports, as it relies on Ginseng [7] to constrain the input and answer quasi-natural language queries by leveraging a multi-level grammar which defines the structure of supported sentences. Similarly, in [8] an unambiguous and controlled subset of natural language with a restricted grammar and a domain-specific vocabulary is used to run queries for static information on source code. On the other hand, more advanced approaches have been developed to support unconstrained natural language queries. In [9], indeed, natural language processing (NLP) techniques are applied to translate free questions to concrete parameters of a third-party query engine.

All the approaches outlined so far are mainly aimed at retrieving static information like specific method calls or write access to certain fields. Our technical contribution describes instead a novel algorithm which brings together NLP and Semantic Web technologies to translate natural language into object-oriented source code. Several research prototypes have been developed to enable the automatic understanding of a natural language description of a program. For instance, Metafor [10], based on concepts from *Programmatic Semantics* [11], is capable of generating class descriptions with attributes and methods. However, its expressiveness is still limited, in the sense that it does not feature the possibility of processing arbitrary English statements. Instead, it can parse a reasonably expressive subset of the English language, to create scaffolding code fragments that can be used to assist the development process. In this sense, it is deeply different from our approach, which aims at mapping any natural language question into the execution of methods extracted from CodeOntology.

More recently, in 2017, SemEval hosted an ambitious challenge [12], aiming at supporting the interaction between users and software APIs, micro-services and applications, using natural language. Most of the work in this area has focused on supervised approaches [13], thereby requiring a dataset mapping natural language to a formal meaning representation. However, this task is different from any previous work related to semantic parsing of natural language commands, as it involves generic programming scenarios and a more comprehensive knowledge base of possible actions. A related problem was also addressed in [14], which targeted the creation of an *if-this-than-that* recipe on the IFTTT[3] platform. The task outlined within the SemEval competition, however, is even more challenging, as it is not limited to *if-then* rules, and it also involves instantiating parameter values. Nevertheless, both approaches are placed in a simplified landscape with respect to our system, which aims at mapping natural language utterances into a real-world and Turing-equivalent programming language.

3   Coarse-Grained Approach

This section describes the coarse-grained approach, which is meant to allow the execution of Java methods, given a natural language description of the intended behavior. The output of such approach is a ranking of the methods in the dataset, based on a metric involving both syntactic and semantic measures. This approach is preliminary to the fine-grained one, which is instead designed to answer more complex questions.

3.1   A Natural Language Interface to OpenJDK

Although CodeOntology already features the possibility of querying source code in a semantic framework powered by the Web of Data, this capability is in fact limited by the complexity of SPARQL queries. Hence, the coarse-grained approach is aimed at providing an easy-to-use and intuitive natural language interface to the entities made available by CodeOntology. We target a RDF repository extracted from OpenJDK 8 [3], containing millions of RDF triples about structural information on source code, actual source code as literals, comments, and semantic links to DBpedia [15] resources.

In particular, we want to allow the end-user to remotely search and execute methods available in the dataset, without necessarily knowing the signature of the method, but only its intended behavior. Thus, we assume that the end-user can provide: *(i)* a natural language description of the method; *(ii)* an unsorted

---

[3] https://ifttt.com/.

list of the actual parameters, optionally including, if the method is not static, the target instance of the method invocation; *(iii)* the expected return type. The system should then run a SPARQL query on the RDF dataset, searching for methods from OpenJDK, whose signature is compatible with the values specified by the user. The retrieved results are subsequently ranked to select the method which most closely matches the natural language description. The selected method is then invoked on the specified input parameters, and the result is then returned to the user, along with the ranking produced by the system. Figure 1 shows the result of the application of the ranking process within the coarse-grained approach.

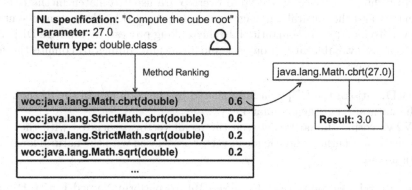

Fig. 1. Example of a simple application of the coarse-grained approach.

3.2 Method Ranking

The ranking of the methods in the dataset relies on the following attributes: *(i)* the name of the method; *(ii)* the Javadoc comment associated with the method; *(iii)* the name of the declaring class; *(iv)* semantic links to DBpedia, already provided by CodeOntology. Several similarity measures are used to produce the final ranking. Such measures are used both at syntactic and semantic level.

Syntactic measures are based on the name of the method, the name of the declaring class and code comments. In particular, the natural language description of the behavior of the method is pre-processed using a standard NLP pipeline which performs sentence splitting, tokenization and lemmatization. Next, we compute the following measures:

- **LS:** normalized Levenshtein similarity against the name of the method;
- **COM:** n-gram overlap against the Javadoc comment related to the method;
- **CN:** n-gram overlap against the name of the declaring class.

More precisely, given two sets S_1 and S_2 of consecutive n-grams from two different sentences, the n-gram overlap is defined as:

$$ngo(S_1, S_2) = 2 \cdot \left(\frac{|S_1|}{|S_1 \cap S_2|} + \frac{|S_2|}{|S_1 \cap S_2|} \right)^{-1} .$$

Thus, the n-gram overlap is computed as the harmonic mean of the degree to which the second sentence covers the first and the degree to which the first sentence covers the second. In practice, for n-grams we set $n = 1$. On the other hand, the Levenshtein distance d_L between two strings is defined as the minimum number of single-character edits, required to change one string into the other. Since we need a similarity value in the range between 0 and 1, we compute the normalized Levenshtein similarity as:

$$s_L(s_1, s_2) = 1 - \frac{d_L(s_1, s_2)}{\max\{|s_1|, |s_2|\}}.$$

Levenshtein distance and n-gram overlap are used to match methods from OpenJDK and the natural language command provided by the user at a syntactic level. To incorporate semantics into the ranking process, we leverage DBpedia links readily available in the dataset and word embeddings to comute the following features:

- **NED:** ratio of the DBpedia links shared by the comment of the method and the natural language command;
- **W2V:** cosine similarity between the mean vector associated with the comment of the method and the mean vector associated with the natural language command.

More precisely, we make use of TagMe [16], to perform Named Entity Disambiguation on the input text and retrieve a set of links to DBpedia resources. Each method available in the dataset provides DBpedia links generated using the same approach, applied on the Javadoc comment. Hence, we use the ratio of the shared links as a measure of semantic relatedness between each retrieved method and the input command.

Moreover, we apply a Word2Vec [17] pre-trained model to retrieve 300-dimensional word vectors from each word in both the natural language specification provided by the user and the comment associated with methods in CodeOntology. The cosine similarity between the mean vector corresponding to the input command and the mean vector associated with each Java method is used as another semantic measure. The final score applied within the ranking process is the average value of the syntactic and semantic measures described so far.

4   Fine-Grained Approach

As we have already mentioned, the fine-grained approach is aimed at dealing with more complex natural language utterances, possibly involving the execution of more methods. Given a question in natural language, this approach is capable of parsing the input question into a Java source code which gets executed to produce the desired answer. This section details how this approach actually works and how it can be used for question answering over source code.

4.1   Dependency Graph Unfolding

Given a natural language question, the fine-grained approach starts by applying Stanford CoreNLP [18] to perform dependency parsing. We assume that the question provided by the user may include primitive literals, such as string literals, integers, Booleans, and parameters of type double. Hence, before parsing the input sentence, care must be taken to replace string literals with a placeholder, in order to prevent the dependency parser from processing also actual arguments. The output of such process is the graph of the dependencies, as shown in Fig. 2.

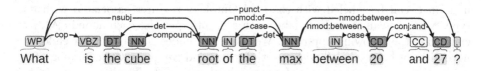

Fig. 2. Result of dependency parsing on a simple input question.

This graph is unfolded into a tree and pruned to remove nodes that are not useful for our purposes. In particular, we allow merging two nodes, depending on the nature of the dependency between the corresponding words. For instance, multiword expressions (MWEs) are merged into a single node, and, similarly, adjectival or adverbial modifiers are joined with the word they refer to. We also allow removing leaf nodes such as conjunctions, determiners and punctuation.

The result is further post-processed, to ensure that all the literal arguments specified by the user correspond to some leaf node of the tree and that no subtree is repeated. Figure 3 shows the result of the application of this approach to the graph depicted in Fig. 2.

4.2   Mapping to a Feasible Execution Tree

The unfolding of the dependency graph results in a tree, such that the set of nodes N can be partitioned into two subsets $\mathcal{L}$ and $\mathcal{M}$, where *(i)* $\mathcal{L}$ is the subset of nodes corresponding to literal actual arguments, *(ii)* $\mathcal{M}$ is the subset of nodes corresponding to natural language utterances denoting a method invocation, *(iii)* each node in $\mathcal{L}$ is a leaf, *(iv)* $N = \mathcal{L} \cup \mathcal{M}$ and $\mathcal{L} \cap \mathcal{M} = \emptyset$.

We want to obtain a tree where each node $i \in \mathcal{M}$ is labeled with a method ranking $\mathcal{R}_i$, that is a sequence $(m_1, s_1) \cdots (m_n, s_n)$, such that *(i)* m_i is a Method for all $i = 1 \ldots n$, *(ii)* $s_i \in [0, 1]$ for all $i = 1 \ldots n$, *(iii)* $i < j \Rightarrow s_i \geq s_j$. To do this, we need to query CodeOntology and rank methods using the coarse-grained approach. However, we only have to select methods whose signature is compatible with the structure of the tree and with the actual arguments provided by the user. Hence, we also label each node with the set of the types it can assume. To this end, we define the set $\mathsf{Types} = \{t \in \mathcal{K} \mid t : T \land T \sqsubseteq \mathsf{Type}\}$, as the set of all types available in our knowledge base $\mathcal{K}$. Next, we define the function $types : N \to 2^{\mathsf{Types}}$, such that:

$$types(i) = \begin{cases} t & \text{if } i \in \mathcal{L} \text{ and } t \text{ is the type of } i \\ returnTypes(\mathcal{R}_i) & \text{if } i \in \mathcal{M} \end{cases},$$

where $returnTypes(\mathcal{R}_i)$ can be computed as:

$$returnTypes(\mathcal{R}) = \begin{cases} \{r\} \cup returnTypes(\mathcal{R}') & \text{if } \mathcal{R} = \mathcal{R}'(m,s) \text{ and} \\ & m \text{ has return type } r \\ \emptyset & \text{if } \mathcal{R} = [] \end{cases}.$$

To assign these type labels to the nodes, we start from the leaves, as each node in $\mathcal{L}$ can be labeled with the CodeOntology resource associated with its type. Next, we can recursively label with a set of types also each node in $\mathcal{M}$, by employing the following approach. We select the nodes such that their children have already been labeled with a set of types and we query CodeOntology for methods that are compatible with the specified arguments. The list of arguments may be unsorted and may also include the target instance of the method invocation. The retrieved methods are then ranked as described in Sect. 3.2 and the corresponding node is labeled with the set of their return types. Algorithm 1 details the described approach.

Algorithm 1. $RankOnTree(i)$

---

1 **if** $i \in \mathcal{L}$ **then**
2 $\quad$ Let t be the type of i
3 $\quad$ $types(i) \leftarrow \{t\}$
4 **else**
5 $\quad$ Let $l = [l_1, \ldots, l_n]$ be the list of the children of i
6 $\quad$ **foreach** $l_j \in l$ **do**
7 $\quad\quad$ $RankOnTree(l_j)$
8 $\quad$ **end**
9 $\quad$ Let $\mathbf{t} = [t_1, \ldots, t_n]$ be a list such that $t_j = types(l_j)$ for each $l_j \in l$
10 $\quad$ Query CodeOntology for methods whose signature is compatible with $\mathbf{t}$
11 $\quad$ Let $\mathcal{R}_i$ be the ranking of the resulting methods, computed using the coarse-grained approach
12 $\quad$ $types(i) \leftarrow returnTypes(\mathcal{R}_i)$
13 **end**

After applying Algorithm 1 on the result of the dependency graph unfolding, we obtain a new tree structure, where each node in $\mathcal{M}$ is labeled with a ranking of methods retrieved from CodeOntology. Figure 3 shows an example of such a tree.

Now, we want to select a method from each ranking, in such a way that the combination of all the selected methods is feasible, meaning that it corresponds to compilable Java source code. At the same time, however, we also need to

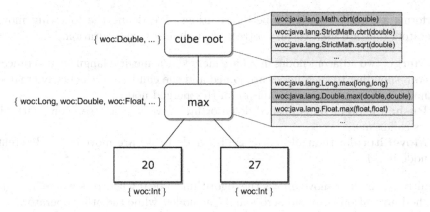

Fig. 3. Mapping to a feasible execution tree.

maximize the total score associated with selected methods. Hence, we have to solve the following integer linear programming problem, where $x_{ij} = 1$ if the j-th method in the ranking $\mathcal{R}_i$ is selected, and $x_{ij} = 0$ otherwise:

$$\text{Maximize} \sum_{i \in \mathcal{M}} \sum_{(m_{ij}, s_{ij}) \in \mathcal{R}_i} x_{ij} \cdot s_{ij} \tag{1}$$

subject to the following constraints: *(i)* $x_{ij} \in \{0,1\}$, *(ii)* $\sum_j x_{ij} = 1$, for all $i \in \mathcal{M}, 1 \le j \le |\mathcal{R}_i|$ and *(iii)* the combination of the selected methods can be compiled.

If a solution to this problem exists, then we can turn the tree into Java source code, which gets executed to answer the original question. Moreover, the average score of selected methods can be interpreted as a measure of the confidence level about the correctness of the solution. The result of such approach is shown in Fig. 3, where selected methods have been highlighted.

4.3 Greedy Search

The algorithmic approach described up to this point may fail to return a correct answer, whenever the tree produced by unfolding the dependency graph cannot be matched to the Java source code corresponding to the input question. In particular, we want to improve the algorithm, so that it is robust to two kinds of situations: *(i)* the tree resulting from the process described in Sect. 4.1 is too detailed, meaning that it has more nodes corresponding to method invocations than needed, or *(ii)* the dependency graph produced by Stanford CoreNLP contains some errors, which can be detected by leveraging knowledge about methods in CodeOntology and typing. There are several ways to extract a tree from the input sentence and, for each tree, several combinations of methods need to be explored. This creates an intractable search space for possible solutions, and, subsequently, we cannot afford an exhaustive search. Thus, we apply a heuristic approach that, starting from the output of the process described in Sect. 4.2, performs a greedy search for better solutions. We define the following move operators that are used to turn a tree into a different configuration:

- **Merge:** two adjacent nodes in $\mathcal{M}$ are merged, the natural language utterances corresponding to such nodes are joined and the children of the newly created node are the union of the children of the merged nodes;
- **Push:** a node in $\mathcal{M}$ is pushed down or up a level in the tree, along with all its children;
- **MoveLiterals:** the children in $\mathcal{L}$ of a node in $\mathcal{M}$ are moved to a different node in $\mathcal{M}$.

Intuitively, the first move allows the algorithm to deal with trees where a single method invocation is spread across multiple nodes, while the other operators are used to handle errors in dependency parsing.

We define the distance between two trees $\mathcal{T}$ and $\mathcal{T}'$, denoted as $TED(\mathcal{T}, \mathcal{T}')$, as the minimum number of moves required to turn one tree into the other. Next, we denote the normalized distance as:

$$NTED(\mathcal{T}, \mathcal{T}') = \frac{TED(\mathcal{T}, \mathcal{T}')}{\max\{|\mathcal{T}|, |\mathcal{T}'|\}},$$

where $|\mathcal{T}|$ is the total number of nodes in $\mathcal{T}$.

Starting from an initial tree $\mathcal{T}_0$, produced as described in Sect. 4.2, the algorithm evaluates all the possible defined moves and applies a greedy search with a *Best-Improvement* strategy, in order to maximize, under the same constraints defined for Eq. 1, the following objective function:

$$z(\mathcal{T}_k) = \frac{1}{|\mathcal{M}_k|} \cdot \sum_{i \in \mathcal{M}_k} \sum_{(m_{ij}, s_{ij}) \in \mathcal{R}_i^k} x_{ij} \cdot s_{ij} - \lambda \cdot NTED(\mathcal{T}_k, \mathcal{T}_0), \qquad (2)$$

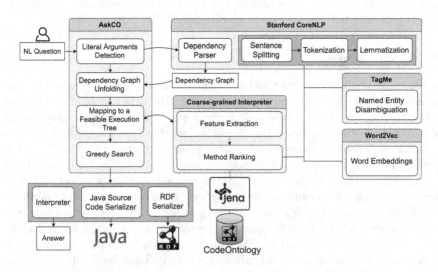

Fig. 4. High-level view of the architecture of the system.

where $\lambda \in [0, 1]$ is a constant, $\mathcal{M}_k$ is the set of non-literal nodes in $\mathcal{T}_k$ and $\mathcal{R}_i^k$ is the method ranking associated with node i in $\mathcal{T}_k$. Overall, Eq. 2 is structured as Eq. 1, with a penalization term which decreases the objective value for trees that are too different from the original tree $\mathcal{T}_0$. In practice, we set λ to 0.5. The algorithm stops when a local optimum is reached and no move can be applied to improve the objective value. Figure 4 shows a high-level view of the architecture of the system, which is available on GitHub at https://github.com/codeontology/question-answering.

5 Experiments

This section provides an evaluation for both the coarse-grained and the fine-grained approaches. Experimental results show that both techniques can be effectively applied on a RDF dataset extracted from OpenJDK 8 [3], with promising results.

5.1 Method Ranking Evaluation

The system implemented for the coarse-grained approach aims at retrieving and ranking Java methods defined within the OpenJDK 8 source code, given a natural language description of the behavior of the method. Providing an evaluation for this coarse-grained ranking of Java methods is challenging, because we are not aware of any dataset pairing natural language commands, with a corresponding set of relevant methods from OpenJDK. Hence, we have extracted a benchmark dataset containing simple questions discussed on StackOverflow[4].

The dataset has been generated by retrieving the most popular questions about the Java programming language, which have been manually filtered to select only the top 122 questions that can be answered with the invocation of a single method from OpenJDK.

For some questions, we may have more than one relevant method, so the dataset has been further manually enriched with missing methods. For instance, the natural language command *"convert a string to an integer"* is associated to two methods, namely the method `java.lang.Integer.parseInt` `(java.lang.String)` and the method `java.lang.Integer.valueOf(java.` `lang.String)`.

Overall, for more than 80% of the questions there is only one relevant method, while some question has even 3 or 4 relevant methods. The dataset is available on figshare[5] under Creative Commons Attribution 4.0 license.

We experiment several combinations of the syntactic and semantic features defined in Sect. 3.2. Table 1 reports the experimental results obtained for the coarse-grained approach. We evaluate the performance of the system based on the Mean Average Precision (MAP) obtained by the produced rankings. However, it is crucial that the first method in the ranking is correct, as it is invoked

---

[4] https://stackoverflow.com/.
[5] https://doi.org/10.6084/m9.figshare.6071663.

by the coarse-grained system. Thus, we also compute the precision at 1 for each ranking, and we report the mean result in Table 1 (MAP@1).

Table 1. Experimental results on method ranking

	Features	MAP@1	MAP
Syntactic features	LS	0.697	0.776
	LS + CN	0.713	0.785
	LS + COM	0.861	0.891
	LS + CN + COM	0.869	0.897
Semantic features	NED	0.607	0.714
	W2V	0.738	0.818
	W2V + NED	0.754	0.822
Syntactic + Semantic features	LS + W2V	0.795	0.852
	LS + W2V + NED	0.803	0.861
	LS + CN + COM + W2V	0.902	0.921
	LS + CN + COM + W2V + NED	**0.902**	**0.923**

As we can see, the best results are obtained by boosting syntactic features with semantics. The coarse-grained approach to the ranking of Java methods, in this case, achieves a Mean Average Precision of 0.923. At the same time, the system is capable of finding and invoking the correct method for the majority of the natural language commands available in the dataset, obtaining a MAP@1 of 0.902.

5.2   Question Answering Evaluation

Experiments on the ranking of Java methods provide a partial evaluation also for the fine-grained approach, as method ranking is the most important step for parsing natural language questions involving the invocation of multiple methods. However, to provide a further evaluation of our fine-grained system, we perform experiments on another benchmark dataset[6] we created, containing 120 questions on mathematical expressions and string manipulation. We can classify each question in the dataset by the number of methods required to provide the correct answer. We obtain that the dataset contains:

- 16 questions requiring the invocation of 1 method;
- 63 questions requiring the invocation of 2 methods;
- 36 questions requiring the invocation of 3 methods;
- 5 questions requiring the invocation of 4 methods.

Hence, the majority of the questions involves the invocation of 2 methods and, on average, 2.25 methods per question are required.

---

[6] available online at: https://doi.org/10.6084/m9.figshare.6071729.

We apply a threshold $t \in [0, 1]$ on the objective value defined by Eq. 2, in order to detect questions that our system is not able to process correctly. When $t = 0$, then the system will provide an answer to all questions in the dataset, while $t = 1$ means that the system basically refuses to process any question. Figure 5 summarizes the performances of the system in response to changes in the value of the threshold. As we can see, when $t = 0$ the system is capable of answering correctly 91% of the questions in the dataset. However, we can increase precision over processed questions using a higher threshold. In particular, setting a threshold $t = 0.15$ allows to get a precision over processed questions of 0.94, while leaving the global result unchanged. When precision over processed questions eventually reaches 1, then global precision equals the rate of processed questions, as clearly shown in Fig. 5.

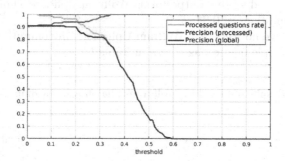

Fig. 5. Performances of the system for different values of the threshold.

It is also interesting to discuss the average size of the rankings, which contain all methods from OpenJDK whose signature is compatible with the actual arguments specified in the natural language question. At this remark, we notice that, on average, the rankings of methods produced by the fine-grained approach on this dataset contain 246.5 methods. The longest ranking includes 677 methods, while the shortest one has 24 methods. Hence, the distribution has a high standard deviation equal to 176.7 methods.

We can compare our approach with the results obtained by the WolframAlpha computational knowledge engine[7]. Of course, our system and WolframAlpha have different capabilities. On the one hand, WolframAlpha can answer a wide range of complex open-domain questions, which cannot be answered by simply invoking methods from OpenJDK. On the other hand, our system is capable of executing natural language commands which are certainly out of the scope of WolframAlpha. However, both approaches should be able to process and answer questions involving mathematical expressions and string manipulation. Table 2 shows the experimental results of the comparison between the systems.

WolframAlpha was able to process 108 out of the 120 questions in the dataset, achieving a global precision of 0.82 and a precision over processed questions of

---

[7] https://www.wolframalpha.com/.

Table 2. Experimental results for the fine-grained approach

	QA over CodeOntology	WolframAlpha
Number of questions	120	120
Processed questions	116	108
Correct answers	109	98
Precision (global)	0.91	0.82
Precision (processed questions)	0.94	0.91

0.91. On the other hand, our approach based on CodeOntology allows processing 116 questions and 109 of such items have been answered correctly. Hence, on this task, the implemented system outperforms WolframAlpha, reaching a precision over processed questions of 0.94.

Interestingly, we noticed that WolframAlpha fails in computing the correct result for some simple queries, as shown in Table 3.

Table 3. Results obtained by WolframAlpha on a set of simple queries

WolframAlpha		
Input	Interpretation	Result
Add 2 to 4	$2 + 4$	6
Add 2 to the max between 3 and 4	$2^{\max\{3,4\}}$	16
Add 2 to the sum of 1 and 3	2^{1+3}	16
What is the uppercase of "abc"?	ToUpperCase["abc"]	"ABC"
Convert "abc" to uppercase	ToUpperCase["Convert \"abc\" to"]	"CONVERT\"ABC\"TO"
What is the length of "abcd"?	StringLength["abcd"]	4
Sum 1 to the length of "string"	-	-

For instance, despite the system is capable of correctly interpreting commands like *"Add 2 to 4"*, it does not parse successfully slightly more complicated sentences such as *"Add 2 to the max between 3 and 4"*.

On the other hand, our approach is able to process correctly the same queries, as shown in Table 4.

Moreover, we can classify questions depending on whether both the systems, only one of them or none of them was able to provide the correct answer. Such categorization is shown in Table 5.

We can use the values reported in Table 5 to perform a McNemar exact test by comparing the case where the two systems provide discordant results (b and c), to a binomial distribution with size parameter $n = b + c$ and $p = 0.5$. The test shows that there exists a statistically significant difference between the two systems, with a confidence level of 99.8%.

Table 4. Results obtained by our approach on a set of simple queries

AskCO		
Input	Interpretation	Result
Add 2 to 4	`Math.addExact(2,4)`	6
Add 2 to the max between 3 and 4	`Math.addExact(2,Math.max(3,4))`	6
Add 2 to the sum of 1 and 3	`Math.addExact(2,Integer.sum(1,3))`	6
What is the uppercase of "abc"?	`"abc".toUpperCase()`	`"ABC"`
Convert "abc" to uppercase	`"abc".toUpperCase()`	`"ABC"`
What is the length of "abcd"?	`"abcd".length()`	4
Sum 1 to the length of "string"	`Long.sum(1,"string".length())`	7

Table 5. Comparison between AskCO and WolframAlpha

	WolframAlpha (correct)	WolframAlpha (failed)	
AskCO (Correct)	$a = 97$	$b = 12$	109
AskCO (Failed)	$c = 1$	$d = 10$	11
	98	22	120

6 Conclusion

This paper introduces two approaches for answering end-user questions on the execution of Java methods. On the one hand, our coarse-grained approach only allows mapping natural language commands to the execution of a single method, but it supports arguments of any arbitrary type, including user-defined classes. On the other hand, the fine-grained approach can handle more complex questions, possibly requiring the execution of multiple methods. However, the input of this approach is a single natural language question which includes the actual arguments as literals, thereby limiting the kinds of the parameters that can be passed by the user. Overall, experimental results show that the approach is promising and, subsequently, it can be effectively used for semantic code search and reuse over CodeOntology.

References

1. Atzeni, M., Atzori, M.: CodeOntology: RDF-ization of source code. In: d'Amato, C., et al. (eds.) ISWC 2017. LNCS, vol. 10588, pp. 20–28. Springer, Cham (2017). https://doi.org/10.1007/978-3-319-68204-4_2
2. Atzeni, M., Atzori, M.: CodeOntology: querying source code in a semantic framework. In: 16th International Semantic Web Conference (Posters & Demo) (2017)
3. Atzeni, M., Atzori, M.: CodeOntology OpenJDK8 dataset. Figshare (2017). https://doi.org/10.6084/m9.figshare.5234878
4. Atzori, M., Mazzeo, G.M., Zaniolo, C.: QA^3: a natural language approach to question answering over RDF data cubes. Semant. Web J. (2018)

5. Devanbu, P.T., Brachman, R.J., Selfridge, P.G., Ballard, B.W.: Lassie - a knowledge-based software information system. In: 12th International Conference on Software Engineering, pp. 249–261. IEEE Computer Society Press (1990)
6. Würsch, M., Ghezzi, G., Reif, G., Gall, H.C.: Supporting developers with natural language queries. In: Proceedings of the 32Nd ACM/IEEE International Conference on Software Engineering, ICSE 2010, vol. 1, pp. 165–174. ACM (2010)
7. Bernstein, A., Kaufmann, E., Kaiser, C., Kiefer, C.: Ginseng: a guided input natural language search engine for querying ontologies. In: 2006 Jena User Conference, Bristol , UK (2006)
8. Panchenko, O., Mller, S., Plattner, H., Zeier, P.D.A.: Querying source code using a controlled natural language. In: The Sixth International Conference on Software Engineering Advances, ICSEA 2011, pp. 369–373, June 2011
9. Kimmig, M., Monperrus, M., Mezini, M.: Querying source code with natural language. In: 26th IEEE/ACM International Conference on Automated Software Engineering (ASE 2011) (2011)
10. Liu, H., Lieberman, H.: Metafor: visualizing stories as code. In: 10th International Conference on Intelligent User Interfaces, IUI. pp. 305–307 (2005)
11. Liu, H., Lieberman, H.: Programmatic semantics for natural language interfaces. In: Extended Abstracts on Human Factors in Computing Systems. CHI (2005)
12. Sales, J.E., Handschuh, S., Freitas, A.: SemEval-2017 task 11: end-user development using natural language. In: 11th International Workshop on Semantic Evaluation, pp. 556–564 (2017)
13. Atzeni, M., Atzori, M.: Towards semantic approaches for general-purpose end-user development. In: 2nd IEEE International Conference on Robotic Computing, IRC, pp. 369–376 (2018)
14. Quirk, C., Mooney, R., Galley, M.: Language to code: learning semantic parsers for if-this-then-that recipes. In: 53rd Annual Meeting of the Association for Computational Linguistics, ACL, pp. 878–888 (2015)
15. Lehmann, J.: DBpedia - a large-scale, multilingual knowledge base extracted from Wikipedia. Semant. Web J. **6**(2), 167–195 (2015)
16. Ferragina, P., Scaiella, U.: TAGME: on-the-fly annotation of short text fragments (by Wikipedia entities). In: 19th ACM International Conference on Information and Knowledge Management, CIKM, pp. 1625–1628 (2010)
17. Mikolov, T., Sutskever, I., Chen, K., Corrado, G.S., Dean, J.: Distributed representations of words and phrases and their compositionality. In: Advances in Neural Information Processing Systems, pp. 3111–3119. Curran Associates, Inc. (2013)
18. Manning, C.D., Surdeanu, M., Bauer, J., Finkel, J., Bethard, S.J., McClosky, D.: The stanford CoreNLP natural language processing toolkit. In: Association for Computational Linguistics (ACL) System Demonstrations, pp. 55–60 (2014)

GraFa: Scalable Faceted Browsing for RDF Graphs

José Moreno-Vega and Aidan Hogan[✉]

IMFD Chile & Department of Computer Science, University of Chile,
Santiago, Chile
ahogan@dcc.uchile.cl

Abstract. Faceted browsing has become a popular paradigm for user interfaces on the Web and has also been investigated in the context of RDF graphs. However, current faceted browsers for RDF graphs encounter performance issues when faced with two challenges: scale, where large datasets generate many results, and heterogeneity, where large numbers of properties and classes generate many facets. To address these challenges, we propose GRAFA: a faceted browsing system for heterogeneous large-scale RDF graphs based on a materialisation strategy that performs an offline analysis of the input graph in order to identify a subset of the exponential number of possible facet combinations that are candidates for indexing. In experiments over Wikidata, we demonstrate that materialisation allows for displaying (exact) faceted views over millions of diverse results in under a second while keeping index sizes relatively small. We also present initial usability studies over GRAFA.

1 Introduction

The Semantic Web community has overseen the publication of a rich collection of datasets on the Web according to a variety of proposed standards [12]. However, current interfaces for accessing such datasets are not generally designed nor intended for end users to interact with directly. The Semantic Web community still lacks effective methods by which end users can interact with such datasets; or as Karger [18] phrases it: *"The Semantic Web's potential to deliver tools that help end users capture, communicate, and manage information has yet to be fulfilled, and far too little research is going into doing so."*

On the other hand, *faceted search* [32][1] has become a familiar mode of interaction for many Web users, popularised in particular by e-Commerce websites like Amazon and eBay. Such interaction is characterised by iteratively refining the active result-set through filter conditions – called *facets* – typically defined to be an attribute (e.g., *type*, *brand*, *country*) and value (e.g., *Toothbrush*, *Samsung*, *India*) that the filtered results should have. Such interaction enables end users to find specific results corresponding to concrete criteria known in advance, or simply to explore and iteratively refine results based on available options.

---

[1] Also known as *"faceted browsing"*, *"faceted navigation"*, etc.

© Springer Nature Switzerland AG 2018
D. Vrandečić et al. (Eds.): ISWC 2018, LNCS 11136, pp. 301–317, 2018.
https://doi.org/10.1007/978-3-030-00671-6_18

While the queries that can be formulated through an iterative selection of facets are generally less expressive than those that can be specified through a structured query language such as SPARQL, faceted browsing is more accessible to a broader range of users unfamiliar with such query languages; furthermore, the end user need not be as familiar with the content or schema of the dataset in question since the facets offered denote the possible filters that can be applied and the number of results to be expected, helping users to avoid empty results.

Adapting faceted search for a Semantic Web context is then a natural idea, where various authors have explored faceted navigation over RDF graphs [7,28] as a potential way to bridge from Semantic Web to end-users. Such works – discussed in more detail in the following section on related work – have explored core themes relating to faceted navigation, including query expressivity, ranking, usability, indexing, performance, reasoning, complexity, etc. However, despite the breadth of available literature on the topic, we argue that more work is required, in particular for faceted browsing over RDF graphs that are *large-scale* (with many triples) and *diverse* (with many properties and classes).

The work presented in this paper was motivated, in particular, by the idea of providing faceted search for Wikidata [29]: a large, collaboratively-edited knowledge-base where users can directly add and curate structured knowledge relating to the Wikipedia project. Though a variety of interfaces exist for interacting with Wikidata[2], including a SPARQL endpoint, query builders, and so forth, none quite cover the main characteristics of a faceted browser (e.g., only displaying options with non-empty results). On the other hand, despite the breadth of works on faceted browsing, we could not find an available system that could load the full ("truthy") Wikidata graph available at the time of writing.

We thus propose a novel faceted browser for diverse, large-scale RDF graphs called GRAFA – GRAph FAcets – that we demonstrate is able to handle the scale and diversity of a dataset such as Wikidata. An initial result set in the system is generated through either keyword search or by selecting an entity type (e.g., *person, building,* etc.). Thereafter, a result set can be refined by selecting a particular property–value (*facet*) that all entities in the next result set should have. A combination of auto-completion and ranking features help ensure that the user is presented with relevant facets and results. Furthermore, at each stage of interaction, only options that lead to non-empty results are returned; this aspect in particular proves the most challenging to implement.

Similar to previous faceted systems [5,31], the GRAFA system is based on Information Retrieval (IR)-style indexes that combines unstructured (text) and semi-structured (facet) information. However, unlike previous such systems, we propose a novel materialisation technique to enable interactive response times at higher levels of scale. The core hypothesis underlying this technique is that although there is a potentially exponential (in the size of the graph) number of combinations of facets that could be considered, few combinations will lead to large result sets that cause slow response times. Hence we propose a technique to

---

[2] https://wikidata.org/wiki/Wikidata:Tools/External_tools; retr. 2018/04/05.

perform an offline analysis of the graph to select facets that are then materialised. Our results show that materialisation can improve worst-case response times by orders of magnitude using a modestly-sized index of precomputed facet views.

To assess the usability of our system, we also present the results of two initial studies. The first user study compares the GRAFA system and the WIKIDATA QUERY HELPER (WQH) interface provided by the Wikidata SPARQL endpoint, asking participants to solve a number of tasks using both systems. Based on the results of this first study, we then made some improvements to the GRAFA system, where in the second study, we asked members of the Wikidata community to use the modified GRAFA system and to answer a questionnaire to rate the usability, usefulness, responsiveness, novelty etc., of the system.

Outline: Section 2 first discusses related work. Section 3 defines the inputs and interactions considered in our faceted browsing framework. Section 4 describes the base indexing scheme used to support these interactions, and Sect. 5 describes the materialisation strategies we use to improve worst-case response times. Turning to evaluation, Sect. 6 focuses on performance, while Sect. 7 focuses on usability. Finally Sect. 8 concludes and discusses future work.

2   Related Work

Various faceted browsers have been proposed for RDF over the years [7,28,32]. Some earlier works include MSPACE [23], ONTOGATOR [20], BROWSERDF [21], /FACET [15], with later proposals including GFACET [13,14], EXPLORATOR [2], RHIZOMER [6], FACETE [25], REVEALD [17], SPARKLIS [8] and HIPPALUS [27]. These works describe evaluations or use-cases involving domain-specific data of low heterogeneity, such as multimedia [20,23,26], suspect descriptions [21], movies [6], cultural heritage [15,20], tweets [1], places [25], biomedicine [17], fish species [27], etc.; furthermore, many of these works delegate data management and query processing to an underlying triple-store/SPARQL engine, and rather focus on issues such as expressiveness, ranking and usability, etc.

Recently Petzka et al. [22] proposed a benchmark for SPARQL systems to test their ability to support faceted browsing capabilities, but again the dataset (referring to transport) contains in the order of tens of classes and properties and we could not find details on the scale of data used for experiments.

A number of later works have explored faceted navigation over more heterogeneous RDF datasets, such as VISINAV [11] operating on RDF data (19 million triples with 21 thousand classes and properties) crawled and integrated from numerous sources on the Web; however, aside from brief discussion of top-k ordering of facets, performance issues were not discussed in detail. Another more scalable proposal is the NEOFONIE [10] system, proposed for faceted search over DBpedia; however, only a small selection of target facets are displayed and no performance results are provided. A more recent scalable approach is that of ELINDA [33], which allows for real-time browsing of DBpedia; however, navigation is not based on facets but rather on interactive bar-charts.

A number of approaches have proposed to use indexing techniques developed for Information Retrieval (IR) to support faceted browsing for RDF. The SEM-PLORE system [31] builds faceted browsing on top of IR-indexes, where facets for the current result set are computed from types, as well as incoming and outgoing relations; a set of top-k facets are constructed by count. Experiments were conducted over DBpedia [19] and LUBM [9] datasets in the order of 100 million triples, showing mean sub-second response times faster than those achievable over selected triple stores. Though this system is along similar lines to what we wish to achieve, the size of the result-sets for which facets are generated in the evaluation is not specified, nor is the value of k for the top-k generation; we could not find materials online to replicate these results, but using a similar implementation later on a more modern version of the same IR engine (Lucene), we find that construction of the full set of facets takes minutes over large result-sets with millions of results. Wagner et al. [30] likewise propose IR-style indexing to support faceted browsing and conduct evaluation over DBpedia, but performance issues are explicitly considered out of scope; however, for their evaluation, we note that the authors mention use of caching to speed-up response times for selected tasks, though no further details are provided.

To the best of our knowledge, the closest published results we found for faceted search over RDF data at the scale of Wikidata was the BROCCOLI system [4,5], which is also based on IR indexes. Though the system has a slightly different focus to ours (semantic search over Wikipedia text enriched with Freebase relations), an index over relations is defined to enable faceted search. The authors propose caching methods to identify and re-use sub-combinations of facets that are frequently required; unlike our approach, this LRU cache is built online from user-queries, whereas we materialise query results offline.[3]

The SEMFACET [3] system addresses a number of issues with respect to faceted browsing for RDF graphs, including reasoning, expressiveness, complexity and efficiency. Though their system can process facets for tens of millions of answers in about 2 s, this requires having all data indexed in memory, which limits scale; hence their evaluation is limited to 20% of DBpedia [19] (3.5 million triples), as well as selected slices of YAGO [16] that fit in memory. Though the system is available for download, we failed to load Wikidata with it. Later work by Sherkhonov et al. [24] discusses the addition of other features to faceted navigation, such as aggregation and recursion, but focuses on studying the complexity of query answering and containment.

3   Faceted Browsing

We now outline the faceted browsing interactions that the GRAFA system currently supports. Beforehand we provide preliminaries for RDF graphs considered as input to the system, mainly to establish notation and nomenclature.

---

[3] We did not find source code for the system to be able to perform tests for Wikidata, though a Freebase demo is available demonstrating interactive runtimes on large result sets: http://broccoli.informatik.uni-freiburg.de/demos/ BroccoliFreebase/; retr. 2018/04/05.

RDF Triples and Graphs: An *RDF triple* (s, p, o) is an element of $\mathbf{IB} \times \mathbf{I} \times \mathbf{IBL}$, where $\mathbf{I}$ is a set of IRIs, $\mathbf{L}$ a set of literals, and $\mathbf{B}$ a set of blank nodes; the sets $\mathbf{I}$, $\mathbf{L}$ and $\mathbf{B}$ are considered pairwise disjoint. The positions of the triple are called *subject*, *predicate*, and *object*, respectively. An *RDF graph* G is a set of triples. Letting $\pi_{\mathrm{S}}(G) = \{s \mid \exists p, o : (s, p, o) \in G\}$ project the ("flat") set of all subjects of G, and letting $\pi_{\mathrm{P}}(G)$ and $\pi_{\mathrm{O}}(G)$ likewise project the set of all predicates and objects of G, we call $\pi_{\mathrm{S}}(G) \cup \pi_{\mathrm{O}}(G)$ the *nodes* of G, $\pi_{\mathrm{S}}(G) \cap \mathbf{I}$ the *entities* of G, and $\pi_{\mathrm{P}}(G)$ the set of *properties* of G. Given an entity s and a property p, we call any o such that $(s, p, o) \in G$ the *value* of property p for entity s.

Keyword Selection: We assume most entities to have values for a label property (e.g., `rdfs:label`, `skos:prefLabel`, `skos:altLabel`) and/or a description property (e.g., `rdfs:comment`, `schema:description`); we also assume that the system is configured with a list of such properties. To generate an initial result-set, users can specify a keyword search, returning a set of entities whose label/description values match the search. Notation-wise, we will denote keyword search as a function $\kappa : 2^G \times \mathbb{S} \to 2^{\pi_{\mathrm{S}}(G)}$, where $\mathbb{S}$ denotes the set of strings (keyword searches). However, to simplify notation, we will consider the input graph as fixed throughout this paper. Hence we abbreviate the function as $\kappa : \mathbb{S} \to 2^{\pi_{\mathrm{S}}(G)}$, taking a string and returning a set of entities according to a keyword-matching function (we discuss implementation of the function in Sect. 4).

Type Selection: To generate an initial set of results, rather than use the keyword search function, a user may prefer to select entities of a given *type* (e.g., *human*, *movie*, etc.). We define a *type* (aka. *class*) to be any value of a *type property* (e.g., `rdf:type`, `wdt:P31`$_{[instance\ of]}$) for any entity; we assume that a fixed set of type properties P_T are preconfigured in the system. We then denote the set of types in a graph G as $T(G) := \{o \mid \exists s, p : (s, p, o) \in G \text{ and } p \in P_T\}$. We denote type selection as $\tau : T(G) \to 2^{\pi_{\mathrm{S}}(G)}$, where $\tau(t) := \{s \mid \exists p \in P_T \text{ such that } (s, p, t) \in G\}$. In summary, $\tau(t)$ returns the set of all entities with the type $t \in T(G)$. Note that we do not currently consider type/class hierarchies.

Facet Selection: Given a current set of results, a user may select a *facet* to further restrict the presented results. Such a facet is here defined to be a property–value pair – e.g., (*director,Kurosawa*) – that each entity in the next result set must have. More formally, given a current result set of entities $E \subseteq \pi_{\mathrm{S}}(G)$, we denote by $E(G) := \{(s, p, o) \in G \mid s \in E\}$ the projection from G of all triples with a subject term in E. Now we can define the facet selection function $\zeta : 2^{\pi_{\mathrm{S}}(G)} \times \pi_{\mathrm{P}}(G) \times \pi_{\mathrm{O}}(G) \to 2^{\pi_{\mathrm{S}}(G)}$ where $\zeta(E, p, o) := \{s \mid (s, p, o) \in E(G)\}$.

Faceted Navigation: We call a sequence of selections of either of the following forms a *faceted navigation*, initiated by keyword or type selection, respectively:

- $\zeta(\zeta(\ldots(\zeta(\kappa(q), p_1, o_1)\ldots, p_{n-1}, o_{n-1}), p_n, o_n)$
- $\zeta(\zeta(\ldots(\zeta(\tau(t), p_1, o_1)\ldots, p_{n-1}, o_{n-1}), p_n, o_n)$

We remark that the ζ function is *commutative*: we can apply the facet selections in any order and receive the same result. Hence, with some abuse of notation, we can unnest and thus more clearly represent the above navigation sequences as a conjunction of criteria, where we use $[\cdot]$ to represent optional criteria:

- $\kappa(q) \, [\wedge \, \zeta(p_1, o_1) \wedge \ldots \wedge \zeta(p_n, o_n)]$
- $\tau(t) \, [\wedge \, \zeta(p_1, o_1) \wedge \ldots \wedge \zeta(p_n, o_n)]$

Type and Facet Interactions: The type selection and facet selection interactions take as input a type t and a facet (p, o) respectively. However, the users may not know the corresponding identifier, hence GRAFA will offer auto-completion search on the labels and aliases of types and the values of facet properties. For example, a user typing al* into the auto-completion box for type selection will receive suggestions such as album, alphabet, military alliance, etc.

Result Display: For each result we display its label, description, and an associated image if available (again we assume that image properties are preconfigured). We further assume that entity identifiers are dereferenceable IRIs, which we can use to offer a link to further information about the entity from the source dataset. We also present the available facets for the current results.

Ranking: We combine three forms of ranking: *frequency*, *relevance* and *centrality*. Frequency indicates the number of results generated by a particular selection. Relevance is particular to keyword-search and uses a TF–IDF style measure to indicate how well a given entity's label(s) and description(s) match a keyword. Centrality measures the importance of a node in the graph, where we use PageRank: we consider each triple $(s, p, o) \in G \cap (\mathbf{I} \times \mathbf{I} \times \mathbf{I})$ in the graph to be a directed edge $s \rightarrow o$ and then apply a standard PageRank algorithm to derive ranks for all nodes. Thereafter, we use these measures in the following way:

- Entities in result pages generated directly from a keyword selection are ranked according to a combination of TF–IDF and PageRank score.
- Entities in result pages generated directly from a type or facet selection are ranked purely according to PageRank score.
- Types suggested by auto-completion are ranked according to PageRank.[4] The count of entities in each type are also displayed.
- Properties displayed in the list of facets are ordered by frequency: the number of entities in the current results with some value for that property.
- Auto-completed facet values are ordered by PageRank.

---

[4] We originally implemented type ranks per frequency (number of results generated), but noted that certain popular types featured undesirably low in this ordering; for example, the type *country* has 207 entities, whereas *third-level administrative country subdivision* has 3792 entities. Hence we changed this ranking to consider PageRank.

Multilingual Support: Where language-tagged labels and descriptions are provided for entities in multiple languages (e.g., ``Denmark''@en, ``Dinamarca''@es), GRAFA can support multiple languages: the user can first select the desired language where search matches text from that language and where labels from that language are used to generate results views. The current online demo of GRAFA supports English and Spanish; language can be switched at any time.

4 Indexing Scheme

The GRAFA system is implemented on top of standard IR-style inverted indexes. More specifically, we base our indexing scheme on Apache Lucene (Core): a popular open source library offering various IR-style indexes, measures, etc.

```
SELECT ?p (COUNT(DISTINCT ?s) AS ?c)
WHERE {
  ?s wdt:P31 wd:Q5.         # human
  ?s wdt:P21 wd:Q6581097. # male
  ?s ?p ?o .
}
GROUP BY ?p
```

```
SELECT ?o (COUNT(?s) AS ?c)
WHERE {
  ?s wdt:P31 wd:Q5.         # human
  ?s wdt:P21 wd:Q6581097. # male
  ?s wdt:P106 ?o .          # occup
}
GROUP BY ?o
```

Fig. 1. Example SPARQL queries to compute facet properties and values over Wikidata; the left query would generate the facet properties and their frequencies for current results representing *male humans*; the right query would generate the facet values and their frequencies if the property *occupation* were then selected

Why not SPARQL? The first reason relates to the features supported, where GRAFA requires keyword search, prefix search (for auto-completion), and ranking primitives; though SPARQL vendors often provide keyword search functionality, these are non-standard and cannot be easily configured; additionally ranking measures based on, for example, PageRank would need to be implemented by reordering (not top-k). Furthermore, to generate, rank and display facet properties and values, our index needs to be able to cope with aggregate queries such as shown in Fig. 1; on the Wikidata Query Service running BLAZEGRAPH, the left query times out, while the right query takes in the order of 37 s. In a locally built index on the same version of Wikidata that we use in our evaluation, VIRTUOSO requires 4 min for the left query and 16 s for the right query. Hence we build custom indexes on top of Lucene, offering us the required features such as keyword search, prefix search, ranking, etc.

Indexing Schemes: We base our search on two (initial) inverted indexes:

- The *entity index* stores an entry (doc.) for each entity. Each entry stores fields to search entities by IRI, labels, description, type IRIs, property IRIs, and property–value pairs. The PageRank value of each entity is also stored.

- The *type index* stores an entry for each type. Each entry stores fields to search types by IRI and labels. The PageRank value of each type is also stored along with its frequency.

Note that types are also entities, and thus types will be included in both indexes. We use a separate types index to quickly find types according to an auto-complete prefix string; furthermore, the types index additionally contains the frequency of (number of entities associated with) a type. We highlight that properties are described by the entity index and are associated with labels, descriptions and defining properties (e.g., *sub-property-of*), etc.

Query Processing: For each type of interaction, we perform the following:

- Keyword selection ($\kappa(q)$): we perform a keyword search on the labels and descriptions fields of the entity index.
- Type selection ($\tau(t)$):
  - Given a user-specified prefix (e.g., "al*") generated by an auto-complete request, we perform a prefix search on label field of the type index and return a list of labels, frequencies and IRIs for matching types.
  - Given a type IRI t selected by the user from the previous auto-complete options, we perform a lookup on the type field of the entity index.
- Facet generation/selection ($\phi \wedge \zeta(p, o)$, where ϕ generates current results E):
  - For the current result set E, we must generate all possible facet properties: their IRIs, labels and frequency *with respect to* E. We thus iterate over E and generate the required information from the property field.
  - Once a p is selected, we must generate all possible facet values: their IRIs, labels, frequency and PageRanks. Let $\epsilon(p)$ denote a query to find all entities with some value for property p executed over the *property* field of the entity index. We thus generate and execute the conjunctive query $\phi \wedge \epsilon(p)$ to find all entities in E with property p, and from these results we generate the list of all pertinent values.
  - Once a (p, o) is selected, we execute the conjunctive query $\phi \wedge \zeta(p, o)$.

To generate the results for any page (for keyword, type or facet selection), the first step of facet generation must be applied to generate the next possible steps.

Performance: Lucene implements efficient intersection algorithms to apply conjunctions. Hence performance issues rather occur when large sets of results are present and the facet selection must find (only) the properties present in E and their frequency with respect to E. For example, given a query $\tau(human)$ in Wikidata, the above process would require scanning 3.6 million results and computing the frequencies of 358 properties. Next, when a property is selected to restrict E with, we may still have to scan many results to compute the available values for p in the set E (and their frequencies). For example, when we execute $\tau(human) \wedge \epsilon(occupation)$, we would now need to scan 3.3 million results to find the values of occupation. Hence the challenge for performance is not due to the difficulty of query processing, but rather the amount of results generated. Under initial experiments with the above indexing scheme, generating the facet properties for type *human* took 135 s; furthermore, such queries are very common as an entry point onto the data. Hence we require optimisations.

5   Materialisation Strategy

To address the aforementioned performance issues, we propose a selective materialisation strategy. This strategy enumerates, off-line, all queries of the form $\tau(t)[\wedge \zeta(p_1, o_1) \wedge \ldots \wedge \zeta(p_n, o_n)]$ that generate greater than or equal to a given threshold α of results. More specifically, the goal is to identify all queries generating a high number ($\geq \alpha$) of results, such as $\tau(human)$, or $\tau(human) \wedge \zeta(gender, male)$, or $\tau(human) \wedge \zeta(gender, male) \wedge \zeta(country, U.S.)$, etc.; the facet properties and values for these queries can then be materialised and indexed.

Choice of Threshold: When selecting α, we are faced with a classical time–space trade-off: we should select a value for α such that queries generating fewer than α results can be processed efficiently using the base indexes, while there are as few as possible queries generating α results to avoid exploding the index. The underlying hypothesis here is that such a value of α exists, which is non-trivial and requires empirical validation (as we will provide in Sect. 6). We say that this is non-trivial since a relatively low value of α can generate a huge number of queries: let $\pi_{\mathrm{PO}}(G) = \{(p, o) \mid \exists s : (s, p, o) \in G\}$ project the property–value facet pairs from G and let $\pi_{\mathrm{PO}}^*(G)$ denote $\pi_{\mathrm{PO}}(G)$ but removing pairs (p, o) where p is a type property. Recall that we denote by $T(G)$ the types of G. For $\alpha = 0$, we would have $|T(G)| \times 2^{|\pi_{\mathrm{PO}}^*(G)|}$ possible queries to contend with containing every combination of type with the powerset of $\pi_{\mathrm{PO}}^*(G)$. For $\alpha = 1$, we could still have the same number (if, e.g., G contains a single subject). More generally:

Lemma 1. *Let $\alpha \geq 1$. Given an RDF graph G with m triples, the total number of queries of the form $\tau(t)[\wedge \zeta(p_1, o_1) \wedge \ldots \wedge \zeta(p_n, o_n)]$ generating more than α results is bounded by the interval $[0, 2^{\lfloor \frac{m}{\alpha} \rfloor} - 1]$.*

Proof. If $|\pi_{\mathrm{S}}(G)| < \alpha$, then no query can generate more than α results, giving the lower bound. Towards the upper-bound, let $\pi_{\mathrm{PO}}^\alpha(G)$ denote the property–value pairs with more than α subjects and let $\Pi_{\mathrm{PO}}^\alpha(G) \subseteq 2^{\pi_{\mathrm{PO}}^\alpha(G)}$ denote all sets of such pairs that cooccur on more than α subjects; these are the queries we need to materialise. We now construct a worst-case G that maximises the value $|\Pi_{\mathrm{PO}}^\alpha(G)|$ with a budget of m triples. To do this, for each subject in G, we will assign the same set of (pairwise distinct) property–value pairs $\{(p_1, o_1), \ldots, (p_k, o_k)\}$. In this case, $|\Pi_{\mathrm{PO}}^\alpha(G)| = 2^k$, representing the powerset of the k property–value pairs. We then need to maximise k; given the inequality $k|\pi_{\mathrm{S}}(G)| \leq m$ for m the budget of triples, we thus need to minimise the number of subjects $|\pi_{\mathrm{S}}(G)|$. But we know that $|\pi_{\mathrm{S}}(G)| \geq \alpha$, otherwise no queries return more than α results; hence we should set $|\pi_{\mathrm{S}}(G)| = \alpha$, which gives us $k = \lfloor \frac{m}{\alpha} \rfloor$ and $|\Pi_{\mathrm{PO}}^\alpha(G)| = 2^{\lfloor \frac{m}{\alpha} \rfloor}$.

With respect to types, note that we can consider this as any other facet by, e.g., setting p_1 to a type property; the only modification required is to not consider the empty set in $\Pi_{\mathrm{PO}}^{\alpha}(G)$, which leads us to the upper bound $2^{\lfloor \frac{m}{\alpha} \rfloor} - 1$.   □

Algorithm: We outline the algorithm to compute the queries generating more than α results. Note that for brevity, we will consider type as a facet. Let $\sigma_{\mathrm{s}=x}(G) := \{(s,p,o) \in G \mid x = s\}$ select the triples in G whose subject is x. In order to compute $\Pi_{\mathrm{PO}}^{\alpha}(G)$ representing the set of all queries with at least α results, a naive algorithm would be to compute from each subject x the powerset of all its property–value pairs $2^{\pi_{\mathrm{PO}}(\sigma_{\mathrm{s}=x}(G))}$ containing at least one type property and then count these sets over all subjects, outputting those with a count of at least α. However, in a dataset such as Wikidata, some subjects have hundreds of property–value pairs, where the powerset for such a subject would be clearly unfeasible to materialise. Instead, we optimise for the fact that a property–value pair with fewer than α subjects can never appear in a conjunctive query with more than α subjects: we compute a restricted powerset $2^{\pi_{\mathrm{PO}}(\sigma_{\mathrm{s}=x}(G)) \cap \pi_{\mathrm{PO}}^{\alpha}(G)}$ that only considers individual (p,o) pairs on each subject x with at least α subjects in G. Thereafter, we can then count the number of subjects for each query and add those with more than α subjects to $\Pi_{\mathrm{PO}}^{\alpha}(G)$. The number of queries generated is still, of course, potentially exponential, and hence it will be important to select a relative high value of α to minimise the set $\pi_{\mathrm{PO}}^{\alpha}(G)$, and thus the exponent.

Indexing: For each query in $\Pi_{\mathrm{PO}}^{\alpha}(G)$ computed in the previous stage, we compute its result set offline, and from that set, we compute the set of facet properties, their frequencies, and the sets of their values. Thus we have precomputed the information needed to generate the results page of each such query (with an index lookup), and to facilitate explorations of the facets on that page.

Keyword Selections: Note that for $\kappa(q)$, given that the number of possible keyword queries q is not bounded, our materialisation approach is not applicable, where we rather simply restrict $\kappa(q)$ to return the top-α results.

6   Performance Evaluation

We now discuss the performance of indexing, materialisation and querying.

Data and Machine: We take the "truthy" dump of Wikidata from 2017/09/13, containing 1.77 billion triples and 74.1 million entities. However, given that we do not consider datatype values, nor labels and descriptions in other languages, the number of Wikidata triples used by GRAFA is 195 million (120 million (p, o) pairs; 75 million labels and descriptions in English and Spanish). The machine used for all experiments has 2× Intel Xeon 4-Core E5-2609 V3 CPUs (@1.9 GHz), 32 GB of RAM, and 2× 2 TB Seagate 7200 RPM 32 MB Cache SATA hard-disks (RAID-1). The code used is available online: https://github.com/joseignm/GraFa/.

Threshold Selection: The selection of the threshold α must find a balance: too high and queries just under the threshold will take too long to run; too low and the number of queries to materialise will explode exponentially. We choose three seconds as a reasonable worst-case response time, which from initial experiments suggested a value of $\alpha = 50,000$. To verify that this would not require materialising too many queries, we counted the subjects associated with each $(p,o) \in \pi_{PO}(G)$ and found that $\frac{149}{10,348,199} \approx 0.001\%$ of (p,o) pairs were associated with more than 50,000 subjects. Ultimately we materialise 141 queries.

Indexing Times: In Table 1, we provide the details of all indexing times. The initial PageRank computation takes 04:30 (hh:mm) and creating the base indexes requires 06:52. Computing the set $\Pi_{PO}^{\alpha}(G)$ for $\alpha =$50,000 took 04:16, while building an index of the properties and their frequency for each such query took 01:13. The most expensive step in the process is materialising the values of such properties, which took 107:18 (4.5 days), where, for each query, we need to build a list of all values for each property. This index of values contains 16,048 query–property keys in total (one for each facet property of a materialised query). An important question is then: is an index on values necessary or could it be optimised? Without indexing values, if a user selects a property on a materialised query with lots of results, where the majority of results have some value for that property, we may still require scanning all the results to generate the value list. For example, if the query is τ(:Human) (3.6 million results) and the user selects ϵ(:occupation) (3.3 million results), without an index for values, all people with some occupation must be scanned to generate all possible values for that property, which would again take minutes. However, some compromise may be possible to reduce this indexing time; one idea is to not materialise values for properties with a low frequency, where of the 358 properties associated with :Human for example, only 31 have more than α results; another idea is to index values for properties independent of the current query, thus potentially suggesting values that may lead to empty results (e.g., on a query for human males, suggesting *first lady* for *occupation*). For now, we simply accept the longer indexing time. On disk, the base index takes up 6 GB of space, the properties index requires 5 MB, while the values index requires 1 GB.

Table 1. Times of all index-creation steps

Process	Time (s)	Time (hh:mm)
Computing PageRank	15,595	004:20
Creating base indexes	24,756	006:52
Identifying queries to materialise	15,382	004:16
Indexing properties	4,356	001:13
Indexing values	386,304	107:18

Query Performance Testing: To test online query performance over these indexes, we created sequential queries simulating user sessions. Each session starts with $\tau(person)$, which offers a lot of results and facet properties; from this initial interaction, the index returns the top 50 ranked results and facet properties for all results. The session then randomly selects a property from the top 20 ordered by frequency $(\epsilon(p))$;[5] the system must then respond with the list of values for that property on the full result set. The session continues by selecting a random value $(\zeta(p,o))$; the system then generates the next results set and list of facet properties for that result set. This process is iterated until there is only one result or there is no further interaction possible, at which point the session terminates.

Query Performance Results: One thousand such sessions were executed. Figure 2 presents the response times for generating results pages with facet properties (τ and ζ queries), while Fig. 3 presents the response times for selecting the values for a property (ϵ queries). These figures show times in milliseconds plotted against the number of results generated (entities or values, resp.); note that the x-axis of Fig. 2 is presented in log-scale and the dashed vertical line indicates the selected value for the α-threshold. In the worst case, a query interaction takes approximately 3 s (for queries just below α), while value selection is possible in all cases under 500 ms. To the right of the α line, we see that materialised queries can be executed in under a second despite large result sizes; without materialisation, these queries took upwards of 2 min to process.

Data: Please note that we make evaluation data, queries, etc., available at https://github.com/joseignm/GraFa/tree/master/misc.

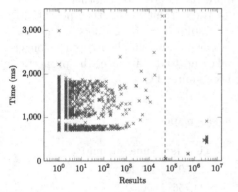

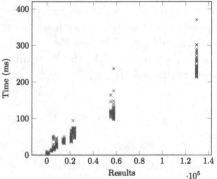

Fig. 2. Times to load result pages (τ, ζ)   **Fig. 3.** Times to load facet values (ϵ)

---

[5] We thus avoid the majority of facet properties with few results, selecting from those with the most results (and thus those more prominently displayed to users).

7   User Evaluation

While the previous section establishes performance results for indexing, materialisation and querying, we now present an initial usability study of the GRAFA system. For this, we implemented a prototype of a user interface as a Java servlet with Javascript enabling interactive client-side features, such as auto-completion. A demo (for Wikidata) is available at http://grafa.dcc.uchile.cl.

User Study Design: We chose a task-driven user study where we give participants ten questions in natural language; for this, we selected the questions and question text from the example queries provided for the Wikidata Query Service (selecting examples answerable as faceted navigations).[6] We list the question text provided to the user and the expected queries they should generate in Table 2; these reflect the SPARQL query and its description in the source.

User Study Baseline: In order to establish a baseline for the tasks, we selected the WIKIDATA QUERY HELPER (WQH) provided on the official Wikidata SPARQL Endpoint[7]; this interface first provides auto-completion on the labels of values and automatically proposes an associated property. For example, a user typing ``mal'' may be suggested *male organism, male,* etc.; upon selecting the latter, the property *sex or gender* is automatically selected, though it can be changed through another auto-completion dialogue. The user can add several property–value pairs in this manner. Suggestions generated through auto-completion are not restricted in a manner that assures non-empty results.

Participants and Instructions: We attained 11 volunteers (students of a Semantic Web course) for a study. Given the question text, we asked the volunteers to use either GRAFA or WQH (switching on every second question) to find the results

Table 2. User study tasks, with question text and expected query to be generated

Question text	Expected query
``Plays''	$\tau(plays)$
``Lakes in Cameroon''	$\tau(lake) \wedge \zeta(country, Cameroon)$
``Lighthouses in Norway''	$\tau(lighthouse) \wedge \zeta(country, Norway)$
``Popes''	$\tau(human) \wedge \zeta(position\ held, Pope)$
``Women born in Wales''	$\tau(human) \wedge \zeta(gender, female) \wedge \zeta(place\text{-}of\text{-}birth, Wales)$
``Papers about Wikidata''	$\tau(scientific\ article) \wedge \zeta(main\ subject, Wikidata)$
``Law & Order episodes''	$\tau(TV\ series\ episode) \wedge \zeta(series, Law\,\&\,Order)$
``Fictional characters from Marvel Universe''	$\tau(fictional\ character) \wedge \zeta(from\ fictional\ universe, Marvel\ Universe)$
``People dying by burning''	$\tau(human) \wedge \zeta(manner\ of\ death, death\ by\ burning)$
``Mosquito species''	$\tau(taxon) \wedge \zeta(parent\text{-}taxon, Culicidae) \wedge \zeta(taxon\text{-}rank, species)$

---

[6] https://wikidata.org/wiki/Wikidata:SPARQL_query_service/queries/examples; retr. 2018/04/05.

[7] https://query.wikidata.org/; retr. 2018/04/05.

and submit the URL, or click skip if they felt unable to find the results; the next task would then be loaded. Half of the participants began with GRAFA and the other half with WQH. They were not instructed on how to use either of the two systems. Afterwards they responded to a brief questionnaire.

User Task Results: We collected results for 55 tasks per system ($\frac{10 \times 11}{2}$). Of these, $\frac{23}{55} \approx 42\%$ were solved correctly in GRAFA, while $\frac{37}{55} \approx 67\%$ were solved correctly in WQH. This was unambiguously a negative result for GRAFA. Investigating the errors further, for GRAFA (32 errors), 10 involved users typing questions directly into the keyword-query text field rather than using type selection as intended; 3 involved selecting incorrect types/facets/values; 19 responses were skipped/left blank/invalid. On the other hand, for WQH (18 errors), 11 responses selected incorrect types/facets/values, while 7 were left blank. Through this study we found a variety of interface issues that we subsequently fixed. We additionally realised that users had a difficult time starting with a type selection; an example is ''Popes'' where users typed ''pope'' into the GRAFA type selection (rather than ''human'' or ''person''); on the other hand, in WQH, typing ''pope'' in the value selection suggested the value *Pope* and, upon selection, the correct property *position held*. On the other hand, in WQH, users sometimes selected the incorrect property, where for a query such as *Women born in Wales*, neither the value *woman* nor *Wales*, when selected, suggests the correct property.

User Questionnaire: After the task, we asked users to answer a brief questionnaire rating the responsiveness and usability of both systems on a Likert 1–7 scale; users rated GRAFA with a mean of 4.5/7 for usability and 4.7/7 for responsiveness; WQH had an analogous mean rating of 5.5/7 for usability and 6.0/7 for responsiveness. Again in this case WQH scored considerably higher than GRAFA. Regarding responsiveness, with subsequent investigation we found that the Javascript libraries for auto-completion were creating lag in the client browser, where we implemented smaller thresholds for suggestions.

Community Questionnaire: Based on the results of this user study, we fixed a number of interface issues in the system, blocking on the selection of a type or

Table 3. Responses to Wikidata community questionnaire

Statement	1	2	3	4	5	6	7	Mean
The system is useful	0	0	0	2	2	4	1	5.4/7
The system offers a novel way to query Wikidata	0	0	0	1	3	4	1	5.6/7
The system is usable	0	1	1	2	2	1	2	4.8/7
Knowledge of Wikidata is not required	0	1	1	2	3	1	1	4.6/7
Load-times did not affect interactivity	0	0	2	2	2	1	2	4.9/7
Ranking of results is intuitive	0	0	1	2	2	4	0	5.0/7
Ranking of facets is intuitive	0	0	3	1	2	2	1	4.7/7

value suggestion, separating type/keyword selection in the interface and so forth. We created a questionnaire that we sent to the Wikidata mailing list asking to try the GRAFA system and then answer a set of 12 questions, where we received nine responses. The results of the questionnaire are presented in Table 3, where most responses were moderately positive about the system. We further asked if they would use the system in future (YES|MAYBE|NO): 4 said YES, while 5 said MAYBE. We made some further improvements based on text comments received, such as to add placeholder examples in the text fields for auto-suggestions.

8   Conclusion

Motivated by the goal of providing users with a faceted interface over Wikidata – and the lack of current techniques and tools by which this could be achieved – in this paper, we have presented methods to enable faceted browsing over large-scale, diverse RDF graphs. A key contribution is our proposed materialisation strategy, which identifies facet queries that are good candidates for indexing. With this technique, worst-case response times drop from minutes to seconds at the cost of increased indexing time. To the best of our knowledge, GRAFA is the only faceted browsing system demonstrated at this level of scale while filtering suggestions that would lead to empty results. With the current system, the faceted browser could be updated for Wikidata on a weekly basis.

On the other hand, the results of our usability experiments were mixed: GRAFA was outperformed by the legacy WQH system in our task-driven user study. Some superficial issues were then fixed, such as blocking auto-complete fields until a selection is made. Though the results were more negative than hoped, we also drew more general conclusions, key amongst which is that, in a diverse graph like Wikidata, users unfamiliar with the dataset may struggle to select types, properties and values corresponding to their intent (e.g., is a *pope* a type or a value?; is *fictional character* a property or a type?). After some improvements to the system, a questionnaire issued to the Wikidata community generated moderately positive results regarding usefulness, novelty, usability, etc.

There are various directions in which this work could be continued. An important aspect for improvement is usability, where based on the aforementioned user study, we conclude that the system should offer more flexible selections; e.g., to automatically detect that *pope* is a value, not a type. The system could also be extended to support more expressive queries, such as ranges on datatype values, value selections, inverses, nested facets, and so forth. Other features – such as reasoning – would yield further challenges at the proposed scale. Furthermore, indexing time is currently prohibitive: investigating incremental indexing schemes would be an important practical contribution. Another important next step would be performing evaluations for other RDF datasets.

In conclusion, although there are various avenues for future work in terms of performance, expressiveness and usability, we hope that by enabling faceted browsing over RDF graphs at new levels of scale, GRAFA already makes a significant step towards making the Semantic Web more accessible to end users.

Acknowledgements. The work was supported by the Millennium Institute for Foundational Research on Data (IMFD) and by Fondecyt Grant No. 1181896.

References

1. Abel, F., Celik, I., Houben, G.-J., Siehndel, P.: Leveraging the semantics of tweets for adaptive faceted search on Twitter. In: Aroyo, L., et al. (eds.) ISWC 2011. LNCS, vol. 7031, pp. 1–17. Springer, Heidelberg (2011). https://doi.org/10.1007/978-3-642-25073-6_1
2. Araújo, S., Schwabe, D.: Explorator: a tool for exploring RDF data through direct manipulation. In: Workshop on Linked Data on the Web, LDOW (2009)
3. Arenas, M., Grau, B.C., Kharlamov, E., Marciuska, S., Zheleznyakov, D.: Faceted search over RDF-based knowledge graphs. J. Web Semant. **37–38**, 55–74 (2016)
4. Bast, H., Bäurle, F., Buchhold, B., Haußmann, E.: Easy access to the freebase dataset. In: International World Wide Web Conference, WWW, pp. 95–98 (2014)
5. Bast, H., Buchhold, B.: An index for efficient semantic full-text search. In: Conference on Information and Knowledge Management, CIKM, pp. 369–378 (2013)
6. Brunetti, J.M., González, R.G., Auer, S.: From overview to facets and pivoting for interactive exploration of semantic web data. IJSWIS **9**(1), 1–20 (2013)
7. Dadzie, A., Rowe, M.: Approaches to visualising linked data: a survey. Semant. Web **2**(2), 89–124 (2011)
8. Ferré, S.: Expressive and scalable query-based faceted search over SPARQL endpoints. In: Mika, P., et al. (eds.) ISWC 2014. LNCS, vol. 8797, pp. 438–453. Springer, Cham (2014). https://doi.org/10.1007/978-3-319-11915-1_28
9. Guo, Y., Pan, Z., Heflin, J.: LUBM: a benchmark for OWL knowledge base systems. J. Web Semant. **3**(2–3), 158–182 (2005)
10. Hahn, R., et al.: Faceted Wikipedia search. In: Abramowicz, W., Tolksdorf, R. (eds.) BIS 2010. LNBIP, vol. 47, pp. 1–11. Springer, Heidelberg (2010). https://doi.org/10.1007/978-3-642-12814-1_1
11. Harth, A.: VisiNav: a system for visual search and navigation on web data. J. Web Semant. **8**(4), 348–354 (2010)
12. Heath, T., Bizer, C.: Linked Data: Evolving the Web into a Global Data Space. Synthesis Lectures on the Semantic Web. Morgan & Claypool Publishers, San Rafael (2011)
13. Heim, P., Ertl, T., Ziegler, J.: Facet graphs: complex semantic querying made easy. In: Aroyo, L., et al. (eds.) ESWC 2010. LNCS, vol. 6088, pp. 288–302. Springer, Heidelberg (2010). https://doi.org/10.1007/978-3-642-13486-9_20
14. Heim, P., Ziegler, J., Lohmann, S.: gFacet: a browser for the web of data. In: International Workshop on Interacting with Multimedia Content in the Social Semantic Web, IMC-SSW (2008)
15. Hildebrand, M., van Ossenbruggen, J., Hardman, L.: /facet: a browser for heterogeneous semantic web repositories. In: Cruz, I., et al. (eds.) ISWC 2006. LNCS, vol. 4273, pp. 272–285. Springer, Heidelberg (2006). https://doi.org/10.1007/11926078_20
16. Hoffart, J., Suchanek, F.M., Berberich, K., Weikum, G.: YAGO2: a spatially and temporally enhanced knowledge base from Wikipedia. Artif. Intell. **194**, 28–61 (2013)
17. Kamdar, M.R., Zeginis, D., Hasnain, A., Decker, S., Deus, H.F.: ReVeaLD: a user-driven domain-specific interactive search platform for biomedical research. J. Biomed. Inform. **47**, 112–130 (2014)

18. Karger, D.R.: The semantic web and end users: what's wrong and how to fix it. IEEE Internet Comput. **18**(6), 64–70 (2014)
19. Lehmann, J., et al.: DBpedia - a large-scale, multilingual knowledge base extracted from Wikipedia. Semant. Web **6**(2), 167–195 (2015)
20. Mäkelä, E., Hyvönen, E., Saarela, S.: Ontogator—A semantic view-based search engine service for web applications. In: Cruz, I., et al. (eds.) ISWC 2006. LNCS, vol. 4273, pp. 847–860. Springer, Heidelberg (2006). https://doi.org/10.1007/11926078_61
21. Oren, E., Delbru, R., Decker, S.: Extending faceted navigation for RDF data. In: Cruz, I., et al. (eds.) ISWC 2006. LNCS, vol. 4273, pp. 559–572. Springer, Heidelberg (2006). https://doi.org/10.1007/11926078_40
22. Petzka, H., Stadler, C., Katsimpras, G., Haarmann, B., Lehmann, J.: Benchmarking faceted browsing capabilities of triplestores. In: International Conference on Semantic Systems, SEMANTICS, pp. 128–135 (2017)
23. schraefel, m.c., Wilson, M., Russell, A., Smith, D.A.: mSpace: improving information access to multimedia domains with multimodal exploratory search. Commun. ACM **49**(4), 47–49 (2006)
24. Sherkhonov, E., Cuenca Grau, B., Kharlamov, E., Kostylev, E.V.: Semantic faceted search with aggregation and recursion. In: d'Amato, C., et al. (eds.) ISWC 2017. LNCS, vol. 10587, pp. 594–610. Springer, Cham (2017). https://doi.org/10.1007/978-3-319-68288-4_35
25. Stadler, C., Martin, M., Auer, S.: Exploring the web of spatial data with facete. In: International World Wide Web Conference, WWW, pp. 175–178 (2014)
26. Tvarožek, M., Bieliková, M.: Generating exploratory search interfaces for the semantic web. In: Forbrig, P., Paternó, F., Mark Pejtersen, A. (eds.) HCIS 2010. IAICT, vol. 332, pp. 175–186. Springer, Heidelberg (2010). https://doi.org/10.1007/978-3-642-15231-3_18
27. Tzitzikas, Y., Bailly, N., Papadakos, P., Minadakis, N., Nikitakis, G.: Using preference-enriched faceted search for species identification. IJMSO **11**(3), 165–179 (2016)
28. Tzitzikas, Y., Manolis, N., Papadakos, P.: Faceted exploration of RDF/S datasets: a survey. J. Intell. Inf. Syst. **48**(2), 329–364 (2017)
29. Vrandecic, D., Krötzsch, M.: Wikidata: a free collaborative knowledgebase. Commun. ACM **57**(10), 78–85 (2014)
30. Wagner, A., Ladwig, G., Tran, T.: Browsing-oriented semantic faceted search. In: Hameurlain, A., Liddle, S.W., Schewe, K.-D., Zhou, X. (eds.) DEXA 2011. LNCS, vol. 6860, pp. 303–319. Springer, Heidelberg (2011). https://doi.org/10.1007/978-3-642-23088-2_22
31. Wang, H., et al.: Semplore: a scalable IR approach to search the web of data. J. Web Semant. **7**(3), 177–188 (2009)
32. Wei, B., Liu, J., Zheng, Q., Zhang, W., Fu, X., Feng, B.: A survey of faceted search. J. Web Eng. **12**(1&2), 41–64 (2013)
33. Yahav, T., Kalinsky, O., Mishali, O., Kimelfeld, B.: eLinda: explorer for linked data. In: International Conference on Extending Database Technology, EDBT, pp. 658–661 (2018)

Semantics and Validation
of Recursive SHACL

Julien Corman[1], Juan L. Reutter[2(✉)], and Ognjen Savković[1]

[1] Free University of Bozen-Bolzano, Bolzano, Italy
[2] PUC Chile and IMFD Chile, Santiago, Chile
jreutter@ing.puc.cl

Abstract. With the popularity of RDF as an independent data model came the need for specifying constraints on RDF graphs, and for mechanisms to detect violations of such constraints. One of the most promising schema languages for RDF is SHACL, a recent W3C recommendation. Unfortunately, the specification of SHACL leaves open the problem of validation against recursive constraints. This omission is important because SHACL by design favors constraints that reference other ones, which in practice may easily yield reference cycles.

In this paper, we propose a concise formal semantics for the so-called "core constraint components" of SHACL. This semantics handles arbitrary recursion, while being compliant with the current standard. Graph validation is based on the existence of an assignment of SHACL "shapes" to nodes in the graph under validation, stating which shapes are verified or violated, while verifying the targets of the validation process. We show in particular that the design of SHACL forces us to consider cases in which these assignments are partial, or, in other words, where the truth value of a constraint at some nodes of a graph may be left unknown.

Dealing with recursion also comes at a price, as validating an RDF graph against SHACL constraints is NP-hard in the size of the graph, and this lower bound still holds for constraints with stratified negation. Therefore we also propose a tractable approximation to the validation problem.

1 Introduction

The success of RDF was largely due the fact that it can be easily published and queried without bounding to a specific schema [4]. But RDF over time has turned into more than a simple data exchange format [2], and a key challenge for current RDF-based applications is checking quality (correctness and completeness) of a dataset. Several systems already provide facilities for RDF validation (see e.g. [12]), including commercial products.[1,2] This created a need for standardizing a declarative language for RDF constraints, and for formal mechanisms to detect and describe violations of such constraints.

---

[1] https://www.topquadrant.com/technology/shacl/.
[2] https://www.stardog.com/docs/.

© Springer Nature Switzerland AG 2018
D. Vrandečić et al. (Eds.): ISWC 2018, LNCS 11136, pp. 318–336, 2018.
https://doi.org/10.1007/978-3-030-00671-6_19

```
:NIAddressShape                 :PolentoneShape
   a sh:NodeShape ;                a sh:NodeShape ;
   sh:property [                   sh:targetClass :Polentone ;
      sh:path :telephone ;         sh:property [
      sh:maxCount 1                   sh:path :address ;
   ] ;                                sh:minCount 1 ;
   sh:property [                      sh:maxCount 1 ;
      sh:path :locatedIn ;            sh:node :NIAddressShape
      sh:minCount 1 ;              ] ;
      sh:maxCount 1 ;              sh:property [
      sh:value :NorthernItaly         sh:path :knows ;
   ] .                                sh:node :PolentoneShape
                                   ] .
```

Fig. 1. Two SHACL shapes, about Polentoni and addresses in Northern Italy

One of the most promising efforts in this direction is SHACL, or Shapes Constraint Language,[3] which has become a W3C recommendation in 2017. SHACL groups constraints in so-called "shapes" to be verified by certain nodes of the graph under validation, and such that shapes may reference each other.

Figure 1 presents two SHACL shapes. The leftmost, named :NIAddressShape, is meant to define valid addresses in Northern Italy, whereas the right one, named :PolentoneShape, defines northern Italians, stereotypically referred to as Polentoni.[4] A node v satisfying the first shape must verify two constraints: the first one states that there can be at most one successor of v via property :telephone. The second one states that there must be exactly one successor (sh:minCount 1 and sh:maxCount 1) of v via property :locatedIn, with value :NorthernItaly.

Validating an RDF graph against a set of shapes is based on the notion of "target nodes", which mandates for each shape which nodes have to conform to it. For instance, PolentoneShape contains the triple :PolentoneShape sh:targetClass :Polentone, stating that its targets are all instances of :Polentone in the graph under validation. But nodes may also have to conform to additional shapes, due to shape references. For instance, in Fig. 1, the shape to the right contains one (non-recursive) shape reference, to :NIAddressShape, stating that every node v conforming to :PolentoneShape must have exactly one :address, which must conform to :NIAddressShape, and one recursive reference, stating that each successor of v via :knows must conform to :PolentoneShape.

By *recursion*, we will always refer to such reference cycles, possibly n-ary (where shape s_1 references s_2, s_2 references s_3,.., s_n references s_1). Unfortunately, the semantics of graph validation with recursive shapes is left explicitly undefined in the SHACL specification: *"... the validation with recursive shapes is not defined in SHACL and is left to SHACL processor implementations. For example, SHACL processors may support recursion scenarios or produce a failure*

---

[3] https://www.w3.org/TR/shacl/.

[4] This example is borrowed from Peter Patel-Schneider: https://research.nuance.com/wp-content/uploads/2017/03/shacl.pdf.

when they detect recursion." The specification nonetheless expresses the expectation that validation of recursive shapes end up being defined in future work. Indeed, shapes references are a core feature of SHACL. Furthermore, in a Semantic Web context, where shapes are expected to be exchanged or reused, reference cycles may naturally appear, intentional or not. Finally, recursion may be viewed as one of the distinctive features of SHACL: without recursion, one ends up with a constraint language whose expressive power is essentially the same as SPARQL.

Another current limitation of the SHACL specification is the lack of a unified and concise formal semantics for the so-called "core constraint components" of the language. Instead, the specification provides a combination of SPARQL queries and textual definitions to characterize these operators. This may be sufficient for reading or writing SHACL constraints, but a more abstract underlying formalization is still missing, in order for instance to devise efficient constraint validation algorithms, identify computational bottlenecks, or to compare SHACL's expressivity with other languages.

Contributions. In this article, we propose a formal semantics for the core constraint components of SHACL, which is robust enough to handle arbitrary recursion, while being compliant with the current standard in the non-recursive case. It turns out that defining such a semantics is far from trivial, due essentially to the combination of three features of the language: recursion, arbitrary negation, and the target-based validation mechanism introduced above. One of the main difficulties is to define in a satisfactory way validation of shapes with so-called *non-stratified* constraints, where negation is used arbitrarily in reference cycles.

To do this, we base our semantic on the existence of a *partial* assignment of shapes to nodes that verifies both constraints and targets, i.e. intuitively a validation of nodes against shapes which may leave *undetermined* whether a given node verifies a shape or violates it. We show that this semantics has desirable formal properties, such as equivalence with classical validation in the presence of stratified constraints.

Recursion, however, comes at a cost, as we show that the problem of validating a graph is worst-case intractable in the size of the graph. Perhaps more surprisingly, we show that this property already holds for stratified constraints, and for a limited fragment of the language, without counting or path expressions. This observation leads us to propose a sound approximation, polynomial in the size of the graph, and whose worst-case execution time can be parameterized.

Organization. Section 2 discusses the problem of recursive SHACL constraints validation, with concrete examples. Then Sect. 3 defines a robust semantics for SHACL, together with a concise abstract syntax, and investigates its formal properties. Section 4 studies computational complexity of the graph validation problem under this semantics, and Sect. 5 proposes a sound approximation algorithm, in order to regain tractability (in the size of the graph under validation). Finally, Sect. 6 reviews alternative languages and formal semantics for graph constraints validation, with an emphasis on RDF.

An extended abstract of this paper has been accepted at the AMW workshop [9]. In addition, an appendix with detailed proofs and a translation from SHACL into our abstract syntax and conversely can be found at [8].

2   Validating a Graph Against SHACL Shapes

This section provides a brief overview of the constraint validation mechanism described in the SHACL specification, and discusses its extension to the case of recursive constraints. We focus here on the problem of deciding whether a graph is valid against a set of shapes. Therefore we purposely ignore the notion of "validation report" defined in the specification, and encourage the interested reader to consult the specification directly.

Checking whether a graph G is valid against a set S of shapes may be viewed as a two-step process. The first step consists in iterating over all shapes $s \in S$, and retrieve their respective target nodes in G. SHACL provides a dedicated language to describe the intended targets of a shape (e.g. the sh:targetClass property in Fig. 1), which is orthogonal to the language used to define constraints. Furthermore, this language has a limited expressivity, allowing all targets of shape s in G to be retrieved in $O(|G| \cdot \log |G|)$, before constraint validation.

```
:SemiPolentoneShape
   a sh:NodeShape ;
   sh:targetNode :Enrico ;
   sh:property [
      sh:path :address ;
      sh:maxCount 1 ;
      sh:node :NIAddressShape
   ] ;
   sh:not [
      sh:path :knows ;
      sh:node :PolentoneShape
   ] .
```

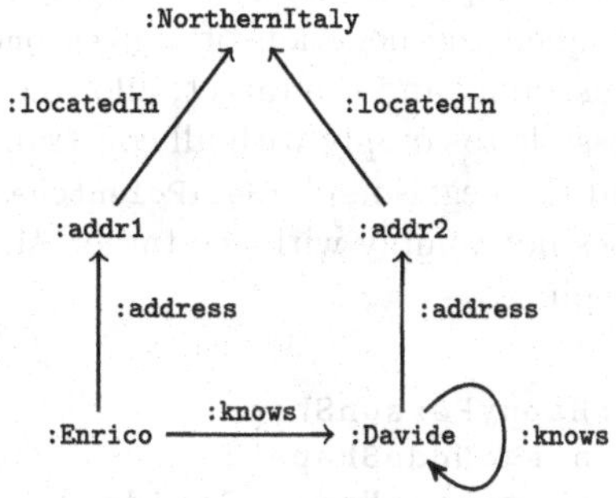

Fig. 2. A SHACL shapes for semi-Polentone, and a graph G to be validated against this shape, together with the shapes of Fig. 1

The second step consists in iterating over each target node v of each shape s, and check whether the node v satisfies s. This check can be represented as a call to a recursive function $validates(s, G, v)$. Some of the constraints for s may be validated by looking locally at the graph, i.e. at the IRI of v and its outgoing paths. But $validates(s, G, v)$ may also trigger a recursive call $validates(s', G, v')$, where s' is a shape referenced by s, and v' is a successor of v in G. It should be noted that v' does not need to be a target node of s'. In turn, $validates(s', G, v')$ may trigger another recursive call, etc.

Another important feature of SHACL is the possibility to declare negated constraints. For instance, shape SemiPolentoneShape in Fig. 2 uses sh:not to describe someone who knows at least one person who is *not* a Polentone (but still lives in Northern Italy). In this case, *validates*(SemiPolentoneShape, G, v) will succeed only if some successor of v via property :knows *violates* the constraints for :PolentoneShape.[5]

2.1 Recursive Constraints with Stratified Negation

Figures 1 and 2, considered together, illustrate a simple case of recursive constraint validation (i.e. constraints with reference cycles). The RDF triple :SemiPolentoneShape sh:targetNode :Enrico indicates that :Enrico is the unique target of shape :SemiPolentoneShape. This is also the only target to be validated in the graph.

To check if :Enrico validates :SemiPolentoneShape, the validation process described in the specification would call *validates*(SemiPolentoneShape, G, :Enrico), triggering an infinite sequence of recursive calls to *validates*(PolentoneShape, G, :Davide). Intuitively, the problems is that *validates* does not keep track of what has been validated (or violated) so far.

A classical solution to ground constraint evaluation in such cases is to define it w.r.t. an *assignment* of (positive and negated) shape labels to nodes. In this example, Enrico can be assigned :SemiPolentoneShape, and :Davide can be assigned the negation of :PolentoneShape. This assignment complies with the constraints and the target, allowing us to validate the graph. Alternatively, it is possible to comply with all constraints by assigning :PolentoneShape to :Davide, and the negation of :SemiPolentoneShape to :Enrico. But this latter assignment does not comply with the target, therefore it would not allow us to validate the graph.

```
: HappyPersonShape                      : NaivePolentoneShape
  a sh: NodeShape ;                       a sh: NodeShape ;
  sh: targetNode : Davide ;               sh: not [
  sh: or (                                  sh: path : knows ;
    [ sh: path : address ;                  sh: node :
      sh: minCount 1 ]                          NaivePolentoneShape
    [ sh: path : knows ;                  ] .
      sh: node :
          NaivePolentoneShape ]
  ) .
```

Fig. 3. Two SHACL shapes which illustrates the need for partial assignments

Several formal frameworks dealing with recursion (such as recursive Datalog [10]) have semantics based on a similar intuition. This notion of assignment is

---

[5] Constraints on node sucessors in SHACL are by default universally quantified. This is why sh:not here requires one successor violating :PolentoneShape to exist.

also used in [7] for ShEx, a constraint language for RDF very similar to SHACL. However, the semantics proposed in [7] would consider the graph of Fig. 2 as invalid, taking only one assignment into consideration, where :Davide is assigned :PolentoneShape, and therefore :Enrico cannot verify :SemiPolentoneShape. The semantics defined in [7] is also restricted to *stratified* constraints, i.e. constraints such that reference cycles have no reference in the scope of a negation (see Definition 8 further below).

2.2 Non-stratified Constraints

Extending assignment-based validation to the non-stratified case raises an interesting question, namely whether such an assignment should be *total*, i.e. assign each shape or its negation to each node of the graph. We illustrate this with validating the graph G of Fig. 2 against the two shapes of Fig. 3.

:Davide is the only target node, for shape :HappyPersonShape. This shape is validated iff :Davide has an address, or knows a naive polentone. Because :Davide has an address, a simple call to *validates*(HappyPersonShape, G, :Davide) would validate the graph. But a total assignment must also assign either :NaivePolentoneShape or its negation to :Davide. And this cannot be done in a consistent manner. If :NaivePolentoneShape is assigned, then :Davide does not verify the corresponding constraint; if the negation of :NaivePolentoneShape is assigned, then :Davide does not violate the constraint. Therefore a semantics based on total assignments would consider the graph invalid.

It should be emphasized that this example is not a limit case: the same problem appears for any (satisfiable) set of shapes containing a reference cycle (of any size), and such that an odd number of references in this cycle are in the scope of a negation. Therefore, if one wants to defines a robust semantics based on assignments for recursive SHACL, it should be based on *partial* assignments, leaving the possibility to assign neither a shape nor its negation to some nodes.

3 Formal Semantics for SHACL

This section provides a formal semantics for recursive SHACL. As explained above, constraint validation is based on *partial* assignment. This semantics *(i)* complies with the current semantics of SHACL for non-recursive constraints, *(ii)* supports arbitrary recursion and negation, and *(iii)* can handle simultaneous validation of multiple targets.

A set of shapes is validated iff *there exists* an assignment (called here *faithful*) complying with it. This is a key difference from query answering, or cautious reasoning in Datalog, interested in *certain answers*, i.e. holding for *all* valid assignments. For instance, in Fig. 2, some faithful assignments assign :PolentoneShape to :Davide, and some do not.

3.1   Notation

Like the SHACL specification, we borrow from SPARQL the notion of *property path*, which describes regular constraints holding over a path in a graph (for the syntax and semantics, we defer to the SPARQL standard [15]). Following [16], if r is a property path and G a graph, we denote with $r(G)$ the evaluation of r, which consists of all pairs (v, v') of nodes in G such that there is a path from v to v' satisfying r.

Similarly, if ψ is a SPARQL query, we denote with $\psi(G)$ the evaluation of ψ in G. Finally, we use $|X|$ to denote the size of structure X.

3.2   Abstract Syntax and Semantics for SHACL Constraints

Syntax. As usual, we find more convenient to work with a logical abstraction of the concrete SHACL language. Our abstraction uses a fragment of first order logic to simulate node shapes, and then unravels so-called SHACL "property shapes" as modal formulas over nodes. Like the SHACL specification, we make the unique name assumption, i.e. we assume that two blank nodes in an RDF graph cannot denote the same individual. We also abstract away from constraints on IRIs and literals (regular expression, datatype, value comparison, etc.), and use a simple constant I instead. Constraints are defined by the following grammar:

$$\phi ::= \top \mid s \mid I \mid \phi_1 \wedge \phi_2 \mid \neg\phi \mid \geq_n r.\phi \mid EQ(r_1, r_2)$$

where s is a shape name, I is an IRI, r is a property path, and $n \in \mathbb{N}^+$. As syntactic sugar, we use $\leq_n r.\phi$ for $\neg(\geq_{n+1} r.\phi)$, and $=_n r.\phi$ for $(\geq_n r.\phi) \wedge (\leq_n r.\phi)$.

Let $\mathcal{L}$ be the language defined by this grammar. A full operator-by-operator translation from SHACL core constraint components to $\mathcal{L}$ and conversely is provided in the online appendix [8] of this article. For non-recursive shape constraints, this is a correct translation, in the sense that a set of constraints in one language and its translation in the other language validate exactly the same graphs, given the same targets. Unfortunately, in the absence of formal semantics for SHACL, this claim cannot be formally proven, but is based on our understanding of the specification. We cannot claim that this also holds for recursive shapes though, because SHACL validation in this case is not defined.

Example 1. We illustrate the syntax with the example from Fig. 1. To express SHACL cardinality constraints (e.g. `sh:maxCount`), we use $\leq_1 r.\phi$, which means that a node can have at most 1 r-successor satisfying ϕ, or $=_1 r.\phi$ for exactly one. Then the constraints for `:NIAddressShape` (abbreviated here as $s_{\texttt{niaddr}}$) can be translated as:

$$(\leq_1 \texttt{telephone}.\top) \wedge (=_1 \texttt{locatedIn.NorthernItaly})$$

where $\top$ is true at every node. In the same way, we can translate the constraints for `:PolentoneShape` (abbreviated here as $s_{\texttt{pol}}$). Both $s_{\texttt{niaddr}}$ and $s_{\texttt{pol}}$ appear in the constraint for $s_{\mathtt{pol}}$. This mimics the SHACL syntax, where both shapes were mentioned:

$$(\leq_0 \mathtt{knows}.\neg s_{\mathtt{pol}}) \wedge (=_1 \mathtt{address}.s_{\mathtt{niaddr}})$$

Semantics. Because shape names may appear in constraint formulas, we define the inductive evaluation of a formula in terms of a node, a graph, and an assignment that mandates which shapes are true or false at each node.

Definition 1 (Assignment). *Let N be a set of shape names, and G a graph. An* assignment σ *for G and N is a total function mapping nodes in G to subsets of $N \cup \{\neg s \mid s \in N\}$, such that s and $\neg s$ cannot be both in $\sigma(v)$.*

Definition 2 (Total assignment). *A assignment σ for G and N is total if either $s \in \sigma(v)$ or $\neg s \in \sigma(v)$, for each node in G and $s \in N$.*

The evaluation $\llbracket \phi \rrbracket^{v,G,\sigma}$ of formula ϕ at node v in graph G given σ is defined in Table 1. In order to evaluate a formula given a partial assignment, we use a 3-valued logic, which, in addition to the usual 1 and 0 for true and false, uses 0.5 to represent an unknown truth value. But if assignments are required to be total, then this third value is not needed:

Observation 1. *Let σ be a total assignment for G and N, and ϕ a constraint formula using shape names in N. Then for each node v of G, either $\llbracket \phi \rrbracket^{v,G,\sigma} = 0$ or $\llbracket \phi \rrbracket^{v,G,\sigma} = 1$.*

The inductive definition of $\llbracket \phi \rrbracket^{v,G,\sigma}$ is standard, aside maybe for the operator $\geq_n r$. Intuitively, $\geq_n r.\phi$ evaluates to true iff at least n r-successors of v validate ϕ, whereas $\geq_n r.\phi$ evaluates to false iff the number of r-successors of v which do or could validate ϕ is strictly inferior to n. This allows the semantics to comply with SHACL cardinality constraints in the non-recursive case.

From SHACL Shapes to $\mathcal{L}$ Constraints. We model a shape as a triple $(s, \phi_s, \mathrm{target}_s)$, where s is a shape name, ϕ_s is a constraint in $\mathcal{L}$, and target_s is a (possibly empty) monadic query retrieving the target nodes of s. If S is a set of shapes, we assume that for each $(s, \phi_s, \mathrm{target}_s) \in S$, if s' appears in ϕ_s, then $(s', \phi_s', \mathrm{target}_s') \in S$. An assignment for G and S is an assignment for G and $\{s \mid (s, \phi_s, \mathrm{target}_s) \in S\}$. Abusing notation, we write "$s \in S$" instead of "$(s, \phi_s, \mathrm{target}_s) \in S$".

3.3 Validation

We finally have all components in place to define graph validation. Intuitively, a graph is valid against a set S of shapes if one can find an assignment σ for G and S complying with targets and constraints. We call such an assignment *faithful*, defined as follows:

Table 1. Inductive evaluation of constraint formula ϕ at node v in graph G given assignment σ

$$[\![\top]\!]^{v,G,\sigma} = 1$$
$$[\![\neg\phi]\!]^{v,G,\sigma} = 1 - [\![\phi]\!]^{v,G,\sigma}$$
$$[\![\phi_1 \wedge \phi_2]\!]^{v,G,\sigma} = \min\{[\![\phi_1]\!]^{v,G,\sigma}, [\![\phi_2]\!]^{v,G,\sigma}\}$$
$$[\![(r_1 = r_2)]\!]^{v,\sigma,G} = \begin{cases} 1, & \text{if } \{v' \mid (v,v') \in r_1(G)\} = \{v' \mid (v,v') \in r_2(G)\} \\ 0 & \text{otherwise} \end{cases}$$
$$[\![I]\!]^{v,\sigma,G} = \begin{cases} 1, & \text{if } v \text{ is the IRI } I, \\ 0 & \text{otherwise} \end{cases}$$
$$[\![s]\!]^{v,G,\sigma} = \begin{cases} 1, & \text{if } s \in \sigma(v) \\ 0, & \text{if } \neg s \in \sigma(v) \\ 0.5 & \text{otherwise} \end{cases}$$
$$[\![\geq_n r.\phi]\!]^{v,\sigma,G} = \begin{cases} 1, & \text{if } |\{v' \mid (v,v') \in r(G) \text{ and } [\![\phi]\!]^{v'G,\sigma} = 1\}| \geq n \\ 0, & \text{if } |\{v' \mid (v,v') \in r(G)\}| - \\ & \quad |\{v' \mid (v,v') \in r(G) \text{ and } [\![\phi]\!]^{v'G,\sigma} = 0\}| < n \\ 0.5 & \text{otherwise} \end{cases}$$

Definition 3 (Faithful Assignment). *A assignment σ for G and S is* faithful *iff $target_s(G) \subseteq \sigma(v)$ for each $(s, \phi_s, target_s) \in S$, and, for each node v in G:*

- *if $s \in \sigma(v)$, then $[\![\phi_s]\!]^{v,G,\sigma} = 1$*
- *if $\neg s \in \sigma(v)$, then $[\![\phi_s]\!]^{v,G,\sigma} = 0$*

Definition 4 (Validation). *A graph G is* valid *against a set S of shapes iff there is a faithful assignment σ for G and S.*

The (online) appendix provides a full translation from SHACL to sets of shapes and conversely, which preserves validation, provided the shapes are non-recursive (i.e. contain no reference cycle). Our notion of validation is more robust though, as it is also well-defined for recursive shapes. In Sect. 4, we study the complexity of the validation problem. But for now, we provide some insight on properties of this semantics.

3.4 Properties of Validation

We introduce some additional notation. First, $\Sigma^{G,S}$ will designate the set of all assignments for G and S. Then we define the "immediate evaluation" operator $\mathbf{T}^{G,S}$ for G and S (or simply $\mathbf{T}$ when obvious from the context). It takes an assignment σ, and returns the assignment $\mathbf{T}(\sigma)$ obtained by evaluating each ϕ_s at each node of G.

Definition 5 (Immediate evaluation operator T).
$\mathbf{T} : \Sigma^{G,S} \to \Sigma^{G,S}$ *is the function defined by*
$s \in (\mathbf{T}(\sigma))(v)$ *iff* $[\![\phi_s]\!]^{v,G,\sigma} = 1$, *and* $\neg s \in (\mathbf{T}(\sigma))(v)$ *iff* $s \in [\![\phi_s]\!]^{v,G,\sigma} = 0$

Finally, we define the preorder $\preceq$ over $\Sigma^{G,S}$ by:

Definition 6 (Preorder $\preceq$).
$\sigma_1 \preceq \sigma_2$ iff $\sigma_1(v) \subseteq \sigma_2(v)$ for each node v in G.

Validation Without Target. The SHACL specification states that a graph G is valid against a set S of shapes if no shape in s has target in G. From Definitions 3 and 4, this also (trivially) holds in the recursive case for our semantics. Somehow surprisingly, validation without target may fail for *total* assignments. For instance, there is no total faithful assignment for the graph of Fig. 2 and the set of shapes containing only shape :NaivePolentoneShape from Fig. 3.

A Stricter Notion of Faithfulness. From Definition 3, a faithful assignment σ is only required to assign s to a node v if ϕ_s is verified by v (given σ), and to assign $\neg s$ to v if ϕ_s is violated by v (given σ). But it is also possible to assign none of these two, even though v verifies of violates ϕ_s (given σ). This may seem counterintuitive, which leads to the following stricter notion of faithfulness:

Definition 7 (Strictly-faithful assignment). *A assignment σ for G and S is strictly faithful iff $target_s(G) \subseteq \sigma(v)$ for each $(s, \phi_s, target_s) \in S$, and, for each node v in G:*

- *if $s \in \sigma(v)$, then $[\![\phi_s]\!]^{v,G,\sigma} = 1$*
- *if $\neg s \in \sigma(v)$, then $[\![\phi_s]\!]^{v,G,\sigma} = 0$*
- *otherwise, $[\![\phi_s]\!]^{v,G,\sigma} = 0.5$.*

We also say that a graph G is strictly valid against a set of shapes S if there is a strictly faithful assignment for G and S.

For instance, there is only one strictly faithful assignment for the graph of in Fig. 2 and the two shapes of Fig. 3. It assigns $\neg$:HappyPersonShape to :addr1, because :addr1 violates the constraint for this shape. There are also several (non-strictly) faithful assignments, some of which assign neither :HappyPersonShape nor its negation to :addr1. So intuitively, non-strict validation allows some form of "lazy" constraint evaluation.

The operator $\mathbf{T}$ provides a more concise definition. Both faithful and strictly faithful assignments must comply with targets for G and S. But in addition, a faithful assignment σ must verify $\sigma \preceq \mathbf{T}(\sigma)$, whereas a strictly faithful assignment σ' must verify $\sigma' = \mathbf{T}(\sigma')$.

Interestingly, these two notions of validation coincide. To prove this, we first need a useful property, the monotonicity of $\mathbf{T}$ w.r.t $\preceq$:

Lemma 1 (monotonicity of T). *For any G, S and $\sigma_1, \sigma_2 \in \Sigma^{G,S}$: if $\sigma_1 \preceq \sigma_2$, then $\mathbf{T}(\sigma_1) \preceq \mathbf{T}(\sigma_2)$.*

We can now state the equivalence:

Proposition 1. *For any G and S, G is valid against S iff G is strictly valid against S.*

Proof (Sketch). The right direction is trivial, because a strictly faithful assignment is faithful. In the other direction, let σ_0 be a faithful assignment for G and S. Define $\Sigma' \subseteq \Sigma^{G,S}$ as all extensions of σ_0, i.e. $\sigma' \in \Sigma'$ iff $\sigma_0 \preceq \sigma'$. From Lemma 1, $\mathbf{T}(\sigma_0) \preceq \mathbf{T}(\sigma')$. And because σ_0 is faithful, $\sigma_0 \preceq \mathbf{T}(\sigma)$. Therefore $\sigma_0 \preceq \mathbf{T}(\sigma')$, i.e. $\mathbf{T}(\sigma') \in \Sigma'$.

Now consider the (meet) semi-lattice $\langle \Sigma', \preceq \rangle$ rooted in σ_0. We just showed that for each $\sigma' \in \Sigma'$, $\mathbf{T}(\sigma') \in \Sigma'$. In addition, from Lemma 1, $\mathbf{T}$ is monotone over $\langle \Sigma', \preceq \rangle$. So from a (weaker version of) the Knaster-Tarski Theorem, $\mathbf{T}$ admits a fixed-point σ_2 over Σ'. And because $\sigma_0 \preceq \sigma_2$, σ_2 complies with all targets for G and S. Therefore σ_2 is strictly faithful for G and S. $\square$

All We Need Is One Target. The following explains why the complexity results provided in Sect. 4 only consider graph validation with a single target node.

Proposition 2. *Given a graph G, set S of shapes and target nodes in G for each $s \in S$, one can construct in linear time a graph G' and set S' of shapes, such that G is valid against S iff G' is valid against S', and S' has a single target in G'.*

Proof. (Sketch). Let $s_1, .., s_n$ be the shapes in S, with respective targets $v_1^1, .., v_1^{m1}, .., v_n^1, .., v_n^{mn}$. Extend G with a fresh node v_0, and an edge (v_o, e_i^j, v_i^j) for each v_i^j, with e_i^j a fresh edge label. Then delete all target expressions in S, and extend S with a fresh shape s_0, with target node v_0, and constraint $\phi_{s_0} \doteq (\geq_1 e_1^{m1}.\top) \wedge .... \wedge (\geq_1 e_1^{mn}.\top)$. $\square$

3.5   Validation and Stratified Negation

Section 2.2 suggested that the need for partial assignments comes from constraints combining circular references with negation, called *non-stratified*. We now make this intuition more precise, showing that we can indeed focus solely on total assignments if the constraints are stratified.

To formalize this idea, we borrow the notion of stratification from Datalog [10] (assuming w.l.o.g that constraints do not contain two consecutive negation symbols).

Definition 8 (stratification). *A set S of shape definitions is stratified if there is a total function str: $S \to \mathbb{N}$ such that:*

- *If s_1 appears in ϕ_{s_2}, then $str(s_1) \leq str(s_2)$*
- *If s_1 appears in ϕ_{s_2} in the scope of a negation then $str(s_1) < str(s_2)$.*

It must be emphasized that the language $\mathcal{L}$ does not include $\leq_n r$ or $=_n r$. If these operators were included, then one would need to redefine the second condition accordingly, as $\leq_n r$ is a form of negation.

The following result confirms that a semantics based on total assignment is sufficient for stratified sets of shapes.

Proposition 3. *Let S be a stratified set of shapes and G a graph. Then there exists a faithful assignment for G and S iff there exists a total faithful assignment for G and S.*

Proof (Sketch). For the right direction, the proof is trivial. For the left direction, to simplify notation, we represent assignments as sets of positive and negative atoms. Let σ be a faithful assignment for G and S, and let $S_1, .., S_n$ be the strata of S, from lowest to highest. The proof constructs an extension σ' of σ, stratum by stratum, initialized with the empty set. For each stratum S_i (starting from S_0), σ' is extended in three steps. First, σ' is extended with σ reduced to atoms with shape names in S_i. Then $\mathbf{T}$ is applied to σ' recursively, until a fixed-point is reached. Finally, σ' is extended with each $s(v)$ such that v is a node in G, $s \in S_i$ and $\neg s(v) \notin \sigma'$. It can be shown by induction on i that this extension of σ' always exists, and complies with all constraints for shapes in $S_0, .., S_i$. So when i reaches n, the last extension of σ' is a total faithful assignment for G and S. $\square$

This result is important for computational reasons. It also implies that 3-valued validation is not easier than 2-valued validation, which may come as a surprise.

4 Complexity

We now study the computational complexity of the validation problem, defined as follows (full proofs are provided in the online appendix):

VALIDATION:
Input: Graph G, set S of shapes
Decide: G is valid against S

Based on Proposition 2, we focused on instances with one target node (for one shape in S). We also assume that this target node is already known. Table 2 summarizes our results. As is customary, since the size of G is likely to be orders of magnitude larger than the size of S, we also study the problems VALIDATION(S) and VALIDATION(G), for a fixed set S of shapes and fixed graph G, called *data complexity* and *constraint complexity* below.

We consider two fragments of the constraint language $\mathcal{L}$: *(i)* $\mathcal{L}_{\geq_1, \neg, \wedge}$ is the fragment defined by the grammar $\phi :: = \ \top \mid I \mid s \mid \phi_1 \wedge \phi_2 \mid \neg \phi \mid \geq_1 p.\phi$, where p is an IRI, and *(ii)* $\mathcal{L}_{\geq_n, \wedge, \vee, r, \mathrm{EQ}}$ is the fragment defined with $\phi :: = \ \top \mid I \mid s \mid \phi_1 \wedge \phi_2 \mid \phi_1 \vee \phi_2 \mid \geq_n r.\phi \mid \mathrm{EQ}(r_1, r_2)$, where r, r_1, r_2 are property paths and $\phi_1 \vee \phi_2$ is interpreted (as expected) as $\neg(\neg\phi_1 \wedge \neg\phi_2)$.

We start by showing an NP upper bound for combined complexity, based on guessing a witnessing faithful assignment. Then we show that this upper bound is tight, even for a fixed set of shapes (data complexity) using stratified negation and basic operators ($\geq_1$, $\neg$ and $\wedge$). We also show that this bound is tight for a fixed graph. Lastly, we show that allowing disjunction but disallowing negation otherwise is sufficient to regain tractability.

Table 2. Computational complexity of VALIDATION. -c stands for *complete*.

Fragment	Data	Constraint	Combined
$\mathcal{L}$ (= SHACL)	NP-c	NP-c	NP-c
Stratified $\mathcal{L}_{\geq 1, \neg, \wedge}$	NP-c	NP-c	NP-c
$\mathcal{L}_{\geq n, \wedge, \vee, r, EQ}$	in P	in P	P-c

Let us start with NP membership. First, all property paths present in S can be materialized in time polynomial in $|G| \cdot |S|$ before validation. In addition, by introducing fresh shape names, S can be transformed in polynomial time into an equivalent set S' of shapes, whose constraints contain at most one operator. Then assuming that we can guess a faithful assignment σ for G and S', we only to check σ is indeed faithful. To do so, it is sufficient to compute the value of $[\![\phi_s]\!]^{v, G, \sigma}$ for each node v in G and $s \in S'$, which is again polynomial in $|G| + |S|$, even with a binary encoding of cardinality constraints. Summing up, we have:

Proposition 4 (Combined – Upper Bound). VALIDATION *is in* NP.

Now for the lower bound, validation is already intractable in data complexity for stratified $\mathcal{L}_{\geq 1, \neg, \wedge}$. This may come as a surprise, considering that data complexity of ground fact entailment in stratified Datalog is in PTIME [10]. We show NP-hardness by a reduction from the satisfiability problem of a propositional circuit: there is a fixed set S of shapes such that every propositional circuit can be transformed (in linear time) into a graph, and this graph is valid against S iff the circuit is satisfiable.

Proposition 5 (Data – Lower Bound). *There is a stratified fixed set S of shapes in $\mathcal{L}_{\geq 1, \neg, \wedge}$ such that* VALIDATION*(S) is* NP*-hard.*

We also show that the problem is NP-hard in constraint complexity for the same fragment (with a reduction from SAT):

Proposition 6 (Constraint – Lower Bound). *There is a fixed graph G such that* VALIDATION*(G) is* NP*-hard, even if S is restricted to stratified sets of shapes in $\mathcal{L}_{\geq 1, \neg, \wedge}$.*

As a more optimistic result, validation is in PTIME if one allows disjunction as a native operator, but disallows negation otherwise. The proof relies on the (unique) minimal fixed-point σ of $\mathbf{T}$ w.r.t. $\preceq$, which can be computed in time polynomial in $|G| + |S|$. Let v_0 be the (unique) target node to validate, against shape s_0. If $\neg s_0 \in \sigma(v_0)$, then G is invalid. Otherwise, it can be shown that there must be an extension of σ (w.r.t. $\preceq$) which is faithful for G and S.

Proposition 7 (Combined – Upper Bound). VALIDATION *is in P for* $\mathcal{L}_{\geq n, \wedge, \vee, r, EQ}$.

Finally, we show PTIME hardness for a sub-fragment of $\mathcal{L}_{\geq n, \wedge, \vee, r, EQ}$ (without property paths and path equality), with a log-space reduction from the problem of evaluating a monotone boolean circuit.

Proposition 8 (Combined – Lower Bound). VALIDATION *is* P-hard *for* $\mathcal{L}_{\geq n, \wedge, \vee, r, EQ}$.

5   Approximation

The above intractability result for data complexity (Proposition 6), and even for a stratified set of shapes, is an important limitation. In order to alleviate this problem, we present in this section an approximation algorithm to decide whether a graph G is valid against a set S of shapes, with an integer parameter k. If k is bounded, then the algorithm is sound, and runs in time polynomial in $|G|$. If k is unbound, then the algorithm is sound and complete, but may run in time exponential in $|G|$. The approximation is sound in that the algorithm returns Valid (resp. Invalid) only if G is valid (resp. not valid) against S.

For readability, from Proposition 2, we focus on validation with a single target node v_0, for shape s_0. Algorithm 1 describes the procedure, composed of two steps. The first step intuitively computes an assignment σ_{minFix} matching all constraints enforced by the graph, regardless of the target. If the validity of G cannot be decided after this (polynomial) step, then σ_{minFix} is extended by assigning s_0 to v_0, and an attempt is made to propagate constraints from v_0 to its successors, in order for v_0 to satisfy ϕ_{s_0}.

Step 1: Minimal Fixed-Point. As a reminder from Sect. 3.3, we use $\Sigma^{G,S}$ to denote the set of all (possibly partial) assignments for G and S. The first step of the algorithm computes the minimal fixed-point σ_{minFix} of the operator $\mathbf{T}$ (see Definition 5) w.r.t. $\preceq$. Because $\langle \Sigma^{G,S}, \preceq \rangle$ is a semi-lattice and $\mathbf{T}$ is monotone w.r.t. $\preceq$ (Lemma 1), σ_{minFix} must exist and be unique. It can also be computed in time polynomial in $|G|$, initializing σ_{minFix} with the empty set, and then applying $\mathbf{T}$ to σ_{minFix} recursively, until a fixed-point is reached. This is performed by procedure COMPUTEMINFIX. If $s_0 \in \sigma_{\mathrm{minFix}}(v_0)$, then the graph is valid, Line 2. Furthermore, any strictly faithful assignment of for G and S must be a fixed-point of $\mathbf{T}$ (see Sect. 3.3), and therefore must extend σ_{minFix}. So from Proposition 1, If $\neg s_0 \in \sigma_{\mathrm{minFix}}(v_0)$, then the graph is invalid, Line 3.

Step 2: Breadth-First Search. The next step consists in searching for a faithful assignment, in a breadth-first fashion, starting from the target node v_0. We abuse notation and use set operators ($\cup, \in$, etc.) to describe the stack. Similarly, for brevity, we represent assignments interchangeably as functions or as sets of (positive and negative) atoms.

Each element of the stack (i.e. each "branch" of this exploration) is a tuple $\langle \sigma, \sigma^P, A, n \rangle$, where:

- σ is the current assignment being constructed, initialized with $\sigma_{\mathrm{minFix}} \cup \{s_0(v_0)\}$.
- $\sigma^P \preceq \sigma$ keeps track of shapes freshly assigned to a node during the previous expansion of σ. For any element of the stack, if σ^P is empty, then no constraint needs to be propagated in this branch, i.e. σ is a faithful assignment, and so the graph is validated, line 7.

- A is a set of atoms of the form $s(v)$, such that $s(v) \notin \sigma$ and $\neg s(v) \notin \sigma$,
- n is the current depth of the exploration, incremented each time σ is extended. When n reaches k, the size of the stack cannot be extended anymore, which triggers a call to REDUCE, line 11, to merge some of the current branches.

Line 8, function EXTEND computes each minimal extensions σ' of σ such that:

- If $s \in \sigma^P(v)$, then $[\![\phi_s]\!]^{v,G,\sigma'} = 1$,
- If $\neg s\sigma^P(v)$, then $[\![\phi_s]\!]^{v,G,\sigma'} = 0$, and
- if $s(v) \in A$, then $\{s, \neg s\} \cap \sigma(v) = \emptyset$.

It can be shown that each call to EXTEND can be executed in time $O(|G|^{|S|})$.

Finally, if the depth n of the exploration reaches k, line 11, then procedure REDUCE prevents the number of elements in the stack to increase. Line 18, function GETCLOSESTPAIR retrieves the two closest assignments σ_1 and σ_2 (in terms of edit distance) in the Stack. Then function GETCONFLICTS 20 retrieves the (possibly empty) set A of atoms which $\sigma_1(v)$ and $\sigma_2(v)$ disagree on, i.e. $s(v) \in A$ if both s and $\neg s$ are in $\sigma_1(v) \cup \sigma_2(v)$, and the procedure REPLACE sets each σ_i to $\sigma_i \setminus \{s(v), \neg s(v)\}$. After this step, either $\sigma_1 \preceq \sigma_2$ or $\sigma_2 \preceq \sigma_1$ must hold, and only the greater of the two (w.r.t $\preceq$) is retained (Line 23) and pushed in the stack.

The number of possible assignments is of $O(2^{|G|})$, but the number of assignments created by EXTEND is $O(|G|^{|S|})$. So if the parameter k is fixed, the reduced stack makes sure that the execution time is $O(|G|^{|S| \cdot k})$.

6   Related Work

Several schema languages have been proposed or implemented for RDF before SHACL, and some of them are closely associated to the design of SHACL. But first, it should be mentioned that RDF Schema (RDFS), contrary to what its name may suggest, is not a schema language in the classical sense, but is primarily used to infer implicit facts.

Among the proposals which do not relate (to our knowledge) to the genesis of SHACL, are proposals for RDF integrity constraints [1,13]. We have not explored a formal comparison between these formalisms and SHACL, but conjecture that they are incomparable with SHACL.

SPIN[6] allows the user to express constraints as SPARQL queries (natively, or using templates) and to declare targets for these constraints, similar to SHACL targets. SPIN became a W3C member submission in 2011, before being explicitly superseded by SHACL in 2017. Being based on SPARQL, it supports negation, but not full recursion.

ShEx has been actively developed since 2012 [6], as a dedicated constraint language for RDF, strongly inspired by XML schema languages. The first version of ShEx did support recursion, but no negation. A formal semantics was provided in [21], based on *regular bag expressions*. Recently, ShEx 2.0[7] incorporated

---

[6] http://spinrdf.org/.
[7] http://shex.io/shex-semantics/.

Algorithm 1. APPROXIMATION

Require: G', S, s_0, v_0, k
1: $\sigma_{\text{minFix}} \leftarrow$ COMPUTEMINFIX(G', S)
2: **if** $s_0 \in \sigma_{\text{minFix}}(v_0)$ **then return** Valid
3: **if** $\neg s_0 \in \sigma_{\text{minFix}}(v_0)$ **then return** Invalid
4: Stack $\leftarrow \langle \sigma_{\text{minFix}} \cup \{s_0(v_0)\}, \{s_0(v_0)\}, \{atoms(G', S)\}, 0 \rangle$
5: **while** NONEMPTY(Stack) **do**
6:     $\langle \sigma, \sigma^P, A, n \rangle \leftarrow$ POP(Stack)
7:     **if** $\sigma^P = \emptyset$ **then return** Valid
8:     **for all** $\sigma' \in$ EXTEND(σ, σ^P, A) **do**
9:        PUSH$(\mathcal{T}, \langle \sigma', \sigma' \setminus \sigma, A, n+1 \rangle)$
10:    **end for**
11:    **if** $n \geq k$ **then** Stack $\leftarrow$ REDUCE(Stack, $|\mathcal{T}|$)
12: **end while**
13: **return** Unknown
14:
15: **procedure** REDUCE(Stack, m)
16:    $i = 0$
17:    **while** $i \leq m$ **do**
18:       $(\langle \sigma_1, \sigma_1^P, A_1, n_1 \rangle, \langle \sigma_2, \sigma_2^P, A_2, n_2 \rangle) \leftarrow$ GETCLOSESTPAIR(Stack)
19:       Stack $\leftarrow$ Stack $\setminus \{\langle \sigma_1, \sigma_1^P, A_1, n_1 \rangle, \langle \sigma_2, \sigma_2^P, A_2, n_2 \rangle\}$
20:       $A \leftarrow$ GETCONFLICTS(σ_1, σ_2)
21:       $\sigma_1 \leftarrow$ REPLACE(σ_1, A)
22:       $\sigma_2 \leftarrow$ REPLACE(σ_2, A)
23:       $\sigma = \max\{\sigma_1, \sigma_2\}$
24:       PUSH(Stack, $\langle \sigma, \sigma_1^P \cup \sigma_2^P, A \cup A_1 \cup A_2, \max\{n_1, n_2\} \rangle$)
25:       $i \leftarrow i + 1$
26:    **end while**
27: **end procedure**

---

negation, and a formal semantics was provided in [7], together with a abstract language called *Shape Schemas*. As highlighted in [5], ShEx and SHACL have lot in common, and the semantics provided in [7] can be directly adapted to SHACL. This proposal is also similar to the one made in this article, in that validation is based on a *typing* verifying target and constraints, similar to our notion of shape assignment. A difference though is that the semantics proposed in [7] is restricted to stratified constraints. Moreover, the (unique) typing used in [7] to define validation favors the validation of shapes in the lowest stratum, so that the graph of Fig. 2 for instance would be considered invalid.

Another line of work is inspired by the Web Ontology Language (OWL), which is based on Description Logics (DLs) [3]. Like RDFS, OWL was not designed as a schema language, but adopts instead the *open-world assumption*, not well-suited to express constraints. Still, proposals have been made to reason with DLs understood as constraints: by introducing *auto-epistemic* operators [11], partitioning DL formulas into regular and constraint axioms [17,22], or reasoning with closed predicates [19]. This last approach was actually proposed as a semantic grounding for SHACL [18], reducing constraint validation to first-order satisfiability with closed binary predicates. But as illustrated with Example Fig. 3, this semantics does not behave well in the presence of targets and non-stratified constraints.

Recursion over negation has been traditionally studied in logical programming (see e.g. [10]), and answer-set programming (see [20] in the context of SPARQL), where stable model semantics (SMS) is one of the most prominent paradigms [14]. But SMS is based on so-called *minimal* models, whereas shape assignments may not be minimal. This makes encoding SHACL into logical programming non trivial, as suggested by complexity results: ground-fact entailment is data-tractable for stratified Datalog, in contrast to our semantics (see Proposition 5). A possible way to relate the two semantics, at least for the stratified case, is to reason about shape "complements" under SMS. Still, our preliminary investigations tend to show that this is not straightforward.

7   Conclusion

The article proposes an abstract syntax and formal semantics for SHACL core constraint components. This semantics is robust enough to handle constraints with arbitrary recursion, which can be expressed in SHACL, but whose validation is left explicitly open in the specification. One of our contributions is to highlight semantic issues related to non-stratified SHACL targets. To address such cases, we adopt a notion of *partial* assignment of (positive and negated) shapes to nodes, and define a semantics with desirable properties, such as monotonicity of forward-chaining, or equivalence with total assignments in the stratified case. We then show that the validation problem is NP-complete for any fragment with at least conjunction, negation and existential quantification, in the size of either graph or constraints, regardless of stratification. Therefore we propose a sound approximation algorithm, parameterized by an integer k, which guarantees termination in time polynomial in the size of the graph.

As a continuation, we plan to investigate other problems, such as (finite) satisfiability of a set of shapes, or SPARQL query containment in the presence of SHACL constraints. We also expect this formalization to be abstract enough to be extended to other constraint languages for graphs, such as ShEx, in order to handle arbitrary recursion.

Acknowledgements. This work was supported by the QUEST, ROBAST and OBATS projects at the Free University of Bozen-Bolzano, and the Millennium Institute for Foundational Research on Data (IMFD), Chile.

References

1. Akhtar, W., Cortés-Calabuig, Á., Paredaens, J.: Constraints in RDF. In: Schewe, K.-D., Thalheim, B. (eds.) SDKB 2010. LNCS, vol. 6834, pp. 23–39. Springer, Heidelberg (2011). https://doi.org/10.1007/978-3-642-23441-5_2
2. Arenas, M., Gutierrez, C., Pérez, J.: Foundations of RDF databases. In: Tessaris, S., et al. (eds.) Reasoning Web 2009. LNCS, vol. 5689, pp. 158–204. Springer, Heidelberg (2009). https://doi.org/10.1007/978-3-642-03754-2_4
3. Baader, F., Calvanese, D., McGuinness, D., Nardi, D., Patel-Schneider, P.: The Description Logic Handbook: Theory, Implementation and Applications. Cambridge University Press, Cambridge (2003)
4. Berners-Lee, T., Hendler, J., Lassila, O.: The semantic web. Sci. Am. **284**(5), 34–43 (2001)
5. Boneva, I.: Comparative expressiveness of ShEx and SHACL (early working draft) (2016)
6. Boneva, I., Labra-Gayo, J.E., Hym, S., Prud'hommeau, E.G., Solbrig, H.R., Staworko, S.: Validating RDF with shape expressions. CoRR, abs/1404.1270 (2014)
7. Boneva, I., Labra Gayo, J.E., Prud'hommeaux, E.G.: Semantics and validation of shapes schemas for RDF. In: d'Amato, C., et al. (eds.) ISWC 2017. LNCS, vol. 10587, pp. 104–120. Springer, Cham (2017). https://doi.org/10.1007/978-3-319-68288-4_7
8. Corman, J., Reutter, J.L., Savkovic, O.: Semantics and validation of recursive SHACL (extended version). Technical report KRDB18-1. KRDB Research Center, Free Univ. Bozen-Bolzano (2018)
9. Corman, J., Reutter, J.L., Savkovic, O.: Validating graph data against recursive constraints: a semantics for SHACL. AMW (2018, to appear)
10. Dantsin, E., Eiter, T., Gottlob, G., Voronkov, A.: Complexity and expressive power of logic programming. ACM Comput. Surv. **33**(3), 374–425 (2001)
11. Donini, F.M., Nardi, D., Rosati, R.: Description logics of minimal knowledge and negation as failure. ACM Trans. Comput. Log. (TOCL) **3**(2), 177–225 (2002)
12. Ekaputra, F.J., Lin, X.: SHACL4P: SHACL constraints validation within Protégé ontology editor. In: ICoDSE (2016)
13. Fischer, P.M., Lausen, G., Schätzle, A., Schmidt, M.: RDF constraint checking. In: Proceedings of the Workshops of the EDBT/ICDT 2015 Joint Conference, EDBT/ICDT, Brussels, Belgium, 27 March 2015, pp. 205–212 (2015)
14. Gelfond, M., Lifschitz, V.: The stable model semantics for logic programming, pp. 1070–1080. MIT Press (1988)
15. Harris, S., Seaborne, A., Prud'hommeaux, E.: SPARQL 1.1 query language. W3C Recomm. **21**(10) (2013)
16. Kostylev, E.V., Reutter, J.L., Romero, M., Vrgoč, D.: SPARQL with property paths. In: Arenas, M., et al. (eds.) ISWC 2015. LNCS, vol. 9366, pp. 3–18. Springer, Cham (2015). https://doi.org/10.1007/978-3-319-25007-6_1
17. Motik, B., Horrocks, I., Sattler, U.: Bridging the gap between OWL and relational databases. Web Semant.: Sci. Serv. Agents World Wide Web **7**(2), 74–89 (2009)
18. Patel-Schneider, P.F.: Using description logics for RDF constraint checking and closed-world recognition. In: AAAI (2015)
19. Patel-Schneider, P.F., Franconi, E.: Ontology constraints in incomplete and complete data. In: Cudré-Mauroux, P., et al. (eds.) ISWC 2012. LNCS, vol. 7649, pp. 444–459. Springer, Heidelberg (2012). https://doi.org/10.1007/978-3-642-35176-1_28

20. Polleres, A., Wallner, J.P.: On the relation between SPARQL1.1 and answer set programming. J. Appl. Non-Class. Log. **23**(1–2), 159–212 (2013)
21. Staworko, S., Boneva, I., Labra-Gayo, J.E., Hym, S., Prud'hommeaux, E.G., Solbrig, H.: Complexity and expressiveness of ShEx for RDF. In: ICDT (2015)
22. Tao, J., Sirin, E., Bao, J., McGuinness, D.L.: Integrity constraints in OWL. In: AAAI (2010)

Certain Answers for SPARQL with Blank Nodes

Daniel Hernández, Claudio Gutierrez, and Aidan Hogan(⊠)

IMFD Chile and Department of Computer Science, University of Chile,
Santiago, Chile
daniel@degu.cl, aidhog@gmail.com

Abstract. Blank nodes in RDF graphs can be used to represent values known to exist but whose identity remains unknown. A prominent example of such usage can be found in the Wikidata dataset where, e.g., the author of Beowulf is given as a blank node. However, while SPARQL considers blank nodes in a query as existentials, it treats blank nodes in RDF data more like constants. Running SPARQL queries over datasets with unknown values may thus lead to counter-intuitive results, which may make the standard SPARQL semantics unsuitable for datasets with existential blank nodes. We thus explore the feasibility of an alternative SPARQL semantics based on certain answers. In order to estimate the performance costs that would be associated with such a change in semantics for current implementations, we adapt and evaluate approximation techniques proposed in a relational database setting for a core fragment of SPARQL. To further understand the impact that such a change in semantics may have on query solutions, we analyse how this new semantics would affect the results of user queries over Wikidata.

1 Introduction

Incomplete information poses a major challenge for data management on the Web. Web data may be incomplete for a variety of reasons: the missing information may be unknown to those who created the dataset, it may be suppressed for privacy reasons, it may not yet have been added to the dataset, it may be a gap left after integrating other datasets, and so forth. A fundamental question for exploiting data on the Web is then how to define the semantics for (and process) queries over incomplete datasets. An important notion of incompleteness is that of *unknown values*. To take a literary example, we know that the poem "Beowulf" was written by *somebody*, but nobody knows who. One option is to simply omit authorship, but we would then lose valuable information that "Beowulf" has *some* author. Various works have then been proposed to deal with incomplete information [10,11,20], amongst which are recent works proposing query rewritings to provide sound answers over databases with unknown values [8,11,12,19]; such works have focused on relational settings.

On the other hand, there is a strong need for methods to deal with incomplete information and unknown values in a Semantic Web setting. In RDF, blank nodes

© Springer Nature Switzerland AG 2018
D. Vrandečić et al. (Eds.): ISWC 2018, LNCS 11136, pp. 337–353, 2018.
https://doi.org/10.1007/978-3-030-00671-6_20

can be used either to represent a resource for which no IRI is defined, or as an existential to represent an unknown value [15]. The RDF standard specifically defines blank nodes with an existential semantics [14]. However, SPARQL [13] does not follow the standard existential semantics of blank nodes in RDF data. As a result, when SPARQL queries are run over datasets where blank nodes are used as existentials to represent unknown values, the results can be unintuitive (or arguably *incorrect* [11]). We now provide such an example taken from the Wikidata [22] knowledge-base, which publishes data associated with Wikipedia as RDF and provides a public SPARQL query interface on the Web.

Example 1. Take the following RDF triples from Wikidata [22][1] and the following two SPARQL queries, which we will denote Q_1 (above) and Q_2 (below):

```
w:NicoleSimpson    w:killedBy  _:b .
w:NicoleSimpson    w:gender    w:Female .
w:ReevaSteenkamp   w:killedBy  w:OscarPistorius .
w:ReevaSteenkamp   w:gender    w:Female .
w:OJSimpson        w:gender    w:Male .
w:OscarPistorius   w:gender    w:Male .
```

```
SELECT ?victim WHERE
{ ?victim w:killedBy ?person .
  ?person w:gender w:Male . }
```

```
SELECT ?victim WHERE
{ ?victim w:killedBy ?person .
  FILTER NOT EXISTS
    { ?person w:gender w:Male . } }
```

In the data, the blank node (_:b) denotes that Nicole Brown Simpson (a victim of homicide) has a killer, but that her killer is unknown. For Q_1, SPARQL dictates a single solution – {?victim/w:ReevaSteenkamp} – which can be considered a *certain answer*; here the (unknown) killer of Nicole Simpson *may* have been male, but *uncertain answers* of this form are not returned. On the other hand, for Q_2, SPARQL again dictates a single solution – {?victim/w:NicoleSimpson} – but this answer is *uncertain* by the same reasoning: we do not know that the killer of Nicole Simpson was not male; _:b could refer to a male in the data. □

These two examples highlight a key problem in the current SPARQL semantics when dealing with unknown values. In the first query only certain answers are returned: answers that hold no matter to whom the unknown value(s) refer(s). In the second query uncertain answers are returned: answers that may or may not hold depending on whom the unknown value(s) refer(s) to. The key premise of this paper is to then ask: should users be offered a *choice* to only return certain answers? Is such a choice important? And what would be its cost?

Regarding cost, unfortunately, query evaluation under certain answer semantics incurs a significant computational overhead; for example, considering queries expressed in the standard relational algebra, if we consider only "complete databases" without unknown values, the data complexity of the standard query evaluation problem is AC^0; on the other hand, the analogous complexity with unknown values under certain answer semantics leaps to coNP-hardness [1].

One can thus hardly blame the design committee of the SPARQL language for choosing to initially overlook the issue of unknown values: not only would the complexity of query evaluation have escalated considerably under something

---

[1] While the example uses real data, for readability, we use fictitious IRIs. In reality, Wikidata uses internal identifiers, such as w:Q268018 to represent Nicole Simpson.

well-founded like a certain-answer semantics, the cost and complexity of correctly implementing the new standard would likewise have jumped considerably. Still, in this paper, we propose that it is time to revisit the issue of evaluating SPARQL queries in the presence of unknown values (blank nodes) in the data.

In terms of need, a recent study of blank nodes suggests that 66% of websites publishing RDF use blank nodes, with the most common use-cases being to represent resources for which no IRI has been defined (e.g., for representing RDF lists), or to represent unknown values [15]. The methods proposed in this paper specifically target datasets using blank nodes in the second sense; such datasets include Wikidata [22], as illustrated in Example 1.

In terms of cost, work by Guagliardo and Libkin [11] offers promising results in terms of the practical feasibility of *approximating* certain answers in the context of relational databases, returning only (but not all) such answers. Performance results suggest that such approximations have reasonable runtimes when compared with standard SQL evaluation. Furthermore, their implementation strategy is based on query rewriting over off-the-shelf query engines, obviating the need to build special-purpose engines, minimising implementation costs.

In this paper, we thus tackle the question: should users be given a choice of certain semantics for SPARQL? Along these lines, we adapt the methods of Guagliardo and Libkin [11] in order to propose and evaluate the first approach (to the best of our knowledge) that guarantees to return only certain answers for a fragment of SPARQL (capturing precisely the relational algebra) over RDF datasets with existential blank nodes, further developing a set of concrete rewriting strategies for the SPARQL setting. We evaluate our rewriting strategies for two popular SPARQL engines – Virtuoso and Fuseki – offering comparison of performance between base queries and rewritten queries (under various strategies), and a comparison of our SPARQL and previous SQL results [11]. We further conduct an analysis of Wikidata user queries to see if a certain answer semantics would really affect the answers over real-world queries and data, performing further experiments to ascertain costs in this setting.

2   Related Work

The conceptual problem of evaluating queries over data with unknown values is well-known in the relational database literature from as far back as the 70's, where Codd presented an extension of the relational model to allow nulls to encode unknown values [7]. Work on querying data with unknown values has continued throughout the decades, mostly in the context of relational databases (e.g., [17,19]), but also in various semi-structured settings (e.g., [2,9]).

A recent milestone has been the development of methods for *approximating* certain answers, where the current work is inspired by the proposal of Guagliardo and Libkin [11] of a method to (under-)approximate certain answers for SQL. Their goal is to trade completeness of certain answers against efficiency, ensuring that only (but not necessarily all) certain answers are returned in the presence of unknown values. We thus adapt their techniques for the SPARQL setting over RDF graphs with unknown values and propose SPARQL-specific rewriting strategies. In the experimental section, we provide a high-level comparison of our results for SPARQL with those published by Guagliardo and Libkin for SQL.

We are not the first work to explore a certain answer semantics for SPARQL. Ahmetaj et al. [2] define a certain answer semantics for SPARQL, but their focus is on supporting OWL 2 QL entailment for queries based on *well-designed patterns* [21], and in particular on complexity results for query evaluation, containment and equivalence. Arenas and Perez [5] also consider a certain answer semantics for SPARQL towards studying conditions for *monotonicity*: a semantic condition whereby answers will remain valid as further data is added to the system; as such, certain answers in their work are concerned with an *open world semantics*. However, determining if a query is *weakly monotonic* – i.e., monotonic disregarding unbound values – is undecidable. Hence Arenas and Ugarte [6] later propose a syntactic fragment of SPARQL that closely captures this notion of weak monotonicity. In contrast to such works, we maintain SPARQL's negation features [3] with a closed world semantics. In many contexts, users are interested in writing SPARQL queries with respect to what the present dataset does/does not contain; where, e.g., as we discuss later, many of the use-case queries for the Wikidata SPARQL service use non-monotonic features, such as difference.

To our knowledge, this is the first work to investigate a certain answer semantics for SPARQL considering existential blank nodes in RDF data.

3   Preliminaries

RDF and Incomplete Information: We assume three pairwise disjoint sets: $\mathbf{B}$ of blank nodes, $\mathbf{C}$ of constants, and $\mathbf{V}$ of attribute names (considered to represent variables when we later speak of queries). We will henceforth refer to blank nodes as simply "blanks". We denote blanks as $\perp_1, \perp_2, \perp_3$, etc.; constants with lowercase a, b, c, etc.; and attribute names with uppercase X, Y, Z, etc.

We define a tuple to be a function (a mapping) $\mu : \mathbf{V} \rightarrow (\mathbf{C} \cup \mathbf{B})$; for simplicity, we will denote the mapping μ with domain $\{X, Y\}$ where $\mu(X) = a$ and $\mu(Y) = b$ simply as $XY \mapsto (a, b)$, or even as the tuple (a, b) if the attributes X and Y are clear from the context. A relation R is determined by a set of tuples with the same domain (we need not consider empty relations).

An RDF graph G (or simply a graph) then corresponds to an instance of a single ternary relation with fixed attributes S, P, O, called *subject*, *predicate* and *object*, where P can only map to $\mathbf{C}$ (i.e. no blanks can occur as predicate), while S and O can map to values from $\mathbf{C} \cup \mathbf{B}$. Here $\mathbf{C}$ represents both IRIs and RDF literals; such a distinction of RDF constants is not exigent for us.

In this model, features for encoding incomplete information – specifically unknown values – are introduced by the semantics of blanks. Here the set $\mathbf{B}$ of blanks appearing in the data are interpreted as existentially-quantified variables in a manner consistent with the RDF semantics [14]; this is how, e.g., Wikidata uses blanks to represent unknown values. Blanks are synonymous with *marked nulls* in a relational setting: nulls that can appear in various locations.

Hence RDF graphs with blanks correspond to (ternary) relations with marked nulls, which have been called *naive tables*, *v-tables*, or *e-tables* by various authors (see [16]); here we will refer to them as *v-tables*. We say a *v*-table/graph is *complete/ground* when no nulls/blanks are used. With this correspondence established, we may henceforth use terms such as graph/table or blank/null interchangeably as best fits the particular context.

The semantics of a *v*-table is then defined as follows. A valuation is a mapping $v : \mathbf{C} \cup \mathbf{B} \to \mathbf{C}$ such that $v(c) = c$ for every $c \in \mathbf{C}$. Valuations are extended to tuples, relations and databases in the natural way. For instance, applying a valuation v to a graph G, denoted $v(G)$, results in the complete graph derived by replacing every blank $\perp_i$ in G by $v(\perp_i)$. The semantics of a *v*-table $[\![R]\!]$ is then given as the ground relations $\{v(R) : v \text{ is a valuation}\}$.

A Core SPARQL Algebra: We now define an algebra for the fragment of SPARQL considered, focusing on set semantics. We first define queries for ground graphs where unknown values will be treated later: A query is a combination of the following algebraic operations: selection (σ_θ), renaming ($\rho_{X/Y}$), projection ($\pi_{\bar{X}}$), natural join ($\bowtie$), union ($\cup$) and difference ($\setminus$). Attribute names ($\mathbf{V}$) are then synonymous with variables in SPARQL. The condition θ of a selection $\sigma_\theta(R)$ is a Boolean combination ($\wedge, \vee, \neg$) of terms of the form $X = Y$ or $X = c$, where X and Y refer to the attributes of R and $c \in \mathbf{C}$. We refer to this fragment as "SRPJUD" capturing the initials of the operations allowed; this fragment corresponds directly to the relational algebra but will be applied to graphs in a SPARQL setting. The correspondence between SRPJUD operators and syntactic SPARQL features is shown in Table 1. Note however that union and difference in SRPJUD follow the relational algebra in that – unlike standard SPARQL – $R \cup S$ and $R \setminus S$ assume that the relations R and S have the same attributes. Thus, SRPJUD does not support generating unbound variables through UNION as supported by SPARQL for attributes not in both R and S. However the difference operator in SRPJUD and the *outer-difference* operator in SPARQL can be mutually expressed using other SRPJUD operators. Taking difference, $R \setminus S$ is a particular case of the outer difference $R - S$ when the attributes of R and S are the same. Conversely, letting $\bar{X}$ denote the set of common attributes of R and S, the outer-difference $R - S$ can be expressed as (1) $R \bowtie (\pi_{\bar{X}}(R) \setminus \pi_{\bar{X}}(S))$ when $\bar{X}$ is non-empty or (2) R when $\bar{X}$ is empty.

Finally, given a query Q expressed in SRPJUD and a graph G, we write $Q(G)$ to denote the result of evaluating Q over the graph G following standard conventions for the relational algebra. Our focus will then be on answering queries in the core SRPJUD fragment over graphs with unknown values. We leave support for the following features of SPARQL for future work: (1) Bag semantics. (2) SPARQL unbounds as created by either SPARQL UNION or OPT. (3) Other features such as property paths, solution modifiers, aggregations, other filter expressions, named graphs, etc. These latter SPARQL features (e.g., aggregation) can, however, be defined *syntactically* on top of the core SRPJUD algebra.

Table 1. Mapping between SPARQL and SRPJUD

$\{\ X\ p\ Y\ \} \leftrightarrow \rho_{S/X}(\rho_{O/Y}(\pi_{S,O}(\sigma_{P=p}(G))))$	
$P\ .\ Q \leftrightarrow P \bowtie Q$	SELECT $\bar{X}$ WHERE $P \leftrightarrow \pi_{\bar{X}}(P)$
P MINUS $Q \leftrightarrow P - Q$	SELECT $(X$ AS $Y)$ WHERE $P \leftrightarrow \rho_{X/Y}(P)$
P UNION $Q \leftrightarrow P \cup Q$	P FILTER $\theta \leftrightarrow \sigma_\theta(P)$

Certain and Possible Answers: We now define a certain- and possible-answer semantics for SPARQL where unknown values are present in the RDF data in the form of blanks. Let Q be a query and G be a graph with blanks. Then a widely used definition of *certain answers* are the answers μ of $Q(G)$ such that $\mu \in Q(v(G))$ for every valuation v. Another more general definition of certain answers – first defined by Lipski [20] and called *certain answers with nulls* by Libkin [18] – states that μ is a certain answer of $Q(G)$ iff $v(\mu) \in Q(v(G))$ for every valuation v; this semantics allows for returning unknown values in answers.

Example 2. Consider an RDF graph G with $\{(a,b,c),(a,d,\perp_1)\}$ and a query $\pi_{\{P,O\}}(G)$. Under *certain answer semantics,* $\{(b,c)\}$ is returned. Under *certain answer semantics with nulls,* $\{(b,c),(d,\perp_1)\}$ is returned; here $\perp_1$ is interpreted as stating that for all valuations, there exists *some* constant there.   □

A complementary notion is that of a *possible answer*: a tuple μ is a possible answer of $Q(G)$ if there exists a valuation v such that $v(\mu) \in Q(v(G))$.

Example 3. Consider again the graph G from Example 2 but instead consider a query $\pi_{\{S,O\}}(\sigma_{P=b}(G)) \setminus \pi_{\{S,O\}}(\sigma_{P=d}(G))$. Under both certain answer semantics an empty result will be returned since there is a valuation μ such that $\mu(\perp_1) = c$. Here (a,c) will be considered a *possible* rather than a certain answer.   □

Given a query Q and an RDF graph G, then we write cert(Q,G) and poss(Q,D) to denote respectively the sets of certain and possible answers, defined as follows:

$$\text{cert}(Q,G) = \bigcap \{\mu \mid v(\mu) \in Q(v(D)) \text{ for all valuations } v\},$$

$$\text{poss}(Q,G) = \bigcup \{\mu \mid v(\mu) \in Q(v(G)) \text{ for all valuations } v\}.$$

Note that the former definition captures certain answers *with nulls,* which we use here; also, note that cert$(Q,G) \subseteq$ poss(Q,G): certain answers are also possible.

4   Approximating Certain Answers

The problem of query evaluation under certain answers is coNP-hard (data complexity); without unknown values the analogous problem is in AC^0. Likewise the definition of the semantics does not directly suggest a practical query answering procedure. Hence in this section we explore an algebra that allows for approximating certain/possible answers based on the notion of maybe tables.

4.1   Unification

We first define the notion of *unification*, which joins tuples with unknown values. We say that μ_1 and μ_2 *unify*, denoted $\mu_1 \Uparrow \mu_2$, iff for every common attribute X that they share, it holds that $\mu_1(X) = \mu_2(X)$ or $\mu_1(X) \in \mathbf{B}$ or $\mu_2(X) \in \mathbf{B}$; in other words, $\mu_1 \Uparrow \mu_2$ holds iff there is a valuation v such that $v(\mu_1(X)) = v(\mu_2(X))$ for every common attribute X. The unification of two tuples $(\mu_1 {}^\frown \mu_2)(X)$ is defined as $\mu_1(X)$ if $\mu_2(X)$ is $\perp$, or $\mu_2(X)$ otherwise. Unification allows to extend the standard operators join, semijoin and anti-semijoin to include the semantics of nulls by replacing the concept of joinable tuples and joins of tuples by the notion of unifiable tuples and unifications of tuples:

$$P \bowtie_{\Uparrow} Q = \{\mu_1 {}^\frown \mu_2 \mid \mu_1 \in P, \mu_2 \in Q, \text{ and } \mu_1 \Uparrow \mu_2\},$$
$$P \ltimes_{\Uparrow} Q = \{\mu_1 \in P \mid \exists \mu_2 \in Q : \mu_1 \Uparrow \mu_2\},$$
$$P \overline{\ltimes}_{\Uparrow} Q = \{\mu_1 \in P \mid \nexists \mu_2 \in Q : \mu_1 \Uparrow \mu_2\}.$$

Such operators will be essential to defining an approximation of certain answers, but they do not appear in SRPJUD and cannot be expressed in this fragment since it does not contain any means to distinguish blanks from constants. Hence to rewrite a SRPJUD query, we need a built-in predicate of the form $bk(X)$ in the target algebra, which evaluates to true for a tuple μ if $\mu(X) \in \mathbf{B}$, or false otherwise. We can now represent $P \ltimes_{\Uparrow} Q$ as $\pi_{\bar{X}}(\sigma_{\theta_{\Uparrow}}(P \bowtie \rho_{\bar{X}/\bar{X}'}(Q)))$ and $P \overline{\ltimes}_{\Uparrow} Q$ as $P - (P \ltimes_{\Uparrow} Q)$, where $\bar{X}$ denotes the attributes/variables of P, $\bar{X}'$ denotes fresh variables, and $\theta_{\Uparrow}$ will be rewritten to $(X = Y \vee bk(X) \vee bk(X'))$.

Translating $P \bowtie_{\Uparrow} Q$ to SRPJUD is more difficult. One option is to use the *active domain*: the set of all possible values to which blanks can be evaluated [17]; however, this would cause obvious practical problems. Hence we will rather use two non-SRPJUD features of SPARQL to implement unification: the ternary conditional operator, which allows for returning one of two values based on a condition (denoted IF$(\cdot, \cdot, \cdot)$ in SPARQL); and the bind operator, which can bind a new value to the relation (denoted BIND$(\cdot, \cdot)$ in SPARQL). From these, we derive a new operator if$_{\theta,X,Y,Z}(\mu)$, which returns $\mu \cup \{Z \mapsto (\mu(X))\}$ if $\mu \models \theta$, or $\mu \cup \{Z \mapsto (\mu(Y))\}$ otherwise. The operator thus creates a new attribute Z and assigns it the value of X if the condition θ is true, otherwise it assigns it the value of Y. We call SRPJUD extended with these unification operators SRPJUD$_{\Uparrow}$. Returning to $P \bowtie_{\Uparrow} Q$, we can first apply a Cartesian product and then unify the results with the ternary conditional operator. More formally, assume that P and Q contain one shared attribute X. We can now express $P \bowtie_{\Uparrow} Q$ as $\rho_{X''/X}(\pi_{X'',\bar{Y}}(\text{if}_{bk(X'),X,X',X''}(\sigma_{\theta_{\Uparrow}}(P \bowtie \rho_{X/X'}(Q)))))$, where $\bar{Y}$ denotes the non-shared attributes of P and Q and $\theta_{\Uparrow}$ is as before. In this case, $P \bowtie \rho_{X/X'}(Q)$ denotes a Cartesian product since there are no shared attributes. This process extends naturally to performing unifications over multiple attributes[2].

---

[2] Note that given two blank nodes on either side, this approach chooses the left blank node arbitrarily and drops the other. This may lead to losing certain answers, which we accept as part of the under-approximation.

4.2   Approximations

We wish to under-approximate certain answers to guarantee that all answers returned are certain while maximising the certain answers returned. But if we consider under-approximating results to a query $P - Q$, intuitively for P we must under-approximate certain answers, while for Q we should *over*-approximate *possible* answers to Q to ensure we remove everything from P that might match under *some* valuation in Q. Note that Q might itself be a query of the form $R - S$; etc. Hence, to under-approximate certain answers for SRPJUD, we need a way to over-approximate possible answers [11]: given a query Q in SRPJUD, we will rewrite it to a pair of queries $(Q^+, Q^?)$ in SRPJUD$_\Uparrow$, under-approximating certain answers and over-approximating possible answers for Q, respectively.

The first operator we define is the selection operator, where we must take care of inequalities involving blanks; we adopt the rewriting proposed by Guagliardo and Libkin [11], and shown by them to have good performance.

Definition 1. *We define the translation of a SRPJUD query Q to a pair of approximation queries $(Q^+, Q^?)$ in SRPJUD$_\Uparrow$ recursively as follows:*

$$G^+ = G, \qquad\qquad\qquad G^? = G,$$
$$(P \cup Q)^+ = P^+ \cup Q^+, \qquad\qquad (P \cup Q)^? = P^? \cup Q^?,$$
$$(P \bowtie Q)^+ = P^+ \bowtie Q^+, \qquad\qquad (P \bowtie Q)^? = P^? \bowtie_\Uparrow Q^?,$$
$$(P - Q)^+ = P^+ \overline{\bowtie}_\Uparrow Q^?, \qquad\qquad (P - Q)^? = P^? - Q^+,$$
$$(\sigma_\theta(P))^+ = \sigma_{\theta^*}(P^+), \qquad\qquad (\sigma_\theta(P))^? = \sigma_{\neg(\neg\theta)^*}(P^?),$$
$$(\pi_{\bar{X}}(P))^+ = \pi_{\bar{X}}(P^+), \qquad\qquad (\pi_{\bar{X}}(P))^? = \pi_{\bar{X}}(P^?),$$
$$(\rho_{X/Y}(P))^+ = \rho_{X/Y}(P^+), \qquad\qquad (\rho_{X/Y}(P))^? = \rho_{X/Y}(P^?),$$

where θ^* denotes the translation defined inductively as follows, noting that X and Y are attributes and a is some constant:

$$(X = Y)^* = (X = Y), \qquad (X \neq Y)^* = (X \neq Y) \wedge \neg\,\mathrm{bk}(X) \wedge \neg\,\mathrm{bk}(Y),$$
$$(X = a)^* = (X = a), \qquad (X \neq a)^* = (X \neq a) \wedge \neg\,\mathrm{bk}(X),$$
$$(\theta_1 \vee \theta_2)^* = \theta_1^* \vee \theta_2^*, \qquad (\theta_1 \wedge \theta_2)^* = \theta_1^* \wedge \theta_2^*.$$

□

4.3   Relation to Certain/Possible Answers

To state formally the relation of certain/possible answers with the corresponding approximation queries given in Definition 1, we require a notion of a subset of answers under unification. Importantly, the following definition is used to ensure that any tuple that unifies with a possible answer (e.g., $(\bot_1, \bot_1)$) will unify with an answer in the over-approximation (e.g., $(\bot_1, \bot_2)$).

Definition 2. *Given P and Q, we state that $P \subseteq_\Uparrow Q$ iff for each tuple $\mu \in P$, there exists $\mu' \in Q$ such that $\nu(\mu') = \mu$ for some valuation ν.*   □

Lemma 1. *Let Q be a SRPJUD query and let $(Q^+, Q^?)$ be the approximation queries for Q as defined in Definition 1. Then, for any RDF graph G, it holds that $Q^+(G) \subseteq \text{cert}(Q, G)$ and $Q^?(G) \supseteq_\Uparrow \text{poss}(Q, G)$.*

Proof. Follows from induction on the structure of the query, following similar techniques as used for Lemmas 1 and 2 of [11].   □

Computing exact certain/possible answers has a high complexity, where Definition 1 directly leads to a rewriting strategy for approximating certain/possible answers. For example, to under-approximate the certain-answers of a SRPJUD query Q, we can rewrite it to the SRPJUD$_\Uparrow$ $Q^?$ and execute that query; furthermore, evaluating queries in SRPJUD$_\Uparrow$ remains tractable in data complexity per the class of base queries SRPJUD (and unlike computing exact certain answers).

5   SPARQL Rewriting Strategies

We now explore alternatives in SPARQL to express the rewriting of Definition 1. All such alternatives are equivalent; in practice however, these strategies can exhibit major performance variations when applied over SPARQL query engines.

The base case in the SPARQL translation is G, which refers to a ternary relation with fixed attributes S, P and O. The basic unit of querying in SPARQL is a *triple pattern*, e.g., XpY ($X \in \mathbf{V}$, $Y \in \mathbf{V}$, $p \in \mathbf{C}$). In RDF, the P attribute cannot take blanks, and hence we do not need to consider unification on that attribute directly. A *basic graph pattern* Q in SPARQL is a join over triple patterns $T_1 \Join \cdots \Join T_k$ where each T_i ($1 \leq i \leq k$) is a triple pattern.

The most complex case to consider is the difference operator $P - Q$, where certain answers are under-approximated by the unification anti-semijoin $P^+ \overline{\Join}_\Uparrow Q^?$ (where $Q^?$ is itself over-approximated). The direct application of the translation rules produces complex queries that can be rewritten to a "friendlier" form for SPARQL engines, as now described. First, given a difference $P - Q$ we say that X is a *correlated* attribute of the difference if X is shared by P and Q. In the following we will assume that $P - Q$ is a difference with at least a correlated variable and that Q is a basic graph pattern.

CNF/DNF Rewritings: In the difference $P - Q$, let $Q = T_1 \Join T_2$ (a common case). The base translation evaluating the required unification in Q is then given as $(T_1 \Join T_2)^? = \beta_{\bar{X}, \bar{X}_1, \bar{X}_2}(\sigma_{\Theta_\Uparrow}(U_1 \Join U_2))$ where U_1 and U_2 are the respective results of replacing shared variables ($\bar{X}$) in $T_1 \Join T_2$ by fresh variables (denoted $\bar{X}_1$ and $\bar{X}_2$), where $\Theta_\Uparrow$ is a conjunction of the standard unifiable condition applied to each pair of renamed variables X_1 and X_2 for X (i.e., $\bigwedge_{X \in \bar{X}}(X_1 = X_2 \vee \text{bk}(X_1) \vee \text{bk}(X_2))$), and where, the operator $\beta_{\bar{X}, \bar{X}_1, \bar{X}_2}$ extends the solution for each $X \in \bar{X}$ using the function if$_{\text{bk}(X_2), X_1, X_2, X}(\cdot)$. These definitions then extend naturally (but verbosely) to the case where Q is $T_1 \Join_\Uparrow \ldots \Join_\Uparrow T_k$. This implies taking the Cartesian product of all triple patterns, filtering by a conjunction of unification conditions $\sigma_{\theta_\Uparrow}$, and then selecting constants over blanks.

The aforementioned unification condition $\Theta_{\Uparrow}$ is in conjunctive normal form (CNF): $\theta_1 \wedge \cdots \wedge \theta_n$ where for $1 \leq i \leq n$, each term θ_i is a disjunctive clause. An alternative solution is to rewrite the unification condition to its equivalent disjunctive normal form (DNF) $\phi_1 \vee \cdots \vee \phi_m$ per a standard conversion. The result is potentially exponential in size; though this does not affect the data complexity, it may have a significant effect on performance in practice. However, this DNF conversion leads to further rewritings that may lead to better performance. First, we can express disjunctions using union ($\cup$) or using disjunctive ($\vee$) selection conditions. Second, since this expression falls on the right-hand side of an anti-semijoin operator, we can also express it as a sequence of such operators. Thus, for the translation of $(P - Q)^+$ into $P^+ \overline{\ltimes}_{\Uparrow} Q^?$, we can consider:

$$P^+ \overline{\ltimes}_{\Uparrow} Q^? = P^+ \overline{\ltimes}_{\Uparrow} \sigma_{\bigwedge_{1 \leq j \leq m} \theta_j}(Q'), \qquad \text{(CNF)}$$

$$P^+ \overline{\ltimes}_{\Uparrow} Q^? = P^+ \overline{\ltimes}_{\Uparrow} \sigma_{\bigvee_{1 \leq j \leq m} \phi_j}(Q'), \qquad \text{(DNF}_1\text{)}$$

$$P^+ \overline{\ltimes}_{\Uparrow} Q^? = P^+ \overline{\ltimes}_{\Uparrow} \bigcup_{1 \leq j \leq m} \sigma_{\phi_j}(Q'), \qquad \text{(DNF}_2\text{)}$$

$$P^+ \overline{\ltimes}_{\Uparrow} Q^? = P^+ \overline{\ltimes}_{\Uparrow} \sigma_{\phi_1}(Q') \ldots \overline{\ltimes}_{\Uparrow} \sigma_{\phi_m}(Q'). \qquad \text{(DNF}_3\text{)}$$

where Q' denotes the rewriting of join variables $\bar{X}$ in Q to produce Cartesian products on all join patterns and the subsequent application of $\beta_{\bar{X}, \bar{X}_1, \ldots, \bar{X}_k}$ to perform unification over those variables. Note, however, that in the cases of DNF_2 and DNF_3, some terms in the disjunction will *not* require a Cartesian product; for example, when we rewrite $P - (T_1 \bowtie T_2)$ to DNF, a disjunctive term on the right of the anti-semijoin will be $(T_1 \bowtie T_2)$ itself (the others will cover the case that join variables in T_1 or T_2 are bound to blanks). This suggests that these options *may* be more efficient despite a *potential* exponential blow-up.

Removing Explicit Unification: Given a base query of the form $P - Q$, if the join variables of Q do not correlate with P, we do not need to perform unification on them. Consider a query $Xpa - (XpY \bowtie Ypb)$. This can be rewritten to $Xpa \overline{\ltimes}_{\Uparrow} (\text{if } _{\text{bk}(Y_2), Y_1, Y_2, Y}(\sigma_{\theta_{\Uparrow}}(XpY_1 \bowtie Y_2pb)))$. However since Y does not appear on the left of the difference, we can simplify to $Xpa \overline{\ltimes}_{\Uparrow} (\sigma_{\theta_{\Uparrow}}(XpY_1 \bowtie Y_2pb))$.

Converting Anti-semijoins to Difference: Given a base query of the form $P - Q$, we can consider cases where the correlating variable(s) of P and Q may or may not yield blanks on either side. In particular, if Q returns a tuple with blanks for all correlating variables, then the entire difference $P - Q$ must be empty. On the other hand, if P returns a tuple with blanks for all correlating variables and Q is non-empty, then that tuple is removed from P. Finally, in cases where we know that the correlating variable(s) of P and Q cannot yield blanks[3], we can convert the anti-semijoin to standard difference. These ideas yield possible optimisations when we know more about which attributes can yield nulls.

---

[3] In standard relational settings, this might be if the correlating variables is a primary key of a table, for example. In RDF, we may detect such a case for subjects or objects of a given property that do not give blanks in a given dataset, for example.

Options for Difference: The SPARQL standard provides several ways for expressing difference. Here we consider two: the operators MINUS and FILTER NOT EXISTS (FNE). The SPARQL standard states that solutions of $(P$ MINUS $Q)$ are the solutions μ_1 of P such that there does not exist a solution μ_2 of Q where $\text{dom}(\mu_1) \cap \text{dom}(\nu_2)$ is not empty and μ_1 is joinable with μ_2. On the other hand, the solutions of P FNE Q are all solutions μ_1 of P such that there does not exist any solution μ_2 for $\mu_1(Q)$, where $\mu_1(Q)$ denotes the result of substituting in Q each variable X in $\text{dom}(\mu_1)$ by $\mu_1(X)$. If $P - Q$ has at least one correlated variable, then P MINUS Q and P FNE Q are equivalent and can be interchanged.

6 Evaluation

Our evaluation presents an initial cost–benefit analysis of a certain answer semantics for SPARQL by addressing the following research questions: RQ1: How do the proposed SPARQL query rewriting strategies compare in terms of performance with the base query, with themselves, with similar results in an SQL setting, and for different SPARQL implementations? RQ2: Does a certain answer semantics significantly change query results in a real-world setting?

6.1 Evaluation Setting

In this section, we describe the SPARQL query engines selected, the machines and configurations used, as well as the datasets and queries. Supporting material can be found online: https://users.dcc.uchile.cl/~dhernand/revisiting-blanks.

Engines and Machines: The query rewriting strategy allows certain answers to be approximated on current SPARQL implementations. We test with two popular engines, with the added benefit of being able to cross-check that the solutions generated by both produce the same answers: Virtuoso (v.7.2.4.2) and Fuseki (v.2.6.0). The machine used is an AMD Opteron Processor 4122, 24 GB of RAM, and a single 240 GB Kingston SUV400S SSD disk; Virtuoso is set with NumberOfBuffers $= 1360000$ and with MaxDirtyBuffers $= 1000000$; Fuseki is initialised with 12 GB of Java heap space.

Rewriting Strategies: We consider various strategies: [B|CNF|DNF$_{1,\ldots,3}$] where B denotes base queries, CNF queries in conjunctive normal form, and DNF queries in disjunctive normal form; we denote these variations as Γ in the following. $[\Gamma^{\nexists} \mid \Gamma^-]$ These queries use either FNE $(\nexists)$ or MINUS $(-)$ in SPARQL. $[\Gamma|\Gamma^*]$ Rather than use isBlank to check if a node is blank or not, in case an engine cannot form an index lookup to satisfy such a condition, we also try adding a triple $(X, a, :Blank)$ to the data for each blank X and a triple pattern to check for that triple in the query (denoted Γ^*); this does not apply to base queries. In total, this leads to 18 possible combinations. Rather than present results for all, we will highlight certain configurations in the results.

6.2   TPC–H Experiments

To address RQ1, we follow the experimental design of Guagliardo and Libkin [11] who provide experiments for PostgreSQL using the TPC-H benchmark. Their results compare the performance of approximations for certain answers with respect to four queries with negations. For this, they modified the TPC-H generator to produce nulls in non-primary-key columns with varying probabilities (1–5%) to generate more/less unknown values. They also use scale factors of 1, 3, 6, and 10, corresponding to PostgreSQL databases of size 1 GB, 3 GB, 6 GB and 10 GB, respectively. We follow their setting as closely as possible to facilitate comparison later. We wrote a conversion tool (similar to the Direct Mapping [4]) to represent TPC-H data as RDF, and convert the TPC-H SQL queries to SPARQL.

Unifications: We first evaluate the proposed rewriting strategies of unifications in the difference operator for SPARQL. The base format of the queries used is $P - (Q \bowtie R)$ where each P, Q, and R is a triple pattern. We then generate between 1,000 and 10,000 triples matching each triple pattern to perform tests at various scales. For the data matching the join variable on Q and R, we generate blanks with a rate of 1, 2, 4 and 8%. These experiments allow us to estimate the costs of unifications in difference without other query operators interfering.

Figure 1 presents performance results. For clarity, we present only a selection of configurations: CNF is equivalent to DNF_1 in this case and we only show the aforementioned $[\cdot^{\#}/\cdot^*]$ variations for the base query and DNF_3 (other variations performed analogously). The first row pertains to Virtuoso while the second pertains to Fuseki. All eight sub-plots are presented with log–log axes (base 10) at the same scale permitting direct comparison across plots (comparing horizontally across engines and comparing vertically across blank rates). The y-axis maximum represents a timeout of 25 min (reached in some cases by Fuseki).

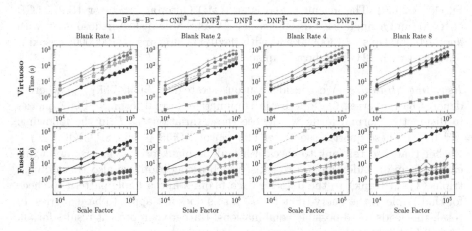

Fig. 1. Unification results for Virtuoso and Fuseki, varying scales and blank rates

(RQ1) The performance of the rewritten queries is (as could be expected) worse than the two base queries for all blank rates, scale factors and engines. In the case of Virtuoso, the base queries generally run in under one second; however, the fastest rewritten queries take at least a second and there is at least an order of magnitude difference between the base query and the fastest rewritten query. Looking at Fuseki, the fastest base query is slower than Virtuoso, but does generally tend to execute within one second (except at the larger scales). However, we see a number of rewriting strategies in the case of Fuseki where the difference is within half-an-order of magnitude of the fastest base case. Otherwise, we see that the choice of strategy is generally not sensitive to the blank rates considered (i.e., lines generally maintain the same ordering across plots), nor is it sensitive to scale (i.e., lines do not generally cross within plots).

Queries: The previous experiments looked at "atomic" unifications. We now run the four TPC-H queries used by Guagliardo and Libkin [11] considering a blank rate of 5%, four scale factors, and two engines. We employ a timeout of 10 min. We also choose one base query ($B^{\#}$) to be compared against the rewritten queries for approximating certain answers. Fuseki repeatedly times out for these experiments hence here we rather focus on the results of Virtuoso.

In Table 2, we present a comparison of the performance results for Virtuoso's fastest rewritten query and the results as presented by Guagliardo and Libkin [11]. More specifically, for a blank rate of 5%, the table shows the range of relative performance between the base query and the best rewritten query execution for that query; since Guagliardo and Libkin do not present absolute runtimes, our comparison is limited to relative performance. Note that due to differences in how SPARQL and SQL treat inequalities over nulls/blanks, Q_3 did not need rewriting for Virtuoso. For Q_2 in PostgreSQL, the actual results drop below the presented numeric precision, returning almost instantaneously for PostgreSQL once a null is found (which confirms that the results are empty).

(RQ1) We see that for Q_1, Virtuoso performs better in relative performance than PostgreSQL, for Q_2 PostgreSQL performs (much) better, for Q_3 there is little difference, while for Q_4 Virtuoso initially performs better than PostgreSQL but then at $\mathbf{SF \geq 3}$, Virtuoso begins to throw an error stating that an internal limit of 2097151 results has been reached (we could not resolve this). Aside from this latter issue, these results show that Virtuoso with our rewriting strategies is competitive with PostgreSQL under SQL-based rewritings for relative performance between base and rewritten queries. Furthermore, unlike in the previous experiments, we observe that in the case of Q_1 and Q_2, Virtuoso is now sometimes faster for the rewritten queries than the base queries: by removing uncertain answers, the number of intermediary solutions to be processed is reduced.

(RQ2) We observe three of the four base queries returning uncertain answers in SPARQL that do not hold under some valuations: for Q_1, 59% of answers are uncertain; for Q_2, all answers are uncertain; whilst for Q_4, 7% of answers are uncertain; we further highlight that these results are present for a blank rate of 5%. These results suggest that for queries with negation, evaluation under standard SPARQL semantics may in some cases return a significant ratio of uncertain/unsound answers even for modest levels of blanks in the dataset; this is to be expected given that, e.g., even a single blank tuple returned from the right-side of a difference can render all results uncertain (as per Q_2).

Table 2. Ranges of average relative performance for scale factor (SF) 1, 3, 6 and 10 on a fixed blank rate of 5%.

Q.	SF = 1	SF = 3	SF = 6	SF = 10
Virtuoso				
Q_1	0.95–0.96	0.95–0.96	0.97–0.99	0.94–0.95
Q_2	0.76–1.07	0.73–0.99	0.89–1.06	0.55–0.77
Q_3	1.00–1.00	1.00–1.00	1.00–1.00	1.00–1.00
Q_4	1.55–1.56	error	error	error
PostgreSQL (G&L [11])				
Q_1	1.01–1.03	0.99–1.01	0.98–1.01	1.00–1.02
Q_2	0.00–0.00	0.00–0.00	0.00–0.00	0.00–0.00
Q_3	1.01–1.04	1.01–1.04	0.99–1.02	1.00–1.06
Q_4	1.75–1.86	1.80–1.93	2.05–2.25	3.54–3.89

Table 3. Numbers of Wikidata use-case queries (from a total of 446) that could be affected by a certain answer semantics

Feature	A	B	C	D
MINUS	13	9	9	2
FILTER NOT EXISTS	23	15	10	1
OPTIONAL w/!BOUND	5	1	0	0
!=	7	5	3	0
Total	47	29	21	3

6.3 Wikidata Survey

Since the previous experiments are based on a synthetic benchmark converted from a relational setting, we performed an analysis of the user-contributed SPARQL queries on the Wikidata Query Service, which offers a more native Semantic Web setting[4]. As previously described, Wikidata uses blanks to represent unknown values; our goal now is to determine whether or not a choice of certain answer semantics could impact a current, real-world setting.

(RQ2) We first inspect the 446 queries to see which could potentially be affected by a certain answer semantics. We provide a summary in Table 3 according to the query features that may cause uncertain answers, with columns helping to indicate why queries with such features do not give uncertain answers in this context: **A** applies no assumptions, counting queries using the pertinent feature; **B** counts the queries that could still give uncertain answers knowing that Wikidata only uses blanks in a single object position; **C** counts the queries that could give uncertain answers further knowing which predicates have blanks; finally, **D** counts the queries whose solutions do change under certain answers. Hence, we see that 10.5% of the queries contain features that could cause uncertain answers, 6.5% of queries could cause uncertain answers even though Wikidata only uses blanks in a single object position, 4.7% of queries could cause uncertain answers knowing which predicates have blank values and which do not, while finally 0.6% of queries actually return uncertain answers.

We provide some statistics on the three Wikidata queries generating uncertain answers in Table 4. First for performance, we run the original query (T_1)

---

[4] https://www.wikidata.org/wiki/Wikidata:SPARQL_query_service/queries/exam ples.

Table 4. Query execution times (ms) for the three Wikidata queries with uncertain answers, and ratio of uncertain answers to total answers

	Local (Virtuoso)			Public (Blazegraph)			Uncertain/total
	T_1	T_2	T_2/T_1	T_1	T_2	T_2/T_1	
Q_1	144464	142989	0.99	53012	TO	–	20/5487
Q_2	7038	521	0.07	1013	2045	2.02	42/42
Q_3	1266326	1419269	1.12	TO	TO	–	12/27221

and a rewritten version for certain answers (T_2) over both a local Virtuoso index of Wikidata, as well as the public Wikidata Query Service (running Blazegraph). (RQ1) While the performance of the first query is comparable under both standard and certain answer semantics for Virtuoso, the latter times out on Blazegraph. On the other hand, the second query is faster on Virtuoso for certain answer semantics, possibly because it is anticipated that all answers will be discarded. This is not the case for Blazegraph, where the rewritten query takes twice the time. In Virtuoso Q_3 takes slightly longer in the rewritten query. Blazegraph times out in all runs of Q_3. We also look at the ratio of uncertain answers the queries would return. (RQ2) The ratio for Q_1 and Q_3 are relatively low, but on the other hand, for Q_2, the ratio is 100%: all answers are uncertain.

For space reasons, we refer to the webpage for further analysis of the Wikidata queries, including more details on the queries returning uncertain answers.

7   Conclusions

In this paper, we have looked at the semantics of SPARQL with respect to RDF graphs that use blank nodes as existential variables encoding unknown values. In particular, we have investigated the feasibility of approximating certain answers in SPARQL, proposing various rewriting strategies. Our initial results suggest that querying for certain/possible answers generally does incur a significant cost, but that at least for Virtuoso, query answering is still feasible (and in some cases faster than under standard semantics). We showed that the relative performance results for Virtuoso under certain answer semantics are competitive with results published for PostgreSQL. In general, we saw that although some queries are executed faster under certain semantics with current SPARQL implementations, for others there can be a significant performance cost. It is important to highlight, however, that experiments were run using off-the-shelf SPARQL implementations; dedicated SPARQL implementations for approximating certain answers may further improve on the performance observed here.

Regarding the question of whether or not offering users a choice of certain answer semantics is important, we performed an analysis of 446 Wikidata queries, where although 10.5% use negation and inequality features that could cause uncertain answers in principle, only 0.6% of the queries return uncertain answers in practice. However, Wikidata only uses unique blanks (acting similar to unmarked nulls) in the object position. It would be interesting to do similar studies for other datasets using existential blanks, though we are not immediately aware of such a dataset that has a set of SPARQL queries to analyse.

In summary, though the results here confirm that certain answers can be effectively approximated using even off-the-shelf SPARQL implementations, the practical motivation for such a SPARQL semantics remains speculative.

Acknowledgements. The work was also supported by the Millennium Institute for Foundational Research on Data (IMFD) and by Fondecyt Grant No. 1181896.

References

1. Abiteboul, S., Kanellakis, P.C., Grahne, G.: On the representation and querying of sets of possible worlds. Theor. Comput. Sci. **78**(1), 158–187 (1991)
2. Ahmetaj, S., Fischl, W., Pichler, R., Simkus, M., Skritek, S.: Towards reconciling SPARQL and certain answers. In: World Wide Web (WWW), pp. 23–33 (2015)
3. Angles, R., Gutierrez, C.: The multiset semantics of SPARQL patterns. In: Groth, P., et al. (eds.) ISWC 2016. LNCS, vol. 9981, pp. 20–36. Springer, Cham (2016). https://doi.org/10.1007/978-3-319-46523-4_2
4. Arenas, M., Bertails, A., Prud'hommeaux, E., Sequeda, J.: A direct mapping of relational data to RDF. W3C Recommendation (2012)
5. Arenas, M., Pérez, J.: Querying semantic web data with SPARQL. In: Principles of Database Systems (PODS), pp. 305–316. ACM (2011)
6. Arenas, M., Ugarte, M.: Designing a query language for RDF: marrying open and closed worlds. In: Principles of Database Systems (PODS), pp. 225–236. ACM (2016)
7. Codd, E.F.: Understanding relations. SIGMOD Rec. **6**(3), 40–42 (1974)
8. Console, M., Guagliardo, P., Libkin, L.: Approximations and refinements of certain answers via many-valued logics. In: Knowledge Representation and Reasoning (KR), pp. 349–358. AAAI Press (2016)
9. David, C., Libkin, L., Murlak, F.: Certain answers for XML queries. In: Principles of Database Systems (PODS), pp. 191–202. ACM (2010)
10. Gheerbrant, A., Libkin, L., Tan, T.: On the complexity of query answering over incomplete XML documents. In: International Conference on Database Theory (ICDT), pp. 169–181 (2012)
11. Guagliardo, P., Libkin, L.: Making SQL queries correct on incomplete databases: a feasibility study. In: Principles of Database Systems (PODS), pp. 211–223. ACM (2016)
12. Guagliardo, P., Libkin, L.: Correctness of SQL queries on databases with nulls. SIGMOD Rec. **46**(3), 5–16 (2017)
13. Harris, S., Seaborne, A., Prud'hommeaux, E.: SPARQL 1.1 query language. W3C Recommendation, March 2013
14. Hayes, P., Patel-Schneider, P.F.: RDF 1.1 semantics. W3C Recommendation, February 2014
15. Hogan, A., Arenas, M., Mallea, A., Polleres, A.: Everything you always wanted to know about blank nodes. J. Web Sem. **27**, 42–69 (2014)
16. Klein, H.-J.: On the use of marked nulls for the evaluation of queries against incomplete relational databases. In: Workshop on Foundations of Models and Languages for Data and Objects. Kluwer (1998)

17. Libkin, L.: Certain answers as objects and knowledge. In: Knowledge Representation and Reasoning (KR). AAAI Press (2014)
18. Libkin, L.: SQL's three-valued logic and certain answers. In: International Conference on Database Theory (ICDT), pp. 94–109 (2015)
19. Libkin, L.: SQL's three-valued logic and certain answers. ACM Trans. Database Syst. **41**(1), 1:1–1:28 (2016)
20. Lipski Jr., W.: On relational algebra with marked nulls preliminary version. In: Principles of Database Systems (PODS), pp. 201–203. ACM (1984)
21. Pérez, J., Arenas, M., Gutierrez, C.: Semantics and complexity of SPARQL. ACM Trans. Database Syst. **34**(3), 16:1–16:45 (2009)
22. Vrandecic, D., Krötzsch, M.: Wikidata: a free collaborative knowledgebase. Commun. ACM **57**(10), 78–85 (2014)

Efficient Handling of SPARQL OPTIONAL for OBDA

Guohui Xiao[1], Roman Kontchakov[2(✉)], Benjamin Cogrel[1], Diego Calvanese[1], and Elena Botoeva[1]

[1] KRDB Research Centre, Free University of Bozen-Bolzano, Italy
{xiao,cogrel,calvanese,botoeva}@inf.unibz.it
[2] Department of Computer Science and Information Systems, Birkbeck, University of London, UK,
roman@dcs.bbk.ac.uk

Abstract. OPTIONAL is a key feature in SPARQL for dealing with missing information. While this operator is used extensively, it is also known for its complexity, which can make efficient evaluation of queries with OPTIONAL challenging. We tackle this problem in the Ontology-Based Data Access (OBDA) setting, where the data is stored in a SQL relational database and exposed as a virtual RDF graph by means of an R2RML mapping. We start with a succinct translation of a SPARQL fragment into SQL. It fully respects bag semantics and three-valued logic and relies on the extensive use of the LEFT JOIN operator and COALESCE function. We then propose optimisation techniques for reducing the size and improving the structure of generated SQL queries. Our optimisations capture interactions between JOIN, LEFT JOIN, COALESCE and integrity constraints such as attribute nullability, uniqueness and foreign key constraints. Finally, we empirically verify effectiveness of our techniques on the BSBM OBDA benchmark.

1 Introduction

Ontology-Based Data Access (OBDA) aims at easing the access to database content by bridging the semantic gap between *information needs* (what users want to know) and their formulation as executable queries (typically in SQL). This approach hides the complexity of the database structure from users by providing them with a high-level representation of the data as an RDF graph. The RDF graph can be regarded as a view over the database defined by a DB-to-RDF mapping (e.g., following the R2RML specification) and enriched by means of an ontology [4]. Users can then formulate their information needs directly as high-level SPARQL queries over the RDF graph. We focus on the standard OBDA setting, where the RDF graph is not materialised (and is called a *virtual RDF graph*), and the database is relational and supports SQL [18].

To answer a SPARQL query, an OBDA system reformulates it into a SQL query, to be evaluated by the DBMS. In theory, such a SQL query can be

© Springer Nature Switzerland AG 2018
D. Vrandečić et al. (Eds.): ISWC 2018, LNCS 11136, pp. 354–373, 2018.
https://doi.org/10.1007/978-3-030-00671-6_21

obtained by (1) translating the SPARQL query into a relational algebra expression over the ternary relation *triple* of the RDF graph, and then (2) replacing the occurrences of *triple* by the matching definitions in the mapping; the latter step is called *unfolding*. We note that, in general, step (1) also includes rewriting the user query with respect to the given (OWL 2 QL) ontology [5,15]; we, however, assume that the query is already rewritten and, for efficiency reasons, the mapping is saturated; for details, see [15,24].

SPARQL joins are naturally translated into (INNER) JOINs in SQL [9]. However, in contrast to expert-written SQL queries, there typically is a *high margin for optimisation* in naively translated and unfolded queries. Indeed, since SPARQL, unlike SQL, is based on a single ternary relation, queries usually contain many more joins than SQL queries for the same information need; this suggests that many of the JOINs in unfolded queries are redundant and could be eliminated. In fact, the semantic query optimisation techniques such as self-join elimination [6] can reduce the number of INNER JOINs [19,21].

We are interested in SPARQL queries containing the OPTIONAL operator introduced to deal with *missing* information, thus serving a similar purpose [9] to the LEFT (OUTER) JOIN operator in relational databases. The graph pattern P_1 OPTIONAL P_2 returns answers to P_1 extended (if possible) by answers to P_2; when an answer to P_1 has no match in P_2 (due to incompatible variable assignments), the variables that occur only in P_2 remain *unbound* (LEFT JOIN extends a tuple without a match with NULLs). The focus of this work is the efficient handling of queries with OPTIONAL in the OBDA setting. This problem is important in practice because (a) OPTIONAL is very frequent in real SPARQL queries [1,17]; (b) it is a source of computational complexity: query evaluation is PSPACE-hard for the fragment with OPTIONAL alone [23] (in contrast, e.g., to basic graph patterns with filters and projection, which are NP-complete); (c) unlike expert-written SQL queries, the SQL translations of SPARQL queries (e.g., [8]) tend to have more LEFT JOINs with more complex structure, which DBMSs may fail to optimise well. We now illustrate the difference in the structure with an example.

Example 1. Let people be a database relation composed of a primary key attribute id, a non-nullable attribute fullName and two nullable attributes, workEmail and homeEmail:

id	fullName	workEmail	homeEmail
1	Peter Smith	peter@company.com	peter@perso.org
2	John Lang	NULL	joe@perso.org
3	Susan Mayer	susan@company.com	NULL

Consider an information need to retrieve the names of people and their e-mail addresses if they are available, with the *preference* given to work over personal e-mails. In standard SQL, the IT expert can express such a preference by means of the COALESCE function: e.g., COALESCE(v_1, v_2) returns v_1 if it is not NULL and v_2 otherwise. The following SQL query retrieves the required names and e-mail addresses:

```
SELECT fullName, COALESCE(workEmail, homeEmail) FROM people.
```

The same information need could naturally be expressed in SPARQL:

```
SELECT ?n ?e { ?p :name ?n  OPTIONAL { ?p :workEmail ?e }
                            OPTIONAL { ?p :personalEmail ?e } }.
```

Intuitively, for each person ?p, after evaluating the first OPTIONAL operator, variable ?e is bound to the work e-mail if possible, and left unbound otherwise. In the former case, the second OPTIONAL cannot extend the solution mapping further because all its variables are already bound; in the latter case, the second OPTIONAL tries to bind a personal e-mail to ?e. See [9] for a discussion on a similar query, which is weakly well-designed [14].

One can see that the two queries are in fact equivalent: the SQL query gives the same answers on the people relation as the SPARQL query on the RDF graph that encodes the relation by using id to generate IRIs and populating data properties :name, :workEmail and :personalEmail by the non-NULL values of the respective attributes.

However, the unfolding of the translation of the SPARQL query above would produce two LEFT OUTER JOINs, even with known simplifications (see, e.g., Q_2 in [8]):

```
SELECT v3.fullName AS n, COALESCE(v3.workEmail,v4.homeEmail) AS e
FROM (SELECT v1.fullName, v1.id, v2.workEmail FROM people v1
LEFT JOIN people v2 ON v1.id=v2.id AND v2.workEmail IS NOT NULL) v3
LEFT JOIN people v4 ON v3.id=v4.id AND v4.homeEmail IS NOT NULL
            AND (v3.workEmail=v4.homeEmail OR v3.workEmail IS NULL),
```

which is unnecessarily complex (compared to the expert-written SQL query above). Observe that the last bracket is an example of a *compatibility filter* encoding compatibility of SPARQL solution mappings in SQL: it contains disjunction and IS NULL.   □

Example 1 shows that SQL translations with LEFT JOINs can be simplified drastically. In fact, the problem of optimising LEFT JOINs has been investigated both in relational databases [12,20] and RDF triplestores [2,8]. In the database setting, *reordering* of OUTER JOINs has been studied extensively because it is essential for efficient query plans, but also challenging as these operators are neither commutative nor associative (unlike INNER JOINs). To perform a reordering, query planners typically rely on simple joining conditions, in particular, on conditions that reject NULLs and do not use COALESCE [12]. However, the SPARQL-to-SQL translation produces precisely the opposite of what database query planners expect: LEFT JOINs with complex compatibility filters. On the other hand, Chebotko et al. [8] proposed some simplifications when an RDBMS stores the *triple* relation and acts as an RDF triplestore. Although these simplifications are undoubtedly useful in the OBDA setting, the presence of mappings brings additional challenges and, more importantly, significant opportunities.

Example 2. Consider Example 1 again and suppose we now want to retrieve people's names, and when available also their work e-mail addresses. We can naturally represent this information need in SPARQL:

```
SELECT ?n ?e { ?p :name ?n OPTIONAL { ?p :workEmail ?e } }.
```

We can also express it very simply in SQL:

```
SELECT fullName, workEmail FROM people.
```

Instead, the straightforward translation and unfolding of the SPARQL query produces

```
SELECT v1.fullName AS n, v2.workEmail AS e
FROM people v1 LEFT JOIN people v2 ON v1.id=v2.id AND
                                      v2.workEmail IS NOT NULL.
```

R2RML mappings filter out NULL values from the database because NULLs cannot appear in RDF triples. Hence, the join condition in the unfolded query contains an IS NOT NULL for the workEmail attribute of v2. On the other hand, the LEFT JOIN of the query assigns a NULL value to workEmail if no tuple from v2 satisfies the join condition for a given tuple from v1. We call an assignment of NULL values by a LEFT JOIN the *padding effect*. A closer inspection of the query reveals, however, that the padding effect only applies when workEmail in v2 is NULL. Thus, the role of the LEFT JOIN in this query boils down to re-introducing NULLs eliminated by the mapping. In fact, this situation is quite typical in OBDA but does not concern RDF triplestores, which do not store NULLs, or classical data integration systems, which can expose NULLs through their mappings.   □

In this paper we address these issues, and our contribution is summarised as follows.

1. In Sect. 3, we provide a succinct translation of a fragment of SPARQL 1.1 with OPTIONAL and MINUS into relational algebra that relies on the use of LEFT JOIN and COALESCE. Even though the ideas can be traced back to Cyganiak [9] and Chebotko *et al.* [8] for the earlier SPARQL 1.0, our translation fully respects *bag semantics* and the *three-valued logic* of SPARQL 1.1 and SQL [13] (and is formally proven correct).
2. We develop optimisation techniques for SQL queries with complex LEFT JOINs resulting from the translation and unfolding: Compatibility Filter Reduction (CFR, Sect. 4.1), which generalises [8], LEFT JOIN Naturalisation (LJN, Sect. 4.2) to avoid padding, Natural LEFT JOIN Reduction (NLJR, Sect. 4.4), JOIN Transfer (JT, Sect. 4.5) and LEFT JOIN Decomposition (LJD, Sect. 4.6) complementing [12]. By CFR and LJN, compatibility filters and COALESCE are eliminated for well-designed SPARQL (Sect. 4.3).
3. We carried out an evaluation of our optimisation techniques over the well-known OBDA benchmark BSBM [3], where OPTIONALs, LEFT JOINs and NULLs are ubiquitous. Our experiments (Sect. 5) show that the techniques of Sect. 4 lead to a significant improvement in performance of the SQL translations, even for commercial DBMSs.

Full version with appendices is available at http://arxiv.org/abs/1806.05918.

2   Preliminaries

We first formally define the syntax and semantics of the SPARQL fragment we deal with and then present the relational algebra operators used for the translation from SPARQL.

RDF provides a basic data model. Its vocabulary contains three pairwise disjoint and countably infinite sets of symbols: IRIs I, blank nodes B and RDF literals L. *RDF terms* are elements of $C = I \cup B \cup L$, *RDF triples* are elements of $C \times I \times C$, and an *RDF graph* is a finite set of RDF triples.

2.1   SPARQL

SPARQL adds a countably infinite set V of *variables*, disjoint from C. A *triple pattern* is an element of $(C \cup V) \times (I \cup V) \times (C \cup V)$. A *basic graph pattern (BGP)* is a finite set of triple patterns. We consider *graph patterns*, P, defined by the grammar[1]

$$P ::= B \mid \text{FILTER}(P,F) \mid \text{UNION}(P_1,P_2) \mid \text{JOIN}(P_1,P_2) \mid$$
$$\text{OPT}(P_1,P_2,F) \mid \text{MINUS}(P_1,P_2) \mid \text{PROJ}(P,L),$$

where B is a BGP, $L \subseteq V$ and F, called a *filter*, is a formula constructed using logical connectives $\wedge$ and $\neg$ from atoms of the form $bound(v)$, $(v = c)$, $(v = v')$, for $v, v' \in V$ and $c \in C$. The set of variables in P is denoted by $var(P)$.

Variables in graph patterns are assigned values by *solution mappings*, which are *partial* functions $s \colon V \to C$ with (possibly empty) domain $dom(s)$. The *truth-value* $F^s \in \{\top, \bot, \varepsilon\}$ of a filter F under a solution mapping s is defined inductively:

- $(bound(v))^s$ is $\top$ if $v \in dom(s)$, and $\bot$ otherwise;
- $(v = c)^s = \varepsilon$ ('error') if $v \notin dom(s)$; otherwise, $(v = c)^s$ is the classical truth-value of the predicate $s(v) = c$; similarly, $(v = v')^s = \varepsilon$ if $\{v, v'\} \not\subseteq dom(s)$; otherwise, $(v = v')^s$ is the classical truth-value of the predicate $s(v) = s(v')$;

$$- (\neg F)^s = \begin{cases} \bot, & \text{if } F^s = \top, \\ \top, & \text{if } F^s = \bot, \\ \varepsilon, & \text{if } F^s = \varepsilon, \end{cases} \quad \text{and} \quad (F_1 \wedge F_2)^s = \begin{cases} \bot, & \text{if } F_1^s = \bot \text{ or } F_2^s = \bot, \\ \top, & \text{if } F_1^s = F_2^s = \top, \\ \varepsilon, & \text{otherwise.} \end{cases}$$

We adopt bag semantics for SPARQL: the answer to a graph pattern over an RDF graph is a multiset (or bag) of solution mappings. Formally, a *bag of solution mappings* is a (total) function Ω from the set of all solution mappings to non-negative integers $\mathbb{N}$: $\Omega(s)$ is called the *multiplicity* of s (we often use $s \in \Omega$ as a shortcut for $\Omega(s) > 0$). Following the grammar of graph patterns, we define respective operations on solution mapping bags. Solution mappings s_1 and s_2 are called *compatible*, written $s_1 \sim s_2$, if $s_1(v) = s_2(v)$, for each $v \in dom(s_1) \cap dom(s_2)$, in which case $s_1 \oplus s_2$ denotes a solution mapping with

---

[1] A slight extension of the grammar and the full translation are given in Appendix A.

domain $dom(s_1) \cup dom(s_2)$ and such that $s_1 \oplus s_2 \colon v \mapsto s_1(v)$, for $v \in dom(s_1)$, and $s_1 \oplus s_2 \colon v \mapsto s_2(v)$, for $v \in dom(s_2)$. We also denote by $s|_L$ the restriction of s on $L \subseteq V$. Then the SPARQL operations are defined as follows:

- FILTER$(\Omega, F) = \Omega'$, where $\Omega'(s) = \Omega(s)$ if $s \in \Omega$ and $F^s = \top$, and 0 otherwise;
- UNION$(\Omega_1, \Omega_2) = \Omega$, where $\Omega(s) = \Omega_1(s) + \Omega_2(s)$;
- JOIN$(\Omega_1, \Omega_2) = \Omega$, where $\Omega(s) = \displaystyle\sum_{\substack{s_1 \in \Omega_1, s_2 \in \Omega_2 \text{ with} \\ s_1 \sim s_2 \text{ and } s_1 \oplus s_2 = s}} \Omega_1(s_1) \times \Omega_2(s_2)$;
- OPT$(\Omega_1, \Omega_2, F) = $ UNION$($FILTER$($JOIN$(\Omega_1, \Omega_2), F), \Omega)$, where $\Omega(s) = \Omega_1(s)$ if $F^{s \oplus s_2} \neq \top$, for all $s_2 \in \Omega_2$ compatible with s, and 0 otherwise;
- MINUS$(\Omega_1, \Omega_2) = \Omega$, where $\Omega(s) = \Omega_1(s)$ if $dom(s) \cap dom(s_2) = \emptyset$, for all solution mappings $s_2 \in \Omega_2$ compatible with s, and 0 otherwise;
- PROJ$(\Omega, L) = \Omega'$, where $\Omega'(s') = \displaystyle\sum_{s \in \Omega \text{ with } s|_L = s'} \Omega(s)$.

Given an RDF graph G and a graph pattern P, the *answer* $[\![P]\!]_G$ to P over G is a bag of solution mappings defined by induction using the operations above and starting from basic graph patterns: $[\![B]\!]_G(s) = 1$ if $dom(s) = var(B)$ and G contains the triple $s(B)$ obtained by replacing each variable v in B by $s(v)$, and 0 otherwise ($[\![B]\!]_G$ is a set).

2.2 Relational Algebra (RA)

We recap the three-valued and bag semantics of relational algebra [13] and fix the notation. Denote by Δ the underlying domain, which contains a distinguished element *null*. Let U be a finite (possibly empty) set of *attributes*. A *tuple over U* is a (total) map $t \colon U \to \Delta$; there is a unique tuple over $\emptyset$. A *relation R over U* is a *bag* of tuples over U, that is, a function from all tuples over U to $\mathbb{N}$. For relations R_1 and R_2 over U, we write $R_1 \subseteq R_2$ ($R_1 \equiv R_2$) if $R_1(t) \leq R_2(t)$ ($R_1(t) = R_2(t)$, respectively), for all t.

A *term v over U* is an attribute $u \in U$, a constant $c \in \Delta$ or an expression $if(F, v, v')$, for terms v and v' over U and a filter F over U. A *filter F over U* is a formula constructed from atoms $isNull(V)$ and $(v = v')$, for a set V of terms and terms v, v' over U, using connectives $\wedge$ and $\neg$. Given a tuple t over U, it is extended to terms as follows:

$$t(c) = c, \text{ for constants } c \in \Delta, \qquad \text{and} \qquad t(if(F, v, v')) = \begin{cases} t(v), & \text{if } F^t = \top, \\ t(v'), & \text{otherwise,} \end{cases}$$

where the *truth-value* $F^t \in \{\top, \bot, \varepsilon\}$ of F on t is defined inductively (ε is *unknown*):

- $(isNull(V))^t$ is $\top$ if $t(v)$ is *null*, for all $v \in V$, and $\bot$ otherwise;
- $(v = v')^t = \varepsilon$ if $t(v)$ or $t(v')$ is *null*, and the truth-value of $t(v) = t(v')$ otherwise;
- and the standard clauses for $\neg$ and $\wedge$ in the three-valued logic (see Sect. 2.1).

We use standard abbreviations $coalesce(v, v')$ for $if(\neg isNull(v), v, v')$ and $F_1 \vee F_2$ for $\neg(\neg F_1 \wedge \neg F_2)$. Unlike Chebotko $et\ al.$ [8], we treat if as primitive, even though the renaming operation with an if could be defined via standard operations of RA.

For filters in positive contexts, we define a weaker equivalence: filters F_1 and F_2 over U are p-$equivalent$, written $F_1 \equiv^+ F_2$, in case $F_1^t = \top$ iff $F_2^t = \top$, for all t over U.

We use standard relational algebra operations: union $\cup$, difference $\setminus$, projection π, selection σ, renaming ρ, extension ν, natural (inner) join $\bowtie$ and duplicate elimination δ. We say that tuples t_1 over U_1 and t_2 over U_2 are $compatible^2$ if $t_1(u) = t_2(u) \neq null$, for all $u \in U_1 \cap U_2$, in which case $t_1 \oplus t_2$ denotes a tuple over $U_1 \cup U_2$ such that $t_1 \oplus t_2 : u \mapsto t_1(u)$, for $u \in U_1$, and $t_1 \oplus t_2 : u \mapsto t_2(u)$, for $u \in U_2$. For a tuple t_1 over U_1 and $U \subseteq U_1$, we denote by $t_1|_U$ the restriction of t_1 to U. Let R_i be relations over U_i, for $i = 1, 2$. The semantics of the above operations is as follows:

- If $U_1 = U_2$, then $R_1 \cup R_2$ and $R_1 \setminus R_2$ are relations over U_1 satisfying $(R_1 \cup R_2)(t) = R_1(t) + R_2(t)$ and $(R_1 \setminus R_2)(t) = R_1(t)$ if $t \notin R_2$ and 0 otherwise;
- If $U \subseteq U_1$, then $\pi_U R_1$ is a relation over U with $\pi_U R_1(t) = \sum_{t_1 \in R_1 \text{ with } t_1|_U = t} R_1(t_1)$;
- If F is a filter over U_1, then $\sigma_F R_1$ is a relation over U_1 such that $\sigma_F R_1(t)$ is $R_1(t)$ if $t \in R_1$ and $F^t = \top$, and 0 otherwise;
- $R_1 \bowtie R_2$ is a relation R over $U_1 \cup U_2$ such that $R(t) = \sum_{\substack{t_1 \in R_1 \text{ and } t_2 \in R_2 \\ \text{are compatible and } t_1 \oplus t_2 = t}} R_1(t_1) \times R_2(t_2)$;
- If v is a term over U_1 and $u \notin U_1$ an attribute, then the $extension$ $\nu_{u \mapsto v} R_1$ is a relation R over $U_1 \cup \{u\}$ with $R(t \oplus \{u \mapsto t(v)\}) = R_1(t)$, for all t. The $extended$ $projection$ $\pi_{\{u_1/v_1, \ldots, u_k/v_k\}}$ is a shortcut for $\pi_{\{u_1, \ldots, u_k\}} \nu_{u_1 \mapsto v_1} \cdots \nu_{u_k \mapsto v_k}$.
- If $v \in U_1$ and $u \notin U_1$ are distinct attributes, then the $renaming$ $\rho_{u/v} R_1$ is a relation over $U_1 \setminus \{v\} \cup \{u\}$ whose tuples t are obtained by replacing v in the domain of t by u. For terms $v_1, \ldots, v_k$ over U_1, attributes $u_1, \ldots, u_k$ (not necessarily distinct from U_1) and $V \subseteq U_1$, let $u'_1, \ldots, u'_k$ be fresh attributes and abbreviate the sequence $\rho_{u_1/u'_1} \cdots \rho_{u_k/u'_k} \pi_{U_1 \cup \{u'_1, \ldots, u'_k\} \setminus V} \nu_{u'_1 \mapsto v_1} \cdots \nu_{u'_k \mapsto v_k}$ by $\rho^V_{\{u_1/v_1, \ldots, u_k/v_k\}}$.
- δR_1 is a relation over U_1 with $\delta R_1(t) = \min(R_1(t), 1)$.

To bridge the gap between partial functions (solution mappings) of SPARQL and total functions (tuples) of RA, we use a $padding$ operation: $\mu_{\{u_1, \ldots, u_k\}} R_1$ denotes $\nu_{u_1 \mapsto null} \cdots \nu_{u_k \mapsto null} R_1$, for $u_1, \ldots, u_k \notin U_1$. Finally, we define the outer union, the (inner) join and left (outer) join operations by taking

$$R_1 \uplus R_2 = \mu_{U_2 \setminus U_1} R_1 \cup \mu_{U_1 \setminus U_2} R_2, \qquad R_1 \bowtie_F R_2 = \sigma_F(R_1 \bowtie R_2),$$
$$R_1 \bowtie\!\!\!\!\bowtie_F R_2 = (R_1 \bowtie_F R_2) \uplus (R_1 \setminus \pi_{U_1}(R_1 \bowtie_F R_2));$$

note that $\bowtie_F$ and $\bowtie\!\!\!\!\bowtie_F$ are $natural$ $joins$: they are over F as well as shared attributes.

---

[2] Note that, unlike in SPARQL, if u is $null$ in either of the tuples, then they are incompatible.

An *RA query* Q is an expression constructed from relation symbols, each with a fixed set of attributes, and filters using the RA operations (and complying with all restrictions). A *data instance* D gives a relation over its set of attributes, for any relation symbol. The *answer to* Q *over* D is a relation $\|Q\|_D$ defined inductively in the obvious way starting from the base case of relation symbols: $\|Q\|_D$ is the relation given by D.

3   Succinct Translation of SPARQL to SQL

We first provide a translation of SPARQL graph patterns to RA queries that improves the worst-case exponential translation of [15] in handling JOIN, OPT and MINUS: it relies on the *coalesce* function (see also [7,8]) and produces linear-size RA queries.

For any graph pattern P, the RA query $\tau(P)$ returns the same answers as P when solution mappings are represented as relational tuples. For a set V of variables and solution mapping s with $dom(s) \subseteq V$, let $ext_V(s)$ be the tuple over V obtained from s by padding it with *nulls*: formally,

$$ext_V(s) = s \oplus \{v \mapsto null \mid v \in V \setminus dom(s)\}.$$

The *relational answer* $\|P\|_G$ to P over an RDF graph G is a bag Ω of tuples over $var(P)$ such that $\Omega(ext_{var(P)}(s)) = [\![P]\!]_G(s)$, for all solution mappings s. Conversely, to evaluate $\tau(P)$, we view an RDF graph G as a data instance *triple*(G) storing G as a ternary relation *triple* with the attributes *sub*, *pred* and *obj* (note that *triple*(G) is a set).

The translation of a triple pattern $\langle s, p, o \rangle$ is an RA query of the form $\pi_{...}\sigma_F\,triple$, where the subscript of the extended projection π and filter F are determined by the variables, IRIs and literals in s, p and o; see Appendix A. SPARQL operators UNION, FILTER and PROJ are translated into their RA counterparts: $\uplus$, σ and π, respectively, with SPARQL filters translated into RA by replacing each *bound*(v) with $\neg isNull(v)$.

The translation of JOIN, OPT and MINUS is more elaborate and requires additional notation. Let P_1 and P_2 be graph patterns with $U_i = var(P_i)$, for $i = 1, 2$, and denote by U their shared variables, $U_1 \cap U_2$. To rename the shared attributes apart, we introduce fresh attributes u^1 and u^2 for each $u \in U$, set $U^i = \{u^i \mid u \in U\}$ and use abbreviations U^i/U and U/U^i for $\{u^i/u \mid u \in U\}$ and $\{u/u^i \mid u \in U\}$, respectively, for $i = 1, 2$. Now we can express the SPARQL solution mapping compatibility:

$$comp_U = \bigwedge_{u \in U} [(u^1 = u^2) \vee isNull(u^1) \vee isNull(u^2)]$$

(intuitively, the *null* value of an attribute in the context of RA queries represents the fact that the corresponding SPARQL variable is not bound). Next, the renamed apart attributes need to be coalesced to provide the value in the representation of the resulting solution mapping; see $\oplus$ in Sect. 2.1. To this end, given an RA filter F over a set of attributes V, terms $v_1, \ldots, v_k$ over V and attributes $u_1, \ldots, u_k \notin V$, we denote by $F[u_1/v_1, \ldots, u_k/v_k]$ the result of replacing each

u_i by v_i in F. We also denote by $coalesce_U$ the substitution of each $u \in U$ with $coalesce(u^1, u^2)$; thus, $F[coalesce_U]$ is the result of replacing each $u \in U$ in F with $coalesce(u^1, u^2)$. We now set

$$\tau(\text{JOIN}(P_1, P_2)) = \rho_{coalesce_U}^{U^1 \cup U^2} \left[\rho_{U^1/U} \tau(P_1) \bowtie_{comp_U} \rho_{U^2/U} \tau(P_2) \right],$$

$$\tau(\text{OPT}(P_1, P_2, F)) = \rho_{coalesce_U}^{U^1 \cup U^2} \left[\rho_{U^1/U} \tau(P_1) \rowtie_{comp_U \wedge \tau(F)[coalesce_U]} \rho_{U^2/U} \tau(P_2) \right],$$

$$\tau(\text{MINUS}(P_1, P_2)) = \pi_{U_1} \rho_{U/U^1} \sigma_{isNull(w)}$$
$$\left[\rho_{U^1/U} \tau(P_1) \rowtie_{comp_U \wedge \bigvee_{u \in U} (u^1 = u^2)} \nu_{w \mapsto 1} \rho_{U^2/U} \tau(P_2) \right],$$

where $w \notin U_1 \cup U_2$ is an attribute and $1 \in \Delta \setminus \{null\}$ is any domain element. The translation of JOIN and OPT is straightforward. For MINUS, observe that $\nu_{w \mapsto 1}$ extends the relation for P_2 by a fresh attribute w with a non-$null$ value. The join condition encodes compatibility of solution mappings whose domains, in addition, share a variable (both u^1 and u^2 are non-$null$). Tuples satisfying the condition are then filtered out by $\sigma_{isNull(w)}$, leaving only representations of solution mappings for P_1 that have no compatible solution mapping in P_2 with a shared variable. Finally, the attributes are renamed back by ρ_{U/U^1} and unnecessary attributes are projected out by π_{U_1}.

Theorem 3. *For any RDF graph G and any graph pattern P, $\|P\|_G = \|\tau(P)\|_{triple(G)}$.*

The complete proof of Theorem 3 can be found in Appendix A.

4   Optimisations of Translated SPARQL Queries

We present optimisations on a series of examples. We begin by revisiting Example 1, which can now be given in algebraic form (for brevity, we ignore projecting away ?p, which does not affect any of the optimisations discussed):

$$\text{OPT}(\text{OPT}(?p \text{ :name } ?n, \ ?p \text{ :workEmail } ?e, \ \top), ?p \text{ :personalEmail } ?e, \ \top),$$

where $\top$ denotes the tautological filter (true). Suppose we have the mapping

$$\text{IRI}_1(\text{id}) \text{ :name fullName} \leftarrow \sigma_{\neg isNull(\text{id}) \wedge \neg isNull(\text{fullName})} \text{people},$$
$$\text{IRI}_1(\text{id}) \text{ :workEmail workEmail} \leftarrow \sigma_{\neg isNull(\text{id}) \wedge \neg isNull(\text{workEmail})} \text{people},$$
$$\text{IRI}_1(\text{id}) \text{ :personalEmail homeEmail} \leftarrow \sigma_{\neg isNull(\text{id}) \wedge \neg isNull(\text{homeEmail})} \text{people},$$

where IRI_1 is a function that constructs the IRI for a person from their ID (an *IRI template*, in R2RML parlance). We assume that the IRI functions are injective and map only *null* to *null*; thus, joins on $\text{IRI}_1(\text{id})$ can be reduced to joins on id, and $isNull(\text{id})$ holds just in case $isNull(\text{IRI}_1(\text{id}))$ holds. Interestingly, the IRI functions can encode GLAV mappings, where the target query is a full-fledged CQ (in contrast to GAV mappings, where atoms do not contain existential variables); for more details, see [10].

The translation given in Sect. 3 and unfolding produce the following RA query, where we abbreviate, for example, $\rho^{\{p^1,p^2\}}_{\{p^4/coalesce(p^1,p^2)\}}$ by $\bar{\rho}_{\{p^4/coalesce(p^1,p^2)\}}$ (in other words, the $\bar{\rho}$ operation always projects away the arguments of its *coalesce* functions):

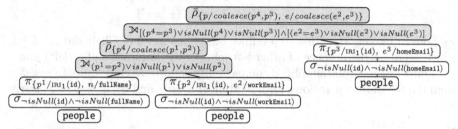

In our diagrams, the white nodes are the contribution of the mapping and the translation of the basic graph patterns: for example, the basic graph pattern ?p :name ?n produces $\pi_{\{p^1/\text{IRI}_1(\text{id}),\, n/\text{fullName}\}}\sigma_{\neg isNull(\text{id})\wedge\neg isNull(\text{fullName})}$ people (we use attributes without superscripts if there is only one occurrence; otherwise, the superscript identifies the relevant subquery). The grey nodes correspond to the translation of the SPARQL operations: for instance, the innermost left join is on $comp_{\{p\}}$ with p renamed apart to p^1 and p^2; the outermost left join is on $comp_{\{p,e\}}$, where p is renamed apart to p^4 and p^3 and e to e^2 and e^3; the two $\bar{\rho}$ are the respective renaming operations with *coalesce*.

4.1 Compatibility Filter Reduction (CFR)

We begin by simplifying the filters in (left) joins and eliminating renaming operations with *coalesce* above them (if possible). First, we can pull up the filters of the mapping through the extended projection and union by means of standard database equivalences: for example, for relations R_1 and R_2 and a filter F over U, we have $\sigma_F(R_1 \cup R_2) \equiv \sigma_F R_1 \cup \sigma_F R_2$, and $\pi_{U'}\sigma_{F'}R_1 \equiv \sigma_{F'}\pi_{U'}R_1$, if F' is a filter over $U' \subseteq U$, and $\rho_{u/v}\sigma_F R_1 \equiv \sigma_{F[u/v]}\rho_{u/v}R_1$, if $v \in U$ and $u \notin U$.

Second, the filters can be moved (in a restricted way) between the arguments of a left join to its join condition: for relations R_1 and R_2 over U_1 and U_2, respectively, and filters F_1, F_2 and F over U_1, U_2 and $U_1 \cup U_2$, respectively, we have

$$\sigma_{F_1} R_1 \bowtie_F R_2 \equiv \sigma_{F_1}(R_1 \bowtie_F R_2), \tag{1}$$

$$\sigma_{F_1} R_1 \bowtie_F R_2 \equiv \sigma_{F_1} R_1 \bowtie_{F \wedge F_1} R_2, \tag{2}$$

$$R_1 \bowtie_F \sigma_{F_2} R_2 \equiv R_1 \bowtie_{F \wedge F_2} R_2; \tag{3}$$

observe that unlike σ_{F_2} in (3), the selection σ_{F_1} cannot be entirely eliminated in (2) but can rather be 'duplicated' above the left join using (1). (We note that (1) and (3) are well-known and can be found, e.g., in [12].) Simpler equivalences hold for inner join: $\sigma_{F_1} R_1 \bowtie_F R_2 \equiv \sigma_{F \wedge F_1}(R_1 \bowtie R_2)$. These equivalences can be, in particular, used to pull up the $\neg isNull$ filters from mappings to eliminate the *isNull* disjuncts in the compatibility condition $comp_U$ of the (left) joins in the translation by means of the standard p-equivalences of the three-valued logic:

$$(F_1 \vee F_2) \wedge \neg F_2 \quad \equiv^+ \quad F_1 \wedge \neg F_2, \tag{4}$$

$$(v = v') \wedge \neg isNull(v) \quad \equiv^+ \quad (v = v'); \tag{5}$$

we note in passing that this step refines Simplification 3 of Chebotko *et al.* [8], which relies on the absence of other left joins in the arguments of a (left) join.

Third, the resulting simplified compatibility conditions can eliminate *coalesce* from the renaming operations: for a relation R over U and $u^1, u^2 \in U$, we clearly have

$$\rho^{\{u^1,u^2\}}_{\{u/coalesce(u^1,u^2)\}} \sigma_{\neg isNull(u^1)} R \quad \equiv \quad \sigma_{\neg isNull(u)} \pi_{U \setminus \{u^2\}} R[u/u^1], \tag{6}$$

where $R[u/u^1]$ is the result of replacing each u^1 in R by u. This step generalises Simplification 2 of Chebotko *et al.* [8], which does not eliminate *coalesce* above (left) joins that contain nested left joins.

By applying these three steps to our running example, we obtain (see Appendix C.1)

4.2 Left Join Naturalisation (LJN)

Our next group of optimisations can remove join conditions in left joins (if their arguments satisfy certain properties), thus reducing them to *natural left joins*.

Some equalities in the join conditions of left joins can be removed by means of attribute duplication: for relations R_1 and R_2 over U_1 and U_2, respectively, a filter F over $U_1 \cup U_2$ and attributes $u^1 \in U_1 \setminus U_2$ and $u^2 \in U_2 \setminus U_1$, we have

$$R_1 \bowtie_{F \wedge (u^1 = u^2)} R_2 \equiv R_1 \bowtie_F \nu_{u^1 \mapsto u^2} R_2. \tag{7}$$

Now, the duplicated u^2 can be eliminated in case it is actually projected away:

$$\pi_{U_1 \cup U_2 \setminus \{u^2\}} (R_1 \bowtie_F \nu_{u^1 \mapsto u^2} R_2) \equiv R_1 \bowtie_F R_2[u^1/u^2] \text{ if } F \text{ does not contain } u^2. \tag{8}$$

So, if F is a conjunction of suitable attribute equalities, then by repeated application of (7) and (8), we can turn a left join into a natural left join. In our running example, this procedure simplifies the innermost left join to

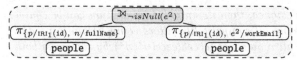

Another technique for converting a left join into a natural left join ($\bowtie$ is just an abbreviation for $\bowtie_\top$) is based on the conditional function *if*:

Proposition 4. *For relations R_1 and R_2 over U_1 and U_2, respectively, and a filter F over $U_1 \cup U_2$, we have*

$$R_1 \bowtie_F R_2 \;\equiv\; \rho^{\{U_2 \setminus U_1\}}_{\{u/if(F,u,null)\,|\,u \in U_2 \setminus U_1\}}(R_1 \bowtie R_2) \quad if \;\; \pi_{U_1}(R_1 \bowtie R_2) \subseteq R_1. \tag{9}$$

Proof. Denote $R_1 \bowtie R_2$ by S. Then $\pi_{U_1} S \subseteq R_1$ implies that every tuple t_1 in R_1 can have at most one tuple t_2 in R_2 compatible with it, and S consists of all such extensions (with their cardinality determined by R_1). Therefore, $\pi_{U_1}(S \setminus \sigma_F S)$ is precisely the tuples in R_1 that cannot be extended in such a way that the extension satisfies F, whence

$$\pi_{U_1}(S \setminus \sigma_F S) \;\equiv\; \pi_{U_1} S \setminus \pi_{U_1} \sigma_F S. \tag{10}$$

By a similar argument, $R_1 \setminus \pi_{U_1} S$ consists of the tuples in R_1 (with the same cardinality) that cannot be extended by a tuple in R_2, and $\pi_{U_1} S \setminus \pi_{U_1} \sigma_F S$ of those tuples that can be extended but only when F is not satisfied. By taking the union of the two, we obtain

$$(R_1 \setminus \pi_{U_1} S) \;\cup\; (\pi_{U_1} S \setminus \pi_{U_1} \sigma_F S) \;\equiv\; R_1 \setminus \pi_{U_1} \sigma_F S. \tag{11}$$

The claim is then proved by distributivity of ρ and μ over $\cup$; see Appendix B.

Proposition 4 is, in particular, applicable if the attributes shared by R_1 and R_2 uniquely determine tuples of R_2. In our running example, id is a primary key in people, and so we can eliminate $\neg isNull(e^2)$ from the innermost left join, which becomes a natural left join, and then simplify the term $if(\neg isNull(e^2), e^2, null)$ in the renaming to e^2 by using equivalences on complex terms: for a term v and a filter F over U, we have

$$if(F \wedge \neg isNull(v), v, null) \;\equiv\; if(F, v, null), \tag{12}$$
$$if(\top, v, null) \;\equiv\; v. \tag{13}$$

Thus, we effectively remove the renaming operator introduced by the application of Proposition 4; for full details, see Appendix C.1.

4.3   Translation for Well-Designed SPARQL

We remind the reader that a SPARQL pattern P that uses only JOIN, FILTER and binary OPT (that is, OPT with the tautological filter $\top$) is *well-designed* [16] if every its subpattern P' of the form $\text{OPT}(P_1, P_2, \top)$ satisfies the following condition: every variable u that occurs in P_2 and outside P' also occurs in P_1.

Proposition 5. *If P is well-designed, then its unfolded translation can be equivalently simplified by* (a) *removing all compatibility filters $comp_U$ from joins and left joins and* (b) *eliminating all renamings $u/coalesce(u^1, u^2)$ by replacing both u^1 and u^2 with u.*

Proof. Since P is well-designed, any variable u occurring in the right-hand side argument of any OPT either does not occur elsewhere (and so, can be projected away) or also occurs in the left-hand side argument. The claim then follows from an observation that, if the translation of P_1 or P_2 can be equivalently transformed to contain a selection with $\neg isNull(u)$ at the top, then the translation of $\textsc{Join}(P_1, P_2)$, $\textsc{Opt}(P_1, P^*, \top)$ and $\textsc{Filter}(P_1, F)$ can also be equivalently simplified so that it contains a selection with the $\neg isNull(u^1)$ or, respectively, $\neg isNull(u^2)$ condition at the top.

Rodríguez-Muro and Rezk [22] made a similar observation. Alas, Example 1 shows that Proposition 5 is not directly applicable to *weakly* well-designed SPARQL [14].

4.4 Natural Left Join Reduction (NJR)

A natural left join can then be replaced by a natural *inner* join if every tuple of its left-hand side argument has a match on the right, which can be formalised as follows.

Proposition 6. *For relations R_1 and R_2 over U_1 and U_2, respectively, we have*

$$\sigma_{\neg isNull(K)} R_1 \bowtie R_2 \equiv R_1 \bowtie R_2, \quad \text{if } \delta\pi_K R_1 \subseteq \pi_K R_2, \text{ for } K = U_1 \cap U_2. \quad (14)$$

Proof. By careful inspection of definitions. Alternatively, one can assume that the left join has an additional selection on top with filters of the form $(u^1 = u^2) \vee isNull(u^2)$, for $u \in K$, where u^1 and u^2 are duplicates of attributes from R_1 and R_2, respectively. Given $\delta\pi_K R_1 \subseteq \pi_K R_2$, one can eliminate the $isNull(u^2)$ because any tuple of R_1 has a match in R_2. The resulting *null*-rejecting filter then effectively turns the left join to an inner join by the outer join simplification of Galindo-Legaria and Rosenthal [12].

Observe that the inclusion $\delta\pi_K R_1 \subseteq \pi_K R_2$ is satisfied, for example, if R_1 has a foreign key K referencing R_2. It can also be satisfied if both R_1 and R_2 are based on the same relation, that is, $R_i \equiv \sigma_{F_i} \pi_{...} R$, for $i = 1, 2$, and F_1 logically implies F_2, where F_1 and/or F_2 can be $\top$ for the vacuous selection. Note that, due to δ, attributes K do not have to uniquely determine tuples in R_1 or R_2. In our running example, trivially, $\delta\pi_{\{p\}}(\pi_{\{p/\text{IRI}_1(\text{id}),\ n/\text{fullName}\}}\text{people}) \subseteq \pi_{\{p\}}(\pi_{\{p/\text{IRI}_1(\text{id}),\ e^2/\text{workEmail}\}}\text{people})$. Therefore, the inner left join can be replaced by a natural inner join, which can then be eliminated altogether because id is the primary key in people (this is a well-known optimisation; see, e.g., [11,21]). As a result, we obtain

$$\sigma_{\neg isNull(p) \wedge \neg isNull(n)}$$
$$\bar{\rho}_{\{e/coalesce(e^2,e^3)\}}$$
$$\Join_{[(e^2=e^3)\vee isNull(e^2)]\wedge \neg isNull(e^3)}$$
$$\pi_{\{p/\text{IRI}_1(\texttt{id}),\ n/\texttt{fullName},\ e^2/\texttt{workEmail}\}} \qquad \pi_{\{p/\text{IRI}_1(\texttt{id}),\ e^3/\texttt{homeEmail}\}}$$
$$\texttt{people} \qquad\qquad \texttt{people}$$

The running example is wrapped up and discussed in detail in Appendices C.1 and C.2.

4.5 Join Transfer (JT)

To introduce and explain another optimisation, we need an extension of relation `people` with a nullable attribute `spouseId`, which contains the `id` of the person's spouse if they are married and `NULL` otherwise. The attribute is mapped by an additional assertion:

$$\text{IRI}_1(\texttt{id})\ \texttt{:hasSpouse}\ \text{IRI}_1(\texttt{spouseId}) \quad \leftarrow \quad \sigma_{\neg isNull(\texttt{id})\wedge \neg isNull(\texttt{spouseId})}\texttt{people}.$$

Consider now the following query in SPARQL algebra:

$$\text{PROJ}(\text{OPT}(\texttt{?p :name ?n}, \text{JOIN}(\texttt{?p :hasSpouse ?s}, \texttt{?s :name ?sn}), \top), \{\texttt{?n, ?sn}\}),$$

whose translation can be unfolded and simplified with optimisations in Sects. 4.1 and 4.2 into the following RA query (we have also pushed down the filter $\neg isNull(sn)$ to the right argument of the join and, for brevity, omitted selection and projection at the top):

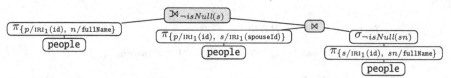

see Appendix C.4 for full details. Observe that the inner join cannot be eliminated using the standard self-join elimination techniques because it is not on a primary (or alternate) key. The next proposition (proved in Appendix B) provides a solution for the issue.

Proposition 7. *Let R_1, R_2 and R_3 be relations over U_1, U_2 and U_3, respectively, F a filter over $U_1 \cup U_2 \cup U_3$ and w an attribute in $U_3 \setminus (U_1 \cup U_2)$. Then*

$$R_1 \Join_F (R_2 \Join \sigma_{\neg isNull(w)} R_3) \equiv$$
$$\rho^{\{U_2 \setminus U_1\}}_{\{u/if(\neg isNull(w),u,null)\ |\ u \in U_2 \setminus U_1\}}((R_1 \Join R_2) \Join_F \sigma_{\neg isNull(w)} R_3),$$
$$if\ \pi_{U_1}(R_1 \Join R_2) \equiv R_1. \quad (15)$$

By Proposition 7, we take sn as the non-nullable attribute w and get the following:

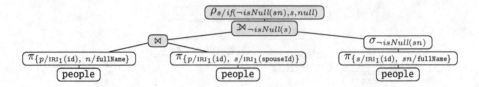

Now, the inner self-join can be eliminated (as id is the primary key of people) and the ρ operation removed (as its result is projected away); see Appendix C.4.

4.6 Left Join Decomposition (LJD): Left Join Simplification [12] Revisited

In Sect. 4.4, we have given an example of a reduction of a left join to an inner join. The following equivalence is also helpful (for an example, see Appendix C.3): for relations R_1 and R_2 over U_1 and U_2, respectively, and a filter F over $U_1 \cup U_2$,

$$\pi_{U_1}(R_1 \bowtie_F R_2) \equiv R_1, \quad \text{if} \quad \pi_{U_1}(R_1 \bowtie R_2) \subseteq R_1. \tag{16}$$

Galindo-Legaria and Rosenthal [12] observe that $\sigma_G(R_1 \bowtie_F R_2) \equiv R_1 \bowtie_{F \wedge G} R_2$ whenever G rejects *nulls* on $U_2 \setminus U_1$. In the context of SPARQL, however, the compatibility condition $comp_U$ does not satisfy the *null*-rejection requirement, and so, this optimisation is often not applicable. In the rest of this section we refine the basic idea.

Let R_1 and R_2 be relations over U_1 and U_2, respectively, and F and G filters over $U_1 \cup U_2$. It can easily be verified that, in general, we can *decompose* the left join:

$$\sigma_G(R_1 \bowtie_F R_2) \equiv (R_1 \bowtie_{F \wedge G} R_2) \uplus$$
$$\sigma_{nullify_{U_2 \setminus U_1}(G)} R_1 \setminus \pi_{U_1}(R_1 \bowtie_{F \wedge nullify_{U_2 \setminus U_1}(G)} R_2), \tag{17}$$

where $nullify_{U_2 \setminus U_1}(G)$ is the result of replacing every occurrence of an attribute from $U_2 \setminus U_1$ in G with *null*. Observe that if G is *null*-rejecting on $U_2 \setminus U_1$, then $nullify_{U_2 \setminus U_1}(G) \equiv^+ \bot$, and the second component of the union in (17) is empty. We, however, are interested in a subtler interaction of the filters when the second component of the difference or, respectively, the first component of the union is empty:

$$\sigma_G(R_1 \bowtie_F R_2) \equiv R_1 \bowtie_{F \wedge G} R_2 \uplus \sigma_{nullify_{U_2 \setminus U_1}(G)} R_1,$$
$$\text{if } F \wedge nullify_{U_2 \setminus U_1}(G) \equiv^+ \bot, \tag{18}$$

$$\sigma_G(R_1 \bowtie_F \sigma_{\neg isNull(w)} R_2) \equiv \sigma_{isNull(w) \wedge nullify_{U_2 \setminus U_1}(G)}(R_1 \bowtie_F \sigma_{\neg isNull(w)} R_2),$$
$$\text{if } F \wedge G \equiv^+ \bot \text{ and } w \in U_2 \setminus U_1. \tag{19}$$

These cases are of particular relevance for the SPARQL-to-SQL translation of OPTIONAL and MINUS. We illustrate the technique in Appendix C.5 on the following example:

FILTER(OPT(OPT(?p a :Product,

FILTER({ ?p :hasReview ?r . ?r :hasLang ?l }, ?l = "en"), ⊤),

FILTER({ ?p :hasReview ?r . ?r :hasLang ?l }, ?l = "zh"), ⊤), *bound*(?r)).

The technique relies on two properties of *null* propagation from the right-hand side of left joins. Let R_1 and R_2 be relations over U_1 and U_2, respectively. First, if $v = v'$ is a left join condition and v is a term over $U_2 \setminus U_1$, then v is either *null* or v' in the result:

$$R_1 \bowtie_{F \wedge (v=v')} R_2 \equiv \sigma_{isNull(v) \vee (v=v')}(R_1 \bowtie_{F \wedge (v=v')} R_2). \qquad (20)$$

Second, non-nullable terms v, v' over $U_2 \setminus U_1$ are simultaneously either *null* or not *null*:

$$R_1 \bowtie_F \sigma_{\neg isNull(v) \wedge \neg isNull(v')} R_2 \equiv$$

$$\sigma_{[\neg isNull(v) \wedge \neg isNull(v')] \vee [isNull(v) \wedge isNull(v')]}(R_1 \bowtie_F \sigma_{\neg isNull(v) \wedge \neg isNull(v')} R_2). \qquad (21)$$

The two equivalences introduce *no new* filters apart from *isNull* and their negations. The introduced filters, however, can help simplify the join conditions of the left joins containing the left join under consideration.

5   Experiments

In order to verify effectiveness of our optimisation techniques, we carried out a set of experiments based on the BSBM benchmark [3]; the materials for reproducing the experiments are available online[3]. The BSBM benchmark is built around an e-commerce use case in which vendors offer products that can be reviewed by customers. It comes with a mapping, a data generator and a set of SPARQL and equivalent SQL queries.

Hardware and Software. The experiments were performed on a t2.xlarge Amazon EC2 instance with four 64-bit vCPUs, 16G memory and 500G SSD hard disk under Ubuntu 16.04LTS. We used five database engines: free MySQL 5.7 and PostgreSQL 9.6 are run normally, and 3 commercial systems (which we shall call X, Y and Z) in Docker.

Queries. In total, we consider 11 SPARQL queries. Queries Q1–Q4 are based on the original BSBM queries 2, 3, 7 and 8, which contain OPTIONAL; we modified them to reduce selectivity: e.g., Q1, Q3 and Q4 retrieve information about 1000 products rather than a single product in the original BSBM queries; we also removed ORDER BY and LIMIT clauses. Q1–Q4 are well-designed (WD). In addition, we created 7 weakly well-designed (WWD) SPARQL queries: Q5–Q7 are similar to Example 1, Q8–Q10 to the query in Sect. 4.6, and Q11 is along the lines of Sect. 4.5. More information is below:

---

[3] https://github.com/ontop/ontop-examples/tree/master/iswc-2018-optional.

Query	Description	SPARQL	Optimisations
Q1	2 simple OPTIONALs for the padding effect (derived from BSBM query 2)	WD	LJN, NLJR
Q2	1 OPTIONAL with a !BOUND filter (encodes MINUS) derived from BSBM query 3	WD	JT
Q3	2 outer-level OPTIONALs, the latter with 2 nested OPTIONALs derived from BSBM query 7	WD	LJN, NLJR
Q4	4 OPTIONALs: ratings from attributes of the same relation derived from BSBM query 8	WD	LJN, NLJR
Q5/6/7	2/3/4 OPTIONALs: preference over 2/3/4 ratings of reviews	WWD	LJN, NLJR
Q8/9/10	2/3/4 OPTIONALs: preference of reviews over 2/3/4 languages	WWD	LJN, LJD
Q11	2 OPTIONALs: country-based preference of home pages of reviewed products	WWD	LJN, NLJR, JT

Data. We used the BSBM generator to produce CSV files for 1M products and 10M reviews. The CSV files (20GB) were loaded into DBs, with the required indexes created.

Evaluation. For each SPARQL query, we computed two SQL translations. The *non-optimised* (N/O) translation is obtained by applying to the unfolded query only the standard (previously known and widely adopted) structural and semantic optimisations [4] as well as CFR (Sect. 4.1) to simplify compatibility filters and eliminate unnecessary COALESCE. To obtain the *optimised* (O) translations, we further applied the other optimisation techniques presented in Sect. 4 (as described in the table above). We note that the optimised Q1 and Q4 have the same structure as the SQL queries in the original benchmark suite. On the other hand, the optimised Q2 is different from the SQL query in BSBM because the latter uses (NOT) IN, which is not considered in our optimisations.

Each query was executed three times with cold runs to avoid any variation due to caching. The size of query answers and their running times (in secs) are as follows:

Query	# answers	PostgreSQL		MySQL		X		Y		Z	
		N/O	O	N/O	O	N/O	O	N/O	O	N/O	O
Q1	19,267	1.79	1.77	0.43	0.38	0.90	0.80	0.56	0.52	29.06	25.09
Q2	6,746	18.75	2.07	19.95	0.36	40.00	16.07	0.44	0.37	27.99	5.97
Q2BSBM			3.88		0.37		20.55		0.38		5.91
Q3	1,355	4.20	0.09	4.70	0.11	5.50	1.60	2.04	0.14	5.45	0.65
Q4	1,174	2.14	0.16	0.86	0.04	3.00	0.60	1.78	0.11	4.38	0.53
Q5	2,294	0.56	0.05	0.01	0.01	1.80	0.30	0.30	0.08	0.51	0.53
Q6	2,294	102.35	0.18	>10 min	0.04	1.90	0.40	4.50	0.14	0.82	0.54
Q7	2,294	102.00	0.17	>10 min	0.04	2.60	0.40	14.57	0.14	1.21	0.53
Q8	1,257	0.07	0.06	0.01	0.01	8.40	1.30	0.08	0.08	295.25	0.40
Q9	1,311	101.20	0.16	>10 min	0.04	>10 min	2.70	4.30	0.11	>10 min	0.43
Q10	1,331	103.30	0.15	>10 min	0.05	>10 min	4.20	5.20	0.14	>10 min	0.43
Q11	3,388	5.26	0.87	3.80	0.21	107.06	2.68	177.95	0.22	7.82	0.13

The main outcomes of our experiments can be summarised as follows.

(a) The running times confirm that the optimisations are effective for all database engines. All optimised translations show better performance in all DB engines, and most of them can be evaluated in less than a second.

(b) Interestingly, our optimised translation is even slightly more efficient than the SQL with (NOT) IN from the original BSBM suite (see Q2BSBM in the table).

(c) The effects of the optimisations are significant. In particular, for challenging queries (some of which time out after 10 min), it can be up to three orders of magnitude.

6 Discussion and Conclusions

The optimisation techniques we presented are intrinsic to SQL queries obtained by translating SPARQL in the context of OBDA with mappings, and their novelty is due to the interaction of the components in the OBDA setting. Indeed, the optimisation of LEFT JOINs can be seen as a form of "reasoning" on the structure of the query, the data source and the mapping. For instance, when functional and inclusion dependencies along with attribute nullability are taken into account, one may infer that every tuple from the left argument of a LEFT JOIN is guaranteed to match (i) at least one or (ii) at most one tuple on the right. This information can allow one to replace LEFT JOIN by a simpler operator such as an INNER JOIN, which can further be optimised by the known techniques.

Observe that, in normal SQL queries, most of the NULLs come from the database rather than from operators like LEFT JOIN. In contrast, SPARQL triple patterns always bind their variables (no NULLs), and only operators like OPTIONAL can "unbind" them. In our experiments, we noticed that avoiding the padding effect is probably the most effective outcome of the LEFT JOIN optimisation techniques in the OBDA setting.

From the Semantic Web perspective, our optimisations exploit information unavailable in RDF triplestores, namely, database integrity constraints and mappings. From the DB perspective, we believe that such techniques have not been developed because LEFT JOINs and/or complex conditions like compatibility filters are not introduced accidentally in expert-written SQL queries. The results of our evaluation support this hypothesis and show a significant performance improvement, even for commercial DBMSs.

We are working on implementing these techniques in the OBDA system Ontop [4].

Acknowledgements. We thank the reviewers for their suggestions. This work was supported by the OBATS project at the Free University of Bozen-Bolzano and by the Euregio (EGTC) IPN12 project KAOS.

References

1. Arias, M., Fernández, J.D., Martínez-Prieto, M.A., de la Fuente, P.: An empirical study of real-world SPARQL queries. In: Proceedings of USEWOD (2011)
2. Atre, M.: Left bit right: for SPARQL join queries with OPTIONAL patterns (left-outer-joins). In: Proceedings of ACM SIGMOD, pp. 1793–1808 (2015)
3. Bizer, C., Schultz, A.: The Berlin SPARQL benchmark. Int. J. Semant. Web Inf. Syst. **5**(2), 1–24 (2009)
4. Calvanese, D., Cogrel, B., Komla-Ebri, S., Kontchakov, R., Lanti, D., Rezk, M., Rodriguez-Muro, M., Xiao, G.: Ontop: answering SPARQL queries over relational databases. SWJ **8**, 471–487 (2017)
5. Calvanese, D., De Giacomo, G., Lembo, D., Lenzerini, M., Rosati, R.: Tractable reasoning and efficient query answering in description logics: the *DL-Lite* family. JAR **39**, 385–429 (2007)
6. Chakravarthy, U.S., Grant, J., Minker, J.: Logic-based approach to semantic query optimization. ACM TODS **15**(2), 162–207 (1990)
7. Chaloupka, M., Nečaský, M.: Efficient SPARQL to SQL translation with user defined mapping. In: Ngonga Ngomo, A.-C., Křemen, P. (eds.) KESW 2016. CCIS, vol. 649, pp. 215–229. Springer, Cham (2016). https://doi.org/10.1007/978-3-319-45880-9_17
8. Chebotko, A., Lu, S., Fotouhi, F.: Semantics preserving SPARQL-to-SQL translation. DKE **68**(10), 973–1000 (2009)
9. Cyganiak, R.: A relational algerba for SPARQL. TR HPL-2005-170, HP Labs Bristol (2005)
10. De Giacomo, G., Lembo, D., Lenzerini, M., Poggi, A., Rosati, R.: Using ontologies for semantic data integration. In: Flesca, S., Greco, S., Masciari, E., Saccà, D. (eds.) A Comprehensive Guide Through the Italian Database Research Over the Last 25 Years. SBD, vol. 31, pp. 187–202. Springer, Cham (2018). https://doi.org/10.1007/978-3-319-61893-7_11
11. Elmasri, R., Navathe, S.: Fundamentals of Database Systems. Addison-Wesley, Boston (2010)
12. Galindo-Legaria, C., Rosenthal, A.: Outerjoin simplification and reordering for query optimization. ACM TODS **22**(1), 43–74 (1997)
13. Guagliardo, P., Libkin, L.: A formal semantics of SQL queries, its validation, and applications. PVLDB **11**(1), 27–39 (2017)
14. Kaminski, M., Kostylev, E.V.: Beyond well-designed SPARQL. In: Proceedings of ICDT (2016)
15. Kontchakov, R., Rezk, M., Rodríguez-Muro, M., Xiao, G., Zakharyaschev, M.: Answering SPARQL queries over databases under OWL 2 QL entailment regime. In: Mika, P., et al. (eds.) ISWC 2014. LNCS, vol. 8796, pp. 552–567. Springer, Cham (2014). https://doi.org/10.1007/978-3-319-11964-9_35
16. Pérez, J., Arenas, M., Gutierrez, C.: Semantics and complexity of SPARQL. ACM TODS **34**(3), 16:1–16:45 (2009)
17. Picalausa, F., Vansummeren, S.: What are real SPARQL queries like? In: SWIM (2011)
18. Poggi, A., Lembo, D., Calvanese, D., De Giacomo, G., Lenzerini, M., Rosati, R.: Linking data to ontologies. J. Data Semant. **10**, 133–173 (2008)
19. Priyatna, F., Corcho, Ó., Sequeda, J.F.: Formalisation and experiences of R2RML-based SPARQL to SQL query translation using morph. In: Proceedings of WWW, pp. 479–490 (2014)

20. Rao, J., Pirahesh, H., Zuzarte, C.: Canonical abstraction for outerjoin optimization. In: Proceedings of ACM SIGMOD, pp. 671–682 (2004)
21. Rodríguez-Muro, M., Kontchakov, R., Zakharyaschev, M.: Ontology-based data access: Ontop of databases. In: Alani, H. (ed.) ISWC 2013. LNCS, vol. 8218, pp. 558–573. Springer, Heidelberg (2013). https://doi.org/10.1007/978-3-642-41335-3_35
22. Rodriguez-Muro, M., Rezk, M.: Efficient SPARQL-to-SQL with R2RML mappings. J. Web Semant. **33**, 141–169 (2015)
23. Schmidt, M., Meier, M., Lausen, G.: Foundations of SPARQL query optimization. In: Proceedings ICDT, pp. 4–33 (2010)
24. Sequeda, J.F., Arenas, M., Miranker, D.P.: OBDA: query rewriting or materialization? In practice, both!. In: Mika, P., et al. (eds.) ISWC 2014. LNCS, vol. 8796, pp. 535–551. Springer, Cham (2014). https://doi.org/10.1007/978-3-319-11964-9_34

Representativeness of Knowledge Bases with the Generalized Benford's Law

Arnaud Soulet[1(✉)], Arnaud Giacometti[1], Béatrice Markhoff[1],
and Fabian M. Suchanek[2]

[1] Université de Tours, LIFAT, Tours, France
{arnaud.soulet,arnaud.giacometti,beatrice.markhoff}@univ-tours.fr
[2] Telecom ParisTech, LTCI, Paris, France
suchanek@telecom-paristech.fr

Abstract. Knowledge bases (KBs) such as DBpedia, Wikidata, and YAGO contain a huge number of entities and facts. Several recent works induce rules or calculate statistics on these KBs. Most of these methods are based on the assumption that the data is a representative sample of the studied universe. Unfortunately, KBs are biased because they are built from crowdsourcing and opportunistic agglomeration of available databases. This paper aims at approximating the representativeness of a relation within a knowledge base. For this, we use the generalized Benford's law, which indicates the distribution expected by the facts of a relation. We then compute the minimum number of facts that have to be added in order to make the KB representative of the real world. Experiments show that our unsupervised method applies to a large number of relations. For numerical relations where ground truths exist, the estimated representativeness proves to be a reliable indicator.

1 Introduction

One of the undisputed successes of the Semantic Web is the construction of huge knowledge bases (KBs). Several recent works use these KBs to derive new knowledge by calculating statistics or deducing rules from the data [7,26,27, 29]. For instance, according to DBpedia, 99% of the places in Yemen have a population of more than 1,000 inhabitants. Thus, we could conclude that Yemeni cities usually have more than 1,000 inhabitants. But is that true in the real world?

Naturally, the reliability of such conclusions depends on the quality of the knowledge base [34] namely its correctness (accuracy of the facts) and its completeness. It is well known that KBs are highly incomplete. This is usually not a problem in statistics and in machine learning, where it is rare to have a complete description of the universe under study. Most approaches work on a sample of the data. In such cases, it is crucial that this sample is representative of the entire universe (or at least, that the bias of this sample is known). For example, it is not a problem if the KB contains only half of the cities of Yemen, if their distribution across different sizes corresponds roughly to the distribution in the

D. Vrandečić et al. (Eds.): ISWC 2018, LNCS 11136, pp. 374–390, 2018.
https://doi.org/10.1007/978-3-030-00671-6_22

real world. Figure 1 illustrates this: there is an ideal knowledge base $\mathcal{K}^*$ divided into two classes A and B that correspond respectively to the places with less than 1,000 inhabitants and other places. The KB $\mathcal{K}_1$ is more complete than the KB $\mathcal{K}_2$. However, $\mathcal{K}_2$ better reflects the distribution between the two classes.

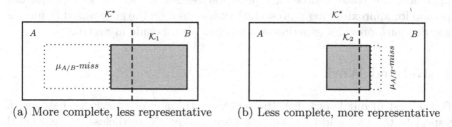

(a) More complete, less representative   (b) Less complete, more representative

Fig. 1. Completeness vs representativeness

Unfortunately, it is not clear whether the data in KBs is representative of the real world. For example, several large KBs, such as DBpedia [2] or YAGO [28], extract their data from Wikipedia. Wikipedia, in turn, is a crowdsourced dataset. In crowdsourcing, contributers tend to state the information that interests them most. As a result, Wikipedia exhibits some cultural biases [6,33]. Inevitably, these biases are reflected in the KBs. For instance, 3,922 entities in DBpedia concern the American company "Disney", which is almost as much as the 4,493 entities concerning Yemen (a country with more than 26 million inhabitants). Wikidata [32], likewise, is the result of crowdsourcing, and may exhibit similar biases. In particular, it is likely that countries such as Yemen are less evenly covered than places such as France – due to the population of contributors. Even if the information in these KBs is correct [13], it is not necessarily representative. If we knew how representative a certain KB is, then we could know whether it is reasonable or not to exploit it for deriving statistics. Such an indication should, for example, prevent us from drawing hasty conclusions about the distribution of the population in the cities of Yemen. But, how to estimate whether a knowledge base is representative or not?

This paper proposes to study the representativeness of knowledge bases by help of the generalized Benford's law. This parameterized law indicates the frequency distribution expected by the first significant digit in many real-world numerical datasets. We use this law as a gold standard to estimate how much data is missing in the KB. More specifically, our contributions are as follows:

- We present a method to calculate a lower bound for the number of missing facts for a relation to be representative. This method works in a supervised context (where the relation is known to satisfy the generalized Benford's law), and in an unsupervised context (where the parameter of the law has to be deduced from the data).
- We prove that, under certain assumptions, the calculated lower bounds are correct both in the supervised and the unsupervised context.

– We show with experiments on real KBs that our method is effective for supervised contexts as well as for unsupervised contexts. The unsupervised method,
  in particular, can audit 63% of DBpedia's facts.

This paper is structured as follows. Section 2 reviews some related work. Section 3
introduces the basic notions of representativeness. In Sect. 4, we propose our
method for approximating representativeness based on the generalized Benford's
law. Section 5 provides experimental results. We conclude in Sect. 6.

2   Related Work

To the best of our knowledge, the representativeness of knowledge bases with
respect to the real world has not yet been studied. Nevertheless, as mentioned
in the introduction, this problem is related to the completeness of KBs.

Completeness. Several recent works have studied the completeness of KBs [25,
34]. Some works propose to manually add information about the completeness
relations [8]. Other approaches mine rules on the data [12] (e.g., people usually
live in the city where they work) and propose to add this information where it
is missing. For this purpose, the work of [12] makes the Partial Completeness
Assumption (PCA): It assumes that, if the KB contains at least one object for
a given relation and a given subject, then it contains all of the objects for this
context. The PCA has been shown to be reasonably accurate in practice [12].
Newer approaches for rule mining take into account the cardinality of the relations, if it is known [30]. Other work aims to determine more generally whether
all objects of a certain relation for a certain subject are present in the KB [11].
For this, the approach uses oracles, such as the PCA and the popularity of the
subject in Wikipedia. Again other work [1,14,17,31] mines class descriptions.
Such approaches are able to determine that a certain attribute is obligatory for
a class – and then allow estimating the number of missing facts per class.

    All of these approaches are concerned with completeness in terms of facts
with respect to the present entities. Our approach, in contrast, also considers
the facts of entities that are missing. Furthermore, none of the above works
studies the representativeness of the KB, i.e., whether or not the distribution of
entities in the KB corresponds to the distribution in the real world.

Representative Sample. Completeness is an important notion for estimating the
quality of a knowledge base, but it is not necessarily the best indicator when
one wants to measure the quality of a distribution. In statistics, several resampling techniques [9] exist to estimate the quality of a sample (median, variance,
quantile), in particular by analyzing the evolution of a measure on a subsample
or by permuting labels. None of these techniques can be used to check whether
a single sample is representative, if the ground truth is unknown – as it is the
case in our scenario.

Benford's Law. When the data is complete, Benford's law [4] is regularly used to detect inconsistencies within the data [22]. If the distribution of the first significant digit of some numerical dataset does not satisfy Benford's law, then the data is assumed to be faulty. For this reason, Benford's law is regularly used to detect frauds in various kind of data: in accounts [23], in elections [19], or in wastewater treatment plant discharge data [3]. However, in all of these cases, Benford's law is used only to estimate the correctness of the data – not its completeness. The work cannot be used, e.g., to decide how many facts are missing in a KB, or whether a KB is representative of the real world.

3   Preliminaries

3.1   Representativeness of Knowledge Bases

For our purposes, a knowledge base (KB) over a set of relations $\mathcal{R}$ and a set of constants $\mathcal{C}$ (representing entities and literals) is a set of *facts* $\mathcal{K} \subseteq \mathcal{R} \times \mathcal{C} \times \mathcal{C}$. We write facts as $r(s, o) \in \mathcal{K}$, where r is the relation, s is the subject, and o is the object. The set of facts for the relation r in $\mathcal{K}$ is denoted by $\mathcal{K}_{|r} = \{r(s, o) \in \mathcal{K}\}$. Given a relation r, $r^{-1}(o, s) \in \mathcal{K}$ means that $r(s, o) \in \mathcal{K}$ where r^{-1} is the inverse relation of r.

In line with the other work in the area [11,17,18,21,24], we denote with $\mathcal{K}^*$ a hypothetical ideal KB, which contains all facts of the real world. Then, the completeness (also called recall) of $\mathcal{K}$, denoted $comp(\mathcal{K})$, is the proportion of facts of $\mathcal{K}^*$ present in $\mathcal{K}$: $comp(\mathcal{K}) = |\mathcal{K} \cap \mathcal{K}^*|/|\mathcal{K}^*|$. For our work, we will make the following assumption:

Assumption 1 (Correctness). *Given a knowledge base $\mathcal{K}$, we assume that all facts of $\mathcal{K}$ are correct i.e., $\mathcal{K} \subseteq \mathcal{K}^*$.*

The correctness assumption is a strong assumption. It has been investigated in [28,34]. In our work, we use it mainly for our theoretical model. Our experiments will show that our method delivers good results even with some amount of noise in the data. Let us now introduce the notion of a *uniform-sampling invariant measure*. A measure μ maps a knowledge base $\mathcal{K}$ to a frequency vector $(f_1, \ldots, f_n) \in \mathbb{R}^n_{\geq 0}$ where each component f_i is the number of observations of the ith characteristic in $\mathcal{K}$. Given a non-zero frequency vector $F = (f_1, \ldots, f_n)$, $\overline{f_i}$ denotes the normalized ith component of F where $\overline{f_i} = f_i / \sum_{i=1}^{n} f_i$. We use the mean absolute deviation (MAD) for comparing two non-zero frequency vectors $F = (f_1, \ldots, f_n)$ and $F' = (f'_1, \ldots, f'_n)$:

$$MAD(F, F') = \frac{1}{n} \sum_{i=1}^{n} \left| \overline{f_i} - \overline{f'_i} \right|$$

F and F' are similar for $\epsilon \ll 1$ iff $MAD(F, F') \leq \epsilon$. In such case, we write $F \sim_\epsilon F'$, or simply $F \sim F'$. A measure μ is uniform-sampling invariant iff for any uniform sample $\mathcal{K}'$ from $\mathcal{K}$ such that $|\mathcal{K}'| \gg 1$, we have $\mu(\mathcal{K}') \sim \mu(\mathcal{K})$.

For instance, in Fig. 1, counting the number of places with less than 1,000 inhabitants (in part A) and more than 1,000 inhabitants (in part B) is a measure with two characteristics (denoted by $\mu_{A/B}$). The measure $\mu_{A/B}$ is uniform-sampling invariant because whatever the uniform sample of a knowledge base $\mathcal{K}$, the proportion of cities with more (or less) than 1,000 inhabitants remains the same. In the following, we consider only uniform-sampling invariant measures.

A knowledge base is *representative* if each measure returns a frequency vector that is proportional to the frequency vector on $\mathcal{K}^*$:

Definition 1 (Representative KB). *A knowledge base $\mathcal{K}$ is representative of $\mathcal{K}^*$ iff $\mu(\mathcal{K}) \sim \mu(\mathcal{K}^*)$ for any uniform-sampling invariant measure μ.*

If a knowledge base $\mathcal{K}$ is unrepresentative, there is at least one measure μ such that $\mu(\mathcal{K}) \nsim \mu(\mathcal{K}^*)$. In this case, since all the facts of $\mathcal{K}$ are correct (Assumption 1), it would be necessary to add new facts to the knowledge base to make it representative for μ. Formally, this number of missing facts of $\mathcal{K}$ for the measure μ, denoted by $\mu\text{-}miss(\mathcal{K})$, is defined as:

$$\mu\text{-}miss(\mathcal{K}) = \min\{|F| : F \subseteq \mathcal{K}^* \wedge \mu(\mathcal{K} \cup F) \sim \mu(\mathcal{K}^*)\}$$

The number of missing facts in $\mathcal{K}$, denoted by $miss(\mathcal{K})$, is the minimum number of facts that have to be added to make the KB representative (whatever the considered measure μ): $miss(\mathcal{K}) = \max_\mu \mu\text{-}miss(\mathcal{K})$. The representativeness of $\mathcal{K}$ estimates whether $\mathcal{K}$ is a representative sample of $\mathcal{K}^*$:

Definition 2 (Representativeness). *The representativeness of $\mathcal{K}$, denoted $rep(\mathcal{K})$, is defined as:*

$$rep(\mathcal{K}) = \frac{|\mathcal{K}|}{|\mathcal{K}| + miss(\mathcal{K})}$$

Interestingly, a KB can be representative without being complete. The representativeness of $\mathcal{K}$ is an upper bound of the completeness: $rep(\mathcal{K}) \geq comp(\mathcal{K})$.

3.2   Problem Statement

The goal of this paper is to approximate the representativeness of a relation r in $\mathcal{K}$ (i.e., the representativeness of $\mathcal{K}_{|r}$) without having a reference knowledge base $\mathcal{K}^*_{|r}$ (which is the most common case in a real-world scenario). This task is ambitious because the calculation of the representativeness of a knowledge base requires to know the distribution of any measure μ on an unknown knowledge base $\mathcal{K}^*_{|r}$. It is obviously not possible to know the distribution $\mu(\mathcal{K}^*_{|r})$ for any measure. In order to calculate an approximation, we propose to use the following observation, which holds for all measures μ:

$$\mu\text{-}miss(\mathcal{K}_{|r}) \leq miss(\mathcal{K}_{|r})$$

This result (which follows from the definition of $miss(\mathcal{K}_{|r})$) means that it is possible to get a lower bound l of the number of missing facts $miss(\mathcal{K}_{|r})$, if some distributions $\mu_i(\mathcal{K}^*_{|r})$ are known. Such a lower bound is useful for calculating an upper bound of the representativeness and the completeness of the knowledge base: $|\mathcal{K}_{|r}|/(|\mathcal{K}_{|r}| + l)$.

Given a knowledge base $\mathcal{K}$ and a relation r, we aim at estimating the representativeness of the relation r in the knowledge base $\mathcal{K}$ by finding a lower bound l such that $l \leq miss(\mathcal{K}_{|r})$.

4 Our Approach

4.1 The Generalized Benford's Law for KBs

The challenge is to find a set of measures whose distribution is known on the ideal knowledge base $\mathcal{K}^*$. To this end, we propose to rely on Benford's law [4]. This law says that, in many natural datasets, the first significant digit of the numbers is unevenly distributed: Around 30% of numbers will start with a "1", whereas only 5% of numbers will start with a "9". This somehow surprising result follows from the fact that many natural numbers follow a multiplicative growth pattern. For example, a city of 1000 inhabitants may grow by 30% each year, thus passing by the values of 1300, 1690, 2197, 2856, 3712, 4826, 6274, 8157, 10604. These values already show a skewed distribution of the first digit, which will repeat itself in the coming years. There are other reasons for such patterns, and Benford's law has since been observed not just for population sizes, but also for prices, stock markets, death rates, lengths of rivers, and many other real-world phenomena [4] – although not all [20]. Technically, Benford's law is a statistical frequency distribution on the first significant digit of a set of numbers, which may or may not apply to a given dataset. In this paper, we use the generalized Benford's law [16], which is parametrized and can thus apply to more datasets.

Definition 3 (Generalized Benford's Law [15]**).** *A set of numbers is said to satisfy a generalized Benford's law (GBL) with exponent $\alpha \neq 0$ if the first digit $d \in [1..9]$ occurs with probability:*

$$B_d^\alpha = \frac{(1+d)^\alpha - d^\alpha}{10^\alpha - 1}$$

The parameter α adds a great flexibility since the choice of this value makes it possible to find Benford's law ($\alpha \to 0$) and the uniform law ($\alpha = 1$). Data that follows a power law ax^{-k} also follows the GBL approximately with $\alpha = -1/k$ [15]. This is, e.g., the case for the out-degree of Web pages [5], with $k = 2.6$.

The GBL can be applied to KBs. Let us look at the relation pop, which links a geographical place to its number of inhabitants (populationTotal in DBpedia, P1082 in Wikidata, and hasNumberOfPeople in YAGO). Figure 2 shows the distribution of first digits of this relation, drilled down to places in the world, in France, and in Yemen. We see that the distribution in the KB roughly follows the GBL. Interestingly, the GBL applies better to the French population than to the Yemeni population. We will now take advantage of this information to measure representativeness.

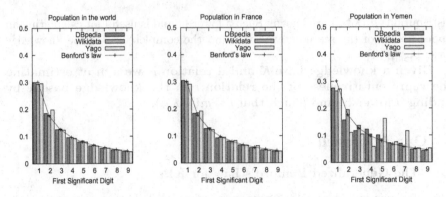

Fig. 2. First significant digit distribution for population

Technically, Fig. 2 presents the frequency vector $(f_1, \ldots, f_9)$ of the first digits of the relation `pop`. Of course, it is not possible to directly calculate the ideal frequency vector $(f_1^*, \ldots, f_9^*)$ of $\mathcal{K}^*$. However, in many cases, we know at least the distribution of the ideal frequency vector (thanks to the GBL). If we do not know the distribution, then our idea is to *learn* the exponent α of the GBL from the observed vector. Once the ideal distribution has been determined, we can use the difference between the observed distribution and the estimated distribution to bound the number of missing facts (Fig. 3).

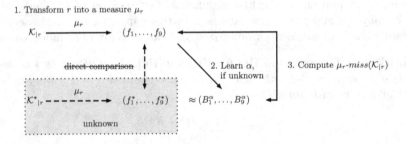

Fig. 3. Overview of the method

More precisely, we propose to proceed as follows:

1. **Transforming a relation into a measure:** Benford's law can only work on numerical datasets. Some relations (such as `pop`) are already numerical. Other relations will have to be transformed into numerical datasets (Sect. 4.2).
2. **Parameterizing the GBL:** To use the GBL, we have to know the parameter α. We distinguish two contexts. In a *supervised* context, the parameter α is known upfront in the real world (as it is the case for the population). Otherwise, in an *unsupervised* context, we learn the parameter α that best

fits the facts in $\mathcal{K}_{|r}$ assuming it is close to the ideal parameter α^* on $\mathcal{K}^*_{|r}$ (Sect. 4.3).

3. **Estimating the number of missing facts:** As the knowledge base is correct, only the addition of new facts would make the frequency vector $(f_1, \ldots, f_9)$ coincide with the distribution of $(B_1^\alpha, \ldots, B_9^\alpha)$ which is (approximately) proportional to $(f_1^*, \ldots, f_9^*)$. The objective of this last step is to calculate the minimum number of facts to add so that $(f_1, \ldots, f_9) \sim (B_1^\alpha, \ldots, B_9^\alpha)$ (Sect. 4.4).

In the following, when we consider a relation r, $\mathcal{K}$ implicitly refers to $\mathcal{K}_{|r}$.

4.2 Transforming Relations into Measures

We show in this section how to transform a relation r into a measure μ_r. The key idea is to transform each relation r into a set of numbers N_r that is a kind of digital signature. Then, we derive a measure μ_r that counts the frequency of each number in N_r having d as first significant digit:

$$\mu_r(\mathcal{K}) = (\#n : \text{the first significant digit of } n \in N_r(\mathcal{K}) \text{ is equal to } d)_{d \in [1..9]}$$

In our example with the relation pop, the measure μ_{pop} counts the number of places that have a population with d as first significant digit. Let us now generalize this principle to two common types of relations:

- **Numerical transformation:** Given a numerical relation r, the numerical transformation keeps all the numbers different from 0:

$$N_r^{num}(\mathcal{K}) = \{number : r(s, number) \in \mathcal{K} \wedge number \neq 0\}$$

Figure 2 illustrates this transformation for relation pop by showing the frequency vector resulting from μ_{pop}.

- **Counting transformation:** Given a relation r, the counting transformation returns for each object o how many facts it has:

$$N_r^{count}(\mathcal{K}) = \{\#s : r(s, o) \in \mathcal{K} \text{ such that } o \text{ is an object of a fact in } \mathcal{K}_{|r}\}$$

For example, for the relation starring, we can count the number of movies for each actor. The left hand-side of Fig. 4 illustrates the resulting frequency vector. We choose to count the number of subjects rather than the number of objects, because relations tend to have more subjects per object than vice versa [12]. However, we can also count the number of objects per subject by applying the above method to r^{-1}. Figure 4 shows two other histograms, one for the relation team (number of players per team) and for birthPlace (number of births per place).

This list of transformations is not exhaustive. For instance, it would be possible to count the number of days since today for a date (e.g. for the birth date relation) or to consider the length of strings. Besides, it is possible to transform the same relation in several ways. In this way, it is possible to obtain more frequency vectors.

4.3   Parameterizing the Generalized Benford's Law

The previous section has given us a measure μ_r that we can apply on the knowledge base $\mathcal{K}$ to calculate a distribution. Now, we want to compare this distribution with the distribution on the ideal KB $\mathcal{K}^*$. This requires knowledge of the parameter α, which depends on the unknown distribution $\mu_r(\mathcal{K}^*)$. We distinguish two settings.

Supervised Setting. In some cases, it is known that $\mu_r(\mathcal{K}^*)$ follows the GBL in the real world with a certain parameter α. For instance, the population of places, the length of rivers, etc. conform to the GBL in the real world with an exponent tending to 0 (see Table 2 below). In that case, the GBL is already parametrized.

Unsupervised Setting. If it is not known whether $\mu_r(\mathcal{K}^*)$ follows the GBL, or if its parameter α is not known, we propose to estimate it from the KB. For this purpose, we make the following assumption:

Assumption 2 (Transferability). *Given a knowledge base $\mathcal{K}$, we assume that if $\mathcal{K}$ conforms to the GBL with exponent α, then the ideal knowledge base $\mathcal{K}^*$ also conforms to the GBL with exponent α.*

This assumption may seem strong: However, it is verified in several cases where we have a ground truth available (see experiments in Sect. 5). The assumption allows us to learn the parameter α that best fits the facts in $\mathcal{K}$. Let us denote by $(f_1, \ldots, f_9)$ the characteristic vector resulting from $\mu_r(\mathcal{K})$ i.e., f_d is exactly the number of occurrences in $N_r(\mathcal{K})$ with d as first significant digit. Let us denote $N = \sum_{d=1}^{9} f_d$. To choose the right parameter α, we use the WLS measure (probability weighted least square or Chi square statistics) as goodness-of-fit measure [15]:

$$WLS_{(f_1,\ldots,f_9)}(\alpha) = \sum_{d=1}^{9} \frac{\left(B_d^\alpha - \frac{f_d}{N}\right)^2}{B_d^\alpha}$$

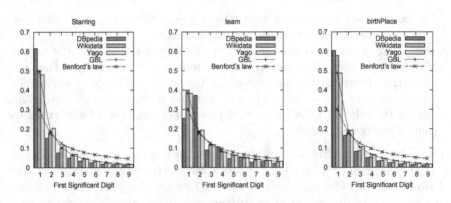

Fig. 4. Examples of measures resulting from counting transformation

Now, choosing the right parameter α means minimizing the WLS measure for the frequency vector $(f_1, \ldots, f_9)$. For this, we use the gradient descent algorithm. For instance, Fig. 4 shows the gap between the GBL and Benford's law for the three relations. For `starring`, α is -1.156 (in DBpedia), -0.759 (in Wikidata) and -0.750 (in YAGO). Once the parameter α has been obtained, we have to assess whether the frequency vector $\mu_r(\mathcal{K})$ conforms to the generalized Benford's law. For this, we use the mean absolute deviation (MAD) defined in Sect. 3.1. To know whether the GBL can be used according to the MAD estimator, we distinguish four cases [16,22]: close conformity (C) when $MAD \leq 0.006$, acceptable conformity (AC) when $0.006 < MAD \leq 0.012$, marginal conformity (MC) when $0.012 < MAD \leq 0.015$, and nonconformity (NC) otherwise. In our running examples, the measure μ_{pop} gives rise to a nonconformity only for Yemeni places in YAGO, because $\alpha = 0.351$ and $MAD(\mu_{\text{pop}}(\mathcal{K}), B^{0.351})$ equals 0.035 (> 0.015). If a measure μ_r leads to a nonconformity, then it is not possible to apply the GBL at all. In all other cases, we can estimate the number of missing facts for the relation r as explained in the next section.

4.4 Estimating the Number of Missing Facts

The purpose of this section is to estimate the number of missing facts for a relation r, knowing that we have an approximation of the expected distribution $(B_1^\alpha, \ldots, B_9^\alpha)$ that is proportional to $(f_1^*, \ldots, f_9^*)$. We assume that all the facts of the knowledge base $\mathcal{K}$ are correct (Assumption 1). Therefore, only the addition of facts can bring the observed distribution of facts $(f_1, \ldots, f_9)$ closer to the expected distribution $(B_1^\alpha, \ldots, B_9^\alpha)$.

Numerical Transformation. When a relation is numerical, the only way to have a number with a given first significant digit is to add a new fact. Intuitively, it is then enough to add facts for each of the digits where the measured frequency is lower than the expected frequency. The following theorem formalizes this idea:

Theorem 1. *Given a knowledge base $\mathcal{K}$ and a measure μ_r^{num} such that $\mu_r^{num}(\mathcal{K}^*)$ satisfies a generalized Bendford's law with exponent α, the number of missing facts for the relation r is:*

$$\mu_r^{num}\text{-}miss(\mathcal{K}) = \max_{d \in [1..9]} \frac{f_d}{B_d^\alpha} - N$$

where $(f_1, \ldots, f_9) = \mu_r(\mathcal{K})$ and $N = \sum_{d=1}^9 f_d$.

This follows from the fact that the expected distribution $f_d/(N + \mu_r^{num}\text{-}miss(\mathcal{K}))$ must be less than B_d^α for each digit d. Table 1 indicates the number of missing facts estimated for the relation `pop` with the unsupervised method, and deduces an approximation of the representativeness. Interestingly, the approximation $\mu_r^{num}\text{-}miss$ for Yemeni places of YAGO is very close to what we obtain in a supervised context (where we know that $\alpha \to 0$) – even though the measure is non-conform for that case. In the supervised context, we calculate that 181 facts are missing, while our estimation tells us that 127 facts are missing. Whatever the KB, our estimation of representativeness confirms our intuition mentioned in the introduction: the population of Yemeni places is less well informed than that of French ones.

Table 1. Representativeness of relations in three KBs (unsupervised context)

Measure	Missing facts			Representativeness		
	DBpedia	Wikidata	YAGO	DBpedia	Wikidata	YAGO
μ_{pop}^{num} in World	15,789	13,720	44,223	0.954	0.961	0.895
μ_{pop}^{num} in France	1,153	1,546	18,829	0.970	0.963	0.918
μ_{pop}^{num} in Yemen	78	4,281	127 (NC)	0.829	0.888	0.577 (NC)
$\mu_{starring}^{count}$	51,179	10,370	2,703	0.892	0.989	0.979
μ_{team}^{count}	41,484	3,373	463	0.980	0.997	0.999
$\mu_{birthPlace}^{count}$	38,664	25,691	470	0.971	0.986	0.998

Counting Transformation. For this transformation, the estimation of the number of missing facts is more complicated, because the addition of a fact for an object can change its first significant digit. For instance, if a number starting with 5 is missing, an object with 5 facts has to be added. One can imagine to add 5 new facts for a new object, to add four new facts for an object that has already 1 fact, to add 3 facts for an object that has already 2 facts, etc. We choose the solution that minimizes the total number of added facts:

Theorem 2. *Given a knowledge base* $\mathcal{K}$ *and a measure* μ_r^{count} *such that* $\mu_r^{count}(\mathcal{K}^*)$ *satisfies a generalized Bendford's law with exponent* α, *the number of missing facts for the relation* r *is:*

$$\mu_r^{count}\text{-}miss(\mathcal{K}) = \sum_{d=1}^{9} ((B_d^\alpha \times m) - f_d) \times d$$

where $m = \max_{d \in [1..9]} \frac{\sum_{i \geq d} f_i}{\sum_{i \geq d} B_i^\alpha}$ *and* $(f_1, \dots, f_9) = \mu_r(\mathcal{K})$.

This follows from the fact that $\sum_{i \geq d} f_i/m \leq \sum_{i \geq d} B_i^\alpha$ for each digit d. For the unsupervised context, Table 1 indicates the number of missing facts estimated for the relations starring/team/birthPlace with our method and deduces an approximation of the representativeness.

Note that for the same relation r, under the two transformations leading to μ_r^{num} and μ_r^{count}, the number of missing facts is bounded by the maximum result: $\max\{\mu_r^{num}\text{-}miss(\mathcal{K}); \mu_r^{count}\text{-}miss(\mathcal{K})\} \leq miss(\mathcal{K})$. Under the same transformation, the missing facts for two distinct relations r_1 and r_2 can be added together: $(\mu_{r_1}\text{-}miss(\mathcal{K}) + \mu_{r_2}\text{-}miss(\mathcal{K})) \leq miss(\mathcal{K})$. We will use these properties in Sect. 5.3 for DBpedia analysis.

4.5   Limitations of Our Approach

Using Theorems 1 and 2, our approach approximates the representativeness of some relation r in the knowledge base $\mathcal{K}$ by finding a lower bound $\mu_r\text{-}miss(\mathcal{K})$ such that $\mu_r\text{-}miss(\mathcal{K}) \leq miss(\mathcal{K}_{|r})$ as requested in Sect. 3.2. This approach works only if Assumption 1 (Correctness) holds. For the unsupervised setting, we also need Assumption 2 (Transferability).

Furthermore, for the GBL to be applicable, the set of numbers N_r has to meet the following two conditions. First, the numbers of N_r have to be distributed across several orders of magnitude: $\log_{10} \max(N_r) - \log_{10} \min(N_r) \geq 1$. For instance, the height of people does not meet this criterion because it is between 100 and 199 cm for most people. In that case, a numerical transformation would lead to a lot of "1" and "2" as first significant digits. For the same reason, it is also not possible to apply the counting transformation to an inverse functional relation r because in that case, each object has only one subject (i.e., $N_r^{count} = \{1, 1, 1, \dots\}$) and then, its prevalence is 0. Second, the cardinality of N_r has to be sufficiently high: $|N_r| \gg 1$. If we do not have enough numbers in N_r, the derived distributions $\mu_r(\mathcal{K})$ will not be reliable enough to learn the parameter α. The next section will show where our method can be applied.

5   Experiments

These experiments answer the following three questions: Is the unsupervised method reliable? Is the representativeness estimated by our method correct? Is the GBL sufficiently effective to be useful for auditing a knowledge base?

All experimental data (the queries, the distributions, the experimental results, and details of the learning method), as well as the source code, are available here: http://www.info.univ-tours.fr/~soulet/prototype/iswc18.

5.1   Verification of the Transferability Assumption

Assumption 2 (Transferability) is a central assumption in the unsupervised approach for learning the GBL parameter. Our first experiment aims to verify if this assumption is true. For this, we compare the parameter α that we obtained by the unsupervised approach to the parameter α of the real world. We found seven relations under the numerical transformation that are known to verify Benford's law in the real world, and that exist in DBpedia and Wikidata. We also found one relation under the counting transformation that exists in our KBs and that is known to follow the GBL in the real world: the out-degree of Wikipedia pages, where $\alpha = -1/2.6 = -0.385$ [5].

Table 2 shows the results obtained for representativeness by Theorem 1 in both supervised and unsupervised contexts. The last column indicates the GBL compliance between the supervised and unsupervised case according to the MAD test (Sect. 4.3). We see that the learned parameter conforms to the ground truth in all cases: it is very close to zero and does not deviate to values that would have

Table 2. Conformity of the unsupervised method with the supervised one

Relation	KB	Sup.		Unsup.		$MAD(B^\alpha, B^{\alpha^*})$
		α^*	Rep.	α	Rep.	
Population of places	DBpedia	0.001	0.949	−0.020	0.954	C
Elevation of places	DBpedia	0.001	0.750	−0.083	0.765	C
Area of places	DBpedia	0.001	0.535	0.143	0.624	AC
Length of water streams	DBpedia	0.001	0.887	0.001	0.887	C
Discharge of water streams	DBpedia	0.001	0.938	−0.105	0.930	AC
Number of deaths	Wikidata	0.001	0.909	−0.106	0.908	AC
Number of injured	Wikidata	0.001	0.883	−0.119	0.875	AC
Out-degree of Wikipedia page	DBpedia	−0.385	0.999	−0.486	0.999	AC

a distorting impact (e.g., $\alpha > 2$, or $\alpha > 5$). For the out-degree of Wikipedia pages, the learned parameter also corresponds well to the real parameter. In addition, the estimator of MAD always indicates a very good conformity (≤ 0.012). This entails that the representativeness that we compute in the unsupervised approach is very similar to the supervised value. In all cases except one, there is less than 1% difference. Even for the least correct prediction (areaTotal) the difference is at most 10%[1].

Finally, we also applied the unsupervised method to numerical relations whose numbers should *not* verify the GBL. In such a situation, the method should have a MAD test that indicates a nonconformity (i.e. >0.015). This is indeed the case for the following relations: Wikipedia page ID (with MAD 0.029), runtime of films (0.077) or albums (0.090), and weight of persons (0.070).

5.2   Validity of Representativeness

In Sect. 3, we postulated that representativeness is an upper bound for completeness. To test this postulation, we simulate an unrepresentative KB as a sample of a known KB. For this purpose, we use the number of inhabitants of French cities from DBpedia as gold standard, because we know that these numbers verify the GBL. We then apply three approaches to degrade this KB:

- **Most-populated:** We removes cities, starting from the least populated to the most populated. This biased sample simulates a KB of Yemeni cities, where only the most populated cities are present.
- **Least-populated:** We remove the most populated cities first. This approach is the opposite of the previous one.
- **Random:** We randomly removes cities. The retained sample of facts is therefore uniformly drawn and it is representative of the original KB.

Our first step is to verify whether our samples conform to Benford's law (Sect. 4.3). This is indeed the case for 100% of samples for the most-populated

---

[1] Different from α, the representativeness varies only between 0 and 1.

approach and the random approach, and for 99% of the samples for the least-populated approach. This validates Assumption 2, and makes our approach applicable. Figure 5 plots the representativeness for the three approaches according to the number of preserved cities in a supervised and unsupervised context. We also plot the real completeness of the sample (w.r.t. the original KB).

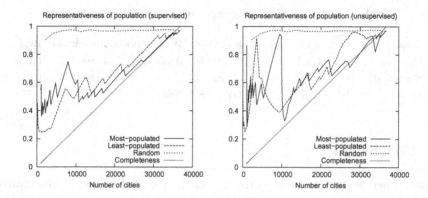

Fig. 5. Impact of incompleteness on French cities using `dbo:populationTotal`

We observe that whatever the approach and the context, representativeness is indeed an upper bound for completeness, as postulated. There is only a single major violation at the point of around 34,000 cities for the most-populated approach, which is due to a wrong approximation of the parameter α in that particular sample. Surprisingly, the representativeness is a very good approximation of completeness for the most-populated and the least-populated approaches. In the case of the supervised context, considering a sample $\mathcal{C} = \mathcal{K}_{|\text{pop}}$ with more than 22,000 cities, the estimated number of cities (i.e., $P = |\mathcal{C} + \mu_{\text{pop}}^{num}\text{-}miss(\mathcal{C})|$) approximates the true number of cities in $\mathcal{K}^*$ (i.e., $T = |\mathcal{K}^*_{|\text{pop}}|$) with less than 5% error: $|P - T|/P \leq 0.05$.

Finally, we observe that as long as the number of cities remains large enough (i.e., greater than 2,500), the representativeness of the random approach is high (around 0.95). This is expected for any large random sample from a complete relation, because a random sample has to be representative in our sense.

5.3 Effectiveness of the GBL for a KB

We considered in DBpedia (France) all the relations with at least 100 facts. We applied the numerical transformation and the counting transformation. We removed all relations whose numbers are not distributed across several orders of magnitude i.e., $\log_{10} \max(N_r) - \log_{10} \min(N_r) < 1$. Table 3 gives a general overview of the resulting 2,920 relations: the number of considered relations, the number of compliant relations (i.e., with $MAD \leq 0.015$), the number of facts, the proportion of facts in DBpedia, the estimated number of missing facts and finally, the estimated representativeness. Clearly, the counting transformation concerns more relations and facts than the numerical transformation. All in all, our analysis covers about 63% of the facts in DBpedia and we estimate its representativeness at 0.719. To make DBpedia's current relations representative, at least 46 million facts would have to be added.

Table 3. Overview of the representativeness of DBpedia (France)

Trans.	# of rel.	# of comp. rel.	# of facts	% of DBpedia	Missing facts	Rep.
Counting	2,920	1,461	117,349,802	0.633	45,869,202	0.719
Numerical	108	43	329,853	0.002	109,603	0.751
Total	2,920	1,487	117,461,855	0.634	45,972,923	0.719

6 Conclusion

In this paper, we have introduced the first method to analyze how representative a knowledge base is for the real world. We believe that representativeness is a dimension of data quality in its own right (along with correctness and completeness), because it is essential for applying statistical or machine learning methods. Our approach quantifies a minimum number of facts that must complement the knowledge base in order to make it representative. Experiments on DBpedia validate our proposal in a supervised and unsupervised context on several relations. Using our method, we estimate that at least 46 million facts are missing for DBpedia to be a representative knowledge base. In future work, we would like to take into account representativeness to correct the result of queries on knowledge bases much like this has been done recently for completeness [10].

References

1. Alam, M., Buzmakov, A., Codocedo, V., Napoli, A.: Mining definitions from RDF annotations using formal concept analysis. In: IJCAI (2015)
2. Auer, S., Bizer, C., Kobilarov, G., Lehmann, J., Cyganiak, R., Ives, Z.: DBpedia: a nucleus for a web of open data. In: Aberer, K., et al. (eds.) ASWC/ISWC -2007. LNCS, vol. 4825, pp. 722–735. Springer, Heidelberg (2007). https://doi.org/10.1007/978-3-540-76298-0_52
3. Beiglou, P.H.B., Gibbs, C., Rivers, L., Adhikari, U., Mitchell, J.: Applicability of Benford's law to compliance assessment of self-reported wastewater treatment plant discharge data. J. Environ. Assess. Policy Manag. (2017). https://doi.org/10.1142/S146433321750017X
4. Benford, F.: The law of anomalous numbers. In: Proceedings of the American Philosophical Society, pp. 551–572 (1938)
5. Broder, A., et al.: Graph structure in the web. Comput. Netw. **33**(1–6), 309–320 (2000)

6. Callahan, E.S., Herring, S.C.: Cultural bias in Wikipedia content on famous persons. J. Assoc. Inf. Sci. Technol. **62**(10), 1899–1915 (2011)
7. de la Croix, D., Licandro, O.: The longevity of famous people from Hammurabi to Einstein. J. Econ. Growth **20**(3), 263–303 (2015)
8. Darari, F., Razniewski, S., Prasojo, R.E., Nutt, W.: Enabling fine-grained RDF data completeness assessment. In: Bozzon, A., Cudre-Maroux, P., Pautasso, C. (eds.) ICWE 2016. LNCS, vol. 9671, pp. 170–187. Springer, Cham (2016). https://doi.org/10.1007/978-3-319-38791-8_10
9. Efron, B.: The Jackknife, the Bootstrap, and Other Resampling Plans, vol. 38. SIAM, Philadelphia (1982)
10. Galárraga, L., Hose, K., Razniewski, S.: Enabling completeness-aware querying in SPARQL. In: Proceedings of the 20th International Workshop on the Web and Databases, pp. 19–22. ACM (2017)
11. Galárraga, L., Razniewski, S., Amarilli, A., Suchanek, F.M.: Predicting completeness in knowledge bases. In: WSDM, pp. 375–383. ACM (2017)
12. Galárraga, L., Teflioudi, C., Hose, K., Suchanek, F.M.: Fast rule mining in ontological knowledge bases with AMIE++. VLDB J. **24**(6), 707–730 (2015)
13. Giles, J.: Internet encyclopaedias go head to head. Nature **438**, 900–901 (2005). https://doi.org/10.1038/438900a
14. Hellmann, S., Lehmann, J., Auer, S.: Learning of OWL class descriptions on very large knowledge bases. Int. J. Semant. Web Inf. Syst. **5**, 25–48 (2009)
15. Hürlimann, W.: A first digit theorem for powers of perfect powers. Commun. Math. Appl. **5**(3), 91–99 (2014)
16. Hürlimann, W.: Benford's law in scientific research. Int. J. Sci. Eng. Res. **6**(7), 143–148 (2015)
17. Lajus, J., Suchanek, F.M.: Are all people married? Determining obligatory attributes in knowledge bases. In: WWW (2018)
18. Levy, A.Y.: Obtaining complete answers from incomplete databases. In: VLDB (1996)
19. Mebane Jr., W.R.: Election forensics: Vote counts and Benford's law. In: Summer Meeting of the Political Methodology Society, UC-Davis, July, pp. 20–22 (2006)
20. Morzy, M., Kajdanowicz, T., Szymański, B.K.: Benford's distribution in complex networks. Sci. Rep. **6**, Article no. 34917 (2016)
21. Motro, A.: Integrity = validity + completeness. TODS **14**, 480–502 (1989)
22. Nigrini, M.: Benford's Law: Applications for Forensic Accounting, Auditing, and Fraud Detection, vol. 586. Wiley, Hoboken (2012)
23. Nigrini, M.J.: A taxpayer compliance application of Benford's law. J. Am. Tax. Assoc. **18**(1), 72 (1996)
24. Razniewski, S., Korn, F., Nutt, W., Srivastava, D.: Identifying the extent of completeness of query answers over partially complete databases. In: SIGMOD (2015)
25. Razniewski, S., Suchanek, F., Nutt, W.: But what do we actually know? In: Proceedings of the 5th Workshop on Automated Knowledge Base Construction, pp. 40–44 (2016)
26. Rebele, T., Nekoei, A., Suchanek, F.M.: Using YAGO for the humanities. In: WHISE workshop (2017)
27. Schich, M., et al.: A network framework of cultural history. Science **345**(6196), 558–562 (2014)
28. Suchanek, F.M., Kasneci, G., Weikum, G.: YAGO: a core of semantic knowledge. In: WWW, pp. 697–706. ACM (2007)
29. Suchanek, F.M., Preda, N.: Semantic culturomics. Proc. VLDB Endow. **7**(12), 1215–1218 (2014)

30. Pellissier Tanon, T., Stepanova, D., Razniewski, S., Mirza, P., Weikum, G.: Completeness-aware rule learning from knowledge graphs. In: d'Amato, C., et al. (eds.) ISWC 2017. LNCS, vol. 10587, pp. 507–525. Springer, Cham (2017). https://doi.org/10.1007/978-3-319-68288-4_30

31. Völker, J., Niepert, M.: Statistical schema induction. In: Antoniou, G., et al. (eds.) ESWC 2011. LNCS, vol. 6643, pp. 124–138. Springer, Heidelberg (2011). https://doi.org/10.1007/978-3-642-21034-1_9

32. Vrandečić, D., Krötzsch, M.: Wikidata: a free collaborative knowledgebase. Commun. ACM **57**(10), 78–85 (2014)

33. Wagner, C., Garcia, D., Jadidi, M., Strohmaier, M.: It's a man's Wikipedia? Assessing gender inequality in an online encyclopedia. In: ICWSM, pp. 454–463 (2015)

34. Zaveri, A., Rula, A., Maurino, A., Pietrobon, R., Lehmann, J., Auer, S.: Quality assessment for linked data: a survey. Semant. Web **7**(1), 63–93 (2016)

Detecting Erroneous Identity Links on the Web Using Network Metrics

Joe Raad[1,3](✉), Wouter Beek[2], Frank van Harmelen[2], Nathalie Pernelle[3],
and Fatiha Saïs[3]

[1] UMR MIA-Paris, INRA, Paris-Saclay University, Paris, France
joe.raad@agroparistech.fr
[2] Department of Computer Science, VU University Amsterdam,
Amsterdam, The Netherlands
{w.g.j.beek,frank.van.harmelen}@vu.nl
[3] LRI, Paris Sud University, CNRS 8623, Paris Saclay University, Orsay, France
{nathalie.pernelle,fatiha.sais}@lri.fr

Abstract. In the absence of a central naming authority on the Semantic Web, it is common for different datasets to refer to the same thing by different IRIs. Whenever multiple names are used to denote the same thing, `owl:sameAs` statements are needed in order to link the data and foster reuse. Studies that date back as far as 2009, have observed that the `owl:sameAs` property is sometimes used incorrectly. In this paper, we show how network metrics such as the community structure of the `owl:sameAs` graph can be used in order to detect such possibly erroneous statements. One benefit of the here presented approach is that it can be applied to the network of `owl:sameAs` links itself, and does not rely on any additional knowledge. In order to illustrate its ability to scale, the approach is evaluated on the largest collection of identity links to date, containing over 558M `owl:sameAs` links scraped from the LOD Cloud.

Keywords: Linked Open Data · Identity · owl:sameAs · Communities

1 Introduction

As the Web of Data continues to grow, more and more large datasets – covering a wide range of topics – are being added to the Linked Open Data (LOD) Cloud. It is inevitable that different datasets, most of which are developed independently of one another, will come to describe (aspects of) the same thing, but will do so by referring to that thing with different names. This situation is not accidental: it is a defining characteristic of the (Semantic) Web that there is no central naming authority that is able to enforce a Unique Name Assumption (UNA). As a consequence, identity link detection, i.e., the ability to determine – with a certain degree of confidence – that two different names in fact denote the same thing, is not a mere luxury but is essential for Linked Data to work. Thanks to identity links, datasets that have been constructed independently

© Springer Nature Switzerland AG 2018
D. Vrandečić et al. (Eds.): ISWC 2018, LNCS 11136, pp. 391–407, 2018.
https://doi.org/10.1007/978-3-030-00671-6_23

of one another are still able to make use of each other's information. The most common predicate that is used for interlinking data on the web is the `owl:sameAs` property. This property denotes a very strict notion of identity that is formalized in model theory. It is defined by Dean et al. [9] as: "an `owl:sameAs` statement indicates that two references actually refer to the same thing". As a result, a statement of the form "x `owl:sameAs` y" indicates that *every* property attributed to x must also be attributed to y, and vice versa.

Over time, an increasing number of studies have shown that `owl:sameAs` is sometimes used incorrectly in practice. For example, Jaffri et al. [15] discuss how erroneous uses of `owl:sameAs` in the linking of DBpedia and DBLP has resulted in several publications being affiliated to incorrect authors. In addition, Ding et al. [10] discuss a number of issues that arise when linking New York Times data to DBpedia. Specifically, they discuss issues that arise when two things are considered the same in some, but not all contexts.

Halpin et al. [13] discuss how the 'sameAs problem', originates from the identity and reference problems in philosophy. In the Semantic Web literature, several approaches have been proposed that focus on limiting this problem. While some approaches consider the introduction of alternative properties that can replace `owl:sameAs` [13], or alternative semantics of the `owl:sameAs` property [1,22], other approaches focus on the (semi-)automatic detection of potentially incorrect `owl:sameAs` statements [5,7,20].

This paper presents a novel approach for the automatic detection of potentially erroneous `owl:sameAs` statements. The approach consists of applying an existing community detection algorithm to an RDF graph that contains solely `owl:sameAs` statements. Based on the communities that are detected, an error degree is calculated for each identity link in the graph. The error degree of an `owl:sameAs` link depends on the density of the community(ies) in which the two terms exist, and whether the identity link is symmetrical or not. It is subsequently used to rank identity links, allowing potentially erroneous links to be identified, and potentially true `owl:sameAs` to be validated.

Since the here presented approach is specifically developed in order to be applied to real-world data, the experiment is run on the largest collection of identity links to date, containing over 558 million `owl:sameAs` links scraped from the LOD Cloud. The evaluation indicates that the calculated error degrees are useful for identifying a large number of correct and erroneous identity links, when applied to this real-world data collection.

The rest of this paper is structured as follows. The next section discusses related work. The approach for detecting potentially erroneous identity links is presented in Sect. 3. The experiments and the evaluation are described in Sect. 4, and Sect. 5 concludes.

2   Related Work

This section will give an overview of the related work on detecting erroneous identity links (Sect. 2.1) and existing approaches for community detection (Sect. 2.2).

We will briefly reflect on why we believe community detection to be a particularly good fit for identity error detection in Sect. 2.3.

2.1 Identity Error Detection

Source Trustworthiness. An early approach for detecting erroneous identity statements in the Web of Data is idMesh [5], a probabilistic and decentralized framework for entity disambiguation. idMesh hypothesizes that links published by trusted sources (e.g., OpenID-based) are more likely to be correct. The approach detects conflicts between `owl:sameAs` and `owl:differentFrom` assertions by using a graph-based constraint satisfaction solver that exploits the symmetric and transitive nature of the `owl:sameAs` relation. The detected conflicts are resolved based on the iteratively refined trustworthiness of the sources from which the assertions originate.

UNA Violations. Several approaches have made use of the hypothesis that individual datasets apply the Unique Name Assumption (UNA) [7,23], and that violations of the UNA that are caused by cross-dataset linking are indicative of erroneous identity links. De Melo [7] applies a linear programming relaxation algorithm that seeks to delete the minimal number of `owl:sameAs` statements such that the UNA is no longer violated. Valdestilhas et al. [23] efficiently detect the resources that share the same equivalence class and that belong to the same dataset, and ranks erroneous candidates based on the number of UNA violations.

Content-Based. Paulheim [21] represents each identity link as a feature vector in a high dimensional vector space, using direct types and in- and/or outgoing properties. They have tested different outlier detection methods in order to assign a score to each link, indicating the likeliness of being an outlier. Cuzzola et al. [6] propose to calculate a similarity score between the names that are involved in a given `owl:sameAs` link, by using the textual descriptions that are associated to these names (e.g., through the `rdfs:comment` property).

Ontology Axiom Violations. Hogan et al. [14] exploit ten OWL 2 RL rules in order to express the semantics of axioms such as *differentFrom* and *complementOf* in order to detect inconsistencies. Whenever an inconsistent equality set is detected, the erroneous links are identified by incrementally rebuilding the equality set in a manner that preserves consistency. Papaleo et al. [20] exploit class disjointness, (inverse) functional properties, locally complete properties, and property mappings in order to detect inconsistencies in an RDF graph made of the subparts of the two RDF descriptions involved in conflicting statements.

Network Metrics. Finally Gueret et al. [12] hypothesizes that the quality of a link can be determined based on how connected a node is within the network in which it appears. The approach is based on the use of three classic network metrics (clustering coefficient, centrality, and degree), and two Linked Data-specific metrics (`owl:sameAs` chains, and description richness). The approach constructs a local network for a set of selected resources by querying the Web of Data. After measuring the different metrics, each local network is first extended by adding new edges and then analyzed again. The result of both analyses is compared to the ideal distribution for the different metrics.

2.2   Community Detection

Despite the absence of a universally agreed upon definition, communities are typically thought of as groups that have dense connections among their members, but sparse connections with the rest of the network. Community detection is a form of data analysis that seeks to automatically determine the community structure of a complex network [18,19]. It has applications in statistical physics, mathematics, computer science, biology, and sociology [24]. Importantly, community detection only requires information that is already encoded in the network topology.

Several community detection algorithms exist, as well as several comparative studies. Lancichinetti et al. [16] analyse 12 community detection algorithms by applying them to the LFR graph benchmark [17]. In a more recent study, Yang et al. [24] have evaluated 8 of the most widely used community detection algorithms, again on the LFR benchmark graphs. From both meta studies, the Louvain algorithm emerges as combining a high accuracy with good computational performance.

The Louvain algorithm [4] is a greedy heuristic method, that starts out by assigning a different community to each node of a given network. It then moves each node over to one of its neighbor communities, specifically, neighbors, the one which results in the highest contribution to a modularity score. In the next step, each community from the previous step is regarded as a single node, and the same procedure is repeated until the modularity (which is always computed with respect to the original graph) no longer increases.

2.3   Discussion

We believe community detection to be a particularly good fit for identity error detection, since it can be applied to the network structure of the owl:sameAs graph itself. Specifically, the approach that we suggest does not require access to resource descriptions, property mappings, or vocabulary alignments. Also, it does not rely on additional assumptions like the UNA that could be false for some dataset (e.g., datasets that are constructed over a longer period of time and/or by a large group of contributors). Finally, current approaches for identity error detection have not always been applied to real-world owl:sameAs links, and no current approach has been evaluated at web scale, i.e., applied to hundreds of millions of links. Since the Louvain algorithm has already been successfully used in other domains, we believe that it can also perform well on the task of detecting owl:sameAs-based communities.

3   Approach

This section presents our approach for detecting erroneous identity links by exploiting the community structure of the identity network itself. This section describes the two main steps that our approach is composed of: firstly, the extraction and compaction of the identity network (Sect. 3.1), and secondly, the ranking of each identity link based on the community structure (Sect. 3.2). Algorithm 1 provides an effective procedure for calculating this ranking.

3.1 Identity Network Construction

The first step of our approach consists of extracting the identity network from a given data graph (Definition 1).

Definition 1 (Data Graph). *A data graph is a directed and labeled graph* $G = (V, E, \Sigma_E, l_E)$. *V is the set of nodes*[1]. *E is the set of node pairs or edges.* Σ_E *is the set of edge labels.* $l_E : E \rightarrow \Sigma_E$ *is the mapping from edges to edge labels.* $l_E(e)$ *denotes the labels of edge e.*

We use e_{ij} to denote the edge between nodes v_i and v_j. From a given data graph G, we can extract the explicit identity network N_{ex} (Definition 2), which is a directed labeled graph that only includes those edges whose labels include `owl:sameAs`.

Definition 2 (Explicit Identity Network). *Given a graph* $G = (V, E, \Sigma_E, l_E)$, *the related explicit identity network is the edge-induced subgraph* $G[\{e \in E \mid \{owl:sameAs\} \subseteq l_E(e)\}]$.

We can reduce the size of the explicit identity network N_{ex} into a more concisely represented undirected and weighted identity network I (Definition 3), without losing any significant information. Since reflexive `owl:sameAs` statements are implied by the semantics of identity, there is no need to represent them explicitly. In addition, since the symmetric statements e_{ij} and e_{ji} make the same assertion: that v_i and v_j refer to the same thing, we can represent this more efficiently, by including only one undirected edge with a weight of 2. A weight of 1 is assigned for edges which either e_{ij} or e_{ji}, but not both, are present in N_{ex}.

Definition 3 (Identity Network). *The identity network is an undirected labeled graph* $I = (V_I, E_I, \{1, 2\}, w)$, *where* V_I *is the set of nodes,* E_I *is the set of edges,* $\{1, 2\}$ *are the edges labels, and* $w : E_I \rightarrow \{1, 2\}$ *is the labeling function that assigns a weight* w_{ij} *to each edge* e_{ij}. *For each explicit identity network* $N_{ex} = (V_{ex}, E_{ex})$, *the corresponding identity network I is derived as follows:*

- $E_I := \{e_{ij} \in E_{ex} \mid i \neq j\}$
- $V_I := V_{ex}[E_I]$, *i.e., the vertex-induced subgraph.*
- $w(e_{ij}) := \begin{cases} 1, & \text{if } e_{ij} \in E_{ex} \\ 2, & \text{if } e_{ji} \in E_{ex} \end{cases}$

---

[1] In RDF, nodes are terms that appear in the subject and/or object position of at least one triple.

3.2   Links Ranking

Given $I = (V_I, E_I, \Sigma_{E_I}, w)$, a partitioning of V_I is a collection of non-empty and mutually disjoint subsets $V_k \subseteq V_I$ that together cover V_I. Since the closure of E_I forms an equivalence set (the semantics of the `owl:sameAs` property states that it is reflexive, symmetric, and transitive), it also induces a unique partitioning. We call members of this partition *identity sets*. These partition members correspond to the connected components of I that we call *equality sets* (Definition 4).

Definition 4 (Equality Set). *Given an identity network* $I = (V_I, E_I, \{1, 2\}, w)$, *an equality set* Q_k *is a connected component of* I.

We want to detect erroneous identity links based on the community structure of each connected component of the identity network. While the number of potential identity links is quadratic in the size of the domain, the representation of equality sets is only linear in terms of the size of the domain. With equality sets, we can implement the following requirements for our algorithm:

- The calculation of erroneous identity links must not have a large memory footprint, since it must be able to scale to very large identity networks, and preferably to all identity statements that appear in the LOD Cloud.
- It must be possible to perform computation in parallel, to allow errors to be detected relatively quickly, preferably directly after the publication of the potential error into the LOD Cloud.
- Calculation must be resilient against incremental updates. Since triples are added to and removed from the LOD Cloud constantly, adding or removing a `owl:sameAs` link must only require a re-ranking of the links within the equality sets that are directly involved in this link.

In order to compute a ranking for the `owl:sameAs` links, we first partition the identity network into different equality sets (several graph partitioning techniques could be applied, such as [2]). Then we detect a set of non-overlapping communities by applying the *Louvain* algorithm [4] for each equality set.

Given an equality set Q_k, the *Louvain* algorithm returns a set of non overlapping communities $C(Q_k) = \{C_1, C_2, \ldots, C_n\}$ where:

- a community C of size $|C|$ (i.e. the number of nodes) is a subgraph of Q_k such that the nodes of C are densely connected (i.e. the modularity of the Q_k is maximized).
- $\bigcup_{1 \leq i \leq n} C_i = Q_k$ and $\forall C_i, C_j \in C(Q_k)$ s.t. $i \neq j, C_i \cap C_j = \emptyset$.

We then evaluate each identity link by relying on its weight and the structure of the communities it occurs in. More precisely, to compute the erroneous degree, we distinguish between two types of links: the *intra-community links* and *inter-community links*.

Definition 5. Intra-Community Link. *Given a community* C, *an intra-community link in* C *noted by* e_C *is a weighted edge* e_{ij} *where* v_i *and* $v_j \in C$. *We denote by* E_C *the set of intra-community links in* C.

Definition 6. Inter-Community Link. *Given two non overlapping communities C_i and C_j, an inter-community link between C_i and C_j noted by $e_{C_{ij}}$ is an edge e_{ij} where $v_i \in C_i$ and $v_j \in C_j$. We denote by $E_{C_{ij}}$ the set of inter-community link between C_i and C_j.*

For evaluating an *intra-community link*, we rely both on the density of the community containing the edge, and the weight of this edge. The lower the density of this community and the weight of an edge are, the higher the *error degree* will be.

Definition 7. Intra-Community Link Error Degree. *Let e_C be an intra-community link of the community C, the intra-community error degree of e_c denoted by $err(e_C)$ is defined as follows:*

$$(a)\; err(e_C) = \frac{1}{w(e_C)} \times \left(1 - \frac{W_C}{|C| \times (|C| - 1)}\right)$$

where $W_C = \sum_{e_C \in E_C} w(e)$

For evaluating an *inter-community link*, we rely both on the density of the inter-community connections, and the weight of this edge. The less the two communities are connected to each other and the lower the weight of an edge is, the higher the *error degree* will be.

Definition 8. Inter-Community Link Error Degree. *Let $e_{C_{ij}}$ be an inter-community link of the communities C_i and C_j, the inter-community error degree of $e_{C_{ij}}$ denoted by $err(e_{C_{ij}})$ is defined as follows:*

$$(b)\; err(e_{C_{ij}}) = \frac{1}{w(e_{C_{ij}})} \times \left(1 - \frac{W_{C_{ij}}}{2 \times |C_i| \times |C_j|}\right)$$

where $W_{C_{ij}} = \sum_{e_{C_{ij}} \in E_{C_{ij}}} w(e)$

4   Experiments

4.1   Dataset

We have tested our approach on the LOD-a-lot dataset [11][2], a compressed data file that contains 28 billion unique triples from the 2015 LOD Laundromat Linked Data crawl [3]. This large subset of the LOD Cloud represents our data graph (Definition 1).

---

[2] http://lod-a-lot.lod.labs.vu.nl.

Algorithm 1. Identity Links Ranking

Input: G: a Data graph
Output: E^{err}: a set of pairs in the from $\{(e_1, err(e_1)), \ldots, (e_m, err(e_m))\}$ with
              m is the number of edges in the identity network extracted from G

1 $I_{ex} \leftarrow ExtractSameAsEdges(G)$; // the explicit identity network
2 $I \leftarrow empty\_graph$; // the identity network
3 **foreach** $(e(v_1, v_2) \in I_{ex}$ and $v_1 \neq v_2)$ **do**
4   **if** $(I.containsEdge(e(v_2, v_1, 1)))$ **then**
5          $I.updateWeight(e(v_2, v_1, 2)$; // set the weight of this edge to 2
6   **else**
7          $I.addEdge(e(v_1, v_2, 1))$; // add this edge to I with a weight $= 1$

8 $P \leftarrow I.partition()$; // partitioning the graph into equality sets
9 **foreach** $(Q \in P)$ **do**
10     $C_{set} \leftarrow LouvainCommunityDetectionAlgorithm(Q)$;
11   **foreach** $(e \in C_{set})$ **do**
12   **if** $(e$ is intra-community-edge(c_i) **then**
13             $err(e) \leftarrow intraCommunityErroneousness(c_i)$;
14   **else**
15   // e is an inter-community edge, c_j is the other community to which
               e is belonging to;
16             $err(e) \leftarrow interCommunityErroneousness(c_i, c_j)$;
17         $E^{err}.add(e, err(e))$;

18 **return** E^{err};

4.2   Quantitative Results

Explicit Identity Network Extraction. We have extracted the explicit identity network (Definition 2) from the data graph described above, by performing a Triple Pattern query of the form $\langle ?, \texttt{owl:sameAs}, ? \rangle$ with the HDT C++ library[3]). This returns a stream of distinct identity pairs, as described in [2]. This extraction process takes around four hours using 1 CPU core, resulting in an explicit identity network of 558.9M edges and 179.73M nodes. The explicit identity network is publicly available at https://sameas.cc/triple.

Identity Network Construction. From the explicit identify network described above, we build an identity network (Definition 3) containing ∼331M weighted edges and 179.67M terms. We leave out ∼2.8M reflexive edges and ∼225M *duplicate* symmetric edges. As a result, we also leave out 67,261 nodes that only appear in such removed edges. This indicates that 68% of the identity network edges are redundantly asserted, with a weight $= 2$.

Graph Partitioning. The next step consists of partitioning the identity network into several equality sets (Definition 4). We have deployed an efficient algorithm described in [2] that partitions the identity network into ∼49M equality

---

[3] https://github.com/rdfhdt/hdt-cpp.

sets, in just under 5 h using 2 CPU cores. The identity sets are publicly available at http://sameas.cc/id.

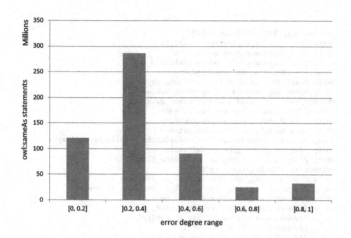

Fig. 1. Error degree distribution of 556M `owl:sameAs` statements

Links Ranking. Once the identity network has been partitioned, we apply the *Louvain* algorithm to detect communities in each equality set. We then assign an error degree to all edges of each equality set. This process takes 80 min[4], resulting an error degree to each irreflexive[5] `owl:sameAs` statement (~556M statements) in the explicit identity network. The error degree distribution of these statements is presented in Fig. 1, showing that around 73% of the statements have an error degree below 0.4. Whilst this distribution is mainly caused by the high number of symmetrical identity statements in the LOD, it also indicates that most equality sets have a rather dense structure. The 179.67M terms of the identity network were assigned into a total of 24.35M communities, with the communities size varying between 2 and 4,934 terms (averaging ~7 terms per community). The Java implementation of the link ranking process is available at http://github.com/raadjoe/LOD-Community-Detection. The erroneous degree of all the `owl:sameAs` statements are available in our identity web service (https://sameAs.cc).

4.3 Community Structure Analysis

In this section we provide a first analysis of the community structure obtained from two equality sets (the largest one and the one about Barack Obama) based on the IRIs contained in the communities. In a 2016 study conducted on the same data collection, de Rooij et al. [8] have shown that the social meaning

---

[4] On an 8 GB RAM Windows 10 machine, using 2 CPU cores.

[5] Reflexive statements were discarded in I, and symmetric ones have the same *err*.

encoded in IRI names significantly coincides with the formal meaning of IRI-denoted resources. Hence, indicating that IRIs can give an idea on the quality of the detected communities.

```
-- Community 258 --  (size = 242)
<http://af.dbpedia.org/resource/Dublin>
<http://am.dbpedia.org/resource/ደብሊን>
<http://an.dbpedia.org/resource/Dublín>
<http://ar.dbpedia.org/resource/دبلن>
<http://ast.dbpedia.org/resource/Ciudá_de_Dublín>
<http://bat-smg.dbpedia.org/resource/Doblėns>
<http://be-x-old.dbpedia.org/resource/Дублін>
<http://br.dbpedia.org/resource/Dulenn>
<http://ca.dbpedia.org/resource/Dublín>
<http://ce.dbpedia.org/resource/Дублин>
<http://commons.dbpedia.org/resource/Dublin_-_Baile_Átha_Cliath>
<http://cs.dbpedia.org/resource/Dublin>
<http://dbpedia.org/resource/Baile_Atha_Cliath>
<http://dbpedia.org/resource/BÁC>
<http://dbpedia.org/resource/Capital_of_Ireland>
<http://dbpedia.org/resource/Capital_of_Republic_of_Ireland>
<http://dbpedia.org/resource/Central_Dublin>
<http://dbpedia.org/resource/City_Center,_Dublin>
<http://dbpedia.org/resource/City_of_Dublin>
<http://dbpedia.org/resource/Dyflin>
<http://dbpedia.org/resource/Europe/Dublin>
<http://dbpedia.org/resource/The_weather_in_Dublin>
<http://dbpedia.org/resource/UN/LOCODE:IEDUB>
<http://dbpedia.org/resource/Visitor_Information_for_Dublin,_Ireland>
<http://dbpedia.org/resource/West_Dublin>
<http://de.dbpedia.org/resource/Dublin>
<http://demo.openlinksw.com/Northwind/Province/ei/Dublin#this>
<http://sws.geonames.org/2964574/>
<http://wordnet.rkbexplorer.com/id/synset-Dublin-noun-1>
<http://www4.wiwiss.fu-berlin.de/flickrwrappr/photos/Dublin>
```

Fig. 2. Excerpt of the 242 terms included in the community containing the IRI http://dbpedia.org/resource/dublin

Community Structure in the Largest Equality Set. The largest equality set Q_{max} contains 177,794 terms connected by 2,849,650 undirected and weighted edges. This equality set is the result of the compaction of 5,547,463 distinct owl:sameAs statements (~1% of the total number of owl:sameAs) and is available at https://sameas.cc/term?id=4073. By looking at the IRIs of this equality set, we can observe that it contains a large number of terms denoting different countries, cities, things and persons (e.g. Bolivia, Dublin, Coca-Cola, Albert Einstein, *Literals*, and so on). Clearly showing that this equality set contains many erroneous owl:sameAs statements.

Applying the *Louvain* algorithm on Q_{max} resulted in 930 non-overlapping communities, with a size varying from 32 to 2,320 terms per community. As a first interpretation on the community structure, we have solely looked at the IRIs. Despite a few exceptions, we can see that this algorithm is able to group related (and possibly identical) terms in the same community, while keeping out unrelated terms in other communities. For instance, the community C_{258}, illustrated in Fig. 2 contains 242 terms. We can see from this excerpt that most of these terms come from the DBpedia dataset and refer to descriptions of Dublin expressed in different languages: `City of Dublin`, `Capital of Ireland`, `Baile Atha Cliath` (Dublin in Irish), `Dyflin` (the old Norse name for The Kingdom of Dublin), etc. However, we can also see that this community contains terms that do not refer to the city of Dublin, but actually refer to the weather in Dublin or visitor information for Dublin.

With this excerpt of the Dublin community, we can see that an `owl:sameAs` statement between two terms in the same community is not necessarily correct, and requires evaluation as well.

Community Structure in the 'Barack Obama' Equality Set. We present here an analysis of the community structure detected on the equality set Q_{obama} which has a reasonable size and thus easier to analyse. The equality set containing the term http://dbpedia.org/resource/Barack_Obama is composed of 440 terms connected by 7,615 undirected and weighted edges. It is built from an explicit identity network of 14,917 `owl:sameAs` statements.

Applying the *Louvain* algorithm on Q_{obama} resulted in 4 non-overlapping communities, with a size varying from 34 to 166 terms per community. This identity set is available at (https://sameas.cc/term?id=5723). The resulting community structure of Q_{obama} is presented in Fig. 3:

- C_0 **(purple)** includes 166 terms, with 98% of the links of this community representing cross-language symmetrical links between DBpedia IRIs (e.g. http://fr.dbpedia.org/resource/Barack_Obama) referring to the person Barack Obama.
- C_1 **(green)** includes 162 terms, mostly DBpedia IRIs of the person Obama in his different roles and political functions (e.g. http://dbpedia.org/resource/President_barack_obama, http://dbpedia.org/resource/senator_obama).
- C_2 **(orange)** includes 78 terms, mostly referring to the presidency and administration of Barack Obama (e.g. http://dbpedia.org/resource/Obama_cabinet, http://dbpedia.org/resource/Barack_Hussein_Obama_administration)
- C_3 **(blue)** includes 34 terms from different datasets denoting various entities such as: Barack Obama the person, his senate career, and a misused literal ("http://dbpedia.org/resource/United_States_Senate_career_of_Barack_Obama", "http://dbpedia.org/resource/Barack_Obama"^^`xsd:string`).

4.4 Links Ranking Evaluation

In order to evaluate the accuracy of our ranking approach, we have conducted several manual evaluations. The judges relied on the descriptions[6] associated to the terms in the *LOD-a-lot* dataset [11], and did not have any prior knowledge about each link's error degree (i.e. whether they are evaluating a well-ranked link or not). In order to avoid any incoherence between the evaluations, the

---

[6] The judges were asked to not consider the `owl:sameAs` statements related to the term.

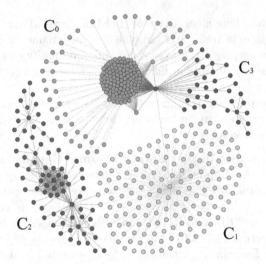

Fig. 3. The communities detected from the equality set containing the term http://dbpedia.org/resource/Barack_Obama using the *Louvain* algorithm. The 4 detected communities are distinguished by their nodes' color. The full figure is available at https://github.com/raadjoe/LOD-Community-Detection/blob/master/Communities-Graph-Obama.svg. (Color figure online)

judges were asked to justify all their evaluations and were given the following instructions: **(a) the same:** if two terms denote the same entity (e.g. Obama and the First Black US President), **(b) related:** not intended to refer to the same entity but closely related (e.g. Obama and the Obama Administration), **(c) unrelated:** not the same nor closely related (e.g. Obama and the Indian Ocean), **(d) can't tell:** in case there are no sufficient descriptions available for determining the meaning of both terms.

A. Accuracy Evaluation in the 'Barack Obama' Equality Set. Firstly, we have relied on the previous observations, made on the community structure presented in Fig. 3, to interpret and evaluate the accuracy of our approach:

 (i) an `owl:sameAs` statement in C_0 has an average error rate of 0.24. The manual evaluation of 30 random `owl:sameAs` statements shows that they are all true identity links.
 (ii) the low density of C_1 has led to several correct `owl:sameAs` statements to have a high error degree (0.9). This is due to the fact that there is only one term linking to all the 161 other terms in this community, with most of these edges being non-symmetrical links.
(iii) the only two `owl:sameAs` statements in this equality set with an error value $\simeq 1$ are the edges in the graph connecting the IRI http://rdf.freebase.com/ns/m.05b6w1g from C_2 to both IRIs http://dbpedia.org/resource/President_Barack_Obama and http://dbpedia.org/resource/President_Obama from C_1. Relying on their descriptions in the *LOD-a-lot*

dataset, we can see that the freebase IRI refers to the presidency of Obama, while the two other IRIs refer to the person Obama, indicating that indeed both statements are incorrect. These two detected incorrect identity statements have led to the false equivalence of the 78 terms of C_2 with the rest of the network's terms.

B. Accuracy Evaluation on a Subset of the Identity Network. In order to evaluate the accuracy over the whole identity network, four of this paper's authors were asked to evaluate a subset of the identity network. The judges were asked to evaluate 200 `owl:sameAs` links (50 links each), representing in an equal manner, each bin of the error degree distribution presented in Fig. 1.

Table 1. Evaluation of 200 `owl:sameAs` links, with each 40 links randomly chosen from a certain range of error degree

Error degree range	0–0.2	0.2–0.4	0.4–0.6	0.6–0.8	0.8–1	Total
same	35(100%)	22(100%)	18(85.7%)	7(77.8%)	15(68.2%)	**97(89%)**
related	0	0	2	2	2	**6**
unrelated	0	0	1	0	5	**6**
related + unrelated	0(0%)	0(0%)	3(14.3%)	2(22.2%)	7(31.8%)	**12(11%)**
can't tell	5	18	19	31	18	**91**
Total	**40**	**40**	**40**	**40**	**40**	**200**

The results presented in Table 1, shows that the higher an error degree is, the more likely that the link is erroneous. More precisely, we may observe that:

- our error degree is able to identify true `owl:sameAs` links with a high accuracy, since 100% of the evaluated links with an error degree ≤ 0.4. are correct (without considering the "*can't tell*" cases).
- when the error degree is between 0.4 and 0.8, 83.3% of the `owl:sameAs` links are correct. However, in 13.3% of the cases, such links might have been used to refer to two different, but related terms.
- an `owl:sameAs` with an error degree >0.8 is an unreliable identity statement, referring in 31.8% of the cases to two different, and mostly unrelated terms.

We have further investigated the 22 evaluated identity links with an error degree over 0.8. Two features were observed from the 7 incorrect identity statements: (i) their error degree is most of the times higher than the true `owl:sameAs` links, and (ii) they all belong to equality sets with a higher number of terms than the true ones. To further investigate these observations, we have evaluated 60 additional links with an error degree >0.9. The first set of links (S1) represents 20 random identity links from the largest equality set. The second set of links (S2) represents 20 random identity links with an error degree $\simeq 1$ (>0.99). The third set of links (S3) represents 20 random links from the largest equality set with an error degree $\simeq 1$. The results presented in Table 2, show that our approach has a

Table 2. Evaluation of 60 `owl:sameAs` links with an error degree >0.9, with the first set of 20 `owl:sameAs` links (S1) randomly chosen from the largest equality set, (S2) randomly chosen from all links with an error degree ≃1, (S3) randomly chosen from the largest equality set with an error degree ≃1

	Largest equality set (S1)	$err \simeq 1$ (S2)	Largest & $err \simeq 1$ (S3)
same	6(50%)	6(60%)	2(11.7%)
related	1	1	2
unrelated	5	3	13
related+unrelated	6(50%)	4(40%)	15(88.2%)
can't tell	8	10	3
Total	20	20	20

high accuracy in detecting erroneous identity links when the threshold is fixed at 0.99 and only equality sets with a high number of terms are considered.

C. Recall Evaluation. In order to calculate the recall of our approach, we have verified how our approach can rank newly introduced erroneous `owl:sameAs` statements. Firstly, we have chosen 40 random terms[7] in the explicit identity network, making sure that all these terms are different, by looking at their descriptions, and that they are not explicitly `owl:sameAs`. From the selected 40 terms, we have generated all the possible 780 undirected edges between them. We added separately, each edge e_{ij} to the identity network with $w(e_{ij}) = 1$, calculated its error degree, and removed it from the identity network before adding the next one. The error degrees of the newly introduced erroneous identity links range from 0.87 to 0.9999. When the threshold is fixed at 0.99, the recall is 93%.

Results Interpretation. The experiments conducted in this paper, on a subset of 28 billion unique triples of the LOD cloud, shows that there exist many false identity statements on the Web. These erroneous `owl:sameAs` statements have led to the false equivalence of many unrelated terms (e.g. Dublin, Coca-Cola, and Albert Einstein), and many related terms (e.g. Barack Obama the person, and his administration). With a total runtime of 11 h, these experiments show that an error degree of every available identity link can be computed in practice. Our manual evaluation of these error degrees suggests that:

1. **our approach can validate a large number of identity links in the LOD:** 73% of the identity links have an error degree of [0-0.4]. All the manually evaluated links in this range were judged as true `owl:sameAs` links (100% accuracy, Table 1).
2. **our approach can detect numerous erroneous identity links in the LOD:** more than 1.2 million `owl:sameAs` links have an error degree of [0.99-1], with a large number of these links coming from large equality sets (e.g.

---

[7] We also made sure to include 5 terms that belong to the same equality set.

~13K links in the largest equality set). Up to 88% of the manually evaluated links with these criteria were judged as false identity statements (Table 2).

3. **refined content-based approaches are needed** for evaluating the remaining `owl:sameAs` links in the LOD (between 50 and 85% were judged as true identity links).

5 Conclusion

We have presented an approach that aims to detect erroneous `owl:sameAs` statements in RDF data sources. Our approach is uniquely based on the topology of the identity network itself. In order to illustrate its ability to scale, we have evaluated our approach over 558 million `owl:sameAs` statements that are scraped from the LOD Cloud. The evaluation shows that the here introduced calculation of an error degree can indeed be used in order to distinguish between correct and incorrect `owl:sameAs` statements. With a total runtime of 11 h, these error degrees can be computed in practice. The erroneous degree of all the evaluated `owl:sameAs` statements are available in our identity web service (https:// sameAs.cc). This will allow others to replicate, check, and hopefully improve upon the here presented results.

The accuracy of the here presented approach could be further improved by combining or comparing results from multiple community detection methods. Since adding a new dataset to the LOD Cloud only requires recalculation of the equivalence sets that are involved in identity assertions within that dataset, it could be useful to test whether the quality of identity links can now be calculated online, e.g., as part of the publication of a dataset into a widely used data catalog.

Acknowledgment. This work was partially conducted within the MaestroGraph project (612.001.553), funded by the Netherlands Organization for Scientific Research (NWO), and was partially supported by the Center for Data Science, funded by the IDEX Paris-Saclay, ANR-11-IDEX-0003-02.

References

1. Beek, W., Schlobach, S., van Harmelen, F.: A contextualised semantics for `owl:sameAs`. In: Sack, H., Blomqvist, E., d'Aquin, M., Ghidini, C., Ponzetto, S.P., Lange, C. (eds.) ESWC 2016. LNCS, vol. 9678, pp. 405–419. Springer, Cham (2016). https://doi.org/10.1007/978-3-319-34129-3_25

2. Beek, W., Raad, J., Wielemaker, J., van Harmelen, F.: sameAs.cc: the closure of 500M `owl:sameAs` statements. In: Gangemi, A., et al. (eds.) ESWC 2018. LNCS, vol. 10843, pp. 65–80. Springer, Cham (2018). https://doi.org/10.1007/978-3-319-93417-4_5

3. Beek, W., Rietveld, L., Schlobach, S.: Lod laundromat (archival package 2016/06) (2016). https://doi.org/10.17026/dans-znh-bcg3

4. Blondel, V., Guillaume, J.-L., Lambiotte, R., Lefebvre, E.: Fast unfolding of communities in large networks. J. Stat. Mech. **2008**(10), P10008 (2008)

5. Cudré-Mauroux, P., Haghani, P., Jost, M., Aberer, K., De Meer, H.: idMesh: graph-based disambiguation of linked data. In: WWW Conference, pp. 591–600 (2009)
6. Cuzzola, J., Bagheri, E., Jovanovic, J.: Filtering inaccurate entity co-references on the linked open data. In: Chen, Q., Hameurlain, A., Toumani, F., Wagner, R., Decker, H. (eds.) DEXA 2015. LNCS, vol. 9261, pp. 128–143. Springer, Cham (2015). https://doi.org/10.1007/978-3-319-22849-5_10
7. de Melo, G.: Not quite the same: identity constraints for the web of linked data. In: des Jardins, M., Littman, M.L. (eds.) AAAI. AAAI Press (2013)
8. de Rooij, S., Beek, W., Bloem, P., van Harmelen, F., Schlobach, S.: Are names meaningful? Quantifying social meaning on the semantic web. In: Groth, P., et al. (eds.) ISWC 2016. LNCS, vol. 9981, pp. 184–199. Springer, Cham (2016). https://doi.org/10.1007/978-3-319-46523-4_12
9. Dean, M., et al.: Owl web ontology language reference. W3C Recommendation, 10 February 2004
10. Ding, L., Shinavier, J., Finin, T., McGuinness, D.L.: owl:sameAs and linked data: an empirical study. In: Proceedings of the Second Web Science Conference (2010)
11. Fernández, J.D., Beek, W., Martínez-Prieto, M.A., Arias, M.: LOD-a-lot – a queryable dump of the LOD cloud. In: d'Amato, C. (ed.) ISWC 2017. LNCS, vol. 10588, pp. 75–83. Springer, Cham (2017). https://doi.org/10.1007/978-3-319-68204-4_7
12. Guéret, C., Groth, P., Stadler, C., Lehmann, J.: Assessing linked data mappings using network measures. In: Simperl, E., Cimiano, P., Polleres, A., Corcho, O., Presutti, V. (eds.) ESWC 2012. LNCS, vol. 7295, pp. 87–102. Springer, Heidelberg (2012). https://doi.org/10.1007/978-3-642-30284-8_13
13. Halpin, H., Hayes, P.J., McCusker, J.P., McGuinness, D.L., Thompson, H.S.: When owl:sameAs isn't the same: an analysis of identity in linked data. In: Patel-Schneider, P.F. (ed.) ISWC 2010. LNCS, vol. 6496, pp. 305–320. Springer, Heidelberg (2010). https://doi.org/10.1007/978-3-642-17746-0_20
14. Hogan, A., Zimmermann, A., Umbrich, J., Polleres, A., Decker, S.: Scalable and distributed methods for entity matching, consolidation and disambiguation over linked data corpora. Web Semant.: Sci. Serv. Agents World Wide Web 10, 76–110 (2012)
15. Jaffri, A., Glaser, H., Millard, I.: URI disambiguation in the context of Linked Data. In: Linked Data on the Web Workshop (LDOW) (2008)
16. Lancichinetti, A., Fortunato, S.: Community detection algorithms: a comparative analysis. Phys. Rev. E 80(5), 056117 (2009)
17. Lancichinetti, A., Fortunato, S., Radicchi, F.: Benchmark graphs for testing community detection algorithms. Phys. Rev. E 78(4), 046110 (2008)
18. Liu, W., Pellegrini, M., Wang, X.: Detecting communities based on network topology. Sci. Rep. 4, 5739 (2014)
19. Newman, M.E.J.: Modularity and community structure in networks. Proc. Natl. Acad. Sci. 103(23), 8577–8582 (2006)
20. Papaleo, L., Pernelle, N., Saïs, F., Dumont, C.: Logical detection of invalid sameas statements in RDF data. In: Janowicz, K., Schlobach, S., Lambrix, P., Hyvönen, E. (eds.) EKAW 2014. LNCS (LNAI), vol. 8876, pp. 373–384. Springer, Cham (2014). https://doi.org/10.1007/978-3-319-13704-9_29
21. Paulheim, H.: Identifying wrong links between datasets by multi-dimensional outlier detection. In: WoDOOM, pp. 27–38 (2014)
22. Raad, J., Pernelle, N., Saïs, F.: Detection of contextual identity links in a knowledge base. In: KCAP (2017)

23. Valdestilhas, A., Soru, T., Ngomo, A.-C.N.: CEDAL: time-efficient detection of erroneous links in large-scale link repositories. In: International Conference on Web Intelligence, pp. 106–113. ACM (2017)
24. Yang, Z., Algesheimer, R., Tessone, C.: A comparative analysis of community detection algorithms on artificial networks. Sci. Rep. **6**, 30750 (2016)

SPgen: A Benchmark Generator for Spatial Link Discovery Tools

Tzanina Saveta[1(✉)], Irini Fundulaki[1], Giorgos Flouris[1],
and Axel-Cyrille Ngonga-Ngomo[2]

[1] Institute of Computer Science - FORTH, Heraklion, Greece
jsaveta@ics.forth.gr
[2] University of Paderborn, Paderborn, Germany

Abstract. A number of real and synthetic benchmarks have been proposed for evaluating the performance of link discovery systems. So far, only a limited number of *link discovery benchmarks* target the problem of linking *geo-spatial* entities. However, some of the largest knowledge bases of the Linked Open Data Web, such as LinkedGeoData contain vast amounts of spatial information. Several systems that manage spatial data and consider the topology of the spatial resources and the topological relations between them have been developed. In order to assess the ability of these systems to handle the vast amount of spatial data and perform the much needed data integration in the Linked Geo Data Cloud, it is imperative to develop benchmarks for geo-spatial link discovery. In this paper we propose the *Spatial Benchmark Generator SPgen* that can be used to test the performance of link discovery systems which deal with topological relations as proposed in the state of the art DE-9IM (Dimensionally Extended nine-Intersection Model). *SPgen* implements all topological relations of DE-9IM between *LineStrings* and *Polygons* in the two-dimensional space. A comparative analysis with benchmarks produced using *SPgen* to assess and identify the capabilities of AML, OntoIdea, RADON and Silk spatial link discovery systems is provided.

1 Introduction

The number of datasets published in the Web of Data as part of the Linked Data Cloud is constantly increasing. The Linked Data paradigm is based on the publication of information by different publishers, and the interlinking of Web resources across knowledge bases. In most cases, the cross-dataset links are not integral to newly created datasets and must be determined automatically, using *link discovery* tools amongst others [1]. The large variety of techniques demands the availability of comparative evaluations to determine which one is best suited for a given use case. Performing such an assessment requires well-defined and widely accepted *benchmarks* to determine the weak and strong points of the proposed techniques and/or tools. A number of real and synthetic benchmarks have been proposed for evaluating the performance of such systems [2].

© Springer Nature Switzerland AG 2018
D. Vrandečić et al. (Eds.): ISWC 2018, LNCS 11136, pp. 408–423, 2018.
https://doi.org/10.1007/978-3-030-00671-6_24

So far, only a limited number of *link discovery benchmarks* target the problem of linking geo-spatial entities. However, some of the largest knowledge bases in the Linked Open Data Web are geo-spatial knowledge bases (e.g., LinkedGeo-Data,[1] with more than 30 billion triples). In particular, considering the topology of the spatial resources and the topological relations between them is of central importance to systems that manage spatial data. We believe that due to the large amount of available geo-spatial datasets employed in various domains, it is critical that benchmarks for geo-spatial link discovery are developed.

In this paper we discuss the *Spatial Benchmark Generator SPgen* that can be used to test the performance of systems that deal with topological relations proposed by the state of the art DE-9IM (Dimensionally Extended nine-Intersection Model) [3]. *SPgen* is developed in the context of the H2020 European project HOBBIT.[2] This benchmark generator implements all topological relations of DE-9IM between *LineStrings* and *Polygons* in the two-dimensional space. *SPgen* follows the choke point-based approach [4] for benchmark design, i.e., it focuses on the technical difficulties of existing systems and implements tests that address those difficulties and "push" systems to resolve them. More specifically we focus on the following choke-points in *SPgen*:

- **Scalability:** produce datasets large enough to stress the systems under test
- **Output quality:** compute *precision, recall* and *f-measure*
- **Time performance:** measure the time the systems need to return the results

To the best of our knowledge such a *generic benchmark generator*, that checks the performance of linking systems for spatial data, does not exist. We also provide a comparative analysis with benchmarks produced using *SPgen* to assess and identify the capabilities of AML [5,6], OntoIdea [7], RADON [8] and Silk [9] spatial link discovery systems.

The outline of the paper is as follows: Sect. 2 discusses related work. We present the Dimensionally Extended nine-Intersection Model and the datasets employed in *SPgen* in Sects. 3 and 4 respectively. *SPgen* is described in detail in Sect. 5 and the experiments we conducted in Sect. 6. We conclude and present future work in Sect. 7.

2   Related Work

SPgen is a generic, schema agnostic and choke-point based [4] benchmark generator that takes as input *trajectories* and checks the performance of linking systems for spatial data. To the best of our knowledge this is the first link discovery benchmark for spatial data. In this section we will discuss the most relevant benchmarks to *SPgen* and more specifically *benchmarks for spatial RDF stores* and *benchmarks for spatial relational databases*.

---

[1] http://linkedgeodata.org/About.
[2] http://www.project-hobbit.eu.

Benchmarks for Spatial RDF Stores. The most relevant benchmark to *SPgen* is Geographica [10] that evaluates RDF stores and consists of micro and macro benchmarks following the approach of Jackpine [11]. Geographica's micro benchmark tests the spatial components of RDF stores using queries that consist of spatial selections, joins and aggregations but it does not address topological relations. Geographica's macro benchmark tests the performance of the RDF stores using reverse geocoding, map search and browsing and a real-world use case from the Earth Observation domain. Kolas [12] proposed a benchmark that extends the LUBM benchmark for RDF stores [13] in order to include spatial entities. In this case, LUBM queries were extended to cover basic types of spatial queries namely location, range, join and nearest neighbor.

Benchmarks for Spatial Relational Databases. The most recent benchmark for spatial databases and the most relevant benchmark to *SPgen* is Jackpine [11] and consists of a micro and a macro benchmark. Jackpine's micro benchmark includes queries based on DE-9IM with queries that focus on spatial analysis. Jackpine's macro benchmark includes queries based on spatial data applications (search and browsing, geocoding, flood risk analysis, etc.). VESPA [14] is a vector-based spatial benchmark that tests the functionality and performance of spatial database systems and includes a set of query and update tasks over synthetic datasets composed of points, lines and polygons. The Á la Carte [15] benchmark produces a synthetic dataset composed only of rectangles and is used to test the performance of different spatial join techniques. Last but not least, one of the first benchmarks that uses real datasets and real queries representative of Earth Science tasks is SEQUOIA [16] and its extension [17]. The queries are related to data loading, raster data management, filtering, spatial joins and path computation over graphs.

3   Dimensionally Extended Nine-Intersection Model (DE-9IM)

The Dimensionally Extended nine-Intersection Model (DE-9IM) [3] or Clementini-Matrix is used for computing the spatial relationships between geometries. It is a topological model, based on the Nine-Intersection Model (9IM), used to describe the spatial relations of geometries in two-dimensional space. The model considers the objects' *interiors*, *boundaries* and *exteriors* and analyzes the intersections of these nine objects parts to determine their relationship.

Spatial relations are *boolean* functions that are used to test the relationships between two geometry objects. The spatial relationships described by DE-9IM are *equals, disjoint, touches, contains, within, intersects, covers, covered by, crosses* and *overlaps* including relations among *LineStrings* and *Polygons*. A *LineString* is a one-dimensional geometric object and consists of a sequence of two or more vertices, along with all points along the linearly interpolated curves (line segments) between each pair of consecutive vertices. The line segments in the line may intersect each other. A *Polygon* is a two-dimensional surface stored as a sequence of points where the first point is connected to the last point defining its exterior bounding ring and zero or more interior rings. In order to better understand the topological relations of DE-9IM it is necessary to define the boundary, interior and exterior of the geometric types. For instance, in the case of *LineString*, the *boundary* (B) are the two end points, the *interior* (I) consists of *points* that are left when the boundary points are removed and the *exterior* (E) are the points not in the interior or boundary. In the case of *Polygon*, the *interior* are the points within the rings, the *boundary* is a set of rings and finally the *exterior* are points not in the interior or boundary.

Given that each geometry is represented by the aforementioned 3 dimensions, all possible relationships between two geometries are represented by a 3×3 matrix of the form:

$$\text{DE9IM}(a,b) = \begin{bmatrix} \dim(I(a) \cap I(b)) & \dim(I(a) \cap B(b)) & \dim(I(a) \cap E(b)) \\ \dim(B(a) \cap I(b)) & \dim(B(a) \cap B(b)) & \dim(B(a) \cap E(b)) \\ \dim(E(a) \cap I(b)) & \dim(E(a) \cap B(b)) & \dim(E(a) \cap E(b)) \end{bmatrix}$$

where dim is the maximum number of dimensions of the intersection ($\cap$) of the interior (I), boundary (B), and exterior (E) of geometries a and b. The dimension of empty sets is equal to -1 or F (false). The dimension of non-empty sets is equal to the maximum number of dimensions of the intersection, specifically, 0 for points, 1 for lines, 2 for areas. Thus, the domain of the model is $\{0, 1, 2, F\}$. A simplified version of $\dim(x)$ values is obtained by mapping the values $\{0, 1, 2\}$ to T (true), so using the boolean domain $\{T, F\}$. The supported spatial relations of DE-9IM are formally described below:

Equals: Two geometries g_1 and g_2 are *equal* if the two geometries are *topologically equal*, that is if their interiors intersect and no part of the interior or boundary of one geometry intersects the exterior of the other. Formally:

$$(I(g_1)I(g_2)) \wedge \neg(I(g_1)E(g_2)) \wedge \neg(B(g_1)E(g_2)) \wedge \neg(E(g_1)I(g_2)) \wedge \neg(E(g_1)B(g_2))$$

Disjoint: Two geometries g_1 and g_2 are *disjoint* if they have no point in common. Formally:

$$\neg(I(g_1)I(g_2)) \wedge \neg(I(g_1)B(g_2)) \wedge \neg(B(g_1)I(g_2)) \wedge \neg(B(g_1)B(g_2))$$

Touches: A geometry g_1 *touches(meets)* a geometry g_2 if they have at least one boundary point in common, but no interior points. Formally:

$$(\neg(I(g_1)I(g_2)) \wedge I(g_1)B(g_2)) \vee$$
$$(\neg(I(g_1)I(g_2)) \wedge B(g_1)I(g_2)) \vee (\neg(I(g_1)I(g_2)) \wedge B(g_1)B(g_2))$$

Contains: A geometry g_1 *contains* a geometry g_2 if g_2 lies in g_1, and the interiors intersect. Another definition is the following: g_1 contains g_2 if no points of g_2 lie in the exterior of g_1, and at least one point of the interior of g_2 lies in the interior of g_1. It is the inverse of *Within*. Formally:

$$(I(g_1)I(g_2)) \wedge \neg(E(g_1)I(g_2)) \wedge \neg(E(g_1)B(g_2))$$

Within: A geometry g_1 is within (inside) geometry g_2 if g_1 lies in the interior of g_2. *Within* is the inverse of *Contains*.

Intersects: A geometry g_1 intersects geometry g_2 if they have at least one point in common.

Covers: A geometry g_1 **covers** geometry g_2 if geometry g_2 lies in g_1. Other definitions: "no points of g_2 lie in the exterior of g_1", or "Every point of g_2 is a point of (the interior or boundary of) g_1". It is the inverse of *CoveredBy*. Formally:

$$((I(g_1)I(g_2)) \wedge \neg(E(g_1)I(g_2)) \wedge \neg(E(g_1)B(g_2)))\vee$$
$$((I(g_1)B(g_2)) \wedge \neg(E(g_1)I(g_2)) \wedge \neg(E(g_1)B(g_2)))\vee$$
$$((B(g_1)I(g_2)) \wedge \neg(E(g_1)I(g_2)) \wedge \neg(E(g_1)B(g_2)))\vee$$
$$((B(g_1)B(g_2)) \wedge \neg(E(g_1)I(g_2)) \wedge \neg(E(g_1)B(g_2)))$$

Covered By: A geometry g_1 is **covered by** geometry g_2 (extends *Within*) if every point of g_1 is a point of g_2, and the interiors of the two geometries have at least one point in common. *Covered by* is the inverse of *Covers*.

Crosses: A geometry g_1 *crosses* geometry g_2 if they share some but not all interior points, and the dimension of the intersection of the two geometries is less than that of at least one of the geometries.

Overlaps: A geometry g_1 *overlaps* geometry g_2 if the geometries share some, but not all points in common, and the intersection has the same dimension as the geometries themselves.

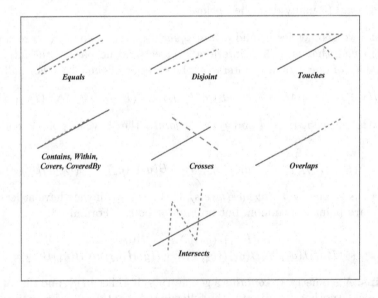

Fig. 1. Examples of DE-9IM topological relations for *LineStrings*

Examples of the DE-9IM relations for *LineStrings* and *Polygons* geometries are shown in Figs. 1 and 2. Figure 1 presents the DE-9IM relations between

LineStrings and Fig. 2 demonstrates the DE-9IM relations between LineStrings and Polygons.

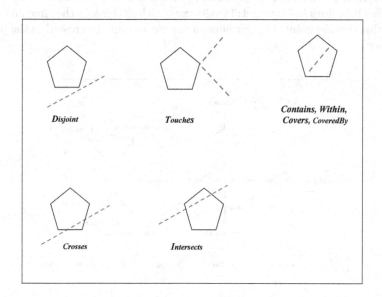

Fig. 2. Examples of DE-9IM topological relations for *LineStrings* and *Polygons*

4   Datasets

In this Section, we present the datasets we experimented with *SPgen*. Recall that the generator is *schema agnostic* and can work, in general, with trajectories, i.e., sequences of *longitude, latitude pairs*. We used two datasets generated from TomTom[3] and Spaten [18].

TomTom Data Generator: TomTom provides a Synthetic Trace Generator[4] developed in the context of the H2020 HOBBIT Project, which facilitates the creation of an arbitrary volume of data from statistical descriptions of vehicle traffic. More specifically, it generates *traces*, with a trace being a list of (longitude, latitude) pairs recorded by one device (phone, car, etc.) throughout one day. The generator uses probability distributions for variables like start and end locations of trips, their starting time or what is the device's update frequency. Using parameters sampled from such distributions, a map is then used to find an appropriate route for the trip and successive points are generated at a regular time interval with typical speeds for each road. TomTom's ontology is shown in Fig. 3. The main class is class `Trace` that contains one or more points

---

[3] https://www.tomtom.com/en_gr/.
[4] https://git.project-hobbit.eu/filipe.teixeira/synthetic-trace-generator/
  container_registry.

(class `Point`) which represents in its turn latitude, longitude pairs. Each point is associated with a *velocity* (class `Velocity`), instances of which have properties `velocityMetric` and `velocityValue`. A point also has attributes `hasTimeStamp` that takes its values in class `xsd:TimeStamp` which designates the time an object was at this specific point. For our benchmark we are only interested in the points of a trace.

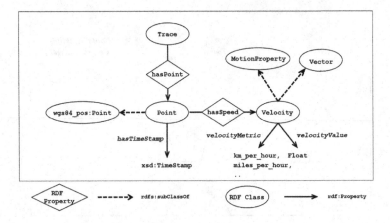

Fig. 3. TomTom schema

Spaten: Spatio-temporal and Textual Big Data Generator: Spaten [18] is an open-source configurable spatio-temporal and textual dataset generator, that can produce large volumes of data based on realistic user behavior. Spaten extracts GPS traces from realistic routes utilizing Google Maps API, and combines them with real POIs and relevant user comments crawled from TripAdvisor. The injection of social properties extracted by existing Twitter graphs to the generated data, along with further parameterization, leads to realistic Geo-Social Network (GeoSN) datasets. Spaten publicly offered GB-size datasets with millions of check-ins and GPS traces.[5] We used the provided trajectories of each user as our second dataset as it takes extremely long time and very powerful computing infrastructure to generate such data - Spaten developers produced those datasets in the period of 2 months. These trajectories consist of timestamps and longitude, latitude pairs (i.e., points) represented in CSV format (Listing 1.1 shows an example). We transformed the given dataset into Turtle format using the TomTom ontology before using it as input dataset in *SPgen*.

```
1   id_1,      timestamp_1,   Point_1
2   id_1,      timestamp_2,   Point_2
3        . . .
4   id_1,      timestamp_n,   Point_n
```

Listing 1.1. Spaten Example Data

---

5   *SPgen*: A Link Discovery Benchmark Generator for Spatial Data

5.1   Overview

In *SPgen*,[6] we focus on relations that follow the DE-9IM (Dimensionally Extended nine-Intersection Model) and determine whether the systems are able to identify those relations between different instances. Each instance is either a *LineString* or a *Polygon*. *SPgen* gets as input traces represented as *LineStrings* and produces a *source* and a *target* dataset. The source dataset is identical to the input traces but is expressed in the Well Known Text format (WKT),[7] whereas the target dataset consists of *LineStrings* or *Polygons* that are generated from the source dataset in such a way that traces in the target dataset have a specific topological DE-9IM relation with the traces of the source dataset.

In *SPgen* we propose a set of test cases whose objective is to test whether link discovery systems for spatial data can identify whether a DE-9IM relation holds between different geometries. *SPgen* implements all topological relations of DE-9IM between LineStrings and Polygons in the two-dimensional space. The *gold standard* is produced after the generation of the source and target using RADON [8]. We discuss in Subsect. 5.4 why we opted for this solution. In the next subsections we will describe *SPgen* in more detail.

5.2   *SPgen* Architecture

The architecture of *SPgen* is shown in Fig. 4. *SPgen* takes a sequence of *traces* as input and a set of *user-defined parameters* such as the (a) number of instances to retrieve from the input dataset, (b) percentage of points to keep for each input trace,[8] (c) geometry of the *target* dataset (note that the target dataset can be either a *LineString* or a *Polygon*) and (d) the DE-9IM topological relation of interest.

The input dataset is processed by the Initialization Module that reads the user-defined parameters and retrieves the input traces by means of SPARQL queries. The retrieved traces are passed to the Resource Generation Module to generate the *source dataset* that transforms each retrieved trace to a *LineString* represented in WKT format. This module interacts with the Resource Transformation Module that generates the target instances represented again in WKT; the module implements the *DE-9IM topological relations* discussed in Sect. 5.3. The relations are implemented as an *extension*[9] of the JTS Topology Suite,[10]

---

[6] https://github.com/hobbit-project/SpatialBenchmark.

[7] WKT is a text markup language for representing vector geometry objects on a map, spatial reference systems of spatial objects and transformations between spatial reference systems. WKT offers a compact machine and human readable representation of geometric objects.

[8] A trace with a possibly huge number of points cannot be processed by systems, hence we would like to give the ability to developers to restrict the trace size.

[9] https://github.com/jsaveta/jtsExtension.

[10] http://svn.code.sf.net/p/jts-topo-suite/code/tags/Version_1.14/.

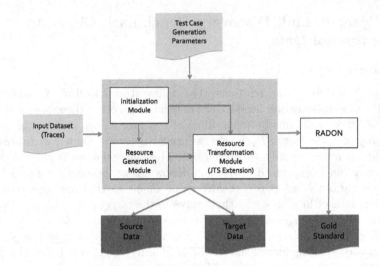

Fig. 4. *SPgen* architecture

a JAVA API that provides a core set of spatial data operations using an explicit precision model and robust geometric algorithms.

The target dataset obtained from the Resource Transformation Module along with the source dataset, is passed as input to RADON.

5.3   Test Cases

SPgen implements all topological relations of DE-9IM between LineStrings and Polygons in the two-dimensional space. Due to space limitations, we only discuss the DE-9IM relation *Disjoint* for LineStrings, and the relation *Within* for LineStrings and Polygons. Other relations are handled in a similar fashion. Our algorithms are based on the idea of the *minimum bounding box* (*bbox*) which is an area defined by two longitudes (in the range $-180\ldots180$) and two latitudes (in the range $-90\ldots90$), such that the resulting bounding box (included within these coordinates) contains the geometry under study.

Disjoint (LineString/LineString): Given a LineString s, and the DE-9IM *Disjoint* relation r, we produce a LineString t disjoint with s, as follows: first, we compute the bounding box $b(s)$ of s, and randomly define longitude, latitude coordinates for a bounding box $b(t)$ that does not intersect with $b(s)$. In order to find $b(t)$, we find sufficiently large (or sufficiently small) coordinates for the minimum (maximum) longitude or latitude coordinates. Second, we generate a random LineString t with the same number of points as s that entirely falls inside $b(t)$, thereby guaranteeing disjointness between s and t.

In the case in which $b(s)$ covers the entire plane (i.e., its longitude, latitude coordinates have the maximum/minimum values), no $b(t)$ can be defined. In these cases, we break s into several smaller LineStrings, say $s_1; \ldots; s_k$, and compute their corresponding bounding boxes $b(s_1); \ldots; b(s_k)$. Then, we use the above process to identify a bounding box $b(t)$ that does not intersect with any of them and create a random target LineString as discussed earlier.

If, despite the partitioning of the bounding box $b(s)$ of LineString s, no appropriate t can be found, then we define a more fine-grained partition and repeat the process which ends when an appropriate disjoint LineString t can be found, or when each pair of consecutive points of s is a partition; if even this fine-grained partition does not allow the definition of an appropriate bounding box, then the original LineString covers the entire plane and no disjoint LineString can be created.

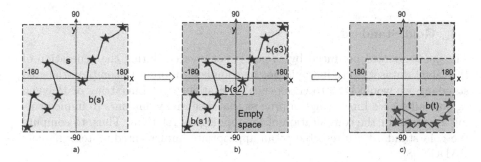

Fig. 5. Example for disjoint (LineString/LineString)

Figure 5 provides an example of the aforementioned process. In subfigure (a) we can see the source LineString s and its bbox $b(s)$. We are in the case where $b(s)$ covers the entire plane, thus we break s into smaller LineStrings and compute their corresponding bounding boxes $b(s_1); b(s_2); b(s_3)$ (subfigure (b)). We do not need to break s more as there is already an empty space where we can generate a bbox $b(t)$ and generate a disjoint to s, target LineString t (subfigure (c)).

Within (LineString/Polygon): Given a LineString s, and DE-9IM *Within* relation r, we produce a Polygon t in which s is *within*, as follows: First, using the JTS API we find the minimum-area convex polygon that contains LineString s. Then, we slightly expand the returned Polygon in order not to cross LineString s and thus we create target Polygon t. In the rare case in which s has one or more points whose longitudes are equal to -180 or 180 or one or more points whose longitudes are equal to -90 or 90, no Polygon that contains s exists. Figure 6 provides an example of the aforementioned process. In subfigure (a) we can see the source LineString s and its bbox $b(s)$ that does not cover the entire plane. Thus, we are able to define a Polygon that contains s (subfigure (b)) and then slightly expand it in order to create a Polygon t that in combination with s follows the definition of DE-9IM *Within* relation (subfigure (c)).

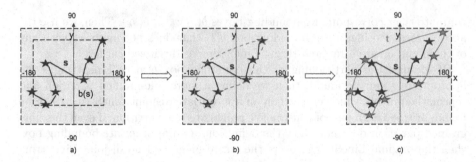

Fig. 6. Example for within (LineString/Polygon)

5.4 Gold Standard

The *gold standard* produced by *SPgen* is not created during the generation of the target dataset, since it would not be complete: in order to compute the gold standard, we would have to check each generated target LineString or Polygon against all source LineStrings, a process that essentially amounts to implementing a system for the computation of the topological relations. Thus, to compute the gold standard, we resorted to an appropriate implemented system, namely RADON [8].

RADON was selected because it is a novel approach for rapid discovery of topological relations among geo-spatial resources. It combines space tiling, minimum bounding box approximation and a sparse index to handle very large datasets. RADON was evaluated with real datasets of various sizes and showed that in addition to being complete and correct, it also outperforms the state of the art spatial link discovery systems by up to three orders of magnitude. Thus, it is appropriate for our purposes.

5.5 Key Performance Indicators

The key performance indicators of a benchmark determine the effectiveness and efficiency of the systems and tools. In *SPgen* we focus on the *output quality* in terms of standard metrics such as *precision, recall* and *f-measure* [19]. We also aim to quantify the *time performance* of the systems measuring the *time* needed by the link discovery system to return results.

6   Experimental Results

In this section we describe the experiments we conducted in order to show how well the various spatial linking systems performed regarding output quality and time performance for datasets of various sizes and for the different DE-9IM topological relations.

DATASETS & TASKS: We ran experiments for all the DE-9IM relations and for LineString/LineString and LineString/Polygon cases for both TomTom and

Spaten datasets ranging from 200 to 2K instances, not exceeding 64 KB per instance due to a limitation of SILK.[11] This is important in order to get a fair comparison for the systems under test. We report here the results for quality output and time performance for all systems.

EXPERIMENTAL SETUP: All the experiments were executed using the HOB-BIT Platform[12] where *SPgen* is integrated and the platform time limit was set to 75 min. Thus, we provide a comparative analysis with benchmarks produced using *SPgen* and were able to assess and identify the capabilities of four systems, namely AgreementMakerLight (AML), OntoIdea, Rapid Discovery of Topological Relations (RADON) and Silk.

TASKS: We divided the experiments into four tasks. In the first two tasks (SLL and LLL), the systems were asked to match LineStrings to LineStrings considering a given relation for 200 and 2K instances for the TomTom and Spaten datasets. In the last two second tasks (SLP, LLP), the systems were asked to match LineStrings to Polygons (or Polygons to LineStrings depending on the relation) again for both datasets. We are only presenting results regarding the time performance and not precision, recall and f-measure as all results from all systems were equal to 1.0 except for OntoIdea (mostly for the Spaten dataset) that were between 0.91 to 0.99.

TASK SLL: SMALL (LINESTRINGS/LINESTRINGS): Fig. 7 presents the time performance for TomTom and Spaten datasets for AML, OntoIdea, Silk and RADON systems for 200 instances. RADON has the best performance in most cases except *Touches* and *Instersects* relations, followed by AML and OntoIdea, while Silk seems to need the most time mainly for the TomTom dataset for *Touches* and *Intersects* relations and for both datasets for *Overlaps*.

TASK LLL: LARGE (LINESTRINGS/LINESTRINGS): Figure 8 presents the time performance for TomTom and Spaten datasets for AML, OntoIdea, Silk and RADON systems for the 2K instances dataset. In contrast to Fig. 7 we have a more clear view of the capabilities of the systems. In this experiment, RADON and Silk have similar behaviour as in the case of the small dataset, but this time it is more clear that the systems need much more time to match instances from the TomTom dataset. RADON has still the best performance in most cases. AML has the next best performance and is able to handle cases better than other systems (e.g. *Touches* and *Intersects*). AML also hits the platform time limit in the case of *Disjoint*. While the time performance of OntoIdea was close to RADON and AML in the smaller dataset, AML is not able to handle the larger dataset.

TASK SLP: SMALL (LINESTRINGS/POLYGONS): Figure 9 presents the time performance for TomTom and Spaten datasets for AML, Silk and RADON for 200 instances (LineStrings/Polygons or Polygons/LineStrings depending on the relation). In contrast to the two first tasks, RADON has the best performance for

---

[11] https://github.com/silk-framework/silk/issues/57.

[12] http://master.project-hobbit.eu.

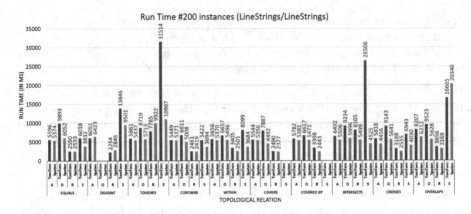

Fig. 7. Time performance for TomTom & Spaten SLL Task for AML(A), OntoIdea(O), Silk(S) and RADON(R) systems

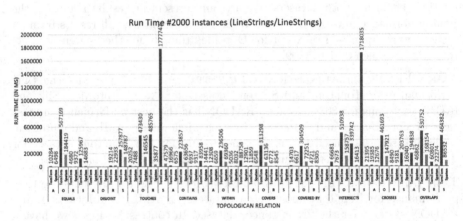

Fig. 8. Time performance for TomTom & Spaten LLL Task and for AML(A), OntoIdea(O), Silk(S) and RADON(R) systems

all relations. AML and Silk have minor time differences and, depending on the case, one is slightly better than the other. All the systems need more time for the TomTom dataset but due to the small size of the instances the time difference is minor.

Task LLP: Large (LineStrings/Polygons): Figure 10 presents the time performance for TomTom and Spaten datasets for AML, Silk and RADON for the 2K instance dataset (LineStrings/Polygons or Polygons/LineStrings depending on the relation). RADON again has the best performance in all cases. AML hits the platform time limit in *Disjoint* relations on both datasets and is better than Silk in most cases except *Contains* and *Within* on the TomTom dataset where it needs an excessive amount of time.

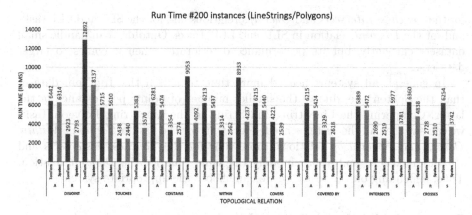

Fig. 9. Time performance for TomTom & Spaten SLP Task and for AML(A), Silk(S) and RADON(R) systems

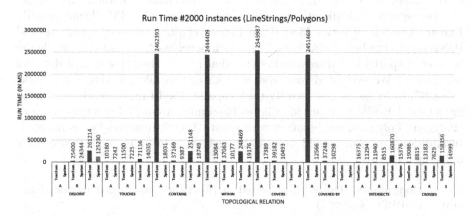

Fig. 10. Time performance for TomTom & Spaten LLP Task and for AML(A), Silk(S) and RADON(R) systems

Discussion

Taking into account the executed experiments we can identify the capabilities of the tested systems as well as suggest some improvements. All the systems participated in most of the test cases except OntoIdea that did not participate in Tasks SLP and LLP and in experiments for the *Disjoint* relation. Also Silk did not participate in *Covers* and *Covered By* experiments.

RADON is the only system that addressed all the tasks, while it can be improved for the *Touches* and *Intersects* relations for the Tasks SLL and LLL and it also has the best performance for the SLP and LLP tasks. AML performs extremely well in most cases. It can be improved in the cases of *Covers/Covered By* and *Contains/Within* when it comes to LineStrings/Polygons Tasks and also in *Disjoint* relations where it hits the platform time limit. Silk can be improved for the *Touches*, *Intersects* and *Overlaps* relations and for the SLL and LLL tasks and for the *Disjoint* relation in SLP and LLP Tasks. OntoIdea can handle small datasets efficiently, but its performance deteriorates when it comes to larger datasets.

In general, all systems needed more time to match the TomTom dataset than the Spaten one, due to the smaller number of points per instance in the latter. Comparing the LineString/LineString to the LineString/Polygon Tasks we can say that all the systems needed less time for the first in *Contains*, *Within*, *Covers* and *Covered by* relations, more time for the *Touches*, *Instersects* and *Crosses* relations, and approximately the same time for the *Disjoint* relation. Thus, depending on the test case we can choose the appropriate system.

7   Conclusions and Future Work

In this paper we presented *SPgen*, a *Spatial Benchmark Generator* that checks whether spatial link discovery systems can identify DE-9IM (Dimensionally Extended nine-Intersection Model) topological relations between LineStrings and Polygons. To the best of our knowledge, such benchmarks do not exist while the number of spatial link discovery systems that identify links for spatial datasets are limited. We evaluated four systems (AML, OntoIdea, RADON and Silk) using *SPgen* to assess and identify their capabilities. In future work, we aim to implement DE-9IM relations for all possible combinations of different geometries (Polygons/Polygons, combination with Points, LineStrings and Polygons, etc.). In addition, we plan to add more data generators in order to test *SPgen* for different use cases.

Acknowledgments. The work presented in this paper was funded by the H2020 project HOBBIT (#688227).

References

1. Ngonga Ngomo, A.-C.: On link discovery using a hybrid approach. J. Data Semant. **1**(4), 203–217 (2012)
2. Saveta, T., Daskalaki, E., Flouris, G., Fundulaki, I., Herschel, M., Ngonga Ngomo, A.-C.: Pushing the limits of instance matching systems: a semantics-aware benchmark for linked data. In: WWW, pp. 105–106. ACM (2015). Poster
3. Strobl, C.: Dimensionally extended nine-intersection model (DE-9IM). In: Shekhar, S., Xiong, H., Zhou, X. (eds.) Encyclopedia of GIS, pp. 240–245. Springer, Cham (2017). https://doi.org/10.1007/978-3-319-17885-1
4. Boncz, P., Neumann, T., Erling, O.: TPC-H analyzed: hidden messages and lessons learned from an influential benchmark. In: Nambiar, R., Poess, M. (eds.) TPCTC 2013. LNCS, vol. 8391, pp. 61–76. Springer, Cham (2014). https://doi.org/10.1007/978-3-319-04936-6_5
5. Cruz, I.F., Antonelli, F.P., Stroe, C.: AgreementMaker: efficient matching for large real-world schemas and ontologies. VLDB Endow. **2**(2), 1586–1589 (2009)

6. Cruz, I.F., et al.: Using agreementmaker to align ontologies for OAEI2011, vol. 814, pp. 114–121 (2011)
7. Khiat, A., Mackeprang, M.: I-Match and OntoIdea results for OAEI 2017. In: OM, p. 135 (2017)
8. Sherif, M.-A., Dreßler, K., Smeros, P., Ngonga Ngomo, A.-C.: RADON - rapid discovery of topological relations. In: AAAI (2017)
9. Smeros, P., Koubarakis, M.: Discovering spatial and temporal links among RDF data. In: LDOW (2016)
10. Garbis, G., Kyzirakos, K., Koubarakis, M.: Geographica: a benchmark for geospatial RDF stores (long version). In: Alani, H., et al. (eds.) ISWC 2013. LNCS, vol. 8219, pp. 343–359. Springer, Heidelberg (2013). https://doi.org/10.1007/978-3-642-41338-4_22
11. Ray, S., Simion, B., Brown, A.D.: Jackpine: a benchmark to evaluate spatial database performance. In: ICDE, pp. 1139–1150. IEEE (2011)
12. Kolas, D.: A benchmark for spatial semantic web systems. In: SSWS (2008)
13. Guo, Y., Pan, Z., Heflin, J.: LUBM: a benchmark for OWL knowledge base systems. Web Semant.: Sci. Serv. Agents World Wide Web **3**(2–3), 158–182 (2005)
14. Paton, N.W., Williams, M.H., Dietrich, K., Liew, O., Dinn, A., Patrick, A.: VESPA: a benchmark for vector spatial databases. In: Lings, B., Jeffery, K. (eds.) BNCOD 2000. LNCS, vol. 1832, pp. 81–101. Springer, Heidelberg (2000). https://doi.org/10.1007/3-540-45033-5_7
15. Gunther, O., Oria, V., Picouet, P., Saglio, J.M., Scholl, M.: Benchmarking spatial joins a la carte. In: SSDM, pp. 32–41. IEEE (1998)
16. Stonebraker, M., Frew, J., Gardels, K., Meredith, J.: The Sequoia 2000 storage benchmark. In: ACM SIGMOD Record, vol. 22, pp. 2–11. ACM (1993)
17. Patel, J., et al.: Building a scaleable geo-spatial DBMS: technology, implementation, and evaluation. In: ACM SIGMOD Record, vol. 26, pp. 336–347. ACM (1997)
18. Doudali, T.D., Konstantinou, I., Koziris, N.: Spaten: a spatio-temporal and textual big data generator. In: IEEE Big Data, pp. 3416–3421 (2017)
19. Goutte, C., Gaussier, E.: A probabilistic interpretation of precision, recall and *F*-score, with implication for evaluation. In: Losada, D.E., Fernández-Luna, J.M. (eds.) ECIR 2005. LNCS, vol. 3408, pp. 345–359. Springer, Heidelberg (2005). https://doi.org/10.1007/978-3-540-31865-1_25

Specifying, Monitoring, and Executing Workflows in Linked Data Environments

Tobias Käfer[1][(✉)] and Andreas Harth[2]

[1] Institute AIFB, Karlsruhe Institute of Technology (KIT), Karlsruhe, Germany
tobias.kaefer@kit.edu
[2] University of Erlangen-Nuremberg (FAU), Nuremberg, Germany
andreas.harth@fau.de

Abstract. We present an ontology for representing workflows over components with Read-Write Linked Data interfaces and give an operational semantics to the ontology via a rule language. Workflow languages have been successfully applied for modelling behaviour in enterprise information systems, in which the data is often managed in a relational database. Linked Data interfaces have been widely deployed on the web to support data integration in very diverse domains, increasingly also in scenarios involving the Internet of Things, in which application behaviour is often specified using imperative programming languages. With our work we aim to combine workflow languages, which allow for the high-level specification of application behaviour by non-expert users, with Linked Data, which allows for decentralised data publication and integrated data access. We show that our ontology is expressive enough to cover the basic workflow patterns and demonstrate the applicability of our approach with a prototype system that observes pilots carrying out tasks in a virtual reality aircraft cockpit. On a synthetic benchmark from the building automation domain, the runtime scales linearly with the size of the number of Internet of Things devices.

1 Introduction

Information systems are increasingly distributed. Consider the growing deployment of sensors and actuators, the modularisation of monolithic software into microservices, and the movement to decentralise data from company-owned data silos into user-owned data pods. The drivers of increasing distribution include:

- Cheaper, smaller, and more energy-efficient networked hardware makes widespread deployment feasible[1].
- Rapidly changing business environments require flexible re-use of components in new business offerings [17].

---

[1] http://www.forbes.com/sites/oreillymedia/2015/06/07/how-the-new-hardware-mo vement-is-even-bigger-than-the-iot/.

© Springer Nature Switzerland AG 2018
D. Vrandečić et al. (Eds.): ISWC 2018, LNCS 11136, pp. 424–440, 2018.
https://doi.org/10.1007/978-3-030-00671-6_25

- Fast development cycles require independent evolution of components [17].
- Privacy-aware users demand to retain ownership of their data[2].

The distribution into components raises the opportunity to create new integrated applications out of the components, given sufficient interoperability. One way to make components interoperable is to equip the components with uniform interfaces using technologies around Linked Data. Consider, e.g. the W3C's Web of Things[3] initiative and the MIT's Solid ("social linked data") project[4], where REST provides an uniform interface to access and manipulate the state of components, and RDF provides an uniform data model for representing component state that allows for using reasoning to resolve schema heterogeneity. While the paradigms for (read-only) data integration systems based on Linked Data are relatively agreed upon [11], techniques for the creation of applications that integrate components with Read-Write Linked Data interfaces are an active area of research [2,4,15,28]. Workflows are a way to create applications, according to Jablonski and Bussler [14], that is highly suitable for integration scenarios, easy to understand (for validation and specification by humans), and formal (for execution and verification by machines). E.g., consider an evacuation support workflow for a smart building (cf. task 4 in our evaluation, Sect. 6), which integrates multiple systems of the building, should be validated by the building management and the fire brigade, verified to be deadlock-free, and executable. Hence, we tackle the research question: *How to specify, monitor, and execute applications given as workflows in the environment of Read-Write Linked Data?*

The playing field for applications in the context of Read-Write Linked Data is big and diverse: As of today, the Linking Open Data cloud diagram[5] lists 1'163 data sets from various domains for read access. The Linked Data Platform (LDP)[6] specifies interaction with Read-Write Linked Data sources. Besides Solid for social networks, a showcase for Read-Write Linked Data is the Web of Things, which is built on sensors and actuators on the Internet of Things. Using such sensors and actuators, we can build applications such as integrated Cyber-Physical Systems, where sensors and actuators provide the interface to Virtual Reality systems (cf. the showcase in the evaluation, Sect. 6.2). Other non-RDF REST APIs provide access to weather reports[7] or building management systems (e.g. Project Haystack[8]) and can be wrapped to support RDF. Using such APIs, we can build applications such as integrated building automation systems (cf. the scenario of the synthetic benchmark in the evaluation, Sect. 6.3).

Traditional environments for workflows are fundamentally different from Read-Write Linked Data. Elmroth et al. argue that the properties of the environment determine the model of computation, which serves as the basis of a

---

[2] "Putting Data back into the Hands of Owners", http://tcrn.ch/2i8h7gp.
[3] http://www.w3.org/WoT/.
[4] http://solid.mit.edu/.
[5] http://lod-cloud.net/.
[6] http://www.w3.org/TR/ldp/.
[7] http://openweathermap.org/.
[8] http://www.project-haystack.org/.

workflow language [5]. Consequently, we have developed ASM4LD [15], a model of computation for the environment of Read-Write Linked Data. In this paper, we investigate an approach for a workflow language consisting of an ontology and operational semantics in ASM4LD. The differences between traditional environments of workflow languages and the environment of Read-Write Linked Data (i.e. RDF and REST) pose the following particular challenges:

Querying and reasoning under the open-world assumption. Ontology languages around RDF such as RDFS and OWL make the open-world assumption (OWA). However, approaches from workflow management operate on relational databases, which make the closed-world assumption (CWA). Closedness allows for testing if something holds for *all* parts of a workflow.

The absence of events in REST. HTTP implements CRUD (the operations create, read, update, delete), but not the subscriptions to events. However, approaches from workflow management use events as change notifications.

While both challenges could be mitigated by introducing assumptions (e.g. negation-as-failure once we reach a certain completeness class [12]) or by extending the technologies (e.g. implement events using Web Sockets[9] or Linked Data Notifications [2]), those mitigation strategies would restrict the generality of the approach, i.e. we would have to exclude components that provide Linked Data, but do not share the assumptions or extensions of the mitigation strategy.

Previous works from Business Process Management, Semantic Web Services, Linked Data, and REST operate on a different model of computation or are complementary: [10,13,19] assume event-based data processing, decision making based on process variables, and data residing in databases under the CWA, whereas our approach relies on integrated state information from the web under the OWA. [28,29] provide descriptions for automated composition or for assisting developers. Currently, we do not see elaborate and correct descriptions available at web scale, which hinders automated composition. We see our approach, which allows for manual composition, as the first step towards automated composition.

The paper is structured as follows: In Sect. 2, we discuss related work. In Sect. 3, we present the technologies on which we build our approach. Next, we present our approach, which consists of two main contributions:

– An ontology to specify workflows models and workflow instances modelled in OWL LD[10] (Sect. 4) that allows for monitoring and execution using querying and reasoning under the OWA. The ontology is strongly related to BPMN, a graphical workflow notation, via the workflow patterns [25].
– An operational semantics for our workflow ontology. We use ASM4LD, a model of computation for Read-Write Linked Data in the form of a condition-action rule language (Sect. 5), which does not require event data and is directly executable. We maintain workflow state in an LDP container.

---

[9] http://www.ietf.org/rfc/rfc6455.txt.
[10] http://semanticweb.org/OWLLD/.

Fast data processing thanks to OWL LD and the executability of ASM4LD allow for the direct application of our approach in practice. In the evaluation (Sect. 6), we present a Virtual Reality showcase, and a benchmark in an Internet of Things setting, specifically in the building automation domain. Moreover, we show correctness and completeness of our approach. We conclude in Sect. 7.

2   Related Work

We now survey related work grouped by field of research.

Workflow Management. Previous work in the context of workflow languages and workflow management systems is based on event-condition-action (ECA) rules, whereas our approach is built for REST, and thus works without events. ECA rules have been used to give operational semantics to workflow languages [13], and to implement workflow management systems [3]. Similar to the case handling paradigm [26], we employ state machines to track the state of activities in a workflow instance.

Web Services. WS-*-based approaches assume arbitrary operations, whereas our approach works with REST resources, where the set of operations is constrained [20,30]. Pautasso et al. proposed extensions to BPEL such that e.g. a BPEL process can invoke REST services [18], and that REST resources representing processes push events [19]. While those extensions make isolated REST calls fit the Web Services processing model of process variable assignments, we propose a processing model based on integrated polled state.

Semantic Web Services. OWL-S and WSMO are mainly concerned with service descriptions and corresponding reasoning for composition. Semantic Web Services build on WS-* technology for workflow execution, e.g. the execution in the context of WSMO, WSMX [10], is entirely event-based. In contrast, our work is based on REST.

Scientific Workflows. Approaches like Taverna [24] and Wings [8] focus on representing the data flow between processing steps. Our approach applies control flow techniques from Workflow Management to REST.

Ontologies for Workflows. Similar to workflows in our ontology, processes in OWL-S are also tree-structured (see Sect. 4) and use lists in RDF. Unlike OWL-S, our ontology also covers workflow instances. Rospocher et al. [22] and the project "Super" developed ontologies that describe process metamodels such as BPMN, BPEL, and EPC. In contrast to our work, their ontologies either require more expressive (OWL) reasoning or do not allow for execution under the OWA.

3   Preliminaries

We next introduce the environment, Read-Write Linked Data, and the model of computation, ASM4LD [15].

Read-Write Linked Data. Linked Data is a collection of practices for data publishing on the web that advocates the use of web standards: HTTP URIs[11] should be used for identifying things. HTTP GET[12] requests to those URIs should be answered using descriptive data, e.g. in RDF[13]. Hyperlinks in the data should enable the discovery of more information[14]. Read-Write Linked Data[15] introduces RESTful write access to Linked Data (later standardised in the LDP specification (see Footnote 6)). Hence, we can access the world's state using multiple HTTP GET requests and enact change using HTTP PUT, POST, DELETE requests.

In the paper, we denote RDF triples using binary predicates[16], e.g. we write for the triple in Turtle notation "`<#wfm> rdf:type :WorkflowModel.`":

$$rdf{:}type(\texttt{<\#wfm>}, \texttt{:WorkflowModel})$$

We abbreviate a class assignment using a unary predicate with the class as predicate name, e.g. $:WorkflowModel(\texttt{<\#wfm>})$. The term $rdf{:}List(\dots)$ is a shortcut, similar to the RDF list shortcut with () brackets in Turtle, and can be regarded as a procedure that (1) takes as argument list elements, (2) adds the corresponding RDF list triples, i.e. with terms $rdf{:}first$, $rdf{:}rest$, and `rdf:nil`, to the current data, and (3) returns the blank node for the RDF list's head.

ASM4LD, A Condition-Action Rule Language. We use a monotonic production rule language to specify both reasoning on RDF data and interaction with Read-Write Linked Data resources [23]. Rule programs in the language consist of initial assertions and rules. The body of all rules is a basic graph pattern query (see Footnote 18) (BGP). We distinguish two types of rules: (1) a derivation rule specifies productions using a BGP in the rule head, and (2) a request rule specifies an interaction using an HTTP request description in the rule head. We assume safe rules and exclude existential variables in rule heads.

As operational semantics for the rule language, we use ASM4LD, an Abstract State Machine-based [9] model of computation for Read-Write Linked Data [15]. In the following, we sketch the operational semantics, where data processing is done in repeated steps, subdivided into the following phases (cf. [15] for details):

(1) The working memory be empty.
(2) Add the initial assertions to the working memory.
(3) Evaluate on the working memory until the fixpoint:
  (a) Request rules that contain GET requests, making the requests and adding the data from the responses to the working memory.

---

[11] http://www.ietf.org/rfc/rfc3986.txt.
[12] http://www.ietf.org/rfc/rfc7230.txt.
[13] http://www.w3.org/TR/rdf11-concepts/.
[14] http://www.w3.org/DesignIssues/LinkedData.html.
[15] http://www.w3.org/DesignIssues/ReadWriteLinkedData.html.
[16] We assume the URI prefix definitions of http://prefix.cc/ The empty prefix denotes http://purl.org/wild/vocab. The base URIs be http://example.org/.

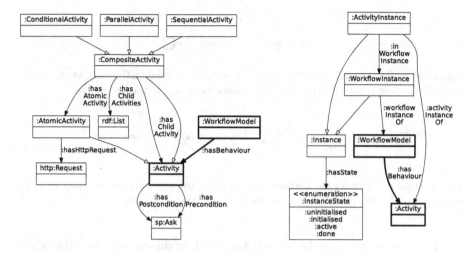

Fig. 1. The ontology to express workflow models and instances as UML Class Diagram. Shared classes between the diagrams are depicted in bold. We use the UML Class Diagram's class, inheritance, association, and enumeration to denote the RDFS ontology language's `rdfs:Class`, `rdfs:subClassOf`, `rdf:Property` with `rdfs:domain` and `rdfs:range`, and instances.

(b) Derivation rules, adding the produced data to the working memory.
   We thus acquire data about the world's current state (from the responses to the GET requests) and reason on this data (using the productions).
(4) Evaluate all request rules that contain PUT/POST/DELETE requests on the working memory and make the corresponding HTTP requests. We thus enact changes on the world's state.

A loop over the phases (1) to (4) implements polling, the way to get information about changes in a RESTful environment. Hypermedia-style link following (to discover new information) can be implemented using request rules, e.g. in the example below.

We use the following rule syntax: In the arguments of the binary predicates, we allow for variables (printed in italics). We print constants in typewriter font. We connect rule head and body using →. The head of a request rule contains one HTTP request with the method as the function name, the target as the first argument, and the RDF payload as the second argument (if applicable). E.g. consider the following rule to retrieve all elements e of a given LDP container:

$$ldp : contains(\texttt{http://example.org/ldpc}, e) \rightarrow \text{GET}(e)$$

4 Activity, Workflow Model and Instance Ontology

To describe workflow models and instances as well as activities, we propose an ontology. We developed the ontology, see Fig. 1[17], with execution based on

---

[17] The ontology can be accessed at http://purl.org/wild/vocab.

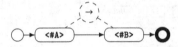

Fig. 2. Workflow (solid: BPMN notation) with sequential activities (`<#A>`, `<#B>`). Dashed: the tree representation with the parent node marked as sequential.

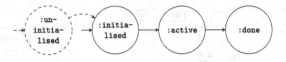

Fig. 3. State machine for the instance resources for the workflow and activity instance resources. The dashed part only concerns workflow instance resources.

querying and reasoning under the OWA in mind. In this section, we define activities, workflows, and instances using the workflow in Fig. 2 as example.

Activities. We regard an atomic activity as a basic unit of work. We characterise an activity by a postcondition represented as a SPARQL ASK query[18], which has to hold in the world's state after the activity has been executed. We use the postcondition (cf. `:hasPostcondition` in Fig. 1) to monitor the execution of activities in workflows. For the execution of an atomic activity, the activity description needs an HTTP request (cf. `:hasHttpRequest` in Fig. 1).

Workflow Models. A workflow model is a set of activities put into a defined order. As notation to describe workflow models, BPMN is a popular choice. The course of action (i.e. control flow) in a BPMN workflow model is denoted using arrows that connect activities and gateways (e.g. decisions and branches). For instance, the middle arrow in the workflow model in Fig. 2 orders activities `<#A>` and `<#B>` sequentially. We call this view on the course of action *flow-based*.

In this paper, instead of a flow-based view on the course of action, we consider a *tree-based* view, as investigated by Vanhatalo et al. [27]. Tree-structured workflow languages include BPEL, a popular language to describe executable workflows. In the tree, activities are leaf nodes. The non-leaf nodes are typed, and the type determines the control flow of the children. The connection between the tree-based (dashed) and the flow-based (solid) workflow representation is depicted in Fig. 2. Flow-based workflows can be losslessly translated to tree-structured workflows and vice versa [21]. We use the tree structure, as checks for completion of workflow parts are easier in a tree. Of the multitude of control flow features of different workflow languages, we support the most basic and common, which have been compiled to the *basic workflow patterns* [25].

---

[18] http://www.w3.org/TR/sparql11-query/.

We now show how to specify workflow models in RDF using Fig. 2's model:

$$:WorkflowModel(\texttt{<\#wfm>}) \land :SequentialActivity(\texttt{<\#root>})$$
$$\land :AtomicActivity(\texttt{<\#A>}) \land :AtomicActivity(\texttt{<\#B>})$$
$$\land :hasBehaviour(\texttt{<\#wfm>}, \texttt{<\#root>})$$
$$\land :hasChildActivities(\texttt{<\#root>}, rdf{:}List(\texttt{<\#A>}, \texttt{<\#B>}))$$

As we assume tree-structured workflows, each workflow model (`<#wfm>`) has a root activity (`<#root>`). If an activity is composite, i.e. a control flow element, then the activity has an RDF list of child activities. Here, `<#root>` is sequential, with the child activities `<#A>`, `<#B>`. The child activities could again be composite, thus forming a tree. Leaves in the tree (here `<#A>` and `<#B>`) are atomic activities. We require child activities to be given in an RDF list, which is explicitly terminated. This termination closes the set of list elements and thus allows for executing workflows under the OWA, which e.g. includes querying whether *all* child activities of a parent activity are `:done`. Yet, for the operational semantics we also need a direct connection between a parent activity and a child activity, which we derive from an RDF list using monotonic reasoning, here:

$$:hasChildActivity(\texttt{<\#root>}, \texttt{<\#A>}) \land :hasChildActivity(\texttt{<\#root>}, \texttt{<\#B>})$$

Instances. Using workflow instances, we can run multiple copies of a workflow model. A workflow instance consequently consists of instances of the model's activities. We model the relation of the instances to their counterparts as shown in Fig. 1. During and after workflow monitoring/execution, the operational semantics maintain the states of instances in an LDP container. At runtime, the instances' states evolve according to the state machine depicted in Fig. 3 (terms from Fig. 1). Section 5 is about the operationalisation of the evolution.

5   Operational Semantics

In this section, we give operational semantics to our workflow language[19] in rules[20]. Before we define the rules, we give an overview of what the rules do.

5.1   Overview

The rules fulfil the following purposes (the numbers are only to guide the reader):

I. Retrieve state[21]
   (1) Retrieve the state of the writeable resources in the LDP container, which maintain the workflow/activity instances' state

---

[19] In a production environment, access control to the instances' LDP container needs to be in place to keep third parties from interfering with the monitoring/execution.
[20] A corresponding Notation3 file can be found at http://purl.org/wild/semantics.
[21] A benefit of using Linked Data throughout is that we can access the workflow/activity instances' state and the world's state in a uniform manner.

(2) Retrieve the relevant world state

II. Initialise workflow instances if applicable

(1) Set the root activity's instance :`active`

(2) Set the workflow instance :`initialised`

(3) Create instance resources for all activities in the corresponding workflow model and set them :`initialised`

III. Finalise workflow instances if their root node is :`done`

IV. Execute and observe :`active` activities

(1) Execution: if an atomic activity turns :`active`, fire the HTTP request

(2) If the postcondition of an :`active` activity is fulfilled, set it :`done`

V. Advance composite activities according to control flow, which includes:

(1) Set a composite activity's children :`active`

(2) Advance between children

(3) Finalise a composite activity

5.2 Condition-Action Rules

We next give the rules for the listed purposes. To shorten the presentation, we factor out those rules that, for workflow execution, fire an activity's HTTP request if the activity becomes :`active`. Those rules are not needed when monitoring. The rules are of the form (the variable *method* holds the request type):

$$AtomicActivity(a) \land hasHttpRequest(a, h) \land http{:}mthd(h, method)$$
$$\land http{:}requestURI(h, u) \land \cdots \rightarrow \text{METHOD}(u, \dots)$$

I. Retrieve State. The following rules specify the retrieval of data where the rule interpreter locally maintains state. Analogously, other rules retrieve the world's state, either by explicitly stating URIs to be retrieved:

$$\text{true} \rightarrow \text{GET}(\text{http://example.org/ldpc})$$

or by following links from data that is already known:

$$ldp{:}contains(\text{http://example.org/ldpc}, e) \rightarrow \text{GET}(e)$$

II. Initialise Workflow Instances. If there is an uninitialised workflow instance (e.g. injected by a third party using a POST request into the polled LDP container), the following rules create corresponding resources for the activity instances and set the workflow instance initialised:

$$WorkflowInstance(i) \land hasState(i, \text{:uninitialised}) \land workflowInstanceOf(i, m)$$
$$\land hasBehaviour(m, a) \rightarrow \text{POST}(\text{server{:}ldpc}, activityInstanceOf(\texttt{<\#it>}, a)$$
$$\land inWorkflowInstance(\texttt{<\#it>}, i) \land hasState(\texttt{<\#it>}, \text{:active}))$$

Also, the workflow instance is set initialised (analogously, we initialise instances for the activities in the workflow model):

$$WorkflowInstance(i) \land hasState(i, \text{:uninitialised}) \land workflowInstanceOf(i, m)$$
$$\rightarrow \text{PUT}(i, WorkflowInstance(i) \land hasState(i, \text{:initialised})$$
$$\land workflowInstanceOf(i, m))$$

III. Finalise Workflow Instances. The done state of the root activity gets propagated to the workflow instance:

$$WorkflowInstance(i) \land hasState(i, \texttt{:active}) \land workflowInstanceOf(i, m)$$
$$\land hasBehaviour(m, a) \land hasState(m, \texttt{:done})$$
$$\rightarrow \textsc{put}(i, WorkflowInstance(i) \land hasState(i, \texttt{:done}) \land workflowInstanceOf(i, m))$$

IV. Monitor Atomic Activities. An activity is done if its postcondition holds.

$$WorkflowInstance(i) \land hasState(i, \texttt{:active}) \land workflowInstanceOf(i, m)$$
$$\land hasDescendantActivity(i, a) \land AtomicActivity(a) \land hasPostcondition(a, p)$$
$$\land ActivityInstance(j) \land activityInstanceOf(j, a) \land hasState(j, \texttt{:active})$$
$$\land sp{:}hasBooleanResult(p, \texttt{true})$$
$$\rightarrow \textsc{put}(j, activityInstanceOf(j, a) \land inWorkflowInstance(j, i) \land hasState(j, \texttt{:done}))$$

To shorten the presentation of the rules in the following, we introduce the following simplifications: We assume that (1) we are talking about an active workflow instance, and (2) that the resource representing an instance coincides with its corresponding activity in the workflow model. (3), the PUT requests in the text do not actually overwrite the whole resource representation but patch the resources by ceteris paribus overwriting the corresponding $hasState(\cdot, \cdot)$ triple.

V. Advance According to Control Flow. We now give the rules for advancing a workflow instance according to the basic workflow patterns (WFPs) [25].

WFP 1: Sequence. If there is an active sequential activity with the first activity initialised, we set this first activity to active:

$$SequentialActivity(s) \land hasState(s, \texttt{:active}) \land hasChildActivities(s, c)$$
$$\land rdf{:}first(c, a) \land hasState(a, \texttt{:initialised}) \rightarrow \textsc{put}(a, hasState(a, \texttt{:active}))$$

We advance between activities in a sequence using the following rule:

$$SequentialActivity(s) \land hasState(s, \texttt{active}) \land hasChildActivity(s, c)$$
$$\land hasState(c, \texttt{done}) \land hasState(n, \texttt{initialised})$$
$$\land rdf{:}first(l, c) \land rdf{:}rest(l, i) \land rdf{:}first(i, n) \rightarrow \textsc{put}(n, hasState(n, \texttt{active}))$$

If we have reached the end of the list of children of a sequence, we regard the sequence as done (the rule is an example of the exploitation of the explicit termination of the RDF list to address the OWA):

$$SequentialActivity(s) \land hasState(s, \texttt{:active}) \land hasChildActivity(s, c)$$
$$\land hasState(c, \texttt{:done}) \land rdf : first(l, c) \land rdf : rest(l, \texttt{rdf:nil})$$
$$\rightarrow \textsc{put}(s, hasState(s, \texttt{:done}))$$

WFP 2: Parallel Split. A parallel activity consists of several activities executed simultaneously. If a parallel activity becomes active, all of its components are set to active:

$$ParallelActivity(p) \land hasState(p, \texttt{active}) \land hasChildActivity(p, c)$$
$$\land hasState(c, \texttt{:initialised}) \rightarrow \textsc{put}(c, hasState(c, \texttt{:active}))$$

WFP 3: Synchronisation. If all the components of a parallel activity are done, the whole parallel activity can be considered done. To find out whether all components of a parallel are done, we mark instances as follows. First, we check whether the first child element of the parallel activity is done and mark the element using the state :doneFromListItemOne:

$$ParallelActivity(p) \wedge hasState(p, \texttt{:active}) \wedge hasChildActivities(p, l)$$
$$\wedge rdf{:}first(l, c) \wedge hasState(c, \texttt{:done}) \rightarrow hasState(c, \texttt{:doneFromListItemOne})$$

Then, starting from the first activity, we go through the list of child activities and propagate the mark between the activities in the list if the activities are done. If the mark reaches the last list element, the whole parallel activity is done:

$$ParallelActivity(p) \wedge hasState(p, \texttt{:active}) \wedge hasChildActivity(p, c)$$
$$\wedge rdf{:}first(l, c) \wedge rdf{:}rest(l, \texttt{rdf:nil}) \wedge hasState(c, \texttt{:doneFromListItemOne})$$
$$\rightarrow \text{PUT}(p, hasState(p, \texttt{:done}))$$

WFP 4: Exclusive Choice. The control flow element choice implements a choice between different alternatives, for which conditions are specified. For the evaluation of the condition, we first have to check whether all child activities are in initialised state, similarly to the rules for WFP 3:

$$ConditionalActivity(a) \wedge hasState(a, \texttt{:active}) \wedge hasChildActivities(a, l)$$
$$\wedge rdf{:}first(l, c) \wedge hasState(c, \texttt{:initialised})$$
$$\rightarrow hasState(c, \texttt{:initialisedFromListItemOne})$$
$$ConditionalActivity(a) \wedge hasState(a, \texttt{:active}) \wedge hasChildActivities(a, l)$$
$$\wedge rdf{:}first(l, c) \wedge hasState(c, \texttt{:initialisedFromListItemOne})$$
$$\wedge rdf{:}rest(l, m) \wedge rdf{:}first(m, d) \wedge hasState(d, \texttt{:initialised})$$
$$\rightarrow hasState(d, \texttt{:initialisedFromListItemOne})$$

If the check succeeded, we can evaluate the conditions and set an activity active:

$$ConditionalActivity(a) \wedge hasState(a, \texttt{:active}) \wedge hasChildActivitiy(a, c)$$
$$\wedge hasState(c, \texttt{:initialisedFromListItemOne}) \wedge hasPrecondition(c, p)$$
$$\wedge rdf{:}first(l, c) \wedge rdf{:}rest(l, \texttt{rdf:nil}) \wedge sp{:}hasBooleanResult(p, \texttt{true})$$
$$\rightarrow \text{PUT}(c, hasState(c, \texttt{:active}))$$

We leave it to the modeller to make sure that the preconditions of the children of a conditional activity are mutually exclusive.

WFP 5: Simple Merge. If one of the children of a conditional activity is done, the whole conditional activity is done:

$$ConditionalActivity(a) \wedge hasState(a, \texttt{:active}) \wedge hasChildActivitiy(a, c)$$
$$\wedge hasState(c, \texttt{:done}) \rightarrow \text{PUT}(a, hasState(a, \texttt{:done}))$$

6   Evaluation

First, we formally show the correctness of our approach to *specifying* workflows
by presenting the relationship of our operational semantics to the formal specification of the basic workflow patterns, which we support completely. Second,
to show the applicability of our approach in a real-world setting, we report on
how we used the approach to do *monitoring* of workflows for human-in-the-loop
aircraft cockpit evaluation in Virtual Reality. Third, we empirically evaluate our
approach to *executing* workflows in a building simulator.

6.1   Mapping to Petri Nets

Van der Aalst et al. use Petri Nets to precisely specify the semantics of the basic
workflow patterns [25]. We now show correctness by giving a mapping of our
operational semantics to Petri Nets. Similar to *tokens* in a Petri Net that pass
between *transitions*, our operational semantics passes the *active* state between
activities using rules (linking to the WFP rules from Sect. 5.2(V)):

- The rule to advance between activities within a :SequentialActivity may
  only set an activity active if its preceding activity has terminated. In the Petri
  Net for the Sequence, a transition may only fire if the preceding transition
  has put a token into the preceding place, see Fig. 4a and the WFP 1 rules.
- Only after the activity before a :ParallelActivity has terminated, the rule
  to advance in a parallel activity sets all child activities active. In the Petri
  Net for the Parallel Split, all places following transition T get a token iff
  transition T has fired, see Fig. 4b and the WFP 2 rules.
- Only if all activities in a :ParallelActivity have terminated, the rules pass
  on the active state. In the Petri Net for the Synchronisation, transition T
  may only fire if there is a place with a token in all incoming arcs (cf. Fig. 4c
  and the WFP 3 rules).
- In the ConditionalActivity, one child activity is chosen by the rule according to mutually exclusive conditions. Similarly, exclusive conditions determine

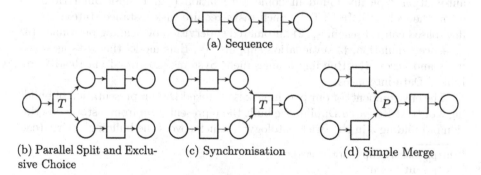

(a) Sequence

(b) Parallel Split and Exclu-   (c) Synchronisation   (d) Simple Merge
sive Choice

Fig. 4. Petri Nets for the basic workflow patterns.

the continuation of the flow after transition T in the Petri Net for the Exclusive Choice, see Fig. 4b and the WFP 4 rules.
- If one child activity of a :ConditionalActivity switches from active to done, the control flow may proceed according to the rule. Likewise, the transition following place P in the Petri Net for the Simple Merge (Fig. 4d) may fire iff there is a token in P, cf. the WFP 5 rules.

Hence, our approach correctly and completely covers the basic workflow patterns.

6.2 Applicability: The Case of Virtual Aircraft Cockpit Design

Together with industry, we successfully applied our approach in aircraft cockpit design [16], where workflow monitoring is used to evaluate cockpit designs regarding Standard Operating Procedures. The monitoring is traditionally done by Human Factors experts using stopwatches in physical cockpits. We built an integrated Cyber-Physical System of Virtual Reality, flight simulation, sensors, and workflows to digitise the monitoring. The challenge was to integrate the different components on both the system interaction and the data level. We built Linked Data interfaces to the components for the interaction integration, and integrated the data using reasoning. Our approach allows for workflow monitoring in the Linked Data setting during runtime. The system's user interface to model workflows has been evaluated by Human Factors experts highly efficient.

6.3 Empirical Evaluation Using a Synthetic Benchmark

The scenario for our benchmark is from the Internet of Things domain, where buildings are equipped with sensors and actuators from different vendors. The devices may be not interoperable, which has been identified by NIST as a major challenge for the building industry [7]. Balaji et al. aim to raise interoperability in Building Management Systems by proposing the Brick ontology [1] to model buildings and Building Management Systems. We thus assume Read-Write Linked Data interfaces to a building's management systems and want to execute building automation tasks. We consider tasks that go beyond rule-based automation typically found in home automation (e.g. Eclipse SmartHome[22]) or on the web (e.g. IFTTT[23]). Such tasks require task instance state, e.g.: (1) flow-based control schemes, (2) automated supervision of cleaning personnel, (3) presence simulation, (4) evacuation support. We thus model the tasks as workflows and access the Building Management Systems integrated via Read-Write Linked Data interfaces.

The environment for our benchmark is a Linked Data representation of building 3 of IBM Research Dublin. We built the representation from a static description of building 3 in the Brick ontology[24], which covers the building's parts (e.g.

---

[22] http://www.eclipse.org/smarthome/.
[23] http://ifttt.com/.
[24] http://github.com/BuildSysUniformMetadata/GroundTruth/blob/2e48662/building_instances/IBM_B3.ttl.

Table 1. Average runtime [s] for workflows Wn in different numbers of buildings.

	W1	W2	W3	W4	W5
1 Building	2	2	6	12	18
10 Buildings	8	9	26	61	75
20 Buildings	12	13	38	80	109
50 Buildings	19	21	61	156	218

rooms) and the building's systems (e.g. lights and switches). We subdivided the description into one-hop RDF graphs around each URI from the building and provide each graph at a corresponding URI. To add state information to the systems, we add writeable SSN[25] properties to the Linked Data interface. To evaluate at different scales, we run multiple copies of the building.

The workload for our benchmark is the control flow of the five representative workflow models proposed by Ferme et al. [6] for evaluating workflow engines, determined by clustering workflows from literature, the web, and industry. We interpreted the five workflow models using the four automation tasks presented above: task 1 corresponds to the first two workflow models; the subsequent tasks to the subsequent workflow models. We assigned the activities in the tasks to two classes: monitoring activities that are checks (e.g. a sensor value), where we attached a postcondition, and execution activities that enact change (e.g. turn on a light), where we attached an HTTP request. For repeatability, the postconditions always hold and the requests do not interfere with the workflow.

The set-up for our evaluation consists of a server with a 32-core Intel Xeon E5-2670 CPU and 256 GB of RAM running Debian Jessie. We deploy the operational semantics and required OWL LD reasoning on Linked Data-Fu 0.9.12[26]. We include reasoning as indicated by the Brick ontology. We maintain building and workflow state in LDP containers, LDBBC 0.0.6[27]. We add workflow instances each 0.2 s after 20 s of warm-up time. The workflow models can be found online[28].

The results of our evaluation can be found in Table 1. Varying the number of activities (W1–W5), and varying the number of devices (proportional to buildings), we observe linear behaviour. The linear behaviour stems from the number of requests to be made, which depends on the number of activities and workflow instances. With no reusable data between buildings, there is no benefit in running the workflows for all buildings on one engine. Instead, we could run one engine per building, thus mirroring the decentralisation of data.

---

[25] http://www.w3.org/TR/vocab-ssn/.
[26] http://linked-data-fu.github.io/.
[27] http://github.com/kaefer3000/ldbbc.
[28] http://people.aifb.kit.edu/co1683/2018/iswc-wild/.

7   Conclusion

We presented an approach to specify, monitor, and execute applications that builds on distributed data and functionality provided as Read-Write Linked Data. We use workflows to specify applications, and thus defined a workflow ontology and operational semantics. We aligned our approach to the basic workflow patterns, reported on an application in Virtual Reality, and evaluated using a synthetic benchmark in an Internet of Things scenario.

The assumptions of the environment of Read-Write Linked Data present peculiar challenges for a workflow system: We work under the OWA and without events as change notifications. Our approach addresses the challenges without adding assumptions to the architecture of the environment, but by modelling a closed world where necessary and by using polling to access the world's state.

We believe that our approach, which brings workflows in a language that is closely related to the popular BPMN notation to Read-Write Linked Data, enables non-experts to engage in the development of applications that can be verified, validated, and executed.

Acknowledgements. We acknowledge helpful feedback on our manuscript from Rik Eshuis and Philip Hake. This work is supported in part by the EU's FP7 (in i-VISION, GA No.@ 605550) and the German BMBF (in AFAP, FKZ 01IS12051).

References

1. Balaji, B., et al.: Brick: towards a unified metadata schema for buildings. In: Proceedings of the 3rd International Conference on Systems for Energy-Efficient Built Environments (BuildSys) (2016)
2. Capadisli, S., Guy, A., Lange, C., Auer, S., Sambra, A., Berners-Lee, T.: Linked data notifications: a resource-centric communication protocol. In: Blomqvist, E., Maynard, D., Gangemi, A., Hoekstra, R., Hitzler, P., Hartig, O. (eds.) ESWC 2017. LNCS, vol. 10249, pp. 537–553. Springer, Cham (2017). https://doi.org/10.1007/978-3-319-58068-5_33
3. Casati, F., Ceri, S., Pernici, B., Pozzi, G.: Deriving active rules for workflow enactment. In: Wagner, R.R., Thoma, H. (eds.) DEXA 1996. LNCS, vol. 1134, pp. 94–115. Springer, Heidelberg (1996). https://doi.org/10.1007/BFb0034673
4. Ciortea, A., Boissier, O., Zimmermann, A., Florea, A.M.: Give agents some REST: hypermedia-driven agent environments. In: El Fallah-Seghrouchni, A., Ricci, A., Son, T.C. (eds.) EMAS 2017. LNCS (LNAI), vol. 10738, pp. 125–141. Springer, Cham (2018). https://doi.org/10.1007/978-3-319-91899-0_8
5. Elmroth, E., Hernández-Rodriguez, F., Tordsson, J.: Three fundamental dimensions of scientific workflow interoperability: model of computation, language, and execution environment. Future Gener. Comput. Syst. **26**(2), 245 (2010)
6. Ferme, V., Skouradaki, M., Ivanchikj, A., Pautasso, C., Leymann, F.: Performance comparison between BPMN 2.0 workflow management systems versions. In: Reinhartz-Berger, I., Gulden, J., Nurcan, S., Guédria, W., Bera, P. (eds.) BPMDS/EMMSAD -2017. LNBIP, vol. 287, pp. 103–118. Springer, Cham (2017). https://doi.org/10.1007/978-3-319-59466-8_7

7. Gallaher, M.P., O'Connor, A.C., Dettbarn Jr., J.L., Gilday, L.T.: Cost analysis of inadequate interoperability in the US capital facilities industry. NIST GCR 04–867 (2004)
8. Gil, Y., Ratnakar, V., Deelman, E., Mehta, G., Kim, J.: Wings for Pegasus. In: Proceedings of the 19th Conference on Innovative Applications of Artificial Intelligence (IAAI) (2007)
9. Gurevich, Y.: Evolving algebras 1993: lipari guide. In: Specification and Validation Methods. Oxford University Press (1995)
10. Haller, A., Cimpian, E., Mocan, A., Oren, E., Bussler, C.: WSMX - a semantic service-oriented architecture. In: Proceedings of the 3rd International Conference on Web Services (ICWS) (2005)
11. Harth, A., Hose, K., Schenkel, R.: Linked Data Management. CRC, Boca Raton (2014)
12. Harth, A., Speiser, S.: On completeness classes for query evaluation on linked data. In: Proceedings of the 26th AAAI Conference on Artificial Intelligence (2012)
13. Hull, R., et al.: Introducing the guard-stage-milestone approach for specifying business entity lifecycles. In: Bravetti, M., Bultan, T. (eds.) WS-FM 2010. LNCS, vol. 6551, pp. 1–24. Springer, Heidelberg (2011). https://doi.org/10.1007/978-3-642-19589-1_1
14. Jablonski, S., Bussler, C.: Workflow Management. International Thomson, London (1996)
15. Käfer, T., Harth, A.: Rule-based programming of user agents for linked data. In: Proceedings of the 11th International Workshop on Linked Data on the Web (LDOW) (2018)
16. Käfer, T., Harth, A., Mamessier, S.: Towards declarative programming and querying in a distributed cyber-physical system: the i-VISION case. In: Proceedings of the 2nd CPSData Workshop (2016)
17. Newman, S.: Building Microservices - Designing Fine-Grained Systems. O'Reilly, Sebastopol (2015)
18. Pautasso, C.: RESTful web service composition with BPEL for REST. Data Knowl. Eng. 68(9), 851 (2009)
19. Pautasso, C., Wilde, E.: Push-enabling RESTful business processes. In: Kappel, G., Maamar, Z., Motahari-Nezhad, H.R. (eds.) ICSOC 2011. LNCS, vol. 7084, pp. 32–46. Springer, Heidelberg (2011). https://doi.org/10.1007/978-3-642-25535-9_3
20. Pautasso, C., Zimmermann, O., Leymann, F.: RESTful web services vs. "Big" web services. In: Proceedings of the 17th International Conference on World Wide Web (WWW) (2008)
21. Polyvyanyy, A., García-Bañuelos, L., Dumas, M.: Structuring acyclic process models. In: Hull, R., Mendling, J., Tai, S. (eds.) BPM 2010. LNCS, vol. 6336, pp. 276–293. Springer, Heidelberg (2010). https://doi.org/10.1007/978-3-642-15618-2_20
22. Rospocher, M., Ghidini, C., Serafini, L.: An ontology for the business process modelling notation. In: Proceedings of the 8th International Conference on Formal Ontology in Information Systems (FOIS) (2014)
23. Stadtmüller, S., Speiser, S., Harth, A., Studer, R.: Data-Fu: a language and an interpreter for interaction with read/write linked data. In: Proceedings of the 22nd International Conference on World Wide Web (WWW) (2013)
24. Turi, D., Missier, P., Goble, C.A., De Roure, D., Oinn, T.: Taverna workflows: syntax and semantics. In: Proceedings of the 3rd International Conference on e-Science and Grid Computing (e-Science) (2007)
25. Van der Aalst, W.M.P., ter Hofstede, A.H.M., Kiepuszewski, B., Barros, A.P.: Workflow patterns. Distrib. Parallel Databases 14(1), 5 (2003)

26. Van der Aalst, W.M.P., Weske, M., Grünbauer, D.: Case handling: a new paradigm for business process support. Data Knowl. Eng. **53**(2), 129 (2005)
27. Vanhatalo, J., Völzer, H., Koehler, J.: The refined process structure tree. In: Dumas, M., Reichert, M., Shan, M.-C. (eds.) BPM 2008. LNCS, vol. 5240, pp. 100–115. Springer, Heidelberg (2008). https://doi.org/10.1007/978-3-540-85758-7_10
28. Verborgh, R., Steiner, T., van Deursen, D., Coppens, S., Vallés, J.G., van de Walle, R.: Functional descriptions as the bridge between hypermedia APIs and the Semantic Web. In: Proceedings of the 3rd International Workshop on RESTful Design (WS-REST) (2012)
29. Zaveri, A.: smartAPI: towards a more intelligent network of web APIs. In: Blomqvist, E., Maynard, D., Gangemi, A., Hoekstra, R., Hitzler, P., Hartig, O. (eds.) ESWC 2017. LNCS, vol. 10250, pp. 154–169. Springer, Cham (2017). https://doi.org/10.1007/978-3-319-58451-5_11
30. Zur Muehlen, M., Nickerson, J.V., Swenson, K.D.: Developing web services choreography standards. Decis. Supp. Syst. **40**(1), 9 (2005)

Mapping Diverse Data to RDF in Practice

Alexandros Chortaras[(✉)] and Giorgos Stamou

National Technical University of Athens, Athens, Greece
{achort,gstam}@cs.ntua.gr

Abstract. Converting data from diverse data sources to custom RDF datasets often faces several practical challenges related with the need to restructure and transform the source data. Existing RDF mapping languages assume that the resulting datasets mostly preserve the structure of the original data. In this paper, we present real cases that highlight the limitations of existing languages, and describe D2RML, a transformation-oriented RDF mapping language which addresses such practical needs by incorporating a programming flavor in the mapping process.

Keywords: RDF mapping · Data integration · Service integration

1 Introduction

An RDF graph is a set of triples, each one of which consists of a subject, predicate, and object. Thus, despite the powerful semantic interpretation of RDF graphs, their machine representation is very simple: a table with a subject, predicate and object column; actually, several relationally-backed RDF stores use such tabular representations. So, when studying how mappings for diverse data formats to RDF graphs can be defined, essentially we have to define how the underlying data can be transformed to a logical tabular representation.

In this framework, much work has been done on transforming data from relational databases (summarized in [8]). Relational data are pretty simple, because they are kept in tables, each row contains a single value for each column, and each row has usually a unique key that can be used to generate unique identifiers. Moreover, SQL is a powerful language that allows the generation of complex custom view by joining tables, selecting data that meet certain conditions, and performing simple data transformations. R2RML, the W3C language for mapping relational databases to RDF [5] is a powerful language, but owes its power mostly to the inherent tabular nature of the source data and the power of SQL which provides almost all needed data manipulation and restructuring.

Beyond relational data, closer to the tabular model are CSV documents and spreadsheets [11,14]. Other formats, such as XML, differ considerably from tabular data owing to their hierarchical structure, and the mapping systems rely

© Springer Nature Switzerland AG 2018
D. Vrandečić et al. (Eds.): ISWC 2018, LNCS 11136, pp. 441–457, 2018.
https://doi.org/10.1007/978-3-030-00671-6_26

on XSLT transformations, XPath and XQuery (e.g. [1,4]). To resolve the poly-morphy of tools and define a uniform way to perform Data-to-RDF mapping, xR2RML [13] and RML [7] extend R2RML to support other data formats.

All such approaches are practical as long as there is no need to alter the struc-ture of the source data, and as long as the underlying source data manipulation languages provide support for data transformations. In a multi-source support-ing language, like RML, this is harder to achieve given that not all sources are backed by powerful languages, such as SQL. In this paper, we propose D2RML, a generic Data-to-RDF Mapping Language, which aims at facilitating the gen-eration of custom RDF data stores by selectively collecting and integrating data from diverse data sources and web services into high quality RDF data stores. D2RML is based on a tabular data representation, on which restructuring, trans-formation and filtering may be applied. D2RML pushes the limits of a mapping language by incorporating 'programming' features. Although a mapping lan-guage cannot substitute a programming language, the real world cases that we discuss demonstrate that such features are essential if such languages aspire to gain acceptance in practice. This paper is an extended version of [3], which refines the restructuring features of D2RML and focuses on real word scenarios.

The rest of the paper has as follows: Sect. 2 gives an overview of R2RML and RML on which our work is mostly based. Section 3 discusses real examples where existing mapping languages turn out to be insufficient. In Sect. 4 we present the simple data model underlying D2RML. In Sect. 5 we describe how several widely used information sources can be cast onto that model, and in Sect. 6 we present the definition of D2RML. Section 7, in place of an evaluation, demonstrates the power of D2RML by describing how it can solve the practical needs outlined in Sect. 3. Section 8 concludes the paper.

2   Related Work, R2RML and RML

RML and xR2RML are two R2RML-based RDF mappings languages that sup-port both relational and non-relational data. As such they share several common features, but differ in some of their focus points. E.g. xR2ML supports mapping from mixed formats (e.g. relational tables with JSON values), and also RDF lists. On the other hand, RML extended with FnO [12] supports interaction with data sources using established vocabularies [6] and interaction with abstract data pro-cessing functions. For both, R2RML is the starting point.

R2RML works with logical tables (`rr:LogicalTable`), which may be base tables, views, or result sets obtained by an SQL query. Each logical table is mapped to RDF triples using one or more triples maps. A triples map is a rule that maps each logical table row to several RDF triples. The rule consists of a subject map that generates the subject of all RDF triples for each row, and several predicate-object maps, that consist of predicate and object or referencing object maps. A predicate map determines predicates for the RDF triples, and an object map their objects. A subject or predicate-object map may include one or more graph maps, which specify a named graph for the resulting triples. Refer-ring object maps allow joining of triples maps. A referring object map specifies a parent triples map (`rr:parentTriplesMap`), the subjects of which will act as objects for the current triples map. RDF terms (i.e. concrete IRIs and literals for the triples) are either declared constants (`rr:constant`), or obtained from the underlying table, view or result set by specifying a column name (`rr:column`) that will supply the values, or generated by a string template (`rr:template`) that includes reference to columns. String templates offer very rudimentary options to manipulate actual database values and generate custom IRIs and literals.

RML extends R2RML by allowing other sources (`rml:LogicalSource`, e.g. JSON or XML files), by defining data iterators (`rml:iterator`) to split the data from such sources into base elements (the equivalent of rows), and by allowing particular references (`rml:reference`), in the form of subelement selectors within the base element, to define the value sources for RDF terms. The iterators and the references depend on the underlying data source, and may be XPath or JSONPath expressions, CSV column names or SPARQL return variable names. Their type is declared using `rml:referenceFormulation`. To describe access to diverse data sources, RML suggests the use of vocabularies, such as DCAT, CSVW, Hydra, and SPARQL-SD. However, these vocabularies in general do not prescribe a way to formulate actual requests (e.g. to a web API that paginates the results using next page access keys).

3   Motivating Examples

Here we present some examples that highlight the need for additional flexibility from an RDF mapping language. All are adapted (to save space) real examples.

Example 1. Consider the following except from a database containing a timeline of modern Greek history events. The database was modeled as a single table.

ID	date	summary	senderA	receiverB	sourceA	senderB	receiverB	sourceB	keywords
304	21/11/ 1940	Palairet calls F.O. saying that ...	Palairet	F.O.	F.O.371/ 24907/ R8517	Palairet	Halifax	F.O.371/ 24907/ R8879	J. Metaxas, Greece - Economy

Each row contains a date, summary and some keywords. Since each event turned out to have in practice at most two references, the modeler of the database included two sets of sender, receiver and source columns. Moreover, keywords are included in a single column, separated by commas, but there may be multi-term keywords in which terms are separated by a dash. xR2RML provides a solution if the keywords entry were in a structured format (eg. JSON). An RDF graph for the above, not inheriting the modeling problems, could be the following:

```
cge:304 [ a cge-t:Event ; cge-t:date "1940-11-21"^^xsd:date ; cge-t:summary "Palairet ..." ;
    cge-t:reference [ a cge-t:Letter ; cge-t:source "F.O.371/24907/R8817" ;
                      cge-t:sender "Palairet" ; cge-t:receiver "F.O." ] ;
    cge-t:reference [ a cge-t:Letter ; cge-t:source "F.O.371/24907/R8879" ;
                      cge-t:sender "Palairet" ; cge-t:receiver "Halifax" ] ;
    cge-t:keyword [ a cge-t:Term ; kvoc-t:text "J. Metaxas" ] ;
    cge-t:keyword [ a cge-t:Term ; kvoc-t:text "Greece", "Economy" ] .
```

The transformation of the keywords, which splits the entry at the commas and then at the dashes to generate a nested structure, is problematic even using the SQL power of R2RML. For other data sources (e.g. CSV files) it would be impossible to do anything more than copy the original data structure. Certainly, this is a not an optimally designed database, but such cases do occur in practice.

Example 2. Consider the following excerpt from the PeriodO (http://perio.do/) gazetteer of historic periods, which is available as a JSON document:

```
{ "periodCollections": {
    "p0339m9": {
      "id": "p0339m9", "type": "PeriodCollection",
      "definitions": { ... ,
        "p0339m9f72b": {
          "id": "p0339m9f72b", "type": "PeriodDefinition", "start": "1204", "stop": "1453",
          "spatialCoverage": [ {"id": "http://dbpedia.org/resource/Greece" } ],
          "localizedLabels": { "eng": [ "Late Byzantine" ] } },
        "p0339m9jq2m": {
          "id": "p0339m9jq2m", ... }, ... } },
    "p08nrfc": { ... }, ... } }
```

Using the SKOS model, we would like to generate the following RDF graph:

```
ark:p0339m9 [ a ark-t:PeriodCollection ] .
ark:p0339m9f72b [ a ark-t:PeriodDefinition ; rdfs:label "Late Byzantine"@en ;
                  ark-t:earliestYear "1204"^^xsd:gYear ; ark-t:latestYear "1453"^^xsd:gYear ;
                  skos:inScheme ark:p0339m9 ; dcterms:spatial dbpedia:Greece ] .
```

Using RML and JSONPath, we could specify a triples map to iterate over the period collections, and then a triples map to iterate over the period definitions to generate the respective triples; but inside a period definition we do not know the enclosing period collection, to generate the `skos:inScheme` triple. Thus we cannot use a referring object map (for period definitions inside period collections). We could possibly specify a third triples map to iterate over the collections and generate triples with object the collection id and subjects the included period ids, but this violates a basic assumption in both R2RML and RML that each iteration over the data should produce a unique subject.

Example 3. Geonames provides its gazetter data as a set of tab-delimited files. Among them, file `admin1Codes.txt` contains top-level administrative regions for all countries, `XX.txt`, where `XX` is a country code, a country's locations, and `alternateNames/XX.txt` alternate location names and links to other resources. Consider line (`GR.ESYE31; Attica; 6692632`) from `admin1Codes.txt`, and lines

```
256601;  Athens; Athinai,Athina; P; PPLC; GR; ESYE31; 445408
445408;  Athens Prefecture; Athena,Athina; A; ADM2; GR; ESYE31; 445408
6692632; Attica; Attica,Attiki; A; ADM1; GR; ESYE31

177543;  264371; el; Athina; 1
1593954; 264371; en; Athens;
2919841; 264371; link; http://en.wikipedia.org/wiki/Athens;
```

from `GR.txt` and `alternatenames/GR.txt`, for Greece, respectively.

The column names are (admin1code, name, geonameid), (geonameid, name, alternate names, feature class, feature code, country code, admin1 code, admin2 code) and (alternateNameid, geonameid, language, alternate name, is Preferred-Name), respectively. From the above, we want to generate the following triples:

```
geo:256601 [ a gn:Feature ; gn:name "Athens" ; gn:featureCode gn:P.PPLC ;
             gn:officialName "Athina"@el ; gn:alternateName "Athens"@en ;
             gn:wikipediaArticle <http://en.wikipedia.org/wiki/Athens> ;
             gn:parentADM1 geo:6692632 ; gn:parentADM2 geo:445408 ] .
geo:445408 [ a gn:Feature ; gn:name "Athens Prefecture" ; gn:featureCode gn:A.ADM2 ;
             gn:parentADM1 geo:6692632 ] .
geo:6692632 [ a gn:Feature ; gn:name "Attica" ; gn:featureCode geo-ont:A.ADM1 ] .
```

To achieve this we need some conditions (e.g. do not include a `gn:parentADM2`) if in a line of `GR.txt` the geonameid and admin2 code coincide). Moreover, in the triples map iterating over `GR.txt`, we need to include a referring object map to perform a join with `admin1Codes.txt`, so as to know that `GR.ESYE31` has geonameid `6692632`. But to do the join we need to concatenate the country code `GR` with the admin1 code `ESYE31`, which in `GR.txt` are provided in distinct columns. In a relational database we could possibly formulate such queries, but with CSV files we have much less flexibility. Even if we overcome somehow the problem of generating a combined key for the join, iteratively joining large CSV files can be inefficient. Instead, we could probably start building the RDF graph by first mapping `admin1Codes.txt`, and then execute the mapping for `GR.txt`, exploiting the contents of the up to then generated RDF graph. Furthermore, in each line `alternateNames/GR.txt`, the value of the language column determines how to interpret the alternate name. If we see `link` we should use `gn:wikipediaArticle`. If we see a language code we should further check the last column: if it is 1 we use `gn:officialName` otherwise `gn:alternateName`. To do this we need conditions and case statements. If the data was relational, we could exploit SQL and define three triples maps, one for each predicate; but with a CSV file this is not possible. Even with relational data, a single triples map, with a conditional statement selecting each time the right predicate, is probably a clearer, more concise, and possibly more efficient, modeling since it requires a unique iteration over the data.

Example 4. Consider the row (4821; 1431–1433 AD; Samothrace; *<inscription>*) of a CSV database of Christian and Byzantine inscriptions provided by the University of Athens that contains an id, a chronology, a location and the actual inscription text. We would like to generate the following RDF graph:

```
bci:4821 [ bci-t:Inscription ; bci-t:chronology "1431-1433 AD" ; bci-t:location "Samothrace" ;
           kvoc-t:date [ a time:DateTimeInterval ;
                         time:hasBeginning tl-t:Y.1431 ; time:hasEnd tl-t:Y.1433 ] ;
           kvoc-t:location geo:734358 ; kvoc-ont:period ark:p0339m9f72b ] .
```

In this case the mapping involves some processing using data analysis services. A geonames linking service that links `Samothrace` to `geo:734358`, the geonames resource for Samothrace, a date recognizer that transforms 1431–1433 AD into a time description using OWL Time vocabulary, and finally a periodO linking service that uses the geonames location and the OWL time date range to classify the inscription to the Late Byzantine period `ark:p0339m9f72b`. The first two services can be seen as data sources that require parameters taking values from the original data; this is supported in RML. But for the third service we need to specify parameters values that are results of the previous two services.

Moreover, we might want to use the results of the two first services only as intermediate results for the first service and not produce any triples for them.

4 Model

The above examples demonstrate that a practical RDF mapping language should include provisions for complex mapping capabilities, that may result from the actual structure or the data, from peculiarities of the data representation choices or models, or from the need for structure altering transformations. To create a general framework for such a language, we define first an abstract tabular data model. Essentially, we assume that data coming from a data source give rise to a tabular structure, which is extensible by the mapping language: new columns may be added by transforming existing ones to generate input data for further transformations. Each cell of the tabular structure may contain a set of values.

Definition 1. *A* set row *of arity k is a tuple $\langle D_1, \ldots, D_k \rangle$, where $D_1, \ldots, D_k$ are sets of values. A* name row *of arity k is a tuple $\langle n_1, \ldots, n_k \rangle$, where $n_1, \ldots, n_k$ are names. A* set table *of arity k with m rows is a tuple $S = \langle N, T \rangle$, where N is a name row and $T = [D_1, \ldots, D_m]$ a list of set rows, all of arity k, such that the i-th elements of $D_1, \ldots, D_m$, for $1 \leq i \leq k$, share all the same domain.*

The names allow us to refer to particular elements of set rows and tables. We denote the set of values that corresponds to name n_i in a set row D by $D[n_i]$ and by $S[n_k]$ (a *column* of S) the list $[D_1[n_k], \ldots, D_m[n_k]]$ of value sets that are obtained from the several set rows of S. For a particular set row D and the several n_i, the sets $D[n_i]$ may have different numbers of values and in general there is no alignment between the individual values among the several sets, and all individual values are equivalent with respect to their relation to the values of the other sets in the same set row.

Definition 2. *A* filter *F over a set table S is a tuple $\langle n, f \rangle$, where n is a column name and f a function, such that $f(D[n]) \subseteq D[n]$ for all set rows D of S.*

We denote the set value $f(D[n])$, obtained by applying F on a set row D by $F(D)$. A filter may be seen as the implementation of a condition.

Data for set tables are acquired from *information sources*. To accommodate several possible information sources in our model, we consider, as in RML, that the information source upon a *request* provides in a reply an *effective data source*, a structure that groups data in several autonomous elements. The division of the effective data source to these autonomous elements is achieved by an *iterator*, which specifies a *logical array*, through whose items the iterator iterates. Each item of a logical array may be a complex structure (another effective data source), so in order to extract from it lists of values to construct set rows values, we need some *selectors*. The selectors transform a logical array into a set table.

Definition 3. *The triple $A = \langle I, t, L \rangle$, where I is an effective source specification, t an iterator, and L a set of selectors, is a* data acquisition pipeline.

Each data acquisition pipeline $\mathcal{A}$ gives rise to a unique set table $\mathcal{S}_\mathcal{A}$. A data acquisition pipeline may be parametric. A parametric data acquisition pipeline $\mathcal{A}'$ that depends on $\mathcal{A}$ is a data acquisition pipeline whose parameters take values from one or more columns of $\mathcal{S}_\mathcal{A}$ and is called a *transformation* of $\mathcal{A}$.

Definition 4. *A series of data acquisition pipelines $\mathcal{A}_0$, $\mathcal{A}_1$, ..., $\mathcal{A}_l$, where each $\mathcal{A}_i$, for $i > 1$, is a transformation that depends on one or more $\mathcal{A}_j$ for $j < i$ is a set table specification. $\mathcal{A}_0$ is the* primary *data acquisition pipeline.*

A set table specification gives rise to a unique set table: $\mathcal{S}_{\mathcal{A}_0}$ extended by columns contributed by $\mathcal{A}_1$, ..., $\mathcal{A}_l$. Each transformation is realized as a series of requests to an information source, after binding the parameters to *all* possible combinations of values obtained from the referred to columns of the set table constructed from the preceding data acquisition pipelines. Thus, a set table specification is evaluated serially. The primary data acquisition pipeline $\mathcal{A}_0$ gives rise to set table $\mathcal{S}_{\mathcal{A}_0}$. Then, for each set row $\mathcal{D}$ of $\mathcal{S}_{\mathcal{A}_0}$, evaluating $\mathcal{A}_1$ gives rise to a set table $\mathcal{S}_{\mathcal{A}_1}(\mathcal{D})$. By flattening all rows of $\mathcal{S}_{\mathcal{A}_1}(\mathcal{D})$ into a single row we obtain a new set row that is appended to $\mathcal{D}$. Doing this for all set rows $\mathcal{D}$ results in $\mathcal{S}_{\mathcal{A}_0\mathcal{A}_1}$. Proceeding this way, eventually $\mathcal{S}_{\mathcal{A}_0}$ is extended to set table $\mathcal{S}_{\mathcal{A}_0\mathcal{A}_1...\mathcal{A}_l}$. More formally, let $n_1, ..., n_k$ be the names, and $[\mathcal{D}_1, ..., \mathcal{D}_m]$ the rows of $\hat{\mathcal{S}} \doteq \mathcal{S}_{\mathcal{A}_0...\mathcal{A}_i}$. Evaluating $\mathcal{A}_{i+1}$ on each row of $\hat{\mathcal{S}}$ produces set tables $\mathcal{S}_{\mathcal{A}_{i+1}}(\mathcal{D}_1), ..., \mathcal{S}_{\mathcal{A}_{i+1}}(\mathcal{D}_m)$. Since all these set tables are produced by the same data acquisition pipeline $\mathcal{A}_{i+1}$, they share the same arity, say k', and let $\hat{s}_1, ..., \hat{s}_{k'}$, be the selectors of $\mathcal{A}_{i+1}$. Thus $\mathcal{S}_{\mathcal{A}_0...\mathcal{A}_{i+1}} = \langle N, \mathcal{T} \rangle$, where $N = \langle n_1, ..., n_k, \mathcal{A}_{i+1}.\hat{s}_1, ..., \mathcal{A}_{i+1}.\hat{s}_{k'} \rangle$, $\mathcal{T} = [\mathcal{D}'_1, ..., \mathcal{D}'_m]$, $\mathcal{D}'_j = [\mathcal{D}_j[n_1], ..., \mathcal{D}_j[n_k], \hat{D}_{j1}, ..., \hat{D}_{jk'}]$ for $1 \leq j \leq m$, and $\hat{D}'_{jl} = \bigcup \mathcal{S}_{\mathcal{A}_{i+1}}(\mathcal{D}_j)[\hat{s}_l]$ for $1 \leq l \leq k'$.

Definition 5. *A triples rule $\mathcal{R}$ over a set table $\mathcal{S} = \langle N, \mathcal{T} \rangle$ is either (a) a triple of filters $\langle \mathcal{F}_s, \mathcal{F}_p, \mathcal{F}_o \rangle$, over $\mathcal{S}$, called the* subject, predicate *and* object filter, *respectively, or (b) a triple $\langle \mathcal{F}_s, \mathcal{F}_p, \hat{\mathcal{R}} \rangle$, over $\mathcal{S}$, where $\mathcal{F}_s$, $\mathcal{F}_p$ are the* subject *and* predicate filter, *respectively, and $\hat{\mathcal{R}}$ another triples rule.*

The implementation of $\mathcal{R}$ is the set of RDF triples $\{(s, p, o) \mid s \in \mathcal{F}_s(\mathcal{D}), p \in \mathcal{F}_p(\mathcal{D}), o \in \mathcal{F}_o(\mathcal{D}), \mathcal{D} \in \mathcal{T}\}$ in case (a), and $\{(s, p, o) \mid s \in \mathcal{F}_s(\mathcal{D}), p \in \mathcal{F}_p(\mathcal{D}), o \in \mathcal{S}_{\hat{\mathcal{R}}}, \mathcal{D} \in \mathcal{T}\}$ in case (b), where $\mathcal{S}_{\hat{\mathcal{R}}}$ are the subjects of the implementation of $\hat{\mathcal{R}}$.

A set of triples rules over some set tables defines a *Data-to-RDF mapping*. The relevant RDF dataset is the implementation of all its triples rules.

5   Retrieving and Interpreting Data

To define general data acquisition pipelines in practice, we consider that an information source, in response to a request, provides some source data. An information source may be either a server (RDBMS, web server/RESTful web service, SPARQL endpoint), the local file system, or a local data model (e.g. an in-memory RDF model). The source data may be an instance of a particular data model (e.g. a SQL result set) or a document (e.g. a JSON or XML document). To obtain source data from an information source we need to specify a request (e.g. an HTTP GET request, an SQL query), as summarized in Table 1. Source data obtained as data models (e.g. SQL result sets) have a unique interpretation, but a document may be interpreted as one of several models. The effective data source is the source data interpreted: SQL and SPARQL results sets denote themselves, a JSON document a JSON Tree, an XML/HTML document a DOM or XDM, and a CSV document a table. In some cases, to obtain an effective data source, we need an intermediate interpretation of the source data as a new information source. This is the case e.g. with an RDF document, which should be seen first as an RDF dataset, from which SPARQL result sets can then be obtained.

Table 1. Information sources, requests and source data

Information source	Request	Source data
RDBMS	SQL SELECT Query	SQL Result Set
Web Server/ RESTful Service	HTTP GET/ POST Request	JSON/XML/CSV/HTML/ RDF/TXT Document
SPARQL Endpoint/ RDF model	SPARQL SELECT Query and Graph IRIs via API	SPARQL Result Set
File System	File Request	JSON/XML/CSV/HTML/ RDF/TXT Document

Table 2. Effective data sources, iterators and selectors

Effective data source	Type	Iterator	Selector
SQL Result Set	Tabular	Row Iterator	Column Name
SPARQL Result Set	Tabular	Row Iterator	Variable Name
JSON Tree	Hierarchical	JSONPath	JSONPath
XDM	Hierarchical	XPath/XQuery	XPath/XQuery
DOM	Hierarchical	CSS Selector	CSS Selector
CSV Document	Tabular	Row Iterator	Column Name/Number
Text Document	Flat	Regular Expression	Regular Expression

The model-specific iterators and selectors needed to convert effective data sources to set tables are shown in Table 2. For example, an SQL (SPARQL result set) obtained through an SQL (SPARQL) SELECT query q that specifies attributes (variables) $n_1, \ldots, n_k$ in the SELECT statement for the returned columns (variables), returns a list of rows $[\langle v_{11}, \ldots, v_{1k} \rangle, \ldots, \langle v_{n1}, \ldots, v_{nk} \rangle]$. Using a row iterator and the column (variable) names $n_1, \ldots, n_k$ as selectors, we obtain the set table $\langle \langle n_1, \ldots, n_k \rangle, [\langle \{v_{11}\}, \ldots, \{v_{1k}\} \rangle, \ldots, \langle \{v_{n1}\}, \ldots, \{v_{nk}\} \rangle] \rangle$. Similarly, a JSONPath (XPath/XQuery, CSS) expression q splits a JSON tree [2] (XDM [15], DOM) interpretation of a JSON (XML/HTML) document $\mathcal{T}$ into a logical array of smaller JSON trees (XDMs, DOMs) $\mathcal{T}_1, \ldots, \mathcal{T}_n$. By executing as selectors JSONPath (XPath/XQuery, CSS) expressions $q_1, \ldots q_k$ over each $\mathcal{T}_1, \ldots, \mathcal{T}_n$ we get the set table $\langle \langle q_1, \ldots, q_k \rangle, [\langle C_{11}, \ldots, C_{1k} \rangle, \ldots, \langle C_{n1}, \ldots, C_{nk} \rangle] \rangle$, where C_{ij} is either the set of values (string values of text or attribute nodes) contained in the array (node set) that results from applying q_j on $\mathcal{T}_i$.

6   D2RML Specification

D2RML draws significantly from R2RML and RML, and follows the same strategy for defining mappings: triples maps, consisting of a subject map and several predicate-object maps. From RML it adopts and extends the interaction with information sources through requests, iterators and selectors. It also extends the expressive capabilities of R2RML and RML by allowing transformations, conditional and case statements, and custom RDF term generation functions. For its semantics, it relies on the data model of Sect. 4. Each triples map corresponds to a set table (Definition 5) and a set of triple rules (Definition 4) with the same subject filter over the common underlying set table. The information source, request and iterator for the original data acquisition pipeline are provided in the triples map definition. Any additional transformations are declared in the order of their application and extend incrementally the underlying set table. The selectors are implicitly declared in the included subject, predicate, object and graph maps.

6.1   Triples Maps

Triples maps are defined as in RML, but tabular data providing information sources are clearly distinguished from non-tabular ones. The inclusion of *Transformation* and *DefinedColumn* lists allow extending the primary set table.

$TriplesMap \leftarrow$ a rr:TriplesMap
    (rr:logicalTable $\langle LogicalTable \rangle$ | dr:logicalSource $\langle LogicalSource \rangle$)?
    (dr:transformations ($\langle Transformation \rangle$+))?
    (dr:definedColumns ($\langle DefinedColumn \rangle$+))?
    (rr:graphMap $\langle GraphMap \rangle$)*
    rr:subjectMap $\langle SubjectMap \rangle$ | rr:subject IRI
    (rr:predicateObjectMap $\langle PredObjMap \rangle$)*

$PredObjMap \leftarrow$ a rr:PredicateObjectMap
    (rr:predicateMap $\langle PredicateMap \rangle$ | rr:predicate IRI)+
    (rr:objectMap ($\langle ObjectMap \rangle$|$\langle RefObjectMap \rangle$) | rr:object (IRI|LIT))+
    (rr:graphMap $\langle GraphMap \rangle$ | rr:graph IRI)*

6.2 Logical Tables and Logical Sources

A *LogicalTable* or *LogicalSource* specifies a data acquisition pipeline (excluding the separators). In the case of query supporting information sources (RDBMSs' and SPARQL services), for compatibility with R2RML, they contain also the query-relevant details of the request. The `is:parameters` predicate helps declare parameters in queries of parametric data acquisition pipelines. For other information sources, the request and any parameters are part of the *InformSource* specification. For non-tabular data information sources, *LogicalSource* should contain the definition of the iterator (`dr:iterator` and `dr:referenceFormulation`) used to split the effective data source.

D2RML introduces two special sources: The first `dr:CurrentModel`, represents the RDF model of the current RDF dataset generated by the D2RML processor, and is interpreted as an *SPARQLTable*. Since the model is constantly updated, we need to define an execution order for the triples maps. Thus, a D2RML document using `dr:CurrentModel` must specify a `is:TriplesMapOrder`, whose `is:mapOrder` defines the execution order of the triple maps as a list. The second source, `dr:SetTable`, allows the generation of a new table from the values of some selected columns (`dr:transferredColumns`) of the dependent on set table. The `dr:SetTable` instance is generated by taking the values of each column of the current row, in order of appearance, and putting them into a new table, by aligning values having the same order. Thus it restructures the data: it converts, one or more sets of values, into a table where each column in each row contains aligned single values. If used as a *LogicalSource*, `dr:SetTable` allows, as in xR2RML, interpreting row values as structured documents (eg. JSON, XML).

LogicalTable$_1$ ← a rr:LogicalTable	a dr:CurrentModel
dr:source ⟨*InformSource*⟩	
SQLTable | *SPARQLTable* | *CSVTable*	*SPARQLTable*
(is:parameters (⟨*DataVariable*⟩+))?	

LogicalTable$_2$ ← a dr:SetTable ; (dr:transferedColumns (⟨*ValueRef*⟩+))?

LogicalSource ← (a dr:LogicalSource ; dr:source ⟨*InformSource*⟩) \| *LogicalTable*$_2$
         dr:iterator LIT ; dr:referenceFormulation IRI

An *SQLTable* is defined as in R2RML. A *SPARQLTable* must specify a query (`dr:sparqlQuery`) and any default or named graphs (`dr:defaultGraph`, `dr:namedGraph`). Finally a, `CSVTable` must specify a delimiter (`dr:delimiter`), whether there is a header line (`dr:headerline`), and possibly a quote, comment, escape, or record separator characters.

6.3 Information Sources

An information source may be a RDBMS, an HTTP server, a SPARQL endpoint, or the file system.

InformSource ← *RDMSSource* | *SPARQLSource* | *HTTPSource* | *FileSource*

An *RDMBSSource* must specify the type of the RDMBS (is:rdbms), the location (is:location) and any needed information for establishing the connection (e.g. username, password). An *HTTPSource* should either specify a single URI (is:uri) or prescribe a full HTTP request (is:request). The latter is specified in using the W3C's 'HTTP Vocabulary in RDF 1.0' [9] and 'Representing Content in RDF 1.0' vocabularies [10]. An *HTTPSource* may contain a list of parameters (is:parameter). To account for result pagination, it may contain a request iterator in the parameter list. A request iterator may be a *KeyRequestIterator* or a *CountRequestIterator*. Both of them should provide the parameter name (is:name) which should appear in the URI of the HTTPRequest, and an initial value (is:initialValue). A key request iterator must specify how to extract each time the new parameter value from the current server results in order to formulate the subsequent request, while a count request iterator must specify an increment (is:increment) and possibly a maximum value (is:maxValue). The set of iteration policies is extensible. A *SPARQLSource* must specify simply the URI of the service (is:uri). A *FileSource* must specify the location of one or more files (is:path) and their encoding (is:encoding). Note that, following the earlier discussion, e.g. a *FileSource* that fetches an RDF file used in conjunction with a *SPARQLTable*, is interpreted as a RDF model information source.

6.4 Transformations and Defined Columns

A *Transformation* extends the underlying set table. Since it is a parametric data acquisition pipeline, its definition includes a *LogicalTable* or *LogicalSource* and one or more *ParameterBindings* to assign parameter values. The latter consists of a reference to a value (*ValueRef*) or a constant value, and the parameter name (dr:parameter) in the corresponding information source the value will be bound to. A *DefinedColumn* allows for in-line set table transformations: to add new columns by applying a series of transformations on particular set table column values without consulting external sources. A *DefinedColumn* should declare the new column name dr:name, the function (dr:function) to generate the values (eg. op:regex, op:replace), and a list of arguments (dr:parameterBinding). It is assumed that the parameter names are provided by the function definition.

Because a *Transformation* may need to work not directly with the value sets of the underlying set table at the level of the selectors (where any alignment between values is lost), but first with higher level iterators that preserve the alignment and then with the value set producing selectors, a transformation may declare one such iterator (dr:bindingIterator) for each transformation that provides parameter bindings. When, executed, the values for parameter bindings will be generated aligned according to the iterator.

Transformation ← a dr:Transformation
          rr:logicalTable ⟨*LogicalTable*⟩ | dr:logicalSource ⟨*LogicalSource*⟩
          (dr:parameterBinding ⟨*ParameterBinding*⟩)+
          (dr:bindingIterator ⟨*BindingIterator*⟩)*

DefinedColumn ← a dr:DefinedColumn ; dr:name LIT ; dr:function IRI
$\qquad$ (dr:parameterBinding ⟨*ParameterBinding*⟩)+

ParameterBinding ← a dr:ParameterBinding
$\qquad$ dr:parameter LIT ; rr:constant LIT | *ValueRef*

BindingIterator ← rr:column LIT | dr:reference LIT
$\qquad$ (dr:transformationReference ⟨*Transformation*⟩)?

6.5 Term Maps and Conditions

The definition of a *TermMap* (*SubjectMap*, *PredicateMap*, *ObjectMap*, *GraphMap* and *LanguageMap*) follows the R2RML specification with the support for filters.

SubjectMap ← a rr:SubjectMap ; *IRIRef* | *BlankNodeRef*
$\qquad$ (*SubjBody CSubjBody**) | *CSubjBody*+

PredicateMap ← a rr:PredicateMap ; (*PredBody CPredBody**) | *CPredBody*+

ObjectMap ← a rr:ObjectMap ; (*ObjBody CObjBody**) | *CObjBody*+

GraphMap ← a rr:GraphMap ; (*GraphBody CGraphBody**) | *CGraphBody*+

LanguageMap ← a rr:LangMap ; (*LangBody CLangBody**) | *CLangBody*+

SubjBody ← (rr:class IRI)* ; (rr:graphMap ⟨*GraphMap*⟩ | rr:graph IRI)*
$\qquad$ (dr:condition ⟨*Condition*⟩)?

[Pred|Graph]Body ← *IRIRef* ; (dr:condition ⟨*Condition*⟩)?

ObjBody ← *IRIRef* | *BlankNodeRef* | *LiteralRef* ; (dr:condition ⟨*Condition*⟩)?

LangBody ← *LiteralRef* ; (dr:condition ⟨*Condition*⟩)?

C[Subj|Pred|Obj|Graph|Lang]Body ← dr:cases (⟨*[Subj|Pred|Obj|Graph|Lang]Body*⟩+)

Condition ← (*ValueRef*)? ; (dr:booleanOperator IRI)|
$\qquad$ (OPERATOR (*ValueRef* | LIT) | dr:operand ⟨*Condition*⟩)+

RefObjectMap ← a rr:RefObjectMap ; rr:parentTriplesMap ⟨*TriplesMap*⟩
$\qquad$ ((rr:joinCondition ⟨*JoinCondition*⟩)+ |
$\qquad$ (dr:parameterBinding ⟨*ParameterBinding*⟩)+)?

JoinCondition ← a rr:Join ; rr:child LIT ; rr:parent LIT

To implement filters, a *TermMap* may contain a condition (dr:condition) and/or a case statement (dr:cases). If it contains a condition, it will be evaluated and the corresponding RDF term will be taken into account only if the condition holds. A condition statement must specify the actual value on which it will operate, and may include several tests which will be jointly evaluated using the operator specified by dr:booleanOperator (op:and or op:or). A test is specified either through an OPERATOR (op:eq, op:le, etc.) and a literal or *ValueRef* which define the value(s) with which the actual value will be compared using OPERATOR, or as a nested condition. Due to the underlying set table model, in general a behaviour of the operators for set arguments should be defined. In the current implementation it is assumed that a condition holds if it holds for a single value pair, but this is a point of further refinement of the language. The operation type (eg. number/string comparison) depends on the operand XSD types. A case statement includes a list of alternative realizations of a *TermMap*, each along with a condition. If the condition evaluates to true, the corresponding *TermMap* is realized, otherwise control flows to the next case.

Finally, a referring object map (*RefObjectMap*) may be defined either as in R2RML or using a *ParameterBinding*, in the case the *LogicalTable* or *LogicalSource* of the referring object's triples map is a parametric data acquisition pipeline: the *ParameterBinding* provides the parameters values to be used in the parametric data acquisition pipeline of the referring object map.

6.6   RDF Terms

RDF terms are specified as in RML, but to account for values coming from transformations, RDF terms are generated through value references, specified by two components: a compulsory `rr:column`, `rr:template` or `dr:reference`, and an optional `dr:transformationReference` to specify the underlying transformation for the `rr:column`, `rr:template` or `dr:reference`. If missing, the primary logical array is assumed. To overcome the limited data manipulation options offered by `rr:template`, a value reference may include local defined columns (`dr:definedColumns`) that are need to generate a particular RDF Term.

IRIRef ← `rr:constant` IRI | *ValueRef* ; (`rr:termType rr:IRI`)?

LiteralRef ← `rr:constant` LIT | *ValueRef* ; (`rr:termType rr:Literal`)?
      ((`dr:languageMap` ⟨*LanguageMap*⟩ | `rr:language` LIT) |
                           `rr:datatype` IRI)?

BlankNodeRef ← *ValueRef* ; (`rr:termType rr:BlankNode`)?

ValueRef ← `rr:column` LIT | `rr:template` LIT | `dr:reference` LIT
      (`dr:transformationReference` ⟨*Transformation*⟩)?
      (`dr:definedColumns` (⟨*DefinedColumn*⟩+))?

7   Evaluation

In Sect. 3 we identified cases where the desired mappings could not be achieved using existing RDF mapping languages. In fact, all cases were real and arose in the context of a project aiming to provide semantic analysis services over a repository of heterogeneous cultural data. This provided a real testbed for the usefulness of D2RML; here we discuss how it helped solving such mappings needs. (To save space we omit the `dr:referenceFormulation` declarations and write `dr:transformationReference` as `dr:tRef`).

Example 1 involved data restructuring: split each joint keyword into a distinct keywords and separate the each keyword's terms. To do this, we first extend the primary set table with a defined column containing the set of split keywords for each row (`op:split` splits its `input` on the provided `separator`). Then in a referring object map, we build a new table (`dr:SetTable`) from each keyword set, and perform there the splitting on the dashes. In this way we achieve a restructuring that preserves the connection of each term with the source keyword:

```
<#EventMap>
    rr:logicalTable [ dr:source <#FOSource> ; rr:tableName "Events" ] ;
    dr:definedColumns ( [
        dr:name "KW" ; dr:function op:split ;
        dr:parameterBinding [ dr:parameter "input" ; rr:column "keywords" ] ;
```

```
        dr:parameterBinding [ dr:parameter "separator" ; rr:constant "," ] ] ) ;
rr:subjectMap [ rr:template  {@tl}{ID}" ; rr:class cge-t:Event ] ;
rr:predicateObjectMap [
    rr:predicate cge-t:keyword ;
    rr:objectMap [
        rr:parentTriplesMap [
            rr:logicalTable [ a dr:SetTable ; dr:transferedColumns ( [ rr:column "KW" ] ) ] ;
            rr:subjectMap [ rr:class cge-t:Term ; rr:termType  rr:BlankNode ] ;
            rr:predicateObjectMap [
                rr:predicate kvoc-t:text ;
                rr:objectMap [
                    dr:definedColumns ( [
                        dr:name "TERM" ; dr:function op:split ;
                        dr:parameterBinding [ dr:parameter "input" ; rr:column "KW" ] ;
                        dr:parameterBinding [ dr:parameter "separator" ; rr:constant "-" ] ] ) ;
                    rr:column "TERM"; rr:termType rr:Literal ] ] ] ] .
```

In Example 2 we needed maps generating more than one subjects for each row to link periods with period collections. The solution is to define first a map to declare each collection as a skos:ConceptScheme, and then a second multi-subject map to link periods with period collections:

```
<#CollectionMap1>
    dr:logicalSource [ dr:source <#PeriodoSource> ; dr:iterator "$.periodCollections.*" ] ;
    rr:subjectMap [ rr:template "{@ark}{$.id}" ;
                        rr:class periodo:PeriodCollection ; rr:class skos:ConceptScheme ] .
<#CollectionMap2>
    dr:logicalSource [ dr:source <#PeriodoSource> ; dr:iterator "$.periodCollections.*" ] ;
    rr:subjectMap [ rr:template "{@ark}{$.definitions.*.id}" ] ;
    rr:predicateObjectMap [ rr:predicate skos:inScheme ;
                        rr:objectMap [ rr:template "{@ark}{$.id}" ; rr:termType rr:IRI ] ] .
```

In Example 3, to avoid joining CSV files, we needed to generate triples linking admin codes to their geonames ids, and then consult these triples to resolve admin code references when generating triples for a country's locations. To achieve this, we first define ADMIN1Map. Because ADMIN1Map should be executed first, we include a triples map order statement. According to that ordering, next is executed CountryMap. This map uses the transformation ADMIN1Trans, which extends the primary set table by a new column with the geonames id of the locations admin code. ADMIN1Trans operates on the dr:CurrentModel logical table, so as to has access to the triples generated by ADMIN1Map. Last comes the AlternateNamesMap which processes the file with the alternate names and contains conditional statements to decide which predicate it should use for each entry:

```
<#Order>
    dr:mapOrder ( <#ADMIN1Map> <#CountryMap> <#AlternateNamesMap> ) .
<#ADMIN1Map>
    rr:logicalTable [ dr:source <#ADMIN1Source> ; dr:delimiter ";" ] ;
    rr:subjectMap [ rr:template "{@geo}{##3}/" ] ;
    rr:predicateObjectMap [ rr:predicate gn:code ;
                        rr:objectMap [ rr:column "##1" ; rr:termType  rr:Literal ] ] .
<#ADMIN1Trans>
    rr:logicalTable [
        a dr:CurrentModel ;
        dr:sparqlQuery "SELECT ?admin1uri WHERE {?admin1uri gn:code \"{@@name@@}\" }" ] ;
    dr:parameterBinding [ dr:parameter "name" ; rr:template "{##6}.{##7}" ] .
<#CountryMap>
    rr:logicalTable [ dr:source <#CountrySource> ; dr:delimiter ";" ] ;
    dr:transformations ( <#ADMIN1Trans> ) ;
```

```
    rr:subjectMap [ rr:template "{@geo}{##1}/" ; rr:class gn:Feature ] ;
    rr:predicateObjectMap [
        rr:predicate gn:parentADM1 ;
        rr:objectMap [ rr:column "admin1uri" ; dr:tRef <#ADMIN1Trans> ; rr:termType  rr:IRI ] ] .
<#AlternateNamesMap>
    rr:logicalTable [ dr:source <#AlternateNamesSource> ; dr:delimiter ";" ] ;
    rr:subjectMap [ rr:template "{@geo}{##2}/" ; rr:termType rr:IRI ] ;
    rr:predicateObjectMap [
        rr:predicateMap [
            dr:cases ( [ rr:constant gn:officialName ;
                         dr:condition [ rr:column "##5" ; op:eq "1" ] ]
                       [ rr:constant gn:alternateName ] ) ] ;
        rr:objectMap [
            rr:column "##4" ; rr:termType  rr:Literal ; dr:languageMap [ rr:column "##3" ] ;
            dr:condition [ rr:column "##3" ; op:neq "link" ] ] ] ;
    rr:predicateObjectMap [
        rr:predicate gn:wikipediaArticle ;
        rr:objectMap [ rr:column "##4" ; rr:termType  rr:IRI ;
                       dr:condition [ rr:column "##3" ; op:eq "link" ] ] ] ] .
```

In Example 4 we needed to define three transformations on the primary set table, the last one of which depended on the first two. The first two (DateMap, LocationMap) are simple triples maps that use a date/location identification information sources. We assume that DateMap returns a JSON array with elements of the form { "start": *start-date-uri*, ''end'': *end-date-uri* }, as it may identify several ranges in the input. Similarly, LocationMap returns an array of { "place": *place-uri* }. Thus, executing these on the primary set table, we get the table extended with two new logical columns. Taking values from the new columns, we can then execute the PeriodoMap transformation. However, we need to match start with end dates; for this we need to specify a *BindingIterator* to declare that we need to iterate on the top level array returned by DateMap:

```
<#DateMap>
    dr:logicalSource [ dr:source <#DatifyService> ; dr:iterator "$" ] ;
    dr:parameterBinding [ dr:parameter "text" ; rr:column "chronology" ] .
<#LocationMap>
    dr:logicalSource [ dr:source <#LocalifyService> ; dr:iterator "$" ] ;
    dr:parameterBinding [ dr:parameter "text" ; rr:column "location" ] .
<#PeriodoMap>
    dr:logicalSource [ dr:source <#PeriodoService> ; dr:iterator "$" ] ;
    dr:bindingIterator [ dr:transformationReference <#DateMap> ; dr:reference "$" ] ;
    dr:parameterBinding [ dr:parameter "start" ;
                          dr:reference "$.start" ; dr:transformationReference <#DateMap> ] ;
    dr:parameterBinding [ dr:parameter "end" ;
                          dr:reference "$.end" ; dr:transformationReference <#DateMap> ] ;
    dr:parameterBinding [ dr:parameter "place" ;
                          dr:reference "$.place" ; dr:transformationReference <#LocationMap> ] .
<#InscriptionsMap>
    rr:logicalTable [ dr:source <#InscriptionsSource> ; dr:delimiter "\t" ] ;
    dr:transformations ( <#DateMap> <#LocationMap> <#PeriodoMap> ) ;
    rr:subjectMap [ rr:template "{@bci}{##1}" ; rr:class bci-t:Inscription ] ;
    rr:predicateObjectMap [
        rr:predicate kvoc-t:location ;
        rr:objectMap [ rr:reference "$.uri" ; dr:tRef <#LocationMap> ; rr:termType rr:IRI ] ] ;
    rr:predicateObjectMap [
        rr:predicate kvoc-t:date ;
        rr:objectMap [
            rr:parentTriplesMap [
                rr:subjectMap [ rr:class time:DateTimeInterval ; rr:termType  rr:BlankNode ] ;
                rr:predicateObjectMap [
                    rr:predicate time:hasBeginning ;
                    rr:objectMap [ dr:reference "$.start" ; dr:tRef <#DateMap> ] ] ;
```

```
    rr:predicateObjectMap [
        rr:predicate time:hasEnd ;
        rr:objectMap [ dr:reference "$.end" ; dr:tRef <#DateMap> ] ] ] ] ;
    rr:predicateObjectMap [
        rr:predicate kvoc-t:period ;
        rr:objectMap [ rr:column "$.uri" ; dr:transformationReference <#PeriodoMap> ] ] .
```

Our D2RML processor is available at http://apps.islab.ntua.gr/d2rml/.

8   Conclusions

Motivated by practical cases of more complex RDF mapping needs not covered by existing languages, we presented D2RML, an extension of R2RML and RML, which, based on an abstract underlying data model, allows the orchestrated retrieval of data from diverse information sources, their transformation using relevant web services, their filtering and manipulation using simple operations, and finally their limited restructuring and mapping to RDF triples.

To offer such capabilities, D2RML adds a programming language flavor to the mapping process, but we claim that this is necessary if such languages are ever going to be widely accepted and used in practice. If in a real mapping problem scenario, the source data do not exactly reflect the structure of the target model, and the modeler needs extended data manipulation capabilities, they usually resort to a programming language, thus invalidating the very usefulness of a mapping language. Our aim was to design a mapping language that would limit the cases where this occurs and where writing custom code turns out to be unavoidable. Further extensions to the language will most probably be needed to accommodate other needs, but the underlying abstract data model provides a solid ground on which to incorporate such extensions.

Acknowledgements. We acknowledge support of this work by 'APOLLONIS' (MIS 5002738), a project implemented under the Action 'Reinforcement of the Research and Innovation Infrastructure', funded by the Operational Programme 'Competitiveness, Entrepreneurship and Innovation' (NSRF 2014-2020) and co-financed by Greece and the European Union (European Regional Development Fund).

References

1. Bischof, S., Decker, S., Krennwallner, T., Lopes, N., Polleres, A.: Mapping between RDF and XML with XSPARQL. J. Data Semant. 1(3), 147–185 (2012)
2. Bourhis, P., Reutter, J.L., Suárez, F., Vrgoc, D.: JSON: data model, query languages and schema specification. In: PODS, pp. 123–135. ACM (2017)
3. Chortaras, A., Stamou, G.: D2RML: integrating heterogeneous data and web services into custom RDF graphs. In: LDOW. CEUR Workshop Proceedings (2018)
4. Connolly, D.: Gleaning resource descriptions from dialects of languages (GRDDL) (2007). https://www.w3.org/TR/grddl/
5. Das, S., Sundara, S., Cyganiak, R.: R2RML: RDB to RDF mapping language (2012). https://www.w3.org/TR/r2rml/

6. Dimou, A., Nies, T.D., Verborgh, R., Mannens, E., de Walle, R.V.: Automated metadata generation for linked data generation and publishing workflows. In: LDOW. CEUR Workshop Proceedings, vol. 1593 (2016)
7. Dimou, A., Sande, M.V., Colpaert, P., Verborgh, R., Mannens, E., de Walle, R.V.: RML: a generic language for integrated RDF mappings of heterogeneous data. In: LDOW. CEUR Workshop Proceedings, vol. 1184 (2014)
8. Hert, M., Reif, G., Gall, H.C.: A comparison of RDB-to-RDF mapping languages. In: I-SEMANTICS, ACM International Conference Proceeding Series, pp. 25–32. ACM (2011)
9. Koch, J., Velasco, C.A., Ackermann, P.: HTTP vocabulary in RDF 1.0 (2017). https://www.w3.org/TR/HTTP-in-RDF10/
10. Koch, J., Velasco, C.A., Ackermann, P.: Representing content in RDF 1.0 (2017). https://www.w3.org/TR/Content-in-RDF10/
11. Langegger, A., Wöß, W.: XLWrap – querying and integrating arbitrary spreadsheets with SPARQL. In: Bernstein, A., et al. (eds.) ISWC 2009. LNCS, vol. 5823, pp. 359–374. Springer, Heidelberg (2009). https://doi.org/10.1007/978-3-642-04930-9_23
12. De Meester, B., Dimou, A., Verborgh, R., Mannens, E.: An ontology to semantically declare and describe functions. In: Sack, H., Rizzo, G., Steinmetz, N., Mladenić, D., Auer, S., Lange, C. (eds.) ESWC 2016. LNCS, vol. 9989, pp. 46–49. Springer, Cham (2016). https://doi.org/10.1007/978-3-319-47602-5_10
13. Michel, F., Djimenou, L., Zucker, C.F., Montagnat, J.: xR2RML: non-relational databases to RDF mapping language (2014). https://hal.inria.fr/hal-01066663v1/document
14. O'Connor, M.J., Halaschek-Wiener, C., Musen, M.A.: M^2: a language for mapping spreadsheets to OWL. In: OWLED. CEUR Workshop Proceedings, vol. 614 (2010)
15. Walsh, N., Snelson, J., Coleman, A.: XQuery and XPath Data Model 3.1 (2017). https://www.w3.org/TR/xpath-datamodel-31/

A Novel Approach and Practical Algorithms for Ontology Integration

Giorgos Stoilos$^{(\boxtimes)}$, David Geleta, Jetendr Shamdasani,
and Mohammad Khodadadi

Babylon Health, London, SW3 3DD, UK
{giorgos.stoilos,david.geleta,jetendr.shamdasani,
mohammad.khodadadi}@babylonhealth.com

Abstract. Today a wealth of knowledge and data are distributed using Semantic Web standards. Especially in the (bio)medical domain several sources like SNOMED, NCI, FMA, and more are distributed in the form of OWL ontologies. These can be matched and integrated in order to create one large medical Knowledge Base. However, an important issue is that the structure of these ontologies may be profoundly different hence using the mappings as initially computed can lead to incoherences or changes in their original structure which may affect applications. In this paper we present a framework and novel approach for integrating independently developed ontologies. Starting from an initial seed ontology which may already be in use by an application, new sources are used to iteratively enrich and extend the seed one. To deal with structural incompatibilities we present a novel fine-grained approach which is based on mapping repair and alignment conservativity, formalise it and provide an exact as well as approximate but practical algorithms. Our framework has already been used to integrate a number of medical ontologies and support real-world healthcare services provided by Babylon Health. Finally, we also perform an experimental evaluation and compare with state-of-the-art ontology integration systems that take into account the structure and coherency of the integrated ontologies obtaining encouraging results.

Keywords: Ontology integration · Ontology matching
Ontology alignment · Conservativity · Mapping repair

1 Introduction

Today a wealth of knowledge and data are distributed using Semantic Web technologies and standards. For example, the Linked Open Vocabularies effort [22] contains more than 600 ontologies for various subjects like geography, multimedia, security, geometry, and more. Especially in the biomedical domain, a large number of ontologies have been developed during the previous decades like SNOMED,[1] NCI [5], UMLS,[2] the Disease ontology [16] and many more, while

---

[1] https://www.snomed.org/.

[2] https://uts.nlm.nih.gov/home.html.

© Springer Nature Switzerland AG 2018
D. Vrandečić et al. (Eds.): ISWC 2018, LNCS 11136, pp. 458–476, 2018.
https://doi.org/10.1007/978-3-030-00671-6_27

Algorithm 1. postProcessKBStructure($\mathcal{O}_1, \mathcal{O}_2, \mathcal{M}$, Config)

Input: Two coherent ontologies $\mathcal{O}_1, \mathcal{O}_2$ and a set of mappings $\mathcal{M}$ between them.

1: $\mathcal{M}_{m\text{-}1} := \{\langle C_i, D\rangle \mid \{\langle C_i, D\rangle, \langle C_j, D\rangle\} \subseteq \mathcal{M} \wedge C_i \neq C_j\}$.
2: $\mathcal{M}' := \mathcal{M} \setminus \mathcal{M}_{m\text{-}1}$
3: **for all** $D \in \mathsf{Sig}(\mathcal{O}_2)$ **do**
4:     $\mathcal{M}' := \mathcal{M}' \cup \mathsf{disambiguate\text{-}m\text{-}1}(\{\langle C_i, D\rangle \mid \langle C_i, D\rangle \in \mathcal{M}_{m\text{-}1}\}, \mathsf{Config})$
5: **end for**
6: Exclusions $:= \emptyset$
7: ConflictSets $:= \{\{m_1, m_2\} \mid \mathcal{O}_1 \cup \mathcal{O}_2 \cup \{m_1, m_2\} \models A \sqsubseteq B, \mathcal{O}_1 \not\models A \sqsubseteq B\}$
8: **for all** $\{\langle A, A'\rangle, \langle B, B'\rangle\} \in$ ConflictSets with $\mathcal{O}_2 \models_{\mathsf{rdfs}} A' \sqsubseteq B'$ **do**
9:     Exclusions $:=$ Exclusions $\cup \{A' \sqsubseteq E \mid A' \sqsubseteq E \in \mathcal{O}_2, \mathcal{O}_2 \models_{\mathsf{rdfs}} E \sqsubseteq B'\}$
10: **end for**
11: **return** $\langle \mathcal{O}_2 \setminus$ Exclusions$, \mathcal{M}'\rangle$

BioPortal [15] is a repository of more than 600 biomedical ontologies. Identifying the common entities between these vocabularies and integrating them is beneficial for building ontology-based applications as one could unify complementary information that these vocabularies contain building a "complete" Knowledge Base (KB).

The problem of computing correspondences (mappings) between different ontologies is referred to as *ontology matching* or *alignment* [18]. A number of ontology matching systems have been developed in the past and have shown to behave well in practice. However, besides classes with their respective labels ontologies usually bring a class hierarchy and depending on how they have been conceptualised they may exhibit significant incompatibilities. For example, in NCI proteins are declared to be disjoint from anatomical structures whereas in FMA proteins are subclasses of anatomical structures. Hence, integrating them without taking into account their logical axioms may lead to many undesired consequences like unsatisfiable classes [11] and/or changes in their initial structure [6]. For these reasons the notions of *conservative alignment* [6,19,20] and *mapping repair* [8,12] have been proposed in the literature. These notions dictate that the mappings should not alter the original ontology structure or introduce unsatisfiable concepts. If they do, then a so-called *violation* occurs which needs to be *repaired* by discarding some of the mappings.

Unfortunately, dropping mappings introduces another problem which is the increase of ambiguity and redundancy. For example, if one drops all mappings between NCI and FMA proteins (due to their structural incompatibilities), then the integrated ontology will contain at least two classes for the same real-world entity. Apart from an unnecessary increase in the size of the integrated ontology this introduces ambiguity and decreases interoperability between services that use classes from the KB. The problem becomes more acute if further sources are integrated in which case we may end up with multiple classes representing the same real-world entity.

Algorithm 2. postProcessNewOntoStructure($\mathcal{O}_1, \mathcal{O}_2, \mathcal{M}$, Config)

Input: Two ontologies $\mathcal{O}_1, \mathcal{O}_2$ and a set of mappings $\mathcal{M}$ computed between them.

1: $\mathcal{M}_{\text{1-m}} := \{\langle C, D_i\rangle \mid \{\langle C, D_i\rangle, \langle C, D_j\rangle\} \subseteq \mathcal{M} \wedge D_i \neq D_j\}$.
2: $\mathcal{M}' := \mathcal{M} \setminus \mathcal{M}_{\text{1-m}}$
3: **for all** $C \in \text{Sig}(\mathcal{O}_1)$ **do**
4:      $\mathcal{M}' := \mathcal{M}' \cup \text{disambiguate-1-m}(\{\langle C, D_i\rangle \mid \langle C, D_i\rangle \in \mathcal{M}_{\text{1-m}}\}, \text{Config})$
5: **end for**
6: ConflictSets $:= \{\{m_1, m_2\} \mid \mathcal{O}_1 \cup \mathcal{O}_2 \cup \{m_1, m_2\} \models A \sqsubseteq B, \mathcal{O}_2 \not\models A \sqsubseteq B\}$
7: **for all** $\{\langle D_1, D_1'\rangle, \langle D_2, D_2'\rangle\} \in \text{ConflictSets}$ **do**
8:      **if** no D such that $\mathcal{O}_2 \models_{\text{rdfs}} D \sqsubseteq D_1' \sqcap D_2'$ exists **then**
9:          **if** $D_1' \sqsubseteq \neg D_2' \in \mathcal{O}_2$ and C exist s.t. $\mathcal{O}_1 \cup \mathcal{O}_2 \cup \mathcal{M}' \models_{\text{rdfs}} C \sqsubseteq D_1' \sqcap D_2'$ **then**
10:             prune($\mathcal{M}' \cup \mathcal{O}_2, \{\{\langle D_1, D_1'\rangle, \langle D_2, D_2'\rangle\}, \{D_1' \sqsubseteq \neg D_2'\}\}$)
11:         **else if** semSim(D_1', D_2') $\leq$ Config.Distance.*thr* **then**
12:             prune($\mathcal{M}', \{\{\langle D_1, D_1'\rangle, \langle D_2, D_2'\rangle\}\}$)
13:         **end if**
14:     **end if**
15: **end for**
16: **return** $\langle \mathcal{O}_2, \mathcal{M}'\rangle$

An effort to construct a large medical KB by integrating existing medical sources recently started in Babylon Health.[3] The KB would serve as the backbone for healthcare services (diagnosis, drug prescription, and more) as well as for other tasks like medical text annotation, understanding, and reasoning. For these purposes a modular and highly configurable ontology integration framework was implemented which is using ontology mapping to discover correspondences between a new medical source and the current KB and enrich the latter with new medical knowledge. There were two major requirements in this effort. First, integrating new sources should not affect the behaviour of the services already functioning with the KB, hence its structure should not change when new sources are integrated. Moreover, the KB should not contain many entities with a large label overlap as this complicates text annotation tasks as well as doctors who are selecting classes form the KB for diagnostic purposes. To address these requirements our framework is using the notion of conservativity for tracking the structural changes, however, in order to repair them we propose a novel fine-grained approach which avoids dropping mappings as much as possible.

First, violations stemming from mappings of higher-multiplicity (i.e., those that map two entities from one ontology to the same entity in the other) are separated from the rest and both are treated differently since they are of different nature. The former are repaired by altering the mappings, however, the latter are repaired by dropping *axioms* from the new ontology. Our motivation is that services have already committed to the structure of the KB and parts of the new ontology that are in disagreement with this conceptualisation can be dropped. In

---

[3] https://www.babylonhealth.com/.

addition, this approach helps reduce ambiguity and duplication of the integrated ontology as much as possible. Regarding violations on the structure of the new ontology, again a distinction between mappings of higher-multiplicity and the rest is made. The former are repaired by dropping mappings, however, the latter can be allowed since, as stated, the hierarchy of the new ontology is given low priority and various heuristics proposed in the literature can be used to guide this process. We formalised our framework using the notion of a (maximal) safe extension of a KB and provided an exact algorithm that is based on computing all repair plans [6].

Unfortunately, detecting all repair plans is known to be computationally very expensive [8,12]. Consequently, we next present a concrete implementation of our framework which is using approximate but efficient algorithms for violation detection (all state-of-the-art systems are based on approximate algorithms). Our implementation has already been used to create a medical KB by integrating SNOMED, NCI, FMA, and CHV, and is currently under use within Babylon. We conclude the paper with an experimental evaluation and a comparison against state-of-the-art mapping repair systems obtaining encouraging results. In more detail, our implementation is currently the only one that can apply a general conservativity-based mapping repair strategy (not only mapping coherency detection) on such a large KB while the created KB contains far less distinct classes with overlapping labels (i.e., less ambiguity and duplication). In addition, it was the only approach for which no conservativity violations could be detected in the integrated KB.

2   Ontologies and Ontology Matching

For brevity reasons, throughout the paper we will use Description Logic notation. For a set of real numbers S we use $\oplus S$ to denote the sum of its elements. For p an ontology prefix and C some class we sometimes write $p{:}C$ to denote that C appears in ontology with prefix p. Hence, for IRI prefixes $p_1 \neq p_2$, $p_1{:}C$ and $p_2{:}C$ denote different classes. For an ontology $\mathcal{O}$ we use $\mathsf{Sig}(\mathcal{O})$ to denote the set of classes that appear in $\mathcal{O}$. Given an ontology $\mathcal{O}$ we assume that all classes C in $\mathcal{O}$ have at least one triple of the form $\langle C \; \mathsf{skos{:}prefLabel} \; v \rangle$ and zero or more triples of the form $\langle C \; \mathsf{skos{:}altLabel} \; v_i \rangle$. For a given class C function $\mathsf{pref}(C)$ returns the string value v in the triple $\langle C \; \mathsf{skos{:}prefLabel} \; v \rangle$. An ontology is called *coherent* if every $C \in \mathsf{Sig}(\mathcal{O})$ with $C \neq \bot$ is satisfiable.

In the literature, the notion of a Knowledge Base is almost identical to that of an ontology, i.e., a set of axioms describing the entities of a domain. In the following, we loosely use the term "Knowledge Base" ($\mathcal{KB}$) to mean a possibly large ontology that has been created by integrating various other ontologies but, formally speaking, a $\mathcal{KB}$ is an OWL ontology.

Ontology matching (or *ontology alignment*) is the process of discovering correspondences (mappings) between the entities of two ontologies $\mathcal{O}_1$ and $\mathcal{O}_2$. To represent mappings we use the formulation presented in [12]. That is, a mapping between $\mathcal{O}_1$ and $\mathcal{O}_2$ is a 4-tuple of the form $\langle C, D, \rho, n \rangle$, where $C \in \mathsf{Sig}(\mathcal{O}_1)$

$D \in \mathsf{Sig}(\mathcal{O}_2)$, $\rho \in \{\equiv, \sqsupseteq, \sqsubseteq\}$ is the *mapping type*, and $n \in (0, 1]$ is the confidence value of the mapping. Moreover, we interpret mappings as DL axioms—that is, $\langle C, D, \rho, n \rangle$ can be seen as the axiom $C \; \rho \; D$ with the degree attached as an annotation. Hence, for a mapping $\langle C, D, \rho, n \rangle$ when we write $\mathcal{O} \cup \{\langle C, D, \rho \rangle\}$ we mean $\mathcal{O} \cup \{C \; \rho \; D\}$ while for a set of mappings $\mathcal{M}$, $\mathcal{O} \cup \mathcal{M}$ denotes the set $\mathcal{O} \cup \{m \mid m \in \mathcal{M}\}$. When not relevant and for simplicity we will often omit ρ and n and simply write $\langle C, D \rangle$. A *matcher* is an algorithm that takes as input two ontologies and returns a set of mappings.

3   An Ontology Integration Framework

Large KBs can be constructed by integrating existing, complementary, and possibly overlapping ontologies. For example, in the biomedical domain, ontologies for diseases, drugs, drug side-effects, genes, and so on, exist that can be integrated in order to build a large medical KB. However, before putting two sources together it would be beneficial to discover their overlapping parts and establish mappings between their equivalent entities.

Example 1. Consider an ontology-based medical application that is using the SNOMED ontology $\mathcal{O}_{\mathsf{snmd}}$ as a KB. Although SNOMED is a large and well-engineered ontology it is still missing medical information like textual definitions for all classes as well as relations between diseases and symptoms. For example, for class the notion of "Ewing Sarcoma" SNOMED only contains the axiom snmd:EwingSarcoma $\sqsubseteq$ snmd:Sarcoma and no relations to signs or symptoms. In contrast, the NCI ontology $\mathcal{O}_{\mathsf{nci}}$ contains the following axiom about this disease:

$$\mathsf{nci{:}EwingSarcoma} \sqsubseteq \exists\mathsf{nci{:}mayHaveSymptom}.\mathsf{nci{:}Fever}$$

We can use ontology matching to establish links between the related entities in $\mathcal{O}_{\mathsf{snmd}}$ and $\mathcal{O}_{\mathsf{nci}}$ and then integrate the two sources in order to enrich our KB. More precisely, using the labels of the aforementioned classes we can identify the following mappings:

$$m_1 = \langle \mathsf{snmd{:}EwingSarcoma}, \mathsf{nci{:}EwingSarcoma}, \equiv \rangle$$
$$m_2 = \langle \mathsf{snmd{:}Fever}, \mathsf{nci{:}Fever}, \equiv \rangle$$

and hence replace our KB with $\mathcal{O}'_{\mathsf{snmd}} := \mathcal{O}_{\mathsf{snmd}} \cup \mathcal{O}_{\mathsf{nci}} \cup \{m_1, m_2\}$. Then, $\mathcal{O}'_{\mathsf{snmd}}$ contains the knowledge that "Ewing sarcoma may have fever as a symptom". $\Diamond$

Unfortunately, it is well-known that integrating ontologies using the initially computed mappings can lead to unexpected consequences like, introducing unsatisfiable classes [11] or structural changes to the input ontologies [6].

Example 2. Consider again the SNOMED and NCI ontologies. Both ontologies contain classes for the notion of "soft tissue disorder" and "epicondylitis". Hence, it is reasonable for a matching algorithm to compute the following mappings:

$$m_1 = \langle \mathsf{snmd{:}SoftTissueDisorder}, \mathsf{nci{:}SoftTissueDisorder}, \equiv \rangle$$
$$m_2 = \langle \mathsf{snmd{:}Epicondylitis}, \mathsf{nci{:}Epicondylitis}, \equiv \rangle$$

Algorithm 3. KnowledgeBaseConstruction($\mathcal{KB}, \mathcal{O}$, Config)

Input: The current KB $\mathcal{KB}$, a new ontology $\mathcal{O}$ and a configuration Config.

1: Mappings := $\emptyset$
2: **for all** matcher : Config.Align.Matchers **do**
3:     **for all** $\langle C, D, \rho, n \rangle \in$ matcher($\mathcal{KB}, \mathcal{O}$) **do**
4:         Mappings := Mappings $\cup \{\langle C, D, \rho, n,$ matcher$\rangle\}$
5:     **end for**
6: **end for**
7: $\mathcal{M}_f := \emptyset$
8: $w = \oplus\{$matcher.$w \mid$ matcher $\in$ Config.Align.Matchers$\}$
9: **for all** $\langle C, D, \rho, \_, \_ \rangle \in$ Mappings such that no $\langle C, D, \rho, n \rangle$ exists in $\mathcal{M}_f$ **do**
10:     $n := \oplus\{n_i \times$ matcher.$w \mid \langle C, D, \rho, n_i,$ matcher$\rangle \in \mathcal{M}\}/w$
11:     **if** $n \geq$ Config.Align.thr **then**
12:         $\mathcal{M}_f := \mathcal{M}_f \cup \{\langle C, D, \rho, n \rangle\}$
13:     **end if**
14: **end for**
15: $\langle \mathcal{O}', \mathcal{M}_f \rangle :=$ postProcessNewOntoStructure($\mathcal{KB}, \mathcal{O}, \mathcal{M}_f$, Config)
16: $\langle \mathcal{O}', \mathcal{M}_f \rangle :=$ postProcessKBStructure($\mathcal{KB}, \mathcal{O}', \mathcal{M}_f$, Config)
17: **return** $\mathcal{KB} \cup \mathcal{O}' \cup \mathcal{M}_f$

However, in NCI we have $\mathcal{O}_{\mathsf{nci}} \models$ nci:Epicondylitis $\sqsubseteq$ nci:SoftTissueDisorder while in SNOMED $\mathcal{O}_{\mathsf{snmd}} \not\models$ snmd:Epicondylitis $\sqsubseteq$ snmd:SoftTissueDisorder. Hence, in the integrated ontology we will have:

$$\mathcal{O}_{\mathsf{snmd}} \cup \mathcal{O}_{\mathsf{nci}} \cup \{m_1, m_2\} \models \text{snmd:Epicondylitis} \sqsubseteq \text{snmd:SoftTissueDisorder}$$

introducing a relation between classes of $\mathcal{O}_{\mathsf{snmd}}$ that did not originally hold and which can have a significant impact on the services of our application which are already based on the structure of $\mathcal{O}_{\mathsf{snmd}}$. $\Diamond$

The amount of such structural changes can be captured by the notion of logical difference [9]. Like in [6] for performance reasons we also use an approximate version of logical difference formalised next.

Definition 1 ([6])**.** *Let A, B be atomic classes (including $\top, \bot$), let Σ be a signature and let $\mathcal{O}$ and $\mathcal{O}'$ be two OWL 2 ontologies. The approximation of the Σ-deductive difference between $\mathcal{O}$ and $\mathcal{O}'$ (denoted $\mathrm{diff}_{\Sigma}^{\approx}(\mathcal{O}, \mathcal{O}')$) as the set of axioms of the form $A \sqsubseteq B$ satisfying: (i) $A, B \in \Sigma$, (ii) $\mathcal{O} \not\models A \sqsubseteq B$, and (iii) $\mathcal{O}' \models A \sqsubseteq B$.*

Using logical difference, the notion of a *conservative alignment* has been proposed in the literature [6,19,20] which dictates that for two ontologies $\mathcal{O}_1$ and $\mathcal{O}_2$ and for $\Sigma_1 = \mathsf{Sig}(\mathcal{O}_1)$ and $\Sigma_2 = \mathsf{Sig}(\mathcal{O}_2)$ the set of mappings $\mathcal{M}$ must be such that $\mathrm{diff}_{\Sigma_1}^{\approx}(\mathcal{O}_1, \mathcal{O}_1 \cup \mathcal{O}_2 \cup \mathcal{M})$ and $\mathrm{diff}_{\Sigma_2}^{\approx}(\mathcal{O}_2, \mathcal{O}_1 \cup \mathcal{O}_2 \cup \mathcal{M})$ are empty. An axiom belonging to either of these sets is called a *(conservativity) violation* and can be "repaired" by removing mappings form the initially computed sets.

Based on the above we have designed a Knowledge Base construction algorithm that is depicted in Algorithm 3. The algorithm accepts as input the current KB $\mathcal{KB}$, a new ontology $\mathcal{O}$ which will be used to enrich $\mathcal{KB}$ and a configuration Config. The configuration is used to tune and change various parameters like thresholds etc., many of which will be described in the rest of the paper. In brief, the algorithm first applies a set of matchers in order to compute a set of mappings between $\mathcal{KB}$ and $\mathcal{O}$ (lines 1–6). The set of matchers to be used is specified in the configuration object (Config.Align.Matchers) and each of them has a different weight assigned (matcher.w). After all matcher have finished, the mappings are aggregated and a threshold is applied (Config.Align.thr) in order to keep only mappings with a high confidence (lines 7–14). As mentioned previously, these mappings are still not the final ones since they may cause conservativity violations. These are handled by two functions, namely postProcessNewOntoStructure and postProcessKBStructure which are discussed in detail in the next section.

4   Safe Ontology Integration

The typical approach to resolve conservativity violations so far has been to remove mappings [6,8,19,20]. However, this approach may introduce other issues like having distinct classes with a large overlap in their labels, hence introducing redundancy and ambiguity. Assume for instance, that in Example 2 we drop mapping m_2. Then, the integrated ontology will contain two different classes for the real-world notion of "epicondylitis" (i.e., nci:Epicondylitis and snmd:Epicondylitis) each with overlapping labels. Subsequently, a service that is using the former class internally cannot interoperate with a service that is using the latter as there is no axiom specifying that the two classes are actually the same.

Instead of removing mappings, another way to repair a violation is by removing axioms from one of the input ontologies.

Example 3. Consider again Example 2 where $\mathcal{O}_{\text{snmd}}$ serves as the current version of the application KB. Instead of computing $\mathcal{KB}_1^{\text{int}} := \mathcal{O}_{\text{snmd}} \cup \mathcal{O}_{\text{nci}} \cup \{m_1, m_2\}$ as in Example 2 assume that we compute the following:

$$\mathcal{KB}_2^{\text{int}} := \mathcal{KB}_1^{\text{int}} \setminus \{\text{nci:Epicondylitis} \sqsubseteq \text{nci:SoftTissueDisorder}\}$$

Then, we have $\mathcal{KB}_2^{\text{int}} \not\models \text{snmd:Epicondylitis} \sqsubseteq \text{snmd:SoftTissueDisorder}$ and hence $\text{diff}_{\text{Sig}(\mathcal{O}_{\text{snmd}})}^{\approx}(\mathcal{O}_{\text{snmd}}, \mathcal{KB}_2^{\text{int}}) = \emptyset$ as desired.   ◇

This approach is reasonable if we assume that an application is already using some Knowledge Base and the role of new ontologies is to enrich and extend it with new information but without altering its structure. Then, parts of the new ontology that cause violations can be dropped.

However, not all violations can be repaired by removing axioms from $\mathcal{O}_2$. This is the case for mappings of higher multiplicity, i.e., those that map two different classes of one ontology to the same class in the other.

Example 4. Consider again ontology $\mathcal{O}_{snmd}$ and $\mathcal{O}_{nci}$. SNOMED contains classes Eczema and AtopicDermatitis whereas NCI contains class Eczema that also has "Atopic Dermatitis" as an alternative label. Hence, a matching algorithm could create two mappings of the form:

$$m_1 = \langle \text{snmd:Eczema}, \text{nci:Eczema}, \equiv \rangle$$
$$m_2 = \langle \text{snmd:AtopicDermatitis}, \text{nci:Eczema}, \equiv \rangle$$

which imply that snmd:Ezcema and snmd:AtopicDermatitis are equivalent although this is not the case in $\mathcal{O}_{snmd}$.   ◇

In these cases it is clear that the only way to repair such violations is by altering the mapping set. One approach would be to drop one of the two mappings or perhaps even change their type from $\equiv$ to $\sqsubseteq$ or $\sqsupseteq$ and we argue that the choice is case dependent. In the previous example, we may decide that SNOMED is more granular than NCI in the sense that Atopic Dermatitis is a type of Eczema whereas the NCI term captures a more general notion. Hence, we may decide to change the mappings to $\langle \text{snmd:Eczema}, \text{nci:Eczema}, \sqsubseteq \rangle$ and $\langle \text{snmd:AtopicDermatitis}, \text{nci:Eczema}, \sqsubseteq \rangle$. However, we may also conclude that the alternative labels in NCI don't strictly denote synonym (alternative) terms for diseases but rather similar ones and hence decide to drop mapping m_2.

Based on the above we introduce the notion of a safe extension of an ontology.

Definition 2. *Let $\mathcal{O}_1$ and $\mathcal{O}_2$ be two ontologies and let $\mathcal{M}$ be a set of mappings computed between them. The* safe extension *of $\mathcal{O}_1$ w.r.t. $\mathcal{O}_2, \mathcal{M}$ is a pair $\langle \mathcal{O}', \mathcal{M}' \rangle$ such that $\mathcal{O}' \subseteq \mathcal{O}_2, \mathcal{M}' \subseteq \mathcal{M}$ and $\text{diff}_{\Sigma}^{\approx}(\mathcal{O}_1, \mathcal{O}_1 \cup \mathcal{O}' \cup \mathcal{M}') = \emptyset$ for $\Sigma = \text{Sig}(\mathcal{O}_1)$.*

The pair of an empty ontology and set of mappings $(\langle \emptyset, \emptyset \rangle)$ is a trivial safe extension but one is usually interested in some maximal safe extension similar to the notion of diagnosis in mapping repair [8].

Definition 3. *Let $\mathcal{O}_1$ and $\mathcal{O}_2$ be two ontologies and let $\mathcal{M}$ be a set of mappings computed between them. A safe extension $\langle \mathcal{O}', \mathcal{M}' \rangle$ of $\mathcal{O}_1$ w.r.t. $\mathcal{O}_2, \mathcal{M}$ is called* maximal *if no safe extension $\langle \mathcal{O}'', \mathcal{M}'' \rangle$ exists s.t. either $\mathcal{O}'' \supset \mathcal{O}'$ or $\mathcal{M}'' \supset \mathcal{M}'$.*

Motivated by the above we have designed Algorithm 4 that accepts as input two ontologies $\mathcal{O}_1, \mathcal{O}_2$ and returns a subset of $\mathcal{O}_2$ and a subset of $\mathcal{M}$ in an attempt to compute a safe extension. The algorithm first processes mappings of higher multiplicity w.r.t. entities in $\mathcal{O}_1$ using function disambiguate-m-1 whose properties are formalised next.

Definition 4. *Given a set of mappings $\mathcal{M} = \{\langle C_1, D \rangle, \langle C_2, D \rangle, \ldots \langle C_n, D \rangle\}$ function* disambiguate-m-1 *returns a set $\mathcal{M}' \subseteq \mathcal{M}$ that satisfies the following property: it contains either a single mapping of the form $\langle C_i, D, \equiv \rangle$ or only mappings of the form $\langle C_i, D, \sqsupseteq \rangle$.*

Afterwards, the algorithm calls algorithm allPlans [6] passing the appropriate parameters in order to compute sets of axioms each of which has the following

---

Algorithm 4. postProcessKBStructure($\mathcal{O}_1, \mathcal{O}_2, \mathcal{M}$, Config)

Input: Two coherent ontologies $\mathcal{O}_1, \mathcal{O}_2$, a set of mappings $\mathcal{M}$, and a Config object.

1: $\mathcal{M}_{\mathrm{m\text{-}1}} := \{\langle C_i, D\rangle \mid \{\langle C_i, D\rangle, \langle C_j, D\rangle\} \subseteq \mathcal{M} \wedge C_i \neq C_j\}$.
2: $\mathcal{M}' := \mathcal{M} \setminus \mathcal{M}_{\mathrm{m\text{-}1}}$
3: **for all** $D \in \mathsf{Sig}(\mathcal{O}_2)$ **do**
4:     $\mathcal{M}' := \mathcal{M}' \cup$ disambiguate-m-1($\{\langle C_i, D\rangle \mid \langle C_i, D\rangle \in \mathcal{M}_{\mathrm{m\text{-}1}}\}$, Config)
5: **end for**
6: P := allPlans($\mathcal{O}_1 \cup \mathcal{O}_2 \cup \mathcal{M}', \mathcal{O}_2, \emptyset, \mathrm{diff}_{\Sigma}^{\approx}(\mathcal{O}_1, \mathcal{O}_1 \cup \mathcal{O}_2 \cup \mathcal{M}'))$ where $\Sigma = \mathsf{Sig}(\mathcal{O}_1)$
7: Pick some $P \in$ P such that no $P' \in$ P exists with $P' \subset P$
8: **return** $\langle \mathcal{O}_2 \setminus P, \mathcal{M}'\rangle$

---

property: if it is removed from $\mathcal{O}_2$ then all violations would be repaired. From all those sets the algorithm picks some minimal one (w.r.t. $\subseteq$) and removes it from $\mathcal{O}_2$.

Lemma 1. *Let $\mathcal{O}_1$ and $\mathcal{O}_2$ be two coherent ontologies and let $\mathcal{M}$ be a set of mappings between them such that every class in $\mathcal{O}_2$ is satisfiable in $\mathcal{O}_1 \cup \mathcal{O}_2 \cup \mathcal{M}$. When applied on $\mathcal{O}_1, \mathcal{O}_2$ and $\mathcal{M}$ Algorithm 4 returns a maximal safe extension of $\mathcal{O}_1$ w.r.t. $\mathcal{O}_2, \mathcal{M}$.*

Although, we are strict with respect to violations that are implied by the mappings to the structure of the KB, we can be more relaxed with respect to violations over the ontology that is being used for the enrichment. Several heuristics have been presented in the literature in order to decide which violations to allow and which to repair. A violation $A \sqsubseteq B \in \mathrm{diff}_{\mathsf{Sig}(\mathcal{O}_2)}^{\approx}(\mathcal{O}_2, \mathcal{O}_1 \cup \mathcal{O}_2 \cup \mathcal{M})$ may be allowed if A and B are somehow semantically related, e.g., if A and B have a common descendant [19]. In contrast, a violation should be repaired if $\mathcal{O}_2 \models A \sqsubseteq \neg B$, i.e., A and B are disjoint [11] or if the assumption of disjointness [13] can be applied to them—that is, if A and B are in different (distant) parts of the hierarchy of $\mathcal{O}_2$ and hence we can assume that they are disjoint.

Motivated by the above we have designed Algorithm 5. Like before mappings of higher multiplicity are treated separately by function disambiguate-1-m. Afterwards, the algorithm iterates over all violations w.r.t. ontology $\mathcal{O}_2$ and uses many of the aforementioned heuristics, like common descendants (line 8), unsatisfiability of classes (lines 9–12) and semantic or taxonomical similarity using function semSim and a pre-defined threshold Config.Distance.thr (lines 13–16) in order to decide to repair them or not. To figure out how to repair a violation algorithm allPlans is utilised again. This time $\mathcal{M}'$ (and possibly $\mathcal{O}_2$) instead of $\mathcal{O}_1$ is passed as a second parameter since we don't want the plans to contain axioms from $\mathcal{O}_1$. Note that in case some class A is unsatisfiable in the integrated ontology either mappings from $\mathcal{M}'$ or axioms from $\mathcal{O}_2$ that lead to this unsatisfiability may be selected to be removed. This choice was motivated by the conceptual differences between NCI and FMA regarding anatomical structures and proteins which are disjoint in one ontology but semantically related in the other. The choice of which plan to pick to remove is again case dependent hence we abstract this away using the function prune which picks at-least one plan.

Algorithm 5. postProcessNewOntoStructure($\mathcal{O}_1, \mathcal{O}_2, \mathcal{M}$, Config)

    Input: Ontologies $\mathcal{O}_1, \mathcal{O}_2$, set of mappings $\mathcal{M}$, and Config object.

1: $\mathcal{M}_{1\text{-m}} := \{\langle C, D_i \rangle \mid \{\langle C, D_i \rangle, \langle C, D_j \rangle\} \subseteq \mathcal{M} \wedge D_i \neq D_j\}$.
2: $\mathcal{M}' := \mathcal{M} \setminus \mathcal{M}_{1\text{-m}}$
3: **for all** $C \in \text{Sig}(\mathcal{O}_1)$ **do**
4:      $\mathcal{M}' := \mathcal{M}' \cup \text{disambiguate-1-m}(\{\langle C, D_i \rangle \mid \langle C, D_i \rangle \in \mathcal{M}_{1\text{-m}}\}, \text{Config})$
5: **end for**
6: LDiff $:= \text{diff}_\Sigma^{\approx}(\mathcal{O}_2, \mathcal{O}_1 \cup \mathcal{O}_2 \cup \mathcal{M}')$ where $\Sigma = \text{Sig}(\mathcal{O}_2)$
7: **for all** $A \sqsubseteq B \in$ LDiff **do**
8:      **if** no C such that $\mathcal{O}_2 \models C \sqsubseteq A \sqcap B$ exists **then**
9:          **if** $B = \bot$ **then**
10:              $\mathcal{O}_2^- := \{C \sqsubseteq \neg D \mid C \sqsubseteq \neg D \in \mathcal{O}_2\}$
11:              P $:= \text{allPlans}(\mathcal{O}_1 \cup \mathcal{O}_2 \cup \mathcal{M}', \mathcal{O}_2^- \cup \mathcal{M}', \emptyset, \{A \sqsubseteq \bot\})$
12:              prune($\mathcal{M}' \cup \mathcal{O}_2$, P)
13:          **else if** semSim(A, B) $\leq$ Config.Distance.*thr* **then**
14:              P $:= \text{allPlans}(\mathcal{O}_1 \cup \mathcal{O}_2 \cup \mathcal{M}', \mathcal{M}', \emptyset, \{A \sqsubseteq B\})$
15:              prune($\mathcal{M}'$, P)
16:          **end if**
17:      **end if**
18: **end for**
19: **return** $\langle \mathcal{O}_2, \mathcal{M}' \rangle$

Lemma 2. *Let $\mathcal{O}_1$ and $\mathcal{O}_2$ be two coherent ontologies and let $\mathcal{M}$ be a set of mappings between them. When applied on $\mathcal{O}_1, \mathcal{O}_2$ and $\mathcal{M}$ Algorithm 5 returns a pair $\langle \mathcal{O}', \mathcal{M}' \rangle$ such that every class in $\mathcal{O}'$ is satisfiable in $\mathcal{O}_1 \cup \mathcal{O}' \cup \mathcal{M}'$.*

Using Lemmas 1 and 2 we can show the main result of our paper.

Theorem 1. *Let $\mathcal{KB}$ and $\mathcal{O}$ be two coherent ontologies, let Config be some configuration and let $\mathcal{KB}'$ be the output of Algorithm 3 when applied on $\mathcal{KB}, \mathcal{O}$ and Config. Then the following hold:*

1. $\text{diff}_\Sigma^{\approx}(\mathcal{KB}, \mathcal{KB}') = \emptyset$ where $\Sigma = \text{Sig}(\mathcal{KB})$.
2. $\mathcal{KB}'$ is coherent.

5 Practical Algorithms

We have provided with concrete implementations of the algorithms and functions presented in the previous section. Regarding matching (lines 2–6 of Algorithm 3), two in-house label-based matchers have been implemented, namely ExactLabelMatcher and FuzzyStringMatcher. The former builds an inverted index of class labels [3] after some string normalisations, like removing possessive cases (e.g., Alzheimer's) and singularisation [10], and matches ontologies using these indexes. The latter is based on the ISub string similarity metric [21]. Since this algorithm does not scale well on large inputs [3] it is mostly used for disambiguating higher-multiplicity mappings or if we wish to re-score subsets of mappings

---

Algorithm 6. planApproximation($\mathcal{O}_1, \mathcal{O}_2, \mathcal{M}$)

Input: Two ontologies $\mathcal{O}_1$ and $\mathcal{O}_2$ a set of mappings between them.

1: Exclusions := $\emptyset$
2: ConflictSets := $\{\{m_1, m_2\} \mid \mathcal{O}_1 \cup \mathcal{O}_2 \cup \{m_1, m_2\} \models_{\mathsf{rdfs}} A \sqsubseteq B, \mathcal{O}_1 \not\models_{\mathsf{rdfs}} A \sqsubseteq B\}$
3: **for all** $\{\langle A, A' \rangle, \langle B, B' \rangle\} \in$ ConflictSets with $\mathcal{O}_2 \models_{\mathsf{rdfs}} A' \sqsubseteq B'$ **do**
4:     Exclusions := Exclusions $\cup \{A' \sqsubseteq E \mid A' \sqsubseteq E \in \mathcal{O}_2, \mathcal{O}_2 \models_{\mathsf{rdfs}} E \sqsubseteq B'\}$
5: **end for**
6: **return** Exclusions

---

with low confidence degrees. In addition to these matchers, the state-of-the-art systems AML [4] and LogMap [7] can also be used in Algorithm 3.

Regarding functions disambiguate-m-1 and disambiguate-1-m the following strategy has been implemented so far:

For a set of mappings $\{\langle C_1, D \rangle, \langle C_2, D \rangle, \ldots, \langle C_n, D \rangle\}$ and some real-value threshold Config.Disamb.th, if $i \in [1, n]$ exists such that the following two conditions hold:
1. $\mathsf{ISub}(\mathsf{pref}(C_i), \mathsf{pref}(D)) > \mathsf{ISub}(\mathsf{pref}(C_j), \mathsf{pref}(D))$ for every $j \neq i$ and
2. $\mathsf{ISub}(\mathsf{pref}(C_i), \mathsf{pref}(D)) \geq$ Config.Disamb.th
then return $\langle C_i, D \rangle$. Similarly in function disambiguate-1-m for sets of mappings of the form $\{\langle C, D_1 \rangle, \langle C, D_2 \rangle, \ldots, \langle C, D_n \rangle\}$.

A major practical consideration of Algorithms 4 and 5 is the call to algorithm allPlans. This algorithm does not scale in practice since it iterates over the power-set of the second parameter (i.e., $\mathcal{O}_2$) [6]. Consequently, as usually done in literature [8,19] we are using approximations of plan computation and violation repair. More precisely, lines 6 and 7 in Algorithm 4 are replaced by the call $P :=$ planApproximation($\mathcal{O}_1, \mathcal{O}_2, \mathcal{M}'$) where the implementation of this function is given in Algorithm 6 and is inspired by the Alcomo repair algorithm [8,12]. This algorithm is based on the assumption that logical differences of the form $A \sqsubseteq B$ stem from *exactly* two mappings which map classes A and B for which $\mathcal{O}_1 \not\models A \sqsubseteq B$ to classes A' and B' in $\mathcal{O}_2$ for which a path of SubClassOf axioms in $\mathcal{O}_2$ exists ($\models_{\mathsf{rdfs}}$), hence implying changes in the structure of $\mathcal{O}_1$. Although in theory this may not always be the case, most violations in practice do follow this pattern. For every such pair of mappings the algorithm picks to remove from $\mathcal{O}_2$ some axiom of the form $A' \sqsubseteq E$, i.e., it tries in some sense to remove the "weakest" axiom from $\mathcal{O}_2$. This choice is motivated by belief revision and the *principle of minimal change* [1].

Example 5. Consider for example the following two ontologies:

$$\mathcal{O}_1 = \{D \sqsubseteq C, C \sqsubseteq B\}$$
$$\mathcal{O}_2 = \{W \sqsubseteq Z, Z \sqsubseteq Y, Y \sqsubseteq X\}$$

and assume the set of mappings $\mathcal{M} = \{m_1, m_2\}$ where $m_1 = \langle D, Y \rangle$ and $m_2 = \langle B, W \rangle$. Clearly, for $\Sigma = \mathsf{Sig}(\mathcal{O}_1)$ we have $B \sqsubseteq D \in \mathsf{diff}_{\Sigma}^{\approx}(\mathcal{O}_1, \mathcal{O}_1 \cup \mathcal{O}_2 \cup \mathcal{M})$ and we can repair this violation by either removing $ax_1 = W \sqsubseteq Z$ or $ax_2 = Z \sqsubseteq Y$.

Algorithm 7. newOntologyApproximate($\mathcal{O}_1, \mathcal{O}_2, \mathcal{M}$)

1: ConflictSets := $\{\{m_1, m_2\} \mid \mathcal{O}_1 \cup \mathcal{O}_2 \cup \{m_1, m_2\} \models_{\mathsf{rdfs}} A \sqsubseteq B, \mathcal{O}_2 \not\models A \sqsubseteq B\}$
2: **for all** $\{\langle D_1, D'_1 \rangle, \langle D_2, D'_2 \rangle\} \in$ ConflictSets **do**
3:     **if** no D such that $\mathcal{O}_2 \models_{\mathsf{rdfs}} D \sqsubseteq D'_1 \sqcap D'_2$ exists **then**
4:         **if** $D'_1 \sqsubseteq \neg D'_2 \in \mathcal{O}_2$ and C exist s.t. $\mathcal{O}_1 \cup \mathcal{O}_2 \cup \mathcal{M}' \models_{\mathsf{rdfs}} C \sqsubseteq D'_1 \sqcap D'_2$ **then**
5:             prune($\mathcal{M}' \cup \mathcal{O}_2, \{\{\langle D_1, D'_1\rangle, \langle D_2, D'_2\rangle\}\}, \{D'_1 \sqsubseteq \neg D'_2\}\}$)
6:         **else if** semSim(D'_1, D'_2) $\leq$ Config.Distance.thr **then**
7:             prune($\mathcal{M}', \{\{\langle D_1, D'_1\rangle, \langle D_2, D'_2\rangle\}\}$)
8:         **end if**
9:     **end if**
10: **end for**

Ontologies $\mathcal{O}_1$ and $\mathcal{O}_2$ as well as KBs $\mathcal{KB}^{int}_{ax_1} = \mathcal{O}_1 \cup \mathcal{O}_2 \setminus \{ax_1\} \cup \mathcal{M}$ and $\mathcal{KB}^{int}_{ax_2} = \mathcal{O}_1 \cup \mathcal{O}_2 \cup \setminus\{ax_1\} \cup \mathcal{M}$ are depicted graphically in Fig. 1, where solid lines denote subclass relations, and dashed lines the two mappings. As we can see, although both integrated ontologies do not exhibit violations over $\mathcal{O}_1$, the two cases differ in the amount of changes they impose on the classes of $\mathcal{O}_1$. More precisely, for $\mathbf{S}(\mathcal{O}_1, \mathcal{O}_2, \mathcal{M}) = \{A \sqsubseteq B \mid A \in \mathsf{Sig}(\mathcal{O}_1), \mathcal{O}_1 \cup \mathcal{O}_2 \cup \mathcal{M} \models A \sqsubseteq B\}$ we have $\mathbf{S}(\mathcal{O}_1, \mathcal{O}_2 \setminus \{ax_2\}, \mathcal{M}) \setminus \mathbf{S}(\mathcal{O}_1, \mathcal{O}_2 \setminus \{ax_1\}, \mathcal{M}) = \{B \sqsubseteq Z, C \sqsubseteq Z, D \sqsubseteq Z, D \sqsubseteq X\}$. Indeed, in this scenario Algorithm 5 will compute Exclusions := $\{ax_1\}$. $\Diamond$

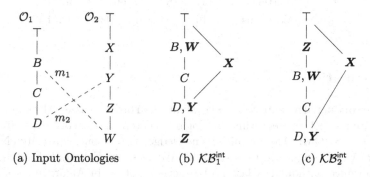

(a) Input Ontologies   (b) $\mathcal{KB}^{int}_1$ (c) $\mathcal{KB}^{int}_2$

Fig. 1. Ontologies of Example 5 and resulting KBs; with bold we denote classes from $\mathcal{O}_2$.

Following a similar approach, the block of lines 6–18 in Algorithm 5 is replaced by the steps depicted in Algorithm 7. Again these steps assume that violations stem from pairs of "conflicting" mappings like those mentioned above. Similarly to Algorithm 5, we are again using the heuristics of common descendants, disjoint classes and class similarity as a guide for repairing the violations; all entailments are checked using RDFS-entailment.

Using Algorithm 3 and the techniques presented above we have started building a large medical KB to be used within Babylon Health. We used the SNOMED

January 2018 release (which contains 340K classes and 511K SubClassOf axioms) as a starting seed KB ($\mathcal{KB}_1$) and have so far iteratively integrated the following ontologies: NCI version 17.12d (which contains 130K classes and 143K SubClassOf axioms), CHV latest version from 2011 (which contains 57 K classes and 0 SubClassOf axioms) and FMA version 4.6.0 (which contains 104K classes and 255K SubClassOf axioms); all ontologies are from the official websites. As a matching algorithm we have so far used our ExactLabelMatcher. Statistics about the KBs that we created after each integration are depicted in Table 1. CHV is a flat list of layman terms of medical concepts. From that ontology we only integrated label information for the classes in CHV that mapped to some class in the Babylon KB; hence only data type properties increased in the KB in that step. The final KB is currently under use by various services within Babylon and we are in the process of also integrating the following resources: SNOMED drug extensions, ICD-10, Read Codes, MeSH, and more.

Table 1. Statistics about the KB after each integration/enrichment iteration.

	SNOMED	+NCI	+CHV	+FMA
Classes	340 995	429 241	429 241	524 837
Properties	93	124	124	219
SubClassOf axioms	511 656	617 542	617 542	713 313
ObjPropAssertions	526 146	664 742	664 742	962 190
DataPropAssertions	543 416	946 801	1 043 874	1 211 459

6   Evaluation

We have conducted an experimental evaluation in order to assess the effectiveness of our approach for integrating ontologies and remedying conservativity viola-tions. Using SNOMED as our initial Knowledge Base we once integrated NCI and then FMA (starting again from scratch). We used our ExactLabelMatcher once with and once without the last post-processing steps in Algorithm 3 (lines 15 and 16). In the following we call the former setting bOWLing and the latter bOWLing_n. We used the latter setting as a "naive" baseline approach.

In addition, we also run Algorithm 3 using AML and two versions of LogMap called LogMap_o and LogMap_c in the following. AML and LogMap_o repair map-pings with respect to coherency, i.e., they only check for conservativity violations that lead to unsatisfiable classes. NCI contains 196 while FMA 33.5K disjoint classes axioms so this mapping repair is relevant. In contrast, LogMap_c also checks for more general conservativity violations using the techniques presented in [19]. For all these systems we disabled the post-processing steps of Algo-rithm 3 in order to assess their mapping repair functionality. On the mapping sets computed by bOWLing_n and LogMap_o we have also run Alcomo [12] as a post-processing step. Alcomo is not a general matcher but a mapping repair system that can be used as a post-processing step. In the following we denote these settings as $\text{bOWLing}_n^{\text{Alc}}$ and $\text{LogMap}_o^{\text{Alc}}$.

Algorithm 3 did not terminate with AML and LogMap_c after running for more than 16 h. As a second attempt we fragmented the ontologies into modules (using a $\sqsubseteq$-reachability based algorithm starting from top-level classes) and fed these one by one to Algorithm 3. For NCI we extracted 53 while for FMA 6 modules. Even in this case AML did not terminate when integrating FMA.

Table 2. Evaluation results

	SNOMED+NCI					
	$|\mathcal{M}|$	$|\mathcal{KB}^{\text{int}}|$	$|\text{LDiff}|$	Time	Loops	ambiguity
bOWLing_n	30 675	677 939	t.o.	12.7	127	16 708
$\text{bOWLing}_n^{\text{Alc}}$	26 825	666 834	0.9m	35.9	100	17 177
bOWLing	19 258	638 702	0	12.2	0	7 810
LogMap_o	27 967	664 837	1.7m	120.9	74	17 632
$\text{LogMap}_o^{\text{Alc}}$	27 763	664 354	1.5m	141.7	71	16 986
LogMap_c	21 838	433 711	897	54.4	0	8,266
AML	32 623	635 876	t.o.	75.0	298	14 353
	SNOMED+FMA					
	$|\mathcal{M}|$	$|\mathcal{KB}^{\text{int}}|$	$|\text{LDiff}|$	Time	Loops	ambiguity
bOWLing_n	8 809	614 728	240k	7.0	3	1 946
$\text{bOWLing}_n^{\text{Alc}}$	7 886	615 291	93k	76.2	1	2 000
bOWLing	8 176	608 060	0	27.9	0	1 440
LogMap_o	7 334	615 252	117k	360.4	1	2 264
$\text{LogMap}_o^{\text{Alc}}$	6 986	615 689	57k	428.4	1	2 253
LogMap_c	6 036	420 424	517	14 004.8	0	1 553

Our results are summarised in Table 2 where we give the number of computed mappings ($|\mathcal{M}|$), the number of SubClassOf axioms in the integrated ontology ($|\mathcal{KB}^{\text{int}}|$), the number of axioms in $\text{diff}_{\text{Sig}(\mathcal{KB})}^{\approx}(\mathcal{KB}, \mathcal{KB}^{\text{int}})$ (denoted by $|\text{LDiff}|$ and with "m" denoting millions), and the time to compute $\mathcal{KB}^{\text{int}}$ (in minutes). Due to the very large size of the KB LDiff cannot be computed by any OWL reasoner so we computed the RDFS-level differences by simply traversing the SubClassOf hierarchy of the KB. In addition, we have also computed the following:

- number of cycles of the form $\{A_1 \sqsubseteq A_2, \ldots, A_n \sqsubseteq A_1\} \subseteq \mathcal{KB}^{\text{int}}$. From a semantic point of view such cycles are not problematic, however, they do complicate graph-based algorithms like hierarchy traversal, extracting paths and depth counting, hence it is a design decision in Babylon to avoid them; input ontologies contain no cycles.

– a notion of "ambiguity" which we defined as the number of times a label appears in two different classes of a given ontology. We also calculated this metric over the original SNOMED, NCI, and FMA ontologies in order to measure their level of ambiguity. We obtained 1055, 4873, and 282, respectively, e.g., in SNOMED 1055 labels appear in multiple classes.

First thing to note from the table is that all systems compute mapping sets of comparable size with the exception of bOWLing on SNOMED+NCI which computes a smaller mapping sets. This is mostly due to functions disambiguate-m-1 and disambiguate-1-m which prune mappings of higher-multiplicity. However, we should note that all mappings computed by this approach are one-to-one mappings, while in all other approaches from the roughly 27k mappings about 17k are actually one-to-one (i.e., fewer than those of bOWLing). The application of Alcomo on the mapping sets does remove some mappings in an attempt to repair the sets while $LogMap_c$ that uses a general conservativity-based repairing approach also computes fewer mappings than $LogMap_o$.

As expected, the ontology produced by bOWLing contains fewer axioms due to the axiom exclusion strategy implemented in line 16 of Algorithm 3 which drops about 30% of NCI axioms and 10% of FMA axioms. However, the gains from this approach are apparent when considering other computed metrics. More precisely, the integrated ontology produced by bOWLing contains no axioms in LDiff in contrast to even more than 1 million new ancestor classes in some of the other approaches. Moreover, there are no cycles and, finally, a very low degree of ambiguity taking also into account the initial ambiguity of these ontologies (see above). The use of Alcomo as a post-processing step on $bOWLing_n$ and $LogMap_o$ does improve the numbers on these metrics, however, as it only focuses on coherency and not general conservativity it does not eliminate them completely. The only comparable approach is $LogMap_c$ which computes a KB without cycles. However, LDiff is still not empty and the approach of dropping mappings increases the ambiguity metric. Recall that we were only able to run $LogMap_c$ on the modules. Had it run on the whole ontology we believe the reported numbers would be higher since as one can note the integrated ontology in this module approach is also much smaller (almost 1/3 smaller). Finally, compared to all other systems our approach is much more scalable requiring a few minutes whereas in all other settings Algorithm 3 could take from one even up to 4h (even when restricted to the modules). Note that in some cases we could not compute LDiff even after 12h (t.o.).

7   Related Work and Conclusions

Constructing large Knowledge Bases (also called Knowledge Graphs these days) is a topic of intensive research and engineering the last years. The works [2,17] focus on extracting medical facts from text and use ontologies like UMLS and SNOMED mostly as flat vocabularies for performing named entity disambiguation, text annotation and information extraction. Hence, the focus is not on merging the medical knowledge and "fusing" the hierarchies. Malacards [14] is an effort for constructing a large disease KB by integrating information from many existing disease ontologies. To identify the overlaps between different sources a label-based unification algorithm is used which is very similar to our Exact-LabelMatcher (labels are normalised, stemmed, singularised, etc. and a hash is created). However, this approach completely discards the hierarchy and the axioms of the original ontologies and the final output is a flat non-ontological structure.

In the current paper we have studied the problem of building large KBs from existing ontologies by integrating them and retaining as much of their initial structure and axioms as possible. Starting with an initial ontology as a seed KB we use new ontologies to extend and enrich it in an iterative way. Overlaps are discovered using ontology matching algorithms and mappings are post-processed in order to preserve properties of the structures of the KB and the new ontology. The algorithm is highly modular as different strategies for handling higher-multiplicity mappings can be implemented and different (or multiple) matchers can be used. Our post-processing steps are based on the notion of conservativity but differently than what is usually done in the literature [6,13,19] we propose to remove axioms from the new ontology in order to repair many of the violations. This is important in order to keep ambiguity low and not have many classes with overlapping labels. We have formalised our framework, designed an exact general algorithm and also presented concrete approximate and practical algorithms. These have already been used in Babylon Health to build a medical Knowledge Base (using SNOMED, NCI, FMA, and CHV) that forms the data and knowledge backbone of various clinical services. Finally, we have conducted an experimental evaluation comparing our conservativity repairing approach to state-of-the-art mapping repair systems obtaining very encouraging results. In summary, our results verify that ambiguity is very-low (almost none introduced compared to the initial ambiguity of the input ontologies), there were no detectable violations (LDiff), no cycles, and our algorithm scales.

8   Proofs

Lemma 1. *Let $\mathcal{O}_1$ and $\mathcal{O}_2$ be two coherent ontologies and let $\mathcal{M}$ be a set of mappings between them such that every class in $\mathcal{O}_2$ is satisfiable in $\mathcal{O}_1 \cup \mathcal{O}_2 \cup \mathcal{M}$. When applied on $\mathcal{O}_1, \mathcal{O}_2$ and $\mathcal{M}$ Algorithm 4 returns a minimal safe extension of $\mathcal{O}_1$ w.r.t. $\mathcal{O}_2, \mathcal{M}$.*

Proof. Let $\langle \mathcal{O}', \mathcal{M}' \rangle$ be the output of Algorithm 4 when applied on $\mathcal{O}_1, \mathcal{O}_2$, and $\mathcal{M}$. $\mathcal{O}' = \mathcal{O}_2 \setminus P$ for some $P \in \mathsf{P}$ and since the second parameter of the call to function allPlans is $\mathcal{O}_2$, then for every such P we have $P \subseteq \mathcal{O}_2$; hence $\mathcal{O}' \subseteq \mathcal{O}_2$.

Next we show that for $\Sigma = \mathsf{Sig}(\mathcal{O}_1)$ we have $\mathsf{diff}^{\approx}_{\Sigma}(\mathcal{O}_1, \mathcal{O}_1 \cup \mathcal{O}' \cup \mathcal{M}') = \emptyset$. Let some $A \sqsubseteq B \in \mathsf{diff}^{\approx}_{\Sigma}(\mathcal{O}_1, \mathcal{O}_1 \cup \mathcal{O}_2 \cup \mathcal{M})$ and assume to the contrary that we have $A \sqsubseteq B \in \mathsf{diff}^{\approx}_{\Sigma}(\mathcal{O}_1, \mathcal{O}_1 \cup \mathcal{O}' \cup \mathcal{M}')$. By definition we have $\mathcal{O}_1 \not\models A \sqsubseteq B$, hence $A \neq B$.

Consider some arbitrary justification $\mathcal{J}$ for $\mathcal{O}_1 \cup \mathcal{O}' \cup \mathcal{M}' \models A \sqsubseteq B$. If $\mathcal{J} \subseteq \mathcal{M}'$ then we have that $\mathcal{O}_1 \cup \mathcal{M}' \models A \sqsubseteq B$. Since $\mathcal{O}_1 \not\models A \sqsubseteq B$ and all mappings relate classes from $\mathcal{O}_1$ to classes in $\mathcal{O}_2$ this can only be the case if these mappings in $\mathcal{J}$ map two classes from $\mathcal{O}_1$ to the same class in $\mathcal{O}_2$: we can use resolution to prove $\mathcal{O}_1 \cup \mathcal{M}' \models A \sqsubseteq B$ in the process of which one of the mappings will introduce a symbol from $\mathcal{O}_2$; since every symbol of $\mathcal{O}_2$ is satisfiable in $\mathcal{O}_1 \cup \mathcal{O}_2 \cup \mathcal{M}$ it can only be removed by resolving it with another mapping that refers it and thus there must be two mappings that map two different classes of $\mathcal{O}_1$ to the same symbol in $\mathcal{O}_2$. However, this is not possible due to the properties of function disambiguate-m-1 which has eliminated from $\mathcal{M}'$ mappings of higher multiplicity.

Hence, every justifications $\mathcal{J}$ of every violation $\alpha \in \mathrm{diff}_\Sigma^{\approx}(\mathcal{O}_1, \mathcal{O}_1 \cup \mathcal{O}_2 \cup \mathcal{M}')$, $\mathcal{J}$ must contain axioms from $\mathcal{O}_2$. But then, by Proposition 10 in [6] we have that P contains all P such that $\mathcal{O}_1 \cup \mathcal{O}_2 \setminus P \cup \mathcal{M} \not\models \alpha$, hence $\mathrm{diff}_\Sigma^{\approx}(\mathcal{O}_1, \mathcal{O}_1 \cup \mathcal{O}' \cup \mathcal{M}') = \emptyset$.

Finally, the maximality condition on $\mathcal{O}'$ is satisfied by selecting some minimal w.r.t. $\sqsubseteq$ plan from P in line 7. $\qquad\square$

Lemma 2. *Let $\mathcal{O}_1$ and $\mathcal{O}_2$ be two coherent ontologies and let $\mathcal{M}$ be a set of mappings between them. When applied on $\mathcal{O}_1, \mathcal{O}_2$ and $\mathcal{M}$ Algorithm 5 returns a pair $\langle \mathcal{O}', \mathcal{M} \rangle$ such that every class in $\mathcal{O}'$ is satisfiable in $\mathcal{O}_1 \cup \mathcal{O}' \cup \mathcal{M}'$.*

Proof. Assume that some arbitrary class A is unsatisfiable in $\mathcal{O}' \cup \mathcal{M}'$. Since $\mathcal{O}_2$ is coherent and $\mathcal{O}' \subseteq \mathcal{O}_2$, then A is also satisfiable in $\mathcal{O}'$. Hence we must have that $A \sqsubseteq \bot \in \mathrm{LDiff}$. Again by coherency of $\mathcal{O}_2$ no C exists such that $\mathcal{O}_2 \models C \sqsubseteq A \sqcap \bot$, hence the algorithm enters block 9–12 and computes a plan for repairing this unsatisfiability ($A \sqsubseteq \bot$). Like in the proof of Lemma 1 a repair plan containing only mappings from $\mathcal{M}'$ exists. Due to the parameters used to call allPlans a repair plan containing disjointness axioms from $\mathcal{O}_2$ may also be present in P. Thus, the call to allPlans with second argument $\mathcal{M}' \cup \mathcal{O}_2$ returns some non-empty repair plan which is removed form $\mathcal{M}' \cup \mathcal{O}_2$ at line 12 repairing this unsatisfiability and contradicting the initial assumption. $\qquad\square$

Theorem 1. *Let $\mathcal{KB}$ and $\mathcal{O}$ be two coherent ontologies, let Config be some configuration and let $\mathcal{KB}'$ be the output of Algorithm 3 when applied on $\mathcal{KB}, \mathcal{O}$ and Config. Then the following hold:*

1. $\mathrm{diff}_\Sigma^{\approx}(\mathcal{KB}, \mathcal{KB}') = \emptyset$ where $\Sigma = \mathrm{Sig}(\mathcal{KB})$.
2. $\mathcal{KB}'$ is coherent.

Proof. By Lemma 2 every class in $\mathcal{O}'$ that is passed as a second parameter in function postProcessKBStructure is satisfiable in $\mathcal{KB} \cup \mathcal{O}' \cup \mathcal{M}'$. Hence, by Lemma 1 it follows that $\langle \mathcal{O}', \mathcal{M}' \rangle$ returned by Algorithm 4 in line 16 is a safe extension of $\mathcal{KB}$ hence item 1. holds.

For item 2, by Lemma 2 and since function postProcessKBStructure only removes axioms from $\mathcal{O}'$ it follows trivially that all classes of $\mathcal{O}'$ are satisfiable in $\mathcal{KB} \cup \mathcal{O}' \cup \mathcal{M}'$. Consider some class A in $\mathcal{KB}$. Since $\mathcal{KB}$ is coherent then A is satisfiable in $\mathcal{KB}$ and by item 1. this class is also satisfiable in $\mathcal{O}'_{KG}$ for otherwise we would have $A \sqsubseteq \bot \in \mathrm{diff}_\Sigma^{\approx}(\mathcal{KB}, \mathcal{KB}')$. $\qquad\square$

References

1. Alchourrón, C.E., Gärdenfors, P., Makinson, D.: On the logic of theory change: partial meet contraction and revision functions. J. Symb. Log. **50**(2), 510–530 (1985)
2. Ernst, P., Siu, A., Weikum, G.: KnowLife: a versatile approach for constructing a large knowledge graph for biomedical sciences. BMC Bioinform. **16**, 157:1–157:13 (2015)
3. Faria, D., Pesquita, C., Mott, I., Martins, C., Couto, F.M., Cruz, I.F.: Tackling the challenges of matching biomedical ontologies. J. Biomed. Semant. **9**(1), 4:1–4:19 (2018)
4. Faria, D., Pesquita, C., Santos, E., Palmonari, M., Cruz, I.F., Couto, F.M.: The AgreementMakerLight ontology matching system. In: Meersman, R., Panetto, H., Dillon, T., Eder, J., Bellahsene, Z., Ritter, N., De Leenheer, P., Dou, D. (eds.) OTM 2013. LNCS, vol. 8185, pp. 527–541. Springer, Heidelberg (2013). https://doi.org/10.1007/978-3-642-41030-7_38
5. Golbeck, J., Fragoso, G., Hartel, F.W., Hendler, J.A., Oberthaler, J., Parsia, B.: The national cancer institute's thésaurus and ontology. J. Web Semant. **1**(1), 75–80 (2003)
6. Jiménez-Ruiz, E., Cuenca Grau, B., Horrocks, I., Berlanga, R.: Ontology integration using mappings: towards getting the right logical consequences. In: Aroyo, L., Traverso, P., Ciravegna, F., Cimiano, P., Heath, T., Hyvönen, E., Mizoguchi, R., Oren, E., Sabou, M., Simperl, E. (eds.) ESWC 2009. LNCS, vol. 5554, pp. 173–187. Springer, Heidelberg (2009). https://doi.org/10.1007/978-3-642-02121-3_16
7. Jiménez-Ruiz, E., Grau, B.C., Zhou, Y.: LogMap 2.0: towards logic-based, scalable and interactive ontology matching. In: Proceedings of the 4th International Workshop on Semantic Web Applications and Tools for the Life Sciences, pp. 45–46 (2011)
8. Jiménez-Ruiz, E., Meilicke, C., Grau, B.C., Horrocks, I.: Evaluating mapping repair systems with large biomedical ontologies. In: Proceedings of the 26th International Workshop on Description Logics, pp. 246–257 (2013)
9. Konev, B., Walther, D., Wolter, F.: The logical difference problem for description logic terminologies. In: Armando, A., Baumgartner, P., Dowek, G. (eds.) IJCAR 2008. LNCS (LNAI), vol. 5195, pp. 259–274. Springer, Heidelberg (2008). https://doi.org/10.1007/978-3-540-71070-7_21
10. Manning, C.D., Surdeanu, M., Bauer, J., Finkel, J.R., Bethard, S., McClosky, D.: The stanford CoreNLP natural language processing toolkit. In: Proceedings of the 52nd Annual Meeting of the Association for Computational Linguistics, ACL, pp. 55–60 (2014)
11. Meilicke, C., Stuckenschmidt, H.: Applying logical constraints to ontology matching. In: Hertzberg, J., Beetz, M., Englert, R. (eds.) KI 2007. LNCS (LNAI), vol. 4667, pp. 99–113. Springer, Heidelberg (2007). https://doi.org/10.1007/978-3-540-74565-5_10
12. Meilicke, C., Stuckenschmidt, H.: An efficient method for computing alignment diagnoses. In: Polleres, A., Swift, T. (eds.) RR 2009. LNCS, vol. 5837, pp. 182–196. Springer, Heidelberg (2009). https://doi.org/10.1007/978-3-642-05082-4_13
13. Meilicke, C., Völker, J., Stuckenschmidt, H.: Learning disjointness for debugging mappings between lightweight ontologies. In: Gangemi, A., Euzenat, J. (eds.) EKAW 2008. LNCS (LNAI), vol. 5268, pp. 93–108. Springer, Heidelberg (2008). https://doi.org/10.1007/978-3-540-87696-0_11

14. Rappaport, N., et al.: MalaCards: an integrated compendium for diseases and their annotation. Database (2013)
15. Salvadores, M., Alexander, P.R., Musen, M.A., Noy, N.F.: Bioportal as a dataset of linked biomedical ontologies and terminologies in RDF. Semant. Web **4**(3), 277–284 (2013)
16. Schriml, L.M.: Disease ontology: a backbone for disease semantic integration. Nucl. Acids Res. **40**(Database–Issue), 940–946 (2012)
17. Shi, L., Li, S., Yang, X., Qi, J., Pan, G., Zhou, B.: Semantic health knowledge graph: semantic integration of heterogeneous medical knowledge and services. BioMed Res. Int. 2017 (2017). (Article ID 2858423)
18. Shvaiko, P., Euzenat, J.: Ontology matching: state of the art and future challenges. IEEE Trans. Knowl. Data Eng. **25**(1), 158–176 (2013)
19. Solimando, A., Jiménez-Ruiz, E., Guerrini, G.: Detecting and correcting conservativity principle violations in ontology-to-ontology mappings. In: Mika, P., et al. (eds.) ISWC 2014. LNCS, vol. 8797, pp. 1–16. Springer, Cham (2014). https://doi.org/10.1007/978-3-319-11915-1_1
20. Solimando, A., Jiménez-Ruiz, E., Guerrini, G.: A multi-strategy approach for detecting and correcting conservativity principle violations in ontology alignments. In: 11th International Workshop on OWL: Experiences and Directions (2014)
21. Stoilos, G., Stamou, G., Kollias, S.: A string metric for ontology alignment. In: Gil, Y., Motta, E., Benjamins, V.R., Musen, M.A. (eds.) ISWC 2005. LNCS, vol. 3729, pp. 624–637. Springer, Heidelberg (2005). https://doi.org/10.1007/11574620_45
22. Vandenbussche, P., Atemezing, G., Poveda-Villalón, M., Vatant, B.: Linked open vocabularies (LOV): a gateway to reusable semantic vocabularies on the web. Semant. Web **8**(3), 437–452 (2017)

Practical Ontology Pattern Instantiation, Discovery, and Maintenance with Reasonable Ontology Templates

Martin G. Skjæveland$^{(\boxtimes)}$, Daniel P. Lupp, Leif Harald Karlsen, and Henrik Forssell

Department of Informatics, University of Oslo, Oslo, Norway
{martige,danielup,leifhka,jonf}@ifi.uio.no

Abstract. Reasonable Ontology Templates (OTTR) is a language for representing ontology modelling patterns in the form of parameterised ontologies. Ontology templates are simple and powerful abstractions useful for constructing, interacting with, and maintaining ontologies. With ontology templates, modelling patterns can be uniquely identified and encapsulated, broken down into convenient and manageable pieces, instantiated, and used as queries. Formal relations defined over templates support sophisticated maintenance tasks for sets of templates, such as revealing redundancies and suggesting new templates for representing implicit patterns. Ontology templates are designed for practical use; an OWL vocabulary, convenient serialisation formats for the semantic web and for terse specification of template definitions and bulk instances are available, including an open source implementation for using templates. Our approach is successfully tested on a real-world large-scale ontology in the engineering domain.

1 Introduction

Constructing sustainable large-scale ontologies of high quality is hard. Part of the problem is the lack of established tool-supported best-practices for ontology construction and maintenance. From a high-level perspective [12], an ontology is built through three iterative phases:

1. Understanding the target domain, e.g., the domain of pizzas
2. Identifying relevant abstractions over the domain, e.g., "Margherita is a particular Italian pizza with only mozzarella and tomato"
3. Formulating the abstractions in a formal language like description logics; here an adapted excerpt taken from the well-known Pizza ontology tutorial:[1]

$$\text{Margherita} \sqsubseteq \text{NamedPizza} \sqcap \exists \text{hasCountryOfOrigin}.\{\text{Italy}\} \tag{1}$$

$$\text{Margherita} \sqsubseteq \exists \text{hasTopping}.\text{Mozzarella} \sqcap \exists \text{hasTopping}.\text{Tomato} \tag{2}$$

$$\text{Margherita} \sqsubseteq \forall \text{hasTopping}.(\text{Mozzarella} \sqcup \text{Tomato}) \tag{3}$$

---

[1] https://protege.stanford.edu/ontologies/pizza/pizza.owl.

© Springer Nature Switzerland AG 2018
D. Vrandečić et al. (Eds.): ISWC 2018, LNCS 11136, pp. 477–494, 2018.
https://doi.org/10.1007/978-3-030-00671-6_28

This paper concerns the third task and targets particularly the large gap that exists between how domain knowledge facts are naturally expressed, e.g., in natural language, and how the same information must be recorded in OWL. The cause of the gap is the fact that OWL at its core supports only unary and binary predicates (classes and properties), and offers no real mechanism for user-defined abstractions with which recurring modelling patterns can be captured, encapsulated, and instantiated. The effect is that every single modelled statement no longer remains a coherent unit but must be broken down into the small building blocks of OWL. And as there is no trace from the original domain statement to the ontology axioms, the resulting ontology is hard to comprehend and difficult and error-prone to manage and maintain.

As a case in point, the Pizza ontology contains 22 different types of pizzas, all of which follow the same pattern of axioms as the encoding of the Margherita pizza seen above. For both the user of the ontology and the ontology engineer this information is opaque. The axioms that make out the instances of the pattern are all kept in a single set of OWL axioms or RDF triples in the same ontology document. Since the pizza pattern is not represented as a pattern anywhere, tasks that are important for the efficient use and management of the ontology, such as finding pattern instances and verifying consistent use of the pattern, i.e., understanding the ontology and updating the pattern, may require considerable repetitive and laborious effort.

In this paper we present *Reasonable Ontology Templates* (OTTR), a language for representing ontology modelling patterns as parameterised ontologies, implemented using a recursive non-cyclic macro mechanism for RDF. A pattern is instantiated using the macro's succinct interface. Instances may be *expanded* by recursively replacing instances with the pattern they represent, resulting in an ordinary RDF graph. Section 2 presents the fundamentals of the OTTR language and exemplifies its use on the pizza pattern. Ontology templates are designed to be practical and versatile for constructing, using and maintaining ontologies; the practical aspects of using templates are covered in Sect. 3. Section 4 concerns the maintenance of ontology template libraries. It presents methods and tools that exploit the underlying theoretical framework to give sophisticated techniques for maintaining template libraries and ultimately the ontologies built from those templates. We define different relations over templates and show how these can be used to define and identify imperfections in a template library, such as redundancy, and to suggest improvements of the library. We believe ontology templates can be an important instrument for improving the efficiency and quality of ontology construction and maintenance. OTTR templates allow the design of the ontology, represented by a relatively small library of templates, to be clearly separated from the bulk content of the ontology, specified by a large set of template instances. This, we believe, supports better delegation of responsibility in ontology engineering projects, allowing ontology experts to build and manage a library of templates and domain experts to provide content in the form of structurally simple template instances. To support this claim we report in Sect. 5 from successful experiments on the use of ontology templates to build and analyse Aibel's large-scale Material Master Data (MMD) ontology. We compare our work with existing approaches in Sect. 6 and present ideas for future work in Sect. 7.

2 Reasonable Ontology Templates Fundamentals

In this section we develop the fundamentals for OTTR templates as a generic macro mechanism adapted for RDF.

An *OTTR template* T consists of a *head*, head(T), and a *body*, body(T). The body represents a *parameterised* ontology pattern, and the head specifies the template's name and its parameters, param(T). A *template instance* consists of a template name and a list of arguments that matches the template's specified parameters and represents a replica of the template's body pattern where parameters are replaced by the instance's arguments. The template body comprises only template instances, i.e., the template pattern is recursively built up from other templates, under the constraint that cyclic template dependencies are not allowed. There is one special *base template*, the TRIPLE *template*, which takes three arguments. This template has no body but represents a single RDF triple in the obvious way. *Expanding* an instance is the process of recursively replacing instances with the pattern they represent. This process terminates with an expression containing only TRIPLE template instances, hence representing an RDF graph.

Example 1. The SUBCLASSOF template is a simple representation of the rdfs:subClassOf relationship. It has two parameters, ?sub and ?super, and a body containing a single instance of the TRIPLE template.

$$
\underbrace{\underbrace{\text{SUBCLASSOF}}_{\text{name}}\underbrace{(\text{?sub, ?super})}_{\text{parameters}}}_{\text{head}} \quad :: \quad \underbrace{\text{TRIPLE}(\text{?sub, rdfs:subClassOf, ?super})}_{\text{instance}} \ .
$$

An example instance of this template is SUBCLASSOF(:Margherita, :NamedPizza); it expands, in one step, to a single TRIPLE instance which represents the (single triple) RDF graph ⟨:Margherita, rdfs:subClassOf, :NamedPizza⟩.

Each template parameter has a *type* and a *cardinality*. (If these are not specified, as in Example 1, default values apply.) The type of the parameter specifies the permissible type of its arguments. The available types are limited to a specified set of classes and datatypes defined in the XSD, RDF, RDFS, and OWL specifications, e.g., xsd:integer, rdf:Property, rdfs:Resource and owl:ObjectProperty. The OWL ontology at ns.ottr.xyz/templates-term-types.owl declares all permissible types and organises them in a hierarchy of *subtypes* and *incompatible* types, e.g., owl:ObjectProperty is a subtype of rdf:Property, and xsd:integer and rdf:Property are incompatible. The most general and default type is rdfs:Resource. This information is used to type check template instantiations; a parameter may not be instantiated by an argument with an incompatible type.

ObjectAllValuesFrom(?X : 1 nonLiteral, ?P : 1 property, ?R : 1 nonLiteral)

:: (?X, rdf:type, owl:Restriction), (?X, owl:onProperty, ?P), (?X, owl:allValuesFrom, ?R) .

SubObjectAllValuesFrom(?X : 1 class, ?P : 1 objectProperty, ?R : 1 class)

:: SubClassOf(?X, _:b1), ObjectAllValuesFrom(_:b1, ?P, ?R) .   (4)

ObjectUnionOf(?X : 1 nonLiteral, ?union : + class)

:: (?X, rdf:type, owl:Class), (?X, owl:unionOf, ?union) .

Fig. 1. Basic OWL OTTR templates

The cardinality of a parameter specifies the number of required arguments to the parameter. There are four cardinalities: *mandatory* (written 1), *optional* (?), *multiple* (+), and *optional multiple* (∗), which is shorthand for ? and + combined. *Mandatory* is the default cardinality. Mandatory parameters require an argument. Optional parameters permit a missing value; *none* designates this value. If *none* is an argument to a mandatory parameter of an instance, the instance is ignored and will not be included in the expansion. A parameter with cardinality *multiple* requires a list as its argument. Instances of templates that accept list arguments may be used together with an *expansion mode*. The mode indicates that the list arguments will in the expansion be used to generate multiple instances of the template. There are two modes: *cross* (written ×) and *zip* (z). The instances to be generated are calculated by temporarily considering all arguments to the instance as lists, where single value arguments become singular lists. In cross mode, one instance per element in the cross product of the temporary lists is generated, while in zip mode, one instance per element in the zip of the lists is generated. List arguments used without an expansion mode behave just like regular arguments. Parameters with cardinality *optional multiple* also accept *none* as a value.

Example 2. Figure 1 contains three examples of OTTR templates that capture basic OWL axioms or restrictions, and exemplify the use of types and cardinalities. The template SubObjectAllValuesFrom represents the pattern ?X ⊑ ∀?P.?R and is defined using the SubClassOf and ObjectAllValuesFrom templates. Note that we allow a Triple instance to be written without its template name. The parameters of SubObjectAllValuesFrom are all mandatory, and have respectively the types class, objectProperty and class. The ObjectUnionOf template represents a union of classes. Here the parameter types are nonLiteral and class, where the latter has cardinality *multiple* in order to accept a list of classes. The type of the first parameter, nonLiteral, prevents an argument of type literal.

Example 3. The pizza pattern presented in the introduction is represented as an OTTR template in Fig. 2(a) together with two example instances. The template takes three arguments: the pizza, its optional country of origin, and its list of toppings. The cross expansion mode (×) on the SubObjectSomeValuesFrom instance causes it to expand to one instance per topping in the list of toppings, e.g., for the first example instance:

- SubObjectSomeValuesFrom(:Margherita, :hasTopping, :Mozzarella) and
- SubObjectSomeValuesFrom(:Margherita, :hasTopping, :Tomato),

creating an existential value restriction axiom for each topping, which results in the set of axioms seen in (2) of the pizza pattern in Sect. 1. By joining SubObjectAllValuesFrom and ObjectUnionOf with a blank node (_:b1), we get the universal restriction to the union of toppings (3). Note that the list of toppings is used both to create a set of existential axioms *and* to create a union class. The optional ?Country parameter behaves so that the SubObjectHasValue instance is not expanded but removed in the case that ?Country is *none*. The first NamedPizza instance in the figure represents exactly the same set of axioms as the listing in Sect. 1.

We conclude this section with the remark that it is in principle possible to choose a "base" other than RDF for OTTR templates, with suitable changes to typing and to which templates are designated as base templates. For instance, we could let templates such as SubClassOf, SubObjectAllValuesFrom, etc. be our base templates, to form a foundation based on OWL. These templates could then be directly translated into corresponding OWL axioms in some serialisation format. (An OTTR template can also be defined as a parameterised Description Logic knowledge base [2].) We have chosen here to base OTTR templates on RDF as this makes a simpler base, and broadens the application areas of OTTR templates, while still supporting OWL.

3   Using Ontology Templates

In this section we present the resources available to enable efficient and practical use of ontology templates: serialisation formats for templates and template instances, tools, formats and specifications that can be generated from templates, and online resources.

Languages. There are currently three serialisation formats for representing templates and template instances: stOTTR, wOTTR, and tabOTTR.

stOTTR[2] is the format used in the examples of Sect. 2 and is developed to offer a compact way of representing templates and instances that is also easy to read and write.

However, to enable truly practical use of OTTR for OWL ontology engineering, we have developed a special-purpose RDF/OWL vocabulary, called *wOTTR*, with which OTTR templates and instances can be formulated. This has the benefit that we can leverage the existing stack of W3C languages and tools for developing, publishing, and maintaining templates. The wOTTR format supports

---

[2] https://gitlab.com/ottr/language/stOTTR/.

writing TRIPLE instances as regular RDF triples. This means that a pattern represented by an RDF graph or RDF/OWL ontology can easily be turned into an OTTR template by simply specifying the name of the template and its parameters with the wOTTR vocabulary. Furthermore, this means that we can make use of existing ontology editors and reasoners to construct and verify the soundness of templates. The wOTTR representation has been developed to closely resemble stOTTR. It uses RDF resources to represent parameters and arguments, and RDF lists (which have a convenient formatting in Turtle syntax) for lists of parameters and arguments. The vocabulary is published at ns.ottr.xyz. A more thorough presentation of the vocabulary is found in [13].

tabOTTR[3] is developed particularly for representing large sets of template instances in tabular formats such as spreadsheets, and is intended for domain expert use.

Generated Queries and Format Specifications. A template may not only be used as a macro, but also, inversely, as a query that retrieves all instances of the pattern and outputs the result in the tabular format of the template head. From a template we can generate queries from both its expanded and unexpanded body. The expanded version allows us to find instances of a pattern in "vanilla" RDF data, while the unexpanded version can be used to collect and transform (in the opposite direction than of expansion) a set of template instances into an instance of a larger template. The latter form is convenient for validating the proper usage of templates within a library, which we present in Sect. 4.

We are also experimenting with generating other specifications from a template, for instance XSD descriptions of template heads, and transformations of these formats, e.g., XSLT transformations. The purpose of supporting other formats is to allow for different data input formats and leverage existing tools for input verification and bulk transformation of instance data to expanded RDF, such as XSD validators and XSLT transformation engines.

Tools and Online Resources. *Lutra*, our Java implementation of the OTTR template macro expander, is available as open source with an LGPL licence at gitlab.com/ottr. It can read and write templates and instances of the formats described above and expand them into RDF graphs and OWL ontologies, while performing various quality checks such as parameter type checking and checking the resulting output for semantic consistency. Lutra is also deployed as a web application that will parse and display any OTTR template available online. The template may be expanded and converted into all the formats mentioned above, including SPARQL SELECT, CONSTRUCT and UPDATE queries, XSD format, and variants of expansions which include or exclude the head or body.

Also available, at library.ottr.xyz, is a "standard" set of ontology templates for expressing common RDF, RDFS, and OWL patterns as well as other example templates. These templates are conveniently presented in an online library that is linked to the online web application.

---

[3] https://gitlab.com/ottr/language/tabOTTR/.

```
NAMEDPIZZA(
    ?Name : 1 class,
    ?Country : ? individual,
    ?Toppings : + class) ·
    ::
    SUBCLASSOF(?Name, :NamedPizza),
    SUBOBJECTHASVALUE(?Name, :hasCountryOfOrigin, ?Country),
    SUBOBJECTALLVALUESFROM(?Name, :hasTopping, _:b1),
    OBJECTUNIONOF(_:b1, ?Toppings),
    x | SUBOBJECTSOMEVALUESFROM(?Name, :hasTopping, ?Toppings) .


NAMEDPIZZA(:Margherita, :Italy, (:Tomato, :Mozzarella))
NAMEDPIZZA(:Grandiosa, none, (:Tomato, :Jarlsberg, :Ham, :SweetPepper))
```

(a) stOTTR serialisation and instances

```
<http://draft.ottr.xyz/pizza/NamedPizza> a ottr:Template ;
ottr:hasParameter
    [ ottr:index 1 ; ottr:classVariable :pizza ] ,
    [ ottr:index 2 ; ottr:individualVariable :country;
            ottr:optional true ] ,
    [ ottr:index 3 ; ottr:listVariable (:toppings) ] .
    ### body:
[] ottr:templateRef t-owl-axiom:SubClassOf ;
    ottr:withValues ( :pizza p:NamedPizza ) .
[] ottr:templateRef t-owl-axiom:SubObjectHasValue ;
    ottr:withValues ( :pizza p:hasCountryOfOrigin :country ) .
[] ottr:templateRef t-owl-axiom:SubObjectAllValuesFrom ;
    ottr:withValues ( :pizza p:hasTopping _:alltoppings ) .
[] ottr:templateRef t-owl-rstr:ObjectUnionOf ;
    ottr:withValues ( _:alltoppings (:toppings) ) .
[] ottr:templateRef t-owl-axiom:SubObjectSomeValuesFrom ;
    ottr:hasArgument [ ottr:index 1; ottr:value :pizza ] ,
        [ ottr:index 2; ottr:value p:hasTopping ] ,
        [ ottr:index 3; ottr:eachValue (:toppings) ] .
```

(b) wOTTR serialisation

```
:pizza  rdfs:subClassOf  p:NamedPizza ,
    [ a                     owl:Restriction ;
      owl:onProperty        p:hasTopping ;
      owl:someValuesFrom    :toppings ] ,
    [ a                     owl:Restriction ;
      owl:onProperty        p:hasTopping
      owl:allValuesFrom     [
          a                  owl:Class ;
          owl:unionOf  ( :toppings ) ] ] ,
    [ a                     owl:Restriction ;
      owl:onProperty        p:hasCountryOfOrigin ;
      owl:hasValue          :country ] .
p:hasTopping            a  owl:ObjectProperty .
p:hasCountryOfOrigin    a  owl:ObjectProperty .
:toppings               a  owl:Class .
```

(c) Expanded RDF graph

```
SELECT *
    { ?param1   rdfs:subClassOf    p:NamedPizza ,
        [ owl:onProperty        p:hasTopping ;
          owl:someValuesFrom    param3item ;
          rdf:type              owl:Restriction ] ,
        [ owl:allValuesFrom   [ owl:unionOf ?param3 ;
                                rdf:type   owl:Class ] ;
          owl:onProperty        p:hasTopping ;
          rdf:type              owl:Restriction ]
    OPTIONAL {
        ?param1   rdfs:subClassOf [
            owl:hasValue         ?param2 ;
            owl:onProperty       p:hasCountryOfOrigin ;
            rdf:type             owl:Restriction ]
    }
        ?param3  (rdf:rest)*/rdf:first ?param3item
    }
```

(d) SPARQL SELECT query

?param1	?param3item	?param2
p:Margherita	p:MozzarellaTopping	
p:Margherita	p:TomatoTopping	
p:Mushroom	p:MozzarellaTopping	
p:Mushroom	p:MushroomTopping	
p:Mushroom	p:TomatoTopping	
p:Napoletana	p:AnchoviesTopping	p:Italy
p:Napoletana	p:CaperTopping	p:Italy
p:Napoletana	p:MozzarellaTopping	p:Italy
p:Napoletana	p:OliveTopping	p:Italy
p:Napoletana	p:TomatoTopping	p:Italy

(e) Excerpt query results

```
#OTTR  prefix
p  http://www.co-ode.org/ontologies/pizza/pizza.owl#
#OTTR  end
#OTTR  template http://draft.ottr.xyz/pizza/NamedPizza
pizza           country     toppings
1               2           3
iri             iri         iri+
p:Margherita    p:Italy     p:Tomato|p:Mozzarella
p:Grandiosa                 p:Tomato|p:Jarlsberg|p:Ham|p:Pepper
#OTTR  end
```

(f) tabOTTR instance serialisation

Fig. 2. NAMEDPIZZA template and example instances in different serialisations

Example 4. Figure 2 contains different representations of the NAMEDPIZZA template. Figure 2(b) contains the published version of the template, available at its IRI address: http://draft.ottr.xyz/pizza/NamedPizza. Figure 2(c) contains the expansion of the template body. Figure 2(d) displays the generated SPARQL query that retrieves instances of the pizza pattern; an excerpt of the results applying the query to the Pizza ontology is given in Fig. 2(e). Figure 2(f) contains a tabOTTR representation of the two instances seen in Fig. 2(a). We encourage the reader to visit the rendering of the template by the web application at osl.ottr.xyz/info/?tpl=http://draft.ottr.xyz/pizza/NamedPizza and explore the various presentations and formats displayed. An example-driven walk-through of the features of Lutra can be found at ottr.xyz/event/2018-10-08-iswc/.

4   Maintenance and Optimisation of OTTR Template Libraries

In this section, we present an initial list and analysis of some of the more central relations between OTTR templates, and discuss their use in template library optimisation. We focus in particular on removing redundancy within a library, where we distinguish two different types of redundancy: a lack of reuse of existing templates, as well as recurring patterns not captured by templates within the library. We present an efficient and automated technique for detecting such redundancies within an OTTR template library.

4.1   OTTR Template Relations

Optimisation and maintenance of OTTR template libraries is made possible by its solid formal foundation. OTTR syntax makes it possible to formally define relations between OTTR templates which can tangibly benefit the optimisation of a template library. Naturally, there are any number of ways templates can be "related" to one another, and the "optimal" size and shape of a template library is likely to be highly domain and ontology-specific. As such, we do not aspire to a best-practice approach to optimising a template library. Instead, we illustrate the point by defining a few central template relations and demonstrating their usefulness for library optimisation and maintenance, independently of the heuristics used. Here, we limit ourselves to template relations defined syntactically in terms of instances, and do not consider, e.g., those defined in terms of semantic relationships between full expansions of templates. We consider the following template relations:

directly depends (DD) S *directly depends on* T if S's body has an instance of T.

depends (D) *depends* is the transitive closure of *directly depends*.

dependency-overlaps (DO) S *dependency-overlaps* T if there exists a template upon which both S and T directly depend.

overlaps (O) S *overlaps* T if there exist template instances i_S, i_T in body(S) and body(T) and substitutions ρ and η of the parameters of S and T resp. such that $\rho(i_S) = i_T$ and $\eta(i_T) = i_S$.

contains (C) S *contains* T if there exists a substitution ρ of the parameters of T such that $\rho(\text{body}(T)) \subseteq \text{body}(S)$.

equals (E) S *is equal to* T if S contains T and vice versa.

Each of the listed relations is, in a sense, a specialisation of the previous one (except for DO, which is a specialisation of DD as opposed to D). For instance, DO imposes no restrictions on the instance arguments, whereas O intuitively requires parameters to occur in compatible positions of i_S and i_T.

NAMEDPIZZA(?Name : 1 class, ?Country : ? individual, ?Toppings : + class)   (cf. Fig. 2(a)) .

ANNOTATEDPIZZA(?Name : 1 class, ?Label : + literal, ?PrefLabel : ? literal, ?Definition : ? literal)

  :: SUBCLASSOF(?Name, :Pizza),

   × | (?Name, rdfs:label, ?Label), (?Name, skos:prefLabel, ?PrefLabel), (?Name, skos:definition, ?Definition) .   (5)

BURGER(?Name : 1 class, ?Condiments : + class, ?Label : + literal, ?PrefLabel : ? literal, ?Definition : ? literal)

  :: SUBCLASSOF(?Name, :Burger),

   × | SUBOBJECTSOMEVALUESFROM(?Name, :hasCondiment, ?Condiments),

   SUBCLASSOF(?Name, _:b2), OBJECTALLVALUESFROM(_:b2, :hasCondiment, _:b3),   (6)

    OBJECTUNIONOF(_:b3, ?Condiments),   (7)

   × | (?Name, rdfs:label, ?Label), (?Name, skos:prefLabel, ?PrefLabel), (?Name, skos:definition, ?Definition) .   (8)

BURGERMEAL(?Name : 1 class, ?Sides : + class)

  :: SUBOBJECTSOMEVALUESFROM(?Name, :hasMain, :Burger),

   SUBOBJECTALLVALUESFROM(?Name, :hasSide, _:b4), OBJECTUNIONOF(_:b4, ?Sides) .   (9)

(a) OTTR template library with redundancies and lack of re-use

NAMEDPIZZA(?Name : 1 class, ?Country : ? individual, ?Toppings : + class)

  :: SUBOBJECTHASVALUE(?Name, :hasCountryOfOrigin, ?Country),

   NAMEDFOOD(?Name, :NamedPizza, ?Toppings, :hasTopping) .   (⋆)

ANNOTATEDPIZZA(?Name : 1 class, ?Label : + literal, ?PrefLabel : ? literal, ?Definition : ? literal)

  :: SUBCLASSOF(?Name, :Pizza),

   ANNOTATION(?Name, ?Label, ?PrefLabel, ?Definition) .   (⋆)

BURGER(?Name : 1 class, ?Condiments : + class, ?Label : + literal, ?PrefLabel : ? literal, ?Definition : ? literal)

  :: NAMEDFOOD(?Name, :Burger, ?Condiments, :hasCondiment),   (⋆)

   ANNOTATION(?Name, ?Label, ?PrefLabel, ?Definition) .   (⋆)

BURGERMEAL(?Name : 1 class, ?Sides : + class)

  :: SUBOBJECTSOMEVALUESFROM(?Name, :hasMain, :Burger),

   SUBOBJECTALLVALUESFROMUNION(?Name, :hasSide, ?Sides) .   (⋆)

New OTTR templates representing previously uncaptured patterns:

NAMEDFOOD(?Name : 1 class, ?Category : 1 class, ?Extras : + class, ?hasExtra : 1 objectProperty)   (10)

  :: SUBCLASSOF(?Name, ?Category),

   × | SUBOBJECTSOMEVALUESFROM(?Name, ?hasExtra, ?Extras),

   SUBOBJECTALLVALUESFROMUNION(?Name, ?hasExtra, ?Extras)

ANNOTATION(?Name : 1 class, ?Label : + literal, ?PrefLabel : ? literal, ?Definition : ? literal)   (11)

  :: × | (?Name, rdfs:label, ?Label), (?Name, skos:prefLabel, ?PrefLabel), (?Name, skos:definition, ?Definition) .

SUBOBJECTALLVALUESFROMUNION(?x : 1 class, ?Property : 1 objectProperty.?RangeList : + class)   (12)

  :: SUBOBJECTALLVALUESFROM(?x, ?Property, _:b1), OBJECTUNIONOF(_:b1, ?RangeList) .

(b) A refactored version of the templates in Fig. 3(a). The refactored templates from Fig. 3(a) are listed first, where a star (⋆) indicates a dependency to a new template, found at the bottom.

Fig. 3. OTTR template library before and after redundancy removal

Example 5. Consider the template library given in Fig. 3(a). All but the BURGERMEAL template contain an instance of SUBCLASSOF, hence all pairs of templates except for (ANNOTATEDPIZZA, BURGERMEAL) have a dependency-overlap. Closer inspection reveals that BURGER contains SUBOBJECTALLVALUESFROM (4, Fig. 1), due to the instances

− SUBCLASSOF(?Name, _:b2)
− OBJECTALLVALUESFROM(_:b2, :hasCondiment, _:b3)

in BURGER (6). (Numbers refer to numbered lines in the figures.) Finally, ANNOTATEDPIZZA and BURGER overlap, since they both directly depend on the same TRIPLE templates (5) and (8). These relationships are depicted in the graph below (dependency relationships omitted for the sake of legibility). Directed/undirected edges depict nonsymmetric/symmetric relations, respectively.

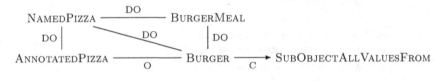

We wish to discuss these relations in the context of *redundancy removal* within an OTTR template library. More specifically, we discuss two types of redundancy:

Lack of reuse is a redundancy where a template S has a contains relationship to another template T, instead of a dependency relationship to T. That is, S duplicates the pattern represented by T, rather than instantiating T. This can be removed by replacing the offending portion of body(S) with a suitable instance of T. A first approach to determining such a lack of reuse makes use of the fact that templates can be used as queries: template S contains T iff T as a query over S yields answers.

Uncaptured pattern is a redundancy where a pattern of template instances is used by multiple templates, but this pattern is not represented by a template. In order to find uncaptured patterns one must analyse in what manner multiple templates depend on the same set of templates. If multiple templates *overlap* as defined above, this is a good candidate for an uncaptured pattern. However, an overlap does not necessarily need to occur for an uncaptured pattern to be present: as demonstrated in the following example, a dependency-overlap can describe an uncaptured pattern that is relevant for the template library.

Example 6. Continuing with our previous example of the library in Fig. 3(a), we find that it contains both an instance of lack of reuse and multiple instances of uncaptured patterns. The containment of SUBOBJECTALLVALUESFROM in BURGER indicates a lack of reuse, and the overlap of BURGER and ANNOTATEDPIZZA is an uncaptured pattern which we refactor into the template ANNOTATION (11). By repairing the lack of reuse in BURGER (6) with an instance of SUBOBJECTALLVALUESFROM, there are two dependency-overlaps that represent uncaptured patterns: the instances (6,7)(9), which are refactored into a new template SUBOBJECTALLVALUESFROMUNION (12), and the dependency-overlap between BURGER and NAMEDPIZZA, which is described by the NAMEDFOOD template (10). These new templates as well as the updated template definitions for the pre-existing ones are given in Fig. 3(b).

4.2   Efficient Redundancy Detection

Naive methods for improving a template library using the relations as described in the previous section quickly become infeasible for large knowledge bases, as they require expensive testing of unification of all template bodies. We have developed an efficient method for finding lack of reuse and uncaptured patterns, which over-approximates the results of unification. The method uses the notion of a dependency pair, which intuitively captures repeated use of templates without considering parameters: a *dependency pair* $\langle I, T \rangle$ is a pair of a multiset of templates I and a set of templates T, such that T is the set of all templates that directly depend on all templates in I, and have at least as many directly depends relationships to each template in I as they occur in I. The idea is that I represents a pattern used by all the templates in T. In order to also detect patterns containing different TRIPLE instances, we will in this section treat a TRIPLE instance (s, p, o) as a template instance of the form $p(s, o)$ and thus treat p as a template. Note that for a set of dependency pairs generated from a template library, the first element in the pair, i.e., the I, is unique for the set, while the T is generally not unique.

Example 7. Three examples of dependency pairs from the library in Fig. 3(a) are

1. $\langle \{\text{SUBCLASSOF}, \text{SUBCLASSOF}, \text{rdfs:label}\}, \{\text{BURGER}\} \rangle$
2. $\langle \{\text{SUBCLASSOF}, \text{OBJECTALLVALUESFROM}\}, \{\text{SUBOBJECTALLVALUESFROM}, \text{BURGER}\} \rangle$
3. $\langle \{\text{skos:definition}, \text{rdfs:label}, \text{skos:prefLabel}\}, \{\text{ANNOTATEDPIZZA}, \text{BURGER}\} \rangle$

The first pair indicates that BURGER is the only template that directly depends on two occurrences of SUBCLASSOF and one occurrence of rdfs:label. Note that BURGER directly depends on other templates too, and these will give rise to other dependency pairs. However there is no other template than BURGER that directly depends on this multiset of templates. The second example shows that SUBOBJECTALLVALUESFROM and BURGER directly depend on the templates SUBCLASSOF and OBJECTALLVALUESFROM.

One can compute all dependency pairs by starting with the set of dependency pairs of the form $\langle \{i : n\}, T \rangle$ where all templates in T have at least n instances of i, and then compute all possible *merges*, where a merge between two clusters

$\langle I_1, T_1 \rangle$ and $\langle I_2, T_2 \rangle$ is $\langle I_1 \cup I_2, T_1 \cap T_2 \rangle$. We have implemented this algorithm with optimisations that ensure we compute each dependency pair only once.

The set of dependency pairs for a library contains all potential lack of reuse and uncaptured patterns in a library. However, note that in the dependency pairs where either I or T has only one element, the dependency pair does not represent a commonly used pattern: If I has only one element then it does not represent a redundant pattern. If T has only one element then the pattern occurs only once. If on the other hand both sets contain two or more elements then the dependency pair might represent a useful pattern to be represented as a template, and we call these *candidate pairs*.

For a candidate pair, there are three cases to consider: 1. the set of instances does not form a pattern that can be captured by a template, as the usage of the set of instances does not unify; 2. the pattern is already captured by a template, in which case we have found an instance of lack of reuse; otherwise 3. we have found one or more candidates (one for each non-unifiable usage of the instances of I) for new templates. The two first cases can be identified automatically, but the third needs user interaction to assess. First, a user should verify for each of the new templates that it is a meaningful pattern with respect to the domain; second, if the template is meaningful, a user must give the new template an appropriate name.

To remove the redundancy a candidate pair $\langle I, T \rangle$ represents, we can perform the following procedure for each template $t \in T$ and $T' = T \setminus \{t\}$. First we check for lack of reuse of t: this may only be the case if t's body has the same number of instances as there are templates in I. We verify the lack of reuse by checking if $t' \in T'$ contains t; this is done by verifying that t used as a query over t''s body yields an answer. If there is no lack of reuse, we can represent the instances of I as they are instantiated in t, as the body of a new template where all arguments are made into parameters. Again, we need to verify that the new template is contained in other templates in T' before we can refactor, and before any refactoring is carried out, a user should always assess the results.

Example 8. Applying the method for finding candidates to the library in Fig. 3(a), gives 19 candidate pairs, two of which are the 2nd and 3rd candidate pair of Example 7. The 1st dependency pair of Example 7 is not a candidate pair since the size of one of its elements ({BURGER}) is one.

By using the process of removing redundancies as described above, we will find that for the 2nd candidate pair of Example 7 we have a lack of reuse of SUBOBJECTALLVALUESFROM in BURGER, as discussed in previous examples. The two instances of SUBCLASSOF and OBJECTALLVALUESFROM in BURGER (see Example 5) can be therefore can be replaced with the single instance: SUBOBJECTSOMEVALUESFROM(?Name, :hasCondiment, _:b3).

From the 3rd candidate pair in Example 7 there is no lack of reuse, but we can represent the pattern as the following template:

```
<NAME>(?x1, ?x2, ?x3, ?x4)
    :: (?x1, rdfs:label, ?x2), (?x1, skos:prefLabel, ?x3), (?x1, skos:definition, ?x4).
```

The template and parameters should be given suitable names and parameters given a type, as exemplified by the ANNOTATION template (11) found in Fig. 3(b). The procedure of identifying dependency pairs and lack of reuse is implemented and demonstrated in the online walk-through at ottr.xyz/event/2018-10-08-iswc/.

For large knowledge bases, the set of candidate pairs might be very large, as it grows exponentially in the number of template instances in the worst case. This means that manually assessing all candidate pairs is not feasible, and smaller subsets of candidates must be automatically suggested. We have yet to develop proper heuristics for suggesting good candidates, but the cases with the most common patterns (the candidates with largest T-sets), the largest patterns (the candidates with the largest I-sets), or large patterns that occur often could be likely sources for patterns to refactor. The latter of the three can be determined by maximising a weight-function, for instance of the form $f(\langle I, T \rangle) = w_1|I| + w_2|T|$. However, these weights might differ from use-case to use-case. Another approach for reducing the total number of candidates to a manageable size, is to let a user group some or all of the templates according to subdomain, and then only present candidates with instances fully contained in a single group. The idea behind such a restriction is that it seems likely that a pattern is contained within a subdomain. We give an example of these techniques in the following section.

5   Use Case Evaluation

In this section we outline an evaluation of OTTR templates in a real-world setting at the engineering company Aibel, and demonstrate in particular our process of finding and removing redundancies over a large, generated template library.

Aibel is a global engineering, procurement, and construction (EPC) service company based in Norway best known for its contracts for building and maintaining large offshore platforms for the oil and gas industry. When designing an offshore platform, the tasks of matching customer needs with partly overlapping standards and requirements as well as finding suitable products to match design specifications are highly non-trivial and laborious. This is made difficult by the fact that the source data is usually available only as semi-structured documents that require experience and detailed competence to interpret and assess. Aibel has taken significant steps to automate these tasks by leveraging reasoning and queries over their *Material Master Data* (MMD) ontology. It integrates this information in a modular large-scale ontology of ∼200 modules and ∼80,000 classes and allows Aibel to perform requirements analysis and matching with greater detail and precision and less effort than with their legacy systems. Since the MMD ontology is considered by Aibel as a highly valuable resource that gives them a competitive advantage, it is not publicly available.

The MMD ontology is produced from 705 spreadsheets prepared by ontology experts and populated by subject matter experts with limited knowledge of

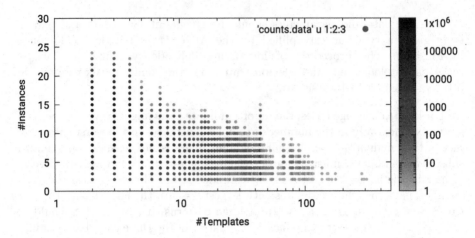

Fig. 4. A scatter plot of the sizes of the two sets for all candidate pairs from Aibel use case. The colour shade denotes the logarithm of the number of candidates at each point.

modelling and semantic technologies. The column headers of the spreadsheets specify how the data is to be converted into an ontology, and the translation is performed by a custom-built pipeline of custom transformations, relational databases, and SPARQL CONSTRUCT transformations. The growing size and complexity of the system, the simple structure of the spreadsheets and lack of common modelling patterns make it hard to keep an overview of the information content of the spreadsheets and enforce consistent modelling across spreadsheets. The absence of overarching patterns also represents a barrier for Aibel's wish to extend the ontology to cover new engineering disciplines, as there are no patterns that are readily available for reuse.

The aim of our evaluation is to test whether OTTR templates and the tools presented in this paper can replace Aibel's current in-house built system and improve the construction and maintenance of the MMD ontology. By exploiting the simple structure of the spreadsheets we automatically generated OTTR templates: one for each spreadsheet (705 templates), one for each unique column header across spreadsheets (476 templates), and one for each axiom pattern used, e.g., existential restriction axiom (4 templates).

To analyse the large template library, we applied the algorithm for finding candidate pairs described in Sect. 4.2, giving a total of 54,795,593 candidate pairs. The scatter plot in Fig. 4 shows the distribution of sizes for the two sets; the largest number of instances and templates for a given candidate is 24 and 474, respectively. The large number of candidates makes it impossible to manually find potential templates, thus we employed the semi-automatic method described in the previous section to suggest possible improvements to the library. In order to demonstrate the process, we selected the candidates that contain a specific template, the template modelling a particular type of *pipe elbows* from the AMSE B16.9 standard, which is an often-used example from the MMD ontology. This template occurs in a total of 12,273 candidates. To reduce the number of candidates further, we removed candidates with instances of a generic character, such as `rdfs:label`, to end up with candidates with domain-specific templates. By using a weight function, we selected the candidate with the largest set of templates and at least 6 instances. From this candidate, with 33 templates and 7 instances, we obtained a template suggestion that we were able to verify is contained by all of the 33 templates in T, by using the template as a query over the templates in T. We added this new template to the library and refactored it into all the 33 templates using its pattern.

Fixing this single redundancy reduced the total number of candidates by over 1.8 million. This great reduction in candidates comes from the fact that fixing a redundancy represented by a candidate C can also fix the redundancies of candidates having a pattern that is contained in, contains, or overlaps C's pattern. This indicates that, despite a very large number of candidates, small fixes can dramatically reduce the overall redundancy. Furthermore, by automatically refactoring all lack of reuse in the entire library, the number of candidates decreases to under 3 million. The average number of instances per template went from 5.6 down to 2.7 after this refactoring. In addition to the redundancies fixed above, we were also able to detect equal templates (pairs of templates both having a lack of reuse of the other). Out of the 931 templates we analysed, only 564 were unique. Thus, we could remove a total of 367 redundant templates from the library. Note that all of the improvements made above should be reviewed by a user, as discussed in Sect. 4.2, to ensure that the new template hierarchy properly represents the domain.

The use case evaluation indicates that OTTR templates and tools can replace Aibel's custom built approach for transforming spreadsheets into ontologies. Indeed, OTTR greatly exceeds the expressivity of Aibel's spreadsheet structure and provides additional formal structure that can be used to analyse and improve the modelling patterns used to capture domain knowledge. As future evaluation, we want to work with Aibel's domain experts in order to identify promising heuristics for finding the best shared patterns. We believe that these new patterns and user requirements from Aibel may foster new ideas for added expressivity and functionality of OTTR languages and tools. Furthermore, we want to evaluate whether we can replace Aibel's hand-crafted queries with queries generated from templates. This would avoid the additional cost of maintaining a large query library, while benefiting from already existing templates and OTTR's compositional nature and tools for building and analysing the generated queries.

6   Related Work

Modularised ontologies, as well as the use and description of ontology design patterns, have attracted significant interest in recent years, as demonstrated by the multitude of languages and frameworks that have emerged. However, a hurdle for the practical large-scale use of ontology design patterns is the lack of a tool supported methodology; see [4] for a discussion of some of the challenges facing ontology design patterns. In this section we present selected work related to our approach that we believe represents the current state of the art.

An early account of the features, benefits and possible use-cases for a macro language for OWL can be found in [14].

The practical and theoretical aspects of OTTR templates were first introduced in [13] and [2]. This paper presents a more mature and usable framework, including formalisation and use of template relations, real-world evaluation, added expressivity in the form of optional parameters and expansion modes, and new serialisation formats.

GDOL [9] is an extension of the Distributed Ontology, Modelling, and Specification Language (DOL) that supports a parametrisation mechanism for ontologies. It is a metalanguage for combining theories from a wide range of logics under one formalism while supporting pattern definition, instantiation, and nesting. Thus it provides a broad formalism for defining ontology templates along similar lines as OTTR. To our knowledge, GDOL has yet to investigate issues such as dependencies and relationships between patterns, optional parameters, and pattern-as-query (the latter being listed as future work). A protege plugin for GDOL is in the works and DOL is supported by Ontohub (an online ontology and specification repository) and Hets (parsing and inference backend of DOL).

Ontology templates as defined in [1] are parameterised ontologies in $\mathcal{ALC}$ description logic. Only classes are parameterised, and parameter substitutions are restricted to class names. This is quite similar to our approach, yet it is not adapted to the semantic web, and nested templates and patterns-as-queries are not considered. Furthermore, it appears this project has been abandoned, as the developed software is no longer available.

OPPL [6] was originally developed as a language for manipulating OWL ontologies. Thus it supports functions for adding and removing patterns of OWL axioms to/from an ontology. It relies heavily on its foundations in OWL-DL and as such can only be used in the context of OWL ontologies. Despite this, the syntax of OPPL is distinct from that of RDF, thus requiring separate tools for viewing and editing such patterns, though a Protégé plugin does exist, in addition to a tool called *Populous* [7] which allows OPPL patterns to be instantiated via spreadsheets. By allowing patterns to return a single element (e.g., a class) OPPL supports a rather restricted form of pattern nesting as compared to OTTR.

Tawny OWL [10] introduces a Manchester-like syntax for writing ontology axioms from within the programming language Clojure, and allows abstractions and extensions to be written as normal Clojure code alongside the ontology. Thus the process of constructing an ontology is transformed into a form of programming, where existing tools for program development, such as versioning, testing frameworks, etc. can be used. The main difference from our approach is that Tawny OWL targets programmers and therefore tries to reuse as much of the standards and tools used in normal Clojure development, whereas we aim to reuse semantic technology standards and tools.

OPLa [5] is a proposal for a language to represent the relationships between ontologies, modules, patterns, and their respective parts. They introduce the OPLa ontology which describes these relationships with the help of OWL annotation properties. This approach does not, however, attempt to mitigate issues arising with the *use* of patterns, but focuses more on the description of patterns, than on practical use.

There are other tools and languages such as XDP [3], built on top of WebProtégé as a convenient tool for instantiating ODPs, the M^2 mapping language [11] that allows spreadsheet references to be used in ontology axiom patterns, and RDF shape languages, such as SHACL [8], that may be used to describe and validate patterns. Although these have similarities with OTTR, we consider these more specialised tools and languages, where for example analysis of patterns is beyond their scope.

7   Conclusion and Future Work

This paper presents OTTR, a language with supporting tools for representing, using and analysing ontology modelling patterns. OTTR has a firm theoretical and technological base that allows existing methods, languages and tools to be leveraged to obtain a powerful, yet practical instrument for ontology construction, use and maintenance.

For future work, the natural next step with respect to template library optimisation is to continue and expand the analysis of Sect. 4, both for existing and new template relations. In particular, it is natural to compare templates both syntactically using their full expansion and in terms of their semantic relationship. The latter would allow us, e.g., to answer questions about consistency and whether a given library is capable of describing a certain knowledge pattern. We also want to develop specialised editors for OTTR templates, such as a plugin for Protégé, and extend support for more input formats, such as accessing data from relational databases.

Acknowledgements. We would like to thank Per Øyvind Øverli from Aibel, and Christian M. Hansen from Acando for their help with the evaluation of OTTR. The second and fourth author were supported by Norwegian Research Council grant no. 230525.

References

1. Blasko, M., Kremen, P., Kouba, Z.: Ontology evolution using ontology templates. Open J. Semant. Web (OJSW) **2**, 15–28 (2015)
2. Forssell, H., et al.: Reasonable macros for ontology construction and maintenance. In: Proceedings of the 30th International Workshop on Description Logics (2017)
3. Hammar, K.: Ontology design patterns in WebProtege. In: Proceedings of the ISWC 2015 Posters and Demonstrations Track (2015)

4. Hammar, K., et al.: Collected research questions concerning ontology design patterns. In: Hitzler, P., et al. (eds.) Ontology Engineering with Ontology Design Patterns, pp. 189–198. IOS Press (2016)
5. Hitzler, P., et al.: Towards a simple but useful ontology design pattern representation language. In: Proceedings of the 8th Workshop on Ontology Design and Patterns (2017)
6. Iannone, L., Rector, A., Stevens, R.: Embedding knowledge patterns into OWL. In: Aroyo, L., et al. (eds.) ESWC 2009. LNCS, vol. 5554, pp. 218–232. Springer, Heidelberg (2009). https://doi.org/10.1007/978-3-642-02121-3_19
7. Jupp, S.: Populous: a tool for building OWL ontologies from templates. BMC Bioinform. **13**(S–1), S5 (2012)
8. Knublauch, H., Kontokostas, D.: Shapes constraint language (SHACL) (2017). W3C Recommendation
9. Krieg-Brückner, B., Mossakowski, T.: Generic ontologies and generic ontology design patterns. In: Proceedings of the 8th Workshop on Ontology Design and Patterns (2017)
10. Lord, P.: The semantic web takes wing: programming ontologies with Tawny-OWL. In: OWLED (2013)
11. O'Connor, M.J., Halaschek-Wiener, C., Musen, M.A.: M2: a language for mapping spreadsheets to OWL. In: OWLED (2010)
12. Ogden, C.K., Richards, I.A.: The Meaning of Meaning. Harvest Book, San Diego (1946)
13. Skjæveland, M.G., et al.: Pattern-based ontology design and instantiation with reasonable ontology templates. In: Proceedings of the 8th Workshop on Ontology Design and Patterns (2017)
14. Vrandečić, D.: Explicit knowledge engineering patterns with macros. In: Proceedings of the Ontology Patterns for the Semantic Web Workshop at the ISWC 2005 (2005)

Pragmatic Ontology Evolution: Reconciling User Requirements and Application Performance

Francesco Osborne$^{(\boxtimes)}$ and Enrico Motta

Knowledge Media Institute, The Open University,
Milton Keynes MK7 6AA, UK
{francesco.osborne,enrico.motta}@open.ac.uk

Abstract. Increasingly, organizations are adopting ontologies to describe their large catalogues of items. These ontologies need to evolve regularly in response to changes in the domain and the emergence of new requirements. An important step of this process is the selection of candidate concepts to include in the new version of the ontology. This operation needs to take into account a variety of factors and in particular reconcile user requirements and application performance. Current ontology evolution methods focus either on ranking concepts according to their relevance or on preserving compatibility with existing applications. However, they do not take in consideration the impact of the ontology evolution process on the performance of computational tasks – e.g., in this work we focus on instance tagging, similarity computation, generation of recommendations, and data clustering. In this paper, we propose the *Pragmatic Ontology Evolution (POE)* framework, a novel approach for selecting from a group of candidates a set of concepts able to produce a new version of a given ontology that (i) is consistent with the a set of user requirements (e.g., max number of concepts in the ontology), (ii) is parametrised with respect to a number of dimensions (e.g., topological considerations), and (iii) effectively supports relevant computational tasks. Our approach also supports users in navigating the space of possible solutions by showing how certain choices, such as limiting the number of concepts or privileging trendy concepts rather than historical ones, would reflect on the application performance. An evaluation of POE on the real-world scenario of the evolving Springer Nature taxonomy for editorial classification yielded excellent results, demonstrating a significant improvement over alternative approaches.

Keywords: Ontology evolution · Domain ontologies · Bibliographic data
Scholarly data · Scholarly ontologies

1 Introduction

Increasingly, organizations are adopting ontologies to describe their large catalogues of items. Indeed, ontologies have proved to be very useful in the context of a variety of tasks [1], including the integration of data from different sources, domain reasoning, classification [2], generation of recommendations [3], cluster analysis [4], community

© Springer Nature Switzerland AG 2018
D. Vrandečić et al. (Eds.): ISWC 2018, LNCS 11136, pp. 495–512, 2018.
https://doi.org/10.1007/978-3-030-00671-6_29

detection [5], sentiment analysis, forecasting [6], and others. Naturally, ontologies need to be regularly maintained and need to evolve according to changes in the domain or new requirements from users or applications [7]. This process is called *ontology evolution* and it is a critical part of the ontology lifecycle. While the literature proposes a variety of frameworks for ontology evolution [8–11], essentially most agree on three fundamental steps in the process: (i) detection of the need for the evolution, (ii) identification of candidate changes, and (iii) validation and assessment of these changes, to ensure that the resulting ontology satisfies the given needs.

Hence, in the first instance, the evolved ontology normally has to comply with a set of requirements, defined to ensure that the ontology remains compatible with the current workflow and usable by the relevant stakeholders.

In the second instance, it is crucial to take into account the impact of the ontology evolution process on relevant applications. Ontologies are often used to enable semantic approaches to data mining, information filtering, trend detection, and other tasks [12], whose performance needs to be taken in consideration when creating a new version of the ontology. Crucially, user needs and applications performance are sometimes in opposition. For example, a very comprehensive representation of items and their features would generally improve the performance of a recommender system, but users may prefer a less complex representation that it is easier to browse, memorize, maintain, and incorporate in their workflow.

In the third instance, domain experts may have preferences about which concepts to privilege that should be considered in the process. For example, they may decide to privilege concepts which are currently trendier rather than historical ones, or those that are more represented in their internal catalogue, rather than considering the full domain.

Finally, users need to be able to understand why a certain concept was selected or discarded and how this relates to the requirements, the user preferences, and the ontology support for some computational tasks.

The motivating scenario for this work concerns the evolution of the internal taxonomy at Springer Nature, which is used for classifying books, journals, and other editorial products. Since this taxonomy is used by a lot of different users and software systems, the evolution process needs to take in consideration both user needs and the impact on applications. For instance, a recommender system for suggesting editorial products described by an ontology [13] would perform differently according to the ontology that it is using. In addition, the process need to be transparent, so that every change can be justified in light of these factors.

Current solutions are not easily applicable to this problem. Most of the methods for selecting the concepts to be included in an evolving ontology address this task by ranking concepts according to a weight derived from information retrieval metrics [14, 15], list of words [16], or online ontologies [11]. These solutions have the advantage of being generic, but present two significant limitations: (i) they do not assess the impact of the new version of the ontology on the performance of the relevant applications, and (ii) they ignore concept synergy, by weighting the relevance of single concepts rather than the overall impact of a combination of concepts. Some approaches do focus on preserving consistency between the ontology and the dependent applications [17–21], however they do not consider the effect of the changes on the performance of computational tasks.

In this paper, we propose the *Pragmatic Ontology Evolution* (POE) framework, a novel approach for selecting from a group of candidates a set of concepts able to produce an ontology that (i) is consistent with the given requirements, (ii) is parametrised with respect to a number of dimensions (e.g., topological considerations), and (iii) supports effectively relevant computational tasks, such as instance tagging, similarity computation, generation of recommendations, and data clustering. POE supports users in navigating the space of possible solutions by showing how certain choices, such as limiting the number of concepts or privileging trendy concepts rather than historical ones, would reflect on the application performance. It also makes it easy to explain why a certain concept was included in the ontology on the basis of its contribution to the performance of a specific task. Finally, it selects the new concepts not only according to individual weights, but also considering their synergy with other concepts.

The rest of the paper is organized as follows. In Sect. 2, we will present a motivating scenario involving the evolution of an editorial taxonomy at Springer Nature. In Sect. 3, we will review the literature regarding ontology evolution and, in particular, the selection of candidate concepts. In Sect. 4, we will discuss POE in details and in Sect. 5, we will evaluate it on a dataset of 1,218 Springer Nature books. Finally, in Sect. 6, we summarize the main conclusions and outline future directions of research.

2 Motivating Scenario: Evolving Springer Nature Market Codes

Springer Nature (SN) is one of the major academic publishing companies and has a vast catalogue of books, journals, and conference proceedings. Like other companies in this space, it has its own editorial classification system, called *Product Market Codes* (PMC). PMC is a taxonomy of research fields that is used to tag editorial items with relevant topics, e.g., "Artificial Intelligence" or "Software Engineering". The resulting metadata are then used for a variety of tasks, such as improving the discoverability of products in digital and physical libraries, supporting marketing decision, and detecting research trends.

It is crucial to keep PMC up to date with the evolution of the research landscape at the right level of granularity. This is particularly challenging in the field of Computer Science, where new areas evolve constantly and taxonomies tend to become obsolete very quickly [22]. In the context of the collaboration between The Open University and Springer Nature [2, 13], we focused on the issue of supporting the evolution of the Computer Science portion of PMC, concentrating in particular on some branches that had become obsolete.

This work builds on our earlier research, which has produced new methods able to generate automatically taxonomies of research areas through large scale-mining of scholarly data. In particular, by applying the Klink-2 algorithm [22] on the Rexplore dataset [23], we generated the Computer Science Ontology (CSO) [24], a large-scale ontology of research topics in Computer Science, which includes about 26K topics linked by about 226K semantic relationships. CSO powers two tools used by SN for tagging and recommending books: Smart Topic Miner [2] and Smart Book Recommender [13].

In accordance with the requirements provided by SN publishing editors, we focused on the evolution of the branches under five concepts of the original PMC taxonomy ("I21017-Artificial Intelligence (incl. Robotics)", "I14029-Software Engineering", and three others – see details in Sect. 5) that we mapped to nine CSO concepts (in the given example "Artificial Intelligence", "Robotics", "Software Engineering", and "Software Design"). We then extracted all their sub-concepts producing 2,451 candidate concepts. However, producing a new version of PMC with all of them, would cause the Computer Science portion of PMC to grow from 89 to 2,540 concepts. This is unfeasible for a variety of pragmatic reasons, including the fact that many books are still manually tagged and curated by editors. We thus needed to find a solution to the evolution of PMC, which ensured that it remained under a certain size. It was also crucial that the new version of the ontology would support effectively tasks such as generation of recommendations, data clustering, and so on. Finally, we would need to be able to produce a justification for the inclusion or the exclusion of a research topic.

This is a typical real-world case in which the first two steps of the ontology evolution process, identifying the need for changes and producing candidate concepts, are relatively easy, since it exists a clear need (new fields in Computer Science are missing) and we already have a good selection of candidate concepts in CSO. On the contrary, there was no clear solution for selecting a set of concepts that would comply with the requirements and support relevant applications.

3   Related Work

Most of the ontology evolution frameworks [7–10] include a phase that regards the verification and selection of candidate changes or concepts to be included in the new version of the ontology. This step is labelled "change validation phase" in the framework of Stojanovic [8], "verification and approval" in Klein and Noy [9], "accepting and rejecting changes" in Noy [10], and it is split in two different phases labelled "validating changes" and "assessing the evolution impact" in the ontology evolution cycle proposed by Zablith [7].

Traditionally, the candidate changes are validated at three different levels [7]: (i) *formal properties-based validation*, which uses formal techniques to preserve the consistency and coherence of the ontology, (ii) *domain base validation*, which exploits domain information to assess the relevance of the candidate changes, and (iii) *application and usage impact*, which measures the effects of the changes on data instances, dependent ontologies, and relevant applications [25]. POE works at the second and third levels, since it assesses the importance of concepts within a certain domain and it evaluates the effect of alternative ontologies on computational tasks.

Approaches to *domain base validation* can be classified according to their focus, which can be either on domain relevance [14–17, 26] or correctness [27]. Text2Onto [14], a well-known system for ontology learning, falls in the first category, since it weights the relevance of the candidate concepts by mean of information retrieval measures, such as Relative Term Frequency (RTF), TF-IDF, and the C-value/NC-value method. SPRAT [15], a tool for automatic pattern-based ontology population, also uses TF-IDF to select the relevant terms that should be included in the ontology. Similarly to POE, they both focus on the inclusion of concepts or terms rather than entire statements. The DINO Framework [16], assesses the relevance of a set of candidate triples according to their Levenshtein distance from a set of wanted or unwanted words, specified by domain experts. The Evolva framework [11] measures the relevance of a statement by generating its ontological context from a set of online ontologies and comparing it to the evolving ontology. The DINAMO-MAS system [26] assesses relationships between terms by means of a confidence score that takes in consideration their lexico-syntactic patterns. Some other systems focus on assessing the correctness of statements. For instance, Sabou et al. [27] verify the correctness of the link between two concepts by exploiting the path connecting the concepts in online ontologies. Similarly to these solutions, POE aims to find the best set of concepts to be included in an evolved ontology. However, it also consider application performance and concept synergy.

Some other approaches focus on assessing the impact of evolution on data instances [28, 29], applications [17–21, 30], and dependent ontologies [25]. Because of lack of space, we will focus our review on the first two categories.

Qin and Atluri [28] propose a method to define and preserve the structural and semantic validity of data instances that are described by an evolving ontology. Similarly, Hartung et al. [29] introduce a generic framework for the study of the evolution of ontologies and ontology-related mappings. We also take into consideration instances and their mapping, but rather than checking their validity, we focus on the impact of their representations on the relevant tasks.

Several approaches address the impact of the resulting ontology on dependent applications, however they focus mainly on preserving consistency and compatibility. For instance, Huang and Stuckenschmidt [17] present MORE, a system that uses temporal logic to detect the consequences of changes. Xuan et al. [18] introduce the floating version model, which preserves compatibility by not allowing a new version of the ontology to falsify axioms that were previously true. Wang et al. [19] propose another technique to maintain the consistency of dependent applications and suggest resolution strategies. Liang et al. [20] present a system that analyses the queries submitted by dependent applications, detects if the relevant entities where changed during the evolution process, and repairs broken queries. Similarly, Kondylakis and Plexousakis [21] propose a formal approach for identifying the impact of ontology evolution on queries and easing query migration. Finally, Groß et al. [30] introduce an approach for measuring the stability of a ontology and show how ontology evolution affected the level of significance of functional enrichment analyses in Biology. Differently from all these systems, POE focuses on the performance of dependent computational tasks rather than on consistency and compatibility, and aims to generate an ontology that can effectively support these tasks.

4 The POE Framework

4.1 Overview of POE

The Pragmatic Ontology Evolution (POE) framework was designed to produce an ontology that complies with the given requirements and performs well on some input tasks, as well as supporting users in exploring the space of solutions. POE takes as input (i) an ontology, (ii) a collection of instances that could be described by the concepts in the ontology, (iii) a set of additional candidate concepts (and their relationships with existing concepts), (iv) a set of requirements, (v) one or more tasks, and, optionally, (vi) four additional parameters defining user preferences. It then finds the combination of candidate concepts that generates the representation of the instances which performs best on the given tasks by first searching in the space of four parameters and then applying a variation of Recursive Feature Elimination [31]. Finally, it returns: (i) a new version of the ontology that complies with the input requirements and effectively supports the relevant tasks, and (ii) a number of statistics that allow users to assess the effect of their preferences (e.g., privileging conservative or novel concepts) on the tasks.

In the PMC scenario, the input ontology is the portion of PMC covering the field of Computer Science, while the instances are the metadata of books published by SN in recent years and tagged with PMC concepts. The set of candidate concepts was built by mapping the PMC concepts that needed to be enriched to relevant concepts in CSO and then selecting all their sub-concepts, as discussed in Sect. 2. The mapping was done semi-automatically by generating candidate mapping with statistical heuristics from Klink-2 [22] and then revising them with the help of SN editors, as described in [2]. This operation yielded 2,451 candidate concepts.

The POE framework is structured in two main steps:

Parameter Optimization. It tests different combinations of four parameters (using grid search) to weigh the candidate concepts. For each combination, it produces an ontology that complies with the requirements, it annotates the instances with it, and it measures the performance of this representation on the tasks. Finally, it returns a ranked list of parameter combinations.

Recursive Concept Elimination. It uses the best parameter combination from previous steps to generate an ontology and applies on it a variation of the Recursive Feature Elimination to iteratively eliminate the least important concepts, until the desired number of concepts is reached.

POE allows users to set *three kind of requirements*: (1) the maximum number of concepts in the ontology, (2) the minimum number of concepts in a branch, and (3) the maximum number of concepts in a branch. Being able to control the dimension of branches is important to produce structurally balanced ontologies. POE also allows the users to define or restrict (within a range) *four parameters* that control the ranking of the candidate concepts.

POE can be used with any task that uses an ontology-derived representation of the instances and whose performance can be evaluated according to an objective metric. In particular, in the current prototype we support four tasks: *instance tagging, similarity computation, generation of recommendations*, and *data clustering*.

In what follows, we will first discuss the basic functions of POE, i.e., the generation of an ontology from a set of parameters (Sect. 4.2), and the evaluation of a ontology on a task (Sect. 4.3). We will then address the two main steps of the POE framework that employ these functionalities: parameter optimization in Sect. 4.4, and Recursive Concept Elimination in Sect. 4.5.

4.2 Topic Ranking

In this phase, we consider the task of selecting a number of concepts to update an ontology as a ranking problem, coherently with the state of the art (e.g., [11, 14–16]). We thus want to assign a weight to every concept and then update the input ontology with the first n concepts that comply with the requirements.

A typical way to do so is assessing a concept importance according to how frequently it is represented in the instances. Intuitively, a concept that is often needed to describe the instances should receive a higher weight than a rarer one. Indeed, previous literature showed that term frequency and TF-IDF perform quite well on this task [14, 15]. We believe however that is possible to have a more comprehensive treatment of this challenge by taking in consideration a number of additional factors. In particular, here we consider four dimensions that can influence the value of a concept in the new ontology and the strategy for mapping it to the instances.

Semantics. As already mentioned, a purely syntactical solution to weigh concepts is to use the frequency of their label in the instances. For example, given the concept *temporal logic*, we could weigh it according to the number of books that contains in the title, abstract, or keyword field the string "temporal logic". Alternatively, we could take a more semantic approach and associate to a concept each instance that contains the label of the concept or of any of its sub-concepts. For example, we could map the concept *temporal logic* to each book that contains one of the alternative labels (e.g., "temporal logics") or sub-concepts (e.g., "temporal operators") in the CSO ontology. This technique has been applied with good results in a variety of fields, such as automatic classification of proceedings [2], technology forecasting [6], recommender systems [3, 13], community detection [5], and others.

Temporal Dimension. It is also useful to consider when the instances were produced. In the scenario of academic publishing, considering recent instances would prioritise the trendiest research topics, which may keep growing and become more popular in the future. However, focusing too much on recent instances, may exclude some significant historical concepts that are still important and may risk prioritising concepts that are experiencing only a transient burst of popularity.

Internal Versus External Instances. The instances can either derive from the catalogue of the organization that has adopted the ontology (e.g., SN books in Computer Science) or they could be generic ones (e.g., all available books in Computer Science). In the first case, the selected concepts will acquire the same biases of the internal dataset. The resulting ontology will be tailored to those specific instances, but may exclude significant concepts that are currently under-represented in the catalogue. Therefore, a company that wants to expand its catalogue and cover new fields may prefer to consider all available instances, while one that is not interested in doing so, may decide to produce a more internally-tailored ontology.

Structural Considerations. Considering only the weight of single concepts may exclude some concepts that are less represented in the instances, but act as good branching point in the ontology and keep the structure easy to browse and explore. Therefore, in some cases it may be advisable to include concepts that are useful from a structural standpoint, even if they appear less frequently in the instances.

We believe that it is useful to take in consideration each of these dimensions when ranking concepts. Therefore, POE takes as input four parameters that can be tuned by the user or optimized on a certain task:

- α (0−1). It controls whether POE uses the syntactic method, the semantic method, or a combination of the two for mapping concepts to instances. If $\alpha = 0$, it will use only the label of a concept, with $\alpha = 1$ it will consider all the sub-topics, otherwise it will use a weighted average.
- $\beta(0 - 1)$. It controls whether the weight will be computed only on instances from an internal dataset or if it will consider also external entities. If $\beta = 0$, POE will use only the internal instances, with $\beta = 1$ only external ones, otherwise it will use a weighted average.
- γ (0 − 1). It modulates the importance of the most recently created instances on the weight. If $\gamma = 0$, POE will weight more recent instances, with $\gamma = 1$ the time dimension will not matter, otherwise it will use a weighted average.
- δ (True, False). It controls whether POE will try to recover structurally important concepts. In the current implementation, a concept is considered structurally important if it has at least three sub-concepts that were selected.

The weight of each concept is computed with the following formula.

$$\left(\left(\log \sum_{y=f}^{l} si_y^{w_y} \right) \beta + \left(\log \sum_{y=f}^{l} se_y^{w_y} \right) (1 - \beta) \right) \alpha + \left(\left(\log \sum_{y=f}^{l} fi_y^{w_y} \right) \beta + \left(\log \sum_{y=f}^{l} fe_y^{w_y} \right) (1 - \beta) \right) (1 - \alpha)$$

Where si_y, se_y, fi_y, fe_y are respectively, for a given year y, the semantic frequency in the internal dataset, the semantic frequency in the external dataset, the syntactic frequency in the internal dataset, and the syntactic frequency in the external dataset; f and l are the first and last year of the analysed period; and $w_y = \frac{1}{(l-y+1)^{2\gamma}}$.

After ranking the concepts, POE selects the first n concepts that comply with the input requirements. The POE framework can adopt any kind of requirements that can be automatically verified by analysing the set of candidate concepts. In the current prototype we take in consideration the minimum and maximum number of concepts for each branch. POE enforces this requirements by first populating each branch with the minimum number of concepts and then inserting the remaining concepts in the branch that are still available until the maximum number of concepts is reached. If δ is true, POE also checks for structurally important concepts and inserts them in place of the ones with lowest weights. Finally, it creates a new version of the ontology which incorporates the selected concepts.

It is also possible to define a list of invalid topics that will not be considered during the selection phase. This option will be used during the Recursive Concept Elimination (Sect. 4.5) to exclude topics that do not perform well on the tasks.

The approach described in this section can be used on its own in alternative to generic methods [14, 15]. The main advantage is that it allows users to explore the space of solutions, possibly with the support of domain experts, and understand how different combination of parameters impact on the resulting ontology. However, it is difficult even for human experts to assess how a new ontology will affect applications. For this reason, we want to take a further step: evaluate the alternative ontologies on the input tasks and suggest the one that yields the best performance.

4.3   Evaluating a Candidate Ontology on a Task

POE evaluates an ontology on some computational tasks by (i) using the ontology for generating a representation of the instances, (ii) running the input tasks on this representation, and (iii) evaluating the performance with the relevant metrics. The instances are represented as a vector in which the elements correspond to the concepts in the ontology and the values weigh the importance of a concept. In the case of PMC, we used the Smart Topic API [13] for representing books as a vectors of research topics in which each topic is assigned a value equal to the number of chapters in which it appears. This is a convenient representation that can support several tasks. The Smart Topic API is a service developed in collaboration with Springer Nature for tagging publications with ontology concepts. It is described in details in [2, 13].

While some tasks (e.g., instance tagging) can be evaluated using simple metrics (e.g., percentage of instances covered), others require a ground truth. For instance, evaluating the performance of a clustering algorithm would usually require a correct set of clusters to compare against. In some cases, such as in the PMC scenario, it is quite expensive to produce a specific gold standard for each task. Therefore, we address this issue by adopting a ground truth ontology that includes all candidate concepts and can be used with every task. The intuition is that we want to select a candidate ontology including no more than n concepts that would perform as well as possible as the full ontology. In the case of PMC, we want to produce an ontology of about 120–200 concepts that can perform as closely as possible to the version which includes all 2,451 candidate concepts from CSO. In the following, we will refer to the candidate ontology as O_c and to the full ontology, which serves as ground truth, as O_f.

It is important to note that if the task in consideration is sensitive to irrelevant or redundant features, the ground truth ontology needs to contain valid concepts and to have been previously evaluated. This is indeed the case with CSO, which was previously tested on several tasks [24], including automatic tagging of scientific publications [1], recommendation generation [2], clustering [5], and technology forecasting [6]. Alternatively, we suggest to pre-filter the candidate concepts [16] or to generate a task-specific gold standard.

The current POE prototype implements four tasks that were developed for the PMC scenario. The implementation of a new task is straightforward since it simple requires to define a representation of the instances, run the task on them, and evaluate the results with a relevant metric. If the input includes several tasks, their overall performance is computed as the average of the resulting metrics.

We will now discuss these tasks and their evaluation.

4.3.1 Instance Tagging

As first task, we consider the automatic tagging that associates each instance to a vector of concepts (via the Smart Topic API [13]). The candidate ontology should enable to generate a relatively granular representation of all the instances. Therefore, we evaluate this task by computing the percentage of instances that are covered by the ontology. Naturally, the definition and quality of the coverage varies according to the scenario and the domain. In the case of PMC, it is important to associate each book to a minimum number of topics, so that they can be browsed and searched with a good granularity. Furthermore, the main topics have to be fairly representative and not appear only in few chapters. We thus consider covered a publication that is associated with at least three concepts that are present in at least three chapters.

4.3.2 Similarity Computation

Computing the similarity of a set of items is a common task that supports more complex tasks such as record linkage, clustering, and so on. We evaluate this task by computing the cosine similarity of each couple of instances according to both O_c and O_f, and then calculating their mean root-mean-squared error.

$$similarity\_performance = 1 - \sqrt{\sum_{i=1}^{n}\sum_{j=1}^{n}\left(\cos(\widehat{c_i}, \widehat{c_j}) - \cos\left(\widehat{f_i}, \widehat{f_j}\right)\right)^2}$$

Where $\cos(\widehat{v_1}, \widehat{v_2})$ is the cosine similarity between vectors $\widehat{v_1}$ and $\widehat{v_2}$, $\widehat{c_i}$ is the vector of instance i produced with the candidate ontology, $\widehat{f_i}$ is the vector of instance i produced with the full ontology, and n is the total number of instances. When the result is near 1 the two ontologies are yielding similar results and thus the candidate ontology is performing well.

4.3.3 Generation of Recommendations

Today several recommender systems use ontologies for enhancing semantically the representation of items or users [3]. In particular, content-based recommenders use feature representations of items to suggest other items that possess similar characteristics. This is the case of *Smart Book Recommender* [13] which suggest SN books relevant to a certain conference.

We generate for each instance, say I, a ranked list of recommendations composed by the 100 instances most similar to I, according to both O_c and O_f. This is realized by computing the cosine similarity of the vector representations derived from the two ontologies. We then assess the agreement of the lists produced by the two ontologies using the Spearman's rank correlation coefficient, a standard metric for evaluating recommender systems. The Spearman's coefficient between two variables equals to the

Pearson correlation between the rank values of those two variables, and it is used when it is important to compare the order of items in a list. It varies between −1 and 1, with 1 (or −1) indicating that the two list exhibit a perfect correlation and 0 indicating that the order of two list is not correlated at all. The performance of O_c on this task is measured according to the following formula:

$$recommender\_performance = \frac{1}{n} \sum_{i=1}^{n} \frac{cov(rc_i, rf_i)}{\sigma_{rc_i} \sigma_{rf_i}}$$

Where σ_{rc_i} and σ_{rf_i} are the standard deviations of the ranked list of items according to O_c and O_f, and $cov(rc_i, rf_i)$ is the covariance of the ranked lists.

4.3.4   Clustering

Cluster analysis is a powerful tool for exploring trends, generating analytics, and informing marketing and political decisions. We first cluster the instances according to both ontologies by using the K-Means++ algorithm and then compare the results with the Rand index, which is a measure of the similarity between two sets of clusters. The Rand index varies between 0 and 1, with 1 indicating that the data are clustered in the same way and 0 indicating that the cluster sets are completely dissimilar.

$$clustering\_performance = \frac{a_i + b_i}{\binom{n}{2}}$$

Where a_i is the number of pairs of instances that are in the same cluster both in the cluster set of O_c and in the cluster set of O_f, and b_i is the number of pairs that are in different clusters.

4.4   Parameter Optimization

Parameter optimization is the first step of the POE approach. In this phase, POE executes a grid search on the space of the four parameters described in Sect. 4.2, produces a candidate ontology for every combination of parameters, and ranks them according to their performance on the tasks, as illustrated in Sect. 4.3. The ontology that performs best is the advisable solution in the space of parameters.

The result of this phase can be used for exploring the space of solutions and assessing the effect of the parameters on the ability of an ontology to perform certain tasks. A simple way to do so is testing if there is any correlation between a parameter and the performance. For instance, Fig. 1 shows the relation between two parameters and the performance obtained on the generation of recommendations task (Sect. 4.3.3) when representing 718 SN books in the 2012–2014 period with the ontology produced by including 40 additional topics to PMC. α is directly correlated with the recommender performance, yielding a Pearson correlation coefficient of 0.69 ($p < 0.0001$). It thus seems that mapping instances with the semantic approach works better when optimizing the ontology for this task. Although, it is interesting to notice that the best results are obtained when $0.5 \leq \alpha \leq 0.75$, therefore a purely semantic approach may be counterproductive. Conversely β exhibits a mild inverse correlation with the performance, yielding a Pearson correlation coefficient of -0.36 ($p < 0.0001$). This indicates that preferring the instances from the internal dataset tends to produce a superior result on this task.

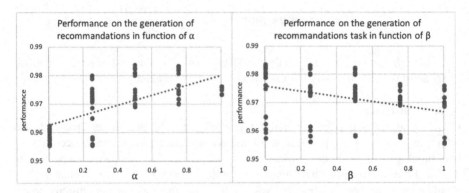

Fig. 1. Performance on the generation of recommendations task in function of α and β.

4.5 Recursive Concept Elimination

The previous step can outperform some more basic methods (see Sect. 5), but still suffers from two main limitations. First, the optimization was limited to the space of parameters, therefore a better solution may exist outside this space. Secondly, the typical strategy of assigning weights to single concepts does not take into consideration concept synergy. Conversely, it is possible that even if concept C_1 has lower weight than C_2, its combination with the other concepts would yield a better overall performance. For instance, two concepts may be redundant (e.g., "Linked Data" and "RDF"), therefore after one of them is selected, adding also the other would yield only a marginal advantage. In this section, we introduce a technique that addresses both limitations.

A comprehensive search outside the space of parameters is computationally intractable since it would need to test all possible permutations. For this reason, performing feature selection in large dimensional input spaces usually involves greedy algorithms. An approach to address this issue in the field of machine learning is the Recursive Feature Elimination algorithm [31], often used with Support Vector Machines and other classifiers. This approach iteratively constructs a model with a set of features, computes their weights, and removes the least important features, until the goal is reached. A crucial advantage of this method is that it takes into account the feature synergy and preserves features whose usefulness requires other features.

We thus adopted a similar procedure, that we label *Recursive Concept Elimination* (*RCE*), as the second step of POE. RCE generates an ontology composed of n concepts by applying the following steps:

1. It produces an ontology with m concepts (where $m > n$) using the best set of parameters detected in the first phase (Sect. 4.4). If no ontology of m concepts complies with the requirements, these are temporarily relaxed.
2. It ranks the concepts according to their importance for the tasks by generating $m - 1$ representations of the instances, each of them lacking a concept, and evaluating them. Each concept is given a weight equal to 1 minus the metric yielded by the evaluation of the representation from which it is absent [32].
3. It discards the j concepts with the smaller weights and returns to step 2, until it reaches n concepts. Finally, it returns the optimized ontology and the ranked set of parameters from the previous phase.

While it is technically possible to directly apply RCE to the full set of candidate concepts, it would not be computationally feasible in most cases. Using the set of best parameters to create an initial ontology of m concepts allows us to obtain a tractable number of RCE iterations.

A further advantage of this method is that it allows users to understand exactly why a concept is there and in which way it relates with the dimensions discussed in Sect. 4.2 and with its performance on a task. Indeed, the ranking order will still be consistent with the set of parameters selected in the first phase and the absence of a concept from the original ranked list would be due to its insufficient performance with regard to the task. The user is thus able to review this information and test different solutions by modulating the input parameters.

5   Evaluation

We tested POE on the task of evolving the PMC taxonomy and used as instances a dataset of Springer Nature publications including 1,218 books in the 2012–2016 period. The evaluation had three aims. First, we wanted to compare POE versus alternative baselines from the state of the art, such as the TF-IDF method adopted in Text2Onto [14] and SPRAT [15]. Secondly, we intended to investigate whether optimizing for a certain task would also yield good performance on related ones. Finally, we intended to assess the effect of training POE on multiple tasks at once.

We thus compared the performance of the ontologies produced by different approaches in supporting the four tasks implemented in POE: automatic tagging (**Task 1**), similarity computation (**Task 2**), generation of recommendations (**Task 3**), and clustering (**Task 4**). In addition to the SN dataset, we adopted the Rexplore dataset [23] as the external source from which to derive statistics, such as the concepts frequencies described in Sect. 4.2 and TF-IDF. The Rexplore dataset is more generic than the SN one and contains 16 million research papers in the field of Computer Science from a variety of academic publishers.

We focused on the evolution of the branches under five concepts of the original PMC taxonomy: *I21017-Artificial Intelligence (incl. Robotics), I23050-Computational Biology/Bioinformatics, I14050-Systems and Data Security, I14029-Software Engineering*, and *I13022-Computer Communication Networks*. These concepts were mapped to nine CSO concepts: Artificial Intelligence, Robotics, Bioinformatics,

Cryptography, Access Control, Software Engineering, Software Design, Computer Networks, and Wireless Telecommunication Systems. Finally, their 2,451 sub-topics were selected as candidate concepts.

We tested fourteen alternative approaches:

- Term Frequency in the SN dataset (**FS**), ranking concepts according to their frequency in SN dataset.
- Term Frequency in the Rexplore dataset (**FR**).
- TF-IDF in the SN dataset (**TS**) (as in [14, 15]), considering the instances under the five branches for the TF and all the instances for the IDF.
- TF-IDF in the Rexplore dataset (**TR**).
- The parameter optimization in POE (Sect. 4.4), yielding the ontology produced from the best combination of parameters for instance tagging (**P1**), similarity computation (**P2**), generation of recommendations (**P3**), clustering (**P4**), and all these tasks together (**P5**).
- The full POE framework returning an ontology optimized for instance tagging (**POE1**), similarity computation (**POE2**), generation of recommendations (**POE3**), clustering (**POE4**), and all these tasks together (**POE5**).

We simulated a realistic situation by training the approaches and computing all the statistics (e.g., TF-IDF) in the 2012–2014 period and then evaluating their performance in the 2015–2016 period. In order to do so, we split the instances dataset in a training set of 718 books and a testing set of 500 books.

We then generated, for each approach, four evolved versions of PMC that included 20, 40, 60, and 80 new concepts and compared their performance using the metrics described in Sect. 4.3. The minimum and maximum number of concepts allowed for each of the five branches was set respectively to 4 and 25. RCE was performed by setting $m = n + 20$ and eliminating one concept at each iteration. POE was implemented in Python and ran on a 2.40 GHz Intel Xeon processor taking between 1 (POE1) and 8 (POE5) hours depending on the task. The computing time is usually not an issue for this kind of task, but if needed it could be cut down by parallelising the parameter optimization and the RCE phase. The existence of statistical differences between the two approaches was explored with the non-parametric Wilcoxon's signed rank test for matched variables.

The material produced during the evaluation and further details about the settings of the approaches are available at http://rexplore.kmi.open.ac.uk/POE.

Tables 1 and 2 show the performance of the approaches on the four tasks. The full version of POE optimized for a task (e.g., POE1 for task 1) obtained the best average result for the task in every case, outperforming both parameter optimization (p = 0.002 with Wilcoxon's rank test), and the other baselines (p = 0.0004). It also obtained the best result for each concept number, with the exception of few cases in which it was outranked by a different version of POE optimized for a similar task. POE5, the version optimized on all tasks at once, proved to be a good compromise by yielding on each task a performance marginally inferior or equal (in case of task 3) to the version of POE specifically optimized for the task (p $\geq$ 0.10). In addition, the parameter optimization step optimized for a task (e.g., P1 for task 1) yielded better results than FS, FR, TS, TR on that same task (p = 0.0004).

Table 1. Performance in task 1 (instance tagging) on the left and task 2 (similarity computation) on the right. In bold the best results. In light grey the version of POE optimized for the task.

	Task 1 (Instance Tagging)					Task 2 (Similarity Computation)				
	20	40	60	80	aver.	20	40	60	80	aver.
FS	0.794	0.946	0.970	0.978	0.922	0.815	0.862	0.882	0.886	0.861
FR	0.818	0.896	0.924	0.972	0.903	0.820	0.852	0.862	0.898	0.858
TS	0.806	0.908	0.944	0.972	0.908	0.821	0.853	0.875	0.882	0.858
TR	0.808	0.916	0.918	0.938	0.895	0.821	0.859	0.867	0.886	0.859
P1	0.858	0.978	**0.990**	0.992	0.955	0.843	0.906	0.933	0.945	0.907
P2	0.858	0.978	**0.990**	0.992	0.955	0.843	0.906	0.934	0.945	0.907
P3	0.872	0.978	0.986	0.992	0.957	0.842	0.906	0.933	0.944	0.906
P4	0.856	0.956	0.978	0.992	0.946	0.842	0.883	0.931	0.945	0.900
P5	0.858	0.982	**0.990**	0.992	0.956	0.843	0.903	0.934	0.945	0.906
POE1	0.910	**0.988**	0.984	**0.994**	**0.969**	0.857	0.899	0.932	0.922	0.903
POE2	0.898	0.986	0.986	0.992	0.966	0.863	**0.913**	**0.941**	**0.947**	**0.916**
POE3	0.906	0.982	**0.990**	0.992	0.968	0.861	**0.913**	0.939	0.946	0.915
POE4	0.834	0.974	0.984	0.988	0.945	0.842	0.899	0.933	0.944	0.904
POE5	0.904	**0.988**	0.984	0.992	0.967	**0.864**	0.907	0.938	0.946	0.914

Table 2. Performance in task 3 (generation of recommendations) on the left and task 4 (clustering) on the right. In bold the best results. In light grey the version of POE optimized for the task.

	Task 3 (Generation of Recommendations)					Task 4 (Clustering)				
	20	40	60	80	aver.	20	40	60	80	aver.
FS	0.940	0.957	0.965	0.966	0.957	0.940	0.866	0.950	0.949	0.926
FR	0.943	0.953	0.957	0.974	0.957	0.943	0.868	0.850	0.962	0.906
TS	0.941	0.954	0.964	0.966	0.956	0.941	0.925	0.887	0.892	0.911
TR	0.941	0.957	0.959	0.971	0.957	0.941	0.894	0.934	0.954	0.931
P1	0.949	0.979	0.989	0.992	0.978	0.937	0.968	0.983	**0.994**	0.970
P2	0.949	0.979	0.990	0.992	0.978	0.937	0.968	0.983	**0.994**	0.970
P3	0.951	0.979	0.989	0.992	0.978	0.939	0.968	0.983	0.982	0.968
P4	0.950	0.970	0.988	0.992	0.975	0.944	0.963	0.977	**0.994**	0.969
P5	0.949	0.978	0.990	0.992	0.977	0.937	0.975	0.983	**0.994**	0.972
POE1	0.958	0.978	0.989	0.980	0.976	0.945	0.977	0.909	0.961	0.948
POE2	0.962	0.981	0.990	0.992	0.981	0.949	0.973	0.982	**0.994**	0.974
POE3	0.963	**0.982**	**0.991**	**0.993**	**0.982**	0.946	**0.979**	0.984	0.987	0.974
POE4	0.953	0.976	0.989	**0.993**	0.978	**0.951**	0.971	**0.985**	**0.994**	**0.975**
POE5	**0.964**	0.981	**0.991**	0.992	**0.982**	0.947	0.968	0.983	0.994	0.973

Furthermore, all the approaches optimized on one the four tasks (including POE5) performed significantly better ($p < 0.0001$) than the ones that simply used statistical techniques. Therefore, it seems that optimizing for one of these tasks holds benefits also on the other ones.

POE3, POE2, and POE5 yielded very good results on all tasks, obtaining the highest average performances, respectively 0.960, 0.959 and 0.959. Interestingly, the performance of POE2 and POE3 on task 4 (clustering) was only slightly inferior to POE4, while the performance of POE4 on task 2 and 3 was not as good. This is probably due to the fact that both task 2 and task 3 concern the similarity between instances, which is also used by K-Means++ for producing the cluster set.

6   Conclusions

We presented the Pragmatic Ontology Evolution (POE) framework, a novel approach that selects concepts to be included in an evolving ontology in accordance with user requirements and their impact on computational tasks. The evaluation showed that the full version of POE outperforms both parameter optimization (p = 0.002) and the other baselines (p = 0.0004).

While POE was initially conceived in the context of tackling a concrete real-world ontology evolution problem, the approach is generally applicable and opens up many interesting avenues of work. In particular, we intend to apply POE on different kinds of ontologies and computational tasks to derive some useful guidelines on how to balance users and application needs. We also intend to further enrich POE by allowing it to handle more complex candidate changes, involving different kinds of semantic relationships. Finally, on the technology transfer side, we will continue our collaboration with Springer Nature, with the aim of supporting its deployment within the editorial team, thus providing a powerful and user-friendly solution to facilitate the process of maintaining and evolving their editorial ontologies.

Acknowledgements. We would like to thank Springer DE for providing us with access to their large repositories of scholarly data.

References

1. Ding, L., Kolari, P., Ding, Z., Avancha, S.: Using ontologies in the semantic web: a survey. In: Sharman, R., Kishore, R., Ramesh, R. (eds.) Ontologies. ISIS, vol. 14, pp. 79–113. Springer, Boston (2007). https://doi.org/10.1007/978-0-387-37022-4_4
2. Osborne, F., Salatino, A., Birukou, A., Motta, E.: Automatic classification of Springer Nature proceedings with smart topic miner. In: Groth, P., et al. (eds.) ISWC 2016. LNCS, vol. 9982, pp. 383–399. Springer, Cham (2016). https://doi.org/10.1007/978-3-319-46547-0_33
3. Middleton, S.E., De Roure, D., Shadbolt, N.R.: Ontology-based recommender systems. In: Staab, S., Studer, R. (eds.) Handbook on Ontologies. INFOSYS, pp. 779–796. Springer, Berlin (2009). https://doi.org/10.1007/978-3-540-24750-0_24
4. Hotho, A., Staab, S., Stumme, G.: Ontologies improve text document clustering. In: Data Mining, ICDM 2003. IEEE (2003)
5. Osborne, F., Scavo, G., Motta, E.: Identifying diachronic topic-based research communities by clustering shared research trajectories. In: Presutti, V., d'Amato, C., Gandon, F., d'Aquin, M., Staab, S., Tordai, A. (eds.) ESWC 2014. LNCS, vol. 8465, pp. 114–129. Springer, Cham (2014). https://doi.org/10.1007/978-3-319-07443-6_9

6. Osborne, F., Mannocci, A., Motta, E.: Forecasting the spreading of technologies in research communities. In: K-CAP 2017, Austin, Texas, USA (2017)
7. Zablith, F., et al.: Ontology evolution: a process-centric survey. Knowl. Eng. Rev. **30**(1), 45–75 (2015)
8. Stojanovic, L.: Methods and tools for ontology evolution (2004)
9. Klein, M., Noy, N.F.: A component-based framework for ontology evolution. In: Workshop on Ontologies and Distributed Systems at IJCAI, vol. 3, p. 4 (2003)
10. Noy, N.F., Chugh, A., Liu, W., Musen, M.A.: A framework for ontology evolution in collaborative environments. In: Cruz, I., et al. (eds.) ISWC 2006. LNCS, vol. 4273, pp. 544–558. Springer, Heidelberg (2006). https://doi.org/10.1007/11926078_39
11. Zablith, F.: Evolva: a comprehensive approach to ontology evolution. In: Aroyo, L., et al. (eds.) ESWC 2009. LNCS, vol. 5554, pp. 944–948. Springer, Heidelberg (2009). https://doi.org/10.1007/978-3-642-02121-3_87
12. Ristoski, P., Paulheim, H.: Semantic web in data mining and knowledge discovery: a comprehensive survey. Web Seman.: Sci. Serv. Agents World Wide Web **36**, 1–22 (2016)
13. Thanapalasingam, T., Osborne, F., Birukou, A., Motta, E.: Ontology-based recommendation of editorial products. In: International Semantic Web Conference 2018, Monterey, CA, USA (2018)
14. Cimiano, P., Völker, J.: Text2Onto. In: Montoyo, A., Muñoz, R., Métais, E. (eds.) NLDB 2005. LNCS, vol. 3513, pp. 227–238. Springer, Heidelberg (2005). https://doi.org/10.1007/11428817_21
15. Maynard, D., Funk, A., Peters, W.: SPRAT: a tool for automatic semantic pattern-based ontology population. In: International Conference for Digital Libraries and the Semantic Web, Trento, Italy (2009)
16. Novacek, V., Handschuh, S.: Semi-automatic integration of learned ontologies into a collaborative framework (2007)
17. Huang, Z., Stuckenschmidt, H.: Reasoning with multi-version ontologies: a temporal logic approach. In: Gil, Y., Motta, E., Benjamins, V.R., Musen, M.A. (eds.) ISWC 2005. LNCS, vol. 3729, pp. 398–412. Springer, Heidelberg (2005). https://doi.org/10.1007/11574620_30
18. Xuan, D.N., Bellatreche, L., Pierra, G.: A versioning management model for ontology-based data warehouses. In: Tjoa, A.M., Trujillo, J. (eds.) DaWaK 2006. LNCS, vol. 4081, pp. 195–206. Springer, Heidelberg (2006). https://doi.org/10.1007/11823728_19
19. Wang, Y., Liu, X., Ye, R.: Ontology evolution issues in adaptable information management systems. In: 2008 IEEE International Conference on e-Business Engineering, ICEBE 2008, pp. 753–758. IEEE (2008)
20. Liang, Y., Alani, H., Shadbolt, N.: Changing ontology breaks queries. In: Cruz, I., et al. (eds.) ISWC 2006. LNCS, vol. 4273, pp. 982–985. Springer, Heidelberg (2006). https://doi.org/10.1007/11926078_79
21. Kondylakis, H., Plexousakis, D.: Ontology evolution: assisting query migration. In: Atzeni, P., Cheung, D., Ram, S. (eds.) ER 2012. LNCS, vol. 7532, pp. 331–344. Springer, Heidelberg (2012). https://doi.org/10.1007/978-3-642-34002-4_26
22. Osborne, F., Motta, E.: Klink-2: integrating multiple web sources to generate semantic topic networks. In: Arenas, M., et al. (eds.) ISWC 2015. LNCS, vol. 9366, pp. 408–424. Springer, Cham (2015). https://doi.org/10.1007/978-3-319-25007-6_24
23. Osborne, F., Motta, E., Mulholland, P.: Exploring scholarly data with rexplore. In: Alani, H., et al. (eds.) ISWC 2013. LNCS, vol. 8218, pp. 460–477. Springer, Heidelberg (2013). https://doi.org/10.1007/978-3-642-41335-3_29
24. Salatino, A.A., Thanapalasingam, T., Mannocci, A., Osborne, F., Motta, E.: The computer science ontology: a large-scale taxonomy of research areas. In: International Semantic Web Conference 2018, Monterey, CA, USA (2018)

25. Klein, M.C., Fensel, D.: Ontology versioning on the Semantic Web. In: SWWS, pp. 75–91 (2001)
26. Sellami, Z., Camps, V., Aussenac-Gilles, N.: DYNAMO-MAS: a multi-agent system for ontology evolution from text. J. Data Semant. 2(2–3), 145–161 (2013)
27. Sabou, M., Fernandez, M., Motta, E.: Evaluating semantic relations by exploring ontologies on the semantic web. In: Horacek, H., Métais, E., Muñoz, R., Wolska, M. (eds.) NLDB 2009. LNCS, vol. 5723, pp. 269–280. Springer, Heidelberg (2010). https://doi.org/10.1007/978-3-642-12550-8_22
28. Qin, L., Atluri, V.: Evaluating the validity of data instances against ontology evolution over the semantic web. Inf. Softw. Technol. 51(1), 83–97 (2009)
29. Hartung, M., Kirsten, T., Rahm, E.: Analyzing the evolution of life science ontologies and mappings. In: Bairoch, A., Cohen-Boulakia, S., Froidevaux, C. (eds.) DILS 2008. LNCS, vol. 5109, pp. 11–27. Springer, Heidelberg (2008). https://doi.org/10.1007/978-3-540-69828-9_4
30. Groß, A., Hartung, M., Prüfer, K., Kelso, J., Rahm, E.: Impact of ontology evolution on functional analyses. Bioinformatics 28(20), 2671–2677 (2012)
31. Guyon, I., Weston, J., Barnhill, S., Vapnik, V.: Gene selection for cancer classification using support vector machines. Mach. Learn. 46(1–3), 389–422 (2002)
32. Kohavi, R., John, G.H.: Wrappers for feature subset selection. Artif. Intell. 97(1–2), 273–324 (1997)

Towards Empty Answers in SPARQL: Approximating Querying with RDF Embedding

Meng Wang[1(✉)], Ruijie Wang[2], Jun Liu[3], Yihe Chen[4], Lei Zhang[5], and Guilin Qi[6]

[1] MOEKLINNS Lab, Xi'an Jiaotong University, Xi'an, China
wangmengsd@stu.xjtu.edu.cn
[2] School of Electronic and Information Engineering, Xi'an Jiaotong University,
Xi'an, China
xjdwrj@stu.xjtu.edu.cn
[3] Guang Dong Xi'an Jiaotong University Academy, Shunde, China
liukeen@xjtu.edu.cn
[4] University of Toronto, Toronto, Canada
[5] FIZ Karlsruhe – Leibniz Institute for Information Infrastructure,
Karlsruhe, Germany
[6] School of Computer Science and Engineering, Southeast University, Nanjing, China

Abstract. The LOD cloud offers a plethora of RDF data sources where users discover items of interest by issuing SPARQL queries. A common query problem for users is to face with empty answers: given a SPARQL query that returns nothing, how to refine the query to obtain a non-empty set? In this paper, we propose an RDF graph embedding based framework to solve the SPARQL empty-answer problem in terms of a continuous vector space. We first project the RDF graph into a continuous vector space by an entity context preserving translational embedding model which is specially designed for SPARQL queries. Then, given a SPARQL query that returns an empty set, we partition it into several parts and compute approximate answers by leveraging RDF embeddings and the translation mechanism. We also generate alternative queries for returned answers, which helps users recognize their expectations and refine the original query finally. To validate the effectiveness and efficiency of our framework, we conduct extensive experiments on the real-world RDF dataset. The results show that our framework can significantly improve the quality of approximate answers and speed up the generation of alternative queries.

Keywords: SPARQL · Empty-answer · RDF · Graph embedding

1 Introduction

With the rapid development of Semantic Web technologies, various knowledge bases are published on the Linked Open Data (LOD) cloud using Resource

© Springer Nature Switzerland AG 2018
D. Vrandečić et al. (Eds.): ISWC 2018, LNCS 11136, pp. 513–529, 2018.
https://doi.org/10.1007/978-3-030-00671-6_30

Description Framework (RDF). To enable users to retrieve the desired data, many RDF datasets provide SPARQL endpoints that allow users to issue basic graph pattern (BGP) queries [11]. However, issuing SPARQL queries requires users to be precisely aware of the structure and schema of the RDF dataset, and this is challenging. Therefore, it is a common scenario for users where an inappropriate query returns an empty set, the so-called empty-answer problem.

Most existing work solves this problem through query relaxation approaches [6–9,12–14] which focus on relaxing RDF terms specified in the original query so that the new relaxed query returns sufficient answers. They find top-k optimal relaxed queries in the exponential search space then evaluate the matching process between the relaxed queries and the RDF graph, which is really time-consuming. **Is it possible to directly retrieve approximate answers for a failing query without changing any parts of the original query? The answer is yes.** In this paper, we stand on recent advances in RDF embedding techniques and address the SPARQL empty-answer problem from the angle of continuous vector space.

Motivating Example: A user wants to find drama films which were released in the United States and directed by Tim Burton. After issuing a SPARQL query over DBpedia [16], the user obtains an empty answer, as shown on the left of Fig. 1. In this example, *Tim Burton, director, country, United States, type,* and *drama films* are specified RDF terms in the SPARQL BGP [11]. Since each specified term must be matched, this is too restrictive for the graph pattern matching. To deal with such failing queries, users usually have no idea which parts of the query should be responsible for the missing possible answers.

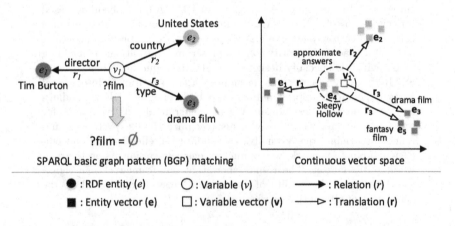

Fig. 1. Failing SPARQL BGP and RDF graph embeddings.

The right of Fig. 1 illustrates the ideal vector representation (i.e., the embedding) of the RDF graph to be queried in a continuous vector space, where entities are represented by vectors (boldface letters), and semantically similar entities are close to each other. For example, e_3 and e_5 are close since *drama film*(e_3) and *fantasy film*(e_5) are similar. The relation between two entities is represented as a translation operation from the head entity to the tail entity, e.g., $\mathbf{e_4} + \mathbf{r_3} \approx \mathbf{e_5}$ when a triple $\langle$*Sleepy Hollow*(e_4), *type*(r_3), *fantasy film*(e_5)$\rangle$ holds. With the translation mechanism, although the expected answer v_1 does not exist in the RDF graph, we can still compute its vector representation $\mathbf{v_1}$ based on specified terms in the SPARQL, e.g., $\mathbf{v_1} \approx \mathbf{e_1} - \mathbf{r_1}$ according to $\langle v_1, r_1, e_1 \rangle$. Then, we can obtain $\mathbf{e_4}$ which is close to $\mathbf{v_1}$ in the space. *Sleepy Hollow*(e_4) is likely to meet the query intention of the original SPARQL query in the RDF graph.

Challenges: Leveraging RDF embeddings is a promising pathway to directly find approximate answers for a failing SPARQL query, but it is also troubling. Our method confronts the following challenges:

- *Limitations of existing embedding models*: We need to project the RDF graph into a continuous vector space, where semantically similar entities are close to each other and the relations among entities are represented by translations. However, none of the existing embedding models (e.g., RDF2vec [20], TransE [3], et al.) meet the requirements.
- *Variety of BGPs*: A BGP may contain multiple different variables and usually contains several triple patterns sharing the same variables or entities. Therefore, how to exploit the interplay between triple patterns and compute approximate answers for each variable is challenging and non-trivial.
- *Comprehensibility of approximate answers*: Obtaining approximate answers without any explanations is inadequate for satisfying users because they may ask why an approximate answer is returned.

Solutions: Given these challenges, we propose a novel framework to address the SPARQL empty-answer problem. The procedure includes the following:

- Firstly, the RDF graph is projected into a continuous vector space by an entity context preserving translational embedding model which is specially designed for SPARQL BGPs.
- Then, given a SPARQL query that returns an empty answer, the SPARQL BGP is partitioned into several parts based on different variables. By leveraging the RDF embeddings and the translation mechanism, approximate answers are further computed based on the vector representations of variables and specified query terms.
- Finally, approximate answers are returned to users. Each returned answer will be attached with an alternative query, which helps users recognize their expectations and refine the original query.

Contributions: Our framework makes the following contributions:

- To the best of our knowledge, we are the first to solve the SPARQL empty-answer problem from a continuous vector space perspective, which improves the quality of the returned answers and speeds up the generation of alternative queries.

- We propose a novel RDF graph embedding model which utilizes the translation mechanism to capture the relations between entities while considering the entity context to make the representations of semantically similar entities close to each other in the vector space.
- We propose efficient algorithms to compute approximate answers attached with alternative queries as the explanations for users.
- We conduct extensive experiments on the real-world dataset to evaluate the effectiveness and efficiency of the framework. These results provide supporting evidence that the framework is powerful in generating effective answers and explanations for failing queries within an acceptable time.

Organization: The remainder of the paper is organized as follows. Section 2 presents the details of the proposed framework. The evaluation of the framework is reported in Sect. 3. The related work is discussed in Sect. 4. Finally, our conclusions and future work are presented in Sect. 5.

2   The Proposed Framework

Before demonstrating the details of our framework, we briefly introduce the notations employed in this paper.

RDF Graph. Let $\mathcal{E}$ be a set of entities, $\mathcal{R}$ be a set of relations between entities. An RDF graph $\mathcal{G} = (\mathcal{E}, \mathcal{R})$ is a finite set of RDF triples in the form $\langle e_h, r, e_t \rangle$, where $e_h, e_t \in \mathcal{E}$ and $r \in \mathcal{R}$. An RDF triple $\langle e_h, r, e_t \rangle$ indicates a directed relation r from the head entity e_h to the tail entity e_t, e.g., $\langle Batman, director, Tim\ Burton \rangle$.

The standard SPARQL [11] contains BGPs and other operations (*UNION*, *OPTIONAL*, *FILTER*, etc.). In this paper, we focus on the SPARQL emptyanswer problem caused by over-constrained BGPs, which already yields a nontrivial problem to study.

BGP. Let $\mathcal{V}$ be a set of entity variables. Each variable $v \in \mathcal{V}$ is distinguished by a leading question mark symbol, e.g., *?film*. A triple pattern is similar to an RDF triple but allows the usage of variables for entities[1], e.g., $\langle$ *?film, director, Tim Burton* $\rangle$. A SPARQL BGP $\mathcal{P} = (\mathcal{E}_\mathcal{P} \cup \mathcal{V}_\mathcal{P}, \mathcal{R}_\mathcal{P})$ is a finite set of triple patterns, where $\mathcal{E}_\mathcal{P} \subseteq \mathcal{E}$, $\mathcal{V}_\mathcal{P} \subseteq \mathcal{V}$ and $\mathcal{R}_\mathcal{P} \subseteq \mathcal{R}$.

SPARQL Query. The official standard [11] defines four different forms of queries on the top of BGPs, namely *SELECT*, *ASK*, *CONSTRUCT*, and *DESCRIBE*. Since *SELECT* is the only form which returns the graph matching results to users, we define a SPARQL query Q as an expression of the form *SELECT S FROM $\mathcal{G}$ WHERE $\mathcal{P}$*, where $\mathcal{P}$ is a BGP, $\mathcal{G}$ is an RDF graph to be queried, and $S \subseteq \mathcal{V}_\mathcal{P}$.

---

[1] To simplify the problem, we do not consider variables on predicates in this paper as such graph patterns are mainly used for exploring RDF schema but rarely used in real-world SPARQL queries [2].

SPARQL Empty-Answer Problem. Given a SPARQL query *SELECT S FROM G WHERE P*, *P* is evaluated to match *G*, and the variables in *P* are substituted by entities in *G*. *SELECT S* employs the matched RDF graphs to provide the final result set *RS*. The SPARQL empty-answer problem refers to the scenario where the final result set is empty, i.e., $RS = \varnothing$.

Given a failing SPARQL query, our goal is to automatically generate top-k answers which approximately meet the original query intention along with corresponding alternative queries. Different from the existing methods [6–9,12–14], our framework solves the empty-answer problem based on a continuous vector space, as introduced in the example of Sect. 1. The proposed framework mainly includes three modules: learning RDF embeddings (described in Sect. 2.1), computing variable embeddings (described in Sect. 2.2), as well as generating approximate answers and alternative queries (described in Sect. 2.3).

2.1 Learning RDF Embeddings

In this module, we aim to embed entities and relations of the underlying RDF graph into a continuous vector space while preserving the inherent structure of the graph. Neural-language-based models, e.g., RDF2vec [20], only generate entity latent representations, and they cannot encode relations between entities. Therefore, we adopt the translation mechanism of TransE [3] to capture the correlations between entities and relations[2]. The translation mechanism in this context represents a relation as a translation operation from the head entity to the tail entity in the continuous vector space. Specifically, if an RDF triple $\langle e_h, r, e_t \rangle \in G$, our objective is to learn embeddings $\mathbf{e_h}$, $\mathbf{r}$ and $\mathbf{e_t}$ which hold $\mathbf{e_h} + \mathbf{r} \approx \mathbf{e_t}$ ($\mathbf{e_t}$ should be a nearest neighbor of $\mathbf{e_h} + \mathbf{r}$). However, directly adopting TransE does not guarantee that semantically similar entities are close to each other in the continuous vector space since it regards an RDF graph as a set of independent triples during the learning process. Triples in the RDF graph are not independent, and semantically similar entities tend to share common context information, e.g., neighboring entities and their associated relations. Therefore, we propose a novel embedding method which considers the entity context information during the translation-based learning process.

Definition 1 (Entity Context). *For an entity $e \in \mathcal{E}$, the context of e is a set $C(e) = \{(r_c, e_c) | r_c \in \mathcal{R}, e_c \in \mathcal{E}, \langle e, r_c, e_c \rangle \in G \vee \langle e_c, r_c, e \rangle \in G\}$, where r_c is the relation between e and its neighbor e_c.*

Given an entity $e \in \mathcal{E}$, our goal is to learn the vector representation $\mathbf{e}$ while preserving its entity context information. To this end, we first define the conditional probability of e given its context $C(e)$ as follows:

$$P(e|C(e)) = \frac{\exp(f_1(e, C(e)))}{\sum_{e' \in \mathcal{E}} \exp(f_1(e', C(e)))}, \tag{1}$$

---

[2] Other translation-based embedding models, such as TransH [21] and TransR [17], can also be easily adopted for the RDF triple encoding.

where $f_1(e', C(e))$ is a score function that measures the correlation between an arbitrary entity e' and the entity context of e. We define $f_1(e', C(e))$ as:

$$f_1(e', C(e)) = -\frac{1}{|C(e)|} \sum_{(r_c, e_c) \in C(e)} f_2(e', r_c, e_c), \tag{2}$$

where $f_2(e', r_c, e_c)$ is the score function of TransE that measures the correlation between e' and $(r_c, e_c) \in C(e)$. $f_2(e', r_c, e_c)$ is formulated as follows:

$$f_2(e', r_c, e_c) = \begin{cases} \|\mathbf{e'} + \mathbf{r_c} - \mathbf{e_c}\|_2^2, & \text{if } \langle e, r_c, e_c \rangle \in \mathcal{G}, \\ \|\mathbf{e_c} + \mathbf{r_c} - \mathbf{e'}\|_2^2, & \text{if } \langle e_c, r_c, e \rangle \in \mathcal{G}, \end{cases} \tag{3}$$

where $\langle e, r_c, e_c \rangle$ and $\langle e_c, r_c, e \rangle$ indicate the directions of r_c. Intuitively, if two entities share more common context information, their embeddings tend to be more similar according to the above equations. By maximizing the joint probability of all entities in $\mathcal{G}$, we define the final objective function as:

$$O = \sum_{e \in \mathcal{E}} \log P(e | C(e)). \tag{4}$$

Considering the over-sized RDF graph, it is impractical to directly compute Eq. (1). Hence, we follow [18] to approximate Eq. (1) based on negative sampling, as formulated in Eq. (5).

$$P(e | C(e)) \approx \sigma(f_1(e, C(e))) \cdot \prod_{e' \in \mathcal{E}_\mathcal{N}^e}^{n} \sigma(f_1(e', C(e))), \tag{5}$$

where n is the number of negative examples, $\sigma(\cdot)$ is the sigmoid function, and e' is the negative entity which is obtained by sampling entities from a uniform distribution over the negative entity set $\mathcal{E}_\mathcal{N}^e$. For each negative entity $e' \in \mathcal{E}_\mathcal{N}^e$, the precondition is $C(e') \bigcap C(e) = \varnothing$.

Based on the above process, entities are encoded into the continuous vector space with their context information such that semantically similar entities are close to each other. The relations between entities are simultaneously captured by the translation mechanism. It is worth mentioning that the generation of embeddings is independent of the rest phases of the framework. Once the RDF embeddings have been learned, we can use them in SPARQL empty-answer problems solving without frequent modification.

2.2   Computing Variable Embeddings

We assume that the initial SPARQL is free of spelling/syntactic errors. In this module, we aim to compute embeddings of variables in the original SPARQL query. Similar to the correlation between an entity and its context, a variable is determined by its neighbors in the BGP of the initial query.

The neighbor of a variable could be a specified entity or any variable else (a BGP may contain multiple variables). Given a failing SPARQL query, we first partition its BGP into several sub-basic graph patterns (sBGPs), each of which contains only one variable connected with a set of specified entities.

Definition 2 (Sub-Basic Graph Pattern). *Given a basic graph pattern* $\mathcal{P} = (\mathcal{E}_\mathcal{P} \cup \mathcal{V}_\mathcal{P}, \mathcal{R}_\mathcal{P})$, *a sub-basic graph pattern (sBGP) for a variable* $v_s \in \mathcal{V}_\mathcal{P}$ *is a set* $\mathcal{SP} = \{\langle v_s, r_s, e_s\rangle | r_s \in \mathcal{R}_\mathcal{P}, e_s \in \mathcal{E}_\mathcal{P}, \langle v_s, r_s, e_s\rangle \in \mathcal{P}\} \cup \{\langle e_s, r_s, v_s\rangle | r_s \in \mathcal{R}_\mathcal{P}, e_s \in \mathcal{E}_\mathcal{P}, \langle e_s, r_s, v_s\rangle \in \mathcal{P}\}$, *where* r_s *is a relation between* v_s *and its neighbor* e_s.

For instance, the left of Fig. 2 illustrates a SPARQL BGP that retrieves American drama films directed by Tim Burton and there is a star actor who was born in New York. We can partition the BGP into two sBGPs, i.e., $sBGP_1$ and $sBGP_2$ for *?film* and *?actor*, respectively.

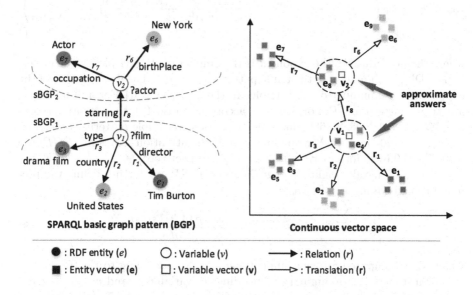

Fig. 2. Failing SPARQL basic graph pattern and RDF graph embeddings.

Then, given a variable v_s and the corresponding sBGP $\mathcal{SP}$, we utilize specified entities in $\mathcal{SP}$ to estimate the embedding of v_s. For a single triple pattern $\langle v_s, r_s, e_s\rangle$ or $\langle e_s, r_s, v_s\rangle$ in $\mathcal{SP}$, we can obtain a preliminary embedding $\tilde{\mathbf{v}}_\mathbf{s}$, computed as follows:

$$\tilde{\mathbf{v}}_\mathbf{s} = \begin{cases} \mathbf{e_s} - \mathbf{r_s}, & \text{if } \langle v_s, r_s, e_s\rangle \in \mathcal{SP}, \\ \mathbf{e_s} + \mathbf{r_s}, & \text{if } \langle e_s, r_s, v_s\rangle \in \mathcal{SP}. \end{cases} \tag{6}$$

For instance, regarding the variable *?film* (v_1) shown in Fig. 2, we can utilize the triple pattern $\langle$*?film, director, Tim Burton*$\rangle$, i.e., $\langle v_1, r_1, e_1\rangle$, to obtain the preliminary embedding of v_1 according to Eq. (6), i.e., $\tilde{\mathbf{v}}_\mathbf{1} = \mathbf{e_1} - \mathbf{r_1}$ in the continuous vector space.

In the sBGP $\mathcal{SP}$, if the variable v_s is involved in multiple triple patterns, we need to jointly consider these triple patterns in the estimation of the variable embedding. Intuitively, different triple patterns may have different impacts on determining the variable embedding. For example, $sBGP_1$ in Fig. 2 consists of three triple patterns, i.e., $\langle ?film,\ director,\ Tim\ Burton\rangle$, $\langle ?film,\ country,\ United\ States\rangle$, and $\langle ?film,\ type,\ drama\ film\rangle$. In the underlying RDF graph, e.g., DBpedia, there are respectively 24, 112,336, and 238 RDF triples matching the three triple patterns. Therefore, the triple pattern $\langle ?film,\ director,\ Tim\ Burton\rangle$ contains more specific information for estimating the variable embedding of $?film$ compared with the other two triple patterns, and it should attract more attention. We define the following attention function $a(v_s, r_s, e_s)$ to measure the attention on a triple pattern $\langle v_s, r_s, e_s\rangle$ when estimating the embedding of v_s.

$$a(v_s, r_s, e_s) = \exp\left(-\frac{|\{e_{v_s}|\langle e_{v_s}, r_s, e_s\rangle \in \mathcal{G} \vee \langle e_s, r_s, e_{v_s}\rangle \in \mathcal{G}\}|}{\sum_{\langle v_s, r'_s, e'_s\rangle \in \mathcal{SP} \vee \langle e'_s, r'_s, v_s\rangle \in \mathcal{SP}} |\{e_{v_s}|\langle e_{v_s}, r'_s, e'_s\rangle \in \mathcal{G} \vee \langle e'_s, r'_s, e_{v_s}\rangle \in \mathcal{G}\}|}\right), \tag{7}$$

where the numerator of the exponent is the number of RDF triples in the underlying RDF graph matching the triple pattern $\langle v_s, r_s, e_s\rangle$. The denominator of the exponent is the number of RDF triples in the underlying RDF graph matching any triple pattern in $\mathcal{SP}$. For instance, according to Eq. (7), the attention scores in $sBGP_1$ are 0.9998, 0.3687, and 0.9979 for the three triple patterns $\langle v_1, r_1, e_1\rangle$, $\langle v_1, r_2, e_2\rangle$, and $\langle v_1, r_3, e_3\rangle$, respectively. And the attention scores in $sBGP_2$ are 0.5967 and 0.6165 for $\langle v_2, r_6, e_6\rangle$, $\langle v_2, r_7, e_7\rangle$, respectively.

By examining all neighbors of v_s in a sBGP $\mathcal{SP}$, we further define the preliminary embedding $\hat{\mathbf{v}}_\mathbf{s}$ of v_s as:

$$\hat{\mathbf{v}}_\mathbf{s} = \frac{\sum_{\langle v_s, r_s, e_s\rangle \in \mathcal{SP} \vee \langle e_s, r_s, v_s\rangle \in \mathcal{SP}} a(v_s, r_s, e_s) \cdot \tilde{\mathbf{v}}_\mathbf{s}}{\sum_{\langle v_s, r_s, e_s\rangle \in \mathcal{SP} \vee \langle e_s, r_s, v_s\rangle \in \mathcal{SP}} a(v_s, r_s, e_s)}, \tag{8}$$

where $\tilde{\mathbf{v}}_\mathbf{s}$ is computed for each single triple pattern according to Eq. (6).

For instance, the preliminary embeddings of variable v_1 and v_2 can be computed as $\hat{\mathbf{v}}_\mathbf{1} = 0.4225 \cdot (\mathbf{e_1} - \mathbf{r_1}) + 0.1558 \cdot (\mathbf{e_2} - \mathbf{r_2}) + 0.4217 \cdot (\mathbf{e_3} - \mathbf{r_3})$ and $\hat{\mathbf{v}}_\mathbf{2} = 0.4918 \cdot (\mathbf{e_6} - \mathbf{r_6}) + 0.5082 \cdot (\mathbf{e_7} - \mathbf{r_7})$, respectively.

For the impacts of relations among variable entities in different sBGPs, we introduce a simple and effective method to compute the final variable embeddings based on the preliminary embeddings. For a variable $v_s \in \mathcal{V_P}$ which is directly linked with other variables in BGP $\mathcal{P} = \{\mathcal{E_P} \cup \mathcal{V_P}, \mathcal{R_P}\}$, we compute its final embedding $\mathbf{v_s}$ as follows:

$$\mathbf{v_s} = \frac{\sum_{\langle v_s, r, v'_s\rangle \in \mathcal{P} \vee \langle v'_s, r, v_s\rangle \in \mathcal{P}} num(v'_s) \cdot f_3(v_s, r, v'_s) + num(v_s) \cdot \hat{\mathbf{v}}_\mathbf{s}}{\sum_{\langle v_s, r, v'_s\rangle \in \mathcal{P} \vee \langle v'_s, r, v_s\rangle \in \mathcal{P}} num(v'_s) + num(v_s)}, \tag{9}$$

where

$$f_3(v_s, r, v'_s) = \begin{cases} \hat{\mathbf{v'_s}} - \mathbf{r}, & \text{if } \langle v_s, r, v'_s\rangle \in \mathcal{P}, \\ \hat{\mathbf{v'_s}} + \mathbf{r}, & \text{if } \langle v'_s, r, v_s\rangle \in \mathcal{P}, \end{cases} \tag{10}$$

and $num(\cdot)$ is the number of triple patterns in the sBGP of a variable. The reason we utilize $num(\cdot)$ is that we assume the preliminary embedding of a variable deserves more attention if it is computed based on more triple patterns. It is worth mentioning that the correlations between variables can be characterized by more sophisticated method, such as an iterative updating algorithm. We will investigate this part in the future.

Finally, we can obtain all variable embeddings of the original query in the continuous vector space. For instance, the final embeddings of variable v_1 and v_2 in Fig. 2 can be computed as $\mathbf{v_1} = 0.4 \cdot (\hat{\mathbf{v}}_2 - \mathbf{r_8}) + 0.6 \cdot \hat{\mathbf{v}}_1 = 0.4 \cdot [0.4918 \cdot (\mathbf{e_6} - \mathbf{r_6}) + 0.5082 \cdot (\mathbf{e_7} - \mathbf{r_7}) - \mathbf{r_8}] + 0.6 \cdot [0.4225 \cdot (\mathbf{e_1} - \mathbf{r_1}) + 0.1558 \cdot (\mathbf{e_2} - \mathbf{r_2}) + 0.4217 \cdot (\mathbf{e_3} - \mathbf{r_3})]$ and $\mathbf{v_2} = 0.6 \cdot (\hat{\mathbf{v}}_1 + \mathbf{r_8}) + 0.4 \cdot \hat{\mathbf{v}}_2 = 0.6 \cdot [0.4225 \cdot (\mathbf{e_1} - \mathbf{r_1}) + 0.1558 \cdot (\mathbf{e_2} - \mathbf{r_2}) + 0.4217 \cdot (\mathbf{e_3} - \mathbf{r_3}) + \mathbf{r_8}] + 0.4 \cdot [0.4918 \cdot (\mathbf{e_6} - \mathbf{r_6}) + 0.5082 \cdot (\mathbf{e_7} - \mathbf{r_7})]$, respectively.

2.3   Generating Approximate Answers and Alternative Queries

In this module, our goal is to discover approximate answers based on embeddings of variables in the continuous vector space. For each approximate answer, we also generate an alternative query as the explanation to help the user recognize his expected information and refine the original query.

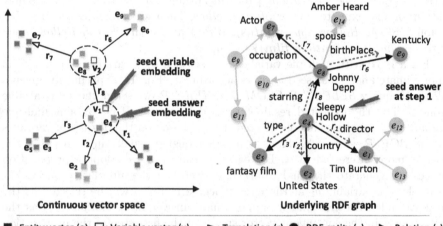

Fig. 3. Approximate answers generation based on the RDF embeddings.

As analyzed in Sect. 2.1, semantically similar entities are close to each other in the continuous vector space. Therefore, given a BGP $\mathcal{P} = (\mathcal{E_P} \cup \mathcal{V_P}, \mathcal{R_P})$, a variable $v_p \in \mathcal{V_P}$, and the embedding $\mathbf{v_p}$ computed through Eq. (9), we can readily find a semantically similar entity e_i of the RDF graph $\mathcal{G} = (\mathcal{E}, \mathcal{R})$ for v_p by computing the distance between $\mathbf{v_p}$ and $\mathbf{e_i}$ in the continuous vector space. We employ the cosine similarity to measure the distance between $\mathbf{v_p}$ and $\mathbf{e_i}$ as follows:

$$sim(\mathbf{v_p}, \mathbf{e_i}) = \frac{\mathbf{v_p} \cdot \mathbf{e_i}}{\|\mathbf{v_p}\| \|\mathbf{e_i}\|}. \tag{11}$$

According to Eq. (11), we can obtain top-k semantically similar entities $\mathcal{E}_\mathcal{K} = \{e_1, ..., e_i, ..., e_k\}$ for v_p.

In a simple case where a failing BGP $\mathcal{P} = (\mathcal{E}_\mathcal{P} \cup \{v_p\}, \mathcal{R}_\mathcal{P})$ contains only one variable v_p, the semantically similar entity set $\mathcal{E}_\mathcal{K}$ implied by Eq. (11) is exactly the set of final approximate answers to v_p. For each approximate answer $e_i \in \mathcal{E}_\mathcal{K}$, we can directly extract a sub-RDF graph $\mathcal{SG}_i$ about e_i from $\mathcal{G}$. The extraction of $\mathcal{SG}_i$ is a searching process based on e_i. Specifically, for each triple pattern $\langle v_p, r_p, e_p \rangle \in \mathcal{P}$ (v_p at the head), its ideal corresponding RDF triple is $\langle e_i, r_p, e_p \rangle$. If $\langle e_i, r_p, e_p \rangle \notin \mathcal{G}$, we figure out the corresponding RDF triple $\langle e_i, r, e' \rangle$ which is most similar to $\langle e_i, r_p, e_p \rangle$ among all the RDF triples in $\mathcal{G}$, formulated as follows:

$$\langle e_i, r, e' \rangle = \underset{\langle e_i, r, e' \rangle \in \mathcal{G}}{\arg \max} \left(\frac{\mathbf{r_p} \cdot \mathbf{r}}{\|\mathbf{r_p}\| \|\mathbf{r}\|} + \frac{\mathbf{e_p} \cdot \mathbf{e'}}{\|\mathbf{e_p}\| \|\mathbf{e'}\|} \right). \tag{12}$$

Analogically, for each triple pattern $\langle e_p, r_p, v_p \rangle \in \mathcal{P}$ (v_p at the tail), we can also compute its corresponding RDF triple. These RDF triples containing e_i form the sub-RDF graph $\mathcal{SG}_i$ which can be utilized to generate a modified BGP $\mathcal{P}'$ that expresses the similar query intention to $\mathcal{P}$. For example, assuming that a BGP $\{\langle \mathit{?film, type, drama\ film} \rangle\}$ returns nothing over an RDF graph, we may obtain an approximate answer *Sleepy Hollow*. Then we can extract a sub-RDF graph $\{\langle \mathit{Sleepy\ Hollow, type, \mathbf{fantasy\ film}} \rangle\}$ and return *Sleepy Hollow* along with $\{\langle \mathit{?film, type, \mathbf{fantasy\ film}} \rangle\}$ for the user.

For a BGP with multiple variables, we select a seed variable and obtain its approximate answers as seed answers according to Eq. (12). Specifically, given a SPARQL query *SELECT $\mathcal{S}$ FROM $\mathcal{G}$ WHERE $\mathcal{P}$*, we select the seed variable from $\mathcal{S}$ (i.e., the expected result expressed by users). If $\mathcal{S}$ contains multiple variables, the seed variable is selected from $\mathcal{S}$ based on the degrees of variables in the BGP $\mathcal{P}$, since a variable at a larger degree usually indicates that the user is more interested in it. For each seed answer, we adopt a propagation process to generate the final returned answer and the alternative query. For example, the variable v_1 at the largest degree in Fig. 3 will be selected as the seed variable. We first find its approximate answer *Sleep Hollow* (e_4) as the seed answer in the continuous vector space. In the first step of the propagation process, we follow Eq. (12) to find the corresponding triples $\{\langle \mathit{Sleep\ Hollow, type,}$ *fantasy film*$\rangle$, $\langle \mathit{Sleep\ Hollow, country, United\ States} \rangle$, $\langle \mathit{Sleep\ Hollow, director,}$ *Tim Burton*$\rangle$, $\langle \mathit{Sleep\ Hollow, starring, Johnny\ Depp} \rangle\}$. After this step, we have determined v_2 as *Johnny Depp*. Then, in the second step, we remove the triple patterns which have already been determined and set *Johnny Depp* (e_8) as a new seed answer for the next step. Repeat the propagation operation until all triple patterns have been determined as illustrated in the right of Fig. 3. Finally, we can generate the final returned answer $\{$ *?film:Sleep Hollow, ?actor:Johnny Depp*$\}$ and the alternative BGP $\{\langle \mathit{?film, type, \mathbf{fantasy\ film}} \rangle$, $\langle \mathit{?film, country,}$ *United States*$\rangle$, $\langle \mathit{?film, director, Tim\ Burton} \rangle$, $\langle \mathit{?film, starring, ?actor} \rangle$, $\langle \mathit{?actor,}$ *occupation, Actor*$\rangle$, $\langle \mathit{?actor, birthplace, \mathbf{Kentucky}} \rangle\}$.

Discussions: We discuss two parts which can be improved during the framework implementation: (1) Given a variable embedding in the vector space, there is no need to traverse all entity embeddings for the top-k similar entities searching. For example, we can partition the vector space into disjoint subspaces, and differentiate entities of the RDF graph according to the subspaces to which they belong. More sophisticated approaches are provided in [15] for this issue. (2) Currently, we assume that a variable at a larger degree is more important in the sub-RDF graph extraction. The preliminary experiments show some effectiveness, but we still need to improve the scalability in the future. For instance, users can be allowed to determine which variable is most important.

3 Experimental Evaluation

To scrutinize the effectiveness and efficiency of the proposed framework, we performed three types of experiments including: (1) Entity context preserving embedding validation; (2) Quality evaluation of approximate answers and alternative queries; (3) Efficiency evaluation. The results demonstrate that our framework significantly outperforms other baselines.

Dataset: DBpedia [16] is extracted from Wikipedia[3] and has become the core dataset of the LOD. In this paper, we employed the English version of DBpedia[4], which consists of 6.7M entities, 1.4K relations and 583M RDF triples.

Queries: According to our investigation, there is no open domain benchmark query set for SPARQL empty-answer problem. Therefore, twenty queries were constructed over DBpedia for the evaluation. The queries were designed to include basic graph patterns with different topological structures (e.g., star, chain, cycle, and complex) based on joins over variables [10].

Baselines: To validate the effectiveness of the consideration of entity context information in the translation-based model, we compared our embedding model with TransE [3]. To evaluate the empty-answer problem solving, we compared our framework with four state-of-the-art query relaxation models, i.e, similarity-based (SB) [7], rule-based (RB) [14], user-preferences-based (UPB) [5], and cooperative-techniques-based (CTB) [9] models. Meanwhile, we also compared our framework to a lite version with directly TransE plugged in.

3.1 Entity Context Preserving Embedding Validation

Our translation-based embedding model leverages the entity context information to encode semantically similar entities and utilizes the translation mechanism to represent relations between entities. The embedding model was implemented in Java, and the following validation was conducted on a Linux server with an Intel Core i7 3.40 GHz CPU and 126 GB memory running Ubuntu-14.04.1. We determined the optimal parameters using a grid search strategy. And the training instances were conducted over 1,000 iterations. The running time per iteration was 562 s. We report the implementation code and detailed parameters in https://github.com/wangmengsd/re.

---

[3] https://www.wikipedia.org.
[4] http://wiki.dbpedia.org/develop/datasets.

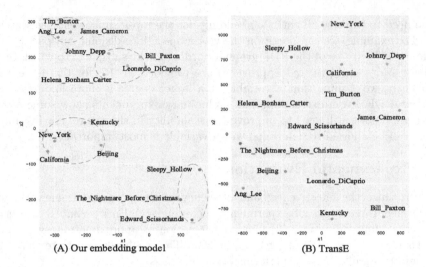

Fig. 4. Visualization of entity embeddings learned by our model and TransE.

For the entity context preserving validation, we projected several sample entity embeddings generated by our embedding method and TransE into two-dimensional spaces using t-SNE[5], as shown in Fig. 4. The result in Fig. 4(a) is consistent with our expectation, where semantically similar entities are close to each other. In contrast, the distribution of entities in Fig. 4(b) does not represent the result we want.

For the translation preserving validation, we followed TransE in [3] and employed *MeanRank* and *Hits@10* as evaluation metrics. Specifically, to test a triple $\langle e_h, r, e_t \rangle$, we removed the head entity e_h or the tail entity e_t. Then we predicted the missing entity e_h or e_t based on $\mathbf{e_t} - \mathbf{r}$ or $\mathbf{e_h} + \mathbf{r}$, and we used the score function Eq. (3) to rank the predictions in a descending order. *Mean-Rank* denotes the average rank of all correct predictions, and *Hits@10* denotes the proportion of correct predictions ranked in the top-*10*. The *MeanRank* of our embedding model is 8.01 and *Hits@10* is 83.98. Whereas, the *MeanRank* of TransE is 8.00 and *Hits@10* is 84.01. Both the results of *MeanRank* and *Hits@10* proved that our embedding model maintained the effectiveness of the translation mechanism.

3.2   Quality of Approximate Answers and Alternative Queries

In this section, we compared our framework with four state-of-the-art query relaxation models [5,7,9,14]. Note that the efficient approach in [9] only computed Maximal Succeeding Subqueries (XSSs) (a kind of relaxed queries), and it didn't support similarity criteria to rank the multiple XSSs and query answers.

---

[5]   https://lvdmaaten.github.io/tsne/.

To evaluate the quality of generated approximate answers and alternative queries, ten evaluators (graduate students in Computer Science, but none of them knows the details of the methods) were asked to evaluate how well an approximate answer satisfies the original query intention (*Sat*) and how similar an alternative query is to the original one (*Sim*). The two metrics *Sat* and *Sim* are rated on a 5-point scale: 0 corresponding to "negative", 1 to "weakly negative", 2 to "neutral", 3 to "weakly positive", and 4 to "positive". For each empty-answer query, we presented top-k[6] approximate answers and alternative queries generated by our framework and four baselines to the evaluators. We also employed *Pearson Correlation Coefficient* to analyze the correlation between the evaluator ratings and similarity scores calculated by the corresponding models. The *Pearson Correlation Coefficient* is a standard measure of the correlation between two variables. The coefficient value ranges from -1 to $+1$, where -1 represents totally negative correlation, 0 represents no linear correlation, and $+1$ represents totally positive correlation. Table 1 reports the average ratings (Avg.Rating) of all five models and the *Pearson Correlation Coefficients* (PCC). We can make the following observations:

Table 1. Results of overall effectiveness.

Top-k		Top-1		Top-3		Top-5		Top-10		Top-20	
Metric		*Sat*	*Sim*	*Sat*	*Sim*	*Sat*	*Sim*	*Sat*	*Sim*	*Sat*	*Sim*
Our method	Avg. rating	**3.5**	**3.25**	**3.15**	**3.03**	**2.72**	**2.61**	1.86	**1.82**	1.41	**1.40**
	PCC	0.54	0.51	0.52	0.48	0.53	0.48	0.52	0.49	0.53	0.49
Lite version with TransE	Avg. rating	0.85	0.33	0.45	0.28	0.3	0.2	0.24	0.09	0.13	0.06
	PCC	0.12	0.07	0.09	−0.03	0.06	−0.05	0.07	0.02	0.05	0.03
SB [7]	Avg. rating	3.2	3.0	3.02	2.75	2.45	2.05	**1.87**	1.77	**1.44**	1.39
	PCC	0.43	0.41	0.43	0.39	0.41	0.36	0.43	0.37	0.43	0.37
RB [14]	Avg. rating	3.05	3.0	2.93	2.68	2.5	2.21	**1.87**	1.74	1.42	1.37
	PCC	0.40	0.39	0.41	0.39	0.42	0.40	0.41	0.38	0.40	0.38
UPB [5]	Avg. rating	2.85	2.75	2.42	2.07	2.11	1.6	1.45	1.21	1.22	0.97
	PCC	0.33	0.29	0.31	0.27	0.32	0.26	0.33	0.27	0.34	0.28
CTB [9]	Avg. rating	2.7	2.75	2.45	2.05	2.16	1.62	1.43	1.18	1.24	0.97
	PCC	-	-	-	-	-	-	-	-	-	-

- Our framework achieved consistently significant improvement on *Sat* and *Sim* compared with all the baselines, which demonstrates the effectiveness of our framework. The reason is that we can directly compare the semantic similarity between expected answers and approximate answers in the continuous vector space. And the entity context preserving embedding model enables our method to generate high quality approximate answers and alternative queries.
- The PCC of our framework is also higher than all other baselines, which indicates that our similarity measuring mechanism (Eqs. (11) and (12)) is more

---

[6] We set $k \in \{1, 3, 5, 10, 20\}$ in this paper.

identical with the perception of users. The reason is that the embeddings are learned based on entity context information of the underlying RDF graph which contains more precise and richer information than the ontological information and statistical language models employed by other models.

- The performances of all methods were affected when increasing k to 20 because more irrelevant answers were generated. However, our method still has its own advantage. Since our method lists approximate answers in a descending order in terms of the similarity, users can obtain the most approximate answers at the top.

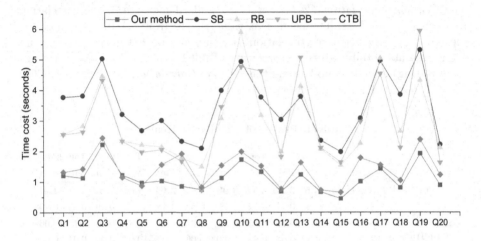

Fig. 5. Time costs of five methods on twenty failing SPARQL queries $Q1 \sim Q20$.

3.3 Efficiency Evaluation

The average time cost of our method to process a failing query is 1.13 s. This amount of time is acceptable for users to obtain approximate answers and alternative queries. We compared the time cost of our framework with other baselines. Figure 5 illustrates the runtime results of solving twenty empty-answer SPARQL queries. We can observe that the time cost of our framework is significantly less than other baselines. The key reason is that our framework is driven and guided by the approximate answer embeddings, which speeds up the generation of possible answers and alternative queries. Another reason is that the similarity computation in the continuous vector space is more efficient than the conventional graph-based computation method over a large RDF graph.

In summary, the evaluation results on effectiveness and efficiency show that our framework can facilitate to generate high-quality approximate answers and alternative queries.

4   Related Work

This section discusses existing related research in the following aspects: RDF query relaxation approaches and RDF graph embedding techniques.

Query relaxation approaches in the RDF context have been proposed to solve the SPARQL empty-answer problem. These methods mainly focus on reformulating the original query into a new relaxed query by removing or relaxing RDF conditions. Four types of models, similarity-based, rule-based, user-preferences-based, and cooperative-techniques-based models are utilized to generate multiple relaxed query candidates. Similarity-based models [6,7] leverage lexical analyses to determine appropriate relaxation candidates. Rule-based models [12–14,19] exploit RDF schema semantics and rewriting rules to perform relaxation. The user-preferences-based model [5] automatically relaxes over-constrained RDF queries based on domain knowledge and user preferences. Cooperative-techniques-based models [8,9] design pruning strategies to reduce the exponential search space of finding Top-k optimal relaxed queries. However, relaxed queries generated by query relaxation approaches may be rather different from initial queries of users because these models cannot consider the expected answers which do not occur in the query results. Over-relaxed queries and irrelevant answers are not effective for the expectation of users.

Existing RDF graphs already include thousands of relation types, millions of entities, and billions of RDF triples [1]. The RDF applications based on conventional graph-based algorithms are compromised by the data sparsity and computational inefficiency. To address these problems, RDF graph embedding techniques [3,4,17,20,21] have been proposed to embed both entities and relations into continuous vector spaces. Among these methods, neural-language-based models [4,20] only generate entity latent representations by training the neural language model of input RDF graphs. As a result, semantically similar entities are close to each other in continuous vector spaces. But we cannot infer relations between entities solely based on entity latent representations. Translation-based models [3,17,21] are effective in modeling relations between entities because of their translation mechanisms. But they do not guarantee that semantically similar entities are close to each other in continuous vector spaces since they regard the RDF graph as a set of independent triples during the learning processes. To sum up, none of the existing models meets the requirements for modeling SPARQL triple patterns in our framework.

5   Conclusions and Future Work

In this paper, we solve the SPARQL empty-answer problem in the continuous vector space. To make semantically similar entities close to each other in the vector space, we propose a novel embedding model which utilizes the translation mechanism to capture the relations between entities while considering the entity context. Then, given a failing SPARQL query, we partition the SPARQL BGP into several parts and compute approximate answers by leveraging RDF embeddings and the translation mechanism. We also generate alternative queries for approximate answers, which helps users recognize their information needs and refine the original query. We conduct extensive experiments on the real-world RDF dataset to validate the effectiveness and the efficiency of our framework.

In future work, we intend to improve the accuracy of variable embedding computation through an iterative updating algorithm. Another development of our research is to address the SPARQL empty-answer problem on graph patterns which contain operators such as *UNION, OPTIONAL, MINUS* and so on.

Acknowledgment. This work was supported by National Key Research and Development Program of China (2018YFB1004500), National Natural Science Foundation of China (61532015, 61532004, 61672419, and 61672418), Innovative Research Group of the National Natural Science Foundation of China (61721002), Innovation Research Team of Ministry of Education (IRT_17R86), Project of China Knowledge Centre for Engineering Science and Technology, Science and Technology Planning Project of Guangdong Province (No. 2017A010101029), and Teaching Reform Project of XJTU (No. 17ZX044).

References

1. Bizer, C., Heath, T., Berners-Lee, T.: Linked data-the story so far. Int. J. Semant. Web Inf. Syst. **5**(3), 1–22 (2009)
2. Bonifati, A., Martens, W., Timm, T.: An analytical study of large SPARQL query logs. Proc. VLDB Endow. **11**(2), 149–161 (2017)
3. Bordes, A., Usunier, N., Garcia-Duran, A., Weston, J., Yakhnenko, O.: Translating embeddings for modeling multi-relational data. In: Advances in Neural Information Processing Systems, pp. 2787–2795 (2013)
4. Cochez, M., Ristoski, P., Ponzetto, S.P., Paulheim, H.: Global RDF vector space embeddings. In: d'Amato, C., et al. (eds.) ISWC 2017. LNCS, vol. 10587, pp. 190–207. Springer, Cham (2017). https://doi.org/10.1007/978-3-319-68288-4_12
5. Dolog, P., Stuckenschmidt, H., Wache, H., Diederich, J.: Relaxing RDF queries based on user and domain preferences. J. Intell. Inf. Syst. **33**(3), 239 (2009)
6. Elbassuoni, S., Ramanath, M., Schenkel, R., Sydow, M., Weikum, G.: Language-model-based ranking for queries on RDF-graphs. In: Proceedings of the 18th ACM conference on Information and Knowledge Management, pp. 977–986. ACM (2009)
7. Elbassuoni, S., Ramanath, M., Weikum, G.: Query relaxation for entity-relationship search. ESWC 2011. LNCS, vol. 6644, pp. 62–76. Springer, Heidelberg (2011). https://doi.org/10.1007/978-3-642-21064-8_5
8. Fokou, G., Jean, S., Hadjali, A., Baron, M.: Cooperative techniques for SPARQL query relaxation in RDF databases. In: Gandon, F., Sabou, M., Sack, H., d'Amato, C., Cudré-Mauroux, P., Zimmermann, A. (eds.) ESWC 2015. LNCS, vol. 9088, pp. 237–252. Springer, Cham (2015). https://doi.org/10.1007/978-3-319-18818-8_15
9. Fokou, G., Jean, S., Hadjali, A., Baron, M.: Handling failing RDF queries: from diagnosis to relaxation. Knowl. Inf. Syst. **50**(1), 167–195 (2017)
10. Görlitz, O., Thimm, M., Staab, S.: SPLODGE: systematic generation of SPARQL benchmark queries for linked open data. In: Cudré-Mauroux, P., et al. (eds.) ISWC 2012. LNCS, vol. 7649, pp. 116–132. Springer, Heidelberg (2012). https://doi.org/10.1007/978-3-642-35176-1_8
11. Harris, S., Seaborne, A., Prud'hommeaux, E.: Sparql 1.1 query language. W3C recommendation **21**(10) (2013)

12. Hogan, A., Mellotte, M., Powell, G., Stampouli, D.: Towards fuzzy query-relaxation for RDF. In: Simperl, E., Cimiano, P., Polleres, A., Corcho, O., Presutti, V. (eds.) ESWC 2012. LNCS, vol. 7295, pp. 687–702. Springer, Heidelberg (2012). https://doi.org/10.1007/978-3-642-30284-8_53

13. Huang, H., Liu, C., Zhou, X.: Approximating query answering on RDF databases. World Wide Web 15(1), 89–114 (2012)

14. Hurtado, C.A., Poulovassilis, A., Wood, P.T.: Query relaxation in RDF. In: Spaccapietra, S. (ed.) Journal on Data Semantics X. LNCS, vol. 4900, pp. 31–61. Springer, Heidelberg (2008). https://doi.org/10.1007/978-3-540-77688-8_2

15. Katayama, N., Satoh, S.: The SR-tree: an index structure for high-dimensional nearest neighbor queries. ACM Sigmod Rec. 26(2), 369–380 (1997)

16. Lehmann, J., et al.: Dbpedia-a large-scale, multilingual knowledge base extracted from Wikipedia. Semant. Web 6(2), 167–195 (2015)

17. Lin, Y., Liu, Z., Sun, M., Liu, Y., Zhu, X.: Learning entity and relation embeddings for knowledge graph completion. In: AAAI, vol. 15, pp. 2181–2187 (2015)

18. Mikolov, T., Sutskever, I., Chen, K., Corrado, G.S., Dean, J.: Distributed representations of words and phrases and their compositionality. In: Advances in Neural Information Processing Systems, pp. 3111–3119 (2013)

19. Poulovassilis, A., Wood, P.T.: Combining approximation and relaxation in semantic web path queries. In: Patel-Schneider, P.F., et al. (eds.) ISWC 2010. LNCS, vol. 6496, pp. 631–646. Springer, Heidelberg (2010). https://doi.org/10.1007/978-3-642-17746-0_40

20. Ristoski, P., Paulheim, H.: RDF2Vec: RDF graph embeddings for data mining. In: Groth, P., et al. (eds.) ISWC 2016. LNCS, vol. 9981, pp. 498–514. Springer, Cham (2016). https://doi.org/10.1007/978-3-319-46523-4_30

21. Wang, Z., Zhang, J., Feng, J., Chen, Z.: Knowledge graph embedding by translating on hyperplanes. In: AAAI, vol. 14, pp. 1112–1119 (2014)

Query-Based Linked Data Anonymization

Remy Delanaux[1]([⊠]), Angela Bonifati[1], Marie-Christine Rousset[2,3],
and Romuald Thion[1]

[1] Université Lyon 1, LIRIS CNRS, Villeurbanne, France
{remy.delanaux,angela.bonifati,romuald.thion}@univ-lyon1.fr
[2] Université Grenoble Alpes, CNRS, INRIA, Grenoble INP, Grenoble, France
marie-christine.rousset@imag.fr
[3] Institut Universitaire de France, Paris, France

Abstract. We introduce and develop a declarative framework for
privacy-preserving Linked Data publishing in which privacy and utility policies are specified as SPARQL queries. Our approach is data-
independent and leads to inspect only the privacy and utility policies in
order to determine the sequence of anonymization operations applicable
to any graph instance for satisfying the policies. We prove the soundness
of our algorithms and gauge their performance through experiments.

1 Introduction

Linked Open Data (LOD) provides access to continuously increasing amounts of
RDF data that describe properties and links among entities referenced by means
of Uniform Resource Identifiers (URIs). Whereas many organizations, institutions and governments participate to the LOD movement by making their data
accessible and reusable to citizens, the risks of identity disclosure in this process are not completely understood. As an example, in smart city applications,
information about users' journeys in public transportation can help re-identify
the individuals if they are joined with other public data sources by leveraging
quasi-identifiers.

The main problem for data providers willing to publish useful data is to determine which anonymization operations must be applied to the original dataset in
order to preserve both individuals' privacy and data utility. For all these reasons,
data providers should have at their disposal the means to readily anonymize their
data prior to publication into the LOD cloud. The majority of the solutions proposed so far and mainly devoted to relational legacy systems rely on variants
of differential privacy as surveyed in [14] or k-anonymity proposed by [18]. Differential privacy offers strong mathematical guarantees of non-disclosure of any
factual information by adding noise to the data, with as a counterpart a low
utility of answers returned by precise queries (as opposed to statistical queries).
In a similar fashion, several k-anonymization methods have been developed that
transform the original dataset into clusters containing at least k records with
indistinguishable values over quasi-identifiers. When taken into account, the utility loss is defined by a metric that measures and minimizes the information loss

© Springer Nature Switzerland AG 2018
D. Vrandečić et al. (Eds.): ISWC 2018, LNCS 11136, pp. 530–546, 2018.
https://doi.org/10.1007/978-3-030-00671-6_31

between the original dataset and the output of the anonymization algorithm. All these approaches fall short in empowering the data providers with the capability of *specifying their own privacy and utility policies*, and in managing anonymization operations as *update operations* on the data.

In this paper, we present a novel declarative framework that *(i)* allows the data providers to specify as queries both the privacy and utility policies they want to enforce, *(ii)* checks whether the specified policies are compatible with each other, and *(iii)*, based on a set of basic update queries, automatically builds candidate sets of anonymization operations that are guaranteed to transform any input dataset into a dataset satisfying the required privacy and utility policies. We provide an algorithm that implements this *data-independent method* starting from privacy and utility policies only, and we prove its soundness.

We believe that our framework is tailored for data publishing in the LOD in which it is important to strike a balance between non-disclosure of information (likely to serve as quasi-identifiers when interconnected with LOD links) and utility preservation for end-users querying the LOD. Our framework is carefully designed to meet all these requirements by leveraging at the same time the expressive power of SPARQL queries and the maturity and effectiveness of SPARQL query engines. It builds on the common wisdom that queries from data providers are more and more available through online SPARQL endpoints [3]. Such queries are a valuable resource to understand the real utility needs of users on publicly available data and to guide the data providers in safe RDF data publishing. For all these reasons, our approach paves the way to a democratization of privacy-preserving mechanisms for LOD data.

The paper is organized as follows. Section 2 reviews related work. Section 3 summarizes preliminaries for explaining the query-based model of privacy and utility that we propose in Sect. 4. Sections 5 and 6 describe respectively our query-based static anonymization method and its experimental assessment. Finally, we conclude the paper in Sect. 7 with further research directions.

2   Related Work

Privacy preserving data publishing (PPDP) has been a long-standing research goal for several research communities, as witnessed by a flurry of work on the topic [7]. A rich variety of privacy models have been proposed, ranging from k-anonymity [18] and l-diversity [15] to t-closeness [13] and ϵ-differential privacy [6]. For each of the aforementioned methods, one or more attack models (such as record linkage, attribute linkage, table linkage and probabilistic attack) are considered, amounting to make two fundamental assumptions: (i) what an attacker is assumed to know about the victim and (ii) under which conditions a privacy threat occurs.

Most of these models, first conceived for relational databases, have been recently extended to the setting of the Semantic Web [12]. Among them, the privacy model that is definitely the closest to our work is k-anonymity that has been recently adapted to RDF graphs in [10,17]. These works focus on defining operations of generalization, suppression or perturbation to apply to values in the range of properties known to be quasi-identifiers for persons identification, along with metrics to measure the resulting loss of information. Our approach is more generic than theirs and also fully declarative since, by leveraging the logical foundations of PPDP for Linked Data in [8], it allows the definition of fine-grained privacy policies specified by queries, and to obtain candidate sets of anonymization operations allowing to practically enforce the requested privacy without loosing the desired utility.

Contrarily to [8], which focuses on the computational complexity of checking whether privacy requirements are fulfilled in Linked Data, we leverage utility policies as queries for which anonymization operations must preserve the original answers, and we employ the interactions of privacy and utility policies in a static analysis method. To the best of our knowledge, our framework is the first to provide practical algorithms for building candidate sequences of atomic operations, described as an open research challenge in [8].

An alternative approach to anonymization for protecting against privacy breaches consists in applying access control methods to Linked Data [11,16,19]. In the Semantic Web setting, when data are described by description logics ontologies, preliminary results on role-based access control have been obtained in [1] for the problem of checking whether a sequence of role changes and queries can infer that an anonymous individual is equal to a known individual. Compared to access control techniques that perform verification at runtime, we focus on a static analysis approach executed only once and guaranteeing that the published datasets do not contain sensitive information.

3 Preliminaries

We introduce the standard notions and concepts for RDF graphs and SPARQL queries. Let $\mathbf{I}$, $\mathbf{L}$ and $\mathbf{B}$ be countably infinite pairwise disjoint sets representing respectively *IRIs*, *literal values* (or *literals*) and *blank nodes*. IRIs (Internationalized Resource Identifiers) are standard identifiers used for denoting any Web resource described in RDF within the LOD. We denote by $\mathbf{T} = \mathbf{I} \cup \mathbf{L} \cup \mathbf{B}$ the set of *terms*, in which we distinguish *constants* (IRIs and literal values) from blank nodes. We also assume an infinite set $\mathbf{V}$ of variables disjoint from the above sets. In the examples, variables in $\mathbf{V}$ are prefixed with a question mark as in the SPARQL language.

Definition 1 (RDF graph). *An RDF graph is a finite set of RDF triples* (s,p,o), *where* $(s,p,o) \in (\mathbf{I} \cup \mathbf{B}) \times \mathbf{I} \times (\mathbf{I} \cup \mathbf{L} \cup \mathbf{B})$.

IRIs appearing in position p into triples denote properties composing the *schema* of the RDF graph.

The queries we consider in our work are built on *graph patterns* that are made of triples with constants and variables (blank nodes are not allowed).

Definition 2 (Graph pattern). *A triple pattern is a triple* $(s,p,o) \in (\mathbf{I} \cup \mathbf{V}) \times (\mathbf{I} \cup \mathbf{V}) \times (\mathbf{I} \cup \mathbf{L} \cup \mathbf{V})$. *A graph pattern is a finite set of triple patterns.*

We can now define the two types of queries under study along with their answers. The first type of queries correspond to the standard notion of *conjunctive queries*, while the second type corresponds to *counting queries* that are the basis for simple analytical tasks.

Definition 3 (Conjunctive query). *A conjunctive query Q is defined by an expression* SELECT $\bar{x}$ WHERE $G(\bar{x}, \bar{y})$ *where $G(\bar{x}, \bar{y})$ is a graph pattern and $\bar{x} \cup \bar{y}$ is the set of its variables, among which $\bar{x}$ are the result (also called the* distinguished*) variables. A conjunctive query Q is alternatively written as $\langle \bar{x}, G \rangle$.*

The evaluation of a query $\langle \bar{x}, G \rangle$ over an RDF graph DB consists in finding mappings μ assigning the variables in G to terms such that the set of triples, denoted $\mu(G)$, obtained by replacing with $\mu(z)$ each variable z appearing in G, is included in DB. The corresponding answer is defined as the tuple of terms $\mu(\bar{x})$ assigned by μ to the result variables.

Definition 4 (Evaluation of a conjunctive query). *Let Q be a conjunctive query defined by $\langle \bar{x}, G \rangle$, and let DB an RDF graph. The answer set of Q over DB is defined by :* $\mathsf{Ans}(Q, DB) = \{\mu(\bar{x}) \mid \mu(G) \subseteq DB\}$.

Definition 5 (Counting query). *Let Q be a conjunctive query. The query* $\mathsf{Count}(Q)$ *is a counting query, whose answer over a graph DB is defined by:* $\mathsf{Ans}(\mathsf{Count}(Q), DB) = |\mathsf{Ans}(Q, DB)|$.

4   Query-Based Policies and Anonymization Operations

Following [8], a privacy policy, represented by a set of conjunctive queries, satisfies the anonymization process if none of the sensitive answers holds in the resulting dataset. This is achieved by letting the privacy queries return no answer or, alternatively, answers with blank nodes, as shown in the remainder. We also model utility policies by sets of queries that can be either conjunctive queries or counting queries useful for data analytics. For satisfying an utility policy, the anonymization process must preserve the answers of all the specified utility queries. We now formally define privacy and utility policies.

Definition 6 (Privacy and utility policies). *Let DB be an input RDF graph, a privacy (resp. utility) policy $\mathcal{P}$ (resp. $\mathcal{U}$) is a set of conjunctive queries (resp. conjunctive or counting queries). Let $\mathsf{Anonym}(DB)$ be the result of an anonymization process of the graph DB by a sequence of anonymization operators.*

A privacy policy $\mathcal{P}$ is satisfied on $\mathsf{Anonym}(DB)$ if for every $P \in \mathcal{P}$ and for any tuple of constants $\bar{c}$, it holds that: $\bar{c} \notin \mathsf{Ans}(P, \mathsf{Anonym}(DB))$. An utility policy $\mathcal{U}$ is satisfied on $\mathsf{Anonym}(DB)$ if for every $U \in \mathcal{U}$ it holds that: $\mathsf{Ans}(U, \mathsf{Anonym}(DB)) = \mathsf{Ans}(U, DB)$.

As usual, we call $|\mathcal{P}|$ (resp. $|\mathcal{U}|$) the *cardinality* of the policy. For a policy $\mathcal{P}$ (resp. $\mathcal{U}$) made of n queries $P_i = \langle \bar{x}_i^P, G_i^P \rangle$ (resp. m queries $U_i = \langle \bar{x}_i^U, G_i^U \rangle$) we call the sum of the cardinalities of their bodies the *size of the policy* defined by $\sum_{i=1}^{n} |G_i^P|$ (resp. $\sum_{i=1}^{m} |G_i^U|$).

The following running example shows that privacy and utility policies might impose constraints on overlapping portions of a dataset.

Example 1. Consider a privacy policy $\mathcal{P} = \{P_1, P_2\}$ on data related to public transportation in a given city, and defined by the two following conjunctive queries written in concrete SPARQL syntax. The first privacy query expresses that travelers' postal addresses are sensitive and shall be protected, and the second privacy query specifies that the disclosure of users identifiers associated with geolocation information (like latitude and longitude as given by the user ticket validation) may also pose a risk (for re-identification by data linkage with other LOD datasets).

```
# Privacy query P1          # Privacy query P2
SELECT ?ad                  SELECT ?u ?lat ?long
WHERE {                     WHERE {
    ?u   a     tcl:User.        ?c   a      tcl:Journey.
    ?u   vcard:hasAddress ?ad.  ?c   tcl:user       ?u.
}                               ?c   geo:latitude   ?lat.
                                ?c   geo:longitude  ?long.
                            }
```

As a consequence, any query displaying either users' addresses or users' identifiers together with their geolocation information would infringe this privacy policy, violating the anonymization of the underlying dataset to be published as open data. The counterpart utility policy is the set of queries $\mathcal{U} = \{U_1, U_2\}$. This set states that users' ages and location related to journeys are to be preserved.

```
# Utility query U1          # Utility query U2
SELECT ?u ?age              SELECT ?c ?lat ?long
WHERE {                     WHERE {
    ?u   a     tcl:User.        ?c   a      tcl:Journey.
    ?u   foaf:age  ?age.        ?c   geo:latitude   ?lat.
}                               ?c   geo:longitude  ?long.
                            }
```

Regarding anonymization operations, we extend the notion of suppression functions considered in [8] that replace IRIs with blank nodes by allowing also triple deletions. The anonymization operations that we consider correspond to **update queries** (Definition 7): when evaluated against an RDF graph DB, the update query DELETE $D(\bar{x})$ INSERT $I(\bar{y})$ WHERE $W(\bar{x}, \bar{z})$ suppresses all occurrences of $D(\bar{x})$ in DB such that $W(\bar{x}, \bar{y})$ can be mapped to a subgraph of DB, and inserts triples corresponding to the pattern $I(\bar{y})$. Note that $\bar{y}$ may contain *existential variables*, i.e., variables that do not appear in $W(\bar{x}, \bar{z})$, and thus cannot be mapped to terms present in DB. This would lead to add triples with blank nodes.

Definition 7 (Update query). *An* update query *(or* update operation*) Q_{upd} is defined by* DELETE $D(\bar{x})$ INSERT $I(\bar{y})$ WHERE $W(\bar{x}, \bar{z})$ *where D (resp. I, W) is a graph pattern which set of variables is $\bar{x}$ (resp. $\bar{y}$, $\bar{x} \cup \bar{z}$). The result of its evaluation over an RDF graph DB is defined by:*

$$\mathsf{Result}(Q_{upd}, DB) = DB \setminus \{\mu(D(\bar{x})) | \mu(W(\bar{x}, \bar{y})) \subseteq DB\}$$
$$\cup \{\mu'(I(\bar{y})) | \mu(W(\bar{x}, \bar{z})) \subseteq DB\}$$

where μ' is an extension of μ to fresh blank nodes, i.e. a mapping such that

$$\mu'(x) = \begin{cases} \mu(x) \text{ when } x \in \bar{x} \cup \bar{z} \\ b_{new} \in \mathbf{B} \text{ otherwise} \end{cases}$$

A deletion query Q_{del} is a particular case of update query where the insertion pattern $I(\bar{y})$ is empty.

In the following section, we will focus on two kinds of **atomic anonymization operations** that correspond respectively to **triple deletions** (i.e., particular case of Definition 7 where $D(\bar{x})$ is reduced to a triple pattern) and **replacement of IRIs by blank nodes** (i.e., particular case of Definition 7 where $D(\bar{x})$ and $I(\bar{y})$ are triple patterns that differ just by the fact that one bound variable of $D(\bar{x})$ is replaced with an existential variable in $I(\bar{y})$).

These two atomic anonymization operations are illustrated in Examples 2 and 3 respectively. From now, by slight abuse of notation w.r.t Definition 7, we will use the SPARQL standard notation [ ] for denoting single existential variables.

Example 2. In the setting of Example 1 related to transportation data, the following query specifies the operation deleting the addresses of users.

```
DELETE  { ?u  vcard:hasAddress   ?ad. }
WHERE   { ?u  a tcl:User.
          ?u  vcard:hasAddress   ?ad.}
```

Example 3. In the same context, this query replaces users' identifiers related to a ticket validation by a blank node.

```
DELETE { ?c  tcl:user    ?u. }
INSERT { ?c  tcl:user    []. }
WHERE  { ?c  a tcl:Journey.
         ?c  tcl:user      ?u.
         ?c  geo:latitude  ?lat.
         ?c  geo:longitude ?long. }
```

5   Finding Candidate Sets of Anonymization Operations

Given privacy and utility policies, the problems of interest that we address in this paper are named COMPATIBILITY and ENUMOPERATIONS. Both problems are generic as they are essentially built on the query-based definition of policy satisfaction, hence they are applicable to larger classes of operations and queries.

Problem 1. The COMPATIBILITY problem.

> **Input** : $\mathcal{P} = \{P_i\}$ a privacy policy and $\mathcal{U} = \{U_j\}$ a utility policy
> **Output:** True if *there exists* a sequence of operations O such that $O(DB)$ satisfies both $\mathcal{P}$ and $\mathcal{U}$ for any DB and False otherwise.

Problem 2. The ENUMOPERATIONS problem.

> **Input** : $\mathcal{P} = \{P_i\}$ a privacy policy and $\mathcal{U} = \{U_j\}$ a utility policy
> **Output:** The set $\mathcal{O}$ of all sequences of operations O such that $O(DB)$ satisfies both $\mathcal{P}$ and $\mathcal{U}$ for any DB.

An algorithm that solves the ENUMOPERATIONS problem solves the COM-PATIBILITY problem as well, by checking whether its output is $\emptyset$.

The rest of this section is devoted to the design of Algorithm 2 that solves the ENUMOPERATIONS problem using update operations (Definition 7) when $\mathcal{P}$ and $\mathcal{U}$ are defined by conjunctive queries (Definition 3). We also define an intermediate step dealing with unitary privacy policies, with Algorithm 1. Note that Algorithm 2 produces a set of *sets* of operations and not a set of *sequences*. As we guarantee that the *sets* of operations hereby computed solve the problem, any sequence obtained by reordering these sets would work as well. Hence, the difference between sets and sequences of operations is fairly immaterial.

If the answer set of Q is preserved by an anonymization process so does its cardinality, implying that any solution for a non-counting query Q is also a solution for its counting counterpart $\mathsf{Count}(Q)$. Similarly, if a utility query Q is satisfied, then its counting counterpart $\mathsf{Count}(Q)$ is also satisfied. Therefore, we focus on non-counting queries in Algorithm 1. However, the opposite implication does not hold, hence we may miss some operation that may guarantee a utility counting query $\mathsf{Count}(Q)$ without guaranteeing a utility non-counting query Q.

5.1   Finding Candidate Sets of Operations for Unitary Privacy Policies

We start with the case where the privacy policy is unitary, i.e. when it is reduced to a singleton $\mathcal{P} = \{P\}$. Intuitively, Algorithm 1 tries to find edges that are in the graph pattern G^P of the privacy policy $\mathcal{P}$ but in none of the utility policy graph patterns G_j^U. For each such an edge, a delete operation is constructed, and possible update operations are considered. Update operations take place in two manners: either the subject of the triple is replaced with a blank node, or its object is replaced with a blank node if it is an IRI. In both cases, the algorithm looks for three alternatives:

- The triple is part of a path of length ≥ 2 in the privacy graph pattern G^P, and therefore the update operation breaks the path thus satisfying the privacy policy P;
- The replaced subject (resp. object) is also the subject (resp. object) of another triple in the privacy query graph G^P and the update operation breaks the link between these triples, hence satisfying the privacy policy P;
- The replaced subject (resp. object) of the triple is also part of the distinguished variables $\bar{x}$ of the privacy policy query, leading to a blank value in the query results.

The soundness of this algorithm is encapsulated in Theorem 1. Due to space constraints, proofs are available in an online appendix.[1] We define the following helper functions that check if update operations are possible:

$$\texttt{check-subject}((s,p,o),G) = \exists (s',p',s) \in G \ \vee$$
$$(\exists (s,p',o') \in G \ \wedge \ \nexists\sigma\, (\sigma(s,p',o') = \sigma(s,p,o)))$$
$$\texttt{check-object}((s,p,o),G) = \exists (o,p',o') \in G \ \vee$$
$$(\exists (s',p',o) \in G \ \wedge \ \nexists\sigma\, (\sigma(s',p',o) = \sigma(s,p,o)))$$

Algorithm 1. Find update operations to satisfy a unitary privacy policy

---

Input : a unitary privacy policy $\mathcal{P} = \{P\}$ with $P = \langle \bar{x}^P, G^P \rangle$
Input : a utility policy $\mathcal{U}$ made of m queries $U_j = \langle \bar{x}_j^U, G_j^U \rangle$
Output: a set of operations O satisfying both $\mathcal{P}$ and $\mathcal{U}$

1 **function** find-ops-unit$(P,\mathcal{U})$:
2   Let H be the graph G^P with all its variables replaced by fresh ones[a];
3   Let $O := \emptyset$;
4   **forall** $(s,p,o) \in H$ **do**
5   Let $c :=$ true;
6   **forall** G_j^U **do**
7   **forall** $(s',p',o') \in G_j^U$ **do**
8   **if** $\exists\sigma\, (\sigma(s',p',o') = \sigma(s,p,o))$ **then**
9                     $c :=$ false;
10   **if** c **then**
11             $O := O \cup \{\texttt{DELETE } \{(s,p,o)\} \texttt{ WHERE } H\}$;
12   **if** check-subject$((s,p,o),H) \vee s \in \bar{x}^P$ **then**
13                 $O := O \cup \{\texttt{DELETE } \{(s,p,o)\} \texttt{ INSERT } \{([\,],p,o)\} \texttt{ WHERE } H\}$;
14   **if** $o \in \mathbf{I} \wedge$ (check-object$((s,p,o),H) \vee o \in \bar{x}^P$) **then**
15                 $O := O \cup \{\texttt{DELETE } \{(s,p,o)\} \texttt{ INSERT } \{(s,p,[\,])\} \texttt{ WHERE } H\}$;
16   **return** O;

---

[a] I.e., with variables that do not appear in any G_j^U.

---

[1] See https://liris.cnrs.fr/~rdelanau/papers/ISWC2018_appx.pdf.

Theorem 1 (Soundness of Algorithm 1). *Let $\mathcal{P}$ be a privacy policy consisting of a single query and let $\mathcal{U}$ be a utility policy. Let $O = $ find-ops-unit$(\mathcal{P},\mathcal{U})$ computed by Algorithm 1. For all $o \in O$, for all RDF graph DB, $\mathcal{P}$ and $\mathcal{U}$ are satisfied by $o(DB)$ obtained by applying the update operation o to DB.*

The behavior of Algorithm 1 is illustrated in the following Example 4.

Example 4 (Example 1 cont'd). Consider the policies $\mathcal{P} = \{P_1, P_2\}$ and $\mathcal{U} = \{U_1, U_2\}$ given in Example 1 with bodies G_1^P, G_2^P, G_1^U and G_2^U, respectively. Let us consider two different runs of Algorithm 1. The call to find-ops-unit$(P_1,\mathcal{U})$ produces the following set O_1 of operations whereas the call to find-ops-unit$(P_2,\mathcal{U})$ produces O_2:

$O_1 = \{$DELETE $\{(?u, \text{vcard:hasAddress}, ?ad)\}$ WHERE G_1^P,

DELETE $\{(?u, \text{vcard:hasAddress}, ?ad)\}$ INSERT $\{([\,], \text{vcard:hasAddress}, ?ad)\}$ WHERE G_1^P,

DELETE $\{(?u, \text{vcard:hasAddress}, ?ad)\}$ INSERT $\{(?u, \text{vcard:hasAddress}, [\,])\}$ WHERE $G_1^P\}$

$O_2 = \{$DELETE $\{(?c, \text{tcl:user}, ?u)\}$ WHERE G_2^P,

DELETE $\{(?c, \text{tcl:user}, ?u)\}$ INSERT $\{([\,], \text{tcl:user}, ?u)\}$ WHERE G_2^P,

DELETE $\{(?c, \text{tcl:user}, ?u)\}$ INSERT $\{(?c, \text{tcl:user}, [\,])\}$ WHERE $G_2^P\}$

Indeed, there is only one way to satisfy P_1, U_1 and U_2: delete or update the address $?ad$ of each user $?u$ as shown in O_1. This goes by either deleting it, replacing the address value by a blank node in the hasAddress triple (possible since $?ad$ is also a distinguished variable), or replacing the user with a blank node (possible since there is another triple originating from the user variable $?u$ in the policy query body). Notice that the update or deletion of the triple {?u a tcl:User} is not authorized, because U_1 would not be satisfied.

The only acceptable operations for P_2, U_1 and U_2 as shown in O_2, are either to delete the link between users and their journeys, or replace each argument of this relation with a blank node. Replacing the subject of the considered triple (the journey variable $?c$) is possible since it is also featured as the subject of other triples in the query body, while replacing the object (the user variable $?u$) is possible since it is a distinguished variable of the privacy query.

5.2   Finding Candidate Sets of Operations for General Privacy Policies

We now extend the previous algorithm to the general case where $\mathcal{P}$ is a set of n queries. The idea is to compute operations that satisfy each P_i using Algorithm 1 and then to distribute the results. The soundness of this algorithm is encapsulated in Theorem 2 and its associated Corollary 1.

Algorithm 2. Find update operations to satisfy policies

> **Input**   : a privacy policy $\mathcal{P}$ made of n queries $P_i = \langle \bar{x}_i^P, G_i^P \rangle$
> **Input**   : a utility policy $\mathcal{U}$ made of m queries $U_j = \langle \bar{x}_j^U, G_j^U \rangle$
> **Output**: a set of sets of operations Ops such that each sequence obtained
>   from ordering any $O \in Ops$ satisfies both $\mathcal{P}$ and $\mathcal{U}$

1 **function** find-ops$(\mathcal{P}, \mathcal{U})$:
2   Let $Ops = \{\emptyset\}$;
3   **for** $P_i \in \mathcal{P}$ **do**
4   Let $ops_i :=$ find-ops-unit$(P_i, \mathcal{U})$;
5   **if** $ops_i \neq \emptyset$ **then** $Ops := \{O \cup \{o'\} \mid O \in Ops \wedge o' \in ops_i\}$;
6   **return** Ops;

Theorem 2 (Soundness of Algorithm 2). *Let $\mathcal{P}$ be a privacy policy and let $\mathcal{U}$ be a utility policy. Let $\mathcal{O} =$ find-ops$(\mathcal{P}, \mathcal{U})$ and let DB be an RDF graph. For any set of operations $O \in \mathcal{O}$, and for any ordering S of O, $\mathcal{P}$ and $\mathcal{U}$ are satisfied by $S(DB)$ obtained by applying to DB the sequence of operations in S.*

Theorem 2 guarantees the soundness of *all* sequences of operations that can be built from the output of Algorithm 2. Corollary 1 leverages this result for the COMPATIBILITY problem.

Corollary 1. *Let $\mathcal{P}$ be a privacy policy and let $\mathcal{U}$ be a utility policy made of counting and non-counting queries. If* find-ops$(\mathcal{P}, \mathcal{U}) \neq \emptyset$ *then the* COMPATI-BILITY *problem has* **True** *as a solution.*

Algorithm 2 guarantees the same robustness to linking attacks as [8]: for any anonymization $DB' = Anonym(DB)$ produced using Algorithm 2, its union with any RDF graph G satisfying the same privacy policy $\mathcal{P}$ will also satisfy $\mathcal{P}$. The reason is that the IRIs possibly common to DB' and G cannot be the images of a mapping from any privacy query. Indeed, the operations that have produced DB' have either deleted triples corresponding to triples in a privacy query or have replaced IRIs involved in mappings from privacy queries to DB with blank nodes (which cannot be joined with blank nodes in G).

The sets of operations produced by Algorithm 2 are *not* equivalent in the sense that they may delete *different sets* of triples in the dataset. Moreover, even for a *given* set of operations, the choice of a possible reordering of its operations may have different effects on the dataset. Indeed, deletions and modifications of triples are not commutative operations but due to the soundness of the algorithm, every obtained solution *satisfies* the privacy and utility policies.

Regarding the complexity of Algorithm 1, its result $O =$ find-ops-unit$(P, \mathcal{U})$ grows linearly with the size of P. Indeed, each triple in the body G^P of P produces at most one delete operation and two update operations. However, regarding the overall complexity of Algorithm 2, if each set O of operations $O \in \mathcal{O} =$ find-ops$(\mathcal{P}, \mathcal{U})$ has cardinality $|\mathcal{P}|$ by construction, the distribution of the results obtained by find-ops-unit on line 4 induces an exponential blowup on the size of $\mathcal{O}$ due to the cartesian product on Line 5. In our experimental assessment (Sect. 6), we will show that in practice the utility and privacy queries in $\mathcal{P}$ and $\mathcal{U}$ oftentimes overlap, thus decreasing drastically the actual number of sequences output by Algorithm 2, possibly to none.

6   Experimental Study

In this section, we present an empirical study devoted to gauge the efficiency of our main algorithm (Algorithm 2) and measure various factors that determine the impact of the overlap and size of the policy queries on its output. The experimental study is organized into three main parts: (1) experimental analysis of the risk of incompatibility between privacy and utility policies; (2) experimental evaluation of the impact of the privacy and utility policies on the number of anonymizations alternatives produced by Algorithm 2; (3) experimental evaluation of Algorithm 2 runtime performance.

Setup and Implementation. We adopted gMark [2], a schema-based synthetic graph and query workload generator, as a benchmark for our experimental study. We used gMark to define the schema of public transportation data, by including types and properties observed in real-world smart city open data platforms[2]. Due to the static nature of our approach, we only need to use such a schema to generate query workloads without the need of generating actual graph instances.

Precisely, we defined a schema with 13 data types and 12 properties capturing information regarding users (including personal data and subscription data for cardholders), ticket validations and user rides (such as geographic coordinates of ticket validations and optional subscription-related data), and information on the transportation network (such as maps). Using gMark, we then built a sample of 500 randomly generated conjunctive queries upon the aforementioned schema, each one containing between 1 and 6 distinguished variables with a size ranging between 1 and 6 triples. As shown in a recent study [3], queries of such size are the most frequent ones in a large corpus of real-world query logs extracted from SPARQL endpoints. This further corroborates our assumption that our query sample is representative of real-world queries formulated by end-users. To account for the structural variability of real-world queries, experiments were performed on workloads using different shapes of queries: chain queries, star queries, star-chain queries and a random mix of star-chain and star queries. For space reasons, we present the results for star-chain queries only. The full list of experiments[3] is available in a notebook at the project's GitHub repository[3].

To generate privacy and utility policies, we fix a number of conjunctive queries to be part of the privacy and utility policies. Then, we randomly pick as many queries as necessary in the query sample to build the policies based on this cardinality, while avoiding duplicates in the same policy and in between both kinds of policies.

---

[2] Notably, the Grand Lyon data website and datasets: https://data.grandlyon.com/.
[3] https://github.com/RdNetwork/Declarative-LOD-Anonymizer.

In all our experiments, we have opted for a balanced cardinality between privacy and utility policies: we have set the policy cardinality equal to 3 for the experiments in Sects. 6.1 and 6.2, whereas Sect. 6.3 features a more extreme case for performance testing with policy cardinality equal to 10. Depending on the experiment, policy size (i.e. the sum of the sizes of the conjunctive queries defining it) may vary since the picked queries have a varying size from 1 to 6.

The *overlap degree* between privacy and utility policies plays an important role in our experiments as a factor likely to impact the results of Algorithm 2. We define it as the ratio between the number of triples appearing in privacy queries that can be mapped to a triple appearing in a utility query and the total size of the privacy policy. More formally, let $\mathcal{P} = \{P_i\}$ and $\mathcal{U} = \{U_j\}$ be privacy and utility policies. The overlap degree between $\mathcal{P}$ and $\mathcal{U}$ is a real number in $[0 \ldots 1]$ defined as:

$$\frac{\sum_{i=1}^{n} |\{t \in G_i^P \mid \exists j \; \exists t' \in G_j^U \; \exists \mu \; \mu(t) = \mu(t')\}|}{\sum_{i=1}^{n} |G_i^P|}$$

Algorithm 2 returns $\emptyset$ as output when it is applied to privacy and utility policies having an overlap degree equal to 1, which are thus incompatible. In Sect. 6.1, we will measure the risk of incompatibility between randomly generated privacy and utility policies by counting the number of cases where this complete overlap occurs.

All our tests have been performed under Windows 10 on a Intel® Core™ i5-6300HQ CPU machine running at 2.30 GHz and 8 GB of RAM. We have implemented our algorithms using Python 2.7. The code of our working prototype along with the datasets and results of our experiments are made open-source and available at the aforementioned project's GitHub repository.

6.1 Measuring Compatibility Between Privacy and Utility Policies

Our goal is to measure the incompatibility rate of privacy and utility policies randomly generated with a fixed cardinality of 3 and a varying size.

We have performed two experiments where we vary the size of the privacy (resp. utility) policy from 6 to 12, which corresponds to privacy (resp. utility) queries having between 2 and 4 triples, while keeping the size of the utility (resp. privacy) policy fixed to 9, which corresponds to utility (resp. privacy) queries with 3 triples. In the first (resp. second) experiment, for each of the 7 privacy (resp. utility) policy sizes, we launch 200 executions of Algorithm 2 and we count the number of executions returning $\emptyset$, which allows to compute the proportion of incompatible policies. For space reasons, we omit the corresponding histograms (available in our online notebook) and we describe the obtained results in the following.

In both experiments, we observed that only **49.2%** and **49.3%** of the 1400 (i.e., corresponding to 200 runs multiplied by 7 data points) executions exhibit compatible policies. This result clearly shows the necessity of design-ing an algorithm which automatically verifies policy incompatibility prior to the anonymization process. It also reveals that even small policy cardinalities (equal to 3 for privacy and utility queries) can already substantially prevent possible anonymizations.

We also noted in the first experiment that the compatibility rate between privacy and utility policies tends to grow with the privacy policy size. This behavior is in clear contrast with the intuition that the more privacy policy is constrained, the less flexibility we have in satisfying them. The explanation however is that increasing the size of the privacy policy decreases the risk that all its triples are mapped with triples in the (fixed size) utility policy, and thus augments the possibilities of satisfying the privacy and utility policies by triple deletions.

We observe the opposite trend in the second experiment: the compatibility rate between privacy and utility policies decreases with the utility policy size. The reason is that requiring more utility for end-users restrains the possibilities of deleting data for anonymization purposes.

6.2   Measuring the Number of Anonymization Alternatives

When applied to compatible privacy and utility policies, Algorithm 2 computes the set of all the candidate sets of update operations that satisfy the input policies. In the worst case, the number of candidate sets corresponds the product of the sizes of the privacy queries. In this experiment, we want to evaluate how this number evolves in practice depending on (1) the overlap between privacy and utility policies, and (2) the total size of the privacy and utility policies.

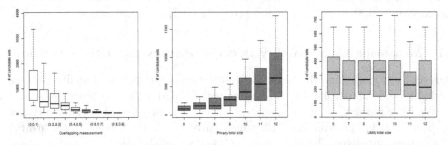

(a) Depending on overlap   (b) Depending on privacy size (c) Depending on utility size

Fig. 1. Candidate set length based on policy overlap, privacy size and utility size

Algorithm 2 has been run on 7000 randomly generated combinations of privacy and utility policies, thus covering a wide spectrum of combinations exhibiting various overlap degrees with various types of queries. For each execution, we compute the overlap degree between the input privacy and utility policies and group results in clusters of 10% before plotting as a boxplot the number of candidate sets in executions featuring the given overlap degree (Fig. 1a). This provides a representation of how many alternatives our algorithm provides for anonymizing a graph, depending on the policies overlap. The boxplot allows to visualize both extreme values and average trends, given that the randomization can easily create extreme cases and outlier values.

We can observe that the number of candidate sets quickly decreases when overlapping grows even slightly. This is easy to understand, given that increasing overlap degree induces that less deletion operations are permitted by the algorithm. As soon as the overlap degree reaches an high value, our algorithm provides very few anonymization alternatives since no possible operation exists to satisfy the given policies.

We use the same experimental settings as in Sect. 6.1 to evaluate how the number of candidate sets evolves as a function of policy size.

Figure 1b displays the results of this experiment when varying privacy size with a fixed utility size of 9 triples. We can observe a steady increase of the number of candidate sets with the privacy size. The explanation for this behavior is that increasing privacy size (with fixed utility size) provide more possible operations for the anonymization.

On the other hand, when varying utility size (with fixed privacy size), the number of candidate sets almost stagnates when increasing the utility size (Fig. 1c). This means that increasing the size of utility queries, without increasing the number of queries itself, does not significantly reduce the anonymization opportunities.

In short, this experiment emphasizes the faint influence of utility policies on possible anonymizations sets, along with the crucial role of privacy policies in shaping possible anonymization operations.

6.3 Runtime Performance

One of the benefits of dealing with a query-driven static method for anonymization is to avoid dealing with the size of an input graph, which could impact performance by increasing runtime. Our static approach only deals with policy size when looking for candidate anonymization sets, which is likely to make the algorithm simple and efficient in general.

To confirm this, we ran the Algorithm 2 for a batch of 100 executions corresponding to input privacy and utility policies of 10 queries each, and we measured the average running time. As a result, we have obtained an average runtime of **0.843 s** over all executions, which turns to be satisfactory in practice.

We can conclude that this static approach provides a fast way to enumerate all the candidate sets of anonymization operations.

7   Conclusion and Future Work

We presented in this paper a novel query-based approach for Linked Open Data anonymization under the form of delete and update operations on RDF graphs. We consider policies as sets of privacy and utility specifications, which can be readily written as queries by the data providers. We further designed a data-independent algorithm to compute sets of anonymization operations guaranteed to satisfy both privacy and utility policies on any input RDF graph. Our proof-of-concept open-source implementation confirms the intuition that (i) the larger is the utility policy, the lesser anonymization operations are available; (ii) the opposite holds for privacy policy but with a stronger impact on the number of candidate anonymization operations; (iii) the more privacy and utility policies are interleaved, the lesser is the number of candidate operations.

Our query-based approach can be combined with ontology-based query rewriting and thus can support reasoning for first-order rewritable ontological languages such as RDFS [20], DL-Lite [5] or EL fragments [9]. More precisely, given a pair of privacy and utility policies made of conjunctive queries defined over an ontology, each set of anonymization operations returned by Algorithm 2 applied to the two sets of their corresponding conjunctive rewritings (obtained using existing query rewriting algorithms [4,5,9]) will produce datasets that are guaranteed to satisfy the policies.

It is also important to emphasize that our approach can be combined with other anonymization approaches (k-anonymity techniques, differential privacy) after the transformation of an input RDF graph by the application of a sequence of operations output by Algorithm 2.

We are planning several orthogonal research directions for future work. A first research direction consists in extending the expressivity of the queries considered in this paper both for defining the policies and the anonymization operations. More expressive privacy and utility queries (with FILTER, NOT EXISTS, aggregate functions) fits in our general framework (Sect. 4) but requires extensions of the Sect. 5 algorithms. In addition to triple deletion and IRI replacement with blank nodes, other anonymization operations can be considered such as value replacement in triples involving datatype properties and IRI aggregation. The point is that each of these operations can be defined with (possibly complex) queries by leveraging SPARQL 1.1 aggregate and update queries as well as calls to built-in functions.

Another future extension is the study of data-dependent solutions, as opposed to the *data-independent* approach introduced in this paper. The proof of Theorem 1 relies on the fact that all instances of the utility queries are completely left unmodified by the deletions operations. However, it may happen that *some* instances common to the privacy and utility queries are suppressed without impacting the answers of the utility queries evaluated over a *given dataset*. An alternative is thus to consider data-dependent solutions, at the cost of running the algorithm on the (possibly huge) dataset. Such an approach could be adopted when no data-independent solution can be found.

Another research direction we envision is to consider an optimization problem that extends the ENUMOPERATIONS problem defined in Sect. 5. The optimization problem consists in finding *optimal* sequences of anonymization operations and not all sequences, where optimality can be defined as minimality w.r.t. a partial order over sequences of anonymization operations (e.g., their size, or a distance between the original and resulting datasets.)

Acknowledgements. This work has been supported by the Auvergne-Rhône-Alpes region through the ARC6 research program for funding Remy Delanaux's PhD, by the LabEx PERSYVAL-Lab (ANR-11-LABX-0025-01), the SIDES 3.0 project (ANR-16-DUNE-0002) funded by the French Program Investissement d'Avenir and the Palse Impulsion 2016/31 programme (ANR-11-IDEX-0007-02) at UDL.

References

1. Baader, F., Borchmann, D., Nuradiansyah, A.: Preliminary results on the identity problem in description logic ontologies. In: Description Logics. CEUR Workshop Proceedings, vol. 1879. CEUR-WS.org (2017)
2. Bagan, G., Bonifati, A., Ciucanu, R., Fletcher, G.H.L., Lemay, A., Advokaat, N.: gMark: schema-driven generation of graphs and queries. IEEE Trans. Knowl. Data Eng. **29**(4), 856–869 (2017)
3. Bonifati, A., Martens, W., Timm, T.: An analytical study of large SPARQL query logs. PVLDB **11**(2), 149–161 (2017)
4. Bursztyn, D., Goadoué, F., Manolescu, I.: Reformulation-based query answering in RDF: alternatives and performance. PVLDB **8** (2015)
5. Calvanese, D., Giacomo, G.D., Lembo, D., Lenzerini, M., Rosati, R.: Tractable reasoning and efficient query answering in description logics: the DL-Lite family. J. Autom. Reason. **39**(3), 385–429 (2007)
6. Dwork, C.: Differential privacy. In: Bugliesi, M., Preneel, B., Sassone, V., Wegener, I. (eds.) ICALP 2006. LNCS, vol. 4052, pp. 1–12. Springer, Heidelberg (2006). https://doi.org/10.1007/11787006_1
7. Fung, B.C.M., Wang, K., Chen, R., Yu, P.S.: Privacy-preserving data publishing: a survey of recent developments. ACM Comput. Surv. **42**(4), 14:1–14:53 (2010)
8. Grau, B.C., Kostylev, E.V.: Logical foundations of privacy-preserving publishing of linked data. In: AAAI, pp. 943–949. AAAI Press (2016)
9. Hansen, P., Lutz, C., Seylan, I., Wolter, F.: Efficient query rewriting in the description logic el and beyond. In: IJCAI (2015)
10. Heitmann, B., Hermsen, F., Decker, S.: k – RDF-neighbourhood anonymity: combining structural and attribute-based anonymisation for linked data. In: PrivOn@ISWC. CEUR Workshop Proceedings, vol. 1951. CEUR-WS.org (2017)
11. Kirrane, S., Mileo, A., Decker, S.: Access control and the resource description framework: a survey. Semant. Web **8**(2), 311–352 (2017)
12. Kirrane, S., Villata, S., d'Aquin, M.: Privacy, security and policies: a review of problems and solutions with semantic web technologies. Semant. Web J. **9**(2), 153–161 (2018)
13. Li, N., Li, T., Venkatasubramanian, S.: t-Closeness: privacy beyond k-Anonymity and l-Diversity. In: ICDE, pp. 106–115. IEEE Computer Society (2007)
14. Machanavajjhala, A., He, X., Hay, M.: Differential privacy in the wild: a tutorial on current practices & open challenges. PVLDB **9**(13), 1611–1614 (2016)
15. Machanavajjhala, A., Kifer, D., Gehrke, J., Venkitasubramaniam, M.: L-diversity: privacy beyond k-anonymity. TKDD **1**(1), 3 (2007)
16. Oulmakhzoune, S., Cuppens-Boulahia, N., Cuppens, F., Morucci, S.: Privacy policy preferences enforced by SPARQL query rewriting. In: ARES, pp. 335–342. IEEE Computer Society (2012)
17. Radulovic, F., García-Castro, R., Gómez-Pérez, A.: Towards the anonymisation of RDF data. In: SEKE, pp. 646–651. KSI Research Inc. (2015)

18. Sweeney, L.: k-Anonymity: a model for protecting privacy. Int. J. Uncertain. Fuzziness Knowl.-Based Syst. **10**(5), 557–570 (2002)
19. Villata, S., Delaforge, N., Gandon, F., Gyrard, A.: An access control model for linked data. In: Meersman, R., Dillon, T., Herrero, P. (eds.) OTM 2011. LNCS, vol. 7046, pp. 454–463. Springer, Heidelberg (2011). https://doi.org/10.1007/978-3-642-25126-9_57
20. W3C: RDF schema 1.1 (2004). http://www.w3.org/TR/rdf-schema/

Answering Provenance-Aware Queries on RDF Data Cubes Under Memory Budgets

Luis Galárraga[1,2](✉), Kim Ahlstrøm[2], Katja Hose[2], and Torben Bach Pedersen[2]

[1] Inria, Rennes, France
luis.galarraga@inria.fr
[2] Aalborg University, Aalborg, Denmark
{kah,khose,tbp}@cs.aau.dk

Abstract. The steadily-growing popularity of semantic data on the Web and the support for aggregation queries in SPARQL 1.1 have propelled the interest in Online Analytical Processing (OLAP) and data cubes in RDF. Query processing in such settings is challenging because SPARQL OLAP queries usually contain many triple patterns with grouping and aggregation. Moreover, one important factor of query answering on Web data is its provenance, i.e., metadata about its origin. Some applications in data analytics and access control require to augment the data with provenance metadata and run queries that impose constraints on this provenance. This task is called provenance-aware query answering. In this paper, we investigate the benefit of caching some parts of an RDF cube augmented with provenance information when answering provenance-aware SPARQL queries. We propose provenance-aware caching (PAC), a caching approach based on a provenance-aware partitioning of RDF graphs, and a benefit model for RDF cubes and SPARQL queries with aggregation. Our results on real and synthetic data show that PAC outperforms significantly the LRU strategy (least recently used) and the Jena TDB native caching in terms of hit-rate and response time.

1 Introduction

In the last years we have seen a steady increase of the amount of Linked Data available on the Web. This data spans over a wide variety of topics ranging from common-sense information to specialized domains such as governmental information, media, life sciences, etc. The data is usually published in RDF [27] and queried with SPARQL [28]. The extended capabilities of SPARQL 1.1—notably the support for aggregation queries—have motivated the publication of multidimensional data, i.e., data cubes, in RDF [1,11,19,31]. Analysis on multidimensional data warehouses and is known as OLAP (On-Line Analytical Processing). The publication of the QB vocabulary [8] has served as a bridge between the Semantic Web and OLAP communities.

© Springer Nature Switzerland AG 2018
D. Vrandečić et al. (Eds.): ISWC 2018, LNCS 11136, pp. 547–565, 2018.
https://doi.org/10.1007/978-3-030-00671-6_32

Imposing constraints on the provenance of query results, a task known as provenance-aware query processing, is crucial in a setting with data coming from multiple independent sources. The provenance of a piece of data is a set of assertions about its origin. If a query must be evaluated over multiple independent data cubes [20], provenance metadata can help restrict OLAP queries to data or sources meeting certain quality constraints [24]. Moreover, provenance metadata allows for the implementation of access control policies on data [4].

The importance of provenance data management motivated the creation of the PROV ontology [23] (PROV-O), the W3C standard to represent provenance information for RDF data. PROV-O provides the data model to describe a set of provenance entities, i.e., RDF resources, which are assigned to the triples in an RDF dataset. There exist multiple representations for provenance-augmented RDF data, such as named graphs and reification. These representations are, though, not exempt from performance issues for very complex queries [25] due to the additional complexity added by the provenance metadata. In this paper we use the named graph representation [5].

A recent formulation for provenance-aware query answering divides the query in two parts: a provenance query and an analytical query [2,35]. The provenance query imposes constraints on the provenance entities of the triples that should be considered to answer the analytical query on the actual data. In a representation of provenance entities using named graphs, this is equivalent to adding a FROM clause to the analytical query for each provenance entity reported by the provenance query. As shown in [2], query response time is seriously affected in frameworks, such as Jena, as the number of FROM clauses increases. This happens because the query engine has to fetch a large number of intermediate results from disk in order to answer the analytical query.

In this paper we propose to alleviate the aforementioned phenomenon by caching some *fragments* of the RDF graph in memory, so that the analytical queries benefit from fast access to the data. Our strategy, called *provenance-aware caching* (PAC), selects the most *beneficial* fragments that fit within a memory budget. Since we are interested in RDF cubes, we assume our queries are OLAP queries, i.e., SPARQL queries with aggregation, grouping and filtering. This assumption reduces the space of possible queries we optimize for, without the need of an explicit query-load. We show that for OLAP analytical queries, it suffices to cache a small percentage of the dataset in order to achieve up to 2x speed-up in query response time. This is particularly convenient for memory-constrained settings. In summary our contributions are:

- A fragmentation scheme tailored for provenance-augmented RDF graphs.
- The formulation of the budgeted provenance-enabled fragment selection problem: The problem of selecting a set of fragments for caching so that as many OLAP queries as possible benefit from fast access to cached data.
- A query rewriting algorithm to answer analytical queries from a set of named graphs and cached fragments.
- A study of the impact of caching on the performance of Jena TDB for provenance-aware SPARQL aggregation queries.

The remainder of this paper is structured as follows. Section 2 introduces the basic concepts of RDF cubes, SPARQL aggregation queries and provenance in RDF. In Sect. 3, we introduce the fragment selection problem with a memory budget for RDF cubes with provenance information. This is followed by an experimental evaluation in Sect. 4 and a discussion of related work in Sect. 5. Section 6 concludes the paper.

2 Preliminaries

2.1 RDF Cubes

In compliance with the official RDF specification [27], we define an *RDF triple t* (or simply a *triple*), as $t = \langle s, p, o \rangle \in (\mathcal{U} \cup \mathcal{B}) \times \mathcal{P} \times (\mathcal{U} \cup \mathcal{B} \cup \mathcal{L})$, where s is the subject, p is the predicate, and o is the object. In this definition, $\mathcal{U}$, $\mathcal{B}$ and $\mathcal{L}$ are countably infinite sets of IRIs, blank nodes and literals. In addition, we define the set of predicates $\mathcal{P} \subseteq \mathcal{U}$ and the set of classes $\mathcal{C} \subseteq \mathcal{U}$. An *RDF dataset* $\mathcal{K}$ is a set of RDF triples. Since RDF defines a graph-like data model, we also refer to RDF datasets as *RDF graphs*. An *RDF cube* $\mathcal{K}_c = \{\mathcal{O}, \mathcal{D}, \mathcal{P_M}, \mathcal{P_D}, \mathcal{P_A}, \Delta\}$ is an RDF graph defined in terms of:

- A set of observations $\mathcal{O} \subseteq \mathcal{U}$.
- A set of measure predicates $\mathcal{P_M} \subseteq \mathcal{P}$, defined between observations and literal numerical values. These predicates are the target of aggregation in OLAP queries.
- A set of dimensions $\mathcal{D}$. Observations are defined by their coordinates in $\mathcal{D}$. Each dimension consists of a hierarchy of classes that describe an observation at different degrees of specificity. Each class defines a *level* in the hierarchy.
- A set of dimension predicates $\mathcal{P_D} \subseteq \mathcal{P}$. These predicates connect the observations with the dimensions in $\mathcal{D}$.
- A set of level attributes $\mathcal{P_A} \subseteq \mathcal{P}$. Level attributes are predicates defined on the class levels of the dimensions. They are often used for grouping and filtering.
- A function $\Delta : \mathcal{D} \rightarrow \mathcal{H}$ that assigns each dimension in $\mathcal{D}$ a *class hierarchy* from the set of hierarchies $\mathcal{H}$. A class hierarchy $H = (L, \prec_L, \gamma, \sigma) \in \mathcal{H}$ consists of a set of class levels $L \subseteq \mathcal{C}$ and a partial order $\prec_L$ on L with a single greatest element. The function $\gamma : L \rightarrow 2^{\mathcal{P_D}}$ assigns each class level in L a set of dimension predicates, whereas the function $\sigma : L \rightarrow 2^{\mathcal{P_A}}$ assigns each class level a set of level attributes.

Example 1. Consider an RDF cube representing a database of air pollution measurements. Each measurement corresponds to an observation in the cube model. A measurement of $12.3 \, \mu\text{g/m}^3$ of the pollutant PM_{10} corresponds to the triple $\langle Obs,\ air{:}pm10,\ 12.3 \rangle$ depicted in Fig. 1. It follows that $Obs \in \mathcal{O}$ and $air{:}pm10 \in \mathcal{P_M}$. The triple $\langle Obs,\ air{:}station,\ St1 \rangle$ defines the coordinates of Obs in the Station dimension. There are three dimensions in our example, i.e., $\mathcal{D} = \{ Year, Station, Sensor \}$. The predicates *air:year, air:station, air:sensor* $\in$

$\mathcal{P}_\mathcal{D}$ are dimension predicates. The function Δ associates each dimension with a class hierarchy. For example, the dimension *Station* is mapped to a hierarchy $H = (L, \prec_L, \gamma, \sigma)$ defined by the classes $L = \{Station, City, Country\}$ and the order *Station* $\prec_L$ *City* $\prec_L$ *Country*. In addition, it holds that $\gamma(Country) = \{air{:}locatedIn\}$. The country's code can be modeled as a level attribute $air{:}ccode \in \mathcal{P}_\mathcal{A}$ of the *Country* class. Thus, $\sigma(Country) = \{air{:}ccode\}$.

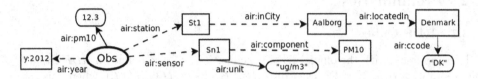

Fig. 1. Observation with 3 dimensions: year, station, and sensor. Measure predicates $\mathcal{P}_\mathcal{M}$ are in solid line style, whereas attribute predicates $\mathcal{P}_\mathcal{A}$ use dotted lines. The dashed edges correspond to the dimension properties $\mathcal{P}_\mathcal{D}$.

2.2   SPARQL Queries

Due to space constraints, we do not provide a rigorous definition of SPARQL queries; instead we resort to the formulation used in [17] to define SPARQL aggregation queries. These are the most common types of OLAP queries. We define a *triple pattern* as a triple $\hat{t} = \langle s, p, o \rangle \in (\mathcal{U} \cup \mathcal{B} \cup \mathcal{V}) \times (\mathcal{P} \cup \mathcal{V}) \times (\mathcal{U} \cup \mathcal{B} \cup \mathcal{L} \cup \mathcal{V})$. The set $\mathcal{V}$ is a set of variables with $(\mathcal{U} \cup \mathcal{B} \cup \mathcal{L}) \cap \mathcal{V} = \emptyset$. A *basic graph pattern* G_p is a set of triple patterns. A SPARQL select query Q is an expression of the form "SELECT V F WHERE $\hat{G}_p$ GROUP BY V' HAVING c" with $V \cup F \neq \emptyset$. In this definition $V \subseteq \mathcal{V}$ is the set of projection variables and F is a set of aggregation expressions of the form $f(g(\hat{V}))$ where $f \in \{COUNT, SUM, AVG, MIN, MAX\}$ and $g(\hat{V})$ is a numerical expression on the set of aggregated variables $\hat{V} \subseteq \mathcal{V}$. $\hat{G}_p$ is an extended basic graph pattern, potentially containing *OPTIONAL* and *FILTER* clauses. The set of grouping variables is a superset of the projection variables ($V' \supseteq V$) whereas c is a Boolean expression on $V \cup F$. The GROUP BY and HAVING clauses are optional.

Example 2. The following SPARQL query computes the maximal concentration of PM_{10} per city in Denmark in 2012 according to the schema in Fig. 1.

```
SELECT ?city (MAX(?ms) as ?max) WHERE {
  ?obs air:pm10 ?ms. ?obs air:year y:2012. ?obs air:station ?st.
  ?st air:inCity ?city. ?city air:locatedIn Denmark.
} GROUP BY ?city
```

2.3   Provenance

There exist multiple provenance models for RDF in the literature [7]; in this paper, we focus on *workflow provenance* [29]: the history of a unit of information from its sources to its current state. This provenance is modeled using RDF by assigning each triple an RDF resource, which we call its *provenance entity*. The set of statements describing the provenance entities of an RDF graph is a *provenance graph*. The PROV Ontology [23] is the W3C specification to model provenance graphs. In this model a provenance entity can represent a data resource such as a file, a web page or the intermediate result of a data transformation process. Any operation on data is modeled as an activity in PROV-O. Those activities can be directly or indirectly carried out by agents: people, organizations or even computer programs.

If $\mathcal{I}$ is a set of *provenance entities* and $f : \mathcal{K} \rightarrow \mathcal{I}$ is a provenance function on the triples of an RDF graph $\mathcal{K}$, a *provenance-augmented RDF graph* $\mathcal{K}_{\mathcal{I}}$ is a set of pairs $\langle t, f(t) \rangle$, which can also be seen as a set of quadruples $\langle s, p, o, i \rangle$, where $i \in \mathcal{I}$. We can also model a provenance-augmented RDF graph as a set of RDF sub-graphs, each containing the triples associated to the same provenance entity. In this view $\mathcal{K}_{\mathcal{I}} = \{\mathcal{K}_{i_1}, \ldots, \mathcal{K}_{i_n}\}$ $(i_1, \ldots, i_n \in \mathcal{I})$ and each RDF sub-graph is a named graph with label i. We define a *provenance-augmented cube* as an RDF cube whose triples have been augmented with provenance entities.

2.4   Provenance-Aware Query Answering

Given a provenance-augmented RDF graph $\mathcal{K}_{\mathcal{I}} = \{\mathcal{K}_{i_1}, \ldots, \mathcal{K}_{i_n}\}$ and a provenance graph $\mathcal{G}_{\mathcal{I}}$ describing the set of provenance entities $i_1, \ldots, i_n \in \mathcal{I}$, a provenance-aware query is a pair of SPARQL queries $\langle q_p, q_a \rangle$ [2,35]. q_p is known as the *provenance query* and is defined on $\mathcal{G}_{\mathcal{I}}$. The provenance query is designed so that it returns a set of provenance entities $I \subseteq \mathcal{I}$. Those provenance entities are used to restrict the scope of the *analytical query* q_a on the RDF graph $\mathcal{K}_{\mathcal{I}}$ to those subgraphs with labels in I. The problem of answering provenance-aware queries on RDF data has been studied in the last years [2,6,34,35]. If provenance information is modeled using named graphs (where the labels are provenance entities), the naive strategy is to augment the analytical query with a FROM clause for every provenance entity reported by the provenance query. In [2] it is shown that this strategy performs poorly in frameworks such as Jena for non-selective provenance queries. A strategy called *full materialization* [35] proposes to first fetch all the triples from the named graphs, and then run the analytical query on the union of those graphs. While this strategy generally outperforms the naive approach, it is not free from performance issues for non-selective analytical queries. Nonetheless, we observe that both strategies require the retrieval of a large number of triples from disk. Therefore, we study the impact of keeping some parts or *fragments* of the dataset in main memory so that queries can benefit from fast access to the data.

3   The Budgeted Provenance-Enabled Fragment Selection Problem

In this section, we define the *budgeted provenance-enabled fragment selection problem*. This is the problem of selecting a set of provenance-enabled RDF data fragments for caching so that we reduce the response time of the analytical query when answering a provenance-enabled query. This is achieved by maximizing the amount of data that is retrieved from the cache. In the following, we describe the three components of our approach, namely the fragmentation strategy, the cost-benefit model, and the query rewriting algorithm. We highlight that our method is query-load agnostic, thus it aims at optimizing for as many queries as possible in the space of analytical queries.

3.1   Fragmentation Strategy

A fragmentation strategy defines how to split a dataset into smaller parts, i.e., fragments. Once the dataset is fragmented, we can decide which parts to put in the cache. We start by defining a fragment for provenance-augmented RDF graphs.

Definition 1. *A fragment signature s_ϕ is a quadruple $\langle s, p, o, i \rangle$ such that each component can be a constant or a variable. We say a quadruple q in a provenance-augmented RDF graph $\mathcal{K}_\mathcal{I}$ matches a fragment signature if there exists an instantiation ρ for the variables in the signature such that $\rho(s_\phi) = q$. The set ϕ of quadruples that match s_ϕ in a provenance-augmented RDF graph $\mathcal{K}_\mathcal{I}$ is a fragment.*

Definition 2. *A provenance-aware fragment tree Φ consists of a set of fragments and a partial order $\sqsupseteq_f$ on those fragments. A fragment ϕ subsumes a fragment ϕ', denoted by $\phi \sqsupseteq_f \phi'$, iff $s_{\phi'} \Rightarrow s_\phi$. If $\phi \sqsupseteq_f \phi'$ then $\phi \supseteq \phi'$.*

Figure 2 shows a provenance-aware fragment tree describing some fragments from the cube introduced in Example 1 with two provenance entities *pr:e1* and *pr:e2*. The root fragment contains the trivial signature, i.e., the signature that matches all quadruples in the dataset. In the second level, we have signatures with restrictions on the provenance entities of the quadruples. The fragments in the third level have restrictions on both the predicate and the provenance entity. Fragments always subsume their children.

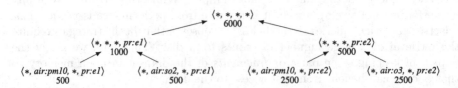

Fig. 2. A provenance-aware fragment tree. The size of each fragment is noted below.

Algorithm 1 describes the method to construct a provenance-aware fragment tree given a provenance-augmented RDF graph. The algorithm first initializes the tree with the trivial signature (line 1). Then for each quadruple in the dataset, the method constructs signatures with (a) bound provenance entity, (b) bound provenance entity and predicate (line 3), and (c) bound provenance entity, predicate and object for the *rdf:type* predicate (lines 4–5). The latter step accounts for the typically large size of the *rdf:type* predicate. If the tree does not contain a signature, the signature is initialized (by setting its size to 1 in line 8) and added to the tree (line 9). Otherwise, the size of the signature is incremented to account for the current quadruple (line 11).

Algorithm 1. BuildProvenanceAwareFragmentTree

Input: a provenance-augmented RDF graph: $\mathcal{K}_\mathcal{I}$
Output: a provenance-aware fragment tree: Φ

1   $\Phi := \{\langle *, *, *, * \rangle\}$
2   **foreach** $q := \langle s, p, o, i \rangle \in \mathcal{K}_\mathcal{I}$ **do**
3     $\Phi' := \{\langle *, *, *, i \rangle, \langle *, p, *, i \rangle\}$
4   **if** $p =$ *"rdf:type"* **then**
5       $\Phi' := \Phi' \cup \{\langle *, p, o, i \rangle\}$
6   **foreach** $s'_\phi \in \Phi'$ **do**
7   **if** $s'_\phi \notin \Phi$ **then**
8         $s'_\phi.size := 1$
9         $\Phi := \Phi \cup s'_\phi$
10   **else**
11        $s'_\phi.size := s'_\phi.size + 1$

12 **return** Φ

When clear from the context, we drop the distinction between a fragment and its signature and refer to both as ϕ. Finally, we highlight that Algorithm 1 produces fragmentation schemes with redundancy. Whether we allow redundancy or not in the set of selected fragments depends on the benefit model. We elaborate on this in the following.

3.2 Cost-Benefit Model

The cost-benefit model quantifies the price we have to pay for caching a fragment, as well as the amount of saved response time induced by using the cached fragment to answer queries. In line with approaches for view materialization [17] we use the number of quadruples matched by the fragment's signature as its cost, i.e., $cost(\phi) = |\phi|$. We say a fragment ϕ is relevant to a provenance-aware query $q = \langle q_p, q_a \rangle$ if at least one of the quadruples in ϕ can be used to answer q—and none of them could lead to a wrong answer. For example, consider a provenance query q_p with result $pr{:}e1$ and the analytical query q_a from Example 2. In this case the analytical query is restricted by the provenance query to the quadruples with provenance $pr{:}e1$. We say that the fragments ϕ and ϕ' with signatures $s_\phi = \langle *, \ air{:}pm10, \ *, \ pr{:}e1 \rangle$ and $s_{\phi'} = \langle *, \ *, \ *, \ pr{:}e1 \rangle$ are relevant to q. In contrast, the fragment ϕ'' with signature $s_{\phi''} = \langle *, \ p{:}unitPrice, \ *, \ pg{:}e2 \rangle$ is not relevant, because q_p does not consider the provenance entity $pg{:}e2$.

Under the assumption that the cost of accessing a quadruple from main memory is insignificant compared to the cost of accessing it from disk, we define the benefit of a cached fragment ϕ as $ben(\phi) = \sum_{q_a \in \mathcal{Q}} |\phi_{q_a} \cap \phi|$. Here $\mathcal{Q}$ is the space of all possible queries and ϕ_{q_a} is the set of quadruples required by the query engine to answer q_a. In other words, the benefit of a cached fragment is the *absolute* number of times one of its quadruples can be fetched to answer a query. It follows that the total benefit of a set of cached fragments $\Phi' \subseteq \Phi$ from a tree Φ, is given by $ben(\Phi') = \sum_{q_a \in \mathcal{Q}} |\phi_{q_a} \cap u(\Phi')|$, with $u(\Phi') = \bigcup_{\phi \in \Phi'} \phi$. Our goal is to find a Φ' with maximal $ben(\Phi')$.

We highlight that in real-world query engines, the benefit of a fragment w.r.t a query q_a may not necessarily depend on the *absolute* number of cached relevant quadruples used to answer q_a, but on the ratio w.r.t the query's relevant set, i.e., $\frac{|\phi_{q_a} \cap u(\Phi')|}{|\phi_{q_a}|}$. For example, it may be more beneficial to retrieve 1000 cached quadruples for a query with $|\phi_{q_a}| = 1000$ than for a query with $|\phi_{q_a}| = 10000$. In the absence of an explicit query load, however, we can only expect to *estimate* the term $|\phi_{q_a} \cap u(\Phi')|$ since the queries q_a as well as ϕ_{q_a} and $|\phi_{q_a}|$ are unknown. In the following we show that the fixed structure of RDF cubes as well as our focus on OLAP queries, both provide hints to guarantee that $ben(\Phi')$ is at least large.

Observation 1: *Distance to observations. The closer to the observations a predicate lies in the schema, the larger the relevance set of its matching fragments is.*

For example, **all OLAP queries** on data cubes involve aggregation of at least one measure. This means that $|\phi_{q_a} \cap \phi| > 0$ for fragments ϕ that match measure quadruples. Furthermore and assuming connected SPARQL queries, filtering or grouping on attributes and dimensions in higher levels always requires to *pass by* the lower levels, hence it is more beneficial to cache quadruples with predicates in the lower levels.

Observation 2: *Diversity. Fragments with larger diversity of predicates have larger relevance sets, i.e., they "touch" more queries.*

Observation 3: *Duplicates. Given a selection of fragments Φ, duplicate quadruples lying in different fragments do not provide additional benefit, because they occupy extra memory without extending the set of relevant queries of Φ.*

Based on these observations and the fragmentation defined by the provenance-aware fragment tree, we devise a selection strategy given a memory budget.

3.3   Fragment Selection

Given a provenance-augmented RDF cube $\mathcal{K}_c = \{\mathcal{O}, \mathcal{D}, \mathcal{P}_\mathcal{M}, \mathcal{P}_\mathcal{D}, \mathcal{P}_\mathcal{A}, \Delta\}$, a maximum budget W in number of quadruples, and a provenance-aware fragment tree Φ, we formulate the *budgeted provenance-enabled fragment selection problem* as an integer linear program (ILP):

$$\text{maximize} \quad \sum_{\phi \in \Phi} d_o(\phi)^{-1} \times d_v(\phi) \times x_\phi$$

$$\text{s.t.} \quad \sum_{\phi \in \Phi} |\phi| \times x_\phi \leq W \quad \text{(budget)}$$

$$\forall p : \phi_{root} \to \cdots \to \phi_k \; : \sum_{\phi \in p} x_\phi \leq 1 \quad \text{(no replication)}$$

$$\forall \phi \in \Phi : x_\phi \in \{0, 1\} \quad \text{(integrality constraints)} \tag{1}$$

Each fragment ϕ in the lattice is assigned a Boolean variable x_ϕ. If $x_\phi = 1$ the fragment is chosen for caching. Hence, the solution to the ILP produces a set of fragments $\Phi_{cached} \subset \Phi$ that will be stored in main memory. Observations 1 and 2 are implemented in the objective function. This function decreases with the distance of the fragment's predicates to the observations (d_o) and increases with the fragment's diversity (d_v)[1]. We define the distance of a predicate to the observations as the number of hops from an observation to the predicate in the schema. In Fig. 1, for example, the predicates *air:pm10* and *air:unit* have distances 1 and 2 respectively. If a fragment ϕ contains quadruples with different predicates, $d(\phi)$ is the smallest distance among all predicates in ϕ. The diversity, on the other hand, is the number of different predicates in quadruples in ϕ. The cost model is encoded in the *budget* constraint. Since duplicate quadruples do not contribute with additional benefit (Observation 3), the *no replication constraint* guarantees that the resulting set Φ_{cached} has no redundancy. Because of this constraint and the partial order encoded in the tree, the ILP solver can pick at most one fragment in a given path from the root to a leaf.

3.4   Query Rewriting

In this section, we describe how to use a selection of cached fragments Φ_{cached} to answer provenance-aware queries $q = \langle q_p, q_a \rangle$. Recall from Sect. 2.4 that in a setting based on named graphs, a provenance-aware query can be answered by adding a FROM clause to the analytical query q_a for each provenance entity $i \in I$ reported by q_p. Each provenance entity corresponds to a named graph that resides in disk. In our setting we count additionally on a set of cached fragments Φ_{cached} that can be accessed from memory. We treat each fragment $\phi \in \Phi_{cached}$ as a memory named graph with label $id(\phi)$, where $id(\phi)$ returns the

---

[1] We omitted the fragment size from the objective function because size benefits too much those fragments with general signatures, i.e., those of the form $\langle *, *, *, i \rangle$.

concatenation of the constant components of s_ϕ (ϕ's signature). Exploiting those memory named graphs to answer the analytical query is the goal of Algorithm 2. The algorithm takes as input an analytical query q_a, a provenance-aware tree Φ_{tree}, the result of the provenance query I, and the set of cached fragments Φ_{cached} reported by our selection strategy in Sect. 3.3. The algorithm returns the graph labels that will be added as FROM clauses to the analytical query.

Algorithm 2. rewriteAnalyticalQuery

Input: Analytical query q_a, provenance-aware tree Φ_{tree}, the provenance query result I, the set of cached fragments Φ_{cached}

Output: A set of graph labels

1   $candidates := relevant := disk := \emptyset$
2   **foreach** $t = \langle s, p, o \rangle \in q_a$ **do**
3   **foreach** $i \in I$ **do**
4          $s := \{\langle *, p, o, i\rangle, \langle *, p, *, i\rangle\} \cap \Phi_{tree}$
5          $relevant := relevant \cup \{\text{most-specific-fragment-in}(s)\}$

6   **foreach** $\phi \in relevant$ **do**
7   **if** $\phi \in \Phi_{cached}$ **then**
8          $candidates := candidates \cup \{\phi\}$
9   **else if** $\exists \, \phi' \in \Phi_{cached} : s_{\phi'} \sqsupseteq_f s_\phi$ **then**
10         $candidates := candidates - \{\hat{\phi} : s_{\phi'} \sqsupseteq_f s_{\hat{\phi}}\} \cup \{\phi'\}$
11   **else**
12         $disk := disk \cup \{i : s_\phi = \langle *, p, *, i\rangle\}$
13         $candidates := candidates - \{\phi : s_\phi \approx \langle -, -, -, i\rangle\}$

14   **return** $\{id(\phi) : \phi \in candidates \cup disk)\}$

Line 1 initializes some intermediate variables. Lines 2–5 compute the most specific fragments in the lattice that are relevant to the analytical query, i.e., fragments whose signatures combine the provenance entities in I with the triple patterns of the analytical query. Then for each relevant fragment ϕ, the algorithm verifies whether ϕ is in the cache (line 9). If so, the fragment is added as a candidate (line 8). Otherwise, the algorithm verifies whether one of ϕ's ancestors (line 9) has been cached. There can be at most of one of such ancestors due to the redundancy constraint discussed in Sect. 3.3. If an ancestor ϕ' is found in the cache, the algorithm adds it to the set of candidates (line 10). Since this addition turns every (possibly) selected descendant of ϕ' redundant, the algorithm removes those descendants from the list of candidates (line 10). If neither ϕ nor any of its ancestors is in the cache, the algorithm takes as candidate the named graph labeled with the provenance entity i in s_ϕ (line 12). This step turns any fragment with signature of the form $\langle -, -, -, i\rangle$ superfluous, and thus unnecessary (line 13). Once the final list of candidates have been computed, Algorithm 2 generates the graph labels that will be used to rewrite the analytical query (line 14).

4   Experiments

4.1   Experimental Setup

Data. We evaluated PAC on several datasets generated with the Star Schema Benchmark (*SSB* [26]) and on the QBOAirbase dataset [11]. The SSB benchmark provides a data generator for a database of line orders processed by a wholesaler. The number of line orders is an argument for the data generator. We converted the SSB dataset into an RDF cube, where each line order corresponds to an observation defined by four dimensions: supplier, part, customer, and date. We generated four datasets with four different numbers of line orders: 80k, 160k, 320k, and 640k. This resulted in 2.3 m, 4.4 m, 7.8 m, and 14.4 m triples respectively. All SSB datasets contain 68 distinct predicates. The QBOAirbase dataset, on the other hand, models air pollution measurements from 36 European countries as an RDF cube augmented with workflow provenance. A measurement corresponds to an observation with coordinates in the time, station (location), and sensor dimensions. We tested our approach on the subset of measurements of Denmark (*qboairbase-dk*) and Great Britain (*qboairbase-gb*). These datasets account for 542k and 4.3 m triples respectively, both with 81 distinct predicates.

Provenance Data and Queries. Since the SSB benchmark does not provide provenance for the data, we augmented each RDF cube with 1000 distinct provenance entities and simulated a set of provenance queries. The provenance entities are assigned to observations in the cube according to two settings: balanced and unbalanced. In the *balanced* setting, each provenance entity is assigned the same number of observations in the cube, whereas in the *unbalanced* setting the ith provenance entity is assigned 2^i triples. We denote the resulting SSB datasets with the prefixes *b*- and *u*- followed by the number of line orders, e.g., *b-ssb-80k* contains 80k line orders with a balanced provenance assignment. We simulated our provenance queries by materializing sets of provenance entities covering from 10% to 90% of the provenance entities in the cube (at intervals of 10%). The datasets qboairbase-dk and qboairbase-gb contain 25.3k and 191.8k different provenance identifiers. For QBOAirbase we constructed a set of 5 provenance queries. These queries impose constraints (a) on whether the data has been quality checked or not (2 queries), (b) on whether we know the data provider or not (2 queries), and (c) on the observation's generation time.

Analytical Queries. The SSB benchmark provides a set of 13 standard OLAP queries [26]. For QBOAirbase [11], we used 8 of the analytical queries available at the project's website[2]. These are the queries where Jena does not time out. For all datasets, we construct provenance-aware queries by combining each analytical query with each of our provenance queries. Each provenance-aware query is executed three non-consecutive times in random order. We averaged the runtimes.

---

[2] http://qweb.cs.aau.dk/qboairbase/.

System Setup and Opponent. We used the Jena (v.3.2) TDB physical database for the in-disk named graphs, and the Jena TDB in-memory store for the cached fragments. All experiments were run in a virtual server with an AMD Opteron 6376 with 8 cores, 128 GB of RAM and 1 TB of disk space running in RAID-5. We tested our queries under two general system settings: (1) after purging the operating system cache and disabling the Jena TDB cache—which we call *cold*, and (2) with the default TDB cache (~50 MB) and a populated OS's cache after having run all the queries at least once. We call this setting *warm*. We compare our approach with the caching provided by Jena TDB and with the LRU caching strategy. The memory budgets are provided as percentages. For Jena TDB a budget of 20% means the engine counts on memory of size 20% the physical database. In contrast, for PAC and LRU the budgets indicate the percentage of triples in the dataset that will be cached. LRU populates the available cache space with the fragments used by the last executed query in a driven-by-size greedy fashion. Jena's standard execution plans timed out with most of the queries, thus we implemented an execution strategy on top of Jena, on which queries are executed on the merge of all relevant in-disk named graphs and cached fragments [2, 35].

Table 1. Performance of different graph filtering strategies (warm setting).

Dataset	PAC			Context index			Naive
	Runtime	Build time	Triples reduction	Runtime	Build time	Triples reduction	Runtime
b-ssb-80k	24.52 s	17.38 s	24.11%	35.97 s	24.45 s	24.10%	33.36 s
u-ssb-80k	25.82 s	17.65 s	22.58%	37.48 s	27.57 s	22.56%	34.79 s
qboairbase-gb	20.04 s	42.72 s	12.00%	103.41 s	134.10 s	−37.51%	24.85 s
qboairbase-dk	1.98 s	5.56 s	13.72%	6.54 s	19.94 s	−35.31%	2.61 s

4.2   Evaluation

Impact of Graph Filtering. We disabled caching and compare PAC's graph filtering and query rewriting with the approach proposed in [2], and a naive query rewriting on the analytical queries. The naive approach rewrites the analytical query by adding a FROM clause for each of the results of the provenance query. In contrast, the approach in [2] defines a *context index* that maps provenance entities to predicate paths, allowing for pruning of the graphs that do not co-occur with predicate paths in the query. In the same spirit, Algorithm 2 filters irrelevant graphs by means of the provenance-aware fragment tree, which encodes the co-occurrences of predicates, object values, and provenance identifiers. Table 1 shows the average runtime and average index built time of the different strategies for four of our datasets. We observe that PAC's filtering outperforms the naive approach in query runtime, because it achieves reductions from 12% to 24% in the total number of materialized triples. While the context index and PAC achieve comparable reductions in the SSB datasets, [2] performs worse for two reasons: (a) it merges all relevant graphs in disk, and (b) it does not handle unions natively. The latter limitation implies that subqueries must be executed independently and their results merged. It also explains why this method sometimes materializes more triples than the naive approach, leading to negative reduction rates. Finally, we highlight that the context index's build time is up to 3x slower than PAC's provenance-aware fragment tree because the context index runs an expensive select query for each predicate path in the index.

Caching vs. In-Memory DB. Table 2 compares the average runtime of PAC, the LRU, and the Jena TDB caching strategies – the two latter with and without PAC's filtering – at budget 20% against full PAC (budget 100%) and the Jena TDB in-memory database in a warm setting. PAC at budget 20% outperforms in total time all caching strategies and the Jena in-memory database. The bottom line is that with PAC's strategic caching, it is not necessary to store everything in main memory for speed-up. In addition, full PAC is 2x faster than the in-memory database thanks to PAC's graph filtering (Algorithm 2).

Table 2. Runtime of a full in-memory database vs. the caching strategies at budget $= 20\%$

Dataset	PAC	LRU+ PAC+F	TDB+ PAC+F	LRU	TDB	Full-PAC	Jena-mem
b-ssb-80k	23.43s	**20.98 s**	23.78 s	35.30 s	33.21 s	**13.03 s**	34.05 s
u-ssb-80k	26.58 s	38.15 s	**26.34 s**	38.15 s	35.84 s	**13.74 s**	35.80 s
airbase-gb	**13.80 s**	20.01 s	17.45 s	22.98 s	25.56 s	**17.86 s**	25.04 s
airbase-dk	1.65 s	2.88 s	**0.02 s**	3.63 s	2.56 s	**2.06 s**	2.75 s
Total	**65.46 s**	82.02 s	67.59 s	100.06 s	117.17 s	**42.69 s**	97.64 s

Impact of the Memory Budget. Figures 3 and 4 show the impact of the memory budget on the average cache hit-rate and the average response time of PAC in a cold setting on four datasets from all our families of datasets. We define the hit-rate as the ratio of graph labels returned by Algorithm 2 that correspond to cached fragments. We observe a monotonically increasing behavior in the hit-rate for all datasets. On the u-ssb-80k and qboairbase datasets, the hit-rate already approaches 80% at budget 10%, contrary to the h-ssb-80k dataset where the increase is more gradual. This phenomenon is mainly caused by the fine granularity of the fragments both in u-ssb-80k and qboairbase. Fine-grained fragments give the selector more flexibility at utilizing the available budget in contrast to very large fragments as the ones found in h-ssb-80k. If a very large fragment does not fit into the remaining cache space, it will not be added, even though it may be relevant to many queries in the query space. The trends in the hit-rate are supported by the runtime behavior in Fig. 4.

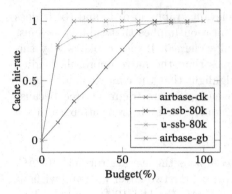

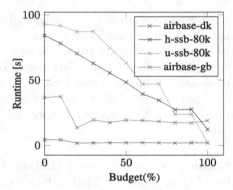

Fig. 3. Budget vs. hit-rate for PAC   **Fig. 4.** Budget vs. runtime for PAC

PAC vs. LRU and TDB. We compare PAC, the LRU caching strategy, and
the Jena TDB native caching for qboairbase-gb on a warm setting in Fig. 5. The
trend is independent of the system setting and is similar for qboairbase-dk. We
first observe that PAC outperforms Jena TDB at all budgets. Only when PAC's
filtering is enabled (TDB+PAC-F), TDB performs comparably to PAC. On the
contrary, LRU seems inadequate for this dataset, even when PAC's filtering is
enabled (LRU+PAC-F). Due to the high diversity of cached fragment signatures
in the qboairbase datasets (approx. 192k), it is unlikely for two consecutive
queries to require the same fragments. This hurts the performance of LRU, which
delivers a hit-rate of 0 for less than 40% budget. LRU+PAC-F does slightly
better, but its maximal hit-rate is no higher than 0.6. The situation is different
for the u-ssb-80k dataset as shown in Fig. 6. While PAC still delivers the best
performance, TDB is outperformed by LRU. The trends are corroborated by the
hit-rate, where PAC is between 0.26 and 0.57 ratio points better than LRU, and
between 0.24 and 0.69 points better than LRU+PAC-F. Our findings in the h-
ssb-80k dataset are alike: PAC is between 0.15 and 0.47 ratio points better than
LRU as displayed in Fig. 7 (between 0.08 and 0.28 points w.r.t. LRU+PAC-F).
All in all, the synergy between graph filtering and a high hit-rate makes PAC
faster than standard caching strategies.

Caching on Bigger Datasets. We also investigate the behavior of the different caching strategies as the number of triples increases in the u-ssb family
of datasets on a warm system setting. We set a budget of 20% and show the
results in Fig. 8. PAC consistently achieves better runtime than its competitors.
In general, all trends observed in the h-ssb-80k and u-ssb-80k datasets remain
constant as the number of triples increases.

Impact of Caching on Queries. We also study the impact of the different
caching strategies on the response time of the individual analytical queries. For
this purpose we compute the area under the curve of response time vs. budget

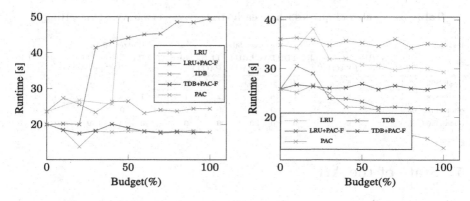

Fig. 5. Runtime on qboairbase-gb   **Fig. 6.** Runtime on u-ssb-80k

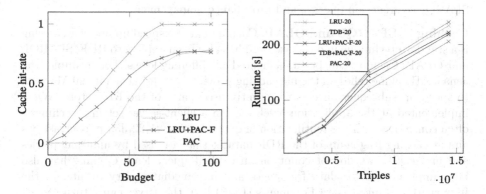

Fig. 7. Hit-rate on h-ssb-80k   **Fig. 8.** Runtime on u-ssb

for each analytical query under the different strategies on the h-ssb-80k, u-ssb-80k, qboairbase-gb, and qboairbase-dk datasets. The runtimes were averaged across all provenance queries. Table 3 shows the number of queries where each strategy wins, that is, the strategy achieves the smallest area under the curve until budgets 20%, 50%, and 100%. We notice that TDB becomes insensitive to the budget argument after a value of 20%. By looking at the winning strategies in each query, we observe that PAC has an almost stable behavior: the set of benefited queries grows monotonically as the budget increases. Despite of being query-load oblivious, PAC with budget 20% wins in 30% of the analytical queries in the SSB datasets, and in 50% of analytical queries in the QBOAirbase datasets.

Table 3. Number of queries where each strategy wins (warm system setting)

Dataset	Budget 20%					Budget 50%					Budget 100%				
	PAC	LRU+PAC+F	TDB+PAC+F	LRU	TDB	PAC	LRU+PAC+F	TDB+PAC+F	LRU	TDB	PAC	LRU+PAC+F	TDB+PAC+F	LRU	TDB
b-ssb-80k	4	0	**5**	0	4	**6**	6	1	0	0	**7**	6	0	0	0
u-ssb-80k	4	1	**5**	2	1	**8**	1	2	1	1	**7**	3	2	0	1
qboairbase-gb	**9**	0	0	0	0	**9**	0	0	0	0	**9**	0	0	0	0
qboairbase-dk	0	0	**9**	0	0	0	0	**9**	0	0	0	0	**9**	0	0

5 State of the Art

This paper studies the impact of caching fragments of a provenance-augmented RDF cube for faster query processing. Therefore, we present the state of the art in terms of three axes: caching in SPARQL and OLAP, query answering on SPARQL aggregation queries, and provenance management.

Caching in SPARQL and OLAP. Caching data to speed up query answering is a standard technique in databases and has also been applied to RDF/SPARQL and OLAP. Caching can be implemented at different levels. For example, the Jena TDB engine relies on the file caching provided by the Java Virtual Machine to speed up subsequent access to recently used parts of the RDF store. When implemented at the application level, e.g., in a client-server setting, caching is often concerned with the reutilization of query results [21,22,33]. In contrast, we aim at caching fragments of the RDF dataset that are used by multiple queries and unlike [21], we do not count on an explicit query load. Caching has also been implemented for data fragments and intermediate query results. In the framework of Linked Data Fragments (LDF) [32], the server can return cached data fragments, leaving the query processing to the client. While PAC's notion of fragments is similar to that of LDF, [32] does not consider provenance and focuses on reducing the server's load for the sake of availability rather than on minimizing response time. Caching has also been applied to OLAP queries [9,16,18]. The system PeerOLAP [18], for example, relies on a P2P network to answer OLAP queries. PeerOLAP reuses the results of queries executed by neighbor peers as data sources. As PeerOLAP, most systems focus on caching recently queried results [3,10]. In [30] a hybrid query engine is proposed; it combines live results with cached data as a trade-off between precision and speed. PAC uses heuristics to strategically pre-cache parts of the data.

SPARQL Aggregation Queries. The interest on optimizing SPARQL queries with aggregation [16,17] started with the publication of SPARQL 1.1 [28]. MARVEL [17] proposes to answer SPARQL aggregation queries on RDF cubes by rewriting the query in terms of a set of views. These views are structured according to a partial order, and selected for query answering based on a cost model as in PAC. Unlike PAC, MARVEL is not a caching approach: it is based on precomputed aggregates –materialized as views– rather than on actual RDF fragments. Moreover, MARVEL does not support provenance. Conversely, [2] supports provenance and assumes that the quadruples produced by the ETL process are all assigned the same provenance entity. The approach proposes a *context index* for graph filtering to speed up the execution of the analytical queries. The index stores the co-occurrence of provenance entities and predicate paths. Albeit not equivalent, PAC's fragment tree supports graph filtering at a better performance without additional assumptions. Besides, [2] does not implement caching.

Provenance Management. The management of provenance is a crucial task for Linked Data and RDF given the decentralized nature of the Web [13,14]. There are several approaches to encode provenance in RDF, such as reification [27], named graphs [5], singleton properties [25], and embedded triples [15]. In this work we focus on workflow provenance [29]. Other approaches study provenance in terms of the lineage of the query results [12,34]: expressions (e.g., a polynomial) that encode the origin of a result w.r.t the triples in the dataset. The TripleProv engine [34] allows for native calculation of lineage for the results of SPARQL queries. Our setting is significantly different, because provenance is encoded as provenance entities described using RDF and the PROV-O ontology [23]. Compared to the notion of lineage for query results, a provenance entity can be seen as the identifier of a precomputed lineage.

6   Conclusions

In this paper, we have presented provenance-aware caching (PAC), an approach to cache fragments of a provenance-augmented RDF graph in order to speed up provenance-aware OLAP queries. Our techniques are query-load agnostic and our experimental evaluation shows that PAC outperforms the Jena TDB native cache and the standard LRU caching strategy in real and synthetic data. The PAC principle can be applied to scenarios where the query-load is unknown, e.g., to bootstrap the cache, or when the workload changes constantly. It is also applicable in settings characterized by locations of "fast" and "slow" access, such as a hybrid drives or remote storage servers. As future work, we envision to integrate explicit dynamic query workloads into our framework, and to extend the fragment definitions beyond equality constraints on the quadruples by, for example, using the provenance graph. All the data and experimental results are available at http://qweb.cs.aau.dk/pac/.

Acknowledgments. This research was partially funded by the Danish Council for Independent Research (DFF) under grant agreement no. DFF-4093-00301 and Aalborg University's Talent Management Programme.

References

1. Ahlstrøm, K., Andersen, A.B., Hose, K., Pedersen, T.B.: Optimizing RDF data cubes for efficient processing of analytical queries. In: COLD (2015)
2. Ahlstrøm, K., Hose, K., Pedersen, T.B.: Towards answering provenance-enabled SPARQL queries over RDF data cubes. In: Li, Y.-F., et al. (eds.) JIST 2016. LNCS, vol. 10055, pp. 186–203. Springer, Cham (2016). https://doi.org/10.1007/978-3-319-50112-3_14
3. Bishop, B., Kiryakov, A., Ognyanov, D., Peikov, I., Tashev, Z., Velkov, R.: A fast track to the web of data. In: SWJ, FactForge (2011)
4. Cadenhead, T., Khadilkar, V., Kantarcioglu, M., Thuraisingham, B.: A language for provenance access control. In: CODASPY (2011)
5. Carroll, J.J., Bizer, C., Hayes, P., Stickler, P.: Named graphs. Provenance and trust. In: WWW (2005)
6. Chebotko, A., Abraham, J., Brazier, P., Piazza, A., Kashlev, A., Lu, S.: Storing, indexing and querying large provenance data sets as RDF graphs in apache HBase. In: SERVICES (2013)
7. Cheney, J., Chiticariu, L., Tan, W.C.: Provenance in databases: why, how, and where. In: Foundations and Trends in Databases (2009)
8. Cyganiak, R., Reynolds, D.: The RDF data cube vocabulary. W3C Recommendation (2014). http://www.w3.org/TR/2014/REC-vocab-data-cube-20140116/
9. Deshpande, P.M., Ramasamy, K., Shukla, A., Naughton, J.F.: Caching multidimensional queries using chunks. SIGMOD Rec. $27(2)$, 259–270 (1998)
10. Erling, O., Mikhailov, I.: RDF support in the virtuoso DBMS. In: Networked Knowledge - Networked Media (2009)
11. Galárraga, L., Ahlstrøm, K., Hose, K.: QBOAirbase: the European air quality database as an RDF cube. In: ISWC, Posters & Demonstrations (2017)
12. Green, T.J., Karvounarakis, G., Tannen, V.: Provenance semirings, In: PODS (2007)
13. Harth, A., Hose, K., Schenkel, R.: Database techniques for linked data management. In: SIGMOD (2012)
14. Harth, A., Hose, K., Schenkel, R.: Linked Data Management. Chapman and Hall/CRC, Boca Raton (2014)
15. Hartig, O.: Foundations of RDF* and SPARQL* - an alternative approach to statement-level metadata in RDF. In: AMW (2017)
16. Ibragimov, D., Hose, K., Pedersen, T.B., Zimányi, E.: Processing aggregate queries in a federation of SPARQL endpoints. In: ESWC (2015)
17. Ibragimov, D., Hose, K., Pedersen, T.B., Zimányi, E.: Optimizing aggregate SPARQL queries using materialized RDF views. In: Groth, P., et al. (eds.) ISWC 2016. LNCS, vol. 9981, pp. 341–359. Springer, Cham (2016). https://doi.org/10.1007/978-3-319-46523-4_21
18. Kalnis, P., Ng, W.S., Ooi, B.C., Papadias, D., Tan, K.L.: An adaptive peer-to-peer network for distributed caching of OLAP results. In: SIGMOD (2002)
19. Kämpgen, B., O'Riain, S., Harth, A.: Interacting with statistical linked data via OLAP operations. In: ILD (2015)
20. Kämpgen, B., Stadtmüller, S., Harth, A.: Querying the global cube: integration of multidimensional datasets from the web. In: Janowicz, K., Schlobach, S., Lambrix, P., Hyvönen, E. (eds.) EKAW 2014. LNCS (LNAI), vol. 8876, pp. 250–265. Springer, Cham (2014). https://doi.org/10.1007/978-3-319-13704-9_20

21. Lorey, J., Naumann, F.: Caching and prefetching strategies for SPARQL queries. In: Cimiano, P., Fernández, M., Lopez, V., Schlobach, S., Völker, J. (eds.) ESWC 2013. LNCS, vol. 7955, pp. 46–65. Springer, Heidelberg (2013). https://doi.org/10.1007/978-3-642-41242-4_5

22. Martin, M., Unbehauen, J., Auer, S.: Improving the performance of semantic web applications with SPARQL query caching. In: Aroyo, L., et al. (eds.) ESWC 2010. LNCS, vol. 6089, pp. 304–318. Springer, Heidelberg (2010). https://doi.org/10.1007/978-3-642-13489-0_21

23. McGuinness, D., Lebo, T., Sahoo, S.: PROV-O: the PROV ontology. W3C Recommendation (2013). http://www.w3.org/TR/2013/REC-prov-o-20130430/

24. Mendes, P.N., Mühleisen, H., Bizer, C.: Sieve: linked data quality assessment and fusion. In: EDBT-ICDT (2012)

25. Nguyen, V., Bodenreider, O., Sheth, A.: Don't like RDF reification?: making statements about statements using singleton property. In: WWW (2014)

26. O'Neil, P., O'Neil, B., Chen, X.: Star schema benchmark. Technical report, UMass/Boston (2009). http://www.cs.umb.edu/~poneil/StarSchemaB.PDF

27. Raimond, Y., Schreiber, G.: RDF 1.1 primer. W3C Recommendation (2014). http://www.w3.org/TR/2014/NOTE-rdf11-primer-20140624/

28. Seaborne, A., Harris, S.: SPARQL 1.1 query language. W3C Recommendation, W3C (2013). http://www.w3.org/TR/2013/REC-sparql11-query-20130321/

29. Theoharis, Y., Fundulaki, I., Karvounarakis, G., Christophides, V.: On provenance of queries on semantic web data. In: IEEE Internet Computing (2011)

30. Umbrich, J., Karnstedt, M., Hogan, A., Parreira, J.X.: Hybrid SPARQL queries: fresh vs. fast results. In: Cudré-Mauroux, P., et al. (eds.) ISWC 2012. LNCS, vol. 7649, pp. 608–624. Springer, Heidelberg (2012). https://doi.org/10.1007/978-3-642-35176-1_38

31. Varga, J., Vaisman, A.A., Romero, O., Etcheverry, L., Pedersen, T.B., Thomsen, C.: Dimensional enrichment of statistical linked open data. J. Web Semant. **40**, 22–51 (2016)

32. Verborgh, R., et al.: Triple pattern fragments: a low-cost knowledge graph interface for the web. J. Web Semant. **37–38**, 184–206 (2016)

33. Williams, G.T., Weaver, J.: Enabling fine-grained HTTP caching of SPARQL query results. In: ISWC (2011)

34. Wylot, M., Cudre-Mauroux, P., Groth, P.: TripleProv: efficient processing of lineage queries in a native RDF store. In: WWW (2014)

35. Wylot, M., Cudre-Mauroux, P., Groth, P.: Executing provenance-enabled queries over web data. In: WWW (2015)

Bash Datalog: Answering Datalog Queries with Unix Shell Commands

Thomas Rebele$^{(\boxtimes)}$, Thomas Pellissier Tanon, and Fabian Suchanek

Telecom ParisTech, Paris, France
thomas.rebele@gmail.com

Abstract. Dealing with large tabular datasets often requires extensive preprocessing. This preprocessing happens only once, so that loading and indexing the data in a database or triple store may be an overkill. In this paper, we present an approach that allows preprocessing large tabular data in Datalog – without indexing the data. The Datalog query is translated to Unix Bash and can be executed in a shell. Our experiments show that, for the use case of data preprocessing, our approach is competitive with state-of-the-art systems in terms of scalability and speed, while at the same time requiring only a Bash shell on a Unix system.

1 Introduction

Motivation. Many data analytics tasks work on tabular data. Such data can take the form of relational tables or TAB-separated files. Even RDF knowledge bases can be seen abstractly as tabular data of a subject, a predicate, an object, and an optional graph id. Quite often, such data has to be preprocessed before the analysis can be made. We focus here on preprocessing in the form of select-project-join-union operations with recursion – removing superfluous columns, selecting rows of interest, recursively finding all instances of a class, etc. The defining characteristic of such a preprocessing step is that it is executed only once on the data in order to constitute the dataset of interest for the later analysis. This one-time preprocessing is the task that we are concerned with.

Databases or triple stores can obviously help. However, loading large amounts of data into these systems may take hours or even days (Wikidata [33], e.g., contains 267GB of data). Another possibility is to use systems such as DLV [18], Souffle [26], or RDFox [22], which work directly on the data. However, these systems load the data into memory. While this works well for small datasets, it does not work for larger ones (as we show in our experiments) Large-scale data processing systems such as BigDataLog [28], Flink [5], Dryad [15], or NoDB [3] can help. However, these require the installation of particular software, the knowledge of particular programming languages, or even a particular distributed infrastructure. Installing and getting to run such systems can take several hours. The user may not have the necessary knowledge and infrastructure to do this (think of a researcher in the Digital Humanities who wants to preprocess a file of census data; or of an engineer in a start-up who has to quickly join log files on a common column; or of a student who wants to extract a subgraph of Wikidata).

© Springer Nature Switzerland AG 2018
D. Vrandečić et al. (Eds.): ISWC 2018, LNCS 11136, pp. 566–582, 2018.
https://doi.org/10.1007/978-3-030-00671-6_33

Our Proposal. In this paper, we propose a method to preprocess tabular data files without installing any particular software. We propose to express the preprocessing steps in Datalog [1]. For example, assume that there is a file `facts.tsv` that contains RDF facts in the form of TAB-separated subject-predicate-object triples. Assume that we want to recursively extract all places located in the United States. The Datalog program in our dialect would be:

```
fact(X, R, Y) :~ cat facts.tsv
locatedIn(X, Y) :- fact(X, "locatedIn", Y) .
locatedIn(X, Y) :- locatedIn(X, Z), fact(Z, "locatedIn", Y) .
main(X) :- locatedIn(X, "USA") .
```

This program prints the file `facts.tsv` into a predicate `fact`. The following two lines are the recursive definition of the `locatedIn` predicate. The `main` predicate is a predefined predicate that acts as the query. Our rationale for choosing Datalog is that it is a particularly simple language, which has just a single syntactic construction, and no reserved keywords. Yet, Datalog is expressive enough to deal with joins, unions, projections, selections, negation, and recursivity. In particular, it can deal with n-ary tables ($n > 3$). If the user deals primarily with RDF data, our approach can also be used with N-Triples files as the A-Box, a subset of OWL 2 RL [21] as the T-Box, and SPARQL [14] for the query.

To execute the Datalog program, we propose to compile it automatically to Unix Bash Shell commands. We offer a Web page to this end: https://www.thomasrebele.org/projects/bashlog. The user can just enter the Datalog program, and click a button to obtain the following Bash code (simplified):

```
awk '$2 == "locatedIn" {print $1 "\t" $3}' facts.tsv > li.tmp
awk '$2 == "USA" {print $0}' li.tmp | tee full.tmp > delta.tmp
while
    join li.tmp delta.tmp | comm -23 - full.tmp > new.tmp
    mv  new.tmp  delta.tmp
    sort -m -o full.tmp  full.tmp delta.tmp
    [ -s delta.tmp ];
do continue; done
cat full.tmp
```

The Bash code can be copy-pasted into a Unix Shell and run. Such a solution has several advantages. First, it does not require any software installation. It just requires a visit to a Web site. The resulting Bash code runs on any Unix-compatible system out of the box. Second, the Bash shell has been around for several decades, and the commands are not just tried and tested, but actually continuously developed. Modern implementations of the `sort` command, e.g., can split the input into several portions that fit into memory, and sort them individually. Finally, the Bash shell allows executing several processes in parallel, and their communication is managed by the operating system.

Contribution. We prose to compile a Datalog program automatically into Bash commands. Our method optimizes the Datalog program with relational algebra optimization techniques, re-uses previously computed intermediate results, and produces a highly parallelized Shell script. For this purpose, our method employs pipes and process substitution. Our experiments on a variety of datasets and preprocessing tasks show that this method is competitive in terms of runtime with state-of-the-art database systems, Datalog engines, and triple stores.

We start with a discussion of related work in Sect. 2. Section 3 introduces preliminaries. Section 4 presents our approach, and Sect. 5 evaluates it.

2   Related Work

Data Processing Systems. Relational Databases such as Oracle, IBM DB2, Postgres, MySQL, MonetDB [4] and NoDB [3] can handle tabular data of arbitrary form, while the triple stores such as OpenLink Virtuoso [8], Stardog, and Jena [6] target RDF data. HDT [11] is a binary format for RDF, which can be used with Jena. RDFSlice [20] can preprocess RDF datasets. Reasoners such as Pellet [23], HermiT [27], RACER [13], Fact++ [30] and Jena [6] can perform OWL reasoning on RDF data. Datalog systems such as DLV [18], Souffle [26], BigDatalog [28], RDFox [22] and others [16,34,35] can efficiently evaluate Datalog queries on large data. Distributed Data Processing systems such as Dryad [15], Apache Tez [25], SCOPE [37], Impala [9], Apache Spark [36], and Apache Flink [5] provide advanced features such as support for SQL or streams.

All of these systems can be used for preprocessing data. However, the vast majority of these systems require the installation of software. The parallelized systems also require a distributed infrastructure. Our approach, in contrast, requires none of these. It just requires a visit to a Web page. The Bash script that we produce runs in a common shell console without any further prerequisites. Interestingly, our approach still delivers comparable performance to the state of the art, as we shall see in the experiments.

Only very few systems do not require a software installation beyond downloading a file (e.g., RDFSlice [20], Stardog, and DLV [18]). Yet, as we shall see in the experiments, these systems do not scale well to large datasets.

Other Work. Linked Data Fragments [32] aim to strike a balance between downloading an RDF data dump and querying it on a server. The method thus addresses a slightly different problem from ours. AI planning with softbots [10] aims to answer queries on an incomplete and evolving database. In our problem setting, however, we have access to all the information. NoSQL Databases such as Cassandra, HBase, and Google's BigTable [7] target non-tabular data. Our method, in contrast, aims at tabular data.

3   Preliminaries

Datalog. A Datalog rule with negation [1] takes the form

$$H :\!— B_1, \ldots, B_n, \neg N_1, \ldots, \neg N_m.$$

Here, H is the head atom, $B_1, \ldots, B_n$ are the positive body atoms, and $N_1, \ldots, N_m$ are the negated body atoms. Each atom is of the form $r(x_1, \ldots, x_k)$, where r is a relation name and $x_1, \ldots, x_k \in \mathcal{V} \cup \mathcal{C}$, where $\mathcal{V}$ is a set of variables and $\mathcal{C}$ is a set of constants. Intuitively, such a rule says that H holds if $B_1, \ldots, B_n$ and none of the $N_1, \ldots, N_m$ holds. We consider only *safe rules*, i.e., each variable in the head or in a negated atom must also appear in a positive body atom. A *Datalog program* is a set of Datalog rules. A set M of atoms is a *model* of a program P, if the following holds: M contains an atom a iff P contains a rule $H :\!\!- B_1, \ldots, B_n, \neg N_1, \ldots, \neg N_m$, such that there exists a substitution $\sigma : \mathcal{V} \to \mathcal{C}$ with $\sigma(B_i) \in M$ for $i \in \{1, \ldots, n\}$ and $\sigma(N_i) \notin M$ for $i \in \{1, \ldots, m\}$ and $a = \sigma(H)$. We consider only *stratified* Datalog programs [1], which entails that there exists a unique minimal model.

OWL RL [21] is a subset of the OWL ontology language. Since every OWL RL ontology can be translated to Datalog [22], we deal with Datalog as the more general case in all of the following.

Relational Algebra. *Relational algebra* [1] provides the semantics of relational database operations on tables. For our purposes, we use the unnamed relational algebra with the operators select σ, project π, join $\bowtie$, anti-join $\triangleright$, and union $\cup$ (see [24] for their definitions). We also use the *least fixed point (LFP) operator* μ [2]. For a function f from a table to a table, $\mu_x(f(x))$ is the least fix point of f for the $\subseteq$ relation. With this, our algebra has the same expressivity as safe stratified Datalog programs, i.e., the translation of safe Datalog with stratified negations to relational algebra is sound and complete [1].

> *Example 2 (Relational Algebra):* The following expression computes the transitive closure of a two-column table *subclass*:
>
> $$\mu_x(\text{subclass} \cup \pi_{1,4}(x \bowtie_{2=1} x))$$
>
> μ computes the least fix point of a function. Here, the function is given by the argument of the μ-operator. To compute the least fix point, we execute the function first with the empty table, $x = \emptyset$. Then the function returns the *subclass* table. Then we execute the function again on this result. This time, the function will join *subclass* with itself, project the resulting table on the first and last column, and add in the original *subclass* table. We repeat this process until no more changes occur.

Unix. Unix is a family of multitasking computer operating systems, which are widely used on servers, smartphones, and desktop computers. One of the characteristics of Unix is that "Everything is a file", which means that files, pipes, the standard output, the standard input, and other resources can all be seen as streams of bytes. Here, we are interested only in *TAB-separated* byte streams, i.e., streams that consist of several *rows* (sequences of bytes separated by a newline character), which each consist of the same number of *columns* (sequences of bytes separated by a tabulator character).

The Bourne-again shell (Bash) is a command-line interface for Unix-like operating systems. A *Bash command* is either a built-in keyword, or a small program. We will only use commands of the POSIX standard. One of them is the awk command, which we use as follows:

```
awk -F$'\t' 'p' b
```

This command executes the program p on the byte stream b, using the TAB as a separator for b. We will discuss different programs p later in this paper.

Pipes. When a command is executed, it becomes a *process*. Two processes can communicate through a pipe, i.e., a byte stream that is filled by one process, and read by the other one. If the producing process is faster than the receiving one, the pipe buffers the stream, or blocks the producing process if necessary. In Bash, a pipe between process p_1 and process p_2 pipes can be constructed by stating $p_1 \mid p_2$. A pipe can also be constructed "on the fly" by a so-called *process substitution*, as follows: $p_1 < (p_2)$. This construction pipes the output of p_2 into the first argument of p_1.

4 Approach

4.1 Our Datalog Dialect

In our Datalog dialect, predicates are alphanumerical strings that start with a lowercase character. Variables start with an uppercase letter. Constants are strings enclosed by double quotes. Constants may not contain quotation marks, TAB characters, or newline characters. For our purposes, the Datalog program has to refer to files or byte streams of data. For this reason, we introduce an additional type of rules, which we call *command rules*. A command rule takes the following form:

$$p(x_1, ..., x_n) \quad :\sim \quad c$$

Here, p is a predicate, $x_1, \ldots, x_n$ are variables, and c is a Bash command. Semantically, this rule means that executing c produces a TAB-separated byte stream of n columns, which will be referred to by the predicate p in the Datalog program. In the simplest case, the command c just prints a file, as in `cat facts.tsv`. However, the command can also be any other Bash command, such as `ls -1`.

Our goal is to compute a certain output with the Datalog program. This output is designated by the distinguished head predicate `main`. An *answer* of the program is a grounded variant of the head atom of this rule that appears in the minimal model of the program. See again Fig. 1 for an example of a Datalog program in our dialect. Our dialect is a generalization of standard Datalog, so that a normal Datalog program can be run directly in our system.

Our approach can also work in "RDF mode". In that mode, the input consists of an OWL 2 RL [21] ontology, a SPARQL [14] query, and an N-Triples file F. We build a `main` predicate for the SPARQL query, and we use a small AWK program that converts F into a TAB-separated byte stream (see our technical report [24] for details). Much like in RDFox [22], we convert the OWL ontology into Datalog rules (see again [24]). For now, we support only a subset of OWL 2 RL: Like RDFox [22], we assume that all classes and properties axioms are provided by the ontology and will not be queried by the SPARQL query. We also do not yet support OWL axioms related to literals. Our SPARQL implementation supports basics graph patterns, property paths without negations, OPTIONAL, UNION and MINUS.

4.2 Loading Datalog

Next, we build a relational algebra expression for the `main` predicate of the Datalog program. Our algorithm is similar to existing approaches [31]. Algorithm 18 takes as input a predicate p, a cache, and a Datalog program P. The algorithm is initially called with $p = \text{main}$, $cache = \emptyset$, and the Datalog program that we want to translate. The cache stores already computed relational algebra plans. This allows us to re-use the same sub-plan multiple times in the final plan, thus the algorithm builds a directed acyclic graph (DAG) instead of a tree.

Our algorithm first checks whether p appears in the cache. In that case, p is currently being computed in a previous recursive call of the method, and the algorithm returns a variable x indexed by p. This is the variable for which we compute the least fix point.

Then, the algorithm traverses all rules with p in the head. For every rule, the algorithm recursively retrieves the plan for the body atoms. The algorithm (anti-)joins the sub-plans, adding selections $\sigma_{j=k}$ if necessary. Finally, it puts the resulting formula into a project-node that extracts the relevant columns.

Algorithm 1. Translation from Datalog to relational algebra

---

1   **fn** mapPred $(p, cache, P)$ **is**
2   **if** $p \in$ cache **then return** x_p ;
3   plan $\leftarrow \emptyset$; newCache $\leftarrow$ cache $\cup \{p\}$
4   **foreach** $p(H_1, ..., H_{n_h}) :\!\!- r_1(X_1^1, ..., X_{n_1}^1), ..., \neg q_1(Y_1^1, ..., Y_{m_1}^1), ...$ in P **do**
5   bodyPlan $\leftarrow \{()\}$
6   **foreach** $r_i(X_1^i, ..., X_{n_i}^i)$ **do**
7   atomPlan $\leftarrow$ mapPred$(r_i, \text{newCache}, P)$
8   **foreach** $(X_j^i, X_k^i) \mid X_j^i = X_k^i, j \neq k$ **do**
9   atomPlan $\leftarrow \sigma_{X_j^i = X_k^i}$(atomPlan)
10   bodyPlan $\leftarrow$ bodyPlan $\bowtie$ atomPlan
11   **foreach** $\neg q_i(Y_1^i, ..., Y_{m_i}^i)$ **do**
12   atomPlan $\leftarrow$ mapPred$(q_i, \emptyset, P)$
13   **foreach** $(Y_j^i, Y_k^i) \mid Y_j^i = Y_k^i, j \neq k$ **do**
14   atomPlan $\leftarrow \sigma_{Y_j^i = Y_k^i}$(atomPlan)
15   bodyPlan $\leftarrow$ bodyPlan $\triangleright$ atomPlan
16   plan $\leftarrow$ plan $\cup \pi_{H_1, ..., H_{n_h}}$(bodyPlan)
17   **foreach** $rule\ p(H_1, ..., H_{n_h}) :\sim c$ in P **do** plan $\leftarrow$ plan $\cup \pi_{H_1, ..., H_{n_h}}(c)$;
18   **return** μ_{x_p}(plan)

Example 3 (Datalog Translation): Assume that there is a two-column TAB-separated file `subclass.tsv`, which contains each class with its subclasses. Consider the following Datalog program P:

(1)   `directSubclass(x,y) :~ cat subclass.tsv`
(2)   `main(x,y) :- directSubclass(x,y).`
(3)   `main(x,z) :- directSubclass(x,y), main(y,z).`

Our algorithm will go through all rules with the head predicate `main`. These are Rule 2 and Rule 3. For Rule 2, the algorithm will recursively call itself and return $\mu_{x_{\text{directSubclass}}}(\emptyset \cup [\texttt{cat subclass.tsv}])$. Since the argument of μ does not contain the variable $x_{\text{directSubclass}}$, this is equivalent to `[cat subclass.tsv]`. For the first body atom in Rule 3, the algorithm returns `[cat subclass.tsv]` just like before. For the second body atom, the algorithm returns x_{main}, because `main` is in the cache. Thus, Rule 3 yields $\pi_{1,4}([\texttt{cat subclass.tsv}] \bowtie_{2=1} x_{\text{main}}))$. Finally, the algorithm constructs

$$\mu_{x_{\text{main}}}([\texttt{cat subclass.tsv}] \cup \pi_{1,4}([\texttt{cat subclass.tsv}] \bowtie_{2=1} x_{\text{main}}))$$

4.3   Producing Bash Commands

The previous step has translated the input Datalog program to a relational algebra expression. Now, we translate this expression to a Bash command by the function b, which is defined as follows:

$b([c]) = c$
An expression of the form $[c]$ is already a Bash command, and hence we can return directly c.

$b(e_1 \cup \ldots \cup e_n)$
We translate a union into a sort command that removes duplicates:

```
sort -u <(b(e_1)) ... <(b(e_n))
```

$b(e_1 \bowtie_{x=y} e_2)$
A join of two expressions e_1 and e_2 on a single variable at position x and y, respectively, gives rise to the command

```
join -t$'\t' -1x -2y
    <(sort -t$'\t' -kx <(b(e_1)))
    <(sort -t$'\t' -ky <(b(e_2)))
```

This command sorts the byte streams of $b(e_1)$ and $b(e_2)$, and then joins them on the common column.

$b(e_1 \bowtie_{x=y,\ldots} e_2)$
The Bash `join` command can perform the join on only one column. If we want to join on several columns, we have to add a new column to each of the byte streams. This new column concatenates the join columns into a single column. This can be achieved with the following AWK program, which we run on both $b(e_1)$ and $b(e_2)$:

```
{ print $0 FS $j_1 s $j_2 s ... s $j_n }
```

Here, the indices $j_1, \ldots, j_n$ are the positions of the join columns in the input byte stream, FS is the field separator, and s is a special separation character (we use ASCII character 2, but any other one can be used as well). Once we have done this with both byte streams, we can join them on this new column as described above. This join will also remove the additional column.

$b(e_1 \rhd_x e_2)$

Just as a regular join, an anti-join becomes a join command. We use the parameter -v1, so that the command outputs only those tuples emerging from e_1 than cannot be joined with those from e_2. We deal with anti-joins on multiple columns in the same way as with multi-column joins.

$b(\pi_{i_1,\ldots i_n}(e))$

A projection becomes the following AWK program, which extracts the given columns from the input byte stream $b(e)$:

```
{ print $i_1 FS ... FS $i_n }
```

$b(\pi_{i:a}(e))$

A constant introduction becomes the following AWK program, which produces a TAB-separated byte stream that inserts the constant in column i of the input byte stream $b(e)$:

```
{ print $1 FS ... $(i-1) FS a FS $i FS ... $n}
```

$b(\sigma_{i=v}(e))$

A selection node gives rise to the following AWK program, which selects the corresponding rows from the input byte stream $b(e)$:

```
$i=="v" { print $0 }
```

This command can be generalized easily to a selection on several columns.

Several of these translations produce process substitutions. In such cases, Bash starts the parent process and the inner process in parallel. The parent process will block while it cannot read from the inner processes. Thus, only the innermost processes run in the beginning. Every process is run asynchronously as soon as input and CPU capacity is available. Thus, our Bash program is not subject to the forced synchronization that appears in Map-Reduce systems.

4.4 Recursion

We have just defined the function b that translates a relational algebra expression to a Bash command. We will now see how to define b for the case of recursion. A node $\mu_x(f(x))$ becomes

```
echo -n > delta.tmp;  echo -n > full.tmp
while
    sort b(f(delta.tmp)) | comm -23 - full.tmp > new.tmp;
    mv  new.tmp  delta.tmp;
    sort -u -m -o full.tmp  full.tmp <(sort delta.tmp);
    [ -s delta.tmp ];
do continue; done
cat full.tmp
```

This code uses 3 temporary files to compute the least fix point of f: `full.tmp` contains all facts inferred until the current iteration. `delta.tmp` contains newly added facts of an iteration. `new.tmp` is used as swap file. The code first creates `delta.tmp` and `full.tmp` as empty files. It then runs f on the delta file. The `comm` command compares the sorted outcome of f to the (initially empty) file `full.tmp`, and writes the new lines to the file `new.tmp`. This file is then renamed to `delta.tmp`. This procedure updates the file `delta.tmp` to contain the newly added facts. The `comm` command cannot write directly to `delta.tmp`, because this file also serves as input to the command produced by $b(f(\text{delta.tmp}))$.

The following `sort` command merges the new lines into `full.tmp`, and writes the output to `full.tmp` (unlike the `comm` command, the `sort` command can write to a file that also serves as input). Now, all facts generated in this iteration have been added to `full.tmp`. The [...] part of the code lets the loop run while the file `delta.tmp` is not empty, i.e., while new lines are still being added. If no new lines were added, the code quits the loop, and prints all facts. Note that, due to the monotonicity of our relational algebra operators, and due to the stratification of our programs, we can afford to run f only on the newly added lines.

4.5   Materialization

Materialization Nodes. To avoid re-computing an algebra expression that has already been computed, we introduce a new type of operator to the algebra, the materialization node. A materialization node $\square(m,(\lambda y : p))$ has two subplans: m is the plan that is used multiple times, and that we will materialize. The lambda function $(\lambda y : p)$ is the main plan, and takes the materialized plan as parameter. The variable y replaces all occurrences of m in the original plan (see [24] for details).

Bash Translation. The translation to Bash is shown on the right. b is the Bash translation function defined in Sect. 4.3. t is a temporary file name. The code first creates a named pipe called `lock_t`. Commands that use t have to wait until $b(m)$ finishes. We ensure this by making these commands read from `lock_t`. Since this pipe contains no data, the commands block. When $b(m)$ finishes, the two `exec` commands close the named pipe, thus unblocking the commands that need t. There can be a rare race condition: $b(m)$ may finish before any process that listens on the pipe was started. In that

```
mkfifo lock_t
(    b(m) > t
     mv lock_t done_t
     cat done_t &
     exec 3> done_t
     exec 3>&-
) &
b_{y→t}(p)
rm t
```

case, the two `exec` commands will try to close a pipe that has no listeners. In such cases, the `exec` command will block. We solve this problem by reading from the pipe with a `cat` command that runs in the background. This way, the pipe has at least one listener, and the `exec` commands will close the pipe. This, however, brings a second problem: If the processes that listen on the pipe were still not started, they will try to listen to a closed pipe. To avoid this problem, we rename the pipe from lock_t to done_t. Such a renaming does not affect any processes that already listen on the pipe, but it will prevent any new processes from listening on the pipe under the old name.

$b_{y \to t}$ extends b as follows: $b_{y \to t}(y)$ generates the bash code cat t, and all plan nodes p_i that have a child y generate the bash code

cat lock_t 2> /dev/null ; $b(p_i)$

As explained above, the cat command blocks the execution until t is materialized. The part "2> /dev/null" removes the error message in case cat is executed when the pipe was already renamed.

4.6 Optimization

Query Optimizations. We apply the usual optimizations on our relational algebra expressions: we push selection nodes as close to the source as possible; we merge unions; we merge projects; we apply a simple join re-ordering. Additionally, we remove an occurrence of a LFP variable, if it cannot contribute new facts; we remove the LFP when there is no recursion; and we extract from a LFP node the non-recursive part of the inner plan (so that it is computed only once at the beginning of the fixed point computation). We also optimize fix point computations in the same way as in the semi-naive Datalog evaluation [1] (see again [24] for details).

Bash Optimizations. We collect different AWK commands that select or project on the same file into a single AWK command. This command runs only once through the file, and writes out all selections and projections into several files, one for each original AWK command. We replace multiple comparisons with constants on the same columns by a hash table lookup. We detect nested sort commands, and remove redundant ones. We run sort -u on the final output to make all results unique. We estimate the number of concurrently running sort commands, and assign each of them an equal amount of memory, if the buffer size parameter is available. Finally, we force all commands to use the same character set and sort order by adding the command export LC_ALL=C to our program.

5 Experiments

We ran our method on several datasets, and compared it to several competitors. All our experiments were run on a laptop with Ubuntu 16.04, an Intel Core i7-4610M 3.00 GHz CPU, 16 GB of RAM, 16 GB of swap space, and 3.8 TB of hard disk space. We used GNU Bash 4.3.48, mawk 1.3.3 for AWK, and GNU coreutils 8.25 for the other POSIX commands.

We emphasize that our goal is not to be faster than each and every system that currently exists. For this, the corpus of related work is simply too large (see Sect. 2). This is also not the purpose of Bash Datalog. The purpose of Bash Datalog is to provide a preprocessing tool that runs without installing any additional software besides a Bash shell. This is an advantage that no competing approach offers. Our experiments then serve mainly to show that our approach is generally comparable in terms of scalability with the state of the art.

Table 1. Runtime for the 14 LUBM queries with 10 universities (155 MB), in seconds. * = no support for querying with a T-Box. We folded the T-Box into the query. ±I = with/without indexes. A dash means that the query is not supported.

Bash	DLV	Souffle	RDFox	Jena TDB	Jena HDT	Stardog	Virtuoso	Postgres*		NoDB*	MonetDB*		RDF-Slice*
								−I	+I		−I	+I	
0.7	9.6	7.8	2.2	25.7	26.4	12.8	11.7	4.8	27.5	>600	1.8	2.6	12.6
1.3	9.3	119.4	2.2	281.3	>600	13.6	11.8	−	−	−	−	−	−
0.9	9.2	8.8	2.2	26.7	27.0	12.7	11.5	7.8	30.5	292.9	1.9	2.7	−
1.9	9.3	11.9	2.2	>600	>600	13.2	12.2	14.7	37.4	>600	2.3	3.1	−
1.4	9.3	10.3	2.2	>600	>600	12.9	−	28.9	51.6	−	2.2	3.0	−
1.9	9.4	11.1	2.4	>600	>600	17.6	−	21.2	43.9	−	3.9	4.7	−
2.4	9.5	56.3	2.2	>600	>600	13.4	−	21.6	44.3	>600	3.0	3.9	−
2.5	9.3	12.9	2.3	>600	>600	15.3	−	−	−	−	−	−	−
3.1	9.4	>600	2.3	>600	>600	13.4	−	71.1	93.8	>600	25.5	26.8	−
2.0	9.3	11.6	2.2	>600	>600	13.5	−	23.0	45.7	>600	5.8	7.1	−
0.9	9.3	7.8	2.2	25.3	35.7	13.0	11.8	−	−	−	−	−	−
1.4	9.2	10.9	2.2	>600	>600	13.1	−	−	−	−	−	−	−
1.4	9.2	10.1	2.2	>600	>600	12.9	−	8.4	31.1	>600	4.3	5.4	−
0.8	9.5	6.7	2.3	34.5	24.8	13.5	12.0	4.8	27.5	19.1	1.9	2.7	3.7
of which loading:				16.8	7.4	11.0	5.9	4.4	24.3		1.7	2.5	

5.1   Lehigh University Benchmark

Setting. The Lehigh University Benchmark (LUBM) [12] is a standard dataset for Semantic Web repositories, which models universities. It is parameterized by the number of universities, and hence its size can be varied. LUBM comes with 14 queries, which are expressed in SPARQL. We compare our approach to Stardog[1], Virtuoso [8], RDFSlice [20], Jena [6], and Jena with the binary triple format HDT [11]. For RDFox [22], Souffle [26], DLV [18], we translated the queries to Datalog in the same way that we translate the queries to Datalog for our own system (Sect. 4.2). For NoDB [3], Postgres[2], and MonetDB [4], we translated the Datalog queries first to an algebra expression, and then to SQL. In this process, we applied the relational algebra optimizations of Sect. 4.6. In this way, the T-Box of the LUBM queries is folded into the SQL query. Not all systems support all types of queries. MonetDB does not support recursive SQL queries. Postgres supports only certain types of recursive queries [24]. The same applies to NoDB. Virtuoso currently does not support intersections. RDFSlice aims at the slightly different problem of RDF-Slicing. It supports only a specific type of join. Also, it does not support recursion.

We ran every competitor on all queries that it supports, and averaged the runtime over 3 runs. We checked whether the query results were correct.

---

[1] https://www.stardog.com/, v. 5.2.0.
[2] https://www.postgresql.org/, v. 10.1.

LUBM10. Table 1 shows the runtimes of all queries on LUBM with 10 universities. The runtimes include the loading and indexing times. For systems where we could determine these times explicitly, we noted them in the last row of the table. Since most systems finished in a matter of seconds, we aborted systems that took longer than 10 min. Among the 4 triple stores (Jena+TDB, Jena+HDT, Stardog, and Virtuoso), only Stardog can finish on all queries in less than 10 min. RDFSlice can answer only 2 queries, and runs a bit faster than Stardog. The 5 database competitors (Postgres, NoDB, and MonetDB – with and without indexes) are generally faster. Among these, MonetDB is much faster than Postgres and NoDB. Postgres and MonetDB are fastest without indexes, which is to be expected when running the query only once. Among the best performing systems are two Datalog systems (Bash Datalog, and RDFox). RDFox shines with a very short and nearly constant time for all queries. We suspect that this time is given by the loading time of the data, and that it dominates the answer computation time. Nevertheless, Bash Datalog is faster than RDFox on nearly all queries on LUBM 10.

Table 2. Runtime for the LUBM queries, in seconds.

Query	LUBM 500 (7.8 GB)							LUBM 1000 (16 GB)					
	Bash	RDFox	Stardog	Virtuoso	MonetDB (no indices)	MonetDB (indices)	RDFSlice	Bash	RDFox	Stardog	MonetDB (no indices)	MonetDB (indices)	RDFSlice
1	**27**	131	582	1577	83	97	229	**75**	273	1955	185	210	1042
2	**53**	132	683	1580				**118**	278	2030			
3	**35**	131	609	1578	88	101		**89**	276	1955	186	217	
4	**95**	129	583	1579	118	131		307	**273**	1962	522	471	
5	**62**	131	498		290	364		**168**	278	1956	894	793	
6	**93**	137	1011		866	797		354	**287**	2361	2066	1934	
7	**122**	134	673		898	753		544	**279**	2005	1809	2016	
8	151	**132**	768					447	**274**	1967			
9	250	**136**	749		2669	3064		712	**283**	2018	3275	3090	
10	**95**	132	678		587	491		334	**277**	1959	1845	1834	
11	**28**	130	498	1576				**64**	273	1957			
12	**56**	130	682					**164**	273	1959			
13	**63**	132	669		312	287		**174**	277	1969	908	955	
14	**28**	136	787	1595	85	99	74	**63**	284	2069	181	217	334
of which load:			489	1575	72	92				1946	160	194	

LUBM500 and LUBM1000. For the larger LUBM datasets, we chose the fastest systems in each group as competitors: RDFox for the Datalog systems, Stardog and Virtuoso for the triple stores, MonetDB for the databases, and RDFSlice as its own group. Table 2 shows the sizes of the datasets and the runtimes of the systems. Our system performs best on more than half of the

queries. The only system that can achieve a similar performance is RDFox. As
before, RDFox always needs just a constant time to answer a query, because it
loads the dataset into main memory. This makes the system very fast. However,
this technique will not work if the dataset is too large, as we shall see next.

5.2 Reachability

Setting. Our next datasets are the social networks LiveJournal [19], com-
orkut [19], and friendster [17]. Table 3 shows the sizes of these datasets. We
used a single query, which asks for all nodes that are reachable from a given
node. As competitors, we chose again RDFox, Stardog, and Virtuoso. We could
not use MonetDB or RDFSlice, because the reachability query is recursive. As
an additional competitor, we chose BigDatalog [28]. This system was already
run on the same LiveJournal and com-orkut graphs in the original paper [28].
We chose 3 random nodes (and thus generated 3 queries) for LiveJournal and
com-orkut. We chose one random node for Friendster.

Results. Table 3 shows the average runtime for each system. Virtuoso was the
slowest system, and we aborted it after 25 min and 50 min, respectively. We did
not run it on the Friendster dataset, because Friendster is 20 times larger than
the other two datasets. Stardog performs better. Still, we had to abort it after
10 h on the Friendster dataset. BigDatalog performs well, but fails with an out
of space error on the Friendster dataset. The fastest system is RDFox. This is
because it can load the entire data into memory. This approach, however, fails
with the Friendster dataset. It does not fit into memory, and RDFox is unable to
run. Bash Datalog runs 50% slower than RDFox. In return, it is the only system
that can finish in reasonable time on the Friendster dataset (4:32h).

Table 3. Runtime for the reachability query, in seconds (OOM = Out of memory;
OOS = Out of space).

Dataset	Nodes	Edges	Bash	RDFox	BigDatalog	Stardog	Virtuoso
LiveJournal	4.8 M	69 M	117	**70**	532	941	>1500
orkut	3.1 M	117 M	225	**121**	1838	1123	>3000
friendster	68 M	2 586 M	**16306**	OOM	OOS	>36000	

5.3 YAGO and Wikidata

Setting. Our final experiments concern the knowledge bases YAGO [29] and
Wikidata [33]. The YAGO data comes in 3 different files, one with the 12 M
facts (814 MB), one with the taxonomy with 1.8 M facts (154 MB), and one with
the 24 M type relations (1.6 GB in size). Wikidata is a single file of 267 GB with
2.1 B triples. We designed 4 queries that are typical for such datasets (Table 1),
together with a T-Box (Table 2).

```
query1(X) :- subClassOf(X, "person") .
query1(X) :- subClassOf(X, Y), query1(Y) .
query2(X) :- hasParent(X, "Louis XIV") .
query3(X) :- hasAncestor(X, "Louis XIV") .
query4(X) :- hasBirthPlace(X, Y), isLocatedIn(Y, "Andorra") .
```

Fig. 1. Knowledge base queries

```
hasParent(X,Y) :- hasChild(Y,X) .
hasAncestor(X,Y) :- hasParent(X,Y) .
hasAncestor(X,Z) :- hasAncestor(X,Y), hasParent(Y,Z) .
isLocatedIn(X,Y) :- containsLocation(Y,X) .
containsLocation(X,Y) :- isLocatedIn(Y,X) .
isLocatedIn(X,Y) :- isLocatedIn(X,Y), isLocatedIn(Y,Z) .
```

Fig. 2. Knowledge base rules

Results. Table 4 shows the results of RDFox and our system on both datasets. On YAGO, RDFox is much slower than our system, because it needs to instantiate all rules in order to answer queries. On Wikidata, the data does not fit into main memory, and hence RDFox cannot run at all. Our system, in contrast, scales to the larger sizes of the data. One may think that a database system such as Postgres may be better adapted for such large datasets. This is, however, not the case. Postgres took 104 s to load the YAGO dataset, and 190 s to build the indexes. In this time, our system has already answered nearly all the queries.

Table 4. Runtime for the Wikidata/YAGO benchmark in seconds. (OOM = out of memory error)

Query	YAGO		Wikidata	
	Bash	RDFox	Bash	RDFox
1	8	483	2259	OOM
2	5	483	2254	OOM
3	293	483	10171	OOM
4	5	481	2270	OOM

Discussion. All of our experiments evaluate only the setting that we consider in this paper, namely the setting where the user wants to execute a single query in order to preprocess the data. Our experiments show that Bash Datalog can preprocess tabular data without the need to install any particular software.

Our approach has some limitations. For example, we could not implement a disk-based hash-join efficiently in Bash commands. Another limitation is the heuristic join reordering. It sometimes introduces large intermediate results, resulting in a less efficient query execution.

Overall, however, our approach is competitive in both speed and scalability to the state of the art. We attribute this to the highly optimized POSIX commands, and to our optimizations described in Sect. 4.6. Furthermore, the startup cost of our system is quite low, as it consists mainly of translating the query to a Bash script.

6   Conclusion

In this paper, we have presented a method to compile Datalog programs into Unix Bash programs. This allows executing Datalog queries on tabular datasets without installing any software. We show that our method is competitive in terms of speed with state-of-the-art systems. Our system can be used online at https://www.thomasrebele.org/projects/bashlog. The Web interface also provides an API which allows to translate Datalog to Bash via an HTTP request. The source code is available at https://github.com/thomasrebele/bashlog. For future work, we aim to explore extensions of this work such as adding support of numerical comparisons to the Datalog language.

Acknowledgments. This research was partially supported by Labex DigiCosme (project ANR-11-LABEX-0045-DIGICOSME) operated by ANR as part of the program "Investissement d'Avenir" Idex Paris-Saclay (ANR-11-IDEX-0003-02).

References

1. Abiteboul, S., Hull, R., Vianu, V.: Foundations of Databases. Addison-Wesley, Boston (1995)
2. Aho, A.V., Ullman, J.D.: The universality of data retrieval languages. In: ACM Symposium on Principles of Programming Languages (1979)
3. Alagiannis, I., Borovica, R., Branco, M., Idreos, S., Ailamaki, A.: NoDB: efficient query execution on raw data files. In: SIGMOD (2012)
4. Boncz, P.A., Kersten, M.L., Manegold, S.: Breaking the memory wall in MonetDB. Commun. ACM **51**(12), 77–85 (2008)
5. Carbone, P., Katsifodimos, A., Ewen, S., Markl, V., Haridi, S., Tzoumas, K.: Apache flink™. IEEE Data Eng. Bull. **38**(4), 28–38 (2015)
6. Carroll, J.J., Dickinson, I., Dollin, C., Reynolds, D., Seaborne, A., Wilkinson, K.: Jena: implementing the semantic web recommendations. In: WWW (2004)
7. Chang, F., et al.: Bigtable: a distributed storage system for structured data. ACM Trans. Comput. Syst. **26**(2), 4 (2008)
8. Erling, O., Mikhailov, I.: RDF support in the virtuoso DBMS. In: Pellegrini, T., Auer, S., Tochtermann, K., Schaffert, S. (eds.) Networked Knowledge. SCI, vol. 221. Springer, Heidelberg (2009). https://doi.org/10.1007/978-3-642-02184-8_2
9. Bittorf, M.K., et al. Impala: a modern, open-source SQL engine for hadoop. In: CIDR (2015)

10. Etzioni, O., Golden, K., Weld, D.S.: Sound and efficient closed-world reasoning for planning. Artif. Intell. **89**(1–2), 113–148 (1997)
11. Fernández, J.D., Martínez-Prieto, M.A., Gutiérrez, C., Polleres, A., Arias, M.: Binary RDF representation for publication and exchange (HDT). Web Semant.: Sci. Serv. Agents World Wide Web **19**, 22–41 (2013)
12. Guo, Y., Pan, Z., Heflin, J.: LUBM: a benchmark for OWL knowledge base systems. J. Web Semant. **3**(2–3), 158–182 (2005)
13. Haarslev, V., Möller, R.: RACER system description. In: Goré, R., Leitsch, A., Nipkow, T. (eds.) IJCAR 2001. LNCS, vol. 2083, pp. 701–705. Springer, Heidelberg (2001). https://doi.org/10.1007/3-540-45744-5_59
14. Harris, S., Seaborne, A., Prud'hommeaux, E.: SPARQL 1.1 query language. W3C Recommendation, March 2013
15. Isard, M., Budiu, M., Yu, Y., Birrell, A., Fetterly, D.: Dryad: distributed data-parallel programs from sequential building blocks. In: EuroSys (2007)
16. Katsogridakis, P., Papagiannaki, S., Pratikakis, P.: Execution of recursive queries in apache spark. In: Rivera, F.F., Pena, T.F., Cabaleiro, J.C. (eds.) Euro-Par 2017. LNCS, vol. 10417, pp. 289–302. Springer, Cham (2017). https://doi.org/10.1007/978-3-319-64203-1_21
17. Kunegis, J.: Konect: the Koblenz network collection. In: WWW (2013)
18. Leone, N., et al.: The DLV system for knowledge representation and reasoning. ACM Trans. Comput. Log. **7**(3), 499–562 (2006)
19. Leskovec, J., Krevl, A.: SNAP datasets: Stanford large network dataset collection. http://snap.stanford.edu/data. Accessed June 2014
20. Marx, E., et al.: Torpedo: improving the state-of-the-art RDF dataset slicing. In: ICSC (2017)
21. Motik, B., et al.: OWL 2 web ontology language profiles. W3C Recommendation, December 2012
22. Motik, B., Nenov, Y., Piro, R., Horrocks, I., Olteanu, D.: Parallel materialisation of datalog programs in centralised, main-memory RDF systems. In: AAAI (2014)
23. Parsia, B., Sirin, E.: Pellet: an OWL DL reasoner. In: ISWC (2004)
24. Rebele, T., Tanon, T.P., Suchanek, F.: Technical report: answering datalog queries with UNIX shell commands. Technical report, Telecom ParisTech (2018). https://www.thomasrebele.org/publications/2018_report_bashlog.pdf
25. Saha, B., Shah, H., Seth, S., Vijayaraghavan, G., Murthy, A.C., Curino, C.: Apache Tez: a unifying framework for modeling and building data processing applications. In: SIGMOD (2015)
26. Scholz, B., Jordan, H., Subotic, P., Westmann, T.: On fast large-scale program analysis in datalog. In: International Conference on Compiler Construction (2016)
27. Shearer, R., Motik, B., Horrocks, I.: HermiT: a highly-efficient OWL reasoner. In: OWLED, vol. 432 (2008)
28. Shkapsky, A., Yang, M., Interlandi, M., Chiu, H., Condie, T., Zaniolo, C.: Big data analytics with datalog queries on spark. In: SIGMOD (2016)
29. Suchanek, F.M., Kasneci, G., Weikum, G.: YAGO: a core of semantic knowledge. In: WWW (2007)
30. Tsarkov, D., Horrocks, I.: FaCT++ description logic reasoner: system description. In: Furbach, U., Shankar, N. (eds.) IJCAR 2006. LNCS (LNAI), vol. 4130, pp. 292–297. Springer, Heidelberg (2006). https://doi.org/10.1007/11814771_26
31. Ullman, J.D.: Principles of Database and Knowledge-Base Systems. W. H. Freeman & Co, New York (1988)
32. Verborgh, R.: Triple pattern fragments: a low-cost knowledge graph interface for the web. J. Web Semant. **37–38**, 184–206 (2016)

33. Vrandecic, D., Krötzsch, M.: Wikidata: a free collaborative knowledgebase. Commun. ACM **57**(10), 78–85 (2014)
34. Wang, J., Balazinska, M., Halperin, D.: Asynchronous and fault-tolerant recursive datalog evaluation in shared-nothing engines. PVLDB **8**(12), 1542–1553 (2015)
35. Wu, H., Liu, J., Wang, T., Ye, D., Wei, J., Zhong, H.: Parallel materialization of datalog programs with spark for scalable reasoning. In: Cellary, W., Mokbel, M.F., Wang, J., Wang, H., Zhou, R., Zhang, Y. (eds.) WISE 2016. LNCS, vol. 10041, pp. 363–379. Springer, Cham (2016). https://doi.org/10.1007/978-3-319-48740-3_27
36. Zaharia, M., Chowdhury, M., Franklin, M.J., Shenker, S., Stoica, I.: Spark: Cluster computing with working sets. In: USENIX Workshop on Hot Topics in Cloud Computing (2010)
37. Zhou, J., Larson, P., Chaiken, R.: Incorporating partitioning and parallel plans into the SCOPE optimizer. In: ICDE (2010)

WORQ: Workload-Driven RDF Query Processing

Amgad Madkour[1](✉), Ahmed M. Aly[2], and Walid G. Aref[1]

[1] Purdue University, West Lafayette, USA
{amgad,aref}@cs.purdue.edu
[2] Google Inc., Mountain View, USA
aaly@google.com

Abstract. Cloud-based systems provide a rich platform for managing large-scale RDF data. However, the distributed nature of these systems introduces several performance challenges, *e.g.*, disk I/O and network shuffling overhead, especially for RDF queries that involve multiple join operations. To alleviate these challenges, this paper studies the effect of several optimization techniques that enhance the performance of RDF queries. Based on the query workload, reduced sets of intermediate results (or reductions, for short) that are common for certain join pattern(s) are computed. Furthermore, these reductions are not computed beforehand, but are rather computed only for the frequent join patterns in an online fashion using Bloom filters. Rather than caching the final results of each query, we show that caching the reductions allows reusing intermediate results across multiple queries that share the same join patterns. In addition, we introduce an efficient solution for RDF queries with unbound properties. Based on a realization of the proposed optimizations on top of Spark, extensive experimentation using two synthetic benchmarks and a real dataset demonstrates how these optimizations lead to an order of magnitude enhancement in terms of preprocessing, storage, and query performance compared to the state-of-the-art solutions.

Keywords: Intermediate results · Basic graph pattern
Distributed SPARQL query processing

1 Introduction

Processing RDF queries involves multiple scans of the same data, *e.g.*, when certain join patterns are frequent and are repeated across multiple queries. This calls for workload-driven mechanisms that cache only the data that is required by the query workload. Network shuffling overhead also degrades query performance in a distributed environment. It occurs when the processing nodes exchange data in order to answer queries. Reducing the network shuffling overhead highly relies on how the data is partitioned across the nodes.

This paper presents *W*orkload-driven *R*DF *Q*uery Processing (WORQ, for short), a system that encapsulates several optimizations that significantly

© Springer Nature Switzerland AG 2018
D. Vrandečić et al. (Eds.): ISWC 2018, LNCS 11136, pp. 583–599, 2018.
https://doi.org/10.1007/978-3-030-00671-6_34

enhance the performance of RDF queries. In particular, WORQ addresses three main issues: (1) how to efficiently *partition* the RDF data in an online fashion, (2) how to *reduce* the intermediate join results of an RDF query in an online fashion, and (3) how to *cache* reusable intermediate join results instead of the final results of an RDF query.

Workload-Driven Partitioning: Data partitioning is common in distributed data management systems. The RDF data is typically divided into several partitions, and then is distributed across the cluster machines. The objective of partitioning is to reduce the query execution time by leveraging parallelism. Data partitioning incurs a preprocessing overhead as it needs to be performed over the whole data. However, for a real workload, only a small fraction of the data is accessed (*e.g.*, see [25]). WORQ adopts a workload-driven approach when partitioning the data. For each query, WORQ identifies each query triple (*i.e.*, an entry consisting of bound and unbound subject, property, and an object) as a subquery. Then, WORQ partitions the data triples by the join attribute of each subquery. The join attribute represents the variable that connects two or more query triples. The join can be between subjects, properties, objects, or a combination of the three attributes. WORQ partitions the data only once for every new query join pattern that is identified.

Join Reductions: Tables are one way of storing RDF data triples. When a single query involves joins between multiple tables that correspond to different query patterns, every binary join operation generates intermediate join results (or *intermediate results*, for short). The intermediate results represent the data that satisfies the binary join and eventually contributes to the final result of the query. However, intermediate results may contain redundant data triples that do not match all the query joins. WORQ minimizes the intermediate results by precomputing join reductions through Bloom-joins [8,16].

Caching: To boost query performance, caching can be employed to improve query response time and increase the throughput of execution. One caching approach is to cache the *results* of each query. However, caching the unique query results incurs significant memory storage overhead. In contrast, WORQ caches (in main memory) the join reductions that correspond to the frequent join patterns. These reductions can be reused by other queries that share the same query patterns.

Queries with Unbound Properties: Some query workloads may have query triples with unbound (*i.e.*, unspecified) properties. For example, the query triple :John ?x :Mary queries all data triples that have a subject :John and an object :Mary, where ?x specifies an unbound property. Answering unbound property queries is challenging for RDF systems that adopt a specific RDF partitioning scheme. Assuming that the data is vertically partitioned [1,13] (VP, for short), the data triples are split into separate files denoted by the property (*i.e.*, predicate) name, where each file contains the subject and object representing the property. Using VP, answering unbound property is challenging because all property files need to be accessed or an index needs to be built on top of each file. In contrast, WORQ utilizes Bloom filters as indexes to efficiently answer unbound property queries.

WORQ is implemented as part of the Knowledge Cubes (KC) proposal [17]. The source code[1] for a Spark-based implementation of WORQ is publicly available for download. Our experimental setup includes two synthetic benchmarks, namely WatDiv [4] and LUBM [10], and a real dataset, namely YAGO2s [7,12]. The purpose of the experiments is to demonstrate three aspects of WORQ : (1) the preprocessing time required given an RDF dataset, (2) the storage overhead incurred to create the RDF database, and (3) the query processing time when answering RDF queries with respect to partitioning and caching. The results illustrate how the presented optimizations provide at least an order of magnitude better results on the three aforementioned aspects when compared to the Hadoop-based state-of-the-art solution.

The contributions of this paper can be summarized as follows:

– We present workload-driven partitioning of RDF triples that can join together in order to minimize the network shuffling overhead based on the query workload.
– We present the use of Bloom filters for computing RDF join reductions online.
– Rather than caching the results of an RDF query, we show that caching the RDF join reductions can boost the query performance while keeping the cache size minimal.
– We study an efficient technique for answering RDF queries with unbound properties using Bloom filters.

The rest of this paper proceeds as follows. Section 2 presents the online reduction of RDF data. Section 3 presents workload-driven partitioning in WORQ . Section 4 presents how WORQ answers unbound-property queries. Section 5 presents the experiments performed over the WatDiv, LUBM, and YAGO datasets. Section 6 presents the related work. Finally, Sect. 7 presents concluding remarks.

2   Online Reduction of RDF Data

WORQ employs Bloom-join [8,16] to compute the reductions between vertical partitions. Many cloud-based systems [13] use vertical partitioning (VP) [1] including the state-of-the-art [27]. VPs can be realized over any relational database system and stored in cloud data sources (e.g. Parquet, ORC2). Bloom-join determines if an entry in one partition qualifies a join condition with another partition. The reductions can be computed in an online fashion using Bloom-join instead of precomputing all possible reductions in an offline fashion (*i.e.*, during the preprocessing phase [27]). Bloom-join utilizes a probabilistic data structure, termed Bloom filter [8]. A Bloom filter does not physically store items, but rather hashes the input against different hash functions. The main functionality of a

---

[1] http://github.com/amgadmadkour/knowledgecubes.

Bloom filter is to determine the existence of an item. Bloom filters can have false-positives, but no false-negatives. Bloom filters are fast to create, fast to probe, and small to store. Also, the false-positives introduce a small percentage of irrelevant rows that eventually are not joined in a Bloom-join. During the evaluation of a join, WORQ uses Bloom filters to probe the join attributes of the query join-patterns. The Bloom filters representing the join attributes filter the rows in both partitions involved in the join, and the results are materialized as a reduction for a specific join pattern, or *reductions*, for short.

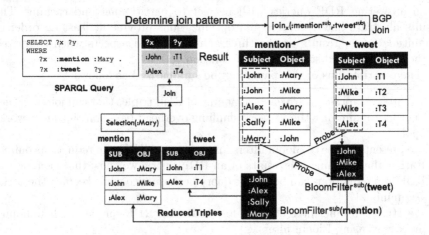

Fig. 1. Evaluating a SPARQL query using Bloom-join between :mention and :tweet

Figure 1 gives an example of using Bloom filters to compute a join reduction. The query has a BGP join between `:mention` and `:tweet` on the Subject attribute. WORQ uses the Bloom filter of $BloomFilter^{sub}(:tweet)$ to compute a reduction for the `:mention` property on the subject column. `:tweet`'s Bloom filter consists of the elements `:John`, `:Mike`, and `:Alex`. Each element in the subject column of the `:mention` partition is probed against the `:tweet` Bloom filter. The reduction for `:mention` represents all the rows that qualify a join between the vertical partitions `:mention` and `:tweet` on the subject attribute. Figure 1 illustrates the entries that qualify the join between `:mention` and `:tweet`, where the vertical partition of `:mention` is reduced from five entries to only three qualifying entries. Similarly, the vertical partition of `:tweet` is reduced from four entries to only two qualifying entries. The reductions for both properties are cached by WORQ in order to be reused by other queries that share the same join patterns. In other words, the `:mention` reduction can be reused by the `:mention` property if it joins with `:tweet` on the subject attribute. Also, the `:tweet` reduction can be reused by the `:tweet` property if `:tweet` joins with the `:mention` property on the subject attribute.

WORQ does not apply selection (*i.e.*, filtering) operations on the original data triples (*i.e.*, VP). Instead, selections are applied on the reductions after the

reductions are computed. For example, the reduction for :mention contains a
selection on the object, namely :Mary. However, the selection has been delayed
until the reductions have been computed from the original data triples. The
advantage of delaying the selection is that the reductions can be reused by other
queries that share the same join patterns. However, if selections are pushed
early on the original data triples, then the reductions will not be representative
of the join operation between the query triples. Finally, the resulting reductions
(including the ones that have been filtered) are joined together based on the
join attribute indicated by the query triples. WORQ does not require a specific
join algorithm to be used. Distributed join algorithms, e.g., broadcast hash join
or sort-merge join that are employed by distributed computational frameworks,
e.g., Spark, can be used [3]. Figure 1 illustrates the final result of the query after
joining both the query triples representing :mention (after the selection) and
:tweet properties, where two entries qualify the join result.

N-ary Join Reductions

WORQ computes the reductions online instead of pre-computing the reductions
offline [27]. In addition, WORQ computes the reductions between all the possible
(n-ary) query-triples instead of computing the reductions in binary form [27].

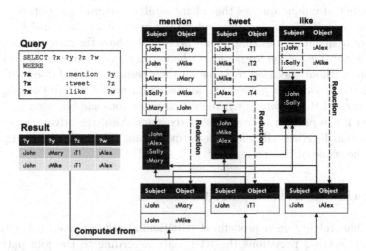

Fig. 2. N-ary join between the reductions of three query triples involving the :mention,
:tweet, and :like VPs

Figure 2 illustrates a SPARQL query with three query-triples that share
the same join attribute (*i.e.*, variable ?x). When the join is computed between
the :mention, :tweet, and :like VPs, only the data triples that are common
amongst the three VPs will qualify as a result. WORQ utilizes Bloom join to
reduce the number of data triples in every VP involved in the join operation,

and hence reduces the intermediate results between the three join operations. WORQ uses the Bloom filters representing the join columns of the three query triples (the subject Bloom filters, in this instance) to reduce the VP entries to the ones that would qualify the join operation. For example, the :mention VP is reduced from five data triples to two triples that have :John as the subject because :John is the only resource that qualifies the :tweet Bloom filter on the subject and the :like Bloom filter on the subject. The same applies to the :tweet VP, where :John is the only resource that qualifies the :mention Bloom filter on the subject and the :like Bloom filter on the subject. Finally, WORQ uses the computed reductions instead of the VPs to evaluate the query. The result of the query includes two rows corresponding to the only resource common across the three property-VPs. The computed reductions are cached to be reused by any other query that contains a join between the three properties on the subject attribute.

Caching of Reductions

Rather than caching portions of the original RDF data or the final query results, WORQ caches (in main-memory) the *reductions* that correspond to the join patterns that are discovered during query processing. Caching intermediate results (*i.e.*, reductions) is suitable in situations where the query workload consists of a high number of unique queries that share similar patterns. In contrast, caching the results is suitable in situations where the query workload consists of a high number of frequent queries that do not necessarily share the same query pattern. WORQ is suitable for the former case where many unique queries can utilize the reductions without the need to cache all their results. WORQ does not assume a specific cache-eviction policy, *i.e.*, any eviction policy. WORQ employs least recently used (LRU) strategy where evicted reductions can be saved to disk and be reused if the pattern they represent reoccurs. Also, the advantage of saving to disk is that filtering will not be performed again. The cache-eviction policy is beyond the scope of this paper.

3   Workload-Driven Partitioning

Rather than relying on a predefined partitioning criteria (*e.g.*, using the subject only), WORQ partitions the RDF data according to the join patterns in the queries received so far. WORQ aims at placing the partitions of the reductions that share the same join attribute on the same machine, which minimizes the shuffling overhead, and more importantly, reduces the query response time. Instead of partitioning the VP, WORQ partitions the *reduction rows* across the machines. After a query is parsed, WORQ identifies the join attributes in the query. Based on the join attributes, the reductions that need to be partitioned are determined. Reduction partitioning is performed only once, and the resulting partitions are reused by any query that has the join pattern that corresponds to the reduction.

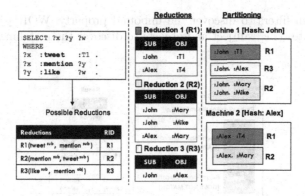

Fig. 3. Workflow for workload-driven partitioning

Figure 3 illustrates a set of query join patterns and their corresponding reductions. The join pattern representing the :tweet property uses the reduction denoted by R1 on the subject. The join pattern representing the :like property uses the reduction denoted by R3 on the subject as well. WORQ partitions the rows of every reduction based on the join attribute (*i.e.*, the subject or object). In Fig. 3, the reductions representing R1, R2, R3 are partitioned using the subject (as the reductions are based on the subject attribute). The reduction rows are hash-distributed across the machines using the join attribute (*i.e.*, subject or object). This partitioning scheme guarantees that all the data triples that are related to the join attributes of the query are co-located on the same machine, and thus allowing the reductions to be computed locally.

4   Queries with Unbound Properties

The performance of unbound-property queries depends on the adopted RDF partitioning scheme. If the data is vertically partitioned, answering unbound-property queries becomes challenging because all the VPs need to be iterated. A straightforward approach to query the unbound properties in a distributed setting is to store the RDF data triples in a single file (*i.e.*, triples file). Distributed file systems, *e.g.*, HDFS, split the files into a set of blocks and distribute the blocks across machines. In this case, RDF query processors can evaluate unbound-property RDF queries in parallel [26], where each machine processes a set of blocks. We refer to this baseline approach as *RDF-Table*. We implement this baseline for evaluation purposes.

WORQ utilizes Bloom filters as cheap indexes to efficiently answer unbound-property queries over data that has been vertically partitioned. WORQ performs two steps to determine the matching properties. The first step is called the *identification* step, where a set of candidate properties are identified. The second step is called the *verification* step, where the candidate properties are verified to eliminate the possibility of false-positives. Given a query, WORQ uses the existing Bloom filters to discover the unbound property. WORQ relies on the *bound* attributes (*i.e.*, subject and object) to discover the matching properties.

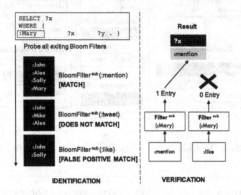

Fig. 4. The identification and verification steps to answer unbound-property queries

Figure 4 illustrates the identification step for answering unbound-property queries. First, the unbound and bound attributes are identified. Then, the bound attributes are used to probe the Bloom filters to determine if the bound values exist for a specific property. If a value exists, the corresponding property is added as a candidate for answering the query. For instance, in Fig. 4, :Mary exists in the :mention property, and is found using a *MATCH* in the corresponding Bloom filter. However, :Mary does not exist in the :tweet property, and hence the Bloom filter returns *DOES NOT MATCH*. Although :Mary does not exist in :like, the Bloom filter returns a *MATCH*, which is a false positive.

Given that Bloom filters can incur false positives, a verification step is needed to ensure the correctness of query evaluation. WORQ verifies the candidate properties by issuing a filter based on the bound attributes with the value indicated in the query triple (*i.e.*, the value that made the candidate property match). If the result-set includes at least one match, then WORQ determines that the candidate property was identified correctly. Otherwise, the candidate property is discarded. Disqualifying data will not happen frequently based on the false-positive rate of the constructed Bloom filters.

5   Experiments

WORQ is compared against S2RDF [27], a Spark-based system that runs over Hadoop. S2RDF [27] proposes an extension to VP, namely ExtVP, where reductions of entries are computed for every vertical partition. S2RDF utilizes *semi-join reductions* [6] to reduce the number of rows in a partition. The reductions represent all RDF query combinations that appear in SPARQL queries (*i.e.*, Subject-Subject, Subject-Object, Object-Subject, Object-Object). However, S2RDF exhibits a substantial preprocessing overhead. Semi-joins are expensive to compute, and generate large network-traffic. In addition, S2RDF generates a large number of files to represent the reductions of the original data. S2RDF translates SPARQL queries to SQL and runs them on Spark SQL. S2RDF has outperformed Hadoop-based systems such as H2RDF+, Sempala, PigSPARQL, SHARD, and other systems such as Virtuoso, where S2RDF has achieved (on average) the best query execution performance [27]. Accordingly, this paper presents a comparison with S2RDF only as S2RDF represents the state-of-the-art Hadoop-based RDF query processing system. WORQ is implemented over Spark (v2.1) where it utilizes Spark DataFrames to represent the reductions. WORQ does not translate the query to SQL. Instead, WORQ implements joins as a series of Spark DataFrame joins. To guarantee a fair setup, all Spark-related parameters are unified for both WORQ and S2RDF. The data for both systems is stored using Parquet[2] columnar-store format. Vertical partitioning has been implemented as a baseline.

5.1 Experimental Setup

The experimental setup datasets and queries proposed by Abdelaziz *et al.* [2] are used. Our experiments are conducted using a real dataset (YAGO2s [7,12]) as well as two synthetic benchmarks (WatDiv [4] and LUBM [10]) that provide widely-adopted query workload generators:

1. **WatDiv** provides a stress-test query workload that allows generating several queries per-pattern. One Billion triples have been generated to demonstrate the query execution performance and preprocessing performance (i.e., the number of files generated, disk space utilization, and loading time). A pre-generated workload provided by WatDiv [4] contains 5000 queries that cover 100 diverse SPARQL patterns, each having 50 variations. A variation represents different bound values for the same query pattern. The variations allow measuring the performance of specific patterns under different selectivities. For the unbound-property queries, we use the query workload provided by Alvarez-Garcia *et al.* [5] that represents 500 queries covering three combinations namely, unbound subject with bound object, unbound object with bound subject, and bound subject and object.
2. **LUBM** provides a query-workload generator, where 1000 queries are generated. Unlike WatDiv, LUBM does not specify the number of patterns.
3. **YAGO2s** consists of 245 million real RDF triples. YAGO2s benchmark queries are used to compare the query execution time [19,27]. There is no publicly available real query workload for YAGO. Generating synthetic queries for YAGO is similar to what WatDiv and LUBM provide while they guarantee generating all possible query shapes.

---

[2] parquet.apache.org.

Our experiments are conducted using an HP DL360G9 cluster with Intel Xeon E5-2660 realized over 5 nodes. The cluster uses Cloudera 5.9 consisting of Spark 2 as a computational framework and Hadoop HDFS as a distributed file-system. Each node consists of 32 GB of RAM, and 52 cores. The total HDFS size is 1 Terabyte. The experiments measure various aspects of WORQ including (1) the number of generated files, (2) the filesystem size, (3) the loading time, (4) the workload query execution performance, (5) the overhead of caching results instead of caching reductions, and (6) the execution performance of unbound properties queries. The data for the 3 benchmarks is loaded into memory before execution.

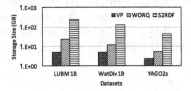

Fig. 5. Disk space utilization   **Fig. 6.** Preprocessing time

5.2   Experimental Results

Preprocessing Performance. Figure 5 gives the disk storage overhead incurred by the three systems over the LUBM, WatDiv, and YAGO2s datasets. VP introduces minimal space overhead across all three systems. The reason is that VP only needs to partition the original triple file based on the property name. Storage in WORQ is composed of the VP and the Bloom filters. S2RDF precomputes all the possible reductions for binary joins ($O(n^2)$, where n is the number of VPs), and stores them on disk along with the original data. Thus, S2RDF introduces the highest disk storage overhead.

Figure 6 gives the preprocessing time for all three systems over the LUBM, WatDiv, and YAGO2s datasets. VP has the smallest loading time due to its simplicity, followed by WORQ, and then S2RDF. The majority of time spent by S2RDF in the preprocessing time involves creating the proposed partitions called ExtVP. The computation involves performing semi-joins between binary partitions in a distributed fashion causing high network shuffling overhead. WORQ incurs a minor overhead compared to VP due to the computation of the Bloom filters.

Query Workload Awareness. For the remaining experiments, the results of VP are omitted due to its low performance. The following experiments demonstrate the query performance of both WORQ and S2RDF across different aspects, *e.g.*, the total execution time, the mean execution time per query pattern, and the mean execution time given the number of join-triples in a

query. WatDiv and LUBM are used due to the availability of workload generators while YAGO2s is omitted as a real query workload is unavailable. However,
a set of benchmark queries [27] are used to measure the performance against
the YAGO2s dataset. In S2RDF, the partitioning is done for every query and
takes place while the queries are being evaluated. S2RDF reports the overall
execution time which includes both the partitioning and the actual execution
time. WORQ follows the same procedure when reporting the overall execution
time.

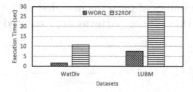

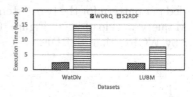

Fig. 7. Mean query execution time   **Fig. 8.** Total query execution time

Figures 7 and 8 give the mean and total execution times based on executing
5000 queries over WatDiv (1 Billion triples) and 1000 queries over LUBM (1
Billion triples). WORQ is consistently better across the two benchmarks. The
difference in performance is attributed to the combination of efficient partitioning and the caching of reduction employed by WORQ as illustrated in later
experiments. WORQ reduces the relations to be joined by computing light-
weight reductions that can fully represent the original data in answering the
RDF queries. Rather than scanning the original (large) data for each query, the
light-weight reductions are used instead.

The difference in performance between LUBM and WatDiv is attributed to
the characteristics of both benchmarks in terms of the number of properties and
the query workload representing each dataset. LUBM consists of 18 properties
while WatDiv consists of 86 properties. The 1 Billion triples for LUBM and
WatDiv are distributed across 18 and 86 properties, respectively. WORQ performs well with the increase in the number of properties. In real datasets, *e.g.*,
YAGO2s [7,12], the number of properties are in hundreds, making WORQ more
appropriate to use than S2RDF.

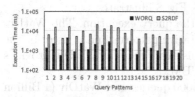

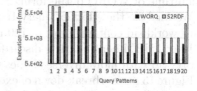

Fig. 9. Mean execution time per query   **Fig. 10.** Mean execution time per query
pattern over WatDiv 1 Billion dataset   pattern over LUBM 1 Billion dataset

Figure 9 gives a break-down of the query execution of 5000 queries over Wat-
Div (1 Billion triples) per query pattern. The x-axis represents the query numbers

and the y-axis represents the execution time. A query pattern represents a set of one or more query triples (*i.e.*, BGP triples) that vary based on the bound and unbound attributes, e.g., one pattern can have two query triples joined by the subject attribute while another pattern would be based on two query triples joined on the object attribute. For every pattern, the mean execution time is recorded for the two systems. Figure 9 shows that WORQ executes each pattern nearly an order of magnitude faster than S2RDF.

Figure 10 gives a break-down of executing 1000 queries over LUBM (1 Billion triples per query pattern). Similar to WatDiv, the mean execution time is recorded per pattern for the two systems. The number of patterns included in the LUBM query workload is 20. Figure 10 shows that all the patterns are executed faster by WORQ than S2RDF.

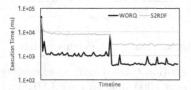

Fig. 11. Execution timeline for two query pattern over WatDiv 1 Billion dataset

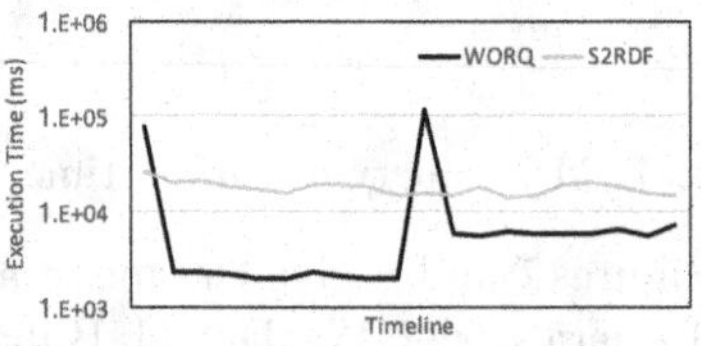

Fig. 12. Execution timeline for two query pattern over LUBM 1 Billion dataset

Figure 11 gives the performance when executing only two patterns over the WatDiv benchmark. The x-axis represents the timeline, where we execute one query pattern first, and then execute another pattern. There are two major spikes in the performance of WORQ that reflect the first time each query pattern was executed. For each pattern, a high query execution overhead is exhibited at the beginning, followed by near-linear performance for the rest of the queries that share the same join pattern.

Figure 12 repeats the same experiment for two patterns over the LUBM benchmark. Similar to Fig. 11, the first time a join pattern is executed, a spike in execution time is exhibited followed by a near-linear performance for the remaining queries. Unlike WatDiv, the computation of the query patterns for the first time over LUBM consumes more time than S2RDF. However, the overall execution time of WORQ outperforms S2RDF as Fig. 8 illustrates.

We analyze the effect of query triples on the query execution. The WatDiv query workload contains a set of 100 representative patterns and is used for the analysis. LUBM benchmark is discarded for this experiment as WatDiv provides a workload with more diverse shapes than LUBM.

Figure 13 gives a break-down of executing 5000 queries over WatDiv (1 Billion triples) given the number of triples per query. From the figure, the number of triples affects the overall query performance, where the query execution time increases as more triples are processed.

Figure 14 gives a break-down of the mean query execution time for 5000 queries over WatDiv (1 Billion triples) based on the number of joins between

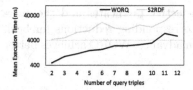

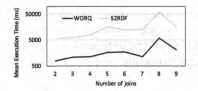

Fig. 13. Mean execution time - number of triples per query over WatDiv 1 Billion

Fig. 14. Mean execution time for joins per pattern over WatDiv 1 Billion

query triples. This is different from the number of query triples experiment, where the number of joins experiment measures the maximum number of joins identified per query, *e.g.*, a query may contain five query triples, but contains a join between two query triples only. To create the experimental setup, every query is first placed in a join group based on the maximum number of joins that it has. Then, the mean execution time is measured for queries within a join group. WORQ achieves nearly an order of magnitude better performance than S2RDF.

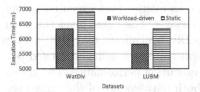

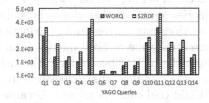

Fig. 15. Mean query execution time using workload-driven and static partitioning

Fig. 16. Execution time of 14 query patterns over YAGO2s dataset

Figure 15 gives a break-down of the mean execution time using workload-driven partitioning and static partitioning of WORQ to illustrate the effect of using the workload-driven component only. Static partitioning is based on subject. In workload-driven partitioning, every query is partitioned based on the join patterns of the query. In contrast, static partitioning is performed based on a pre-specified criteria, *e.g.*, partitioning by subject. Static partitioning was performed on the subject column. Figure 15 demonstrates that workload-driven partitioning contributes positively towards the overall query execution performance over the two datasets. The partitioning time is dependent on where data is originally stored on the cluster and generally incurs a minor cost. The query evaluation time given where data is partitioned dominates the execution time.

Figure 16 gives a break-down of the query execution over 14 benchmark YAGO2s queries [27]. The x-axis represents the query numbers and the y-axis represents the execution time. The queries were designed to take into consideration various query shapes, *e.g.*, star-shaped, and resources selectivities. Each query was executed 5 times using different selective predicates and the average time was reported. For every query, the corresponding reductions for both WORQ and S2RDF were loaded into memory in advance. WORQ achieves better query execution performance over all queries.

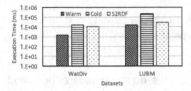

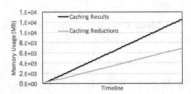

Fig. 17. Mean query execution time given warm and cold cache for WORQ over WatDiv and LUBM

Fig. 18. Memory usage based on caching results and caching reductions

Caching of Reductions. Figure 17 demonstrates the effect of caching on the query performance. *Cold* queries are those with patterns that have not been executed before, i.e., that have no corresponding reductions in the cache. *Warm* queries are those that share the same pattern as queries that executed before, i.e., that have corresponding reductions in the cache. The figure gives the mean execution time of 5000 queries from the WatDiv benchmark and 1000 queries from the LUBM benchmark. The figure demonstrate how utilizing cached patterns (*i.e.*, reductions) achieves better query execution performance. The reason LUBM cold cache is worse is because while both datasets are of the same overall size (1B triples), one contains 18 files/predicates (LUBM) in contrast to 87 files/predicates in WatDiv so the filtering time is higher for LUBM queries. Also, WORQ pays a price only once when a query pattern is seen for the first time. However, the cold-start cost is minor.

Figure 18 gives a break-down of the memory usage over 5000 unique queries covering 100 patterns. Using a workload of 5000 unique queries, the figure demonstrates how the size of the cached queries grows over time and surpasses the size of cached reductions. The memory usage for caching the query results can reach more than 10 GB over 5000 queries while caching reductions exhibits a slower memory usage curve. The conclusion is that caching the reductions is more suitable than caching the query results in situations where there are many unique queries that share common patterns.

Performance of Unbound-Property Queries. Schatzle *et al.* [27] do not evaluate the performance of S2RDF for unbound-property queries as it is out of the scope of their current work. In addition, S2RDF adopts a VP structure to answer queries, leading to degraded query performance over unbound-property queries. Therefore, we use the RDF-Table approach described in Sect. 4 as a baseline. We evaluate three query patterns based on the attributes of an RDF triple, namely a bound subject and object, a bound subject, and a bound object.

Table 1. Unbound property results - (BSO) Bound Subject and Object, (BS) Bound Subject, (BO) Bound Object

System	BSO-Mean	BSO-Sum	BS-Mean	BS-Sum	BO-Mean	BO-Sum
WORQ	1.25 ms	10.49 min	4.18 ms	34.84 min	3.52 ms	29.34 min
RDF-Table	5.3 ms	44.44 min	3.80 ms	31.67 min	4.35 ms	36.26 min

Table 1 gives the result of processing 500 queries with bound subject and object (BSO), bound subject (BS), and bound object (BO) over WatDiv (1 Billion triples). For bound subject and object (BSO), the mean execution time per query is nearly five times better than the baseline. This is attributed to the Bloom filter usage, where the number of false-positives is reduced by evaluating the properties against two bound values instead of one bound value, *e.g.*, queries with bound subject only or a bound object only. For bound subject (BS), the mean execution time of WORQ is comparable to that of RDF-Table. This performance is due to two main reasons. The first is the efficiency of RDF-Table within Spark as RDF-Table performs predicate pushdown filtering in parallel and the result is aggregated back to the driver (*i.e.*, master node). The second is that the data of the RDF-Table is sorted by the subject, allowing the predicate-pushdown to work efficiently. For bound object, the mean execution time is also comparable to that of RDF-Table. The overall execution time of WORQ is better than RDF-Table. The reason for the better result is attributed to the lack of sorting on the object column for the RDF dataset. This gives WORQ performance advantage when executing bound object queries.

6 Related Work

Graph-based partitioning is an NP-complete problem [14], and hence hash partitioning heuristics [21,31] are employed instead of graph-based partitioning in order to partition RDF data efficiently. However, sophisticated partitioning techniques [11,15,22,28] cannot guarantee that no data will be shuffled when processing complex queries with multiple joins. Several techniques [23,29] utilize the query workload to enhance the partitioning of RDF data. In addition, one study [4] demonstrates the need to continuously adapt to workloads in order to guarantee consistent performance. Characteristic sets [18] capture the set of properties that occur together for a given subject. However, characteristic sets are data-driven and are tied to star-shaped queries only. Castilo *et al.* [9] perform evaluation of SPARQL queries using (offline) materialized results coined RDF-MatView indexes. In contrast to materialized views, WORQ does not materialize results but instead identifies reductions that can be reused across queries that share the same join patterns. H2RDF+ [20] provides a result-based workload-aware RDF caching engine that manages to dynamically index frequent workload subgraphs in real time. However, caching the final results of RDF queries incurs significant storage overhead and cannot generalize to a broader query workloads. Yang *et al.* [30] propose caching the intermediate results of basic graph patterns in SPARQL queries. However, the proposed approach is tied to the join orders that would result in different intermediate results. Alvarez-Garcia *et al.* [5] introduce a compressed index called k2-triples for answering unbound-property queries. However, the proposed index is not applicable in a distributed setting. Ravindra *et al.* [24] uses a non-relational algebra based on a TripleGroup data model to answer unbound-property queries.

7   Concluding Remarks

This paper presents several optimizations for RDF query processing over vertically partitioned triples. First, we present how to use Bloom join to compute reduced sets of intermediate results (or reductions, for short) that are common for certain join pattern(s) in an online fashion. Second, we study the effect of caching these reductions instead of caching the final results of each query. Third, we present how to partition the RDF data triples using the join attributes of the query instead of using a predefined partitioning criteria. Fourth, we present how to efficiently answer queries with unbound properties using Bloom filters. Extensive experimentation using the WatDiv, LUBM, and YAGO2s demonstrate how a realization of these optimizations can lead to an order of magnitude enhancement in terms of preprocessing time, storage, and query performance. Bloom filters/join is one case study. N-ary filtering can utilize any set membership structure (*e.g.*, Bloom, Cuckoo, Roaring Bitmaps) so long as we can add and check elements in a set. The novelty is in how membership structures (*e.g.*, Bloom Filter) are used to filter data and answer unbound property queries efficiently in a distributed setting. For future work, we will investigate further query processing enhancements including load-balanced partitioning of reductions, generalized filtering (exact vs. approximate structures), and spatio-temporal RDF filtering.

References

1. Abadi, D.J., Marcus, A., Madden, S.R., Hollenbach, K.: Scalable semantic web data management using vertical partitioning. In: VLDB (2007)
2. Abdelaziz, I., Harbi, R., Khayyat, Z., Kalnis, P.: A survey and experimental comparison of distributed SPARQL engines for very large RDF data. PVLDB **10**(13), 2049–2060 (2017). Article no. 9
3. Agathangelos, G., Troullinou, G., Kondylakis, H., Stefanidis, K., Plexousakis, D.: RDF query answering using apache spark : review and assessment. In: DESWEB (2018)
4. Aluç, G., Hartig, O., Özsu, M.T., Daudjee, K.: Diversified stress testing of RDF data management systems. In: ISWC, pp. 197–212 (2014)
5. Álvarez-García, S., Brisaboa, N., Fernández, J.D., Martínez-Prieto, M.A., Navarro, G.: Compressed vertical partitioning for efficient RDF management. Knowl. Inf. Syst. **44**(2), 439–474 (2015)
6. Bernstein, P.A., Chiu, D.-M.W.: Using semi-joins to solve relational queries. J. ACM **28**, 25–40 (1981)

7. Biega, J., Kuzey, E., Suchanek, F.M.: Inside YAGO2s. In: WWW, pp. 325–328. ACM Press, New York (2013)
8. Bloom, B.: Space/time trade-offs in hash coding with allowable errors. Commun. ACM **13**(7), 422–426 (1970)
9. Castillo, R., Leser, U.: Selecting materialized views for RDF data. In: ICWE (2010)
10. Guo, Y., Pan, Z., Heflin, J.: LUBM: a benchmark for OWL knowledge base systems. Web Semant. **3**, 158–182 (2005)
11. Gurajada, S., Seufert, S., Miliaraki, I., Theobald, M.: TriAD: a distributed shared-nothing RDF engine based on asynchronous message passing. In: SIGMOD, pp. 289–300 (2014)
12. Hoffart, J., Suchanek, F.M., Berberich, K., Weikum, G.: YAGO2: a spatially and temporally enhanced knowledge base from Wikipedia. Artif. Intell. **194**, 28–61 (2013)
13. Kaoudi, Z., Manolescu, I.: RDF in the clouds: a survey. VLDB J. **42**, 67–91 (2015)
14. Karypis, G., Kumar, V.: A fast and high quality multilevel scheme for partitioning irregular graphs. SIAM **20**, 359–392 (1998)
15. Lee, K., Liu, L.: Scaling queries over big RDF graphs with semantic hash partitioning. VLDB **6**, 1894–1905 (2013)
16. Mackert, L.F., Lohman, G.M.: R* optimizer validation and performance evaluation for local queries. ACM SIGMOD Record **15**(2), 84–95 (1986)
17. Madkour, A., Aref, W.G., Basalamah, S.: Knowledge cubes - a proposal for scalable and semantically-guided management of Big Data. In: BigData, pp. 1–7. IEEE, October 2013
18. Neumann, T., Moerkotte, G.: Characteristic sets: accurate cardinality estimation for RDF queries with multiple joins. In: ICDE, pp. 984–994 (2011)
19. Neumann, T., Weikum, G.: The RDF-3X engine for scalable management of RDF data. VLDB **19**, 91–113 (2010)
20. Papailiou, N., Tsoumakos, D., Karras, P., Koziris, N.: Graph-aware, workload-adaptive SPARQL query caching. In: SIGMOD, pp. 1777–1792 (2015)
21. Papailiou, N., Konstantinou, I., Tsoumakos, D., Karras, P., Koziris, N.: H2RDF+ : high-performance distributed joins over large-scale RDF graphs. In: BigData (2013)
22. Peng, P., Zou, L., Chen, L., Zhao, D.: Query workload-based RDF graph fragmentation and allocation. In: EDBT, pp. 377–388 (2016)
23. Rabl, T., Jacobsen, H.-A.: Query centric partitioning and allocation for partially replicated database systems. In: Proceedings of the 2017 ACM International Conference on Management of Data - SIGMOD 2017 (2017)
24. Ravindra, P., Anyanwu, K.: Scaling unbound-property queries on big RDF data warehouses using MapReduce. In: EDBT, pp. 169–180 (2015)
25. Rietveld, L., Hoekstra, R., Schlobach, S.: Structural properties as proxy for semantic relevance in RDF graph sampling. ISWC **8797**, 81–96 (2014)
26. Rohloff, K., Schantz, R.E.: High-performance, massively scalable distributed systems using the MapReduce software framework. In: PSIEtA, pp. 1–5 (2010)
27. Schätzle, A., Przyjaciel-Zablocki, M., Skilevic, S., Lausen, G.: S2RDF: RDF querying with SPARQL on spark. In: VLDB, vol. 9, pp. 804–815 (2016)
28. Wu, B., Zhou, Y., Yuan, P., Liu, L., Jin, H.: Scalable SPARQL querying using path partitioning. In: ICDE, pp. 795–806 (2015)
29. Yan, D., et al.: Quegel: a general-purpose query-centric framework for querying big graphs. In: VLDB (2016)
30. Yang, M., Wu, G.: Caching intermediate result of SPARQL queries. In: WWW (2011)
31. Zeng, K., Yang, J., Wang, H., Shao, B., Wang, Z.: A distributed graph engine for web scale RDF data. In: VLDB, pp. 265–276, February 2013

Canonicalisation of Monotone SPARQL Queries

Jaime Salas and Aidan Hogan[✉]

IMFD Chile and Department of Computer Science,
University of Chile, Santiago, Chile
jsalas@dcc.uchile.cl, aidhog@gmail.com

Abstract. Caching in the context of expressive query languages such as
SPARQL is complicated by the difficulty of detecting equivalent queries:
deciding if two conjunctive queries are equivalent is NP-complete, where
adding further query features makes the problem undecidable. Despite
this complexity, in this paper we propose an algorithm that performs syntactic canonicalisation of SPARQL queries such that the answers for the
canonicalised query will not change versus the original. We can guarantee that the canonicalisation of two queries within a core fragment
of SPARQL (monotone queries with select, project, join and union) is
equal if and only if the two queries are equivalent; we also support other
SPARQL features but with a weaker soundness guarantee: that the (partially) canonicalised query is equivalent to the input query. Despite the
fact that canonicalisation must be harder than the equivalence problem,
we show the algorithm to be practical for real-world queries taken from
SPARQL endpoint logs, and further show that it detects more equivalent queries than when compared with purely syntactic methods. We
also present the results of experiments over synthetic queries designed to
stress-test the canonicalisation method, highlighting difficult cases.

1   Introduction

SPARQL endpoints often encounter performance problems in practice: in a survey of hundreds of public SPARQL endpoints, Buil-Aranda et al. [2] found
that many such services have mixed reliability and performance, often returning errors, timeouts or partial results. This is not surprising: SPARQL is an
expressive query language that encapsulates and extends the relational algebra,
where even the simplified decision problem of verifying if a given solution is contained in the answers of a given SPARQL query for a given database is known
to be PSPACE-complete [17] (combined complexity). Furthermore, evaluating
SPARQL queries may involve an exponential number of (intermediate) results.
Hence, rather than aiming to efficiently support all queries over all database
instances for all users, the goal is rather to continuously improve performance:
to increase the throughput of the most common types of queries answered.

   An obvious means by which to increase throughput of query processing is to
re-use work done for previous queries when answering future queries by *caching*

© Springer Nature Switzerland AG 2018
D. Vrandečić et al. (Eds.): ISWC 2018, LNCS 11136, pp. 600–616, 2018.
https://doi.org/10.1007/978-3-030-00671-6_35

results. In the context of caching for SPARQL, however, there are some significant complications. While many engines may apply low-level caches to avoid, e.g., repeated index accesses, generating answers from such data can still require a lot of higher-level query processing. On the other hand, caching at the level of queries or subqueries is greatly complicated by the fact that a given abstract query can be expressed in myriad equivalent ways in SPARQL.

Addressing the latter challenge, in this paper we propose a method by which SPARQL queries can be *canonicalised*, where the canonicalised version of two queries Q_1 and Q_2 will be (syntactically) identical if Q_1 and Q_2 are *equivalent*: having the same results for any dataset. Furthermore, we say that two queries Q_1 and Q_2 are *congruent* if and only if they are equivalent modulo variable names, meaning we can rewrite the variables of Q_2 in a one-to-one manner to generate a query equivalent to Q_1; our proposed canonicalisation method then aims to give the same output for queries Q_1 and Q_2 if and only if they are congruent, which will allow us to find additional queries useful for applications such as caching.

Example 1. Consider two queries Q_A and Q_B asking for names of aunts:

```
SELECT DISTINCT ?z WHERE {
  ?x :sister ?y . ?y :name ?z .
  { ?w :mother ?x . }
    UNION { ?w :father ?x. } }
```

```
SELECT DISTINCT ?n WHERE {
  { ?a :name ?n . ?c :mother ?p . ?p :sister ?a . }
  UNION
  { ?a :name ?n . ?c :father ?p . ?p :sister ?a . } }
```

Both queries are congruent: if we rewrite the variable ?n to ?z in Q_B, then both queries are equivalent and will return the same results for any RDF dataset. Canonicalisation aims to rewrite both queries to the same syntactic form.   □

Our main use-case for canonicalisation is to improve *caching* for SPARQL endpoints: by capturing knowledge about query congruence, canonicalisation can increase the hit rate for a cache of (sub-)queries [16]. Furthermore, canonicalisation may be useful for *analysis of SPARQL logs*: finding repeated/congruent queries without pair-wise equivalence checks; *query processing*: where optimisations can be applied over canonical/normal forms; and so forth.

A fundamental challenge for canonicalising SPARQL queries is the high computational complexity that it entails. More specifically, the QUERY EQUIVALENCE PROBLEM takes two queries Q_1 and Q_2 and returns true if and only if they return the same answers for any database instance. In the case of SPARQL, this problem is NP-complete even when simply permitting joins (conjunctive queries). Even worse, the problem becomes undecidable when features such as projection and optional matches are combined [18]. Canonicalisation is then at least as hard as the equivalence problem, meaning it will likewise be intractable for even simple fragments and undecidable when considering the full SPARQL language.

We thus propose a canonicalisation procedure that does not change the semantics of an input query (i.e., is *correct*) but may miss congruent queries (i.e., is *incomplete*) for certain features. We deem such guarantees to be sufficient for use-cases where completeness is not a strong requirement, as in the case of caching where missing a congruent query will require re-executing the query (which would have to be done in any case). For *monotone queries* [19] in a core SPARQL fragment, we provide both correctness and completeness guarantees.

The procedure we propose is based on first converting SPARQL queries to a graph-based (RDF) algebraic representation. We then initially apply canonical labelling to the graph to consistently name variables, thereafter converting the graph back to a SPARQL query following a fixed syntactic ordering. The resulting query then represents the output of a baseline canonicalisation procedure for SPARQL. To support further SPARQL features such as UNION, we extend this procedure by applying normal forms and minimisation over the intermediate algebraic graph prior to its canonicalisation. Currently we focus on canonicalising SELECT queries from SPARQL 1.0. However, our canonicalisation techniques can be extended to other types of queries (ASK, CONSTRUCT, DESCRIBE) as well as the extended features of SPARQL 1.1 (including aggregation, property paths, etc.) while maintaining correctness guarantees; this is left to future work.

Extended Version: An online version of this paper provides additional definitions, proofs, and experimental results [20].

2   Preliminaries

RDF: We first introduce the RDF data model, as well as notions of isomorphism and equivalence relevant to the canonicalisation procedure discussed later.

Terms and Graphs. RDF assumes three pairwise disjoint sets of terms: *IRIs*: **I**, *literals* **L** and *blank nodes* **B**. An *RDF triple* (s, p, o) is composed of three terms – called *subject*, *predicate* and *object* – where $s \in \mathbf{IB}$, $p \in \mathbf{I}$ and $o \in \mathbf{ILB}$.[1] A finite set of RDF triples is called an *RDF graph* $G \subseteq \mathbf{IB} \times \mathbf{I} \times \mathbf{ILB}$.

Isomorphism. Blank nodes are defined as existential variables [10] where two RDF graphs differing only in blank node labels are thus considered *isomorphic* [7]. Formally, let $\mu : \mathbf{IBL} \rightarrow \mathbf{IBL}$ denote a mapping of RDF terms to RDF terms such that μ is the identity on **IL** ($\mu(x) = x$ for all $x \in \mathbf{IL}$); we call μ a *blank node mapping*; if μ maps blank nodes to blank nodes in a one-to-one manner, we call it a *blank node bijection*. Let $\mu(G)$ denote the image of an RDF graph G under μ (applying μ to each term in G). Two RDF graphs G_1 and G_2 are defined as isomorphic – denoted $G_1 \cong G_2$ – if and only if there exists a blank node bijection μ such that $\mu(G_1) = G_2$. Given two RDF graphs, the problem of determining if they are isomorphic is GI-complete [11], meaning the problem is in the same complexity class as the standard GRAPH ISOMORPHISM PROBLEM.

Equivalence. The *equivalence* relation captures the idea that two RDF graphs entail each other [10]. Two RDF graphs G_1 and G_2 are *equivalent* – denoted $G_1 \equiv G_2$ – if and only if there exists two blank node mappings μ_1 and μ_2 such that $\mu_1(G_1) \subseteq G_2$ and $\mu_2(G_2) \subseteq G_1$ [8]. A graph may be equivalent to a smaller graph (due to redundancy). We thus say that an RDF graph G is *lean* if it does not have a proper subset $G' \subset G$ such that $G \equiv G'$; otherwise

---

[1] We use, e.g., **IBL** as a shortcut for $\mathbf{I} \cup \mathbf{B} \cup \mathbf{L}$.

we can say that it is *non-lean*. Furthermore, we can define the *core* of a graph G as a lean graph G' such that $G \equiv G'$; the core of a graph is known to be unique modulo isomorphism [8]. Determining equivalence between RDF graphs is known to be NP-complete [8]. Determining if a graph G is lean is known to be coNP-complete [8]. Finally, determining if a graph G' is the core of a second graph G is known to be DP-complete [8].

Graph Canonicalisation. Our method for canonicalising SPARQL queries involves representing the query as an RDF graph, applying canonicalisation techniques over that graph, and mapping the canonical graph back to a SPARQL query. As such, our query canonicalisation method relies on an existing graph canonicalisation framework for RDF graphs called BLABEL [12]; this framework offers a sound and complete method to canonicalise graphs with respect to isomorphism ($\text{ICAN}(G)$) or equivalence ($\text{ECAN}(G)$). Both methods have exponential worst-case behaviour; as discussed, the underlying problems are intractable.

SPARQL. We now provide preliminaries for the SPARQL query language [9]. For brevity, our definitions focus on SPARQL *monotone queries* (MQs) [19] – permitting selection $(=, \wedge, \vee)^2$, join, union and projection – for which we can offer sound and complete canonicalisation.

Syntax. Let **V** denote a set of query variables disjoint with **IBL**. We define the abstract syntax of a SPARQL MQ as follows:

1. A *triple pattern* t is a member of the set **VIB** $\times$ **VI** $\times$ **VIBL** (i.e., an RDF triple allowing variables in any position). A triple pattern is a *query pattern*.
2. If both Q_1 and Q_2 are query patterns, then $[Q_1 \text{ AND } Q_2]$, and $[Q_1 \text{ UNION } Q_2]$ are also query patterns.
3. If Q is a query pattern and V is a set of variables such that for all $v \in V$, v appears in some triple pattern contained in Q, then $\text{SELECT}_V(Q)$ is a *query*.[3]

Blank nodes in SPARQL queries are considered to be non-distinguished query variables where we will assume they have been replaced with fresh query variables. Per the final definition, we currently do not support subqueries and assume, w.l.o.g., that all queries have a projection $\text{SELECT}_V(Q)$.

Algebra. We will now define an algebra for such queries. A *solution* μ is a partial mapping from variables in **V** appearing in the query to constants from **IBL** appearing in the data. Let $\text{dom}(\mu)$ denote the variables for which μ is defined. We say that two mappings μ_1 and μ_2 are *compatible*, denoted $\mu_1 \sim \mu_2$, when $\mu_1(v) = \mu_2(v)$ for every $v \in \text{dom}(\mu_1) \cap \text{dom}(\mu_2)$. Letting M, M_1 and M_2 denote sets of solutions, we define the algebra as follows:

$$M_1 \bowtie M_2 := \{\mu_1 \cup \mu_2 \mid \mu_1 \in M_1, \mu_2 \in M_2, \mu_1 \sim \mu_2\}$$
$$M_1 \cup M_2 := \{\mu \mid \mu \in M_1 \text{ or } \mu \in M_2\}$$
$$\pi_V(M) := \{\mu' \mid \exists \mu \in M : \mu' \subseteq \mu, \text{dom}(\mu') = V \cap \text{dom}(\mu)\}$$

---

[2] This is expressed by placing constants in triple patterns.

[3] Note that **SELECT** $*$ is equivalent to returning all variables (or omitting the feature).

Union is defined here in the SPARQL fashion as a union of mappings, rather than relational algebra union: the former can be applied over solution mappings with different domains, while the latter does not allow this.

Semantics. Letting Q denote an MQ pattern in the abstract syntax, we denote the evaluation of Q over an RDF graph G as $Q(G)$. Before defining $Q(G)$, first let t denote a triple pattern; then by $\mathbf{V}(t)$ we denote the set of variables appearing in t and by $\mu(t)$ we denote the image of t under a solution μ. Finally, we can define $Q(G)$ recursively as follows:

$$t(G) := \{\mu \mid \mu(t) \in G, \mathrm{dom}(\mu) = \mathbf{V}(t)\}$$
$$[Q_1 \text{ AND } Q_2](G) := Q_1(G) \bowtie Q_2(G)$$
$$[Q_1 \text{ UNION } Q_2](G) := Q_1(G) \cup Q_2(G)$$
$$\text{SELECT}_V(Q)(G) := \pi_V(Q(G))$$

Set vs. Bag. The previous definitions assume a *set semantics* for query answering, meaning that no duplicate mappings are returned as solutions [17]. However, the SPARQL standard, by default, considers a *bag* (aka. *multiset*) *semantics* for query answering [9], where the cardinality of a solution in the results captures information about how many times the query pattern matched the underlying dataset [1]. We thus use the extended syntax $\text{SELECT}_V^{\Delta}(Q)$, where $\Delta = \text{true}$ indicates set semantics and $\Delta = \text{false}$ indicates bag semantics.

Containment and Equivalence. Query containment asks: *given two queries Q_1 and Q_2, does it hold that $Q_1(G) \subseteq Q_2(G)$ for all possible RDF graphs G?* If so, we say that Q_2 *contains* Q_1, which we denote by the relation $Q_1 \sqsubseteq Q_2$. On the other hand, query equivalence asks, *given two queries Q_1 and Q_2, does it hold that $Q_1(G) = Q_2(G)$ for all possible RDF graphs G?* In other words, Q_1 and Q_2 are equivalent if and only if Q_1 and Q_2 contain each other. If so, we say that $Q_1 \equiv Q_2$. In this paper, we relax the equivalence notion to ignore labelling of variables; more formally, let $\nu : \mathbf{V} \to \mathbf{V}$ be a one-to-one mapping of variables and, slightly abusing notation, let $\nu(Q)$ denote the image of Q under ν (rewriting variables in Q wrt. ν); we say that Q_1 and Q_2 are *congruent* (denoted $Q_1 \cong Q_2$) if and only if there exists ν such that $Q_1 \equiv \nu(Q_2)$. An example of such query congruence was provided in Example 1.

The complexity of query containment and equivalence vary from NP-complete when just AND is allowed (with triple patterns), upwards to UNDECIDABLE once, e.g., projection and optional matches are added [18]. For MQs, containment and equivalence are NP-complete for the related query class of *Unions of Conjunctive Queries* (UCQs) [19], which allow the same features as MQs but disallow joins over unions. Interestingly, though MQs and UCQs are equivalent query classes – i.e., for any UCQ there is an equivalent MQ and vice-versa – containment and equivalence for MQs jumps to Π_2^P-complete [19]. Intuitively this is because MQs are more succinct than UCQs; for example, to find a path of length n where each node is of type A or B, we can create an MQ of size $O(n)$, but it requires a UCQ of size $O(2^n)$. We consider MQs since real-world SPARQL queries may arbitrarily nest joins and unions (canonicalisation will rewrite them to UCQs).

Most of the above results have been developed under set semantics. In terms of bag semantics, we can consider an analogous containment problem: that the answers of Q_1 are a *subbag* of the answers of Q_2, meaning that the multiplicity of an answer in Q_1 is always less-than-or-equals the multiplicity of the same answer in Q_2. In fact, the decidability of this problem remains an open question [4]; on the other hand, the equivalence problem is GI-complete [4], and thus in fact probably *easier* than the case for set semantics (assuming GI $\neq$ NP): under bag semantics, conjunctive queries cannot have redundancy, so intuitively speaking we can test a form of isomorphism between the two queries.

3   Related Work

Various works have presented complexity results for query containment and equivalence of SPARQL [5,13,14,18,23,24]. With respect to implementations, only one dedicated library has been released to check whether or not two SPARQL queries are equivalent: SPARQL Algebra [14]. The problem of determining equivalence of SPARQL queries can, however, be solved by reductions to related problems, where Chekol et al. [6] have used a μ-calculus solver and an XPath-equivalence checker to implement SPARQL equivalence checks. Recently Saleem et al. [22] compared these SPARQL query containment methods using a benchmark based on real-world query logs; we use these same logs in our evaluation. These works do not deal with canonicalisation; using an equivalence checker would require quadratic pairwise checks to determine all equivalences in a set or stream of queries; hence they are impractical for a use-case such as caching.

To the best of our knowledge, little work has been done specifically on canonicalisation of SPARQL queries. In analyses of logs, some authors [3,21] have proposed some syntactic canonicalisation methods – such as normalising whitespace or using a SPARQL library to format the query – that do manage to detect some duplicates, but not more complex cases such as per Example 1. Rather the most similar work to ours (to the best of our knowledge) is the SPARQL caching system proposed by Papailiou et al. [16], which uses a canonical labelling algorithm (specifically Bliss) to assign consistent labels to variables, allowing to recall isomorphic graph patterns from the cache for SPARQL queries. However, their work does not consider factoring out redundancy caused by query operators (aka. *minimisation*), and hence they would not capture equivalences as in the case of Example 1. In general, our work focuses on canonicalisation of queries whereas the work of Papailiou et al. [16] is rather focused on caching; compared to them we capture a much broader notion of query equivalence than their approach based solely on canonical labelling of query variables. It is worth noting that we are not aware of similar methods for canonicalising SQL queries.

4 Query Canonicalisation

Our approach for canonicalising SPARQL MQs involves representing the query as an RDF graph, performing a canonicalisation of the RDF graph (including the application of algebraic rewritings, minimisation and canonical labelling), ultimately mapping the resulting graph back to a final canonical SPARQL UCQ.

4.1 Representational Graph for UCQs

The MQ class is closed under join and union (see Q_A, Example 1). As the first query normalisation step, we will convert MQ queries to UCQs of the form $\mathsf{select}_V^\Delta(\mathsf{union}(\{\mathsf{and}(\{Q_1^1,\ldots Q_m^1\}),\ldots,\mathsf{and}(\{Q_1^k,\ldots Q_n^k\})\}))$ following a standard DNF-style expansion (we refer to the extended version for more details [20]). The output UCQ may be exponential in size. Thereafter, given such a UCQ, we define its *representational graph* (or R-graph for short) as follows.

Definition 1. *Let $\beta()$ denote a function that returns a fresh blank node and $\beta(x)$ a function that returns a blank node unique to x. Let $\iota(\cdot)$ denote an id function such that if $x \in \mathbf{IL}$, then $\iota(x) = x$; otherwise if $x \in \mathbf{VB}$, then $\iota(x) = \beta(x)$. Finally, let Q be a UCQ; we define $\mathrm{R}(Q)$, the R-graph of Q, as follows:*

- *If Q is a triple pattern (s,p,o), then $\iota(Q)$ is set as $\beta()$ and*

$$\mathrm{R}(Q) = \{(\iota(Q),:\mathsf{s},\iota(s)),(\iota(Q),:\mathsf{p},\iota(p)),(\iota(Q),:\mathsf{o},\iota(o)),(\iota(Q),\mathsf{a},:\mathsf{TP})\}$$

- *If Q is $\mathsf{and}(\{Q_1,\ldots,Q_n\})$, then $\iota(Q)$ is set as $\beta()$ and*

$$\mathrm{R}(Q) = \{(\iota(Q),:\mathsf{arg},\iota(Q_1)),\ldots,(\iota(Q),:\mathsf{arg},\iota(Q_n)),(\iota(Q),\mathsf{a},:\mathsf{And})\}$$
$$\cup\,\mathrm{R}(Q_1)\cup\ldots\cup\mathrm{R}(Q_n)$$

- *If Q is $\mathsf{union}(\{Q_1,\ldots,Q_n\})$, then $\iota(Q)$ is set as $\beta()$ and*

$$\mathrm{R}(Q) = \{(\iota(Q),:\mathsf{arg},\iota(Q_1)),\ldots,(\iota(Q),:\mathsf{arg},\iota(Q_n)),(\iota(Q),\mathsf{a},:\mathsf{Union})\}$$
$$\cup\,\mathrm{R}(Q_1)\cup\ldots\cup\mathrm{R}(Q_n)$$

- *If Q is $\mathsf{select}_V^\Delta(Q_1)$, then $\iota(Q)$ is set as $\beta()$ and*

$$\mathrm{R}(Q) = \{(\iota(Q),:\mathsf{arg},\iota(Q_1)),(\iota(Q),:\mathsf{distinct},\Delta),(\iota(Q),\mathsf{a},:\mathsf{Select})\}$$
$$\cup\,\{(\iota(Q),:\mathsf{var},\iota(v))\mid v\in V\}\cup\mathrm{R}(Q_1)$$

where "a" abbreviates $\mathsf{rdf}:\mathsf{type}$ and Δ is a boolean datatype literal.   □

Example 2. Here we provide an example of the R-graph for query Q_A and Q_B in Example 1: the R-graph has the same structure for both queries assuming that a UCQ normal form is applied beforehand (to Q_A in particular). For clarity, we embed the types of nodes into the nodes themselves; e.g., the uppermost node expands to 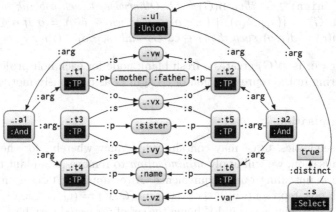.

Due to the application of UCQ normal forms, we have a projection, over a union, over a set of joins, where each join involves one or more triple patterns.   □

We also define the inverse $R^-(R(Q))$, mapping an R-graph back to a UCQ query, such that $R^-(R(Q))$ is congruent to the Q [20].

4.2 Projection with Union

Unlike the relational algebra, SPARQL MQs allow unions of query patterns whose sets of variables are not equal. This may give rise to existential variables, which in turn can lead to further equivalences that must be considered [19].

Example 3. Returning to Example 1, consider a query $Q_C \equiv Q_B$, a minor variant of Q_B using different non-projected variables in the union:

```
SELECT DISTINCT ?n WHERE { { ?a :name ?n . ?c :mother ?m . ?m :sister ?a . }
        UNION { ?a :name ?n . ?c :father ?f . ?f :sister ?a . } }
```

Such unions are permitted in SPARQL. Likewise we could rename both occurrences of ?a on the left of the union in Q_C without changing the solutions since ?a is not projected. Any correspondences between non-projected variables across a union are thus syntactic and do not affect the semantics of the query.   □

We thus distinguish the blank node representing every non-projected variable in each CQ of the R-graph produced previously. Letting G denote $R(Q)$, we define the CQ roots of G as $cq(G) = \{y \mid (y, a, : And) \in G\}$. Given a term r and a graph G, we define $G[r]$ as the sub-graph of G rooted in r, defined recursively as $G[r]_0 = \{(s, p, o) \in G \mid s = r\}$, $G[r]_i = \{(s, p, o) \in G \mid \exists x, y : (x, y, s) \in G[r]_{i-1}\} \cup G[r]_{i-1}$, with $G[r] = G[r]_n$ such that $G[r]_n = G[r]_{n+1}$ (the fixpoint).

We denote the blank nodes representing variables in G by $var(G) = \{v \in \mathbf{B} \mid \exists(s, p) : (s, p, v) \in G \land p \in \{: s, : p, : o\}\}$, and we denote the blank nodes representing unprojected variables in G by $uvar(G) = \{v \in var(G) \mid \nexists s : (s, : var, v) \in G\}$. Finally we denote the blank nodes representing projected variables in G by $pvar(G) = var(G) \setminus uvar(G)$. We can now define how variables are distinguished.

Definition 2. *Let G denote $\mathrm{R}(Q)$ for a* UCQ *Q. We define the variable distinguishing function $\mathrm{D}(G)$ as follows. If there does not exist a blank node x such that $(x, \mathsf{a}, \mathsf{:Union}) \in G$, then $\mathrm{D}(G) = G$. Otherwise if such a blank node exists, we define $\mathrm{D}(G) = \{(s, p, \delta(o)) \mid (s, p, o) \in G\}$, where $\delta(o) = o$ if $o \notin \mathsf{uvar}(G)$; otherwise $\delta(o) = \beta(r, o)$ such that $r \in \mathsf{cq}(G)$ and $(s, p, o) \in G[r]$.*   □*

In other words, $\mathrm{D}(G)$ creates a fresh blank node for each non-projected variable appearing in the representation of a CQ in G as previously motivated.

4.3   Minimisation

Under set semantics, UCQs may contain redundancy whereby, for the purposes of canonicalisation, we will apply *minimisation* to remove redundant triple patterns while maintaining query equivalence. After applying UCQ normalisation, the R-graph now represents a UCQ of the form $(Q, V) := (Q_1 \cup \ldots \cup Q_n, V)$, with each $Q_1, \ldots, Q_n$ being a CQ and V being the set of projected variables. Under set semantics, we then first remove *intra-CQ redundancy* from the individual CQs; thereafter we remove *inter-CQ redundancy* from the overall UCQ.

Bag Semantics. We briefly note that if projection with bag semantics is selected, the UCQ can only contain one (syntactic) form of redundancy: exact duplicate triple patterns in the same CQ. Any other form of redundancy mentioned previously – be it intra-CQ or inter-CQ redundancy – will affect the multiplicity of results [4]. Hence if bag semantics is selected, we do not apply any redundancy elimination other than removing duplicate triple patterns in CQs.

Set-Semantics/CQs. We now minimise the individual CQs of the R-graph by computing the core of the sub-graph induced by each CQ independently. But before computing the core, we must ground projected variables to avoid their removal during minimisation. Along these lines, let G denote an R-graph $\mathrm{D}(\mathrm{R}(Q))$ of Q. We define the grounding of projected variables as follows: $L(G) = \{(s, p, \lambda(o)) \mid (s, p, o) \in G\}$, where if o denotes a projected variable, $\lambda(o) = \mathsf{:}o$ for $\mathsf{:}o$ a fresh IRI computed for o; otherwise $\lambda(o) = o$. We assume for brevity that variable IRIs created by λ can be distinguished from other IRIs. Finally, let $\mathsf{core}(G)$ denote the core of G. We can then minimise each CQ as follows.

Definition 3. *Let G denote $\mathrm{D}(\mathrm{R}(Q))$. We define the* CQ-*minimisation of G as $\mathrm{c}(G) = \{\mathsf{core}(L(G[x])) \mid x \in \mathsf{cq}(G)\}$. We call $C \in \mathrm{c}(G)$ a CQ core.*   □

Example 4. Consider the following query, Q_D:

```
SELECT DISTINCT ?z WHERE {
  { ?w :mother ?x . } UNION { ?w :father ?x. ?x :sister ?y . }
    UNION { ?c :mother ?d . ?d :sister ?y . }
  ?d ?p ?e . ?e :name ?f . ?x :sister ?y . ?y :name ?z }
```

This query is congruent to the previous queries Q_A, Q_B, Q_C. After applying UCQ normal forms, we end up with the following R-graph for Q_D:

```
SELECT DISTINCT ?z WHERE {
  { ?w1 :mother ?x1 . ?d1 ?p1 ?e1 . ?e1 :name ?f1 .
    ?x1 :sister ?y1 .  ?y1 :name ?z .  }
  UNION { ?w2 :father ?x2 . ?x2 :sister ?y2 . ?d2 ?p2 ?e2 .
    ?e2 :name ?f2 . ?x2 :sister ?y2 . ?y2 :name ?z .  }
  UNION { ?c3 :mother ?d3 . ?d3 :sister ?y3 . ?d3 ?p3 ?e3 .
    ?e3 :name ?f3 . ?x3 :sister ?y3 . ?y3 :name ?z .  } }
```

We then replace the blank node for the projected variable ?z with a fresh IRI, and compute the core of the sub-graph for each CQ (the graph induced by the CQ node with type : And and any node reachable from that node in the directed R-graph). Figure 1 depicts the sub-R-graph representing the third CQ (omitting the : And-typed root node for clarity since it will not affect computing the core). The dashed sub-graph will be removed from the core per the map: {_:vx3/_:vd3, _:t35/_:t32, _:t33/_:t32, _:vp3/: sister, _:ve3/_:vy3, _:t34/_:t36, _:vf3/: vz, ...}, with the other nodes mapped to themselves. Observe that the projected variable : vz is now an IRI, and hence it cannot be removed from the graph.

If we consider applying this core computation over all three conjunctive queries, we would end up with an R-graph corresponding to the following query:

```
SELECT DISTINCT ?z WHERE {
  { ?w1 :mother ?x1 . ?x1 :sister ?y1 . ?y1 :name ?z }
  UNION { ?w2 :father ?x2 . ?x2 :sister ?y2 . ?y2 :name ?z . }
  UNION { ?c3 :mother ?d3 . ?d3 :sister ?y3 . ?y3 :name ?z . } }
```

We see that the projected variable is preserved in all CQs. However, we are still left with (inter-CQ) redundancy between the first and third CQs.   □

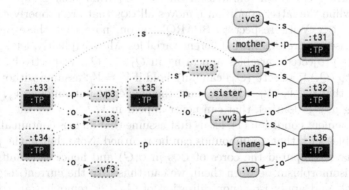

Fig. 1. R-graph of a CQ showing minimisation by leaning

Set Semantics/UCQs. After minimising individual CQs, we may still be left with a union containing redundant CQs as highlighted by Example 4. Hence we must now apply a higher-level minimisation of redundant CQs. While it may be tempting to simply compute the core of the entire R-graph – as would work for Example 4 and, indeed, as would also work for unions in the relational algebra – unfortunately SPARQL union again raises some non-trivial complications [19].

Example 5. Consider the following (unusual) query:

```
SELECT DISTINCT ?n WHERE { { ?m :cousin ?n . } UNION { ?x ?y ?n . } }
```

If we were to compute the core over the R-graph for the entire UCQ, we would remove the second CQ as follows:

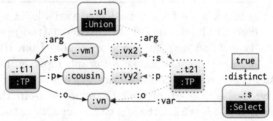

This would leave us with the following query:

```
SELECT DISTINCT ?n WHERE { ?m :cousin ?n . }
```

But this has changed the query semantics where we lose non-cousin values.   □

Instead, we must check containment between pairs of CQs [19]. Let $(Q, V) := (Q_1 \cup \ldots \cup Q_n, V)$ denote the UCQ under analysis. We need to remove from Q:

1. all Q_i $(1 \leq i \leq n)$ such that there exists Q_j $(1 \leq j < i \leq n)$ such that $\mathsf{select}_V(Q_i) \equiv \mathsf{select}_V(Q_j)$; *and*
2. all Q_i $(1 \leq i \leq n)$ where there exists Q_j $(1 \leq j \leq n)$ such that $\mathsf{select}_V(Q_i) \sqsubset \mathsf{select}_V(Q_j)$ (i.e., proper containment where $\mathsf{select}_V(Q_i) \not\equiv \mathsf{select}_V(Q_j)$);

The former condition removes all but one CQ from each group of equivalent CQs while the latter condition removes all CQs that are properly contained in another CQ. With respect to SPARQL union, note that these definitions apply to cases where CQs have different variables. More explicitly, let $V_1, \ldots, V_n$ denote the projected variables appearing in $Q_1, \ldots, Q_n$, respectively. Observe that $\mathsf{select}_{V_i}(Q_i) \sqsubseteq \mathsf{select}_{V_j}(Q_j)$ can only hold if $V_i = V_j$: assume without loss of generality that $v \in V_i \setminus V_j$, where v must then generate unbounds in V_j, creating a mapping μ, $v \in \mathrm{dom}(\mu)$, that can never appear in V_i.[4]

To implement condition (1), let us first assume that all CQs contain all projection variables such that no unbounds can be returned. Note that in the previous step we have computed the cores of CQs in $c(G)$ and hence it is sufficient to check for isomorphism between them; we can thus take the current R-graph G_i for each Q_i and apply iso-canonicalisation of G_i [12], removing any other Q_j $(j > i)$ whose G_j is isomorphic. Thereafter, to implement condition (2), we can check if there exists a blank node mapping μ such that $\mu(G_j) \subseteq G_i$, for $i \neq j$ (which is equivalent to checking *simple entailment*: $G_i \models G_j$ [8]).

---

[4] We assume that CQs without variables may generate an empty mapping ($\{\mu\}$ with $\mathrm{dom}(\mu) = \emptyset$) if the CQ is contained in the data, or no mapping ($\{\}$) otherwise. This means we will not remove such CQs (unless they are precisely equal to another CQ) as they will generate a tuple of unbounds in the results if and only if the data match.

Now we drop the assumption that all CQs contain all variables in V, meaning that we can generate unbounds. To resolve such cases, we can partition $\{Q_1, \ldots, Q_n\}$ into various sets of CQs based on the projected variables they contain, and then apply equivalence and containment checks in each part.

Definition 4. *Let* $C(G) = \{C_1, \ldots, C_n\}$ *denote the* CQ *cores of* $G = D(R(Q))$. *A* CQ *core* C_i *is in* $E(G)$ *iff* $C_i \in C(G)$ *and there does not exist a* CQ *core* $C_j \in C(G)$ *($i \neq j$) such that:* $\mathsf{pvar}(C_i) = \mathsf{pvar}(C_j)$; *and* $C_i \cong C_j$ *with* $j < i$ *or* $C_j \models C_i$. □

Definition 5. *Let* $E(G) = \{C_1, \ldots, C_n\}$ *denote the minimal* CQ *cores of* $G = D(R(Q))$. *Let* $P = \{(s, p, o) \in G \mid \exists(s, \mathsf{a}, \mathsf{:Select}) \in G\}$ *and* $U = \{(s, p, o) \in G \mid \exists(s, \mathsf{a}, \mathsf{:Union}) \in G, \text{ and } p = \mathsf{:arg} \text{ implies } \exists C \in E(G) : \{o\} = \mathsf{cq}(C)\}$. *We define the minimisation of* G *as* $M(G) = \bigcup_{G' \in E(G)} L^-(G') \cup P \cup U$, *where* $L^-(G')$ *denotes the replacement of variable IRIs with their original blank nodes.* □

The result is an R-graph representing a redundancy-free UCQ.

4.4 Canonical Labelling and Query Generation

We take the minimal R-graph $E(G)$ generated by the previous methods and apply the iso-canonicalisation method $\mathrm{ICAN}(E(G))$ to generate canonical labels for the blank nodes in $E(G)$; having normalised the UCQ algebra and removed redundancy, applying this process will finally abstract away the naming of variables in the original query from the R-graph. Then we are left to map from the R-graph back to a query, which we do by applying $R^-(\mathrm{ICAN}(E(G)))$; in $R^-(\cdot)$, we order triple patterns in CQs, CQs in UCQs and variables in the projection lexicographically. The result is the final canonicalised UCQ in SPARQL syntax. Soundness and completeness results for MQs are given in the extended version [20].

4.5 Other Features

We can represent other (non-MQ) features of SPARQL (e.g., filters, optional, etc.) as an R-graph in an analogous manner to that presented here; thereafter, we can apply canonical labelling over that graph without affecting the semantics of the underlying query. However, we must be cautious with UCQ rewriting and minimisation techniques. Currently in queries with non-UCQ features, we detect subqueries that are UCQs (i.e., use only join and union) and apply normalisation only on those UCQ subqueries considering any variable also used outside the UCQ as a virtual projected variable. Combined with canonical labelling, this provides a cautious (i.e., sound but incomplete) canonicalisation of non-MQ queries.

4.6 Implementation

We implement the described canonicalisation procedure using two main libraries: JENA for parsing and executing SPARQL queries; and BLABEL for computing the core of RDF graphs and applying canonical labelling. The containment checks

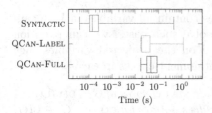

Fig. 2. Runtimes for LSQ queries

Table 1. High-level results for canonical-ising LSQ queries, including the total time taken and (max) duplicates (**D.**) found

Algorithm	Time (s)	D.	Max.D.	Queries
SYNTACTIC	211	3,960	12	768,618
QCAN-LABEL	28,066	10,722	40	768,618
QCAN-FULL	77,022	10,722	40	768,618

over CQs are implemented using SPARQL ASK queries (with Jena). In the following, we refer to our system as QCAN: Query CANonicalisation. Source code is available at https://github.com/RittoShadow/QCan, while a simple online demo can be found at http://qcan.dcc.uchile.cl/.

5   Evaluation

We now evaluate the proposed canonicalisation procedure for monotone SPARQL queries. In particular, the main research questions to be empirically assessed are as follows. RQ1: *How is the performance of canonicalisation?* RQ2: *How many additional duplicate queries can the canonicalisation process expect to find versus baseline syntactic methods in a real-world setting?* To address these questions, we present two experimental settings. In the first setting, we apply our canonicalisation method over queries from the Linked SPARQL Queries (LSQ) dataset [21], which contains queries taken from the logs of four public SPARQL endpoints. In the second setting, we create a benchmark of more difficult synthetic queries designed to stress-test the process. All experiments were run on a single machine with two Intel Xeon E5-2609 V3 CPUs and 32 GB of RAM running Debian v.7.11.

5.1   Real-World Setting

In the first setting, we perform experiments over queries from endpoint logs taken from the LSQ dataset [21], where we extract the unique strings for SELECT queries that could be parsed successfully by JENA (i.e., that were syntactically valid), resulting in 768,618 queries (see the extended version [20] for details). Over these queries, we then apply three experiments for increasingly complete and expensive canonicalisation, as follows. SYNTACTIC: We pass the query through the Jena SPARQL parser and serialiser, parsing the query into an abstract algebra and then writing the algebraic query back to a SPARQL query. QCAN-LABEL: We parse the query, applying canonical labelling to the query variables and reordering triple patterns according to the order of the canonical labels. QCAN-FULL: We apply the entire canonicalisation procedure, including parsing, labelling, UCQ rewriting, minimisation, etc. We can now address our research questions.

(RQ1:) Per Table 1, canonicalising with QCAN-LABEL is 127 times slower than the baseline SYNTACTIC method, while QCAN-FULL is 365 times slower than SYNTACTIC and 2.7 times slower than QCAN-LABEL; however, even for the slowest method QCAN-FULL, the mean canonicalisation time per query is a relatively modest 100 ms. In more detail, Fig. 2 provides boxplots for the runtimes over the queries; we see that most queries under the SYNTACTIC canonicalisation generally take around 0.1–0.3 ms, while most queries under QCAN-LABEL and QCAN-FULL take 10–100 ms. We did, however, find queries requiring longer: approximately 2.5 s in isolated worst cases for QCAN-FULL.

(RQ2:) Canonicalising with QCAN-LABEL finds 2.7 times more duplicates than the baseline SYNTACTIC method. On the other hand, canonicalising with QCAN-FULL finds no more duplicates than QCAN-LABEL: we believe that this observation can be explained by the relatively low ratio of true MQ queries in the logs [20], and the improbability of finding redundant patterns in real queries. The largest set of duplicate queries found was 12 in the case of SYNTACTIC and 40 in the case of QCAN-LABEL and QCAN-FULL.

5.2 Synthetic Setting

Many queries found in the LSQ dataset are quite simple to canonicalise. In order to see how the proposed canonicalisation methods perform for more complex queries, we propose two categories of synthetic query: the first category is designed to test the canonicalisation of CQs, particularly the canonical labelling and intra-CQ minimisation steps; the second category is designed to test the canonicalisation of UCQs, particularly the UCQ rewriting and inter-CQ minimisation steps. Both aim at testing performance rather than duplicates found.

Synthetic CQ Setting. In order to test the minimisation of CQs, we select difficult cases for the canonical labelling and core computation of graphs [12]. More specifically, we select the following three (undirected) graph schemas:

2D GRIDS: For $k \geq 2$, the k-2D-grid contains k^2 nodes, each with a coordinate $(x, y) \in \mathbb{N}^2_{1\ldots k}$, where nodes with distance one are connected; the result is a graph with $2(k^2 - k)$ edges.

3D GRIDS: For $k \geq 2$, the k-3D-grid contains k^3 nodes, each with a coordinate $(x, y, z) \in \mathbb{N}^3_{1\ldots k}$, where nodes with distance one are connected; the result is a graph with $3(k^3 - k^2)$ edges.

MIYAZAKI: This class of graphs was designed by Miyazaki [15] to enforce a worst-case exponential behaviour in NAUTY-style canonical labelling algorithms. For k, each graph has $20k$ nodes and $30k$ edges.

To create CQs from these graphs, we represent each edge in the undirected graph by a pair of triple patterns $(v_i, : \mathsf{p}, v_j)$, $(v_j, : \mathsf{p}, v_i)$, with $v_i, v_j \in \mathbf{V}$ and $:\mathsf{p}$ a fixed IRI for all edges. In order to ensure that the canonicalisation involves CQ minimisation, we enclose the graph pattern in a `SELECT DISTINCT` v query, which provides the most challenging case for canonicalisation: applying set semantics and projecting (and thus "fixing") a single query variable v. We then run the FULL canonicalisation feature, which for CQs involves computing the core

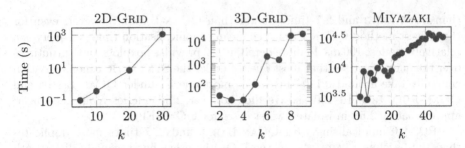

Fig. 3. Runtimes for threes types of synthetic CQs

of the R-graph and applying canonical labelling. Note that under minimisation, 2D-GRID and 3D-GRID graphs collapse down to a core with a single undirected edge, while MIYAZAKI graphs collapse down to a core with a 3-cycle.

In Fig. 3 we present the runtimes of the canonicalisation procedure, where we highlight that the y-axis is presented in log scale. We see that instances of 2D-GRID for $k \leq 10$ can be canonicalised in under a second. Beyond that, the performance of canonicalisation lengthens to seconds, minutes and even hours.

Synthetic MQ Setting. We also performed tests creating MQs in CNF (joins of unions) of the form $(t_{1,1} \cup \ldots \cup t_{1,n}) \bowtie \ldots \bowtie (t_{m,1} \cup \ldots \cup t_{m,n})$, where m is the number of joins, n is the number of unions, and $t_{i,j}$ is a triple pattern sampled (with replacement) from a k-clique of triples with a fixed predicate (such that $k = m + n$) to stress-test the performance of the canonicalisation procedure, where each such query will be rewritten to a query of size $O(n^m)$. Detailed results are available in the extended paper [20]; in summary, QCAN-FULL succeeds up to $m = 4$, $n = 8$, taking about 7.4 h, or $m = 8$, $n = 2$, taking 3 min; for values of $m = 8$, $n = 4$ and beyond, canonicalisation fails.

6   Conclusions

This paper describes a method for canonicalising SPARQL (1.0) queries considering both set and bag semantics. This canonicalisation procedure – which is sound for all queries and complete for monotone queries – obviates the need to perform pairwise containment/equivalence checks in a list/stream of queries and rather allows for using standard indexing techniques to find congruent queries. The main use-cases we foresee are query caching, optimisation and log analysis.

Our method is based on (1) representing the SPARQL query as an RDF graph, over which are applied (2) algebraic UCQ rewritings, (3 – in the case of set semantics) intra-CQ and inter-CQ normalisation, (4) canonical labelling of variables and ordering of query syntax, before finally (5) converting the graph back to a canonical SPARQL query. As such, by representing the query as a graph, our method leverages existing graph canonicalisation frameworks [12].

Though the worst-case complexity of the algorithm is doubly-exponential, experiments show that canonicalisation is feasible for a large collection of real-world SPARQL queries taken from endpoint logs. Furthermore, we show that the number of duplicates detected doubles over baseline syntactic methods. In more challenging experiments involving synthetic settings, however, we quickly start to encounter doubly-exponential behaviour, where the canonicalisation method starts to reach its practical limits. Still, our experiments for real-world queries suggests that such difficult cases do not arise often in practice.

In future work, we plan to extend our methods to consider other query features of SPARQL (1.1), such as subqueries, property paths, negation, and so forth; we also intend to investigate further into the popular OPTIONAL operator.

Acknowledgements. The work was supported by the Millennium Institute for Foundational Research on Data (IMFD) and by Fondecyt Grant No. 1181896.

References

1. Angles, R., Gutierrez, C.: The multiset semantics of SPARQL patterns. In: Groth, P., et al. (eds.) ISWC 2016. LNCS, vol. 9981, pp. 20–36. Springer, Cham (2016). https://doi.org/10.1007/978-3-319-46523-4_2
2. Buil-Aranda, C., Hogan, A., Umbrich, J., Vandenbussche, P.-Y.: SPARQL web-querying infrastructure: ready for action? In: Alani, H., et al. (eds.) ISWC 2013. LNCS, vol. 8219, pp. 277–293. Springer, Heidelberg (2013). https://doi.org/10.1007/978-3-642-41338-4_18
3. Arias Gallego, M., Fernández, J.D., Martínez-Prieto, M.A., de la Fuente, P.: An empirical study of real-world SPARQL queries. In: Usage Analysis and the Web of Data (USEWOD) (2011)
4. Chaudhuri, S., Vardi, M.Y.: Optimization of real conjunctive queries. In: Principles of Database Systems (PODS), pp. 59–70. ACM Press (1993)
5. Chekol, M.W., Euzenat, J., Genevès, P., Layaïda, N.: SPARQL query containment under SHI axioms. In: AAAI Conference on Artificial Intelligence (2012)
6. Wudage Chekol, M., Euzenat, J., Genevès, P., Layaïda, N.: Evaluating and benchmarking SPARQL query containment solvers. In: Alani, H., et al. (eds.) ISWC 2013. LNCS, vol. 8219, pp. 408–423. Springer, Heidelberg (2013). https://doi.org/10.1007/978-3-642-41338-4_26
7. Cyganiak, R., Wood, D., Lanthaler, M.: RDF 1.1 Concepts and Abstract Syntax. W3C Recommendation, February 2014. http://www.w3.org/TR/rdf11-concepts/
8. Gutierrez, C., Hurtado, C.A., Mendelzon, A.O., Pérez, J.: Foundations of semantic web databases. J. Comput. Syst. Sci. **77**(3), 520–541 (2011)
9. Harris, S., Seaborne, A., Prud'hommeaux, E.: SPARQL 1.1 Query Language. W3C Recommendation, March 2013. http://www.w3.org/TR/sparql11-query/
10. Hayes, P., Patel-Schneider, P.F.: RDF 1.1 Semantics. W3C Recommendation, February 2014. http://www.w3.org/TR/rdf11-mt/
11. Hogan, A.: Skolemising blank nodes while preserving isomorphism. In: World Wide Web Conference (WWW), pp. 430–440. ACM (2015)
12. Hogan, A.: Canonical forms for isomorphic and equivalent RDF graphs: algorithms for leaning and labelling blank nodes. ACM TWeb **11**(4), 22:1–22:62 (2017)

13. Kaminski, M., Kostylev, E.V.: Beyond well-designed SPARQL. In: International Conference on Database Theory (ICDT), pp. 5:1–5:18 (2016)
14. Letelier, A., Pérez, J., Pichler, R., Skritek, S.: Static analysis and optimization of semantic web queries. ACM Trans. Database Syst. **38**(4), 25:1–25:45 (2013)
15. Miyazaki, T.: The complexity of McKay's canonical labeling algorithm. In: Groups and Computation, II, pp. 239–256 (1997)
16. Papailiou, N., Tsoumakos, D., Karras, P., Koziris, N.: Graph-aware, workload-adaptive SPARQL query caching. In: ACM SIGMOD International Conference on Management of Data, pp. 1777–1792. ACM (2015)
17. Pérez, J., Arenas, M., Gutierrez, C.: Semantics and complexity of SPARQL. ACM Trans. Database Syst. **34**(3), 16:1–16:45 (2009)
18. Pichler, R., Skritek, S.: Containment and equivalence of well-designed SPARQL. In: Principles of Database Systems (PODS), pp. 39–50 (2014)
19. Sagiv, Y., Yannakakis, M.: Equivalences among relational expressions with the union and difference operators. J. ACM **27**(4), 633–655 (1980)
20. Salas, J., Hogan, A.: Canonicalisation of monotone SPARQL queries. Technical report. http://aidanhogan.com/qcan/extended.pdf
21. Saleem, M., Ali, M.I., Hogan, A., Mehmood, Q., Ngomo, A.-C.N.: LSQ: the linked SPARQL queries dataset. In: Arenas, M., et al. (eds.) ISWC 2015. LNCS, vol. 9367, pp. 261–269. Springer, Cham (2015). https://doi.org/10.1007/978-3-319-25010-6_15
22. Saleem, M., Stadler, C., Mehmood, Q., Lehmann, J., Ngomo, A.N.: SQCFramework: SPARQL query containment benchmark generation framework. In: Knowledge Capture Conference (K-CAP), pp. 28:1–28:8 (2017)
23. Schmidt, M., Meier, M., Lausen, G.: Foundations of SPARQL query optimization. In: International Conference on Database Theory (ICDT), pp. 4–33. ACM (2010)
24. Theoharis, Y., Christophides, V., Karvounarakis, G.: Benchmarking database representations of RDF/S stores. In: Gil, Y., Motta, E., Benjamins, V.R., Musen, M.A. (eds.) ISWC 2005. LNCS, vol. 3729, pp. 685–701. Springer, Heidelberg (2005). https://doi.org/10.1007/11574620_49

Cross-Lingual Classification of Crisis Data

Prashant Khare[1](✉), Grégoire Burel[1](✉), Diana Maynard[2](✉),
and Harith Alani[1](✉)

[1] Knowledge Media Institute, The Open University, Milton Keynes, UK
{prashant.khare,g.burel,h.alani}@open.ac.uk
[2] Department of Computer Science, University of Sheffield, Sheffield, UK
d.maynard@sheffield.ac.uk

Abstract. Many citizens nowadays flock to social media during crises to share or acquire the latest information about the event. Due to the sheer volume of data typically circulated during such events, it is necessary to be able to efficiently filter out irrelevant posts, thus focusing attention on the posts that are truly relevant to the crisis. Current methods for classifying the relevance of posts to a crisis or set of crises typically struggle to deal with posts in different languages, and it is not viable during rapidly evolving crisis situations to train new models for each language. In this paper we test statistical and semantic classification approaches on cross-lingual datasets from 30 crisis events, consisting of posts written mainly in English, Spanish, and Italian. We experiment with scenarios where the model is trained on one language and tested on another, and where the data is translated to a single language. We show that the addition of semantic features extracted from external knowledge bases improve accuracy over a purely statistical model.

Keywords: Semantics · Cross-lingual · Multilingual
Crisis informatics · Tweet classification

1 Introduction

Social media platforms have become prime sources of information during crises, particularly concerning rescue and relief requests. During Hurricane Harvey, over 7 million tweets were posted about the disaster in just over a month[1], while over 20 million tweets with the words #sandy and #hurricane were posted in just a few days during the Hurricane Sandy disaster.[2] Sharing such vital information on social media creates real opportunities for increasing citizens' situational awareness of the crisis, and for authorities and relief agencies to target their efforts more efficiently [23]. However, with such opportunities come real challenges, such as the handling of such large and rapid volumes of posts, which

---

[1] https://digital.library.unt.edu/ark:/67531/metadc993940/.
[2] Mashable: Sandy Sparks 20 Million Tweets http://mashable.com/2012/11/02/hurricane-sandy-twitter.

© Springer Nature Switzerland AG 2018
D. Vrandečić et al. (Eds.): ISWC 2018, LNCS 11136, pp. 617–633, 2018.
https://doi.org/10.1007/978-3-030-00671-6_36

renders manual processing highly inadequate [7]. The problem is exacerbated by the findings that many of these posts bear little relevance to the crisis, even those that use the dedicated hashtags [11].

Because of these challenges, there is an increasingly desperate need for tools capable of automatically assessing crisis information relevancy, to filter out irrelevant posts quickly during a crisis, and thus reducing the load to only those posts that matter. Recent research explored various classification methods of crisis data from social media platforms, which aimed at automatically categorising them into *crisis-related* or *not related* using supervised [10,13,21,25] and unsupervised [18] machine learning approaches. Most of these methods use statistical features, such as n-grams, text length, POS, and hashtags.

One of the problems with such approaches is their bias towards the data on which they are trained. This means that classification accuracy drops considerably when the data changes, for example when the crisis is of a different type, or when the posts are in a different language, in comparison to the crisis type and language the model was trained on. Training the models for all possible crisis types and languages is infeasible due to time and expense.

In our previous work, we showed that adding semantic features increases the classification accuracy when training the model on one type of crisis (e.g. floods), and applying it to another (e.g. bushfires) [11]. In this paper, we tackle the problem of language, where the model is trained on one language (e.g. English), but the incoming posts are in another (e.g. Spanish). We explore the role of adding semantics in increasing the multilingual fitness of supervised models for classifying the relevancy of crisis information.

The main contributions of this paper can be summarised as follow:

1. We build a statistical-semantic classification model with semantics extracted from BabelNet and DBpedia.
2. We experiment with classifying relevancy of tweets from 30 crisis events in 3 languages (English, Spanish, and Italian).
3. We run relevancy classifiers with datasets translated into a single language, as well as with cross-lingual datasets.
4. We show that adding semantics increases cross-lingual classification accuracy by 8.26%–9.07% in average F_1 in comparison to traditional statistical models.
5. We show that when datasets are translated into the same language, only the model that uses BabelNet semantics outperforms the statistical model, by 3.75%.

The paper is structured as follows: Sect. 2 summarises related work. Sections 3 and 4 describe our approach and experiments on classifying cross-lingual crisis data using different semantic features. Results are reported in Sects. 4.2 and 4.3. Discussion and conclusions are in Sects. 5 and 6.

2 Related Work

Classification of social media messages about crises and disasters in terms of their relevancy has been addressed already by a number of researchers [2,3,9–12,22,

25]. The classification types can differ, however. Some classify simply as relevant (related) or not; some include a partly relevant category; while others include the notion of informativeness (where informative is taken to mean providing useful information about the event). For example, Olteanu et al. [16] use the categories *related and informative, related but not informative,* and *not related.* Others treat relevance and informativeness as two separate tasks [5].

Methods for this kind of classification use a variety of supervised machine learning approaches, usually relying on linguistic and statistical features such as POS tags, user mentions, post length, and hashtags [8–10,19,21]. Approaches range from traditional classification methods such as Support Vector Machines (SVM), Naive Bayes, and Conditional Random Fields [9,17,21] to the more recent use of deep learning and word embeddings [3].

One of the drawbacks to these approaches is their lack of adaptability to new kinds of data. [9] took early steps in this area by training a model on messages about the Joplin 2011 tornado and applying it to messages about Hurricane Sandy, although the two events are still quite similar. [12] took this further by using semantic information to adapt a relevance classifier to new crisis events, using 26 different events of varying types, and showed that the addition of semantics increases the adaptability of the classifier to new, unseen, types of crisis events. In this paper, we develop that approach further by examining whether semantic information can help not just with new events, but also with events in different languages.

In general, adapting classification tools to different languages is a problem for many NLP tasks., since it is often difficult to acquire sufficient data to train separate models on each language. This is especially true for tasks such as sentiment analysis, where leveraging information from data in different languages is required. In that field, two main solutions have been explored: either translating the data into a single language (normally English) and using this single dataset for training and/or testing [1]; or training a model using weakly-labelled data without supervision [6]. Severyn et al. [20] improved performance of sentiment classification using distant pre-training of a CNN, consisting of inferring weak labels (emoticons) from a large set of multilingual tweets, followed by additional supervised training on a smaller set of manually annotated labels. In the other direction, annotation resources (such as sentiment lexicons) can be transferred from English into the target language to augment the training resources available [14]. A number of other approaches rely on having a set of correspondences between English and the target language(s), such as those which build distributed representations of words in multiple languages, e.g. using Wikipedia [24].

We test two similar approaches in this paper for the classification of information relevancy in crisis situations: (a) translate all datasets into a single language; (b) make use of high-quality feature information in English (and other languages) to supplement the training data of our target language(s).

As far as we know, while these kinds of language adaptation methods have been frequently applied to sentiment analysis, they have not been applied to crisis classification methods. Our work extends mainly on the previous work using hierarchical semantics from knowledge graphs to perform crisis-information classification through a supervised machine learning approach [11,12], by generating statistical and semantic features for all relevant languages and then using this to train the models, regardless of which language is required.

3   Experiment Setup

Our aim is to train and validate a binary classifier that can automatically differentiate between *crisis-related* and *not related* tweets in cross-lingual scenarios. We generate the statistical and semantic features of tweets from different languages and then train the machine learning models accordingly. In the next sections we detail: (i) the datasets used in our experiments; (ii) the statistical and semantic sets of features used; and (iii) the classifier selection process.

3.1   Datasets

For this study, we chose datasets from multiple sources. From the CrisisLex platform[3] we selected 3 datasets: CrisisLexT26, ChileEarthquakeT1, and SOSItalyT4. CrisisLexT26 is an annotated dataset of 26 different crisis events that occurred between 2012 and 2013. Each event has 1000 labeled tweets, with the labels *'Related and Informative'*, *'Related but not Informative'*, *'Not Related'* and *'Not Applicable'*. These events occurred around the world and hence covered a range of languages. ChileEarthquakeT1 is a dataset of 2000 tweets in Spanish (from the Chilean earthquake of 2010), where all the tweets are labeled by relatedness (relevant or not relevant). The SOSItalyT4 set is a collection of tweets spanning 4 different natural disasters which occurred in Italy between 2009 and 2014, with almost 5.6k tweets labeled by the type of information they convey ("damage", "no damage", or "not relevant"). Based on the guidelines of the labeling, both *"damage"* and *"no damage"* indicate relevance.

We chose all the labeled tweets from these 3 collections. Next, we converged some of the labels, since we aim to generate a binary classifier. From CrisisLexT26, we merged *'Related and Informative'* and *'Related but not Informative'* into the *Related* category, and merged *Not Related* abd *Not Applicable* into the *Not Related* category. For SOSItalyT4 we add the tweets labeled as "damage" and "no damage" to the "Related" category, and the "not relevant" to the "Not Related" category.

Finally, we removed all duplicate instances from the individual datasets to reduce content redundancy, by comparing the tweets in pairs after removing the special characters, URLs, and user-handles (i.e., '@' mentions). This resulted in 21,378 *Related* and 2965 *Not Related* documents in the CrisisLexT26 set, 924 *Related* and 1238 *Not Related* in the Chile Earthquake set, and 4372 *Related* and 878 *Not Related* in the SOSItalyT4 set.

---

[3] crisislex.org/.

Next, we applied 3 different language detection APIs: detectlanguage[4], langdetect[5], and TextBlob[6]. We labeled the language of the tweet where there was agreement by at least 2 of the APIs. The entire data constituted more than 30 languages, where English (en), Spanish (es), and Italian (it) comprised almost 92% of the collection (29,141 out of 31755). Considering this distribution, we focused our study on these 3 languages. To this end, we first created an unbalanced set (in terms of language) for training the classifier (see Table 1-*unbalanced*). In order to reduce the imbalance between *Related* and *Not Related* tweets, we thus only selected 8,146 tweets in total out of the 29,141 tweets. Next, we create a *balanced* version of the corpus where we split the data into a training and test set for each language, with equal distribution throughout, to remove any bias (Table 1- *balanced*).

Table 1. Data size for English(*en*), Spanish(*es*), and Italian(*it*)

Language	Unbalanced		Balanced			
			Train		Test	
	Not related	Related	Not related	Related	Not related	Related
English (*en*)	2060	2298	612	612	201	200
Italian (*it*)	813	812	612	612	201	200
Spanish (*es*)	1039	1124	612	612	201	200
Total	3912	4234	1836	1836	603	600

We also provide, in Table 2, a breakdown of all the original datasets to give an overview of the language distribution within each crisis event set.

3.2 Feature Engineering

We define two types of feature sets: *statistical* and *semantic*. Statistical features are widely used in various text classification problems [8–10,13,21,25] and so we consider these as our baseline approach. These capture the quantifiable statistical properties and the linguistic features of a textual post, whereas semantic features determine the named entities and associated hierarchical semantic information.

Statistical Features were extracted for each post in the dataset, following previous work, as follows:

– *Number of nouns:* nouns refer to entities occurring in the posts, such as people, locations, or resources involved in the crisis event [8,9,21].
– *Number of verbs:* these indicate actions occurring in a crisis event [8,9,21].

---

[4] https://detectlanguage.com.
[5] https://pypi.python.org/pypi/langdetect.
[6] http://textblob.readthedocs.io/en/dev/.

Table 2. Language distribution (in %) in crisis events data

Event	Language (%)				Event	Language (%)			
	en	it	es	Other		en	it	es	Other
Colorado Wildfire	99.30	0	0.09	0.61	CostaRica Quake	45.67	1.96	44.03	8.33
Guatemala Quake	23.84	1.20	69.56	5.40	Italy Quake	18.53	71.10	9.70	0.77
Philippines Flood	91.31	0	0.98	7.71	Typhoon Pablo	81.22	0.22	4.40	14.17
Venezuela Refinery	8.93	0.22	89.8	1.06	Alberta Flood	99.48	0	0	0.52
Australia Bushfire	98.94	0.0	0.10	0.97	Bohol E'quake	86.5	0.12	0.12	13.25
Boston Bombing	93.22	0.21	2.12	4.34	Brazil Club Fire	31.6	0	1.79	66.61
Colorado Floods	99.67	0	0.11	0.22	Glasgow Helicopter	99.86	0	0.11	0.03
LA Airport Shoot	97.07	0.11	1.30	1.52	LacMegantic Train	52.57	0.21	1.16	46.06
Manila Flood	72.40	0.22	0.22	27.16	NY Train Crash	99.86	0.14	0	0
Queensland Flood	99.56	0.09	0	0.35	Russia Meteor	87.56	0.64	2.56	9.24
Sardinia Flood	10.93	88.49	0.12	0.46	Savar Building	86.90	0.82	5.19	7.09
Singapore Haze	97.47	0.0	0	2.53	Spain Train Crash	43.13	0	54.67	2.20
Typhoon Yolanda	91.59	0.11	1.83	6.47	Texas Explosion	94.99	0	3.00	2.01
L'Aquila Quake	4.89	88.58	1.43	5.10	Emilia Quake	1.02	87.99	0.34	10.65
Genova Flood	2.09	95.12	0	2.79	Chile Quake	10.82	0.19	82.00	6.99

- *Number of pronouns:* similar to nouns, pronouns include entities such as people, locations, or resources.
- *Tweet Length:* total number of characters in a post. The length of a post could indicate the amount of information [8,9,19].
- *Number of words:* similar to Tweet Length, number of words may also be an indicator of the amount of information [9,10].
- *Number of Hashtags:* these reflect the themes of a post, and are manually generated by the authors of the posts [8–10].
- *Unigrams:* the entire data (text of each post) is tokenised and represented as unigrams [8–10,13,21,25].

The spaCy library[7] is used to extract the Part Of Speech (POS) features (e.g., *nouns, verbs, pronouns*). Unigrams for the data are extracted with the regexp tokenizer provided in NLTK.[8] We removed stop words using a dedicated list.[9] Finally, we applied the TF-IDF vector normalisatiton on the unigrams in order to weight the importance of tokens in the documents according to their relative importance within the dataset, and represent the entire data as a set of vectors. This results in a vocabulary size in unigrams (for each language in the balanced data, combining test and train data) of *en*-7495, *es*-7121, and *it*-4882.

Semantic Features are curated to generalise the information representation of the crisis situations across various languages. Semantic features are designed to be broader in context and less crisis-specific, in comparison to the actual text of the posts, thereby helping to resolve the problem of data sparsity. To this

---

[7] SpaCy Library, https://spacy.io.

[8] http://www.nltk.org/_modules/nltk/tokenize/regexp.html.

[9] https://raw.githubusercontent.com/6/stopwords-json/master/stopwords-all.json.

end, we use the Named Entity Recognition (NER) service Babelfy[10], and two different knowledge bases for creating these features: BabelNet[11] and DBpedia[12]. Note that the semantics extracted by these tools are in English, and hence they bring the multilingual datasets a bit closer linguistically. The following semantic information is extracted:

- *Babelfy Entities*: Babelfy extracts the entities from each post in different languages (e.g., *news*, *sadness*, *terremoto*), and disambiguates them with respect to the BabelNet [15] knowledge base.
- *BabelNet Senses (English)*: for each entity extracted from Babelfy, the English labels associated with the entities are extracted (e.g. *news→news*, *sadness→sadness*, *terremoto→earthquake*).
- *BabelNet Hypernyms (English)*: for each entity, the direct hypernyms (at distance-1) are extracted from BabelNet and the main sense of each hypernym is retrieved in English. From our original entities, we now get *broadcasting*, *communication*, and *emotion*).
- *DBpedia Properties*: for each annotated entity we also get a DBpedia URI from Babelfy. The following properties associated with the DBpedia URIs are queried via SPARQL: `dct:subject`, `rdfs:label` (only in English), `rdf:type` (only of the type http://schema.org and http://dbpedia.org/ontology), `dbo:city`, `dbp:state`, `dbo:state`, `dbp:country` and `dbo:country` (the location properties fluctuate between `dbp` and `dbo`) (e.g., `dbc:Grief`, `dbc:Emotions`, `dbr:Sadness`).

The inclusion of semantic features such as hypernyms has been shown to enhance the semantic and contextual representation of a document by correlating different entities, from different languages, with a similar context [12]. For example, the entities *policeman*, *policía* (Spanish for police), *fireman*, and *MP (Military Police)* all have a common hypernym (English): *defender*. By generalising the semantics in one language, English, we avoid the sparsity that often results from having various morphological forms of entities across different languages (see Table 3 for an example). Similarly, the English words *floods* and *earthquake* both have *natural disaster* as a hypernym, as does *inondazione* in Italian, ensuring that we know the Italian word is also crisis relevant. Adding the semantic information, through *BabelNet Semantics*, results in a vocabulary size in unigrams of: *en*-12604, *es*-11791, and *it*-8544.

Finally, we extract DBpedia properties of the entities (see Table 3) in the form of subject, label, and location-specific properties. This semantic expansion of the dataset forms the *DBpedia Semantics* component, and results in a vocabulary size in unigrams of: *en*-21905, *es*-15388, *it*-10674. The two types of semantic features (*BabelNet* and *DBpedia*) are used both individually and also in combination, to develop the binary classifier.

---

[10] http://babelfy.org.
[11] http://babelnet.org.
[12] http://dbpedia.org.

Table 3. Semantic expansion with BabelNet and DBpedia semantics

Feature	Post A	Post B
	'#WorldNews! 15 feared dead and 100 people could be missing in #Guatemala after quake http://t.co/uHNST8Dz'	'Van 48 muertos por terremoto en Guatemala http://t.co/nAGG3SUi via @ejeCentral'
Babelfy entities	feared, dead, people, missing, quake	muertos, terremoto
BabelNet sense (English)	fear, dead, citizenry, earthquake	slain, earthquake
BabelNet hypernyms (English)	geological_phenomenon, natural disaster, group	geological_phenomenon, natural disaster, dead
DBpedia properties	dbr:Death, dbc:Communication, dbr:News, dbc:Geological_hazards, dbc:Seismology, dbr:Earthquake	dbc:Geological_hazards, dbc:Seismology, dbr:Earthquake, dbr:Death
Google translation	To es-'#Noticias del mundo! 15 muertos temidos y 100 personas podrían estar desaparecidas en #Guatemala después terremoto http://t.co/uHNST8Dz'	To en-'48 people killed by earthquake in Guatemala http://t.co/nAGG3SUi via @ejeCentral'

3.3   Classifier Selection

In order to address the binary classification problem, the high dimensionality resulting from unigrams of tweets and semantic features, and the need to avoid overfitting, were taken into consideration. The training data instances (which varied between 1200–4500 under different experimental setups) were much smaller in size than the large dimensionality of the features (ranging between 9000–20000). We therefore opted for a Support Vector Machine (SVM) with a Linear Kernel [4] as the classification model. As discussed in [3,11], SVM performs better than other common approaches such as classification and regression trees (CART) and Naive Bayes in similar classification problems. The work by [3] also shows almost identical performance (in terms of accuracy) of SVM and CNN models in classification of the tweets. In [11], we showed the appropriateness of SVM Linear kernel over RBF kernel, Polynomial kernel, and Logistic Regression in such a classification scenario.

4   Cross-Lingual Classification of Crisis-Information

We demonstrate and validate our classification models through multiple experiments designed to test various criteria and models. We experiment on the models created with the following combinations of statistical and semantic features, thereby enabling us to assess the impact of each classification approach:

- *SF*: uses only the statistical features; this model is our baseline.
- *SF+SemBN*: combines statistical features with semantic features from *Babel-Net* (entity sense, and their hypernyms in English, as explained in Sect. 3.2).
- *SF+SemDB*: combines statistical features with semantic features from *DBpedia* (label in English, type, and other DBpedia properties).
- *SF+SemBNDB*: combines statistical features with semantic features from *BabelNet* and *DBpedia*.

We apply and validate the models above in the following three experiments:

Monolingual Classification with Monolingual Models: In this experiment, we train the model on one language and test it on data in the same language. This tests the value of adding semantics to the classifier over the baseline when the language is the same.

Cross-lingual Classification with Monolingual Models: Here we evaluate the classifiers on crisis information in languages that were not observed in the training data. For example, we evaluate the classifier on Italian when the classifier was trained on English or Spanish.

Cross-lingual Classification with Machine Translation: In the third experiment, we evaluate the classifier when the model is trained on data in a certain language (e.g. Spanish), and used to classify information that has been automatically translated from other languages (e.g. Italian and English) into the language of the training data. The translation is performed using the Google Translate API.[13] To perform this experiment, we first translate the data from each of our three languages in turn into the other two languages.

All experiments are performed on both (i) the *unbalanced* dataset, to adhere to the natural distribution of these languages; and (ii) the *balanced* dataset, to remove bias towards any particular language which is caused by the uneven distribution of these languages in our datasets. By default, we refer to results from the balanced dataset unless we specifically mention the unbalanced one. Results are reported in terms of *Precision* (P), *Recall* (R), F_1 score (F_1), and ΔF_1 (% change over baseline $\frac{(semantic\ model\ F_1 - SF\ F_1) * 100}{SF\ F_1}$, where $SF\ F_1$ is the F_1 score in SF model).

---

[13] https://cloud.google.com/translate/.

4.1 Results: Monolingual Classification with Monolingual Models

For the monolingual classification, a 5-fold cross validation approach was adopted and applied to individual datasets of *English*, *Italian*, and *Spanish*. Results in Table 4 show that adding semantics has no impact compared with the baseline (SF model) when the language of training and testing is the same.

Table 4. Monolingual Classification Models – 5-fold cross-validation (best F_1 score is highlighted for each model). *en*, *it*, and *es* refer to English, Italian, and Spanish respectively.

		SF			SF+SemBN				SF+SemDB				SF+SemBNDB			
Test	Size	P	R	F_1	P	R	F_1	ΔF_1	P	R	F_1	ΔF_1	P	R	F	ΔF_1
en	4358	0.833	0.856	0.844	0.84	0.858	0.849	0.59	0.826	0.844	0.835	-1.07	0.829	0.845	0.836	-0.95
it	1625	0.703	0.721	0.711	0.712	0.714	0.713	0.28	0.696	0.706	0.701	-1.4	0.702	0.715	0.708	-0.42
es	2163	0.801	0.808	0.804	0.812	0.809	0.810	0.75	0.799	0.795	0.797	-0.87	0.798	0.798	0.798	-0.75
Avg.				0.786			0.791	0.54			0.778	-1.1			0.781	-0.71

Unbalanced Data (from Table 1-*unbalanced*)

		SF			SF+SemBN				SF+SemDB				SF+SemBNDB			
Test	Size	P	R	F_1	P	R	F_1	ΔF_1	P	R	F_1	ΔF_1	P	R	F	ΔF_1
en	1224	0.832	0.830	0.831	0.835	0.805	0.820	-1.32	0.835	0.799	0.816	-1.80	0.829	0.808	0.818	-1.56
it	1224	0.690	0.729	0.709	0.703	0.722	0.712	0.42	0.689	0.716	0.702	-0.99	0.708	0.718	0.712	0.42
es	1224	0.798	0.765	0.781	0.794	0.783	0.789	1.02	0.779	0.754	0.766	-1.92	0.780	0.773	0.776	-0.64
Avg.				0.774			0.774	0.04			0.761	-1.57			0.769	-0.59

Balanced Data (from Table 1-*balanced*)

4.2 Results: Cross-Lingual Classification with Monolingual Models

This experiment involves training on data in one language and testing on another. Results, shown in Table 5, indicate that when using the statistical features alone (SF - the baseline), average F_1 is 0.557. When semantics are included in the classifier, average classification performance improvement (ΔF_1) is by 8.26%–9.07%, with a standard deviation (SDV) between 10.9%–13.86% across all three semantic models, for all the test cases. Similarly, when applied to *unbalanced* datasets, performance increases by 7.44%–9.78%.

While the highest gains are observed in SF+SemBNDB, the SF+SemBN seems to exhibit a consistent performance by improving over the SF baseline in 5 out of 6 cross-lingual classification tests, while SF+SemDB and SF+SemBNDB each show improvement in 4 out of 6 tests.

4.3 Results: Cross-Lingual Crisis Classification with Machine Translation

The results from cross-lingual classification after language translation are presented in Table 6. For each training dataset, we translate the test data into the

Table 5. Cross-Lingual Classification Models (best F_1 score is highlighted for each model).

	Size	SF			SF+SemBN				SF+SemDB				SF+SemBNDB			
Unbalanced Data (from Table 1- *unbalanced*)																
Train Test		P	R	F_1	P	R	F_1	ΔF_1	P	R	F_1	ΔF_1	P	R	F	ΔF_1
en	4358															
it	1625	0.576	0.522	0.417	0.598	0.562	0.518	24.2	0.595	0.576	0.553	32.6	0.609	0.588	0.568	36.2
es	2163	0.674	0.633	0.604	0.663	0.654	0.645	6.79	0.653	0.649	0.643	6.46	0.649	0.641	0.633	4.8
it	1625															
en	4358	0.469	0.474	0.449	0.547	0.545	0.538	19.82	0.508	0.508	0.504	12.25	0.516	0.516	0.516	14.9
es	2163	0.635	0.610	0.586	0.643	0.627	0.612	4.43	0.601	0.60	0.596	1.70	0.625	0.620	0.614	4.78
es	2163															
en	4358	0.633	0.62	0.604	0.60	0.572	0.532	-11.9	0.623	0.618	0.610	0.99	0.606	0.592	0.571	-5.46
it	1625	0.536	0.533	0.521	0.529	0.529	0.528	1.34	0.526	0.526	0.526	0.96	0.539	0.539	0.539	9.78
Avg.				0.530			0.562	7.44			0.572	9.16			0.573	9.78
SDV				0.082			0.053	13.08			0.053	12.3			0.044	14.47
Balanced Data (from Table 1-*balanced*)																
Train Test	Size	P	R	F_1	P	R	F_1	ΔF_1	P	R	F_1	ΔF_1	P	R	F	ΔF_1
en	1224															
it	401	0.539	0.515	0.429	0.588	0.571	0.549	28	0.569	0.568	0.568	32.4	0.578	0.576	0.572	33.3
es	401	0.689	0.688	0.688	0.669	0.668	0.668	-2.9	0.647	0.644	0.641	-6.8	0.666	0.661	0.659	-4.2
it	1224															
en	401	0.521	0.521	0.521	0.581	0.581	0.580	11.3	0.558	0.552	0.539	3.5	0.550	0.546	0.538	3.3
es	401	0.655	0.646	0.640	0.672	0.655	0.647	1.1	0.638	0.636	0.635	-0.78	0.637	0.633	0.631	-1.4
es	1224															
en	401	0.609	0.593	0.578	0.657	0.620	0.597	3.3	0.667	0.666	0.665	15	0.660	0.653	0.650	12.4
it	401	0.529	0.522	0.489	0.534	0.534	0.532	8.8	0.551	0.546	0.533	9	0.555	0.551	0.543	11
Avg.				0.557			0.596	8.26			0.597	8.71			0.599	9.07
SDV				0.096			0.053	10.94			0.057	13.86			0.054	13.6

language of the training data. For example, when the training data is in English (*en*), the Italian data is translated to English, and is represented in the table as *it2en*. We aim to analyse two aspects here: (i) how semantics impacts the classifier on the translated content; and (ii) how the classifiers perform over the translated data in comparison to cross-lingual classifiers, as seen in Sect. 4.2.

From the results in Table 6, we see that based on average % change ΔF_1 of all translated test cases (*en2it,es2it*, etc.), SF+SemBN outperforms the statistical classifier (SF) by 3.75% (balanced data) with a standard deviation (SDV) of 4.57%. However, the other two semantic feature models (SF+SemDB and SF+SemBNDB) do not improve over the statistical features when the test and training data are both in the same language (after translation). The SF+SemBN shows improvement in 4 out of 6 translated test cases, except when trained on Spanish (*es*).

Comparing the best performing model from translated data, i.e. SF+SemBN, and overall baseline (SF model from cross-lingual classification Table 5-*balanced*), the SF+SemBN (translation) has an average F_1 gain (ΔF) across each translated test case over the baseline of 15.23% (with a SDV 12.6%). For example, compare

Table 6. Cross-Lingual Crisis Classification with Machine Translation (best F_1 score is highlighted for each event).

Unbalanced Data (from Table 1- *unbalanced*)

Train Test	Size	SF P	SF R	SF F_1	SF+SemBN P	SF+SemBN R	SF+SemBN F_1	ΔF_1	SF+SemDB P	SF+SemDB R	SF+SemDB F_1	ΔF_1	SF+SemBNDB P	SF+SemBNDB R	SF+SemBNDB F	ΔF_1
en	4358															
it2en	1625	0.644	0.613	0.591	0.635	0.611	0.593	0.34	0.582	0.568	0.548	-7.27	0.597	0.580	0.561	-5.0
es2en	2163	0.681	0.681	0.681	0.667	0.667	0.667	-2.0	0.669	0.661	0.659	-3.2	0.664	0.661	0.660	-3.1
it	1625															
en2it	4358	0.609	0.601	0.588	0.636	0.618	0.597	1.53	0.570	0.570	0.569	-3.2	0.575	0.574	0.571	-2.9
es2it	2163	0.647	0.629	0.612	0.675	0.636	0.607	-0.81	0.609	0.595	0.578	-5.5	0.620	0.603	0.583	-4.7
es	2163															
en2es	4358	0.643	0.626	0.609	0.661	0.634	0.610	0.16	0.654	0.654	0.653	7.2	0.649	0.648	0.646	6.07
it2es	1625	0.585	0.584	0.583	0.590	0.590	0.589	1.03	0.581	0.580	0.580	-0.51	0.586	0.585	0.584	0.17
Avg.				0.611			0.611	0.03			0.598	-2.1			0.60	-1.6
SDV				0.036			0.029	1.3			0.046	5.1			0.04	4.2

Balanced Data (from Table 1-*balanced*)

Train Test	Size	SF P	SF R	SF F_1	SF+SemBN P	SF+SemBN R	SF+SemBN F_1	ΔF_1	SF+SemDB P	SF+SemDB R	SF+SemDB F_1	ΔF_1	SF+SemBNDB P	SF+SemBNDB R	SF+SemBNDB F	ΔF_1
en	1224															
it2en	401	0.624	0.583	0.546	0.622	0.598	0.577	5.7	0.561	0.558	0.554	1.46	0.594	0.588	0.581	6.4
es2en	401	0.675	0.671	0.669	0.704	0.696	0.693	3.6	0.701	0.671	0.658	-1.6	0.695	0.674	0.664	-0.74
it	1224															
en2it	401	0.583	0.578	0.572	0.639	0.631	0.625	9.3	0.547	0.546	0.545	-4.7	0.551	0.551	0.551	-3.6
es2it	401	0.638	0.621	0.609	0.703	0.668	0.653	7.2	0.619	0.603	0.590	-3.1	0.610	0.596	0.582	-4.4
es	1224															
en2es	401	0.686	0.678	0.675	0.691	0.670	0.661	-2.0	0.691	0.691	0.691	2.3	0.683	0.683	0.683	1.2
it2es	401	0.594	0.594	0.593	0.586	0.586	0.586	-1.2	0.580	0.576	0.570	-3.9	0.579	0.576	0.571	-3.7
Avg.				0.610			0.633	3.75			0.601	-1.59			0.605	-0.83
SDV				0.052			0.045	4.57			0.059	2.9			0.054	4.14

ΔF between *it-en2it* in SF+SemBN in the translated model and *it-en* in SF in the cross-lingual model, similarly for the other 5 test cases. Based on an average of ΔF across all the test cases, the SF+SemBN (from translation) and SF+SemBN (from cross-lingual models), both perform well over the baseline (SF from cross-lingual model), by 8.26% and 15.23% respectively.

4.4   Cross-Lingual Ranked Feature Correlation Analysis

To understand the impact of the semantics and the translation on the discriminatory nature of the cross-lingual data from different languages, we analysed the correlation between ranked features of each dataset under different models. For this, we considered the *balanced* datasets across each language and took the entire data by merging the training and test data for each language. Next, we calculated Information Gain over each dataset (English (*en*), Spanish, (*es*), and Italian (*it*)), across all 4 models (SF, SF+SemBN, SF+SemDB, SF+SemBNDB). We also calculated Information Gain over the translated datasets (*en2it, en2es, es2it, es2en, it2en,* and *it2es*). This provides the ranked list of features, in terms of their discriminatory powers in the classifiers, in each selected dataset.

Table 7. Spearman's Rank Order Correlation between ranked informative features (based on IG) across models and languages

Data/Model	SF	SF+SemBN	SF+SemDB	SF+SemBNDB	Translation	
en − es	0.573	0.385	0.349	0.373	0.515(en-es2en)	0.449(es-en2es)
en − it	−0.179	0.402	0.111	0.315	0.266(en-it2en)	0.594(it-en2it)
es − it	0.418	0.222	0.503	0.430	0.678(es-it2es)	0.612(it-es2it)

For each pair of datasets, such as English (*en*) - Spanish (*es*), we consider the common ranked features with $IGscore > 0$, and calculate the Spearman's Rank Order Correlation (ranges between $[-1, 1]$) across the two ranked lists. For the translated data, we analysed pairs where one dataset is translated to the language of another dataset, such as *en-it2en* and *it-en2it*.

Table 7 shows how the correlation varies across the data. These variations can be attributed to a number of aspects. The overlap of crisis events while sampling the data is a crucial parameter, as the data was sampled based on language, and the discreteness of the source events (Table 2) was not taken into consideration. This can particularly be observed in the *en-es* correlation, where the highest correlation is without the semantics. This also explains the better performance of the SF model over the *semantic* models when trained on *en* and evaluated on *es* (Table 5-*balanced*). The correlation between *en-it* is ˜−0.179, which indicates nearly 'no correlation'. The increase in discriminative-feature correlations between datasets once semantics are added is in part due to the extraction of semantics in English (see Sect. 3.2), thus bringing the terminologies closer semantically as well as linguistically.

Translating the data to the same language shows an increment in the correlation. This is expected for multiple reasons. Firstly, having the data in the same language enables the identification of more similar features such as verbs and adjectives across the datasets. Secondly, given the similarity in the different types of events covered under the three languages, such as *floods* and *earthquakes*, the nature of the information is likely to have a high contextual overlap.

5   Discussion and Future Work

Our aim is to create hybrid models, by mixing semantic features with the statistical features, to produce a crisis data classification model that is largely language-agnostic. The work was limited to English, Spanish, and Italian, due to the lack of sufficient data annotations in other languages. We are currently designing a CrowdFlower annotation task to expand our annotations to several other languages.

We ran our experiments on both *balanced* and *unbalanced* datasets. However, performance over the balanced dataset provides a fairer comparison, since biases towards the dominant languages are removed. We also experimented with classifying data in their original languages, as well as automatically translating the data into the language of the training data. Results show that with balanced datasets, translation improves the performance of all classifiers, and reduces the benefits of using semantics in comparison to the statistical classifier (SF; the baseline). One could conclude that if the data is to be translated into the same language that the model was trained on, then the statistical model (SF) might be sufficient, whereas if translation is not viable (e.g., data arriving in unpredicted languages, or where translations are too inaccurate or untrustworthy) then the model that mixes statistical and semantics features is recommended, since it produces higher classification accuracies.

In this work, the classifiers were trained and tested on data from various types of crisis events. It is natural for some nouns to be identical across various languages, such as names of crises (e.g. Typhoon Yolanda), places, and people. In future work we will measure the level of terminological overlap between the datasets of different languages.

We augmented all datasets with semantics in English (Sect. 3.2). This is mainly because BabelNet (version 3.7) is heavily biased towards English[14]. Most existing entity extractors are skewed heavily towards the English language, and hence as a byproduct of adding their identified semantics, more terms (concepts) in a single language (English) will be added to the datasets. As a consequence, this will bring the datasets of different languages closer together linguistically, thus giving an advantage to semantic models over purely statistical ones in the context of cross-lingual analysis. We performed a comparison of vocabulary similarity between the language datasets, before and after the addition of semantics, to also comprehend the overlapping of the vocabulary. For instance, the cosine similarity between (without semantics) *en-it* is 0.311, *en-es* is 0.536, and *it-es* is 0.32. Adding semantics increased the cosine similarity across all the datasets. In current experiments, we had 6 test cases in each classification model; despite the consistency observed across 6 cross-lingual test cases, we would need more observations to establish that the gain achieved by the semantic models over the baseline models is statistically significant. Repeating these experiments over more languages should help in this; alternatively, creating multiple train and test splits for each test case could also complement such analysis, which was not feasible in this study due to insufficient data to create multiple splits for each dataset. However, we did perform a 10 iteration of 5-fold cross validation over the entire dataset across all the feature sets and found that SF+SemBN(*BabelNet Semantics*) model outperformed all others (particularly baseline with a statistically significant value of p = 0.0192, on a two-tailed t-test).

In this work, we experimented with training the model on one language at a time. Another possibility is to train the model on multiple languages, thus increasing its ability to classify data in those languages. However, generating such a multilingual model is not always feasible, since it requires annotated data in all the languages it is intended to analyse. Furthermore, the need for models that can handle other languages is likely to remain, since the language of data shared on social media during crises tends to differ substantially, depending on

---

[14] Almost 17 million word senses in English, next highest is French, with 7 million senses http://live.babelnet.org/stats.

where these crises are taking place. Therefore, the ability of a model to classify data in a new language will always be a clear advantage. The curated data (with semantics) and code, in this work, is being made available for research purposes.[15]

6 Conclusion

Determining which tweets are relevant to a given crisis situation is important in order to achieve a more efficient use of social media, and to improve situational awareness. In this paper, we demonstrated the ability of various models to classify crisis related information from social media posts in multiple languages. We tested two approaches: (1) adding semantics (from *BabelNet* and *DBpedia*) to the datasets; and (2) automatically translating the datasets to the language that the model was trained on. Through multiple experiments, we showed that all our semantic models outperform statistical ones in the first approach, whereas only one semantic model (using BabelNet) shows an improvement over the statistical model in the second approach.

Acknowledgment. This work has received support from the European Union's Horizon 2020 research and innovation programme under grant agreement No. 687847 (COMRADES).

References

1. Araujo, M., Reis, J., Pereira, A., Benevenuto, F.: An evaluation of machine translation for multilingual sentence-level sentiment analysis. In: Proceedings of the 31st Annual ACM Symposium on Applied Computing, pp. 1140–1145. ACM (2016)
2. Burel, G., Saif, H., Alani, H.: Semantic wide and deep learning for detecting crisis-information categories on social media. In: d'Amato, C., et al. (eds.) ISWC 2017. LNCS, vol. 10587, pp. 138–155. Springer, Cham (2017). https://doi.org/10.1007/978-3-319-68288-4_9
3. Burel, G., Saif, H., Fernandez, M., Alani, H.: On semantics and deep learning for event detection in crisis situations. In: Workshop on Semantic Deep Learning (SemDeep) at ESWC (2017)
4. Cristianini, N., Shawe-Taylor, J.: An Introduction to Support Vector Machines and Other Kernel-Based Learning Methods. Cambridge University Press, New York (2000)
5. Derczynski, L., Meesters, K., Bontcheva, K., Maynard, D.: Helping crisis responders find the informative needle in the tweet haystack. arXiv preprint arXiv:1801.09633 (2018)
6. Deriu, J., et al.: Leveraging large amounts of weakly supervised data for multilanguage sentiment classification. In: Proceedings of the 26th International Conference on World Wide Web, International World Wide Web Conferences Steering Committee, pp. 1045–1052 (2017)

---

[15] https://github.com/pkhare/iswc_codebase.

7. Gao, H., Barbier, G., Goolsby, R.: Harnessing the crowdsourcing power of social media for disaster relief. IEEE Intell. Syst. **26**(3), 10–14 (2011)
8. Imran, M., Elbassuoni, S., Castillo, C., Diaz, F., Meier, P.: Extracting information nuggets from disaster-related messages in social media. In: ISCRAM (2013)
9. Imran, M., Elbassuoni, S., Castillo, C., Diaz, F., Meier, P.: Practical extraction of disaster-relevant information from social media. In: Proceedings of the 22nd International Conference on World Wide Web, pp. 1021–1024. ACM (2013)
10. Karimi, S., Yin, J., Paris, C.: Classifying microblogs for disasters. In: Proceedings of the 18th Australasian Document Computing Symposium, pp. 26–33. ACM (2013)
11. Khare, P., Burel, G., Alani, H.: Classifying crises-information relevancy with semantics. In: Gangemi, A., et al. (eds.) ESWC 2018. LNCS, vol. 10843, pp. 367–383. Springer, Cham (2018). https://doi.org/10.1007/978-3-319-93417-4_24
12. Khare, P., Fernandez, M., Alani, H.: Statistical semantic classification of crisis information. In: Workshop on HSSUES at ISWC (2017)
13. Li, R., Lei, K.H., Khadiwala, R., Chang, K.C.C.: TEDAS: a twitter-based event detection and analysis system. In: 2012 IEEE 28th International Conference on Data Engineering (ICDE), pp. 1273–1276. IEEE (2012)
14. Mihalcea, R., Banea, C., Wiebe, J.: Learning multilingual subjective language via cross-lingual projections. In: Proceedings of the 45th Annual Meeting of the Association of Computational Linguistics, pp. 976–983 (2007)
15. Navigli, R., Ponzetto, S.P.: BabelNet: the automatic construction, evaluation and application of a wide-coverage multilingual semantic network. Artif. Intell. **193**, 217–250 (2012)
16. Olteanu, A., Vieweg, S., Castillo, C.: What to expect when the unexpected happens: social media communications across crises. In: Proceedings of the 18th ACM Conference on Computer Supported Cooperative Work and Social Computing, pp. 994–1009. ACM (2015)
17. Power, R., Robinson, B., Colton, J., Cameron, M.: Emergency situation awareness: twitter case studies. In: Hanachi, C., Bénaben, F., Charoy, F. (eds.) ISCRAM-med 2014. LNBIP, vol. 196, pp. 218–231. Springer, Cham (2014). https://doi.org/10.1007/978-3-319-11818-5_19
18. Rogstadius, J., Vukovic, M., Teixeira, C., Kostakos, V., Karapanos, E., Laredo, J.A.: CrisisTracker: crowdsourced social media curation for disaster awareness. IBM J. Res. Dev. **57**(5), 4-1–4-13 (2013)
19. Sakaki, T., Okazaki, M., Matsuo, Y.: Earthquake shakes twitter users: real-time event detection by social sensors. In: Proceedings of the 19th International Conference on World Wide Web, pp. 851–860. ACM (2010)
20. Severyn, A., Moschitti, A.: UNITN: training deep convolutional neural network for twitter sentiment classification. In: Proceedings of the 9th International Workshop on Semantic Evaluation (SemEval 2015), pp. 464–469 (2015)
21. Stowe, K., Paul, M.J., Palmer, M., Palen, L., Anderson, K.: Identifying and categorizing disaster-related tweets. In: Proceedings of The Fourth International Workshop on Natural Language Processing for Social Media, pp. 1–6 (2016)
22. Tonon, A., Cudré-Mauroux, P., Blarer, A., Lenders, V., Motik, B.: ArmaTweet: detecting events by semantic tweet analysis. In: Blomqvist, E., Maynard, D., Gangemi, A., Hoekstra, R., Hitzler, P., Hartig, O. (eds.) ESWC 2017. LNCS, vol. 10250, pp. 138–153. Springer, Cham (2017). https://doi.org/10.1007/978-3-319-58451-5_10

23. Vieweg, S., Hughes, A.L., Starbird, K., Palen, L.: Microblogging during two natural hazards events: what twitter may contribute to situational awareness. In: Proceedings of the SIGCHI Conference on Human Factors in Computing Systems, pp. 1079–1088. ACM (2010)
24. Wick, M., Kanani, P., Pocock, A.C.: Minimally-constrained multilingual embeddings via artificial code-switching. In: AAAI, pp. 2849–2855 (2016)
25. Zhang, S., Vucetic, S.: Semi-supervised discovery of informative tweets during the emerging disasters. arXiv preprint arXiv:1610.03750 (2016)

Measuring Semantic Coherence
of a Conversation

Svitlana Vakulenko[1], Maarten de Rijke[2], Michael Cochez[3,4,5],
Vadim Savenkov[1], and Axel Polleres[1,6,7(✉)]

[1] Vienna University of Economics and Business, Vienna, Austria
{svitlana.vakulenko,vadim.savenkov,axel.polleres}@wu.ac.at
[2] University of Amsterdam, Amsterdam, The Netherlands
derijke@uva.nl
[3] Fraunhofer FIT, 53754 Sankt Augustin, Germany
[4] Informatik 5, RWTH University Aachen, Aachen, Germany
[5] Faculty of Information Technology, University of Jyvaskyla, Jyvaskyla, Finland
michael.cochez@fit.fraunhofer.de
[6] Complexity Science Hub Vienna, Vienna, Austria
[7] Stanford University, Stanford, CA, USA

Abstract. Conversational systems have become increasingly popular
as a way for humans to interact with computers. To be able to provide intelligent responses, conversational systems must correctly model
the structure and semantics of a conversation. We introduce the task
of measuring semantic (in)coherence in a conversation with respect to
background knowledge, which relies on the identification of semantic
relations between concepts introduced during a conversation. We propose and evaluate graph-based and machine learning-based approaches
for measuring semantic coherence using knowledge graphs, their vector
space embeddings and word embedding models, as sources of background
knowledge. We demonstrate how these approaches are able to uncover
different coherence patterns in conversations on the Ubuntu Dialogue
Corpus.

1 Introduction

Conversational interfaces are seeing a rapid growth in interest. Conversational
systems need to be able to model the structure and semantics of a human conversation in order to provide intelligent responses. The requirement conversations be
coherent is meant to improve the probability distribution over possible dialogue
states and candidate responses.

A conversation is an information exchange between two or more participants.[1]
An essential property of a conversation is its *coherence*; De Beaugrande and
Dressler [9] describe it as a "continuity of senses." Coherence constitutes the out-
come of a cognitive process, and is, therefore, an inherently subjective measure.

---

[1] We use the terms "dialog" and "conversation" interchangeably, while "dialog" refers
specifically to a two-party conversation.

© Springer Nature Switzerland AG 2018
D. Vrandečić et al. (Eds.): ISWC 2018, LNCS 11136, pp. 634–651, 2018.
https://doi.org/10.1007/978-3-030-00671-6_37

It is always relative to the background knowledge of participants in the conversation and depends on their interpretation of utterances. Coherence reflects the ability of an observer to perceive meaningful relations between the concepts and to be critical of the new relations being introduced. Meaning emerges through the interaction of the knowledge presented in the conversation with the observer's stored knowledge of the world [24]. In other words, a conversation has to be assigned an interpretation, which depends on the knowledge available to the agent.

In this paper we focus on analyzing semantic relations that hold within dialogues, i.e., relations that hold between the concepts (entities) mentioned in the course of the same dialogue. We call this type of relation *semantic coherence.* We focus on semantic relations but ignore other linguistic signals that make a text coherent from a grammatical point of view. A classic example to illustrate the difference is due to Chomsky [4]: "Colorless green ideas sleep furiously" – a syntactically well formed English sentence that is semantically incoherent.

Our hypothesis is that, apart from word embeddings, recognizing concepts in the text of a conversation and determining their semantic closeness in a background knowledge graph can be used as a measure for coherence. To this end, we propose and evaluate several approaches to measure semantic coherence in dialogues using different sources of background knowledge: both text corpora and knowledge graphs. The contributions that we make in this paper are threefold: (1) we introduce a dialogue graph representation, which captures relations within the dialogue corpus by linking them through the semantic relations available from the background knowledge; (2) we formulate the semantic coherence measuring task as a binary classification task, discriminating between real dialogues and generated adversary samples,[2] and (3) we investigate the performance of state-of-the-art and novel algorithms on this task: top-k shortest path induced subgraphs and convolutional neural networks trained using vector embeddings.

The main challenge in applying structural knowledge to natural language understanding becomes apparent when we do not just try to differentiate between genuine conversations and completely random ones, but create adversarial examples as conversations that have similar characteristics compared to the positive examples from the dataset. Then, the results achieved using word embeddings are usually best and suggest that knowledge graph (KG) embeddings would potentially be an efficient way to harness the structure of entity relations. However, KG embedding-based models rely on entity linking being correct and cannot easily recover from errors made at the entity linking stage compared to other graph-based approaches that we use in our experiments.

2 Related Work

Several lines of research are relevant to our work: discourse analysis, dialogue systems and knowledge graphs.

---

[2] As there is no standard corpus available for this task, we test against 5 ways to generate artificial negative samples.

2.1 Discourse Analysis

Previous work on discourse analysis demonstrates good results in recognizing discourse structure based on lexical cohesion for specific tasks such as topic segmentation in multi-party conversations [12]. Term frequency distribution on its own already provides a strong signal for topic drift. A more sophisticated approach to assess text coherence is based on the entity grid representation [2], which represents a text as a matrix that captures occurrences of entities (columns) across sentences (rows) and indicates the role entity plays in the sentence (subject, object, or other). This approach relies on a syntactic (dependency) parser to annotate the entity roles and is, therefore, also targeted at measuring lexical cohesion rather than semantic relations between concepts. The de facto standard testbed for discourse coherence models is the information (sentence) ordering task [16]; it was recently extended to a convolutional neural network-based model for coherence scoring [22]. The best results to date were demonstrated by incorporating a fraction of semantic information from an external knowledge source (entity types classification) into the original entity grid model [10]. Cui et al. [7] push the state-of-the-art on the sentence ordering task by incorporating word embeddings at the input layer of a convolutional neural network instead of the entity grid.

In summary, background knowledge has been found to be able to provide a strong signal for measuring coherence in discourse.

2.2 Dialogue Systems

In contrast to previous research focused on measuring coherence in a monologue, we consider the task of evaluating coherence in a written dialogue setting by analyzing the largest multi-turn dialogue corpus available to date, the Ubuntu Dialogue Corpus [17].

Research in dialogue systems focuses on developing models able to generate or select from candidate utterances, based on previous interactions. Lowe et al. [18] evaluated several baseline models on the Ubuntu Dialogue Corpus for the next utterance classification task. Their error analysis suggests that the models can benefit from an external knowledge of the Ubuntu domain, which could provide the missing semantic links between the concepts mentioned in the course of the conversation. This work motivated us to consider evaluating whether relations accumulated in large knowledge graphs could provide missing semantics to make sense of a conversation.

2.3 Knowledge Graphs

Knowledge graphs (KGs) were successfully applied for disambiguating natural language text in a variety of tasks, such as information retrieval [3,13] and textual entailment [26]. They serve an important role by providing additional relations that help to bridge the lexical gap and gain a more complete understanding of the context in comparison with shallow approaches based on lexical features alone.

There was also a recent surge in development of question answering interfaces to KGs [1,19,28].

Our work is orthogonal to these lines of work, as it seeks to discover the potential and limitations of KGs to support natural language *understanding* beyond single search queries or factoid question answering towards a holistic interactive experience, which recognizes and supports the natural (coherent) flow of a conversation.

3   Measuring Semantic Coherence

In this section, we describe several approaches to modeling a conversation and measuring its coherence. We use dialogues, i.e., a two-party conversation to illustrate our approaches. Our approaches could also be applied to multi-party conversations.

We propose to measure dialogue coherence with a numeric score that indicates more coherent parts of a conversation and provides a signal for topic drift. Our approach is based on the assumption that naturally occurring human dialogues, on average, exhibit more coherence than their random permutations.

3.1   Dialogue Graph

We model a dialogue as a graph D, which contains 4 types of nodes P, U, W, C and edges E between them. P refers to the set of conversation participants, U – the set of utterances, W – the set of words and C – the set of concepts.

The words w in a conversation are grouped into utterances $\forall w \in W, \exists (u, w) \in E$ such that $u \in U$,[3] which belong to one of the conversation participants $\forall u \in U \exists (p, u) \in E$ such that $p \in P$. Every utterance can belong to only one of the participants, while the same words can be re-used in different utterances by the same or different participants. Words may refer to concepts from the background knowledge $(w, c) \in E$, where $w \in W, c \in C$. Several words may refer to a single concept, while the same concept may be represented by different sets of words. The sequence in which words appear in a conversation is given by the consecutive set of edges $T = \{(w_1, w_2), (w_2, w_3), \ldots\}$ such that $T \subset E$, indicating the dialogue flow.

The first three types of nodes P, U, and W together with their relations are available from the dialogue transcript itself, while the set of concepts C and relations between them constitute the semantic representation (meaning) of a dialogue. The meaning is not directly observable, but is constructed by an observer (one of the dialogue participants or a third party) based on the available background knowledge. The background knowledge supplies additional links, which we refer to as *semantic relations*. They link words to concepts they may refer to: (w, c) (see footnote 3) and different concepts to each other (c_i, c_j). These

---

[3] For simplicity, we ignore the role of word order; it can be re-constructed from the order within the conversation T if needed, see below.

external relations provide the missing links between words, which explain and justify their co-occurrence. The absence of such links gives an important signal to the observer, and may indicate a topic switch or discourse incoherence. However, some of the valid links may also be missing from the background knowledge.

An example dialogue graph is illustrated in Fig. 1. The dialogue consists of four utterances represented by nodes $u1$–$u4$. In the graph we also illustrate a subgraph extracted from the background knowledge, which links the concepts $c1$ dbr:Gedit and $c4$ dbr:Ubuntu(OS) to the concept $c*$ dbr:GNOME, which was not mentioned in the conversation explicitly. This link represents semantic relation between the dialogue turns: $(u1, u2)$ and $(u3, u4)$, indicating semantic coherence in the dialogue flow. In this example, the semantic relation extracted from the background knowledge corresponds to the shortest path of length 2, i.e., the distance between the concepts mentioned in the dialogue was two relations introducing one external concept from the background knowledge. $c*$ can consist of more than one entity, but encompass a whole subgraph summarizing various relations, which hold between entities, and are represented via alternative paths between them in a knowledge graph. In the next section, we describe our approach to empirically learn semantic relations that are characteristic for human dialogues, using different sources of background knowledge and different knowledge representation models.

3.2   Semantic Relations

We collect semantic relations between concepts referenced in a dialogue from our background knowledge. We consider two common sources of background knowledge: (1) unstructured data: word co-occurrence statistics from text corpora; (2) semi-structured data: entities and relations from knowledge graphs. In order to be able to use a KG as a source of background knowledge we need to perform an entity linking step, which maps words to semantic concepts (w, c), where concepts refer to entities stored in KG. We consider two approaches to retrieve relations between the entities mentioned in a dialog, namely vector space embeddings and subgraph induction via the top-k shortest paths algorithm.

Embeddings. Embeddings are generated using the distributional hypothesis by representing an item via its context, i.e., its position and relations it holds with respect to other items. Embeddings are multi-dimensional vectors (of a fixed size), which encode the distributional information of an item (a word in the a or a node in a graph), i.e., its position and relations to other items in the same space. This is achieved by computing vector representations towards an optimality criteria defined with a certain output function, which depends on the embedding vectors being trained. Thus, embeddings efficiently encode (compress) an original sparse representation of the relations (e.g., an adjacency matrix) for each of the items. It provides an easy and fast way to access this information (relationship structure). Following this approach, every concept c_i in our dialogue graph (Fig. 1) is assigned to an n-dimensional vector, which

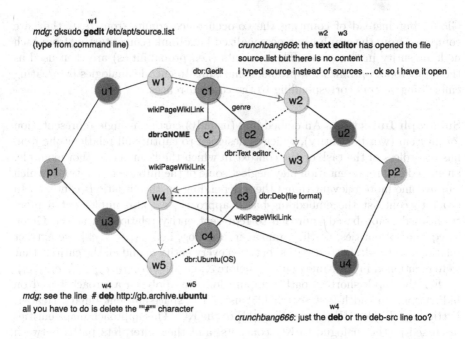

Fig. 1. Dialogue graph example along with the annotated dialog. We focus specifically on the layer of concepts in the middle $[c_1, \ldots, c_4]$ attempting to bridge the semantic gap in the lexicon of a conversation using available knowledge models: word embeddings and a knowledge graph.

encodes its location in the semantic space, and loses all the edges, which explicitly specified its relations to other concepts in the space.

We consider two types of embeddings to represent concepts mentioned in a dialog, one for each of our background knowledge sources: word embeddings trained on a text corpus, and entity embeddings trained on a KG. For word embeddings, we use word2vec [20], in particular the skip-gram variant, which aims to create embeddings such that they are useful for predicting words which are in the neighborhood of a given word. GloVe [23] is a word embedding method, with the explicit goal of embedding analogies between entities. This method does not work directly on the text corpus, but rather on co-occurrence counts which are derived from the original corpus.

For graph embeddings, we use two methods that can be scaled to large graphs, such as DBpedia and Wikidata: biased RDF2Vec [5] (using random walks) and Global RDF Vector Space Embeddings [6]; we refer to the latter ones as *KGlove* embeddings. RDF2Vec is based on word2vec. It works by first generating random walks on the graph, where the edges have received weights which influence the probability of following these edges. During the walk, a sentence is generated consisting of the identifiers occurring on the nodes and edges traversed. For each entity in the graph, many walks are performed and hence a large text is generated. This text is then used for training word2vec. KGlove is based on

GloVe, but instead of counting the co-occurrence counts from text, they are computed from the graph using personalized PageRank scores starting from each node or entity in the graph. These counts (i.e., probabilities) are then used as the input to an optimization problem that aims to encode analogies by creating embedding vectors corresponding to the co-occurrences.

Subgraph Induction. An embedding (usually) carries a single representation for an item (word or entity), which is designed to capture all relations the item has regardless of the task or the context in which the item occurs. For example, an embedding representation may neglect some of the infrequent relations, which can become more relevant than others depending on the situation (context). In order to contrast the embedding-based approach, we also implement a more traditional graph-based approach to represent entity relations in a KG. Given a sequence of entities, as they appear in a dialog, i.e., $[c_1, c_2 \ldots c_n]$, we extract relations, as top-k shortest paths, between every entity c_i and all the entities that were mentioned in the same dialogue before c_i, i.e., $(c_1, c_i), (c_2, c_i), \ldots, (c_{i-1}, c_i)$.

For the top-k shortest path computation, we apply an approach based on bidirectional breadth-first search [25] using the space-efficient binary Header, Dictionary, Triples (HDT) encoding [11] of the KG. This approach maps entities discussed in the dialogue to KG concepts, and then interprets paths between concepts in the KG as semantic relations between the respective entities. Many such relations are never mentioned in the conversation and only become explicit through the path enumeration over the KG. By increasing the number of desired shortest paths k and the maximum path length ℓ, one can discover more relations, including those that might be omitted or obscured in the entity embedding representation in the case of a random walk or frequency-based embedding algorithms. An obvious downside of this increase in recall is reduced efficiency.

3.3   Dialogue Classification

We measure semantic coherence by casting the task into a classification problem. The score produced by the classifier corresponds to our measure of semantic coherence.

Since human dialogues are expected to exhibit a certain degree of incoherence due to topic drift and since relations are missing from our background knowledge, we cannot assume every concept in our dialogue dataset to be coherent with respect to the other concepts in the same dialog. However, it is reasonable to assume that on average a reasonably large set of concepts extracted from a human dialogue exhibits a higher degree of coherence than a randomly generated one. We build upon this assumption and cast the task of measuring semantic coherence as a binary classification task, in which real dialogues have to be distinguished from corrupted (incoherent) dialogues. We consider positive and negative examples for whole conversations, represented as a sequence of words or entities, which constitute the input for the binary classifier. Effectively, these examples provide a supervision signal for measuring and aggregating distances between words/concepts by learning the weights for the neural network classifier.

Negative Sampling. To produce negative (adversarial) examples for the binary classification task we propose five sampling strategies:

- RUf: Random uniform. For every positive example we choose a sequence of entities (or words for training on word embeddings) of the same size from the vocabulary uniformly at random; so, we double the size of the dataset effectively by supplementing it with completely randomly generated (i.e., presumably incoherent) counterexamples.
- SqD: Sequence disorder. Randomly permute the original sequence, which is similar in spirit to the sentence ordering task for evaluating discourse coherence [16]. The key difference is that we rearrange the order of words (entities), which may also occur within the same sentence (utterance), rather than permuting whole sentences.
- VoD: Vocabulary distribution. For every positive example choose a sequence of entities of the same length from the vocabulary using the same frequency distribution as in the original corpus; so, VoD is very similar to RuF, but tries to emulate "structure" to some extent by choosing similar term frequencies.
- VSp: Vertical split. Create a negative example by permuting two positive examples replacing utterances of one of the conversation participants with utterances of a participant from a different conversation.
- HSp: Horizontal split. Create a negative example by permuting two positive examples merging the first half of one conversation with the second half of a different conversation.

Convolutional Neural Network. To solve the binary classification task we train a classifier using a convolutional neural network architecture, which is applied to sequences of words and entities to distinguish irregular semantic drift, which was deliberately injected into conversations, from smooth drift which occur within real conversations.

It is a standard architecture previously employed for a variety of natural language tasks, such as text classification [14]. The network consists of (1) an input layer, which appends the pre-trained embeddings for each of the word (entity) from the dialogue sequence; (2) a convolutional layer, which consist of filters (arrays of trainable weights) sliding over and learning predictive local patterns in the previous layer of the input embeddings; (3) a max pooling layer, which aggregates the features learned by the neighboring filters; (4) the hidden layer, a fully connected layer, which allows combining features from all the dimensions with a non-linear function; and (5) the output layer is a fully connected layer, which aggregates the scores to make the final prediction. See also Sect. 4.2 for details.

4 Evaluation Setup

The source code of our implementation and evaluation procedures is publicly accessible.[4] We also release our dataset used in the evaluation, which contains

---

[4] https://github.com/vendi12/semantic_coherence.

dialogue annotations with DBpedia entities and shortest paths, for reproducibility and further references.[5]

4.1   Dataset

Dialogues. Our experiments were performed on a sample of dialogues from the Ubuntu Dialogue Corpus[6] [17], which contains 1,852,869 dialogues in total, with one dialogue per file in TSV format, and is the largest conversational dataset to date. There are multiple challenges related to using this corpus, however. The dialogues were automatically extracted from a public chat using several heuristics selecting two user handles and segmenting based on the timestamps. The dialogues cannot be considered as perfectly coherent since some of the related utterances are missing from the dialogues; there can be several different topics discussed within the same conversation and the asynchronous nature of on-line communication often results in semantic mismatch in the dialogue sequence. While we cannot guarantee local coherence of the real dialogues, we expect them to be on average more coherent, when comparing to the dialogues randomly generated by sampling entities (words) from the vocabulary or merging entities (words) from different dialogues, which we refer to as negative samples, or adversaries, in our binary classification task.

We proceed by annotating a sample of 291,848 dialogues from the Ubuntu Dialogue Corpus with the DBpedia entities using the DBpedia Spotlight public web service[7] [8]. The input to the entity linking API is the text for each utterance in a conversation. Next, we considered only the dialogues where both participants contribute at least 3 new entities each, i.e., every dialogue in our dataset contains minimum 6 entities shared between the dialogue partners. The threshold for entities per conversation was chosen to ensure there is enough semantic information for measuring coherence. This way, we end up with a sample of 45,510 dialogues, which we regard as true positive examples of coherent dialogue. It contains 17,802 distinct entities and 21,832 distinct words that refer to these. The maximum size of a dialogue in this dataset is 115 entities or 128 words referring to them. We shuffled the dialogues and selected 5,000 dialogues for our test set. While this procedure means we cannot test our approach on short conversations, with fewer entities, we consider 45K dialogues to be a representative dataset for evaluating our approach.

The negative samples for both training and test set were generated using five different sampling strategies described in Sect. 3.3. Each development set consists of 81,020 samples (50% positive and 50% negative). We further split it into a training and validation set: 64,816 and 16,204 (20%) samples, respectively. Our test set comprises the remaining 5,000 positive examples, and 5,000 generated negative samples.

---

[5] https://github.com/vendi12/semantic_coherence/tree/master/data.

[6] https://github.com/rkadlec/ubuntu-ranking-dataset-creator.

[7] http://model.dbpedia-spotlight.org/en/annotate.

Knowledge Models. We compared the performance on our task across two types of embeddings models trained on two different knowledge source types: GloVe [23] and Word2Vec [20] for the word embeddings, and biased RDF2vec [5] and KGloVe [6] for the knowledge graph entity embeddings.

We utilise two publicly available word embedding models: GloVe embeddings pre-trained on the Common Crawl corpus (2.2M words, 300 dimensions)[8] and Word2Vec model trained on the Google News corpus (3M words, 300 dimensions).[9] 1,578 words from our dialogues (7%) were not found in the GloVe embeddings dataset and received a zero vector in our embedding layer. Thus, GloVe embeddings cover 20,254 words from our vocabulary (93%). Word2Vec embeddings cover only 73% of our vocabulary.

For RDF2Vec and KGloVe we used publicly available pre-trained global embeddings of the DBpedia entities (see [5,6], respectively). For KGlove we used all different embeddings, while for RDF2Vec we experimented with the embeddings that gave the best performance in [5]. KGlove embeddings cover 17,258 entities from our vocabulary (97%), while Rdf2Vec provides 62–77% due to different importance sampling strategies of the embedding approaches.

The shortest paths used were extracted from dumps of DBpedia (April 2016, 1.04 billion triples) and Wikidata (March 2017, 2.26 billion triples).[10]

4.2   Implementation

Our neural network model is implemented using the Keras library with a TensorFlow backend. The one-dimensional (temporal) convolutional layer contains 250 filters of size 3 and stride (step) 1. The max pooling layer is global, the hidden layer is set to 250 dimensions. There are two activation layers with rectified linear unit (ReLU) after the convolutional and the hidden layers to capture also non-linear dependencies between input and output, and two dropout layers with rate 0.2 after the embeddings and hidden layers to avoid overfitting. The last ReLU activation is projected onto a single-unit output layer with a sigmoid activation function to obtain a coherence score on the interval between 0 and 1.

The network is trained using the Adam optimizer with the default parameters [15] to minimize the binary cross-entropy loss between the predicted and correct value. All models were trained for 10 epochs in batches of 128 samples and early stopping after 5 epochs if there is no improvement in accuracy on the validation set.

To compute the shortest paths we merged the dumps of DBpedia and Wikidata into a single 36 GB binary file in HDT format [11] (DBpedia+Wikidata HDT), with an additional 21 GB index on the subject and the object components of triples. We set the parameters of the algorithm in our experimental evaluation as follows: k for the number of shortest paths to be retrieved from the graph to 5, the maximum length ℓ of a path to 9 edges (relations) and a

---

[8] https://nlp.stanford.edu/projects/glove/.

[9] https://code.google.com/archive/p/word2vec/.

[10] http://www.rdfhdt.org/datasets/.

```
PREFIX ppf: <java:at.ac.wu.arqext.path.>
PREFIX dbr: <http://dbpedia.org/resource/>
SELECT * WHERE {
?X ppf:topk ("--source" dbr:Directory_service
                       dbr:Gnome dbr:GNOME
                       dbr:Desktop_environment
            "--target" dbr:Desktop_computer
```

Fig. 2. k-shortest path query (cf. [25] to extract relevant connections between entities from the knowledge graph

Table 1. The top 5 most common entities and relations in the Ubuntu Dialogue dataset: mentioned entities – from linking dialogue utterances to DBpedia entities via Dbpedia Spotlight Web service; context entities and relations – from the shortest paths between the mentioned entities in DBpedia.

Top #	Mentioned entities		Context entities		Relations	
	Label	Count	Label	Count	Label	Count
1	Ubuntu(philosophy)	1605	Ubuntu(OS)	1058	WikiPageWikiLink	51014
2	Sudo	708	Linux	725	Gold/hypernym	319
3	Booting	676	Microsoft_Windows	208	Ontology/genre	178
4	APT(Debian)	405	FreeBSD	175	OperatingSystem	140
5	Live_CD	314	Smartphone	171	rdf-schema#seeAlso	116

timeout terminating the query after 2 s to cope with the scalability issues of the algorithm. Our top-k shortest paths algorithm implementation is available via a SPARQL endpoint[11] using the syntax shown in Fig. 2.

The function `at.ac.wu.arqext.path.topk` is a user defined extension available as a Jena ARQ extension.[12]

5   Evaluation Results

Table 1 reports the most common entities and relations, which while not being mentioned in the course of a dialogue, were on the shortest paths (in the KG) between other entities that were explicitly mentioned in the dialogue, i.e., which constitute an implicit dialogue context. While Dbpedia Spotlight dereferenced "Ubuntu" mentions to the concept related to philosophy rather than to the popular software distribution, the graph-based approach succeeds in recovering the correct meaning of the word by extracting this concept from the shortest paths that lie between the other entities mentioned in dialogues. Almost all relations obtained from the KG correspond to the links between the corresponding Wikipedia web pages (wikiPageWikiLink).

---

[11] http://wikidata.communidata.at.
[12] https://bitbucket.org/vadim_savenkov/topk-pfn.

5.1 Semantic Distance

The length of the shortest path (number of edges, i.e., relations on the path) is a standard measure used to estimate semantic (dis)similarity between entities in a knowledge graph [21]. We observe how it correlates with a standard measure to estimate similarity between vectors in a vector space, *cosine distance*, defined as: $1 - \cos(x, y) = 1 - \frac{xy^\mathsf{T}}{\|x\|\|y\|}$. Figure 3 showcases different perspectives on semantic similarity (coherence) between the entities in real and generated dialogues as observed in different semantic spaces (w.r.t. the knowledge models), alignments and differences between them. The barplots reflect the distributions of the semantic distances between entities in dialogues. The semantic distances are measured using cosine distances between vectors in the vector space for word (Word2Vec and GloVe) and KG (RDF2Vec) embeddings, and in terms of the shortest path lengths in the DBpedia+Wikidata KG. We observe that the real dialogues (True positive) tend to have smaller distances between entities: 1–2 hops or at most 0.3 cosine distance, while randomly generated sequences are skewed further off. Embeddings produce much more fine-grained (continuous) representation of semantic distances in comparison with the shortest path length metric. Distributions produced by different word embeddings are very similar in shape, while the one from KG embeddings is steeper and skewed more to the center, there are only a few entities further than 0.7, while this is the top for the random distances in word embeddings.

We also discover the bottleneck of our shortest path algorithm at length 5. Since the set of relevant entities for which the paths are computed grows proportionally to the dialogue length, depending on the degree of the node the number of expanded nodes quickly reaches the limit on the memory size. In our case, the algorithm retrieved the paths of length at most 5 due to the 2-s timeout, while the parameter for the maximum length of the path ℓ was set to 9.

5.2 Classification Results

Our evaluation results from training a neural network on the task of measuring (in)coherence in dialogues are listed in Table 2. It summarizes the outcomes of models trained on different embeddings using different types of adversarial samples (negative sampling strategies are described in Sect. 3.3). For the KG embeddings, we report only the approaches that performed best across different test splits.[13]

From the results we observe that the easiest task was to distinguish real dialogues from randomly generated sequences. When the model was trained with randomly generate dialogues, accuracies often reach close to 100%. However, this same model performs poorly when used for any other type of non-genuine messages we created. In the best case (KGloVe Uni), still only 10% of messages randomly sampled from the vocabulary distribution were correctly detected.

---

[13] The full result table is available on-line: https://github.com/vendi12/semantic_coherence/blob/master/results/results.xls.

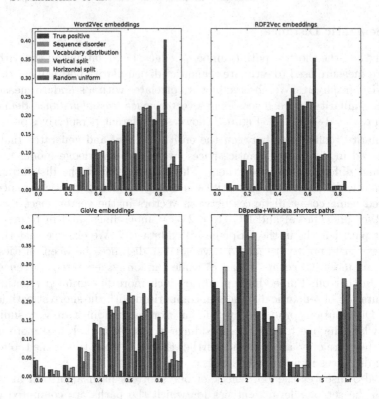

Fig. 3. Distribution of cosine distances for different data splits using Word2Vec and GloVe word embeddings (left), and RDF2Vec KG embeddings (top right), compared with the distribution of shortest path lengths in DBpedia+Wikidata KG (bottom right). Words in real dialogues (True positive) are more related than frequent domain words (Vocabulary distribution), and much more than a random sample (Random uniform).

This indicates that there is a need to experiment with the other types as well. We also observe that the models that are trained with specific adversarial examples are best in separating that type. However, even when the model is not explicitly trained to recognize a specific type of dialogue, but instead trained on other types of adversarial examples, it is sometimes still able to classify messages correctly. This happens, for example, in the case of KGloVe Uniform where the adversarial messages are sampled from the Vocabulary distribution and the model is still able to detect around 70% of randomly generated messages.

The dialogues generated by permuting the sequence of entities (words) in the original dialogues (the sequence ordering task) were harder to distinguish (The best performing model resulted in an accuracy of 0.79). Finally, the hardest task was to discriminate the adversarial examples generated by merging two different dialogues together (vertical and horizontal splits). This was expected as these dialogues have short sequences of genuine dialogue inside, making them hard to classify.

Table 2. Accuracy on the test set across different embedding and sampling approaches. The table shows for 7 different embedding strategies (4 types), how the embedding performs when trained with data from different generated adversarial examples. For example, the underlined value in the table (0.92), means that GloVe word embeddings, when trained with genuine and Vertical split (VSp) adversarial examples, is able to correctly find 92% of the Vocabulary distribution (VoD) adversarial examples in the test set. In the same row, in the TPos column, it can be seen that 60% of the genuine messages were correctly identified. Hence, this results in an average accuracy of 0.76. In blue highlight, we indicate the results where the adversarial examples for training the model where of the same type as for testing the model. In bold, we indicate the best result for each adversarial example type. Abbreviations: TPos – True Positive, TNeg – True Negative, RUf – Random uniform, VoD – Vocabulary distribution, SqD – Sequence disorder, VSp – Vertical split, HSp – Horizontal split, Avg – Average, PRS – PageRank Split, PR – PageRank, Uni – Uniform, PrO – Predicate Object.

Embeddings	Data split	TPos	RUf	Avg	VoD	Avg	SqD	Avg	VSp	Avg	HSp	Avg	Avg
						Accuracy							
						TNeg							
Word2Vec	RUf	0.99	0.99	**0.99**	0.02	0.50	0.02	0.50	0.01	0.50	0.01	0.50	0.60
	VoD	0.89	0.62	0.75	0.90	0.89	0.53	0.71	0.18	0.54	0.20	0.54	**0.69**
	SqD	0.75	0.65	0.70	0.88	0.81	0.81	0.78	0.27	0.51	0.29	0.52	0.66
	VSp	0.59	0.50	0.55	0.82	0.71	0.41	0.50	0.59	0.59	0.61	0.60	0.59
	HSp	0.62	0.39	0.50	0.71	0.66	0.38	0.50	0.55	0.58	0.63	0.63	0.58
GloVe	RUf	0.99	0.99	**0.99**	0.00	0.50	0.01	0.50	0.00	0.50	0.00	0.50	0.60
	VoD	0.93	0.38	0.66	0.93	**0.93**	0.39	0.66	0.19	0.56	0.08	0.51	0.66
	SqD	0.76	0.71	0.73	0.91	0.84	0.82	**0.79**	0.16	0.46	0.15	0.45	0.66
	VSp	0.60	0.25	0.42	<u>0.92</u>	0.76	0.43	0.51	0.65	0.62	0.66	0.63	0.59
	HSp	0.71	0.34	0.52	0.81	0.76	0.30	0.50	0.55	**0.63**	0.66	**0.68**	0.62
rdf2vec PRS	RUf	0.98	0.99	**0.99**	0.02	0.50	0.02	0.50	0.02	0.50	0.01	0.50	0.60
	VoD	0.79	0.68	0.73	0.83	0.81	0.34	0.57	0.36	0.57	0.35	0.57	0.65
	SqD	0.59	0.48	0.54	0.72	0.66	0.67	0.63	0.43	0.51	0.40	0.50	0.56
rdf2vec PR	HSp	0.57	0.59	0.58	0.72	0.64	0.43	0.50	0.59	0.58	0.67	0.62	0.58
KGloVe Uni	RUf	0.92	0.97	**0.94**	0.11	0.51	0.09	0.50	0.08	0.50	0.07	0.50	0.59
	VoD	0.54	0.88	0.71	0.73	0.64	0.61	0.58	0.51	0.52	0.52	0.53	0.60
	SqD	0.55	0.62	0.58	0.64	0.59	0.63	0.59	0.47	0.51	0.45	0.50	0.56
KGloVe PrO	HSp	0.31	0.81	0.56	0.75	0.53	0.69	0.50	0.77	0.54	0.70	0.51	0.53
KGloVe PR	HSp	0.47	0.69	0.58	0.61	0.54	0.54	0.50	0.57	0.52	0.65	0.56	0.54

The best performance across all test settings was achieved using the word embeddings models, especially GloVe performed well. KG embeddings, while performing reasonably well on the easier tasks (RUf and VoD), fell short to distinguish more subtle changes in semantic coherence. For the KG embedding weighting approaches, we noticed that the ones which performed well in earlier work, also worked better in this task. In particular, it was noticed that the weighting biased by PageRank computed on the Wikipedia links graph results in better results in machine learning tasks.

As discussed in Sect. 4, RDF2vec has fewer entity embeddings than KGloVe, when trained from the same original graph (DBpedia). KGloVe will provide an embedding, even when not much is known about a specific entity. In case of a node that does not have any edges, KGloVe will assign a random vector to it. In contrast, RDF2Vec will prune infrequent nodes. Another problem that affects KG embeddings are incorrectly recognized entities. There is no linking required for needed word embeddings since it represents different meanings of the word in a single vector.

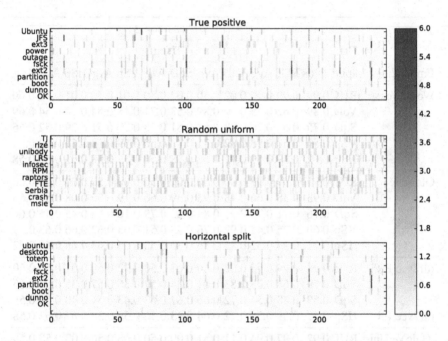

Fig. 4. Heatmap of the activations on the output of the word embeddings layer. Notice the vertical-bar pattern indicating a stronger semantic relation between the words in a real dialogue (top) in comparison with a random word sequence (middle). The topic drift effect can be observed when two different dialogues are concatenated (horizontal split – bottom): the bars at the top are shifted in comparison with the bars in the second half of the conversation, comparing to the coherence patterns observed in the real dialogue (top).

Overall, we want to be able not only to tell to which degree a dialogue is (in)coherent but also to identify the regions in the dialogue where coherence was disrupted, or to partition the dialogue into coherent segments indicating the shifts between different topics. Visualization of the activations in the output of the convolutional layer of the Glove word embeddings-based model exhibits distinct vertical activation patterns, which can be interpreted as traces of local coherence the model is able to recognize (See Fig. 4).

6   Conclusion

We considered the task of measuring semantic coherence of a conversation, which introduces an important and challenging problem that requires operating vast amounts of heterogeneous knowledge sources to infer implicit relations between the utterances, i.e., bridging the semantic gap in understanding natural language. We proposed and evaluated several approaches to this problem using alternative sources of background knowledge, such as structured (knowledge graph) and unstructured (text corpora) knowledge representations. These approaches detect semantic drift in conversations by measuring coherence with respect to the background knowledge. Our models were trained for dialogs but the approach does not restrict the number of conversation participants. The model's performance depends to a large extent on the choice of background knowledge source, with respect to the conversation domain. The conversation needs to contain a sufficient number of recognized entities to signal its position within the semantic space.

Our results indicate promising directions as well as challenges in applying structural knowledge to analyse natural language. We show that the use of word embeddings in text classification is superior to some existing knowledge graph embeddings. This is an important insight, advancing research by uncovering limitations of state-of-the-art knowledge graph embeddings and indicating directions for improvements.

Knowledge graph embeddings constitute a potentially powerful method to efficiently harness entity relations for tasks that require estimates of semantic similarity. However, their use relies on the correctness of the entity linking performance. Errors made at this stage in the pipe-line approach do propagate into the classification results, but we noticed that they are rather consistent, which partially mitigates the problem. Our experiments showed that graph-based approaches are more robust to errors in entity linking than knowledge graph embeddings, which is an important insight for future work: this effect can likewise be expected with other existing entity linking approaches.

More research is needed on how to make a knowledge graph embeddings-based model more robust to uncertainty in entity linking, such as end-to-end learning on graphs [29]. Also, combining evidence from both structured (knowledge graphs) and unstructured (text) data sources has a great potential to mitigate knowledge sparsity, increase support and interpretability of semantic relations [27]. We provide a test bed for the semantic coherence task, which can be used to compare word- and entity-based representation approaches, and their combinations, whereupon others can build.

Acknowledgments. This work is supported by the project 855407 "Open Data for Local Communities" (CommuniData) of the Austrian Federal Ministry of Transport, Innovation and Technology (BMVIT) under the program "ICT of the Future." Svitlana Vakulenko was supported by the EU H2020 programme under the MSCA-RISE agreement 645751 (RISE_BPM). Axel Polleres was supported under the Distinguished Visiting Austrian Chair Professors program hosted by The Europe Center of Stanford University. Maarten de Rijke was supported by Ahold Delhaize, Amsterdam Data Science, the Bloomberg Research Grant program, the China Scholarship Council, the Criteo Faculty Research Award program, Elsevier, the European Community's Seventh Framework Programme (FP7/2007-2013) under grant agreement nr 312827 (VOX-Pol), the Google Faculty Research Awards program, the Microsoft Research Ph.D. program, the Netherlands Institute for Sound and Vision, the Netherlands Organisation for Scientific Research (NWO) under project nrs CI-14-25, 652.002.001, 612.001.551, 652.001.003, and Yandex. All content represents the opinion of the authors, which is not necessarily shared or endorsed by their respective employers and/or sponsors.

References

1. Athreya, R.G., Ngonga, A., Usbeck, R.: Enhancing community interactions with data-driven chatbots - the DBpedia chatbot. In: WWW 2018 Companion. ACM (2018)
2. Barzilay, R., Lapata, M.: Modeling local coherence: an entity-based approach. Comput. Linguist. **34**(1), 1–34 (2008)
3. Blanco, R., Ottaviano, G., Meij, E.: Fast and space-efficient entity linking for queries. In: WDSM 2015, pp. 179–188. ACM (2015)
4. Chomsky, N.: Syntactic Structures. Mouton and Co., The Hague (1957)
5. Cochez, M., Ristoski, P., Ponzetto, S.P., Paulheim, H.: Biased graph walks for RDF graph embeddings. In: WIMS 2017, pp. 21:1–21:12 (2017)
6. Cochez, M., Ristoski, P., Ponzetto, S.P., Paulheim, H.: Global RDF vector space embeddings. In: d'Amato, C., et al. (eds.) ISWC 2017. LNCS, vol. 10587, pp. 190–207. Springer, Cham (2017). https://doi.org/10.1007/978-3-319-68288-4_12
7. Cui, B., Li, Y., Zhang, Y., Zhang, Z.: Text coherence analysis based on deep neural network. In: CIKM 2017, pp. 2027–2030. ACM (2017)
8. Daiber, J., Jakob, M., Hokamp, C., Mendes, P.N.: Improving efficiency and accuracy in multilingual entity extraction. In: I-SEMANTICS 2013, pp. 121–124 (2013)
9. De Beaugrande, R., Dressler, W.: Textlinguistics. Longman, Harlow (1981)
10. Elsner, M., Charniak, E.: Extending the entity grid with entity-specific features. In: ACL 2011, pp. 125–129. ACL (2011)
11. Fernández, J.D., Martínez-Prieto, M.A., Gutiérrez, C., Polleres, A., Arias, M.: Binary RDF representation for publication and exchange (HDT). JWS **19**, 22–41 (2013)
12. Galley, M., McKeown, K., Fosler-Lussier, E., Jing, H.: Discourse segmentation of multi-party conversation. In: ACL 2003, pp. 562–569 (2003)
13. Hasibi, F., Balog, K., Garigliotti, D., Zhang, S.: Nordlys: a toolkit for entity-oriented and semantic search. In: SIGIR 2017, pp. 1289–1292 (2017)
14. Kim, Y.: Convolutional neural networks for sentence classification. CoRR abs/1408.5882 (2014)
15. Kingma, D.P., Ba, J.: Adam: a method for stochastic optimization. CoRR abs/1412.6980 (2014)

16. Lapata, M.: Probabilistic text structuring: experiments with sentence ordering. In: ACL 2003, pp. 545–552 (2003)
17. Lowe, R., Pow, N., Serban, I., Pineau, J.: The ubuntu dialogue corpus: a large dataset for research in unstructured multi-turn dialogue systems. In: SIGDIAL 2015, pp. 285–294 (2015)
18. Lowe, R.T., Pow, N., Serban, I.V., Charlin, L., Liu, C., Pineau, J.: Training end-to-end dialogue systems with the ubuntu dialogue corpus. D&D 8(1), 31–65 (2017)
19. Lukovnikov, D., Fischer, A., Lehmann, J., Auer, S.: Neural network-based question answering over knowledge graphs on word and character level. In: WWW 2017, pp. 1211–1220. ACM (2017)
20. Mikolov, T., Sutskever, I., Chen, K., Corrado, G.S., Dean, J.: Distributed representations of words and phrases and their compositionality. In: NIPS 2013, pp. 3111–3119 (2013)
21. Mohammad, S., Hirst, G.: Distributional measures as proxies for semantic relatedness. CoRR abs/1203.1 (2012)
22. Nguyen, D.T., Joty, S.R.: A neural local coherence model. In: ACL 2017, pp. 1320–1330 (2017)
23. Pennington, J., Socher, R., Manning, C.D.: Glove: global vectors for word representation. In: EMNLP 2014, pp. 1532–1543. ACL (2014)
24. Petöfi, J.S.: Semantics, pragmatics, text theory. Università di Urbino (1974)
25. Savenkov, V., Mehmood, Q., Umbrich, J., Polleres, A.: Counting to k or how SPARQL1.1 property paths can be extended to top-k path queries. In: SEMANTICS 2017, pp. 97–103 (2017)
26. Silva, V.S., Freitas, A., Handschuh, S.: Recognizing and justifying text entailment through distributional navigation on definition graphs. In: AAAI 2018 (2018)
27. Thoma, S., Rettinger, A., Both, F.: Towards holistic concept representations: embedding relational knowledge, visual attributes, and distributional word semantics. In: d'Amato, C., et al. (eds.) ISWC 2017. LNCS, vol. 10587, pp. 694–710. Springer, Cham (2017). https://doi.org/10.1007/978-3-319-68288-4_41
28. Usbeck, R., Ngomo, A.N., Haarmann, B., Krithara, A., Röder, M., Napolitano, G.: 7th open challenge on question answering over linked data (QALD-7). In: 4th SemWebEval Challenge at ESWC 2017, pp. 59–69 (2017)
29. Wilcke, X., Bloem, P., de Boer, V.: The knowledge graph as the default data model for learning on heterogeneous knowledge. Data Sci. 1(1–2), 39–57 (2017)

Combining Truth Discovery and RDF Knowledge Bases to Their Mutual Advantage

Valentina Beretta[1(✉)], Sébastien Harispe[1], Sylvie Ranwez[1],
and Isabelle Mougenot[2]

[1] LGI2P, IMT Mines Ales, Univ Montpellier, Ales, France
`valentina.beretta@mines-ales.fr`
[2] UMR 228 Espace Dev UM, Maison de la Télédétection, Montpellier, France

Abstract. This study exploits knowledge expressed in RDF Knowledge Bases (KBs) to enhance Truth Discovery (TD) performances. TD aims to identify *facts* (true claims) when conflicting claims are made by several sources. Based on the assumption that true claims are provided by reliable sources and reliable sources provide true claims, TD models iteratively compute value confidence and source trustworthiness in order to determine which claims are true. We propose a model that exploits the knowledge extracted from an existing RDF KB in the form of rules. These rules are used to quantify the evidence given by the RDF KB to support a claim. This evidence is then integrated into the computation of the confidence value to improve its estimation. Enhancing TD models efficiently obtains a larger set of reliable facts that *vice versa* can populate RDF KBs. Empirical experiments on real-world datasets showed the potential of the proposed approach, which led to an improvement of up to 18% compared to the model we modified.

Keywords: Truth discovery · RDF KBs · Rule mining
Source trustworthiness · Value confidence

1 Introduction

Several popular initiatives, such as DBpedia [2], Yago [17] and Google Knowledge Vault [5], automatically populate Knowledge Bases (KBs) with Web data. The performance of this Knowledge Base Population (KBP) process is critical to ensuring the quality of the KB. In particular, it requires dealing with complex cases in which several conflicting data are extracted from different sources, e.g. different automatic extractors will provide different birth places for Pablo Picasso. Approaches based on voting or naive strategies that only consider the most frequently provided data value are *de facto* limited. Such approaches are unable to deal with spam-based attacks or duplicated errors, which are common on the Web. Dealing with this problem therefore requires distinguishing values

D. Vrandečić et al. (Eds.): ISWC 2018, LNCS 11136, pp. 652–668, 2018.
https://doi.org/10.1007/978-3-030-00671-6_38

according to their sources. In this study, we propose an approach that serves KBP integrating potentially conflicting data provided by multiple sources; it relies on a general framework that can be used to address conflict resolution problems by exploiting prior knowledge defined in existing KBs.

Several techniques based on Knowledge Fusion have been proposed in order to automatically obtain reliable information. Most of them suppose that information veracity strictly depends on source reliability. Intuitively, the more reliable a source is, the more reliable the information it provides. In turn they also assume that source reliability depends on information veracity, i.e. reliable information is provided by reliable sources. Truth Discovery (TD) methods are unsupervised approaches based on these assumptions aimed at identifying the most reliable of a set of conflicting triples – for *functional* predicates, i.e. when there is a single true value for a property of a real-world entity. This study aims to enhance the TD framework using knowledge extracted from an existing RDF KB to obtain a larger set of correct *facts* that could be used to populate RDF KBs. More precisely, it makes the following contributions:

- A novel approach that can be used to enrich traditional TD models by incorporating additional information given by recurrent patterns extracted from a KB. A state-of-the-art rule mining system is used to extract rules that represent these patterns. A method is proposed for selecting the most useful rules to be used to evaluate veracity of triples. Moreover, since each rule contributes to TD performances according to its quality, a function that aggregates the existing rule quality metrics is also defined. High-quality rules will have a higher weight than low-quality rules;
- An extensive evaluation of the proposed approach; interestingly, it shows that the TD framework can benefit from information derived by rules. As a consequence, we point out how the creation of high quality RDF KBs may benefit from the use of highly reliable TD models. The datasets and source code proposed in this study are open-source and freely accessible online.[1]

The paper is organized as follows. Section 2 presents an overview of the TD framework and how it can be applied in the RDF KB context. It also describes the state-of-the-art rule mining techniques that are used in our work to detect interesting recurrent patterns. Section 3 explains how additional information extracted from KBs is integrated into the TD framework. The proposed approach is evaluated and discussed in Sect. 4. Finally, Sect. 5 reports the main findings and discusses perspectives.

2 Related Work and Preliminaries

In this section we introduce the formal aspect of TD, its goal and the key elements required to achieve it. We then formally present rules and their quality metrics. We will then be able to use them to exploit identified recurrent patterns to increase confidence in certain triples.

---

[1] https://github.com/lgi2p/TDwithRULES.

In this study, we assume that sources provide their claims in the form of RDF triples $\langle subject, predicate, object \rangle \in I \times I \times (I \cup L)$ where I is the set of Internationalized Resource Identifiers (IRIs) and L the set of literals.

The following definition introduces all TD components (source, data items and values). Since the TD and Linked Data (LD) fields use different notations, this definition aims at clarifying the correspondence between terms belonging to each field.

Definition 1 (Truth Discovery). *Let $D \subseteq I \times I$ be a set of data items where each $d \in D$ is a pair $(subj, pred)$ that refers to a functional property $(pred \in I)$ of an entity $(subj \in I)$. Let $V \subseteq I \cup L$ be a set of values that can be assigned to these data items and S be the set of sources. Each source $s \in S$ can associate a value $v \in V$ (corresponding to $obj \in I \cup L$) to a data item $d \in D$, hence providing a claim v_d that corresponds to the RDF triple $\langle subj, pred, obj \rangle$. **Truth Discovery** associates a value confidence to each claim and a trustworthiness score to each source. It then iteratively estimates these quantities to identify the true value v_d^* for each data item.*

Several TD approaches have been proposed, as detailed in recent surveys [4,10]. The models differ from one another in the way they compute the value confidence of claims and the trustworthiness of sources. Some of them use no additional information, while others attempt to improve TD performances using external support such as extractor information (i.e. the confidence associated with extracted triples), the temporal dimension, hardness of facts, common sense reasoning or correlations. Models that take correlations into account can be divided according to the kinds of correlations they consider: source correlations, value correlations or data item correlations. To the best of our knowledge, no existing work takes advantage of data item correlations in the form of recurrent patterns to improve TD results. The idea is that the confidence of a certain claim can increase when recurrent patterns occur which are associated with the considered data item. This kind of correlation can be used to enhance existing TD models. In this study, a rule mining procedure is used to identify patterns in data. We specify the major aspects of the rule mining below.

2.1   Recurrent Pattern Detection from RDF KBs

Several techniques can be used to identify regularities in data. For instance, link mining models are often used for that purpose in knowledge base completion [13]. In this study we prefer to use rule mining techniques because they are easily interpretable [1]. Rules generalize patterns in order to identify useful suggestions that can be used to generate new data or correct existing data [6]. We therefore propose to exploit these suggestions in order to solve conflicts among triples provided by different sources. Given our problem setting, where rules are used to reinforce the confidence of a claim, we are particularly interested in Horn rules. Considering Datalog-style, a Horn rule $r : B_1 \wedge B_2 \wedge \cdots \wedge B_n \rightarrow H$, i.e. $r : \widehat{B} \rightarrow H$, is an implication from a conjunction of atoms called the body to a single atom called the head [12]. An atom is usually denoted $pred(subj, obj)$, where $subj$ and obj can be variables or constants. Considering that an instantiation of an atom is a substitution of its variables with IRIs, an atom a holds under an instantiation σ in a KB K if $\sigma(a) \in K$. Moreover, a body $\widehat{B}$ holds under σ in K, if each atom in $\widehat{B}$ holds [7]. Note that in our setting each instantiated atom $pred(subj, obj)$ can also be represented as an RDF triple $\langle subj, pred, obj \rangle$.

Rule extractors rely on the Closed World Assumption (CWA). This means that when a fact is not known (does not belong to the KB) it is considered to be false. This assumption is more often appropriate when KBs are complete. On the contrary, RDF KBs are based on the Open World Assumption (OWA). When dealing with incomplete information the OWA is preferable. If information is missing we need to distinguish between false and unknown information. A triple that does not appear in the KB is not systematically false. In this context, methods have recently been proposed that mine rules from RDF KBs such as DBpedia or Yago, taking the OWA into account [15]. An example of a rule mining system that considers the OWA is AMIE [8]. It is based on the Partial Completeness Assumption (PCA): if a KB contains some object values for a given pair *(subject, predicate)*, it is assumed that all object values associated with it are known. This assumption can generate counter-examples, required for rule mining models, but do not appear in RDF KBs, which often contain only positive facts. Alternative assumptions and metrics have been proposed to extract rules under the OWA [9,13,18]. In this study, we use AMIE because it is a state-of-the-art system and its source code is freely available online.

2.2 Rule Quality Metrics

Any rule, independently of the system used to extract it, can be evaluated by several quality metrics; among them the most well-recognized measures are *support* and *confidence* [1,11,19]. Support represents the frequency of a rule in a KB, while confidence is the percentage of instantiations of a rule in the KB, compared to the instantiations of its body. Based on the formal definition given in [8], for the sake of coherence and clarity, we present how these metrics are computed below. In the rest of the paper we do not make a comparison of the different quality metrics because it is out of the scope of this study. The primary aim here is to evaluate the potential of integrating knowledge extracted from an RDF KB into a TD process. However, since we are aware that robust metrics could have an impact on TD results, we plan to study such a comparison in future studies.

Considering a Horn rule $r : \widehat{B} \rightarrow H$ where H is composed of a single atom $p(x, y)$, its support is defined by:

$$supp(\widehat{B} \rightarrow p(x, y)) := \#(x, y) : \exists z_1, \ldots, z_n : \widehat{B} \wedge p(x, y) \tag{1}$$

where $z_1, \ldots, z_n$ are the variables contained in the atoms of the rule body $\widehat{B}$ apart from x and y, and $\#(x, y)$ is the number of different pairs x and y.

Its confidence is computed using the following formula:

$$conf(\widehat{B} \to p(x,y)) := \frac{supp(\widehat{B} \to p(x,y))}{\#(x,y) : \exists z_1, \ldots, z_n : \widehat{B}} \qquad (2)$$

This formula was introduced to evaluate the quality of rules using the CWA. It is too restrictive when dealing with the OWA. For this reason Galarraga et al. defined a new confidence, called $conf_{PCA}$ [8]. It makes a distinction between false and unknown facts based on PCA. In this setting, if a predicate related to a particular subject, never appears in the KB, then it can neither be considered as true nor false. This new confidence based on PCA is evaluated as follows:

$$conf_{PCA}(\widehat{B} \to p(x,y)) := \frac{supp(\widehat{B} \to p(x,y))}{\sum_j supp(\widehat{B} \to p(x,j))} \qquad (3)$$

where j's are all instantiations of the object variable related to predicate p and having subject x. Using PCA, $conf_{PCA}$ normalizes the support by the set of true and false facts that does not include the unknown ones.

In the next section, we describe how these quality measures are combined into a single measure. Having a more robust metric is important because it is the quality of each rule that will determine its contribution to the computation of the overall evidence that supports a certain claim.

3   Incorporating Rules into the Truth Discovery Framework

This section presents how extracted rules are integrated into truth discovery models. To that end, we define the concepts of *eligible* and *approving* rules, which will be used to identify the most useful rules that need to be taken into account when evaluating the confidence of a claim. Then we describe how information associated with these rules is quantified to further introduce the new confidence estimation formulas used by our TD framework.

3.1   Eligible and Approving Rules

It may not be useful to consider the entire set of extracted rules (denoted R) in order to improve value confidence. For instance, some rules could have a body that is not related to a given data item. Therefore, given a claim $\langle d, v \rangle$, i.e. v_d, where $d = (subj, pred)$, only *eligible* rules are used as potential evidence to improve its confidence estimation. They are defined in the following way.

Definition 2 (Eligible Rule). *Given a KB K, a set of rules $R = \{r : \widehat{B} \to H\}$ extracted from K where $H = p(x,y)$ and a claim $\langle d, v \rangle$ where $d = (subj, pred)$, a rule $r \in R$ is an **eligible rule** when its body holds, i.e. all of its body atoms appear in K when all rule variables are instantiated w.r.t. the data item subject. Moreover, its head predicate has to correspond to the one in the claim under examination, i.e. $(\sigma(\widehat{B}) \in K) \wedge (H = pred(subj, y))$.*

In our context, the eligibility of a rule depends on the subject and the predicate that compose a data item d. Thus, all claims related to the same data item $d = (subj, pred)$ have the same set of eligible rules, denoted $R_d = \{r \in R \mid (\sigma(\widehat{B}) \in K) \wedge (H = pred(subj, y))\}$.

Once eligible rules for a claim v_d have been collected, the proposed approach checks how many of these rules endorse (approve) v_d, i.e. how many rules support v_d.

Definition 3 (Approving Rule). *Given a KB K, a set of eligible rules $R_d = \{r : \widehat{B} \to H\}$ where $H = pred(subj, y)$ and a claim $\langle d, v \rangle$ where $d = (subj, pred)$, a rule $r \in R_d$ is an **approving rule** when the value predicted by r corresponds to the claimed value v, i.e. $(\sigma(\widehat{B}) \in K) \wedge (H = pred(subj, v))$.*

The set of approving rules for v_d is represented by $R_d^v \subseteq R_d$ where d indicates that the rules are eligible for a certain data item d and v indicates that the rules predict/support value v. Formally, we obtain $R_d^v = \{r \in R_d \mid (\sigma(\widehat{B}) \in K) \wedge (H = pred(subj, v))\}$.

Example. Given a KB K, reported in Table 1, and the rules:

- $r_1 : speaks(x, z) \wedge officialLang(y, z) \to bornIn(x, y)$
- $r_2 : residentIn(x, w) \wedge cityOf(w, y) \to bornIn(x, y)$

Given the following claims about the birth location of some painters $\langle Picasso, bornIn, Spain \rangle$, $\langle Picasso, bornIn, Málaga \rangle$ and $\langle Monet, bornIn, France \rangle$, the set of eligible rules for data item $d_A = (Picasso, bornIn)$ is $R_{d_A} = \{r_1, r_2\}$. The predicate in the head corresponds to the predicate in the claim and when all occurrences of variable x are replaced by $Picasso$ in r_1's and r_2's body, they are both verified. However, when $d_B = (Monet, bornIn)$ the set of eligible rules is $R_{d_B} = \{r_2\}$ because, even though the head and claim predicate are the same using both rules, if the x variable is substituted by $Monet$ the body of r_1 is not verified.

The set of approving rules for the first, second and third claims are respectively $R_{d_A}^{Spain} = \{r_1\}$, $R_{d_A}^{Málaga} = \emptyset$ and $R_{d_B}^{France} = \{r_2\}$.

Before explaining how additional information related to approving and eligible rules is quantified and then incorporated into the TD framework, we describe a function used to integrate the two quality aspects we are interested in, for each rule. This enables better weighting of each rule's contribution during the evaluation of a claim.

Table 1. Illustrative set of triples.

predicate	subject	object	predicate	subject	object
officialLang	(Spain,	Spanish)	residentIn	(Picasso,	Paris)
speaks	(Picasso,	Spanish)	cityOf	(Paris,	France)
residentIn	(Monet,	Vétheuil)	cityOf	(Vétheuil,	France)

3.2   Combining Rule Quality Measures

Support and conf_{PCA} represent different aspects of a rule, see Sect. 2.2. We propose an aggregate function to combine them into a single quality metric since, in our context, it is important to take both aspects into account. It may happen that two rules r_1 and r_2 have the same confidence, but different supports. For instance, if $\text{conf}_{PCA}(r_1) = \text{conf}_{PCA}(r_2) = 0.8$, $supp(r_1) = 5$ and $supp(r_2) = 500$, then r_2 deserves a higher level of *credibility* than r_1 since r_2 has been observed more often than r_1.

To address this issue, a function *score* : $R \rightarrow [0,1]$ is defined. It is based on Empirical Bayes (EB) methods [16]. EB adjusts estimations resulting from a limited number of examples that may happen by chance. Estimations are modified in function of available examples and prior expectations. When many examples are available, estimation adjustments are small. On the contrary, when there are only few examples, the adjustments are greater. They are corrected w.r.t. the average value that is expected by *a priori* knowledge. Given a family of the prior distribution of available data, EB is able to directly estimate its hyper parameters from the data. Then, it updates the prior belief with new evidence. In other words, the estimation that can be computed from the new examples is modulated w.r.t. prior expectation. The new estimation corresponds to the expected value of a random variable following the updated distribution. In our case, a more robust conf_{PCA}, i.e. the proportion of positive examples among all examples considered, needs to be estimated. The prior expectation on our data can be modelled using a *Beta* distribution that is characterized by parameters α and β. Once the model has estimated them, it uses this distribution as prior to modulate each individual estimate. This estimation will be equal to the expected value of the updated distribution $Beta(\alpha + X, \beta + (N - X))$, where X is the number of new positive examples and N is the total number of new examples. The new expected value is $(\alpha + X)/(\alpha + \beta + N)$. This value is returned by the aggregation function. In summary, given the hyper parameters α_S and β_S, the value returned by *score* for a rule $r : \widehat{B} \rightarrow p(x,y)$ is computed as follows:

$$score(r) = \frac{\alpha_S + supp(r)}{\alpha_S + \beta_S + \sum_j supp(\widehat{B} \rightarrow p(x,j))} \tag{4}$$

where $supp(r)$ is the support of r and $\sum_j supp(\widehat{B} \rightarrow p(x,j))$ is the number of triples containing data item (x,p). The returned score appears to be similar to conf_{PCA}, but it takes the cardinality of the examples into account.

Once this score has been estimated for each rule, the proposed approach sums up all this new information and integrates it into the value confidence estimation formula.

3.3   Assessing a Rule's Viewpoint on Claim Confidence

All the evidence provided by rules for a claim v_d is summarized in a *boosting factor* that can be seen as the confidence that is assigned by these rules to

v_d. More precisely, it represents the proportion of eligible rules that confirm a given claim v_d. In other words it evaluates the percentage of approving rules out of the entire set of eligible rules, i.e. $|R_d^v|/|R_d|$. It is returned by a function $boost : D \times V \rightarrow [0,1]$. As anticipated, the proposed model weights each rule differently w.r.t. its quality $score$. The higher the $score$ of a rule, the stronger its impact should be on computing the $boosting\ factor$. Intuitively, given a claim v_d where $d = (subj, pred)$ and a set of rules R extracted from a KB K, the proposed model evaluates the $boosting\ factor$ in the following way:

$$boost(d, v_d) \approx \frac{\sum\limits_{r \in R_d^v} score(r)}{\sum\limits_{r \in R_d} score(r)} \tag{5}$$

where R_d^v is the set of approving rules, R_d is the set of eligible rules and $score :$ $R \rightarrow [0,1]$ represents the quality score associated with a rule (as detailed in Sect. 3.2). Since the $boosting\ factor$ consists in evaluating a proportion, EB is used also in this case to obtain a better estimation, less likely to be the result of chance. As explained in Sect. 3.2, when applying EB, initially the parameters α_b and β_b of a $Beta$ distribution are estimated from the available data using methods of moments. Then this prior is updated based on evidence associated with a specific v_d. Thus, the $boosting\ factor$, corresponding to the expected value of the updated prior, is equal to:

$$boost(d, v_d) = \frac{\alpha_b + \sum\limits_{r \in R_d^v} score(r)}{\alpha_b + \beta_b + \sum\limits_{r \in R_d} score(r)} \tag{6}$$

where α_b and β_b are the hyper parameters of the Beta distribution representing the available examples. Since AMIE does not consider any $a\ priori$ knowledge such as the partial order of values to extract rules, we decided to use it to further exploit rule information and compute a more refined boosting factor. More precisely, considering a partial order $V = (V, \preceq)$, when a rule r explicitly predicts a value v, we assume that it implicitly supports all more general values v' such that $v \preceq v'$. In other words, the evidence provided as support by a rule to a value is propagated to all its generalizations. Therefore, in this case the boosting factor $boost_{PO}(d, v_d)$ indicates the percentage of approving rules out of all eligible rules, for both the value under examination and all of its more specific values. The subscript PO in the name of the boosting factor underlines the fact that the Partial Order among values is taken into account. The set R_d^v in Eq. 6 is replaced by the set $R_d^{v+} = \{r \in R_d \mid \widehat{B} \wedge H = p(x, v'), v' \preceq v\}$.

3.4 Integrating Rules' Viewpoints into Confidence Computation

All the elements required to integrate information given by recurrent patterns into TD models have been defined. Since the $boosting\ factor$ depends on the claim, only the confidence formula has been updated. As proof of concept, in this study we modified *Sums* [14] whose estimation formulas are:

$$t^i(s) = \frac{1}{\max\limits_{s' \in S} \sum\limits_{v'_d \in V^{s'}} c^{i-1}(v'_d)} \sum\limits_{v_d \in V^s} c^{i-1}(v_d) \tag{7}$$

$$c^i(v_d) = \frac{1}{\max\limits_{v'_d \in V} \sum\limits_{s' \in S^{v'_d}} t^i(s')} \sum\limits_{s \in S^{v_d}} t^i(s) \tag{8}$$

We modified Eq. 8 proposing *Sums$_{RULES}$*. This new model integrates the additional information given by rules into the confidence formulas as follows:

$$c^i_{rules}(v_d) = \frac{1}{norm_{v_d}} \left[(1 - \gamma)c^i(v_d) + \gamma \, boost(d, v_d) \right] \tag{9}$$

where $\gamma \in [0, 1]$ is a weight that calibrates the influence assigned to sources and KB for estimating value confidences. For the sake of coherence, when using *boost$_{PO}$* we considered the partial order also for the computation of the confidence formula, as suggested in a previous study [3]. We refer to the model that uses confidence formula $c^i_{PO}(v_d)$, taking the partial order into account, as *Sums$_{PO}$*. It computes the confidence of v_d considering all the trustworthiness of sources that provide the value v for the data item d, i.e. the claim v_d under examination, or a more specific value than v. Indeed as highlighted above when claiming a value, we also consider that a source implicitly supports all its generalizations. Similarly, the model that integrates both the *boost$_{PO}$* and rules is indicated as *Sums$_{RULES\&PO}$* and is defined as follows:

$$c^i_{RULES\&PO}(v_d) = \frac{1}{norm_{v_d}} \left[(1 - \gamma)c^i_{PO}(v_d) + \gamma \, boost_{PO}(d, v_d) \right] \tag{10}$$

Note that, while *Sums* and *Sums$_{RULES}$* return a true value for each data item selecting the value with the highest confidence, *Sums$_{RULES\&PO}$* and *Sums$_{PO}$* required a more refined and greedy procedure to select the most informative true value. Indeed, considering the partial order of values, the highest confidence is always assigned to the most general value (it is implicitly supported by all the others). Thus, since systematically returning the most general value each time is not worthwhile, the selection procedure leverages the partial order to identify the expected value. Starting from the root, at each step it selects the closest specialization of the value with the highest confidence. The procedure stops when there are no more specific values, or when the confidence of the selected values is lower that a given threshold θ defining the minimal confidence score required to be consider as a true value. For further details see [3].

4 Experiments and Results

In order to obtain an extended overview of the proposed approach, several experiments were carried out on synthetic and real-world datasets. First of all, experiments were conducted using synthetic datasets to determine the improvement obtained by $Sums_{RULES}$ (Eq. 9) and $Sums_{RULES\&PO}$ w.r.t. their baseline, i.e. $Sums$ [14] (Eq. 8) and $Sums_{PO}$ (Eq. 10) considering different scenarios. Note that, in both cases, the baseline corresponds to set $\gamma = 0$ in the new confidence formula of the proposed models. A second set of experiments was conducted using a real-world dataset to test the proposed approach in a realistic scenario. A comparison with existing models is also presented.

The rules used in the experiments, as well as their support and $conf_{PCA}$ were extracted from DBpedia by AMIE. To ensure that the rules considered are abstractions of a sufficient number of facts, we selected those with the highest head coverage. We selected 62 rules for the predicate *birthPlace*. Examples of these rules are reported in Table 2.

Table 2. Examples of rules extracted by AMIE from DBpedia for *birthplace* predicate.

@prefix db: <http://dbpedia.org/resource/>.
@prefix db-owl: <http://dbpedia.org/ontology/>.

?a db-owl:deathPlace ?b	→?a db-owl:birthPlace ?b
?a db-owl:country ?b	→?a db-owl:birthPlace ?b
?a db-owl:deathPlace ?b ∧ ?b db-owl:language db:English_language	→?a db-owl:birthPlace ?b

4.1 Experiments on Synthetic Data

The synthetic datasets were used to evaluate the proposed model on various scenarios depending on the granularity of the true values provided. Experts usually provide specific true values. Non-expert users provide general values, which remain true. To evaluate the performance in these contexts, we measured the expected value rate/recall (returned values that correspond to expected ones), the true but more general value rate (returned values that are more general than the expected ones) and the erroneous value rate (values that are neither expected nor general) obtained by different model settings.

Generation. The main elements required to generate these datasets are: a ground truth, a partial order and a set of claims provided by several sources on different data items [3]. The ground truth was generated by selecting a subset of 10000 DBpedia instances having the birthPlace property, considering the related value as the true one. Also the partial order of values was constructed using the DBpedia ontology. Partial order relationships were added between all classes subsumed, i.e. rdfs:subClassOf, by dbpedia-dbo:Place class and between those classes and their instances. Moreover, the relationships were added to all instances for which the property dbpedia-dbo:isPartOf or dbpedia-dbo:country exists. Since dbpedia-owl:Thing is the most abstract concept in DBpedia, all the values belonging to the partial order graph were rooted to it. In order to obtain a partial order of values respecting the properties of a Directed Acyclic Graph, all cycles induced by incorrectness on the part-of property were removed.[2] For the generation of the claims, 1000 sources and 10000 data items were considered. Table 3 reports all the features regarding the generation of the claim set. The main feature is related to the distribution used to select the granularity of the true values provided. Based on this feature, three types of dataset were generated: EXP, LOW_E and UNI figuring, respectively, the behaviors of experts, a mix of experts and non-experts, and non-expert users. Considering that Picasso was born in Málaga, for example, in the case of EXP datasets, the sources tend to provide true values such as Málaga, Andalusia, Spain, while in the case of UNI datasets they will also provide general values such as Europe or the Continent. For each scenario, 20 synthetic datasets were generated.

Table 3. Features of synthetic datasets.

Feature	Description
Source coverage	Each source provides a number of claims that is exponentially distributed.
Source trustworthiness	The trustworthiness distribution is Gaussian with average 0.6 and standard deviation 0.4. This means that the sources are mostly reliable and only a few of them are always or never correct.
# of true claims per source	Each source provide a true value w.r.t. its trustworthiness level.
# of distinct true values per data item	$1..V_d^{true}$ where $V_d^{true} = \{v \in V : v_d^* \preceq v\}$
Granularity of the true value provided	Each source provides a true value having a granularity that approaches the granularity of the expected true value w.r.t. a high decay-rate exponential distribution (EXP), a low decay-rate exponential distribution (LOW_E) and a uniform distribution (UNI).
# of distinct false values per data item	$1..30$ values belonging to $V_d^{false} = V_d^{true} \setminus \{v|v \preceq v_d^*\}$

Results. The results, summarized in Fig. 1, show that the proposed approach enables the definition of TD models that benefit from the use of *a priori* knowledge given by an external RDF KBs. Indeed, the number of correct *facts* identified by the proposed model usually increases w.r.t. the baseline. Intuitively, since

---

[2] We assumed that abstract concepts should have higher out-degree than less abstract ones. Thus, for each cycle, the edge whose target is the node with the highest out-degree was removed. Analysing the discarded edges, the heuristic works.

the number of correct *facts* increases, a new KB that is populated with the true claims identified by the improved TD will be of higher quality.

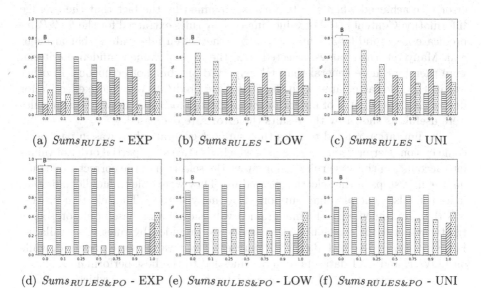

(a) $Sums_{RULES}$ - EXP   (b) $Sums_{RULES}$ - LOW   (c) $Sums_{RULES}$ - UNI

(d) $Sums_{RULES\&PO}$ - EXP (e) $Sums_{RULES\&PO}$ - LOW (f) $Sums_{RULES\&PO}$ - UNI

Fig. 1. Expected (horizontal line bars), true but more general (diagonal line bars) and erroneous values (dotted bars) obtained by $Sums_{RULES}$ and $Sums_{RULES\&PO}$ on different datasets with several γ. The letter B indicates the baseline model results.

The improvement obtained by considering both $Sums_{RULES}$ and $Sums_{RULES\&PO}$ was always greater for UNI datasets than for EXP or LOW_E ones. Since identifying true values in UNI settings was harder than in the other cases (the highest disagreement among sources on the true values was modeled by UNI), the baseline obtained the lowest recall. Using additional information tackles the high level of disagreement among sources and thus enables full exploitation of the higher scope for improvement that was available in the case of the UNI setting.

Considering $Sums_{RULES}$ the best recall was obtained with different γ values. For UNI datasets, the optimal configuration was when $\gamma = 1$. In such a case, it was considered that no information provided by sources was useful and that only rules should be used to solve conflicts among claims (when rules are available). This was true only for the extreme situation represented by UNI datasets where disagreement among sources was so high that the recall obtained by baseline model remained under 10%. Indeed, in the other cases it was advantageous to take both source trustworthiness and rule information into account. For EXP datasets, the optimal γ value was 0.1, while for LOW_E it was 0.9. Low γ values were preferred in EXP settings because in this case sources that provide true values are quite sure about the expected one, and it is thus less useful to consider the rules' viewpoints. Moreover, this setting was the only situation where considering external knowledge was damaging in terms of recall. Nevertheless, the error rate obtained by $Sums_{RULES}$ when $0 < \gamma < 1$ was always lower than the error rate achieved when $\gamma = 0$. This is explained by the fact that the average Information Content[3] (IC) of values inferred by rules extracted for the *birthPlace* predicate is around 0.53. This means that they often infer values that are general. Many returned values, selected with the highest value confidence criteria, were therefore more general than the expected one but not erroneous. In other words, the rules associated with the *birthPlace* predicate were more effective for discovering the country of birth than the expected location. However using rules were useful, as shown by the results the error rate decreased.

The limitation related to rules that support general values was in part overcome by considering $Sums_{RULES\&PO}$, which also takes the partial order of values into account. In this case rules can improve the selection of the correct value during the first steps of the selection procedure. They were able to handle and dominate the false general values supported by many sources. The selection process was then continued with the fine-grained values evaluated based only on source trustworthiness information since no evidence provided by rules was available. For $Sums_{RULES\&PO}$ tested on EXP datasets, low γ values were preferred, while on LOW_E and UNI datasets high γ values led to the best performance.

The best overall recall was obtained by $Sums_{RULES\&PO}$, which considers both kinds of *a priori* knowledge: extracted rules and partial order of values.

4.2   Experiments on Real-World Data

These experiments were conducted to test the proposed model in a realistic scenario. Since the results of experiments on synthetic data showed that the most interesting results were obtained by considering both extracted rules and the partial order of values, we compared the results obtained in this case with those obtained by existing TD methods[4] [20]. The evaluation protocol consisted in counting the number of values returned by a model that are equal to the expected values. In this setting, the number of general values returned were not analyzed since the main aim of TD models is to return the expected value, not its generalizations.

Generation. We collected a set of claims related to the predicate dbo:birthPlace, i.e. people's birth location. As data item subject, we randomly selected a subset of 480 DBpedia instances of type dbo:Person having the property birthPlace and having at least one eligible rule. For each data item we collected a set of webpages (up to 50) containing at least one occurrence of the

---

[3] Information Content indicates the degree of abstraction/concreteness of a concept w.r.t. an ontology. It monotonically increases from the most abstract concept (its $IC = 0$) to the most concrete ones discriminating the granularity of different values.

[4] For these models we used the implementation available at http://www.github.com/daqcri/DAFNA-EA.

subject's full name and the words "was born", i.e. the natural language expression that is usually used to introduce the birth location of a person. Given a webpage and its data item, we defined two procedures for extracting the provided claim. Procedure A selects, as claimed value, the location (identified by DBpedia-spotlight API) that co-occurs in the same sentence and is nearest to the word "born". Procedure B adds a constraint to procedure A: a value can be selected only if it appears after the first occurrence of the subject's full name in the text. Two different datasets were created based on procedures A and B, respectively DataA and DataB. For building our ground truth, we assumed that the values defined in DBpedia as birth location for each data item were the true ones. Since in the collected claims, values that were more specific than the expected one (contained in the ground truth) were provided, we manually checked if these specifications were true. For 20 instances that we manually checked, 10 were found to be true specifications. Note that as partial order we considered the same one as for the experiments on synthetic data. The procedures, source code and datasets obtained are available online at https://github.com/lgi2p/TDwithRULES.

Results. We can observe that for both datasets DataA and DataB we improved the performance by 18% and 14% respectively compared to the baseline, i.e. *Sums* – the approach we decided to modify. Table 4 shows the results obtained by the best configuration of parameters where both extracted rules and partial order were considered.

When comparing the proposed approach to existing TD models, it did not outperform the others, see Table 5. Note that our study focused on modifying *Sums* which is considered to be one of the most well studied models, but not necessarily the most effective one. After investigating the errors, we found out that it was mainly due to a limitation of *Sums*: it rewards sources having high coverage and, meanwhile, penalizes those with low coverage. Indeed *Sums* computes the trustworthiness of a source by summing up all the confidence of the claims it provides. Thus the higher the number of claims a source provides, the higher the trustworthiness of the source. The problem is that *Sums* does not distinguish between sources always providing true values, but having different coverage. While Wikipedia.org is correctly considered as a high reliable source, an actor's fan club website is incorrectly considered as unreliable. Even if the information it provides is correct, because it covers only one data item its trustworthiness will be lower than the one of Wikipedia.org (source having a high coverage). In real-world datasets very few sources have high coverage, and most of them have low coverage – power law phenomenon. In this scenario the sources having high coverage dominate the specialized ones. Therefore, no extraction errors from high coverage sources are allowed. Indeed if an incorrect value is extracted from Wikipedia.org (for instance when the sentence refers to another person), this will be incorrectly considered as the true one. Since this cannot be guaranteed (the extraction procedures we defined are voluntarily naive), we propose a post-processing procedure that alleviates this problem. Before selecting the true value, it sets equal to 0 all the confidence of those values that are

Table 4. Recall obtained using *Sums* and its modifications on DataA and DataB.

Model	DataA	DataB
Sums	0.448	0.473
Sums$_{PO}$ ($\gamma = 0.0, \theta = 0.05$)	0.517	0.566
Sums$_{RULES\&PO}$ ($\gamma = 0.3, \theta = 0.0$)	0.527	0.548
Sums$_{RULES\&PO}$ ($\gamma = 0.3, \theta = 0.05$)	0.565	0.590
Sums$_{RULES\&PO}$ +post-proc. ($\gamma = 0.3, \theta = 0.1$)	**0.631**	**0.614**

Table 5. Recall obtained using existing models on DataA and DataB.

Existing model	DataA	DataB
Voting	0.640	0.625
TruthFinder	**0.646**	0.622
2-Estimates	0.631	0.635
3-Estimates	0.008	0.612
Cosine	0.636	0.635
AccuCopy	0.638	0.640
Accu	0.638	**0.660**
Depen	0.431	0.494
AccuSim	0.413	0.448
SimpleLCA	0.631	**0.660**
GuessLCA	0.644	0.646

provided by only a single source. We assume that it is highly improbable that the same extraction error occurs, i.e. the erroneous value should therefore be provided only once. This solution, indicated as *Sums*$_{RULES\&PO}$ + post-proc., obtained performances comparable with existing models for DataA and DataB. While it enables to avoid some of the extraction errors (occurring more with the most naive procedure A), it is still not capable of assigning lower trustworthiness levels to specialized sources.

Given these observations, in real-world settings it is very important to consider the power law phenomenon. The results show that *Sums* is not efficient in this kind of situation. Nevertheless, using additional information (partial order and extracted rules) improved the results w.r.t. the baseline approach, and this is promising for the principles introduced in this study. As shown in Table 4, the improvement due to taking this information into account was 18% for DataA and of 14% for DataB. Moreover, through this study we also show that correctness and the granularity of values in DBpedia can be improved using TD models. Claims on data items can easily be collected on the Web. When more specific values than the one contained in DBpedia are found, they can be verified using TD model.

5   Conclusion

Solving information conflicts in an automated fashion is critical for the development of large RDF KBs populated by heterogeneous information extraction systems. In this study, we suggest using TD models as unsupervised techniques to populate RDF KBs. In order to create high quality KBs and exploit current ones, we propose improving an existing TD model (*Sums*) using knowledge extracted from an external RDF KB in the form of rules. Several experiments that show the validity of the proposed model were conducted. The performances of the proposed model show higher recall than baseline methods (up to 18% of improvement). The datasets, source code and procedures are all available online. We plan to apply the rationale of the proposed model to other TD models in order to outperform them all. In addition, we envisage extending the evaluation methodology in order to consolidate our results by considering other predicates and non-functional ones such as those used in ISWC Semantic Web Challenge 2017. Currently, we do not consider as negative evidence the fact that a rule predicts a different value than the one contained in a claim. In the future, we envisage studying how to incorporate this information, as well as explicit axioms, subjectivity information and contextual dependencies (such as diachronicity).

References

1. Agrawal, R., Imieliński, T., Swami, A.: Mining association rules between sets of items in large databases. In: SIGMOD 1993, vol. 22, pp. 207–216. ACM (1993)
2. Auer, S., Bizer, C., Kobilarov, G., Lehmann, J., Cyganiak, R., Ives, Z.: DBpedia: a nucleus for a web of open data. In: Aberer, K., et al. (eds.) ASWC/ISWC -2007. LNCS, vol. 4825, pp. 722–735. Springer, Heidelberg (2007). https://doi.org/10. 1007/978-3-540-76298-0_52
3. Beretta, V., Harispe, S., Ranwez, S., Mougenot, I.: How can ontologies give you clue for truth-discovery? An exploratory study. In: WIMS 2016, p. 15. ACM (2016). https://doi.org/10.1145/2912845.2912848
4. Berti-Équille, L., Borge-Holthoefer, J.: Veracity of Data: From Truth Discovery Computation Algorithms to Models of Misinformation Dynamics. Synthesis Lectures on Data Management. Morgan & Claypool Publishers (2015). https://doi. org/10.2200/S00676ED1V01Y201509DTM042
5. Dong, X., et al.: Knowledge vault: a web-scale approach to probabilistic knowledge fusion. In: KDD 2014, pp. 601–610. ACM (2014). https://doi.org/10.1145/2623330. 2623623
6. Galárraga, L.: Interactive rule mining in knowledge bases. In: Actes des 31e Conférence sur la Gestion de Données (BDA 2015), Île de Porquerolles (2015)
7. Galárraga, L., Suchanek, F.M.: Towards a numeric rule mining language. In: Proceedings of Automated Knowledge Base Construction workshop (2014)
8. Galárraga, L., Teflioudi, C., Hose, K., Suchanek, F.M.: Fast rule mining in ontological knowledge bases with AMIE+. VLDB J. **24**(6), 707–730 (2015). https:// doi.org/10.1007/s00778-015-0394-1
9. Lehmann, J., Völker, J. (eds.): Perspectives On Ontology Learning, vol. 18. IOS Press (2014). https://doi.org/10.3233/978-1-61499-379-7-i
10. Li, Y., et al.: A survey on truth discovery. SIGKDD Explor. Newsl. **17**(2), 1–16 (2016). https://doi.org/10.1145/2897350.2897352
11. Maimon, O., Rokach, L.: Data Mining and Knowledge Discovery Handbook, vol. 2. Springer, Heidelberg (2005). https://doi.org/10.1007/b107408
12. Nebot, V., Berlanga, R.: Finding association rules in semantic web data. Knowl.-Based Syst. **25**(1), 51–62 (2012)
13. Nickel, M., Murphy, K., Tresp, V., Gabrilovich, E.: A review of relational machine learning for knowledge graphs. Proc. IEEE **104**(1), 11–33 (2016)

14. Pasternack, J., Roth, D.: Knowing what to believe (when you already know something). In: COLING 2010, pp. 877–885. Association for Computational Linguistics, Stroudsburg, PA, USA (2010)
15. Quboa, Q.K., Saraee, M.: A state-of-the-art survey on semantic web mining. Intell. Inf. Manag. **5**(01), 1–10 (2013)
16. Robbins, H.: An empirical Bayes approach to statistics. In: Proceedings of the Third Berkeley Symposium on Mathematical Statistics and Probability Volume 1: Contributions to the Theory of Statistics, pp. 157–163. University of California Press, Berkeley, California (1956)
17. Suchanek, F.M., Kasneci, G., Weikum, G.: YAGO: a core of semantic knowledge. In: WWW 2007, pp. 697–706. ACM (2007). https://doi.org/10.1145/1242572. 1242667
18. Pellissier Tanon, T., Stepanova, D., Razniewski, S., Mirza, P., Weikum, G.: Completeness-aware rule learning from knowledge graphs. In: d'Amato, C., et al. (eds.) ISWC 2017. LNCS, vol. 10587, pp. 507–525. Springer, Cham (2017). https:// doi.org/10.1007/978-3-319-68288-4_30
19. Ventura, S., Luna, J.M.: Quality measures in pattern mining. In: Ventura, S., Luna, J.M. (eds.) Pattern Mining with Evolutionary Algorithms, pp. 27–44. Springer, Cham (2016). https://doi.org/10.1007/978-3-319-33858-3_2
20. Waguih, D.A., Berti-Equille, L.: Truth discovery algorithms: an experimental evaluation. CoRR abs/1409.6428 (2014)

Content Based Fake News Detection
Using Knowledge Graphs

Jeff Z. Pan[1(✉)], Siyana Pavlova[1], Chenxi Li[1,2], Ningxi Li[1,2], Yangmei Li[1,2],
and Jinshuo Liu[2(✉)]

[1] University of Aberdeen, Aberdeen, UK
jeff.z.pan@abdn.ac.uk
[2] Wuhan University, Wuhan, China
liujinshuo@whu.edu.cn

Abstract. This paper addresses the problem of fake news detection.
There are many works already in this space; however, most of them
are for social media and not using news content for the decision making. In this paper, we propose some novel approaches, including the
B-TransE model, to detecting fake news based on news content using
knowledge graphs. In our solutions, we need to address a few technical
challenges. Firstly, computational-oriented fact checking is not comprehensive enough to cover all the relations needed for fake news detection.
Secondly, it is challenging to validate the correctness of the extracted
triples from news articles. Our approaches are evaluated with the Kaggle's 'Getting Real about Fake News' dataset and some true articles from
main stream media. The evaluations show that some of our approaches
have over 0.80 F1-scores.

1 Introduction

With the widespread popularization of the Internet, it becomes easier and more
convenient for people to get news from the Internet than other traditional media.
Unfortunately, open Internet fuels the spread of a great many fake news without
effective supervision. *Fake news* are news articles that are intentionally and
verifiably false, and could mislead readers [AG17a]. With characteristics of low
cost, easy access, and rapid dissemination, fake news can easily mislead public
opinion, also disturb the social order, damage the credibility of social media,
infringe the interests of the parties and cause the crisis of confidence [VRA18,
SCV+17]. We all know how it has occurred and exerted an influence in the past
2016 US presidential elections [AG17a]. Hence, it is important and valuable to
develop methods for detecting fake news.

Most existing works on fake news detection are based on styles, focusing
on capturing the writing style of news content as features to classify news
articles [GM17, Gil17, Wan17, JLY17]. Although they can be effective, these
approaches cannot explain what is fake in the target news article. On the
other hand, knowledge based (or content based) fake news detection, which

D. Vrandečić et al. (Eds.): ISWC 2018, LNCS 11136, pp. 669–683, 2018.
https://doi.org/10.1007/978-3-030-00671-6_39

is also known as fact checking [SSW+17], is more promising, as the detection is based on content rather than style. Existing content based approaches focus on path reachability trying to find a path in an existing knowledge graph [PVGPW17,PCE+17] for a given triple [LCSR+15,SFMC17,SW16]. However, there are a few limitations of the existing content-based approaches, which lead to the following research questions:

RQ1: What happens if we do not have a knowledge graph in the first place, but only have articles? For a fake news topic, it is likely that at the beginning we do not have the knowledge graph to rely on for fact-checking. Our idea is either to construct knowledge graphs based on (true and fake) news articles bases, or to utilize related sub-graphs from open knowledge graphs. In the former case, we can also construct two knowledge graphs on the same topic: one is based on fake news articles and the other one based on true news articles. It should be noted that fake news articles are also available in online fake news web sites, such as 'the Onion', which often provide different categories of fake news articles. In the latter case, we could extract the sub-graph centered on the background topic of news articles from the open knowledge graph. Hence, we construct an external knowledge graph for these news articles based on facts related to the background topic in DBpedia dataset[1].

RQ2: Can we use incomplete and imprecise knowledge graphs for fake news detection? All computational knowledge-based approaches mainly focus on simple common relations between entities, such as *"country"*, *"child"*, *"employerOf"*. And the knowledge graphs they use are too incomplete and imprecise to cover the complex relations that appeared in fake news articles. For example, the triple *(Anthony Weiner, cooperate with, FBI)* extracted from a news article has the entities of *"Anthony Weiner"* and *"FBI"*, and the relation of *"cooperate with"*. The entities are easily found in open knowledge but the relation is not. In this paper, our idea is to make use of knowledge graph embedding for computing semantic similarities, so as to accommodate incomplete and imprecise knowledge graphs. As far as we know, this is the first work on this direction. We use a basic knowledge graph embedding model, namely TransE [BUGD+13] , to test the potential of knowledge graph embedding methods in content based fake news detection.

RQ3: How can we use Knowledge Graph Embedding for content based fake news detection? We firstly propose an approach to utilizing TransE [BUGD+13] to train a single model on a given knowledge graph, such as a subset of an open knowledge graph, or one that is constructed based on some news articles. Secondly, we propose a approach to generating a binary TransE model (B-TransE) which combines a negative model with a positive single model. Furthermore, in order to improve the performance, we also propose a hybrid approach to using a fusion strategy to combine the feature vectors produced by the models above.

Our major contributions of this paper are summarized as follows:

---

[1] http://wiki.dbpedia.org/.

- To the best of our knowledge, we are the first to propose the approach of content based fake news detection by making use of incomplete and imprecise knowledge graphs.
- We proposed a few approaches to exploit knowledge graph embedding to facilitate content based fake news detection.
- Our experiments show that our binary model approach outperforms our single model approach, and that our hybrid approach improve the performance of fake news detection.
- Our experiments show that our approaches outperform the Knowledge Stream approach in the test datasets.

2   Related Works

2.1   Fake News Detection

An effective approach is of prime importance for the success of fake news detection that has been a big challenge in recent years. Generally, those approaches can be categorized as knowledge-based and style-based.

Knowledge-Based. The most straightforward way to detect fake news is to check the truthfulness of the statements claimed in news content. Knowledge-based approaches are also known as fact checking. The expert-oriented approaches, such as Snopes[2], mainly rely on human experts working in specific fields to help decision making. The crowdsourcing-oriented approaches, such as Fiskkit[3] where normal people can annotate the accuracy of news content, utilize the wisdom of crowd to help check the accuracy of the news articles. The computational-oriented approaches can automatically check whether the given claims have reachable paths or could be inferred in existing knowledge graphs. Ciampaglia et al. [LCSR+15] take fact-checking as a problem of finding shortest paths between concepts in a knowledge graph; they propose a metric to assess the truth of a statement by analyzing path lengths between the concepts in question. Shiralkar et al. [SFMC17] propose a novel method called"Knowledge Stream(KS)" and a fact-checking algorithm called Relational Knowledge Linker that verifies a claim based on the single shortest, semantically related path in KG. Shi et al. [SW16] view fake news detection as a link prediction task, and present a discriminative path-based method that incorporates connectivity, type information and predicate interactions.

Style-Based. Style-based approaches attempt to capture the writing style of news content. Mykhailo Granik et al. [GM17] find that there are some similarity between fake news and spam email, such as they often have a lot of grammatical mistakes, try to affect reader's opinion on some topics in manipulative way and use similar limited set of words. So they apply a simple approach for fake news detection using naive Bayes classifier due to those similarity. Gilda [Gil17]

---

[2] http://www.snopes.com/.
[3] http://fiskkit.com.

applies term frequency-inverse document frequency (TF-IDF) of bi-grams and probabilistic context free grammar (PCFG) detection and test the dataset on multiple classification algorithms. Wange [Wan17] investigates automatic fake news detection based on surface-level linguistic patterns and design a novel, hybrid convolutional neural network to integrate speaker related metadata with text. Jiang *et al.* [JLY17] find that some key words tend to appear frequently in the micro-blog rumor. They analyze the text syntactical structure features and presents a simple way of rumor detection based on LanguageTool.

2.2 Knowledge Graph Embedding

Bordes *et al.* [BUGD+13] propose a method, named TransE, which models relationships by interpreting them as translations operating on the low-dimensional embeddings of the entities. TransE is very efficient while achieving state-of-the-art predictive performance, but it does not perform well in interpret such properties as reflexive, one-to-many, many-to-one, and many-to-many. So, Wang *et al.* [WZFC14] propose TransH which models a relation as a hyperplane together with a translation operation on it. Lin *et al.* [LLS+15] propose TransR to build entity and relation embeddings in separate entity space and relation spaces. TransR learns embeddings by first projecting entities from entity space to corresponding relation space and then building translations between projected entities. Ji *et al.* [JHX+15] propose a model named TransD, which uses two vectors to represent a named symbol object (entity and relation), and the first one represents the meaning of a(n) entity (relation), the other one is used to construct mapping matrix dynamically.

3 Basic Notions

In this section we introduce some basic notions related to content-based classification of news articles with external knowledge.

A *knowledge graph* KG describes entities and the relations between them. It can be formalised as $KG = \{E, R, S\}$, where E denotes the set of entities, R the set of relations and S the triple set. An *article base* AB is a set of news articles for each of which we have a title, a full content text and an annotation of *true* or *fake*. A knowledge graph may be a readily available for fact checking, such as DBpedia, or one needs to construct one from an article base.

In this paper, we use the knowledge graph embedding (KGE) method TransE to facilitate fake news detection. Typical knowledge graph completion algorithms are based on knowledge graph embedding (KGE). The idea of embedding is to represent an entity as a k-dimensional vector h (or t) and defines a scoring function f_r (h, t) to measure the plausibility of the triplet (h, r, t) in the embedding space. The representations of entities and relations are obtained by minimising a global loss function involving all entities and relations. Different KGE algorithms often differ in their scoring function, transformation and loss function. When a knowledge graph is converted into vector space, more semantic computations can be applied than just reasoning and querying.

The task of *fact checking* is to check if a target triple (h, r, t) is true based on a given knowledge graph. The task of *content based fake news detection* (or simply fake news detection), is to check if a target news article is true based on its title and content, as well as some related knowledge graph.

4 Our Approach

4.1 Framework Overview

To detect whether a news article is true or not, and to answer our research questions as outlined in Sect. 1, we propose a solution which uses, a tool to produce knowledge graphs (KG), a single B-TransE model, a binary TransE model and finally hybrid approaches. Firstly, we generate background knowledge by producing three different KG. This part addresses RQ1 and RQ2. Then we use a B-TransE model to build entity and relation embedding in low-dimensional vector space and detect whether the news article is true or not. We test a single TransE model and a binary TransE model and thus answer RQ3. Finally, we use some hybrid approaches to improve detection performance.

For the task of background knowledge generation, we consider three types of KGs: one is based on fake news article base; one is based on open KG, such as DBpedia, a crowd-sourced community effort to extract structured information from Wikipedia; one is based on true news article base from reliable news agencies.

The external KG extracted from open knowledge graph includes two parts: $KG_1 = \{E_1, R_1, S_1\}$ based on entities from fake article base and $KG_2 = \{E_2, R_2, S_2\}$ centered on the topic of news articles. These are further described in Sect. 5.2.

External KGs such as DBpedia are excellent for general knowledge facts, such as *(Barack Obama, birthPlace, Hawaii)*. However, they are incomplete and imprecise as such KGs do not contain enough relations to represent current events, as the latter are generated daily. An example of such a relation is *(Anthony Weiner, cooperate with, FBI)*, which is not contained in DBpedia. Despite this, in Sect. 5, we show that an incomplete and imprecise external open KG can perform well on the task of fake news detection.

The entities and the relation from the example above, however, can easily be extracted from an article on the topic. In order to be able to assess news items as true or fake, we propose an approach which uses external knowledge generated from real world news articles. We propose using a set of true and a set of fake articles to generate two models: $\mathbf{M}$ and $\mathbf{M'}$ as described in Sect. 3. We summarize these articles, as using the full article text causes redundancies and increases runtime. We further explore the performance of our approach, using only external knowledge from article bases, in order to answer the question what happens when we do not have a KG in the first place, but only news articles. In Fig. 1 we outline the methods used to generate a KG from an article base.

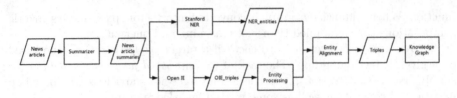

Fig. 1. Triple extraction from an article base

To construct KG from news articles, we start with a set of news articles and use OpenIE[4] to extract triples first. However, OpenIE does not perform well in triple extraction of news, so we propose some methods to improve the quality of the triples, including Stanford NER[5] and others which are further discussed in Sect. 5.2. We then perform entity alignment and obtain the triples which constitute our article based KG.

Once we have generated our three external KG, we use TransE to train a single model on each of them and compare their performance. To the best of our knowledge, this is the first work to apply knowledge graph embedding for fake news detection. Thus we use the basic TransE model. Since all the translation-based models aim to represent entities and relations in a vector space and there is no great difference between these models on our dataset, we choose the most basic model TransE. The single model is further described in Sect. 4.2 and an outline of its usage can be seen in Fig. 2. Our results, presented in Sect. 5, show that the external open KG has the best performance.

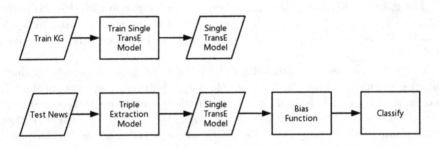

Fig. 2. Single TransE model

Then, we explore what happens when we combine a negative single model and a positive single model. The binary TransE (B-TransE) model is further described in Sect. 4.3 and an outline of its usage can be seen in Fig. 3. In Sect. 5 we then show that binary models perform better than single ones.

---

[4] https://nlp.stanford.edu/software/openie.html.
[5] https://nlp.stanford.edu/software/CRF-NER.html.

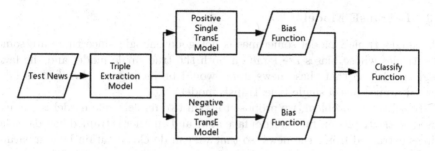

Fig. 3. Binary TransE model

Finally, we use a hybrid approach using an early fusion strategy that combines the feature vectors produces by the models above in order to improve detection performance. Further details of this approach are in Sect. 4.4.

4.2 Single TransE Model

To judge whether a given news article is true or fake through a knowledge graph, we extract triples from the news article and represent the triples in vector space, so that we can judge whether the news article is true or fake by the vectors. We use a Knowledge Graph to train a TransE model, which represents triples as vectors, and we name our method Single TransE Model.

In the Single TransE Model, we define TransE model as $\mathbf{M}$, and a triple based on $\mathbf{M}$ as $(\mathbf{h}, \mathbf{t}, \mathbf{r})$. We denote the triples extracted from one news item as $\mathbf{TS}$, so each triple is defined as $triple_i = (\mathbf{h}_i, \mathbf{t}_i, \mathbf{r}_i)$, where i means the index of the triple in $\mathbf{TS}$. We represent one news item as $\mathbf{N} = \{\mathbf{TS}, \mathbf{M}\}$.

To classify one news item, we calculate the bias of each triple in $\mathbf{TS}$. The bias of $triple_i$ is defined as

$$f_b(triple_i) = ||\mathbf{h}_i + \mathbf{r}_i - \mathbf{t}_i||_2^2 \tag{1}$$

Then we use these biases to classify the news item through a classifier. There are two ways we use these biases to do classification.

Avg Bias Classification. For the first one, we use average bias of a triple set to classify the news item through a classifier and name it Avg Bias Classification. The average bias of a triple set is defined as

$$f_{avgB}(TS) = \frac{\sum_{i=1}^{n} f_b(triple_i)}{|\mathbf{TS}|} \tag{2}$$

where the $|\mathbf{TS}|$ refers to the size of the triple set.

Max Bias Classification. The second one, the Max Bias Classification uses the max bias of a triple set to judge whether a news item is true or fake. The max bias of a triple set is defined as

$$f_{maxB}(TS) = f_b(triple_{max}) \tag{3}$$

Where max refers to the index of the triple whose bias is the maximum.

4.3   B-TransE Model

The single TransE model sometimes is not good enough, since there are some true triples whose biases are large on both the true single model and the fake single model, so that these news items would be incorrectly classified as fake news if we use just a single true TransE model.

To solve this problem, we propose to train two models, one model is trained based on the triples extracted from fake news and another is trained based on the triples extracted from true news, so that we can do classification by comparing the biases of the true model and the biases on the fake model. We name it B-TransE model:

– the model based on true news is defined as $\mathbf{M}$, and a triple based on $\mathbf{M}$ is defined as $(\mathbf{h}, \mathbf{t}, \mathbf{r})$
– the model based on fake news is defined as $\mathbf{M'}$, and a triple based on $\mathbf{M'}$ is defined as $(\mathbf{h'}, \mathbf{t'}, \mathbf{r'})$

In the B-TransE Model, we represent one news item as $\mathbf{N} = \{\mathbf{TS}, \mathbf{TS'}, \mathbf{M}, \mathbf{M'}\}$, $\mathbf{TS}$ refers to triple set extracted from the news based on $\mathbf{M}$ and each triple is defined as $triple_i = (\mathbf{h}_i, \mathbf{t}_i, \mathbf{r}_i)$, and $\mathbf{TS'}$ refers to triple set based on $\mathbf{M'}$ and each triple is $triple'_i = (\mathbf{h'}_i, \mathbf{t'}_i, \mathbf{r'}_i)$, where i refers to the index of the triple in each triple set.

We define the bias of $triple_i$ and $triple'_i$ as

$$f_b(triple_i) = ||\mathbf{h}_i + \mathbf{r}_i - \mathbf{t}_i||_2^2 \tag{4}$$

$$f_b(triple'_i) = ||\mathbf{h'}_i + \mathbf{r'}_i - \mathbf{t'}_i||_2^2 \tag{5}$$

To judge whether a news item is true or fake, we propose two classify functions and do some experiments to verify the efficiency of each method. **Max Bias Classify** The first way, we use max bias on true single model and max bias on fake single model to do classification. And the Max Bias Classify function is defined as

$$f_{mc}(N) = 0, if f_b(triple_{max}) < f_b(triple'_{max}) \tag{6}$$

$$f_{mc}(N) = 1, otherwise \tag{7}$$

where $f_{mc}(N) = 0$ means the news item is true, and $f_{mc}(N) = 1$ means it is fake. **Avg Bias Classify** The another way, we use average bias on true single model and average bias on fake single model to do classification. And the Avg Bias Classify function is defined as

$$f_{ac}(N) = 0, if f_{avgB}(TS) < f_{avgB}(TS') \tag{8}$$

$$f_{ac}(N) = 1, otherwise \tag{9}$$

where $f_{ac}(N) = 0$ means the news item is true, and $f_{ac}(N) = 1$ means it is fake.

4.4   Hybrid Approaches

To improve the detection performance, we need a fusion strategy to combine the feature vectors from different models. The fusion strategy we use is known as early (feature-level) fusion, which means integrating different features first and using those integrated-features do classification.

In this part, we use the bias vector of the triple, whose bias is the maximum, rather than bias to do classifiction. The bias vector is defined as

$$\mathbf{v}_i = \mathbf{h}_i + \mathbf{r}_i - \mathbf{t}_i \tag{10}$$

The max bias vector is defined as Vec_{max}. We use two different feature vectors:

1. max bias vectors from the model based on true news is defined as Vec_{max};
2. max bias vectors from the model based on fake news is defined as Vec'_{max}.

The integrated vector is defined as V, so that:

$$V = (Vec_{max}, Vec'_{max}) \tag{11}$$

which means we concatenate two different max bias vectors to get an integrated vector, and we use this vector to do classification.

5   Experiments and Analysis

5.1   Data

Fake and True News Article Bases. We use two article bases for our experiments: one with fake news and one with news that we regard as true. We use Kaggle's 'Getting Real about Fake News' dataset, which contains news articles on the 2016 US Election, and we select 1,400 of this dataset as our Fake News Article Base (FAB). These articles have been manually labeled as *Bias, Conspiracy, Fake, Bull Shit*, which we regard as *fake*. Our True News Article Base (TAB) was produced by using the BBC News, Sky News and The Independent websites to scrape 1,400 news articles, which were on the topic of US Election and were published between 1st January and 31st December 2016. These articles have not been manually labeled, however, for the purposes of our experiments, we regard them as *true*. The statistics of two article bases are shown in Table 1. We divide each article base into two parts, 1,000 are for training a model and 400 are for testing.

Knowledge Graphs. We produce three knowledge graphs for our experiments: one named FKG based on FAB, one named D4 (DBpedia 4-hop) from DBpedia, and one named NKG based on TAB.

Table 1. Statistics of fake and true news article bases.

Article base	Label	Source	Quantity
FAB	*fake*	Kaggles Getting Real about Fake News	1,400
TAB	*true*	BBC, Sky, Independent news	1,400

FKG. FKG= $\{E_0, R_0, S_0\}$ is constructed using the training set of FAB. FKG has the following characteristics: $|E_0| = 4K$ entities, $|R_0| = 1.2K$ relations, $|S_0| = 8K$ triples.

D4. To build our KG from DBpedia with 4 hops, we use SPARQL query endpoint interface[6] to interview DBpedia dataset online. We selected 4 hops as it provides a good trade-off between coverage and noise level. There is a public SPARQL endpoint over the DBpedia dataset[7]. DB4 includes two parts, they are KG_1 and KG_2. $KG_1 = \{E_1, R_1, S_1\}$ based on entities from FAB. It has the following characteristics: $|E_1| = 215K$ entities, $|S_1| = 760K$ triples. $KG_2 = \{E_2, R_2, S_2\}$ centered on 2016 US election. We take the entity "United States presidential election 2016" as h_0, extract all triples within four hops. It has the following characteristics: $|E_2| = 132K$ entities, $|R_2| = 5,211$ relations, $|S_2| = 312K$ triples. The reason we extract 4-hop subgraph is that one more hop produces lots of repetitive triples, and most appear in the 4-hop one. We just need to make sure that we get triples related to the topic even some are not related tightly, which also makes the KG construction easier and general.

NKG. We produce NKG= $\{E_3, R_3, S_3\}$ using the training set of TAB. NKG has the following characteristics: $|E_3| = 15K$ entities, $|R_3| = 3,751$ relations, $|S_3| = 19k$ triples.

5.2 Experiment Setup

Article Summarization. We use the titles and the first two sentences of each article to produce the summaries. We did some small-scale experiment, and found that the above summarisation works better than other choices. The intuition behind is that the main message of a news article is often contained in the title and the first two sentences.

Knowledge Extraction. We use an extraction model to extract train triples from 1k fake news, which is used to train FML, and extract train triples from 1k true news, which is used to train FML. Simultaneously, we use an extraction model to extract test triple sets from 400 fake news and 400 true news, which means translating each news item into a triple set with a fake or true label. We use OpenIE to perform triple extraction. However, OpenIE does not perform

---

[6] https://rdflib.github.io/sparqlwrapper/.
[7] http://dbpedia.org/sparql.

very well on triple extraction from news articles. Thus, we use the following four methods to improve the quality of the entities and relations in the triples extracted:

- We disambiguate pronouns so that a text such as "The man woke up. He took a shower." would be transformed to "The man woke up. The man took a shower". We use Neuralcoref to do this.
- We use NLTKs WordNetLemmatizer to transform any verbs in the triples to their present tense.
- We shorten the length of the entities, which is extracted though OpenIE and is named OpenIEEntity. We find out the word which is real entity in the entity extracted though OpenIE and remove other words. Such as "western mainstream media like John Kerry" is shortened to "western mainstream media".
- We use Stanford NER to extract entities from news, which is named NEREntity. Then align the OpenIEEntities to NEREntities.

To produce the two parts of the external KG from an open knowledge graph, as outlined in Sect. 4.1, we use the following steps:

1. $KG_1 = \{E_1, R_1, S_1\}$ based on entities from a fake article base. Firstly, to obtain the set of entities E_1 from triples in fake article base. And then, to extract all triples S_1 from the open knowledge graph with these entities as subjects and objects respectively.
2. $KG_2 = \{E_2, R_2, S_2\}$ centered on the topic of news articles. This sub-KG reflects true statements about the news topic in the real world. We take the entity h_0 that is the most related to the topic as the center, and extract all triples S_2 within a certain number of hops. As shown in Fig. 4, it is a simplified three-hop sub-graph example. Supposing the node "0" to be h_0, firstly, to extract all triples denoted as T_1 that has the formula as (h_0, r, t). Secondly, to extract all triples denoted as T_2 that has the formula as (h_1, r, t), where h_1 refers to an entity in T_1, also one of the nodes "1" in the figure. And the rest can be done by analogy.

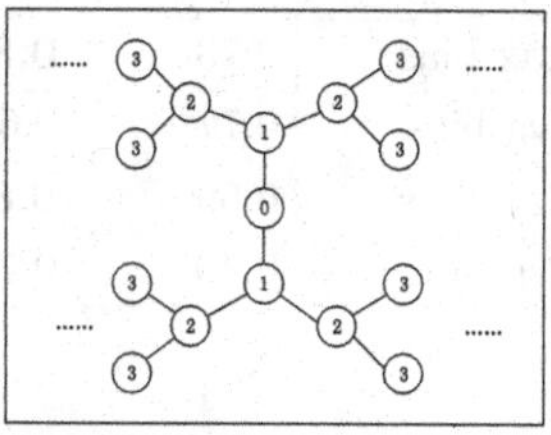

Fig. 4. A simplified three-hop sub-graph example

Model Generation. We generate three single trained models based on TransE for our experiments: the first model FML using the negative knowledge graph

FKG; the second model TML-D4 using the positive knowledge graph D4; the third model TML-NKG using the positive knowledge graph NKG.

FML. The TransE model gets the input of S_0 and automatically produces the trained model FML.

TML-D4. The TransE model gets the input of $S_1 + S_2$ and automatically produces the trained model TML-D4.

TML-NKG. The TransE algorithm gets the input of S_3 and automatically produces the trained model TML-D4.

5.3   Fake News Detection

Using Single Models. The results of the single TransE model with different bias function are shown in Table 2, which shows that: (1) TML-D4 model performs the best (in terms of F score) for the fake news detection task. It suggests that using incomplete knowledge graph can still be effective for fake news detection task. (2) FML and TML-NKG model also perform pretty well, which suggests that using imprecise knowledge graphs can also be effective for fake news detection. This also suggests that, if we do not have knowledge graph in the first place, but only have articles, contracting a knowledge graph from articles is a effective method. (3) Max Bias significantly outperforms Avg Bias in terms of F Score. Maybe there are a few true triples in the triple set of one true news, so that the average bias of the triple set becomes smaller. Since not all the triples extracted from one fake news is false, max bias is more useful in fake news detection task. (4) TML-D4 performs a little better than TML-NKG and FML. The results may correlate with the training data of the TransE model: There are 1 K training news of TML-NKG and FML, but there are 132 K entities and 312 K triples of the training data set of TML-D4.

Table 2. Performance of single TransE model.

Models	Bias function	Precision	Recall	F1 score
FML	Max bias	0.75	0.78	0.77
	Avg bias	**0.80**	0.65	0.72
TML-D4	Max bias	0.73	**0.86**	**0.79**
	Avg bias	0.77	0.68	0.72
TML-NKG	Max bias	0.69	**0.86**	0.77
	Avg bias	0.79	0.71	0.75

Using B-TransE Model. The results of B-TransE Model with different bias function are shown in Table 3, from which we observe that the B-TransE Model is better than the Single TransE Model. This suggests that es the approach based on one related knowledge graph is not enough, and that one should combine related knowledge graph with external knowledge graphs.

Table 3. Performance of different models.

Models	Bias function	Precision	Recall	F1 score
FML + TML-D4	Max bias	**0.85**	**0.80**	**0.83**
	Avg bias	0.80	0.78	0.79
FML + TML-NKG	Max bias	0.75	0.79	0.77
	Avg bias	0.81	0.72	0.76

Hybrid Approaches. In this section, we do experiments on the test sets using the hybrid approach described in Sect. 4.4. Experimental results of combining different models are shown in Table 4. We use vectors from a single TransE model and integrated vectors from a B-TransE Model. The classification we use is SVM [Joa98, SS02], and we choose 'poly', 'linear' and 'rbf' as kernel functions. From Table 4, we can draw a conclusion that: the hybrid approach can further improve the single and binary model approaches.

Table 4. Performance of different models.

Approaches	Kernel	Precision	Recall	Accuracy
FML	poly	0.22	**0.90**	0.63
TML-D4	poly	0.82	0.87	0.85
FML + TML-D4	poly	**0.83**	0.88	**0.89**
FML	linear	0.61	0.91	0.75
TML-D4	linear	0.79	0.88	0.86
FML + TML-D4	linear	**0.81**	**0.92**	**0.87**
FML	rbf	0.74	**0.79**	0.81
TML-D4	rbf	0.94	0.77	0.80
FML + TML-D4	rbf	**0.95**	0.74	**0.81**

Knowledge Stream. Finally, we test Knowledge Stream approach [SFMC17] on the 400 true articles and 400 fake articles required that a file exists for each article which contains all of the triples extracted from the given article and with IDs for each entity and relation accordingly. Once these files existed, they were run in Knowledge Stream to produce scores for each triple in each file. Table 5 shows the results of the comparison of the performance of the TransE FML (which is not even the best single model from our approach, as discussed above) and that of Knowledge Stream. From the table we observe that while Knowledge Stream has a very high recall value, TransE outperforms it significantly. Therefore, we conclude that: Our single TransE model is better than Knowledge Stream on the task of fake news detection when the background knowledge graph is constructed from real news articles.

Table 5. Performance of different models.

Method	Function	Precision	Recall	F1 score
Knowledge stream	Max	0.50	0.99	0.66
	Avg	0.47	**1.0**	0.64
TransE FML	Max bias	0.75	0.78	**0.77**
	Avg bias	**0.80**	0.65	0.72

6   Conclusion and Future Work

In this paper, we tackle the problem of content based fake news detection. We have proposed some novel approaches of fake news detection based on incomplete and imprecise knowledge graphs, based on the existing TransE model and our B-TransE model. Our findings suggest that even incomplete and imprecise knowledge graph can help detect fake news.

As for future work, we will explore the following directions: (1) To combine our content based approaches with style-based approaches. (2) To provide explanations for the results fake news detection, even with incomplete and imprecise knowledge graphs. (3) To explore the use of the schema of knowledge graphs as well as approximate reasoning [PRZ16] and uncertain reasoning [PTRT12, SFP+13, JGC15] in fake news detection.

Acknowledgements. The work is supported by the Aberdeen-Wuhan Joint Research Institute.

References

[AG17a]   Allcott, H., Gentzkow, M.: Social media and fake news in the 2016 election. J. Econ. Perspect. **31**(2), 211–236 (2017)

[BUGD+13] Bordes, A., Usunier, N., Garcia-Duran, A., Weston, J., Yakhnenko, O.: Translating embeddings for modeling multi-relational data. In: Advances in Neural Information Processing Systems, pp. 2787–2795 (2013)

[Gil17]   Gilda, S.: Evaluating machine learning algorithms for fake news detection. In: 2017 IEEE 15th Student Conference on Research and Development (SCOReD), pp. 110–115. IEEE (2017)

[GM17]   Granik, M., Mesyura, V.: Fake news detection using Naive Bayes classifier. In: 2017 IEEE First Ukraine Conference on Electrical and Computer Engineering (UKRCON), pp. 900–903. IEEE (2017)

[JGC15]   Fokoue, A., Sycara, K., Tang, Y., Garcia, J., Pan, J.Z., Cerutti, F.: Handling uncertainty: an extension of DL-Lite with subjective logic. In: Proceedings of 28th International Workshop on Description Logics, DL 2015 (2015)

[JHX+15] Ji, G., He, S., Xu, L., Liu, K., Zhao, J.: Knowledge graph embedding via dynamic mapping matrix. In: Proceedings of the 53rd Annual Meeting of the Association for Computational Linguistics and the 7th International Joint Conference on Natural Language Processing (Volume 1: Long Papers), vol. 1, pp. 687–696 (2015)

[JLY17]   Jiang, Y., Liu, Y., Yang, Y.: LanguageTool based university rumor detection on Sina Weibo. In: 2017 IEEE International Conference on Big Data and Smart Computing (BigComp), pp. 453–454. IEEE (2017)

[Joa98]   Joachims, T.: Making large-scale SVM learning practical. Technical report, SFB 475: Komplexitätsreduktion in Multivariaten Datenstrukturen, Universität Dortmund (1998)

[LCSR+15]   Ciampaglia, G.L., Shiralkar, P., Rocha, L.M., Bollen, J., Menczer, F., Flammini, A.: Computational fact checking from knowledge networks. PloS one 10, e0128193 (2015)

[LLS+15]   Lin, Y., Liu, Z., Sun, M., Liu, Y., Zhu, X.: Learning entity and relation embeddings for knowledge graph completion. In: AAAI, vol. 15, pp. 2181–2187 (2015)

[PCE+17]   Pan, J.Z., et al. (eds.): Reasoning Web 2016. LNISA, vol. 9885. Springer, Cham (2017). https://doi.org/10.1007/978-3-319-49493-7

[PRZ16]   Pan, J.Z., Ren, Y., Zhao, Y.: Tractable approximate deduction for OWL. Artif. Intell. 235, 95–155 (2016)

[PTRT12]   Pan, J.Z., Thomas, E., Ren, Y., Taylor, S.: Tractable fuzzy and crisp reasoning in ontology applications. IEEE Comput. Intell. Mag. 7, 45–53 (2012)

[PVGPW17]   Pan, J.Z., Vetere, G., Gomez-Perez, J.M., Wu, H. (eds.): Exploiting Linked Data and Knowledge Graphs in Large Organisations. Springer, Heidelberg (2017). https://doi.org/10.1007/978-3-319-45654-6

[SCV+17]   Shao, C., Ciampaglia, G.L., Varol, O., Flammini, A., Menczer, F.: The spread of fake news by social bots. arXiv preprint arXiv:1707.07592 (2017)

[SFMC17]   Shiralkar, P., Flammini, A., Menczer, F., Ciampaglia, G.L.: Finding streams in knowledge graphs to support fact checking. arXiv preprint arXiv:1708.07239 (2017)

[SFP+13]   Sensoy, M., et al.: Reasoning about uncertain information and conflict resolution through trust revision. In: Proceedings of the 12th International Conference on Autonomous Agents and Multiagent Systems, AAMAS 2013 (2013)

[SS02]   Schölkopf, B., Smola, A.J.: Learning with Kernels: Support Vector Machines, Regularization, Optimization, and Beyond. MIT Press, Cambridge (2002)

[SSW+17]   Shu, K., Sliva, A., Wang, S., Tang, J., Liu, H.: Fake news detection on social media: a data mining perspective. ACM SIGKDD Explor. Newslett. 19(1), 22–36 (2017)

[SW16]   Shi, B., Weninger, T.: Fact checking in heterogeneous information networks. In: Proceedings of the 25th International Conference Companion on World Wide Web, pp. 101–102. International World Wide Web Conferences Steering Committee (2016)

[VRA18]   Vosoughi, S., Roy, D., Aral, S.: The spread of true and false news online. Science 359(6380), 1146–1151 (2018)

[Wan17]   Wang, W.Y.: "liar, liar pants on fire": A new benchmark dataset for fake news detection. arXiv preprint arXiv:1705.00648 (2017)

[WZFC14]   Wang, Z., Zhang, J., Feng, J., Chen, Z.: Knowledge graph embedding by translating on hyperplanes. In: AAAI, vol. 14, pp. 1112–1119 (2014)

Author Index

Printed in the United States
by Bookmasters

Printed in the United States
By Bookmasters